FOUNDATIONS OF

Climatology

FOUNDATIONS OF

Climatology

AN INTRODUCTION TO PHYSICAL, DYNAMIC, SYNOPTIC, AND GEOGRAPHICAL CLIMATOLOGY

E. T. STRINGER

UNIVERSITY OF BIRMINGHAM

W. H. FREEMAN AND COMPANY

San Francisco

Printed in the United States of America.

Library of Congress Catalog Card Number: 72-81920

International Standard Book Number: 0-7167-0242-8

1 2 3 4 5 6 7 8 9

To the memory of my parents,

Edward H. Stringer (1888–1957),
Lilian M. Stringer (1891–1954),

without whose encouragement and self-sacrifice this book
would never have been commenced;
and to

Gloria, my wife,

without whose devotion and hard work it
would never have been completed.

Contents

Preface

The foundations of climatology lie in the realms of physics, mathematical statistics, and synoptic meteorology. Many excellent textbooks deal with these subjects, others with the basic facts of climate, but no textbook covers all these fields in a manner both easy to understand and sufficiently rigorous for the serious student. Such a textbook I have endeavoured to provide, on the basis of nineteen years' experience in teaching climatology and meteorology to a wide variety of students.

Foundations of Climatology is intended as a text, both for students in university geography departments where climatology is taken seriously as a subject in its own right, and for teachers of physics, mathematics, and geography who wish to interest their pupils in atmospheric science. I also hope that the book will prove useful in the training of new entrants, with only modest mathematical backgrounds, to the assistant, forecaster, and climatologist grades of state meteorological services, and that it will aid physical and biological scientists, medical and public health specialists, engineers, architects, planners, and others who wish to apply climatic information to their own fields in a scientific manner. I have written it purely from the view point of classical physics. Other climatologies are, of course, possible: written, for example, from the biological, human, economic, or historical standpoints; but the physical approach is the one most neglected in geographical literature, the one, in my experience, most likely to be difficult for the students, and the one most likely to advance climatology, both as a science and as a service to mankind.

In numerous elementary books on climatology, atmospheric phenomena are treated purely descriptively. However valuable these descriptions may be for the nonspecialist, they are of no real use to the serious student of climatology, whether he or she is in a university, college, or other organization, or is working as an amateur. Climatology deals with numbers, and one of the first things a person with a scientific mind will want to do, on coming upon a set of climatic figures, is to see if he can use these figures to predict another set. He will find no guidance toward this end in the elementary texts. The only alternative has been provided by standard meteorological texts, which are written for persons with a good background in mathematics and physics.

Foundations of Climatology is intended to help the nonmathematician make the difficult transition from elementary, qualitative climatology to advanced, quantitative climatology. Climatology is becoming more and more an exact science, and the intending climatologist must be numerate as well as literate. Essential mathematical concepts and physical relationships, although introduced into the text when necessary, are there kept to a minimum, but their proofs and formulae are given in detail in the appendixes at the end of the book. Only knowledge of very elementary algebra, trigonometry, and physics is assumed in the text. Calculus and vector notation are used occasionally in the text, and extensively in the appendixes, but one can read the body of the text without a knowledge of these subjects.

From logical statements and deductive principles, expressed in mathematical terms, one can build up an intellectual picture of world and local climates without making a single observation. Such a picture provides the geographer with a stimulating problem to be investigated by field observation. The creating of such a theoretical picture is the *raison d'etre* for climatology as a science in its own right, because not only does the theory provide answers to the problems that arise from actual observations of weather and climate, but it also enables us to make predictions that have scientific validity. Most existing textbooks of climatology completely ignore this theoretical picture of climate, mainly because they have been written by geographers, who usually know little, and care less, about mathematics and physical science. To remedy this deficiency, I have devoted much time during the past decade to an investigation of the mathematical and physical foundations of climatology, of which this book is one result. The picture of climate it presents is completely different from those presented in most texts. Many topics usually given a prominent place, climatic regions and air-mass climatology, for example, here occupy relatively insignificant places; instead, I have emphasized those aspects of climatology that depend on fundamental principles and therefore have distinct value in predictive systems. Purely factual descriptions of the distributions of the climatic elements on the continents are readily available elsewhere.

Climatology is ultimately concerned with actual places and regions, however abstruse its theory becomes, and is thus part of the broad field of geography, as well as being an integral part of the science of meteorology. In addition, climatology is not just the study of "average weather," in terms of long-period means. It is also concerned with meteorological phenomena averaged over periods of a few months, a few days or hours, even a few seconds. It is "normal" for the atmosphere to behave irregularly, and it is more realistic to focus attention on the irregularities (the "weather") than on the smooth long-period mean (the "climate"), which may or may not exist in nature.

I prefer to link weather and climate in a meteorological spectrum, comprising all atmospheric phenomena whether periodic or irregular. The shorter wavelengths of this spectrum include microvariations of the meteorological elements with periods of a few minutes or seconds, and the hour-by-hour variations of ordinary weather. In its middle portion, the spectrum covers climatic variations with periods varying from a few months to tens or hundreds of years; at its long-wave end, the spectrum includes geological variations in climate, with periods of thousands of millions of years. The only difference between curves representing the time or space fluctuations about their respective climatic means of, say, microtemperature changes, everyday weather variations, and decade-by-decade climatic fluctuations, is one of scale. Because of this, it is logical to study them all by one technique, which permits direct comparison between one scale and another; such a technique is the mathematical theory of turbulence. Turbulence concepts are fundamental for microclimatology, the study of climates in the lowest layers of the atmosphere; and in turn, microclimatology forms the basis of much of applied climatology, which is becoming increasingly important in the solution of specific practical problems in agriculture, engineering, industry, transportation, architecture, and public health.

Present-day climatology is an exciting science, with a future full of promise. Electronic computing machines now allow us to solve, in a reasonable time, the complex equations of motion and energy balance that explain how climates are formed: accurate long-range forecasting—the domain of the climatologist rather than the weatherman—is now a distinct possibility. Earth satellites enable global weather patterns to be actually observed, not merely inferred; and the possibility of direct observations of the planets nearest to the Earth will permit real comparative studies of "world" climates to be made for the first time. Increasing knowledge of climate may enable us to modify or even control it, with the help of the new sources of energy becoming available.

ACKNOWLEDGMENTS

I have first of all to thank my former teachers in the Meteorological Office, almost two decades ago, who kindled my interest in the rigorous approach to climatology. In particular, I thank Professor A. F. Jenkinson for considerable guidance in the field of mathematical statistics, and Mr. P. J. Meade and Mr. K. H. Smith for stimulating thoughts in the fields of dynamic meteorology and synoptic meteorology, respectively. Special thanks are due to Professor David L. Linton, Head of the Department of Geography in the University of Birmingham, for his interest in the book at all stages in its preparation. For many years, Dr. Gordon T. Warwick, Reader in the Department of Geography, has been a most valued and constructive critic. The late Professor R. H. Kinvig, and Mrs. B. Eckstein, helped me in the initial phases of my climatological career.

My climatological ideas have profited from discussions held with American and Canadian climatologists during a period I spent at McGill University in 1964 as Visiting Professor of Climatology. For providing the opportunities for these discussions, I have to thank Dr. F. Kenneth Hare, Dr. Trevor Lloyd, Professor Theo. L. Hills, Dr. H. E. Landsberg, Mr. Morley K. Thomas, and Dr. John R. Mather.

No little word of thanks is due my publishers, for their patience amid the innumerable delays that have attended the writing of the book, and for their care in preparing it for publication; the efforts of Mr. Aidan A. Kelly, of the Editorial Department, are worthy of special mention. I must also acknowledge with gratitude the care with which Professor Gordon Manley read and commented on the book in its early stages.

The labors of typing the book have been shared between a number of ladies: special thanks are due to Mrs. R. Priestman, Mrs. M. L. Stones, Mrs. J. Needham, Miss D. Morgan, and my wife. Indeed, I must record my sincere appreciation of my wife's forebearance of the last six years, during which the demands of two small sons and a book-finishing husband have been, to say the least, somewhat trying.

Birmingham, England,
July 1971

E. T. STRINGER

FOUNDATIONS OF
Climatology

Introduction

Weather is a phenomenon that varies enormously over the face of the Earth, and hence is of great geographical significance. Travelers and writers have from the earliest times described the almost endless varieties of weather to be found from place to place, and also from time to time at the same place. But scientific explanation of the geographical pattern of weather is very recent indeed.

Weather is the sum total of atmospheric conditions at a given place at a given time; climate is a generalization or integration of weather conditions for a given period of time within a given area. Weather is an everyday experience; climate is an abstract concept.

Meteorological phenomena are essentially turbulent (see Figure I.1): on the mean pattern or flow, i.e., the climate, are superimposed the "eddies," the irregularities of the actual occurrences, i.e., the weather. In the past, climatology has studied the mean pattern of atmospheric behavior during long periods (usually 30 years), and meteorology has been concerned with the study of day-to-day conditions. However, the mean pattern or flow shown in Figure I.1 can be interpreted as representing very different time intervals, say, from 5 minutes (microclimates) to 1,000 years (planetary climate). Similarly, the irregularities may refer to second-by-second or day-by-day fluctuations, or to the variations exhibited during periods of 30 years or 3,000 years. Meteorology and climatology should not be defined as completely separate fields, since they are both interested in the same phenomena.

Today meteorology is recognized as atmospheric science in its widest sense. Its concerns include the physics, chemistry, and dynamics of the atmosphere, and the direct effects of those dynamics on the Earth's surface, on the oceans, and on life in general.[1]* Its goals are the complete understanding, accurate prediction, and artificial control of atmospheric phenomena. It is one of the most complex fields of both natural and applied science.

Climatology is the scientific study of climate; its concerns include the practical applications of such study. It uses the same basic data as meteorology, and its results are of great use to meteorologists, particularly in weather prediction, and in applications of meteorology to problems in industry, agriculture, transport, architecture, biology, and medicine. The aim of climatology is to discover, explain, and exploit for the benefit of man the normal behavior of atmospheric phenomena, bearing in mind that irregularities in atmospheric behavior are the rule, not the exception: the irregularities at one scale of climatic behavior define the norms for smaller-scale phenomena.

Insofar as climatology explains the norm in general† terms, it is clearly part of meteorology, but insofar as its emphasis is on specific climatic conditions at particular points on the Earth's surface, it is clearly part of geography. The variations in the Earth's surface have profound effects on the interchange of heat, moisture, and momentum between land, water, and atmosphere, and are vital in determining specific climatic conditions; here local empirical observation as well as meteorological theory is absolutely necessary. Climatology thus does not belong wholly within the fields of either meteorology or geography. It is a science—really an applied science—whose methods are strictly meteorological, but whose aims and results are geographical. In

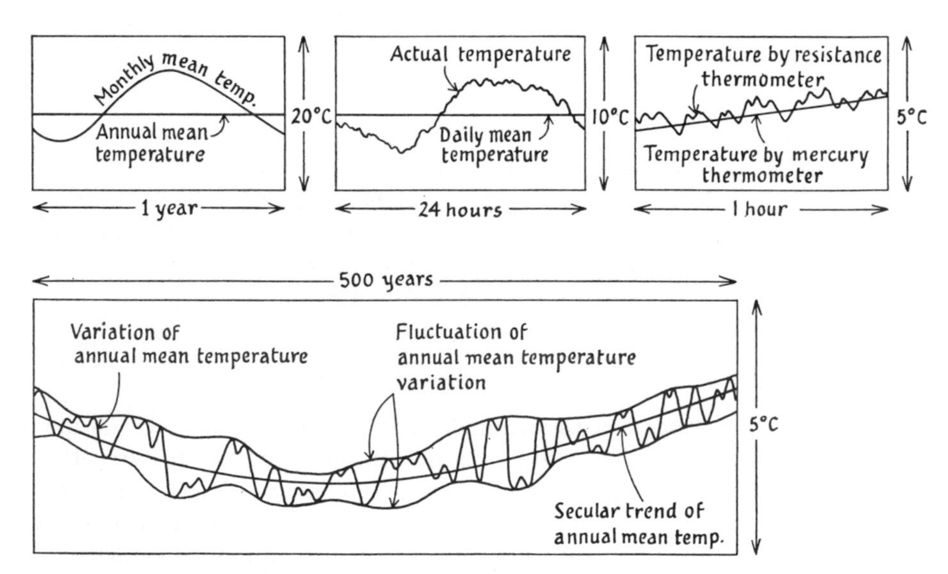

FIGURE I.1.
Hypothetical examples of turbulent fluctuations in air temperature.

* The numbered notes will be found at the end of each chapter. The abbreviations used in the notes are listed, beginning on p. 559.

† I.e., by means of mathematical and physical principles, which have universal application. "General" is not intended to mean " vague."

practice, the climatologist needs to learn as many meteorological techniques as possible, and needs to steep himself in geography to determine the problems that will be most immediately worth his attention.

Very frequently, the climatologist finds, when investigating a problem, that the basic meteorological groundwork has not been done, and that he must do it himself before he can solve his problem. He must therefore keep well abreast of current developments in meteorology, no branches of which he can afford to ignore.

Basically, there are three methods of working within the field of climatology. *Climatography* is the presentation of climatic data, either verbally or cartographically, or in the form of tables, graphs, or statistical parameters.[2] *Physical and dynamic climatology* is the explanation of climatic phenomena in terms of physics and dynamics. Sometimes physical climatology is defined as concerned mainly with radiation and vertical exchange processes between the atmosphere and the Earth's surface, in contrast to dynamic climatology, which is defined as concerned with the movements of the atmosphere in terms of hydrodynamics. Or physical climatology may be regarded as the application of practical—even experimental—physics to climatic phenomena,

A Functional Classification of Some of the Main Divisions of Climatology

Division	*Purpose*
Descriptive climatology	The provision of climatological information in an easily understood form.
Statistical climatology	The reduction of the mass of climatic records to a form both concise and precise.
Mathematical climatology	The isolation of those aspects of climate that can be given an exact mathematical representation.
Synoptic climatology	The isolation of those aspects of climatology that are of use in weather forecasting.
Microclimatology	The discovery of the features of climate that are characteristic of the lowest few meters of the atmosphere, and the isolation of the factors controlling them.
Macroclimatology	The study of climate on a worldwide scale.
Mesoclimatology	The isolation of the atmospheric entities that control the climates of areas a few square miles in extent.
Local climatology	The study of the climates of areas a few square miles in extent.
Topoclimatology	The study of the climate of specific places.

and dynamic climatology as the application of mathematical concepts. Whatever the definitions adopted, however, physical and dynamic climatology both involve scientific explanation, whereas climatography is pure description.[3]

The third method is that of *applied climatology*, which is the scientific analysis of climatic data in order to apply it to solving specific design or operational problems within such fields as industry and technology, agriculture, forestry, or medicine.[4] Applied climatology is distinct from *applied meteorology*, which applies current weather data, analyses, and forecasts to specific practical problems.

Climatology may be subdivided according to function, instead of according to method of approach. Many of these subdivisions are studied by all three methods. Thus *synoptic climatology* uses the concepts of physical and dynamic climatology to derive its chief tool, the synoptic chart, and its products are either prediction tools (applied climatology) or climatographies.

The History of Climatology and Meteorology[5]

The earliest records of meteorological knowledge we have are generally supposed to be in the weather lore of the Greek poets and the Hebrew writers of the Old Testament, but there are even older sources—for example, the Vedas of India, and the cuneiform tablets of the Tigris-Euphrates area—that deserve more attention than they have received. The Greek philosophers speculated on the causes of various phenomena of weather and climate, and Aristotle's *Meteorologica* is believed to be the earliest meteorological treatise, dated about 350 B.C. Hippocrates' *Airs, Waters, and Places* (ca. 400 B.C.) was probably the first climatography.

Climatology originated in ancient Greece. To the early Greek philosophers, the term "climate" (κλιμα) meant "slope," and referred to the curvature of the Earth's surface. They believed that significant regional differences in weather occurred only from north to south, that the sloping of the Earth's surface, up to the south and down to the north, gave rise to the torrid, temperate, and frigid zones. The work of the Alexandrian philosophers—Eratosthenes and Aristarchus in particular—included astronomical and mathematical geography of an order that was not surpassed until modern times; it is possible that their climatological knowledge was much more sophisticated than they have been given credit for.

During the great period of geographical exploration, which began in A.D. 1450, it was recognized that climates were not simple latitudinal belts, but covered very irregular areas, and were affected by the general circulation of the atmosphere as well as by the distribution and configuration of the continents and oceans. By the middle of the seventeenth century, scientific method had begun to appear. The barometer was invented by Torricelli in 1643, and the thermometer by Galileo in 1593. Bacon (1561–1626) began applying the inductive method to both weather and climate, which idea was also followed by Descartes (1596–1650). The first foundations of atmospheric thermodynamics were laid when Boyle (1627–1691) discovered the relation between the pressure and volume of a gas.

The central problem of climatology in the eighteenth century was conceived to be the summarizing of weather observations in order to give the probabilities that the various meteorological phenomena would occur at given phases of the annual and daily cycles. The development of new instruments resulted in an abundance of data, and it was considered that local variations should be eliminated as far as possible, leaving only broad scale differences in climate. Thus began standardization of exposure and instrumentation.

The annual and daily marches of the climatic elements were analyzed by orthodox methods of astronomy, in particular by use of the arithmetic mean, which had been introduced by Cotes and Simpson in the early eighteenth century as a device for eliminating chance errors in astronomical and physical measurement. The daily and annual cycles were obviously related in some way to the astronomical motions of the Earth, and it was hoped to discover some pattern—possibly akin to the regularity expressed by Kepler's laws for the planets—in the apparently irregular fluctuations shown by the daily observations. However, analysis by arithmetic averaging did not lead to such satisfactory results with climatic data as it did in astronomy.

No meteorological theory was available in the early eighteenth century on which to base explanations of the apparent climatic phenomena and relationships which analysis was bringing to light. This was the time of pure empiricism, and since climatic maps were not yet available, description was purely mathematical. Equations were set up to describe, for example, the distribution of observed mean annual temperature as a function of latitude, or as a function of angular distance from certain assumed cold poles. None of these investigations gave any real understanding of the physical processes underlying climate.

The cartographic approach to climatology originated in 1817 with von Humboldt's treatise on temperature, which contained a map of mean annual isotherms for much of the northern hemisphere.[6] Temperatures were reduced to sea level to eliminate local influences, which procedure has continued to the present day, despite the fact that these local influences are frequently just what the geographer needs to study. The basic concept behind the cartographic approach to meteorology, that the wind brings the weather, was employed by von Buch in 1820. Dove in 1827 used essentially synoptic concepts to explain local weather in terms of ideal or model polar and equatorial air currents, and Howard in 1820 described the climate of London in terms of alternating cold and warm masses of air both near ground level and aloft, but they had no charts to help them. Dove also was the first to set forth, in 1828, the kinetic theory of cyclone energy. A theoretical approach to climate was made by Schmidt in 1830. He derived an equation for the heat gain per unit area of the Earth's surface exposed to solar radiation and the heat loss by earth radiation.[7]

In 1841, Espy first set forth the convection theory of cyclone energy; its basis, as later formulated by von Helmholtz, is that the main source of energy is latent heat of water vapor. In 1845, Berghaus produced the first world map of precipitation, using shading, and in 1848, Dove published the first maps of monthly mean temperature. Land observations of surface winds were considered to be unrepresentative, and so the first wind charts were for the oceans, produced between 1848 and 1860 by Maury from the records of sailing ships. In 1857 was published what is probably the first major climatography, Blodget's description of North America. Also in 1857, Meech was the first to calculate from basic radiation theory the amount of solar radiation received at the outer limits of the atmosphere in all latitudes. The first world map showing areas of constant seasonal precipitation was published by Mühry in 1862.[8]

With the development of the synoptic chart between 1860 and 1865, bound up with Buys-Ballot's discovery of the empirical relationship between wind and pressure, and the development of wireless telegraphy, a new type of study began: investigation of the geometry of mean sea-level pressure patterns and their relation to weather.[9] The first national meteorological services of the world were organized between 1855 and 1880 to work along these lines. Renou in 1864 produced the first map of annual mean pressure for France and adjacent regions, and the first world maps of monthly mean pressure were produced by Buchan in 1868.[10]

Unfortunately, the idea that oval or radially symmetric cyclones—which were very easy to draw—were the main carriers of bad weather was just elegant enough to discourage more careful investigation. Progress was very slow at this time, not only because of the lack of upper-air data—and hence a lack of knowledge of the three-dimensional structure of the atmosphere—but also because of a lack of theoretical knowledge on the part of the weathermen.

There were a few exceptions to this general stagnation. The British Meteorological Office published, between 1869 and 1880, a series of meteorograms, which were graphs giving the detailed time-variations of pressure, temperature, wind, humidity, and precipitation from automatic recording instruments set up in seven United Kingdom observatories.[11] If the data in these graphs had been correlated and followed up, frontal models could have been developed fifty years before the Polar Front theory was set up.

Until the middle of the nineteenth century, the greater part of meteorological study had been climatological, but with the introduction of the synoptic chart, which allows the representation of simultaneous conditions over a large area, meteorology and climatology began to move away from each other. Meteorology became a scientific study of current weather and its prediction, based largely on instantaneous pressure distributions, and climatology a historical study of the normal pattern of atmospheric behavior, based on arithmetic means. Köppen clearly recognized in 1874 that the arithmetic mean was an unsatisfactory value, but it has continued until today as the basic unit in some backwaters of climatology. Other statistical techniques have been more successful, in particular the method of least squares, frequency analysis, the correlation coefficient, and harmonic analysis of periodic phenomena, which was first applied to atmospheric pressure.

Synoptic climatology can be said to have begun with Köppen, who in the 1870's grouped two years' weather observations for St. Petersburg according to pressure patterns, using the empirical relationship between wind and pressure discovered in 1860. The first world wind charts, covering both land and sea areas, were drawn up by Coffin in 1876. Following Meech, Wiener in 1877 (and Angot in 1883) also calculated from radiation theory the amount of solar radiation received at the outer limits of the atmosphere in all latitudes. These investigations could yield only relative amounts, because the actual intensity of solar radiation outside the Earth's atmosphere was not known until the end of the nineteenth century.[12]

The empirical cyclone model of Ley in 1878 gave the essential features of the three-dimensional structure of a frontal low, based on 8,000 observations of the movement of cirriform cloud. The idea was taken up by Abercromby and Marriott in their cyclone model of 1883, in which weather lore was analyzed in terms of isobaric geometry. The cold front was conceived by Ley in 1878 and Köppen in 1879 as a cold-air wedge invading warmer air; unfortunately, they thought it was a shallow phenomenon (600 m thick) of only local importance. The first world map of temperature regions was published by Supan in 1879, and the first world map of precipitation using mean annual isohyets by Loomis in 1882.[13]

In 1883, L. Teisserenc de Bort produced the first mean-pressure maps showing the great seasonal anticyclones and cyclones, the "centers of action" on which the study of the general circulation of the atmosphere was based. In 1886, he published the first world maps of annual and monthly cloudiness. In 1887, he published the first chart of the pattern of pressure distribution in the upper air, which he based on indirect evidence and computations. In 1889, he published the first world map of January mean pressures for the lowest four km of the atmosphere. This map exhibited the zone of intense westerly winds between latitudes 30° and 50° N, forming two stationary long waves in their present positions. He explained this phenomenon in terms of differential thermal effects at the Earth's surface.[14] In 1888 von Helmholtz successfully explained billow clouds in terms of meso scale gravity waves and deduced theoretically the existence of a global belt of atmospheric discontinuity. The squall line was first deduced by Durand-Greville in 1890, as an independent linear organism of cloud and wind change extending right across Europe.[15]

The oldest approach to climatology is the verbal description which will always have its place if climatic data are to be made palatable to the nonspecialist. The classic work at this period was Hann's *Handbook of Climatology*, in three volumes; the first, published in 1883, was on general climatology, and the two later volumes, on regional climatology, included noninstrumental data, and literary and eyewitness descriptions of local weather and climate. Earlier maps of monthly mean temperature and pressure were revised for Bartholomew and Herbertson's *Atlas of Meteorology* (1889), which remained standard for many years.[16]

During the final years of the nineteenth century, direct aerology from upper-air soundings was begun on a small scale, for example, by Teisserenc de Bort and Assmann in Europe, and by Rotch in the United States.[17] The real breakthrough began with the application of hydrodynamic concepts to the synoptic chart by the Norwegian V. Bjerknes and his co-workers. Ever since Laplace had written in 1799 that his goal was to reduce all known phenomena to the law of gravity by means of strict mathematical deduction, weather and climate had been considered to be Laplacian problems, i.e., events of inanimate Nature that ought to be predictable from the laws of mechanics. Bjerknes was continuing the program implied by the Laplacian goal when he developed his theories of atmospheric circulation between 1897 and 1902 in Stockholm, as was Margules in Vienna, who in 1901 showed that the potential energy of a pressure distribution generally amounts to less than 10 per cent of the kinetic energy of the winds balancing it, and that isobaric geometry was thus "a mere cogwheel in the atmospheric machine."[18]

In applying dynamics to synoptic meteorology, the general idea was to search for two phenomena in the atmosphere, waves and vortices, whose behavior under laboratory conditions was well-understood, and whose movements and development could be predicted from thermo-hydrodynamic theory. A variety of vortices occur in climatic phenomena, from eddies a few feet across to large-scale circulations of continental extent. Usually, there is first a wave, which then turns into a vortex; at the time of transition, potential energy is converted into kinetic energy and heat, and cannot be turned into potential energy again within the life cycle of the same vortex. The motion of waves consists simply of a periodic redistribution of already existing energy. Vortices (for example, cyclones) therefore pose more difficult problems than waves.

Of the two main theories of cyclone energy then in existence, the convection theory and the kinetic theory, the kinetic theory was favored by the Vienna school of meteorology and developed further by it during the beginning years of this century. The Vienna school consisted of very able men—notably Margules, Exner, von Ficker, Defant, and W. Schmidt—but lacking the balance between empiricism and theory that would be found in the other later schools, it missed some important opportunities. Thus, Exner understood the theory of short gravity-inertia waves, and von Ficker was familiar with the Moazagotl wave clouds of the mountain regions of central Europe, but the theory and the clouds were not fitted together as cause and effect until Queney, Küttner, and Lyra did so in the concept of lee standing waves, between 1936 and 1943.[19]

Hydrodynamics was applied to meteorology by two main techniques: *Lagrangian*, stressing the trajectory of air particles, and *Eulerian*, stressing their field or distributional aspect. Shaw and Lempfert in 1906 provided a Lagrangian proof of the existence of Dove's two model air currents, and produced a new cyclone model in 1911.[20] Without a knowledge of fronts, however, the model could not be developed or the life history of the cyclone outlined. The synoptic chart was a Eulerian concept.

V. Bjerknes moved from Stockholm to Leipzig in 1913, and there set up the first genuine school of meteorology, founded on his new system of applying physical hydrodynamics to meteorology and oceanography. This was the real beginning of physical and dynamic meteorology. Bjerknes believed in a rational, physical approach to all weather phenomena, and in the possibility—indeed, necessity—of applying physical laws to atmospheric states at any moment; "case studies" therefore became important in the new meteorology.[21]

In 1917, Bjerknes moved to Bergen, and during the next twelve years inspired very important developments in meteorology. The Bergen school developed dense networks of observing stations in southwest Norway, and designed realistic models of atmospheric structure that fitted all the surface observations as far as possible and that, for the first time in synoptic meteorology, took the dynamic effects of mountains into account. Its members also began using indirect evidence of air movements, temperatures, and humidities aloft—provided by clouds and other hydrometeors—to form an indirect aerology.[22]

The first major achievement of the Bergen school was the discovery of the warm front by J. Bjerknes and Solberg in 1918. The second was the discovery of the occlusion process by Bergeron in 1919, and the third was the formulation of the well-known Polar Front theory of cyclone formation in 1922. Calwagen, Schinze, and Spinnangr proved the existence of cyclone series, and in 1926 Calwagen used aircraft soundings to form the first direct aerology.[23]

At the time of the formulation of the Polar Front theory, emphasis was on the traveling depressions and anticyclones of middle latitudes. The enthusiastic disciples of the Bergen school believed that all local weather could ultimately be explained in terms of air masses and fronts, once enough local varieties of these had been recognized. They also believed that depressions and anticyclones would have some effect on the upper atmosphere, but, lacking observations, they did not emphasize such effect.

Köppen produced in 1918 the first detailed classification of world climates, based on the vegetation cover of the land, which was assumed to be determined by the combination of temperature and rainfall (see p. 22 for a detailed discussion). In 1921 the field of *complex climatology* was developed by the Russian geographer Federov. It attempted to describe a local climate in terms of daily weather observations, i.e., in terms of the relative frequencies of various weather types or groups of weather types, instead of mean values. His work did not become well-known to western climatologists until 1959. Federov's system was really complex climatography, and was similar to the synoptic climatology that would be developed by western meteorologists during the Second World War.[24]

In 1930 Bergeron outlined the field of *dynamic climatology*, which he defined as the statistical treatment of stable weather types (air masses and fronts) as complete phenomena or dynamic-thermodynamic processes (in current terminology, we would call this field synoptic climatology). Such statistical treatment would have required more knowledge of atmospheric dynamics and thermodynamics than was then available; as a result, his suggestion was not followed up by the qualitative-minded geographical climatologists of the period, even though its possible application to long-range weather forecasting promised great benefits for mankind. Apart from Bergeron's outlining of the field, dynamic climatology began in 1932 with Hesselberg's attempt to define conditions under which the hydrodynamic equations of motion defining instantaneous conditions can be applied to mean states of the atmosphere.[25]

Detailed theoretical work on insolation was published by Milankovitch in 1920 and 1930, and on the radiation balance of the Earth by Simpson in 1928–29 and by Baur and Phillips from 1934 to 1936. Simpson calculated values of the radiation balance over the world in ten-degree squares, from radiation theory, and his values were used for radiation maps until very recently. Baur and Phillips gave values for latitudinal zones in the northern hemisphere.[26]

Between 1930 and 1940, Köppen and Geiger published their five-volume *Handbuch der Klimatologie*, which was designed to replace that of Hann. Their first volume was general, but the other four volumes were regional and were written by local experts, since the data were too voluminous for any two men to master. In general, there was more emphasis on tables and maps, and less on personal description, than in Hann. The second detailed classification of climatcs was published by Thornthwaite from 1931 to 1933; Thornthwaite took evaporation into account and thus considered the effectiveness of precipitation as well as its amount (see p. 24 for a detailed discussion).[27]

Between 1935 and 1940, a school that developed at Frankfurt under R. Scherhag introduced a combination of direct and indirect quantitative aerology that showed how the movements of surface weather systems are controled by the lower stratosphere. Their concept of "stratospheric coupling," which considered adjustments between weather systems at ground level and movements of the upper atmosphere to be mutual, thus tended to disprove the ideas of the Bergen school. The Frankfurt school also included the work of F. Baur on extended-range weather forecasting, a truly climatological study, in which sunspots and other solar features are used as the basis for forecasting general weather conditions six months or more ahead.[28]

Bjerknes and his co-workers of the Bergen school had concentrated on small synoptic systems, because they were unable to deal quantitatively with the interaction between several weather systems in the general circulation. This latter problem was successfully taken up at the University of Chicago under C.-G. Rossby. In 1937, Rossby, a Swedish hydrodynamicist, commenced studies of the general circulation of the atmosphere at the Massachusetts Institute of Technology. He later moved to Chicago, and his first followers, H. R. Byers, J. Namias, H. Wexler, and H. C. Willett, later became leaders in the dynamic study of both weather and climate. In Scandinavia, E. Palmén became a collaborator of the Chicago school.[29]

The Chicago school developed the technique of isentropic analysis, introduced successful simplified models of the large-scale atmospheric circulation, in particular, the models of the long "Rossby waves" and of the jetstream, and, for the first time in meteorology, used the electronic computer. (If the computer facilities and the upper-air data used by the Chicago school had been available between 1911 and 1912, then the early Laplacian attempt of the British mathematician Richardson to predict the atmospheric state over central Europe six hours ahead might have succeeded.[30])

The central theme of the Chicago school was Lagrangian. Its adherents focused their attention on the movements of the air, from which they obtained pressure as a by-product, not vice versa as with the Bergen school. Whereas the work of the Bergen school was purely meteorological, that of the Chicago school was in essence climatological, since it applied hydrodynamic concepts to mean features of the atmospheric circulation, by averaging charts, usually over periods of five or thirty days, and by assuming the atmosphere to be barotropic, which allowed radiation, friction, thermodynamic solenoids, the atmospheric water-vapor cycle, and the production and dissipation of kinetic energy to be ignored.

Considering the approximations used, the success of the Chicago school was remarkable. Its main achievement was that it proved, by means of the vorticity-conservation theorem, that large-scale features of weather and climate could be deduced from the inherent dynamic properties of the atmosphere, without reference to the variations in the Earth's surface. In 1940 the Chicago school proved that the mid-troposphere was dominant in control of weather. This line of thought culminated in the concept of "thermal steering" developed by the British meteorologist R. C. Sutcliffe between 1939 and 1947 as the outcome of his development theory, in which the wind, pressure, and temperature fields of the atmosphere were linked in terms of both hydrodynamics and thermodynamics. The result was a completely new approach to the study of both weather and climate, by means of *thickness patterns*.[31]

The field of *microclimatology*, the study of the climate of the lowest layers of the atmosphere, was first outlined in detail by Geiger in 1927, but its development did not begin until the Second World War. Physical techniques could be applied here; solution of many practical problems in agriculture, forestry, biology, air pollution, and chemical warfare provided an immediate goal; and progress was rapid. Outstanding contributions in this field have been made by Thornthwaite, the British meteorologists Sutton and Pasquill, and the Australian meteorologist Priestley. Microclimatology showed the way climatology as a whole was to develop in the future, as a broad field in which the work of many different types of scientist overlaps, not as the province of any one group of workers.[32]

By 1942, weather data on a "probability risk" basis for months, even years, ahead had become necessary for planning military operations, and clearly could not be provided by the day-to-day forecasting techniques of the synoptic meteorologist. In order to provide such data, W. C. Jacobs in the United States began developing *synoptic climatology*. He characterized it as both a practical and a research field, concerned with the description and explanation of local climates in terms of the large-scale atmospheric circulation. In Britain, C. S. Durst also recognized that climate is the synthesis of weather, and that the purpose of synoptic climatology is to describe the average weather of a given locality in a given synoptic situation.[33]

Thornthwaite in 1948 published a completely new classification of climates, based on a concept from physics, that of potential evapotranspiration.[34] For the first time, a rational, scientific basis for classification was available, but the classification was complex, and indicated that the climates of no two points on the Earth's surface are identical. At their best, climatic classifications provide only pigeonholes for convenient storage or presentation of geographical information; they do not come to grips with the real problems of climatology.

Giao in 1949 defined the object of dynamic climatology to be deducing the mean properties of the atmosphere, and the mean atmospheric circulation over specific areas, from hydrodynamic theory, the fixed influences of the Earth's surface—its heat sources (warm ocean currents) and sinks (large ice caps), and its varying degrees of aerodynamic roughness, wetness, and reflectivity (albedo)—being incorporated into the mathematical models. This view of the scope of dynamic climatology concurs with the current opinion of meteorologists.[35] That there is need of a rigorous dynamic climatology is shown by the fact that, even after a hundred years or more of discussion, it is still impossible to decide by dialectic methods whether the last ice age was caused by the sun's being hotter (or cooler) than it is today.[36]

Theoretical climatology, termed *Climatonomy* by Lettau in 1950, has as its object the precise, inductive explanation of climatic phenomena. Commencing with basic concepts of local energy balance over simple surfaces, it builds these up into theoretical models of heat sources and sinks on a planetary scale (*energetic climatology*), and finally integrates these models into an explanation of the general circulation of the atmosphere (*circulation climatology*). How significant its techniques are for even elementary climatological study is shown by the discovery that monsoons are dynamic features of the atmosphere, and that explanation of them in terms of thermal convection is unnecessarily complicated.[37]

The business of refining and improving climatic maps is still going on today. The main standard sources before the production of modern national climatic atlases to World Meteorological Organization specifications[38] were Bartholomew and Herbertson's *Atlas* and Shaw's *Manual of Meteorology*. Modern climatic maps tend more and more to take processes into account. Statistical averages of raw data are not really meaningful in map form, because a map showing the geographical distribution of one element is really exhibiting the sum total of different distributions produced by several meteorological processes. Hence the trend now is to concentrate on maps showing the observed results of definable meteorological processes, for example, heat and moisture balance. One reason such maps are needed is that the standardization of radiation-measuring instruments lagged behind that of other meteorological instruments because of the complexities involved, and because the theoretically calculated radiation distributions could not be compared with observed distributions until 1955, when Budyko published maps for the world and Houghton, in more detail, for North America.[39]

Early descriptions of the geographical pattern of weather and climate were purely verbal or cartographic. Modern descriptions tend to be increasingly statistical, sometimes employing quite refined mathematics, and their results, although absolutely essential for climatology, are not of obvious interest to persons who wrongly think that climate is a simple phenomenon. A good example of this trend is provided by the study of climatic change. Early studies of climatic change were based on simple qualitative inspection of long observational series, for example, of Stockholm temperatures by Ehrenheim in 1824 and of Edinburgh temperatures by Buchan in 1867. Later studies have ranged from studies involving literary research and simple statistics to detailed statistical investigations by power-spectrum techniques which require computers. These rigorous mathematical investigations have shown that many of the early conclusions were erroneous.[40]

There have, of course, been numerous derivative descriptive accounts, of which that of Kendrew is the best-known.[41] Unfortunately, this type of work has been identified with climatology by many readers, although it is only climatography. Descriptive climatography is the task of encyclopaedists or lexicographers, not scientists, however useful or interesting its results may be. The tendency among climatologists today is to avoid such description, if for no other reason that that it has been all but completed.

The Education of the Climatologist

Because climatology is a science, the climatologist should be educated as a scientist. Only a few curricula for climatology in the English-speaking world make a conscious effort both to produce climatological scientists, and to give due weight to both the meteorological and the geographical side of the science.[42] Training in climatology in the western world has usually taken place in four types of organization: in the meteorology departments of universities, particularly in the United States; in the geography departments of universities; in various agricultural, forestry, biological, and medical colleges; and in government meteorological departments. Almost none of these organizations provide a really satisfactory climatological education.[43]

Unfortunately, the brilliance of the Bergen school has had a bad influence on climatological education. The basis of the Bergen approach was the belief that weather can be completely explained by atmospheric pressure and its derivatives. This idea was taken up and extended by Petterssen in his classic manual on *Weather Analysis and Forecasting*.[44] Publicists and popularizers of the Polar Front theory are responsible for the widespread but quite erroneous belief that frontal and air-mass analysis *is* meteorology, instead of being only one particular model that meteorologists and climatologists employ. The Polar Front theory was a boon to the writers of elementary-school textbooks, and most people are familiar with the daily synoptic chart, but very few know how much mathematics and physics goes into making that simple-looking chart.

Although, to the uncritical, the daily weather map may appear to provide enough material for understanding the geographical pattern of weather, it does not. The features on which the map is based—pressure, winds, temperature, humidity, clouds, and so on—are themselves part of the weather, and the argument is circular. The scientist would much prefer his daily maps to portray such fundamental quantities as energy, momentum, and inertia, because these are the ultimate terrestrial quantities of which the weather is the objective manifestation, and well-tried techniques are available for dealing with these quantities. Atmospheric science can not yet completely explain weather from such basic considerations, and so dispense entirely with the empirical synoptic chart, but it can at least apply concepts from physics that would render the chart less empirical. Some of the broad-scale geographical patterns of weather can now be explained at least qualitatively without recourse to synoptic charts.

EXPLANATION AND SYNTHESIS IN CLIMATOLOGY

In modern weather science, weather is not considered "explained" unless it has been shown to be a logical consequence of certain well-established properties of the atmosphere, e.g., its density, weight, heat content, and energy balance. The meteorologist as a scientist, as distinct from the professional weather forecaster, will have achieved one of his ultimate aims if, by logical deduction from the atmosphere's inherent properties, he can establish global weather patterns that correspond perfectly with the observed patterns. Establishing such patterns is an immense scientific problem; its solution is entirely feasible on a regional, chorographic basis, but the minute variations of actual weather and climate which the observations bring to light mean that solution on a local, topographic scale is not likely to be possible. This is the real justification for the existence of climatology as a distinct science, separate and apart from meteorology.

Complete scientific explanation—and hence prediction—of the world and chorographic patterns of weather would be of great practical use to mankind, even though such prediction would need to be supplemented by local "account book" checks of current weather by the existing regional forecasting offices, in order to add detail to the general forecasts. Explanation and prediction of instantaneous weather patterns is the job of the meteorologist, explanation and prediction of the mean weather pattern that of the climatologist. By comparison of "actual" with "predicted" conditions, the former on a local and the latter on a chorographic scale, the exact role of the surface features of the Earth, both large and small, in the formation of weather and climate could be assessed. Assessing their role in such formation is one of the standing problems of geography. It was expressed particularly well by H. R. Mill, himself both geographer and meteorologist, who regarded geography as the study of the effect of the form of the Earth's surface on all fluid distributions.[45]

Weather, of course, is an element of landscape (in the broadest sense); better, it is the "skyscape" that complements the "landscape" and completes our external physical environment. But the landscape, both physical and human, is also a factor in the formation of weather, and microclimatology clearly must take it into account. Local topography, with its variation in aspect, slope, soil type, altitude, vegetation, buildings, man-made surfaces, artificial pools, and so on, influences the local climate, and quite often the local weather. But the really interesting fact is that the processes which create a microclimate, i.e., radiation exchange processes when there is little or no wind, and turbulent eddying when the air is moving faster, also create weather systems on a vastly greater scale. The little gusts or eddies of wind we feel when the air blows past our faces have their analogies in the great cyclones and anticyclones that move across the face of the Earth, forming, increasing in size and strength, and then ultimately disappearing. Dust devil and hurricane, in spite of their widely different scales, have the same turbulent nature; it is clearly feasible to hope that the physical theory of *turbulence* should be able to explain them both. The explanation is very difficult, in terms of both the observations and the mathematics, but the success so far achieved indicates that the theory of turbulence is indeed one link between microfeatures and macrofeatures of weather and climate.

One result of weather observations since the Second World War is that it is now possible to study weather patterns globally or at least hemispherically. The atmosphere is a whole; a change in the weather at one point on the Earth's surface may be due to processes operating in the atmosphere thousands of miles away. This is proved in weather-analysis offices every day, and it is now recognized that to explain completely the weather at a particular place at a particular time, it may be necessary to trace back to processes operating in the atmosphere over areas of thousands of square miles on the other side of the world. This global approach has been made scientifically possible by C.-G. Rossby's discovery of the long waves in the upper atmosphere and by his application of the concept of vorticity conservation to the study of the atmosphere. With the introduction of Rossby's ideas, it became obvious that the upper air contains the real key to atmospheric processes, and hence surface weather, over a large area.

Thus climatologists are now concerned with two different levels of the atmosphere, and with phenomena on two different scales. On the macroscale, attention is focused on the whole atmosphere, from top to bottom, over a region at least the size of a continent. On the microscale, attention is focused on the few feet immediately above and below the ground, where the exchange of heat, moisture, and momentum takes place under the influences of the processes of radiation and turbulent diffusion. The traveling depressions and anticyclones are recognized as results of interaction between processes on these two scales; although such pressure systems are responsible for much (but by no means all) of our everyday weather, and are therefore useful in everyday forecasting, they are not the ultimate terrestrial causes of our weather.

Although the concerns with these two different levels of the atmosphere must ultimately merge to explain not only our weather and climate but also our atmosphere as a whole, such merging has not yet begun. The techniques useful for the many practical problems on the two different scales are so specialized that, in practice, they are the province of different groups of workers; no climatologists are now really engaged in synthesizing the two areas of concern.

Even when the theory of turbulence is perfected mathematically, there will still be the problem of describing everyday weather synoptically in terms of it. Macrostudies are largely theoretical, the province of the dynamic climatologist and the long-range weather forecaster. Microstudies are mainly empirical, involving detailed, accurate observations covering a short period of time. The gap between the two is bridged effectively only when a case study is made of actual weather in some specific locale or region, for example, studies of the relationships between weather in North America and the northern hemisphere atmospheric circulation that appear regularly in *The Monthly Weather Review* of the U.S. Weather Bureau, Miller's account of San Francisco summer fogs, and Patton's study of the effect of snow cover on climate in the Sierra Nevada of California.[46] Despite the great value of these contributions (and they are very much better than the "climatic region" or even "air-mass climatology" approaches to regional weather description), they are not complete climatologies. The circulation studies are made by meteorologists who have been trained essentially in the dynamic approach and who have little interest in the microaspects of the balance between local meteorological processes in specific landscapes. The California studies are essentially topoclimatological and microclimatological, not much concerned with explaining empirical ideas by means of dynamic global concepts. Yet microfeatures and global concepts must both be dealt with if weather is to be described and properly explained.

The geographer has a great part to play in this desirable synthesis. He has always described weather as a differentiator or integrator of landscapes in the broad sense, but in recent years he has fought shy of synthesis because of the apparent divergence of modern meteorology from his particular interests. I hope to show in this book that this divergence is only a matter of techniques, not of subject matter, and that modern meteorological literature contains a vast amount of material of great value and interest to the geographer.

THE ANALYTICAL VIEWPOINT IN CLIMATOLOGY

Scientifically speaking, there are at least two ways of looking at nature. The empiricist—for example, the field geographer or geologist, or the naturalist—takes nature as it actually is (or, rather, as he thinks it is), catalogues and classifies the observations he makes, and, finally, accounts for the patterns he thus discovers by a process of simple logic. The theoretician—for example, the physicist, the astronomer, or the biochemist—uses observations (his own or the empiricist's) to set up a working model (experimental or mathematical) of the situation. Essentially, he conceives the observations to be evidence of the working of a system that he can describe by means of a mathematical structure. The system is a theoretical one, established by the processes of scientific inference, and based on axioms or self-evident principles.[47] By means of mathematical reasoning—i.e., pure deduction—he can then make predictions, in other words, derive information about events that have not yet been observed. The advantage of such an analytical approach to the climatologist will be obvious. Climate is not a matter of direct observation; it must be inferred. To state that such-and-such an area has such-and-such a climate is to make a scientific prediction about the meteorological phenomena that will likely be observed in that area.

The climatographer's viewpoint is empirical; the climatonomer's approach is theoretical. The climatographer's method is to take all his meteorological data, and to analyze the figures statistically. He can then produce (*a*) a general climatological model constructed from long-period averages; and (*b*) a series of synoptic-climatological models, each depicting the average weather conditions associated with certain specified airflow types or general wind directions. The only general principle such models illustrate is the relationship between altitude and certain climatic elements. Their disadvantage is that they are all constructed from *averages*, which iron out the day-to-day irregularities so important to the student of weather. The pattern of, for example, rainfall for any one day will most probably not fit any one of these models accurately. One could conclude that the models do not describe features of nature, but are merely very convenient (and useful) ways of summarizing observations.

Furthermore, examination of a year's daily rainfall maps may show that on some days when the wind blew from, say, WSW, the rain fell in the same geographical pattern, a pattern that is completely unrelated to topography, and that clearly cannot be explained by the altitude-rainfall relation already established empirically. The pattern would be ironed out by the averaging, but it clearly exists, and should not be ignored. This is where the theoretical approach comes in, for concepts from physics show how under certain conditions, lee waves may cause air to rise in just the pattern needed to explain the rainfall.

Lee waves occur when air is forced to rise over a mountain ridge, and then performs similar rises and descents in the lee of the ridge, each one of which may produce the same "orographic weather" as the mountain. It is knowledge of the reaction of the airflow to this "invisible orography," plus knowledge of the transfer of heat, moisture, and momentum by atmospheric turbulence, that provides the basis for a really scientific study of the geographical pattern of weather. The simple relationships between topography and weather that could be deduced from synoptic-climatological models could not provide such a basis, as study of almost any sequence of daily weather maps will show. The empirical approach that such models depend on must assume that all movements of the atmosphere result from either orographic lifting or simple convection. But the meteorologists have proved from dynamic theory (i.e. mathematically) that there are numerous "properties" of the atmosphere that must be manifested as vertical and obliquely vertical air currents, given certain combinations of temperature, pressure, humidity, and airspeed. Therefore, certain weather patterns, such as those caused by lee waves, must depend on these properties. Since we know that these properties exist, and since we know that they must be manifested whenever the right combination of temperature, pressure, and airspeed occurs, we should not be surprised when the rainfall distribution they cause appears, and our analysis would be simplified accordingly.

The Elementary Facts of Weather and Climate

However one chooses to define weather and climate, there is a simple practical distinction between them. Weather is a matter of everyday experience, whereas climate is a statistical generalization of that experience. Thus if we know that two places, or two regions, differ in climate, then they must also differ in weather, but from the climatic details alone it would be impossible to decide just how the "weathers" would differ on any particular day. A survey of the variation of the climatic elements over the Earth can thus be a useful preliminary to a geographical study of weather patterns, if it is approached in the correct frame of mind.

To use a homely analogy, a person desiring weather information who consults maps or tables of climatic data is like one who wants to listen to a particular radio program, and consults a station log to find the station that will broadcast it. Weather, like the radio program, is the interesting reality, but to understand it, we must first "tune in" to the appropriate climate; otherwise we shall have no yardstick for comparison. In other words, climate is the "carrier wave" on which the irregular "signals" of weather are superimposed. We have to study the "normal" distribution of the meteorological elements before we can begin to understand the departures from "normality" which constitute weather.

Of all the climatic elements, *solar radiation* is the most basic. The intensity and duration of sunshine measures part of this radiation, but there are also invisible parts of the radiation that have not by any means been as closely studied. *Cloud cover* is possibly the next most important, because it determines both how much solar radiation reaches the ground, and how much is radiated back into space from the Earth's surface. *Temperature*, depending on radiation, influences cloud cover and controls humidity, and is obviously important in itself. *Winds* and *pressure* are interdependent, the latter representing the weight of the atmosphere, the former representing the air movement that ensues when adjacent vertical columns of the atmosphere have different weights. *Humidity* indicates the amount of water in vapor form in the atmosphere; *precipitation* measures the amount of water in its liquid or solid phase.

In addition to these primary elements, which, with the exception of precipitation, may be regarded as having continuous distributions in both space and time, the climatic elements also include various *weather systems*. These can be as complex and large as depressions and anticyclones, or as small as squall lines; they can release enormous amounts of energy, like monsoons, typhoons, and tornadoes, or can be purely local phenomena, like thunderstorms. All these weather systems may be regarded as disturbances of the mean motion (general circulation) of the atmosphere.

Apart from studying world distribution patterns of the various meteorological elements and weather systems, we can study the geographic patterns revealed by combinations of the elements and systems; these geographical patterns are clearly more important in the study of real weather and climate than the patterns of the elements themselves. However, much less work has been done on the combinations than on single elements. The best-known climate classification systems that use several elements, the Köppen and Thornthwaite systems, still do not take all the important ones into consideration.

Figures I.2 to I.5 present world maps of the average distribution of surface air

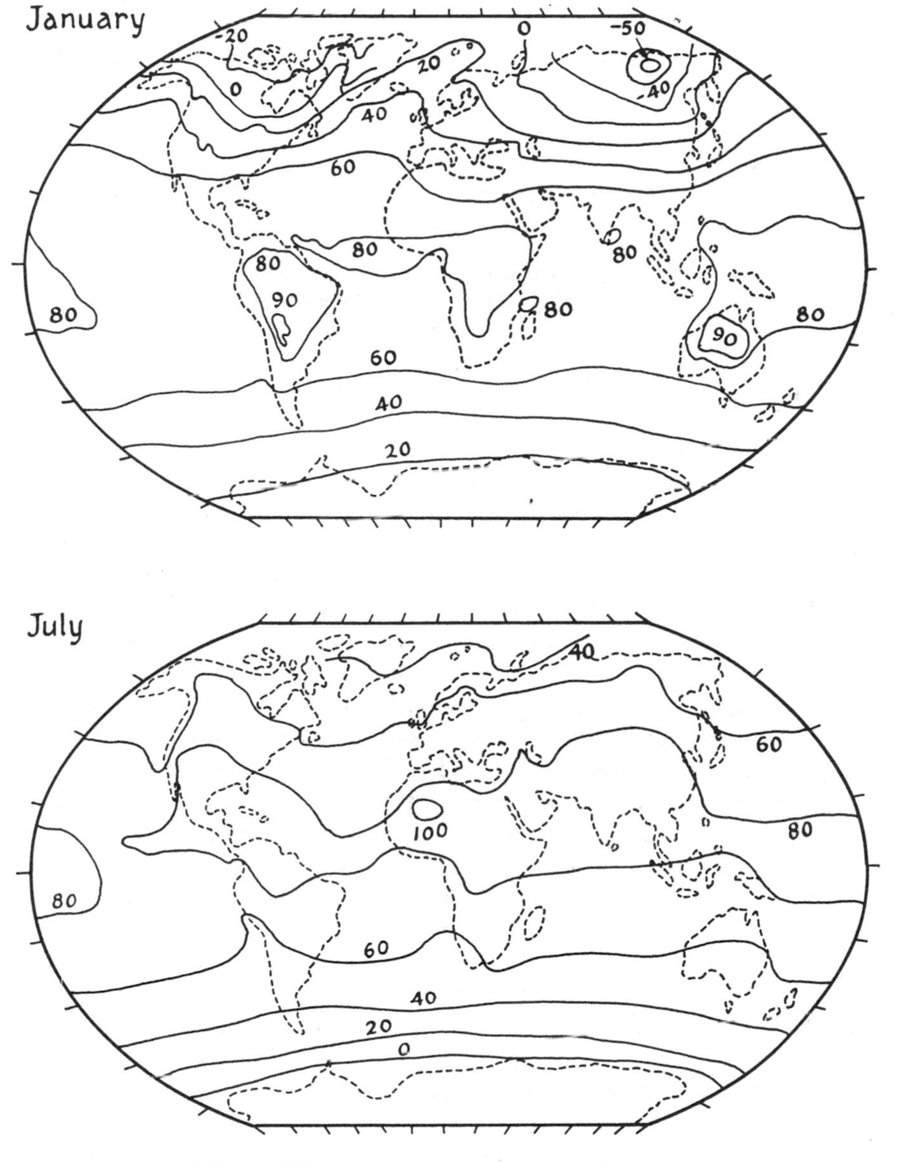

FIGURE I.2.
Annual variation in mean sea-level temperature (°F).

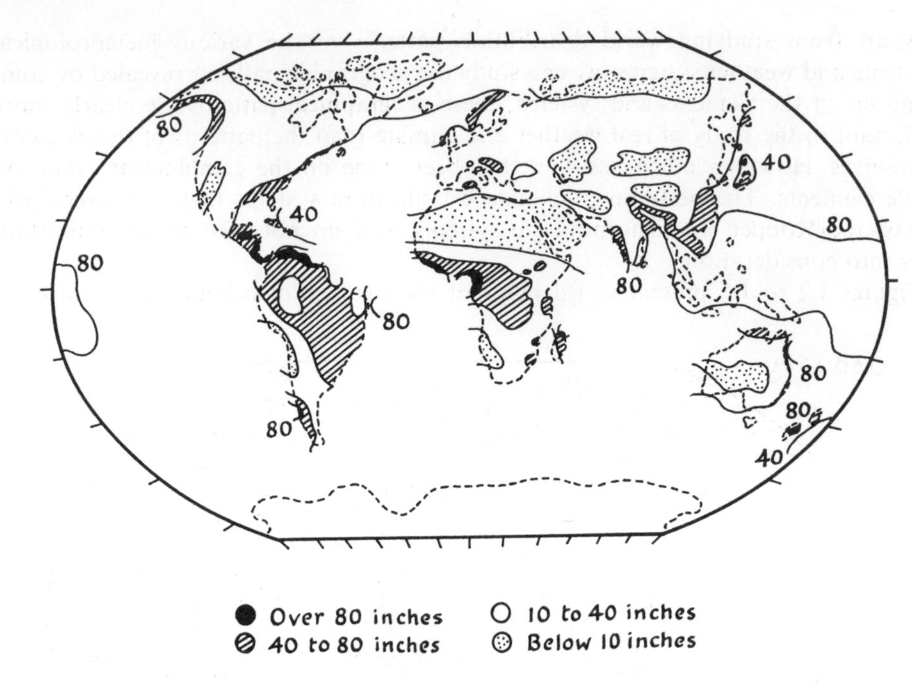

FIGURE I.3
Annual precipitation totals (after Landsberg)

temperatures,* precipitation, vapor pressure, and total solar radiation (direct plus diffuse). The maps were constructed on very different bases. Temperature maps are based on measurements of a continuous quantity with simple, reliable instruments, and can therefore be regarded as reasonably accurate, since interpolation between observing stations is relatively easy. Precipitation maps are based on imperfect sampling of a discontinuous quantity, and hence are subject to an error that is practically impossible to determine. Unlike temperature maps, therefore, they can be relied on for only relative information.

Humidity maps are based on vapor pressure, a quantity that cannot be measured directly, but must be calculated by physical theory from observations of wet- and dry-bulb temperatures and atmospheric pressure.[48] Maps of total radiation are indirect, based on the Savinov-Angström formula with parameters found from actual observation.[49] No simple radiation instrument has yet been invented that can successfully integrate all the known factors that determine how much solar radiation reaches the ground. Solar radiation covers so many wavelengths that no one instrument can respond to all of them.

Maps of resultant winds (see Figure I.6) show a statistical concept, necessary because the apparently simple phenomenon we call "wind" has two attributes, direction and strength, which cannot be combined in a simple measurement without elaborate instrumentation. The usual compromise solution, separate maps of wind direction and windspeed, the former showing wind roses and the latter isopleths of windspeed, is not

---

* I.e., temperatures recorded by thermometers mounted 3 to 4 feet above the ground under standard conditions of exposure.

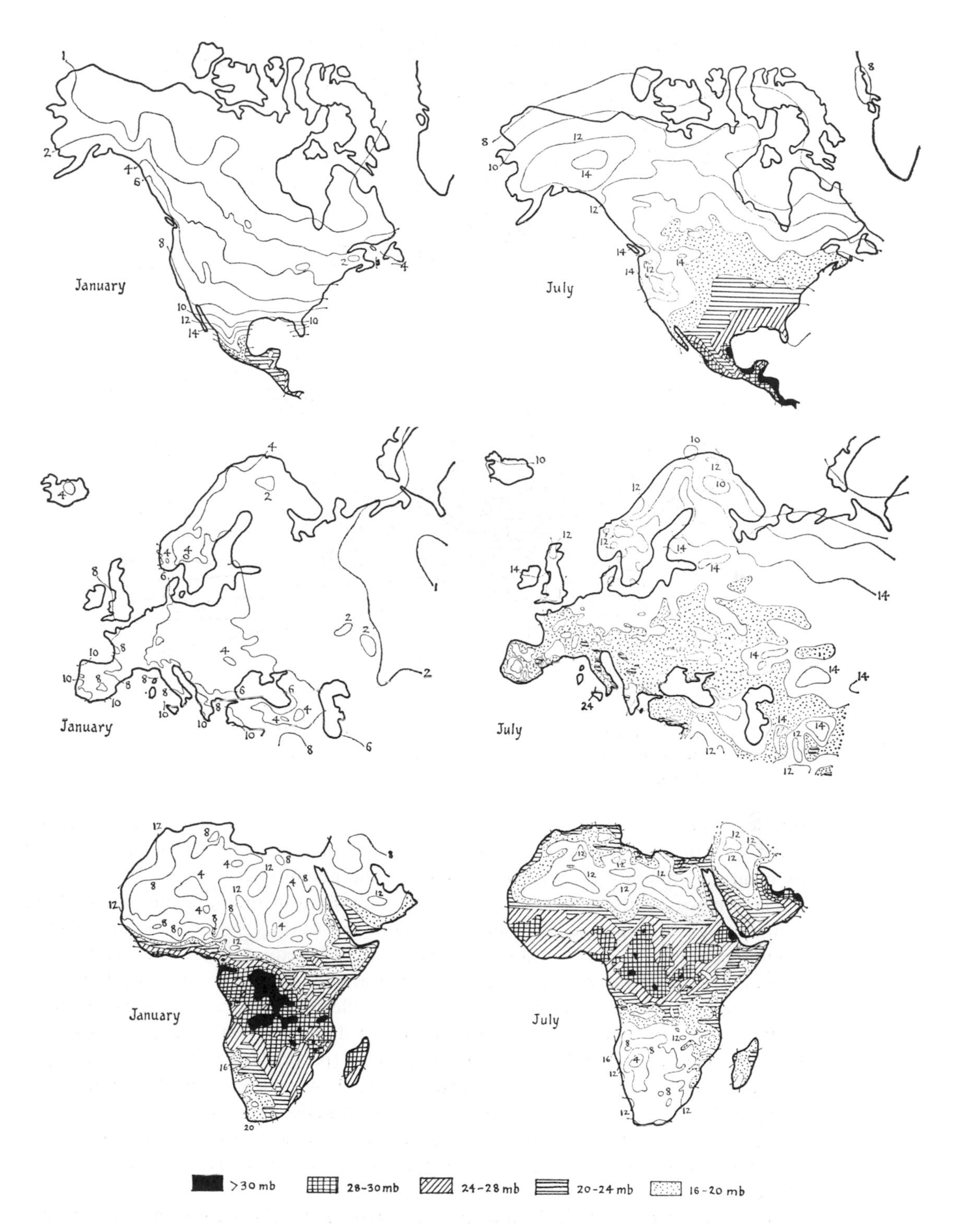

FIGURE I.4.
Mean monthly vapor pressures, reduced to mean sea level (after G. A. Tunnell).

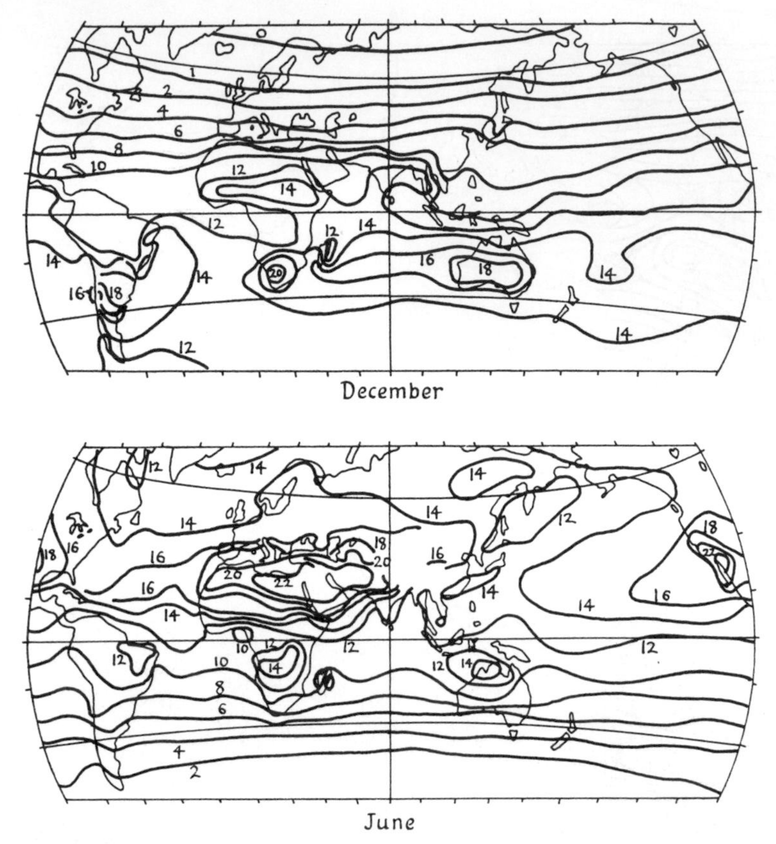

FIGURE I.5.
Total radiation received at the Earth's surface. Complex areas (e.g., high mountains) are omitted. Units are kg-cal per cm² per month. (After Budyko.)

very valuable because the two quantities cannot be combined as they are in nature. Wind is a form of momentum, and it is incomplete to speak of the movement of momentum without specifying both direction and magnitude.

These maps are also one-sided, in that they only show incoming quantities. Without certain physical quantities moving away from the Earth's surface, the atmosphere would cease to function as it does. No world maps based on direct measurement are available showing the broad pattern of, for example, the return of invisible radiant energy to the atmosphere and to space (i.e., the reflection of ultraviolet solar radiation by the atmosphere and the emission of infrared radiation by both the atmosphere and the Earth's surface), and the return of latent energy (particularly water vapor, in evapotranspiration) to the atmosphere. Furthermore, these processes are functions of so many different variables that it is doubtful if small-scale maps, showing their average global values, would have any significance, but they certainly must be taken into account—usually on a local or regional basis—in the study or explanation of climate.

It is obviously very difficult to combine analytic maps like these with their very different properties, but there is no avoiding the fact that climate is an integration of weather elements. Köppen's method of integration was to let nature do the synthesis,

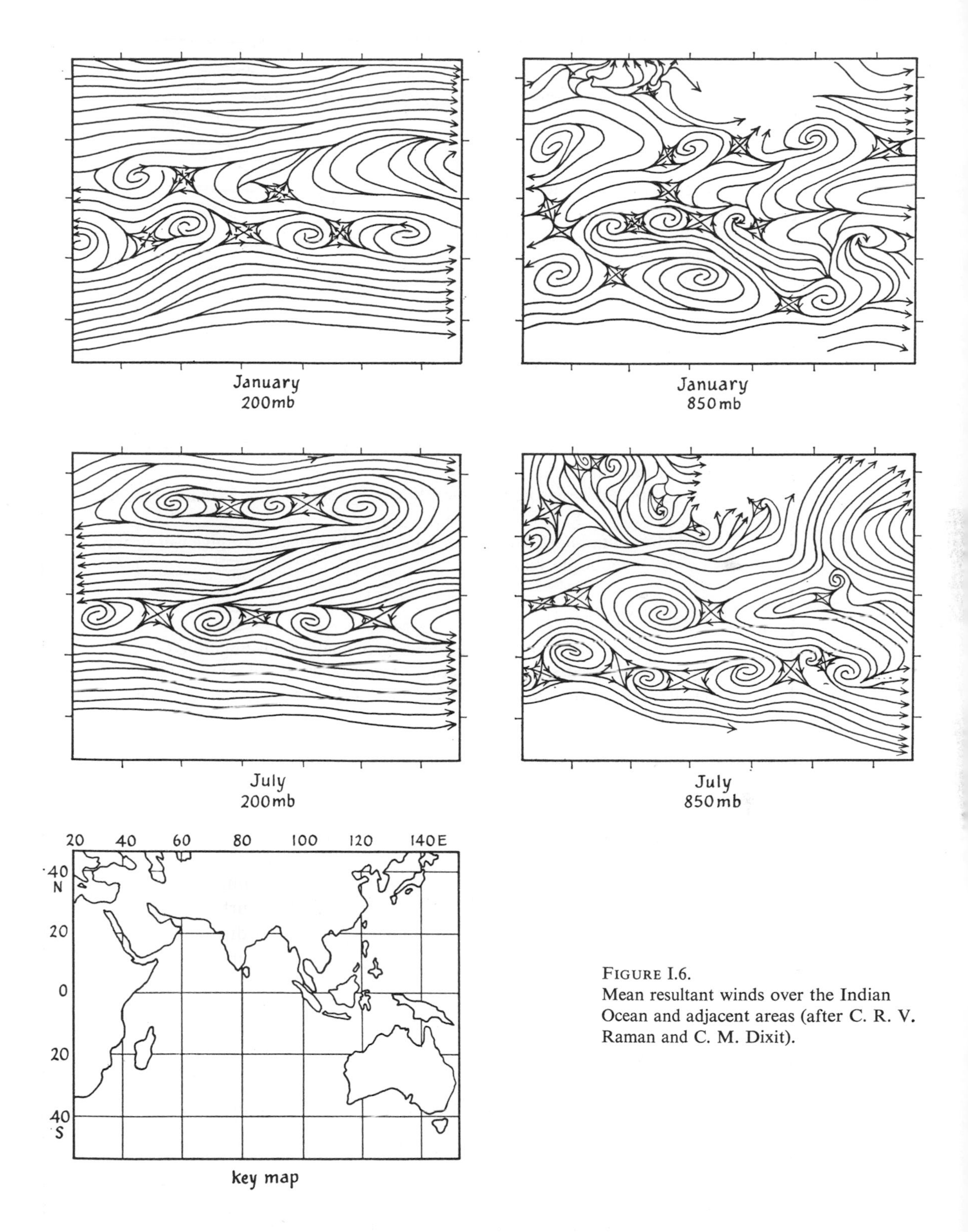

FIGURE I.6.
Mean resultant winds over the Indian Ocean and adjacent areas (after C. R. V. Raman and C. M. Dixit).

Köppen's Classification of Climates

Main Climatic Groups

Tropical rainy climates	*A*
Dry climates	*B*
Warm temperate rainy climates	*C*
Cold snow forest climates	*D*
Polar climates	*E*

"A" Climates: Tropical Rainy

Temperature of the coldest month greater than 64.4°F (18°C).

SUBDIVISIONS

Af —Regions with no dry season. At least 2.4 inches (6 cm) precipitation in driest month.

Aw—Distinct dry season. One month with precipitation less than 2.4 inches (6 cm).

Am—Monsoon. Short dry season, but sufficient moisture to give wet ground throughout the year.

"B" Climates: Dry

Arid regions where evaporation exceeds precipitation. Distinguished from A, C, and D climates by combination of precipitation and temperature. Climate is dry if rainfall is less than $0.44t - 8.5$, where t is annual mean temperature in °F.

SUBDIVISIONS

BS —Steppe.

BW—Desert. Annual rainfall less than $\frac{0.44t - 8.5}{2}$.

Limits for steppe and desert climates may need adjustment according to distribution of precipitation.

Precipitation	Arid/nonarid boundary	BW/BS boundary
Maximum in summer	$r = 0.44t - 3$	$r = \frac{0.44t - 3}{2}$
Maximum in winter	$r = 0.44t - 14$	$r = \frac{0.44t - 14}{2}$

FURTHER SUBDIVISIONS

h—Subtropical deserts with average temperature greater than 64.4°F.

k—Middle-latitude deserts.

"C" Climates: Warm Temperate Rainy

Average temperature of coldest month less than 64.4°F and greater than 26.6°F. Average temperature of warmest month over 50°F.

SUBDIVISIONS

Cf —At least 1.2 inches precipitation in driest month, difference between wettest month and driest month less than for Cw and Cs.
Cw—Winter dry season. At least ten times as much precipitation in wettest month of summer as in driest month of winter.
Cs —Summer dry season. At least three times as much rain in wettest month of winter as in driest month of summer, the latter having less than 1.2 inches precipitation.

"D" Climates: Cold Temperate Snow

Average temperature of warmest month above 50°F and that of coldest month below 26.6°F. Chief divisions, Df and Dw, as defined for C climates.

"E" Climates: Polar

Average temperature of warmest month under 50°F.

SUBDIVISIONS

ET—Tundra. Average temperature of warmest month above 32°F.
EF—No month with temperature above 32°F.

Further subdivisions which may be found

a —Hot summer. Average temperature of warmest month over 71.6°F.
b —Cool summer. Average temperature of warmest month below 71.6°F.
c —Cool, short summer. Less than four months over 50°F.
d —Average temperature of coldest month below − 36.4°F.
H —Polar climate due to high altitude.
h —Hot, dry climate.
g —Ganges subtype. Hottest month before summer solstice.
i —Isothermal subtype. Annual range of temperature less than 9°F.
k —Cool, dry climate (middle-latitude deserts).
k′ —Temperature of warmest month under 64.4°F.
m —Mixed or monsoon type of tropical climate.
n —Dry climate with frequent fog.
n′ —Dry climate with high humidity.
w′—Rainfall maximum in autumn.
w″—Two distinct rainfall maxima.

Example of Classification

The British Isles: Cfb in the south, Cfc in the Scottish Highlands.

Thornthwaite's First Classification of Climates

The scheme is based on the following factors.

(a) Precipitation effectiveness ratio: P/E ratio $= \frac{P}{E}$, where $P =$ total monthly precipitation, $E =$ total monthly evaporation (inches).
(b) Precipitation effectiveness index: P/E index = sum of the twelve P/E ratios for the year.
(c) Thermal efficiency ratio: T/E ratio $= \frac{T - 32}{4}$, where T is the average temperature for a particular month (°F).
(d) Thermal efficiency index: T/E index = sum of the twelve T/E ratios for the year.

Determination of P/E index for stations without evaporation data:

$$P/E \text{ index} = \sum_{1}^{12} 115 \left(\frac{P}{T - 10}\right)^{10/9}.$$

Humidity Provinces	P/E index	Characteristic vegetation
A (wet)	128 and above	rainforest
B (humid)	64–127	forest
C (subhumid)	32–63	grassland
D (semiarid)	16–31	steppe
E (arid)	under 16	desert

r = rainfall abundant in all seasons, s = summer dry season, w = winter dry season, d = rainfall deficient all year.

Temperature Provinces	T/E index
A′ (tropical)	128 and above
B′ (mesothermal	64–127
C′ (microthermal)	32–63
D′ (taiga)	16–31
E′ (tundra)	1–15
F′ (frost)	0

There are 120 possible combinations of P/E, seasonal rainfall concentration, and T/E, theoretically, but only 32 occur on Thornthwaite's original world map.

Example of Classification

The British Isles: BC′r on the whole, but CC′r in East Anglia.

by using vegetation zones. The disadvantage of this procedure was that the vegetation boundaries were imperfectly known in Köppen's time, and the resulting climatic divisions were purely conventional, and not in any sense functional.

Thornthwaite's first classification took evaporation into account, whereas Köppen used only temperature and precipitation. His moisture categories were based on an empirical relationship established in the arid southwest United States, and there is no guarantee that it applies elsewhere. The second Thornthwaite classification (see Figure I.7) was based on a physical concept, the total evaporating power of the air when an unlimited supply of moisture is available, which must directly control vegetation growth. The disadvantage of this scheme is that the soil-moisture storage at any place must be known before the potential evapotranspiration value can be interpreted, and certain physical assumptions must be made if maps of the former quantity are to be constructed. The resulting climatic categories are extremely numerous—as indeed climates are in nature—and it is impossible to produce a simple world map based on them.

An alternative approach to climatic integration on a global scale is to base a system of classification on the atmospheric circulation, which is obviously an important determinant of climatic differences. Figure I.8 is produced on this basis. The obvious disadvantage is that circulation classification cannot differentiate local climates.

A major drawback to any attempt at global climatic classification is that it presupposes climates to be stable phenomena, which is by no means the case. The only stability shown by climates is in their "normal" range of fluctuation. Figure I.9 illustrates the range of fluctuation of some of Köppen's climatic boundaries. More impressively, Figure I.10 demonstrates deviations from the thirty-year "normal" pressure patterns, and Figure I.11 shows some deviations from mean monthly patterns. Clearly, these deviations are the really significant features for both scientific and practical purposes.

Study of Figures I.11 to I.13 inclusive shows that there must be a phase of climatology rather analogous to the investigation of daily weather situations in meteorology, the difference being that the patterns studied are five-day, thirty-day, thirty-year, hundred-year, or almost any other, instead of patterns for only one day. This phase of climatology is the key to studies of regional and world climate, and is based on fundamental concepts of dynamic meteorology, as well as on averages of synoptic observations.

An entirely different phase of climatology is the study of local climate, in particular, of *topoclimate*, the climate of a specific place.[50] Basically, this study need not involve many synoptic considerations, but it does require the use of theoretical concepts from micrometeorology and physical meteorology. Since accurate measurements are required of many climatic elements not usually included in either the synoptic or the climatological networks, it also demands a fair knowledge of problems of instrumentation.

Concentrating attention on a small area, say, one square mile, involves making climatic measurements under nonstandard conditions of exposure. The local geographical patterns of temperature, humidity, and so on, will change with wind direction, and the microclimatic vertical profiles of heat, moisture, and air movement will differ above such landscape features as woods, meadows, arable fields, or small gardens. It is therefore very difficult to produce accurate topoclimatic maps, and many such maps are necessary to describe adequately the climate of a small locality.

On the smallest scale of all, it is quite possible to produce accurate maps of

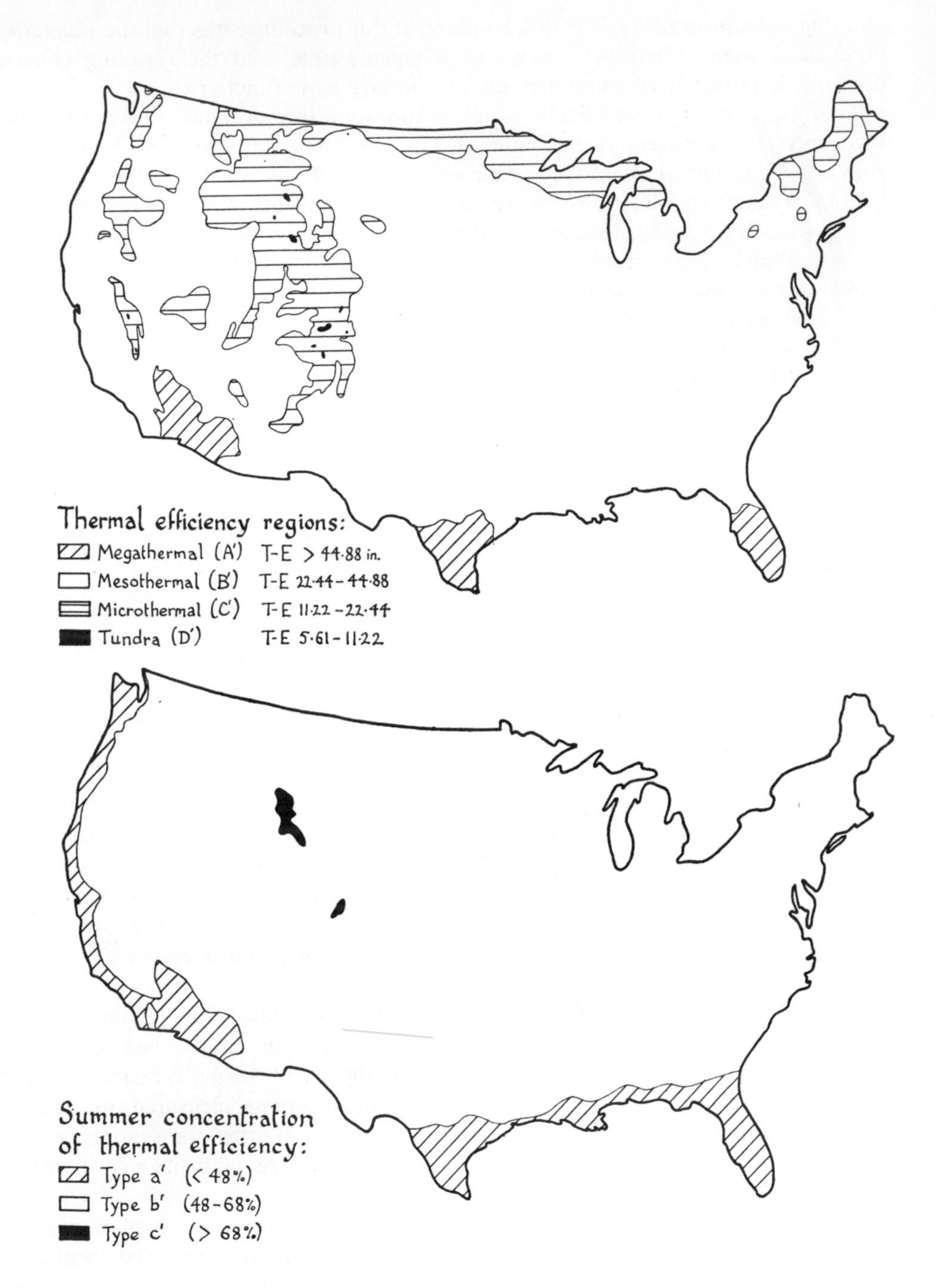

FIGURE I.7.
The climatic regions of the United States according to the 1948 Thornthwaite system (after C. W. Thornthwaite).

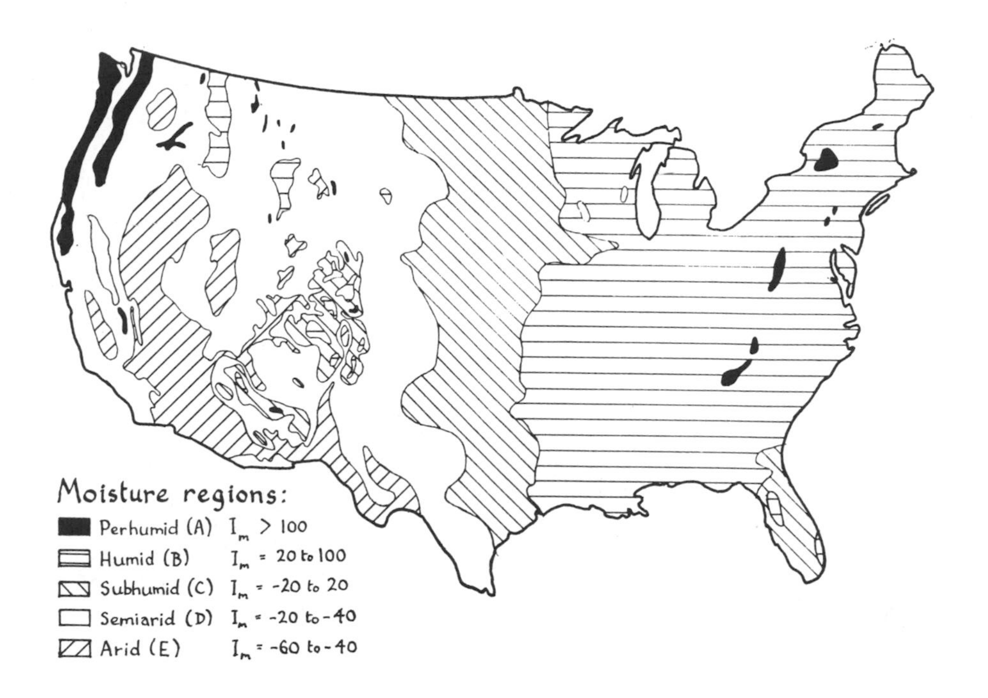
Moisture regions:
Perhumid (A) I_m > 100
Humid (B) I_m = 20 to 100
Subhumid (C) I_m = -20 to 20
Semiarid (D) I_m = -20 to -40
Arid (E) I_m = -60 to -40

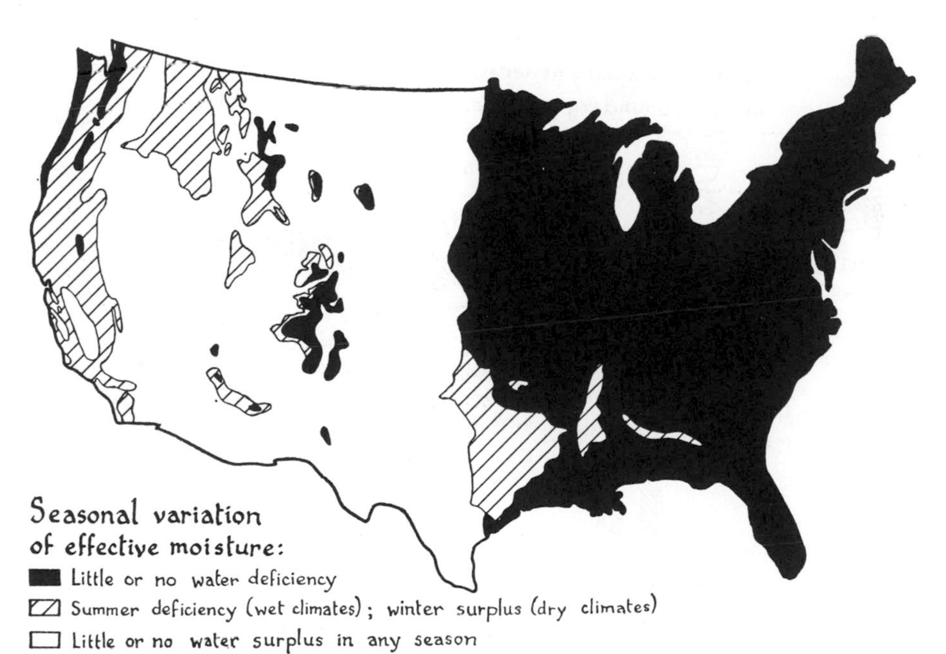
Seasonal variation
of effective moisture:
Little or no water deficiency
Summer deficiency (wet climates); winter surplus (dry climates)
Little or no water surplus in any season

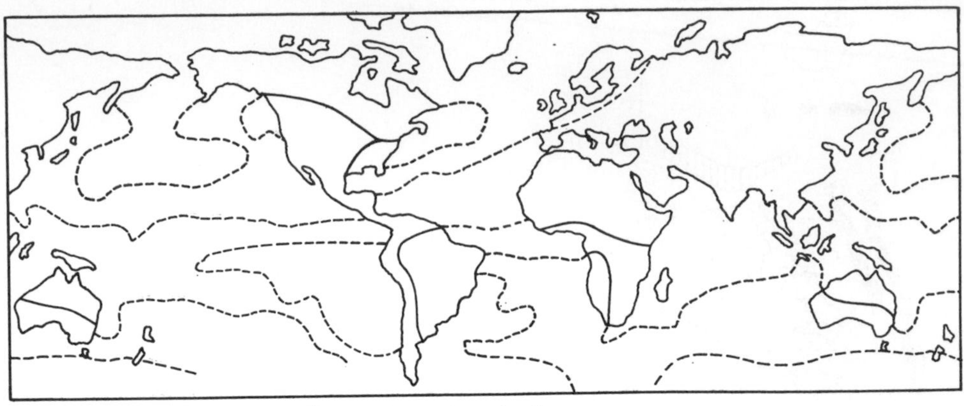

Dynamic regions in January

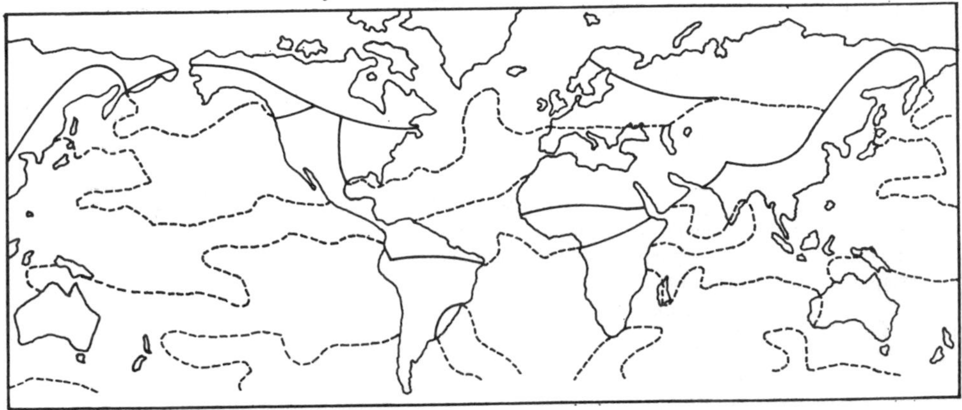

Dynamic regions in July

——— Regional boundary generated by airstreams from unlike sources.

- - - - Regional boundary generated by divergence of airstreams.

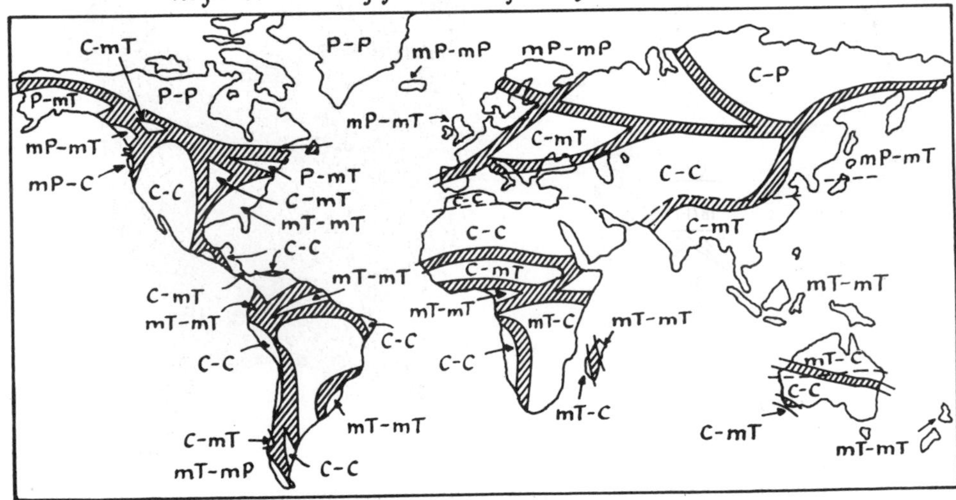

Major static climatic regions

//// Major air mass boundary on continents. Air masses: first letter indicates type of air mass prevailing in January, second letter, type prevailing in July.

--- Equatorial margin of winter westerlies.

FIGURE I.8.
Climatic regions, as defined in terms of the atmospheric circulation (after J. R. Borchert).

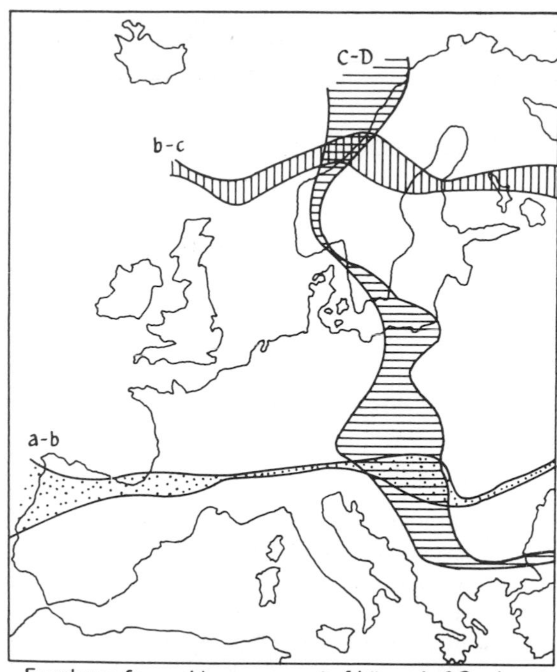

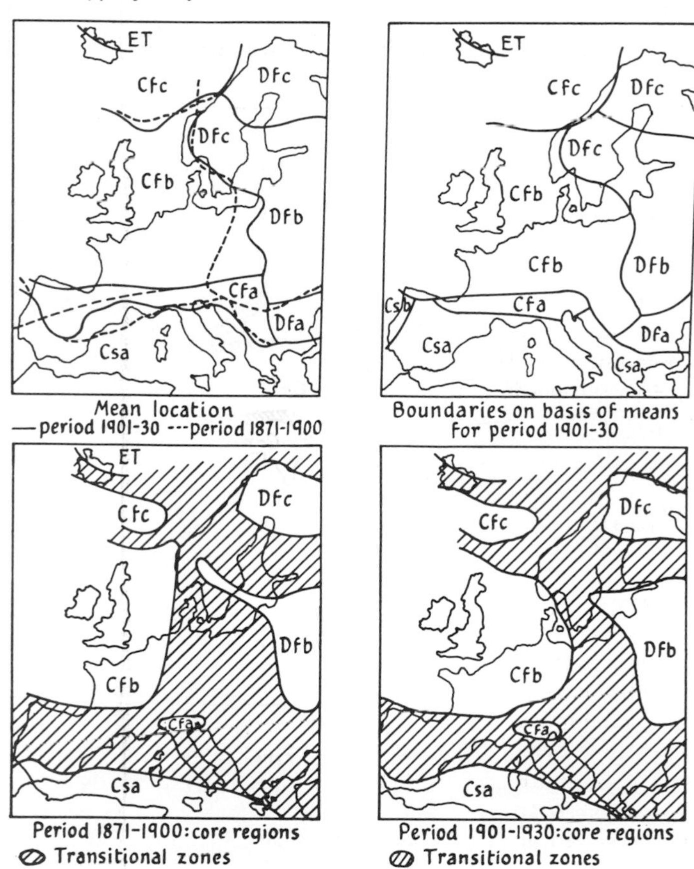

FIGURE I.9.
Dynamic nature of Köppen's climatic regions. Maps B, C, and E are based on "climatic years," i.e., on annual mean values for each year during periods specified. (After S. Gregory.)

Thornthwaite's Second Classification of Climates

The scheme is based on the following factors:

(a) Average annual potential evapotranspiration (i.e., water need) expressed in terms of a moisture index,

$$I_m = \frac{1}{e}(100s - 60d),$$

where e is the water need, s is the water surplus, and d is the water deficiency; e is determined from

$$e = 1.6\,(10t/I)^a$$

where t is the mean monthly temperature (°C), a is a geographical constant, I is the heat index, and e is in centimeters per month. A water surplus occurs whenever precipitation exceeds potential evaporation. A water deficiency occurs whenever potential evaporation exceeds precipitation, provided the effect of water stored in the soil is allowed for.

(b) Heat index, I, determined as the sum over the twelve months of the monthly heat index, i, defined by $i = (t/5)^{1.514}$, where t is again mean monthly temperature.

(c) Thermal efficiency, defined in terms of a T/E index that in practice equals e.

Moisture Provinces	Moisture Index, I_m
A —Perhumid	100 and above
B_4—Humid	80–100
B_3—Humid	60–80
B_2—Humid	40–60
B_1—Humid	20–40
C_2—Moist subhumid	0–20
C_1—Dry subhumid	−20 to 0
D —Semiarid	−40 to −20
E —Arid	−60 to −40

Temperature Provinces	T/E index (inches)
E'—Frost	Less than 5.61
D'—Tundra	5.61–11.22
C'_1—Microthermal	11.22–16.83
C'_2—Microthermal	16.83–22.44
B'_1—Mesothermal	22.44–28.05
B'_2—Mesothermal	28.05–33.66
B_3—Mesothermal	33.66–39.27
B_4—Mesothermal	39.27–44.88
A' —Megathermal	More than 44.88

SUBDIVISIONS

(i) For moist climates (i.e., moisture provinces A, B, and C_2), the subdivisions are defined in terms of an index of aridity, $I_a = 100d/e$.

	I_a
r = little or no water deficiency	0–16.7
s = moderate summer water deficiency	16.7–33.3
w = moderate winter water deficiency	16.7–33.3
s_2 = large summer water deficiency	more than 33.3
w_2 = large winter water deficiency	more than 33.3

(ii) For dry climates (i.e., moisture provinces, C_1, D, and E), the subdivisions are defined in terms of an index of humidity, $I_h = 100s/e$.

	I_h
d = little or no water surplus	0–10
s = moderate winter water surplus	10–20
w = moderate summer water surplus	10–20
s_2 = large winter water surplus	more than 20
w_2 = large winter water surplus	more than 20

Example of Thornthwaite's Second Classification

Salisbury, N.Y., $AC_2'rb_1'$.

FIGURE I.10.
'Normal' pressure maps for various periods. Each map represents mean conditions for a 30-year period: 1825 represents 1811–1840, etc. (After Fay.)

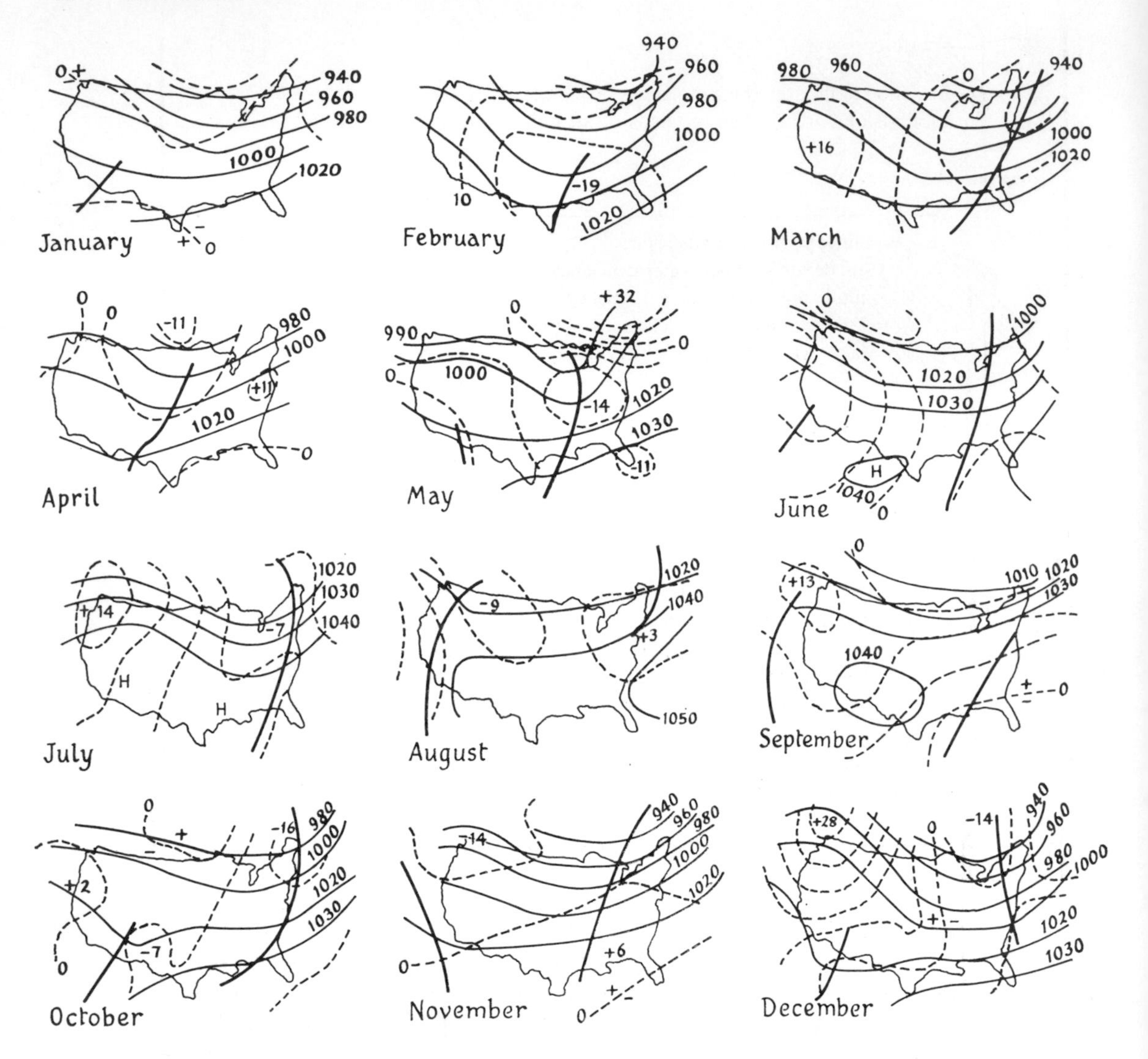

——— Contours of height of 700-mb pressure surface in tens of feet.

- - - - - Departures from normal of contours in tens of feet.

▬▬▬ Trough lines

FIGURE I.11.
Mean monthly 700-mb circulation patterns
for the United States in 1960 (after the U.S. Weather Bureau).

crypto-climate, the climate of enclosed spaces (Figure I.14), because the influence of extraneous elements can be eliminated or minimized by appropriate instrumentation. Such micro-geographical maps find many uses in applied climatology. (They are not, strictly speaking, microclimatological maps, since this term refers essentially to the *vertical* distribution of the climatic elements in the lower layers of the atmosphere.)

It will be obvious that no simple, all-inclusive geographical expression of climate is

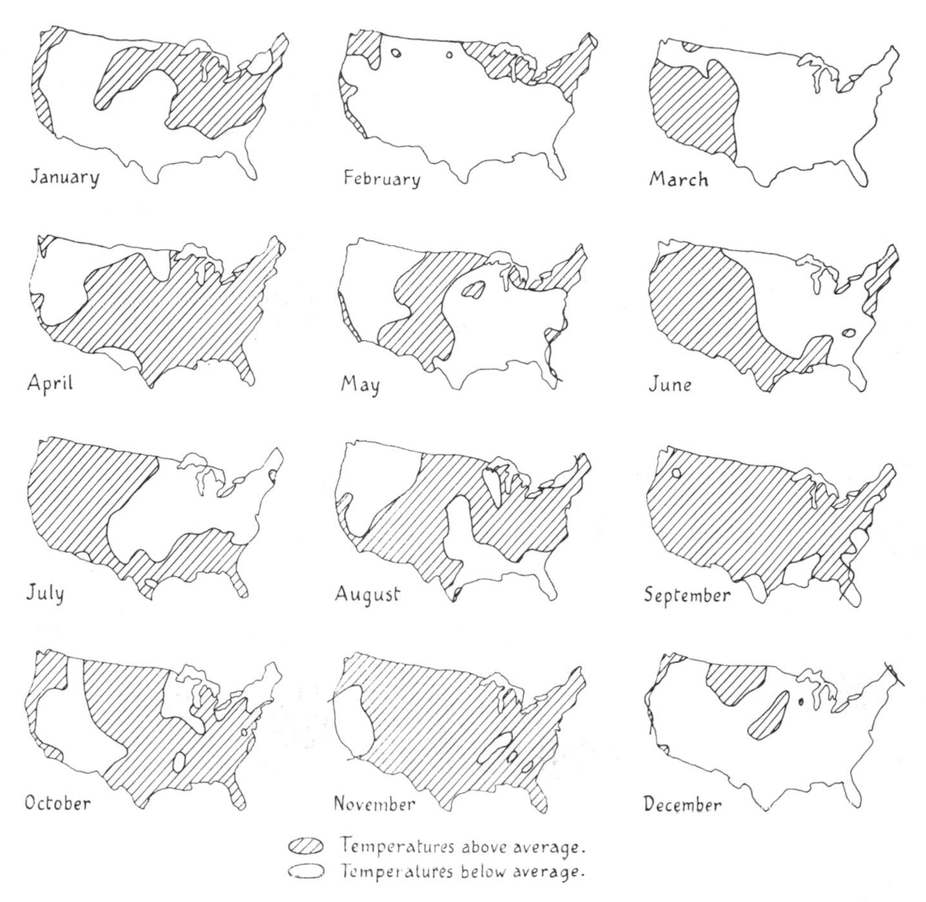

FIGURE I.12.
Departures from normal of monthly mean temperatures for the United States in 1960 (after the U.S. Weather Bureau).

possible at any scale. Only at the largest (macroclimate) and smallest (cryptoclimate) scales is it possible to produce a map that both depicts the distribution of a climatic element and is based on measurement of the element itself. At other scales, there are too many complicating influences, and the concept of the geographical distribution of an element is meaningful only in either an instantaneous or a synoptic sense. Some of these influences are geographical; these include especially the character of the landscape, and the local heat, moisture, and momentum sources and sinks, which affect the local radiation balance. Others are meteorological, the most important being the turbulence of the lower atmosphere, which affects the rate of both vertical and horizontal diffusion of atmospheric properties. Köppen presented a classification of macroclimate on a single map, but Thornthwaite later showed that four or even five maps are necessary to describe adequately just *one* of the relationships that Köppen assumed was simple.

The end-product of geography is, of course, description, which may be cartographic, verbal, or mathematical. Cartographic description in map form (for horizontal geographical distributions) or sectional or profile form (for vertical geographical distributions) is the most usual, and is taken here to be the normal form of geographical

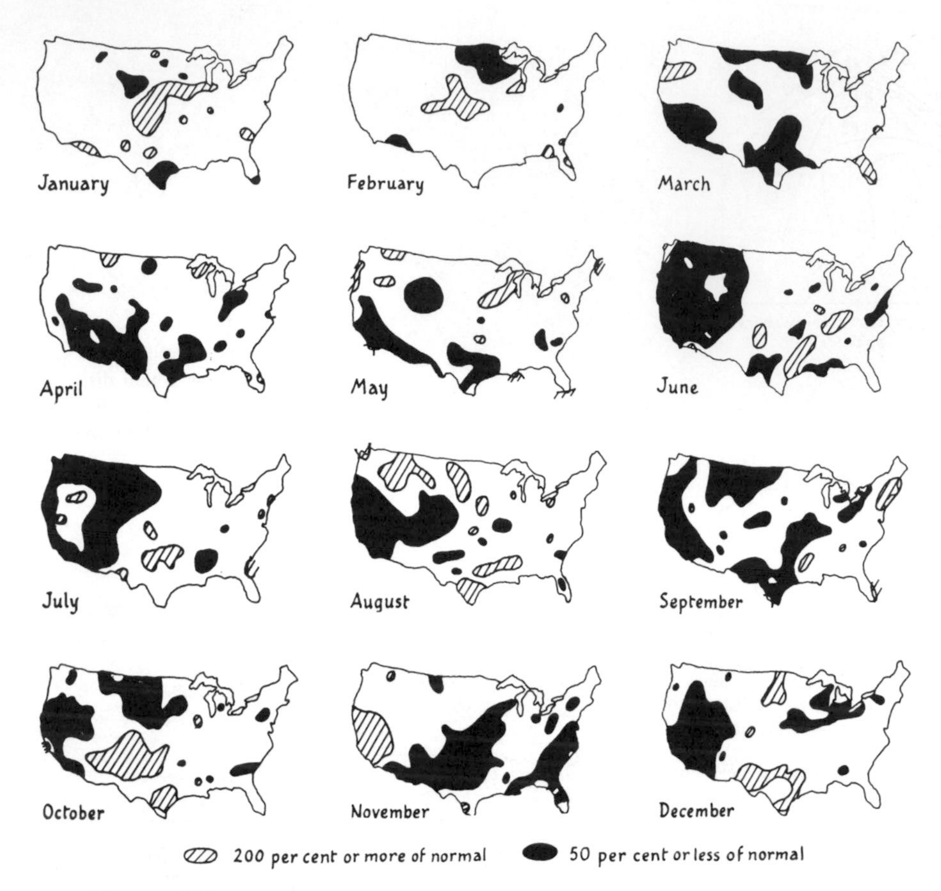

FIGURE I.13.
Departures from normal of monthly precipitation totals for the United States in 1960 (after the U.S. Weather Bureau).

expression. It should not be forgotten, however, that mathematical notation is the most precise and economical form of description, and that many of the mathematical relationships found in the appendixes to this book are really geographical descriptions.* The term *mapping* has a distinct meaning to the mathematician, and if the maps constructed by the climatologist or geographer are to be valid, they must be consistent with the properties of maps as demonstrated by the mathematician. In other words, although a climatic map seems to the empiricist to be a much easier and more attractive way to present climatic facts than the mathematical equation, this ease is misleading, for in fact, a map is only one form of mathematical structure, and a quite complex one at that.[51]

In conclusion, the functions of various types of scientist in the field of climatology should be considered. The physical scientists, including physicists, chemists, and physically oriented biologists, have the task of discovering the properties of the atmosphere,

* For example, the hydrostatic relationship $\partial p/\partial z = -\rho g$ perfectly describes the vertical stratification of pressure and density, and hence temperature, above the Earth's surface, in relation to gravity variations.

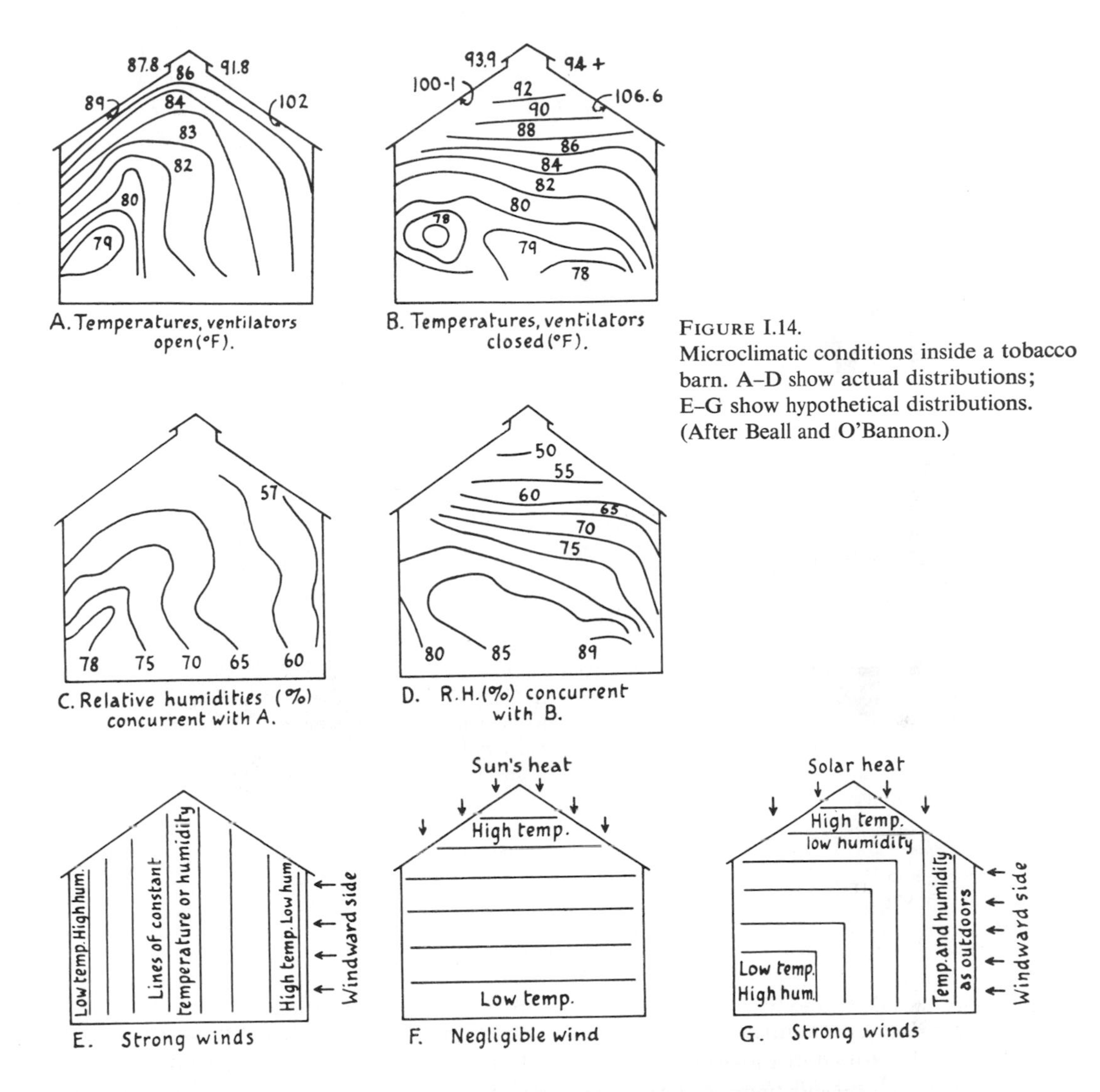

FIGURE I.14.
Microclimatic conditions inside a tobacco barn. A–D show actual distributions; E–G show hypothetical distributions. (After Beall and O'Bannon.)

and presenting these in mathematical models that depict idealized situations. Obviously these models will be refined and redesigned as progress is made in mathematics. The engineers—aeronautic, chemical, civil, electrical, mechanical, and so on—and the physiologists are occupied with specific problems in which meteorological influences are often important; it is to such workers that we may look for many of the "constants" in the theoreticians' equations. The same applies to forestry and agricultural scientists, systems-analysis specialists, and cyberneticists.

The professional "weather scientist," the synoptic meteorologist, has as his climatological function the task of building up a knowledge of how weather usually behaves in a given locality. In many cases, he will develop his own forecasting rules by a process of intuition, and these rules will require the labors of the theoretical meteorologist to explain them.

The aim of the geographer in the field of climatology is to assess the extent to which a specific phenomenon of weather or climate is tied to one locality or region and is thus a "unique" event, and to what extent it is purely dynamic, i.e., a consequence of an atmospheric property (or properties) being manifested at that place and time because of physical processes operating in the atmosphere and completely unrelated to the nature of the Earth's surface at the locality in question. In almost every meteorological event, there are elements of both dynamics and geography, but their relative importance varies with time.

Notes to Introduction

1. The definitions adopted are those in R. E. Huschke, ed., *Glossary of Meteorology* (Boston, Mass., 1959).

2. For an example of climatographic techniques, see *MC*, pp. 307-31.

3. A good example of modern physical and dynamic climatology is provided by *DC*; the best-known climatography is probably W. G. Kendrew's *The Climates of the Continents* (Oxford, 5th ed., 1961).

4. See W. C. Jacobs, *Wartime Developments in Applied Climatology*, *AMM*, vol. 1, no. 1 (1947); also H. E. Landsberg and W. C. Jacobs, "Applied Climatology," *CM*, pp. 976-90.

5. Basic sources on the history of climatology are: N. Shaw, *Manual of Meteorology*, vol. I (Cambridge, 1926); C. W. Thornthwaite and J. Leighly, "Status and prospects of climatology," *Sci. Monthly*, 57 (1943), 457; J. Leighly, "Climatology since the year 1800," *TAGU*, 30 (1949), 658; J. Leighly, "Climatology," in P. E. James and C. F. Jones, eds., *American Geography, Inventory and Prospect* (Syracuse, N.Y., 1954), pp. 335-57; T. Bergeron, "Methods in Scientific Weather Analysis and Forecasting," *ASM*, pp. 440-70.

For recent accounts of the work of persons mentioned in this section, see: H. H. Frisinger, *W*, 21 (1966), 443, on Descartes; J. Munzar, *W*, 22 (1967), 360, on von Humboldt; M. Bôcher, *BAMS*, 46 (1965), 448, on Dove and Espy; B. J. R. Blench, *W*, 18 (1963), 83, on Luke Howard; H. E. Landsberg, *BAMS*, 45 (1964), 268, on Blodget; P. H. Putnins, *BAMS*, 48 (1968), 408, on Baur; and G. W. Platzman, *BAMS*, 49 (1968), 496, on Richardson. See also *BAMS*, 49 (1968), 802, for Scherhag's work at the Free University of West Berlin since 1949. For details of persons not mentioned in the text, but who nevertheless are worthy of note, see: Sir G. Sutton, *W*, 20 (1965), 270, and A. F. Sealey, *W*, 22 (1967), 391, on Admiral Fitzroy, who published in 1863 what amounted to an approach to a frontal theory of depressions; H. H. Frisinger, *BAMS*, 48 (1967), 265, on Isaac Greenwood, the holder of the first chair of mathematics and experimental philosophy in North America, who took weather observations in Boston in the eighteenth century, and whose attempts to organize world observations of marine meteorology were ended by his unfortunately early death; B. Sinclair, *BAMS*, 46 (1965), 779, on G. A. Hyde, who was involved in the early days of systematic weather observation in the United States; and M. Bôcher, *BAMS*, 46 (1965), 448, on the work of W. C. Redfield in the United States, who realized, from his observations of a New England gale in 1821, that a cyclonic storm was a gigantic whirlwind.

6. A. von Humboldt, *Mem. Phys. Chim. Soc. d'Arcueil*, 3 (1817), 462-602; trans. by D. Brewster, *Edinburgh Phil. J.*, 3 (1820), 256; 4 (1821), 23; 5 (1821), 29. Map reproduced in N. Shaw, *Manual of Meteorology*, I, 260-61.

7. For references to Dove's original papers, see *ASM*, p. 472. See also H. W. Dove, *Meteorologische Untersuchungen* (Berlin, 1837). L. Howard, *The Climate of London, Deduced from Meteorological Observations, Made at Different Places in the Neighborhood of the Metropolis* (London: 2 vols., 1820; 3d ed. of vol. I, 1833). See also L. von Buch, *Abhandl. Akad. Wiss. Berlin, 1818-19* (1820), p. 104. J. C. E. Schmidt, *Lehrbuch der Mathematischen und Physischen Geographie* (Göttingen, 1830), II, 351-56.

8. J. P. Espy, *Philosophy of Storms* (Boston, 1841). H. Berghaus, *Physikalischer Atlas*, vol. I (Gotha, 1845). H. W. Dove, *Temperaturtafeln nebst Bemerkungen über die Verbreitung der Märme auf der Oberfläche der Erde und Ihre Jährlichen Periodischen*

Veränderungen (Berlin, 1848). M. F. Maury, *Wind and Current Charts* (Washington, D.C., 1848-60). L. Blodget, *Climatology of the United States and of the Temperate Latitudes of the North American Continent* (Philadelphia, 1857). L. M. Meech, *Smithsonian Contributions to Knowledge*, vol. 9, no. 2 (1857). A. Mühry, *Klimatographische Übersicht der Erde* (Leipzig and Heidelberg, 1862; separate supplement, 1865).

9. C. H. D. Buys-Ballot, *Eenige Regeelen voor Anstaande Weersveranderingen in Nederland* (Utrecht, 1860). This empirical law states, in effect, that if one stands with one's back to the wind in the northern (southern) hemisphere, then low (high) pressure will lie to one's left. The same law was deduced mathematically by C. M. Guldberg and H. Mohn, *Études sur les mouvements de l'atmosphere* (Oslo: vol. I, 1876; vol. II, 1880).

10. E. Renou, *Ann. Soc. Météorol. France*, 12 (1864), 240; A. Buchan, Trans. Roy. Soc. Edinburgh, 25 (1869), 575.

11. Published in the British Meteorological Office's *Quarterly Weather Report* for the years 1869 to 1880.

12. W. Köppen, *Reports in Meteorol.*, vol. 4, no. 4 (1874). J. H. Coffin, *Winds of the Earth, Smithsonian Contributions to Knowledge*, vol. 20 (1876). C. Wiener, *Z. Math. Phys.*, 22 (1877), 341. A. Angot, *Ann. Bur. Cent. Météorol. France, 1883*, part 1 (1885), p. 121.

13. W. C. Ley, "Clouds and Weather Signs," *Modern Météorology* (London, 1879). R. Abercromby and W. Marriot, *QJRMS*, 9 (1883), 27. W. C. Ley, *Symon's Meteorol. Mag.*, 13 (1878), 33. W. Köppen, *Ann. Hydrodyn. Maritime Meteorol.*, 7 (1879), 324. A. Supan, *Petermanns Mitteil.*, 25 (1879), 349. E. Loomis, *Am. J. Sci.*, 23 (series 3, 1882), 1.

14. L. Teisserenc de Bort, *Ann. Bur. Cent. Météorol. France: 1881*, part 4 (1883), p. 1; *1884*, part 4 (1886), p. 27; *1885*, part 4 (1887), p. 35; and *1887*, part 4 (1889).

15. H. von Helmholtz, *Sitzber. Preuss. Akad. Wiss.* (1888), p. 647, and *Meteorol. Z.*, 13 (1888), 329. E. Durand-Greville, *Ann. Bur. Cent. Météorol. France, 1890*, part 1 (1892), p. 249.

16. J. Hann, *Handbuch der Klimatologie* (Stuttgart, 1883; 2d ed., 3 vols., 1897; 3d ed., 1908-11; 4th ed. of vol. I, rev. by K. Knoch, 1932).

17. For reference to Assmann's original paper, see *ASM*, p. 470. L. Teisserenc de Bort, *Mem. Acad. Sci. Paris*, 134 (1902), 987. A. L. Rotch, *QJRMS*, 24 (1898), 250.

18. P. S. Laplace, *Traité de Mécanique céleste* (Paris, vol. I, 1799); V. Bjerknes, *Meteorol. Z.*, 19 (1902), 97. For reference to Bjerknes' 1898 paper, see *ASM*, p. 470. M. Margules, *Denkschr. Akad. Wiss. Wien*, 73 (1901), 329.

19. Examples of the work of the Vienna school include a general text, F. M. Exner's *Dynamische Meteorologie* (Leipzig and Berlin, 1917), and various studies on local wind circulations, for example: A. Defant, *Meteorol. Z.*, 27 (1910), 161; H. von Ficker, *Denkschr. Akad. Wiss. Wien*, 78 (1906), 38, and 85 (1910), 113; W. Schmidt, *Wiss. Veröff Deutsch-Öst. Alpenvereins* (Innsbruck, 1930). On lee waves, see: P. Queney, *La Météorol.*, 12 (1936), 334 and 453; J. Küttner, *Beitr. Phys. Freien Atmos.*, 25 (1938), 79; G. Lyra, *Z. Angew. Math. Mech.*, 23 (1943), 1. For a fairly recent advocate of the convection theory, see A. Refsdal, *Geofys. Publ.*, vol. 5, no. 12 (1930), and vol. 9, no. 12 (1932).

20. N. Shaw and R. G. K. Lempfert, *Life History of Surface Air Currents* (London, 1906).

21. In *Meteorol. Z.*, 21 (1904), 1, Bjerknes outlined his program for investigations in meteorology and oceanography on the basis of his new system of physical hydrodrynamics. Details were later developed in the following classic works: V. Bjerknes and J. W. Sandström, *Dynamic Meteorology and Hydrography, Vol. I: Statics, Publ. Carnegie Inst.*, vol. 88 (Washington, 1910); V. Bjerknes, O. Devik, and T. Hesselberg, *Dynamic Meteorology and Hydrography, Vol. II: Kinematics, Publ. Carnegie Inst.*, vol. 88 (Washington, 1911); V. Bjerknes, J. Bjerknes, H. Solberg, and T. Bergeron, *Physikalische Hydrodynamik* (Berlin, 1933).

22. The original version of the Bergen school model was given in J. Bjerknes, *Geofys. Publ.*, vol. 1, no. 2 (1919). The first outline of air-mass analysis was given by T. Bergeron, *Geofys. Publ.*, vol. 5, no. 6 (1928).

23. J. Bjerknes and H. Solberg, *Geofys. Publ.*, vol. 2, no. 3 (1921), and vol. 3, no. 1 (1922). T. Bergeron, *Three-Dimensionally Combining Synoptic Analysis*, part 2 (Moscow, 1934); for the complete reference (in Russian), see *ASM*, p. 471. E. G. Calwagen, *Geofys. Publ.*, vol. 3, no. 10 (1926). G. Schinze, *Untersuchungen zur Aerologischen Synoptik* (Hamburg, 1932).

24. W. Köppen, *Petermanns Mitteil*, 64 (1918), 193 and 243 (the term "classification of climate" first appeared in W. Köppen, *Geogr. Z.*, 6, 1900, 593 and 657). E. E. Federov, "A new method in view in climatography," report (in Russian) at *Congress of Russian Amateurs of World Knowledge* (1921); trans. by P. E. Lydolf, *AAAG*, 49 (1959), 120-41.

25. T. Bergeron, *Meteorol. Z.*, 47 (1930), 246; trans. by H. C. Willett, *MWR*, 59 (1931), 219. T. Hesselberg, *Beitr. Phys. Freien Atmos.*, 19 (1932), 291.

26. M. Milankovitch, *Théorie mathématique des phénomènes thermiques produits par la radiation solaire* (Paris, 1920), and "*Mathematische Klimalehre und Astronomische Theorie der Klimischwangungen*," in Köppen and Geiger, eds., *Handbuch der Klimatologie* (Berlin, 1930), vol. I. G. C. Simpson, "Some studies in terrestrial radiation," *Mem. Roy. Meteorol. Soc.*, 2 (1928), 69; "Further studies in terrestrial radiation," *loc. cit.*, 3 (1928), 1; "The distribution of terrestrial radiation," *loc. cit.*, 3 (1929), 53. Simpson's values are used, for example, in Landsberg, *Handbook of Meteorology* (New York, 1945), pp. 935-36. F. Baur

and H. Phillips, *Beitr. Geophys.*, 42 (1934), 160; 45 (1935), 81; 47 (1936), 218.

27. W. Köppen and R. Geiger, eds., *Handbuch der Klimatologie* (Berlin, 1930-40). C. W. Thornthwaite, *GR*, 21 (1931), 633, and *GR*, 23 (1933), 433.

28. R. Scherhag, *Das Wetter*, 51 (1934), 111, and *Neue Methoden der Wetteranalyse und Wetterprognose* (Berlin, 1948). For an outline of Baur's system of climatological forecasting, see F. Baur, *CM*, pp. 814-31, and *Physikalisch-Statistische Regeln als Grundlagen für Wetter- und Witterungsvorhersagen*, part I (Frankfurt a.M., 1956).

29. The basic discoveries of the Chicago School were first published in: C.-G. Rossby, *J. Marine Res.*, 1 (1937-38), 15 and 239; C.-G. Rossby *et al.*, "Application of fluid mechanics to the problem of the general circulation of the atmosphere," *BAMS*, 18 (1937), 201; C.-G. Rossby, *QJRMS*, 66 (1940), Suppl., p. 68, and *J. Marine Res.*, 2 (1939), 38.

30. L. F. Richardson, *Weather Prediction by Numerical Process* (Cambridge, 1922).

31. R. C. Sutcliffe, *QJRMS*, 64 (1938), 495; *QJRMS*, 65 (1939), 518; *QJRMS*, 73 (1947), 370.

32. R. Geiger, *Das Klima der Bodennahen Luftschicht* (1927); trans. by M. N. Steward *et al.*, *The Climate Near the Ground* (Cambridge, Mass., 1950). For Thornthwaite's contributions, see, *inter alia*: C. W. Thornthwaite and B. Holzman, "The determination of evaporation from land and water surfaces," *MWR*, 67 (1939), 4; C. W. Thornthwaite, *TAGU*, 22 (1941), 429, on the chemical absorption hygrometer as a meteorological instrument; C. W. Thornthwaite and B. Holzman, *Measurement of Evaporation from Land and Water Surfaces*, *U.S. Dept. Agric. Tech. Bull.*, no. 817 (1942); C. W. Thornthwaite and J. C. Owen, *MWR*, 68 (1941), 315, on a dew-point recorder for measuring atmospheric moisture; C. W. Thornthwaite *et al.*, *On the Measurement of Vertical Winds near the Ground*, *PC*, no. 14 (1941). O. G. Sutton, *Atmospheric Turbulence* (London, 1949); see also *Proc. Roy. Soc.*, vol. 146, series A (1934), on wind structure and evaporation in a turbulent atmosphere. F. Pasquill, *Atmospheric Diffusion* (London, 1962); see also *Proc. Roy. Soc.*, vol. 198, series A (1949), for eddy diffusion of water vapor and heat near the ground. C. H. B. Priestley, *Turbulent Transfer in the Lower Atmosphere* (Chicago, 1959); see also C. H. B. Priestley and W. C. Swinbank, *Proc. Roy. Soc.*, 189, series A (1947), 543, for vertical transport of heat by turbulence in the atmosphere.

33. W. C. Jacobs, *BAMS*, 27 (1946), 306, and *AMM*, vol. 1, no. 1 (1947); C. S. Durst, *CM*, pp. 967-74.

34. C. W. Thornthwaite, *GR*, 38 (1948), 55.

35. A. Giao, *GPA*, 15 (1949), 114; A. Giao and M. Ferreira, *GPA*, 34 (1956), 101. For current opinion, see *DC*.

36. J. von Neumann, in *DC*, p. 11.

37. H. Lettau, quoted by Landsberg, *AMM*, vol. 3, no. 12 (1957).

38. See *BWMO* for details of progress and publications concerning these atlases.

39. See the maps in *HBES*; see also C. W. Thornthwaite, J. R. Mather, and D. B. Carter, *PC*, 9 (1958), 1, and H. G. Houghton, *JM*, 11 (1954), 1.

40. F. W. Ehrenheim, *Om Climaternas Rörlighet* (Stockholm, 1824); A. Buchan, *J. Scot. Meteorol. Soc.*, 13 (1867), 4. For a literary study, see, for example, G. Manley, *QJRMS*, 79 (1953), 242, on the mean temperature of central England, 1698-1952. For a relatively simple example of modern techniques, see D. H. McIntosh, *QJRMS*, 79 (1953), 262, on annual recurrences in Edinburgh temperature. For the more advanced techniques, see pp. 85–101 in *Techniques*.

41. W. G. Kendrew, *The Climates of the Continents* (Oxford, 1922, and many editions since).

42. Curricula have been developed at, e.g.: the Laboratory of Climatology (C. W. Thornthwaite Associates, Centerton, New Jersey), specializing in on-the-job graduate training in microclimatology and topoclimatology; McGill University, Montreal, where the Geography Department provides graduate training in microclimatology at research centers in Labrador, Quebec, Guyana, and Barbados; and the University of Birmingham, England, where the department of Geography provides a formal graduate course in climatology and applied meteorology (including weather forecasting).

43. Anyone interested in the current and future education of climatologists should look at existing trends in the education of meteorologists. On the philosophy and practice of meteorological education in the United States, see W. A. Baum, *BAMS*, 48 (1967), 746. On the facilities for meteorological education in North America, see *WW*, 20 (1967), 96-135. For examples of meteorological education at particular universities, see A. Wiin-Nielsen, *BAMS*, 48 (1967), 758, for the University of Michigan, and R. C. Sutcliffe, *W*, 23 (1967), 382, for the University of Reading, England. For details of the McGill University climatology program in the Barbados, see B. J. Garnier, *BAMS*, 49 (1968), 636. For a climatology curriculum for WMO Class I personnel specializing in climatology, see R. A. Bryson, *BAMS*, 48 (1967), 752. The universities with meteorology departments include the really scientific parts of climatology in their courses in meteorology proper, and the climatological courses are usually concerned mainly with climatography. The climatology courses given in university geography departments usually suffer from the handicap that few geographers are experienced in the mathematical side of climatology. The courses given in agricultural, biological, and other institutions are usually under the direction of

scientists who have had to educate themselves in climatology because they needed to solve practical problems of a climatological nature in their own fields, and whose outlook is therefore too specialized. Climatological courses given in state meteorological services are usually severely biased toward the concept of climatology as routine date processing.

44. S. Petterssen, *Weather Analysis and Forecasting* (New York, 1940; 2d ed., 2 vols., 1956).

45. This view of geography was expressed in Mill's article "Geography" in the *Encyclopaedia Britannica* (14th ed., 1929). Although Mill stated in this authoritative article (reprinted for many years, e.g., in 1951) that geography is "the exact and organised knowledge of the distribution of phenomena on the surface of the Earth," he also stated that the characteristic task of geography "is to investigate the control exercised by the forms and vertical relief of the surface of the lithosphere directly or indirectly on the various mobile distributions." Mill's essentially pragmatic view of the task of geography may not be the same as the conclusion reached by R. Hartshorne, *The Nature of Geography*, *AAAG*, vol. 29 (1939), and *Perspective on the Nature of Geography* (London, 1959), who argued, on philosophical grounds, that geography is essentially concerned with areal differentiation. However, Mill's view has the merit, to the climatologist at least, that it provides the geographer with a problem that not only is capable of solution, but also has definite applications in the fields of applied climatology and atmospheric science generally.

46. See D. H. Miller, *UCPG*, vol. 11 (1955), and C. P. Patton, *UCPG*, vol. 10, no. 3 (1956).

47. See H. Jeffreys, *Scientific Inference* (Cambridge, 1937), for a review of the procedure involved.

48. For details, see G. A. Tunnell and E. J. Sumner, *MM*, 78 (1949), 258 and 295; see also G. A. Tunnell, *MM*, 82 (1953), 103.

49. For details of the Savino-Angström formula, see Appendix 4.24 in *Techniques*.

50. For details of the concept of *topoclimatology*, see C. W. Thornthwaite, *BWMO*, 2 (1953), 40, and 6 (1957), 2. See also *AAAG*, 51 (1961), 345.

51. For details concerning the mathematics of one aspect of map analysis, i.e., the analysis of scalar fields and their rates of change, see W. J. Saucier, *Principles of Meteorological Analysis* (Chicago, 1955), pp. 96-131 and 138-44.

CHAPTER 1

The Atmosphere

Geographical exploration, the examination by man of his external world in terms of place, has undergone many changes in emphasis as man's concept of his world has expanded, and as the general mood of human thought has evolved. Originally thought of only two-dimensionally, with time regarded as significant only in special cases or for specific reasons, man's physical world has now acquired a new dimension, since the development of aviation in the twentieth century has led to exploration of the atmosphere and outer space. The Earth's surface must now be thought of as including both the atmosphere and the oceans, as any photograph from an artificial Earth satellite or a rocket will demonstrate.

Until the nineteenth century, explorers were almost always merchants, missionaries, adventurers, or military or naval personnel, rarely professional geographers, who were usually advisers to governments or royalty, or were academics, and who subjected the reports of the explorers to critical assessment. Today the geographical explorers of the atmosphere are airmen, astronauts, and meteorologists; again, very few are professional geographers.

It is universally recognized among geographers that exploration of at least one's local surroundings is essential training for the budding geographer. Hopefully, in the future, it will become increasingly recognized that some experience in exploring the atmosphere is necessary for the budding climatologist, if he hopes to carry out original research. In the past, there have been geographers who added to our knowledge of the geography of the atmosphere from personal experience.[1]

The only way to see the Earth's surface in perspective is to view it from high above: one then becomes convinced that the face of the Earth is literally very much clouded, or at least covered by a semitransparent haze one is not aware of from the ground, and that the concept of landscape is meaningless without its associated skyscapes or weatherscapes in all their multiplicity. Mountains are useful as observation platforms, but they generate local weather of their own, and so tend to give an unrepresentative picture. Aircraft and balloons are much more useful to the climatologist who wishes to obtain first-hand knowledge of his element. Experienced pilots provide definite evidence that, although blue skies look much the same everywhere in the world to the "groundhog," they are actually quite varied.[2] In fact, there is a true geography of the atmosphere, more elusive and powerful than the geography of the Earth's surface.

At the other end of the scale of observation, the literary naturalists, such as White and Hudson, regarded weather as a part of the local landscape, and faithfully described the meteorological characteristics of their region or parish, as well as the details of its natural and human history. Their method of verbal description was very interesting, but hardly concise, and could not of course be expected to yield information accurate enough for modern scientific purposes.[3]

Unlike most geographical phenomena, weather changes very rapidly in most parts of the world. Hence geographical description of weather must emphasize process and causation much more than with other geographical features. Normally, it is more useful for the geographer to describe the conditions (i.e., atmospheric properties) under which specific weather phenomena are probable for a given locale or area, since such conditions can indicate cause or process, than to describe the phenomena themselves from records of past occurrences. For the form of the description, mathematics is the most appropriate, and the most concise, when the operative processes are fairly well-understood, and cartography is best when they are not. Verbal description is necessary only to emphasize or explain critical points in the mathematics or cartography.

Just as other mobile features of the Earth's surface are best-described by reference to certain fixed features of topography, so the infinitely complex phenomena of weather are conveniently described in relation to relatively permanent features of the topography of the atmosphere. It is the task of the climatologist to discover and chart these quasipermanent features in the confusing mass of instantaneous patterns with which the meteorologist confronts him.

The first step in scientific geographical description of the Earth's solid surface is the construction of contour maps showing the amount of land above or below mean sea level. Other distributions can then be described in terms of the contour maps. The first step in scientific geographical description of the atmosphere is to prepare maps depicting the amount of air above or below mean sea level or a surface roughly parallel thereto (see Appendix 1.1).* These maps of the atmosphere's relative topography (see Figures 1.1 to 1.4) indicate the large-scale flow of air, and allow estimates to be made of various physical properties (vorticity, for example) which we can demonstrate to be significant in bringing about weather changes. Unlike topographic maps for the solid Earth, they present patterns which change seasonally, secularly, and cyclically.

Thus the atmosphere may be regarded in effect as a series of surfaces superimposed

---

* The Appendixes begin on p. 509.

one above another. All these surfaces, at some time or place, are important in deciding weather conditions at ground level. That the influence of atmospheric topography on air movement is not completely represented by these maps, however, is indicated by work on a trade-wind island, where it was found convenient to postulate an "equivalent mountain," an imaginary mountain which, had it actually been present, would have caused the air to move in the observed manner.[4]

The atmospheric contour charts in Figures 1.1 to 1.4 describe its "permanent" features, its mountains, valleys, and hills, although their permanence varies from several weeks to only a few hours. In addition, the atmosphere exhibits analogies to large rivers in its jetstreams (see the atmospheric cross sections in Figures 1.5 and 1.6*), and to gigantic waterfalls in its mountain waves. But no mobile phenomena on the Earth's solid surface are comparable to the great circumpolar whirls of the atmosphere, or the vast westerly airstreams of middle latitudes. Some of these features of atmospheric topography are also visible on maps of instantaneous pressure values. Jetstreams, cyclones, and anticyclones appear on daily weather maps; other features, the upper-air long waves in the westerlies, for example, must usually be brought to light by averaging the daily maps over periods of days.

If, in addition to the amount of air below or above a given height, the temperature and moisture content of the atmosphere at various heights is also considered (Figures 1.7 to 1.13), it becomes clear that the atmosphere has its humid and desert regions, its tropical and glacial zones, just like the solid surface of the Earth, but the locations of some of them are surprising.† The different regions and zones are characterized by certain cloud sequences and types of sky, which are often of great complexity. Some of these cloud sequences and sky types depend on the underlying topography of the solid Earth. More often, however, they are purely the consequence of certain dynamic processes—hydrodynamic convergence or divergence, in particular—acting on the general properties of the atmosphere in that latitudinal belt.

Maps depicting atmospheric geography in fair detail have become feasible only since 1950, although geographical exploration of the atmosphere has been going on sporadically for a long time. Before discussing the details of atmospheric geography, we should review some historical aspects of atmospheric exploration.[5]

A Brief History of the Exploration of the Atmosphere

The first instrumental observation of a change with height of a property of the atmosphere was made by Pascal in 1643. Pascal found that the reading on his barometer decreased as he ascended the Tour St. Jacques (52 m high) in Paris. Five years later, Périer confirmed the fact that barometric pressure decreases with height, when he climbed the Puy-de-Dôme (1,460 m) to carry out an experiment involving the balancing of liquids. Mountaineering was a very difficult activity at that time (see Appendix 1.2), and so the success of Périer's work did not herald the real age of atmospheric exploration.

---

* In these figures, positive values indicate westerly winds and negative values easterly. The blacked-out curves along the bottom of each graph represent the general orography of the Earth's surface.

† Notice that whereas Figures 1.7 to 1.9 give air temperatures as measured, in Figures 1.10 and 1.11 the temperatures have been reduced adiabatically to 1,000 mb, so that these last two figures show temperature in an atmosphere that is everywhere at the same pressure.

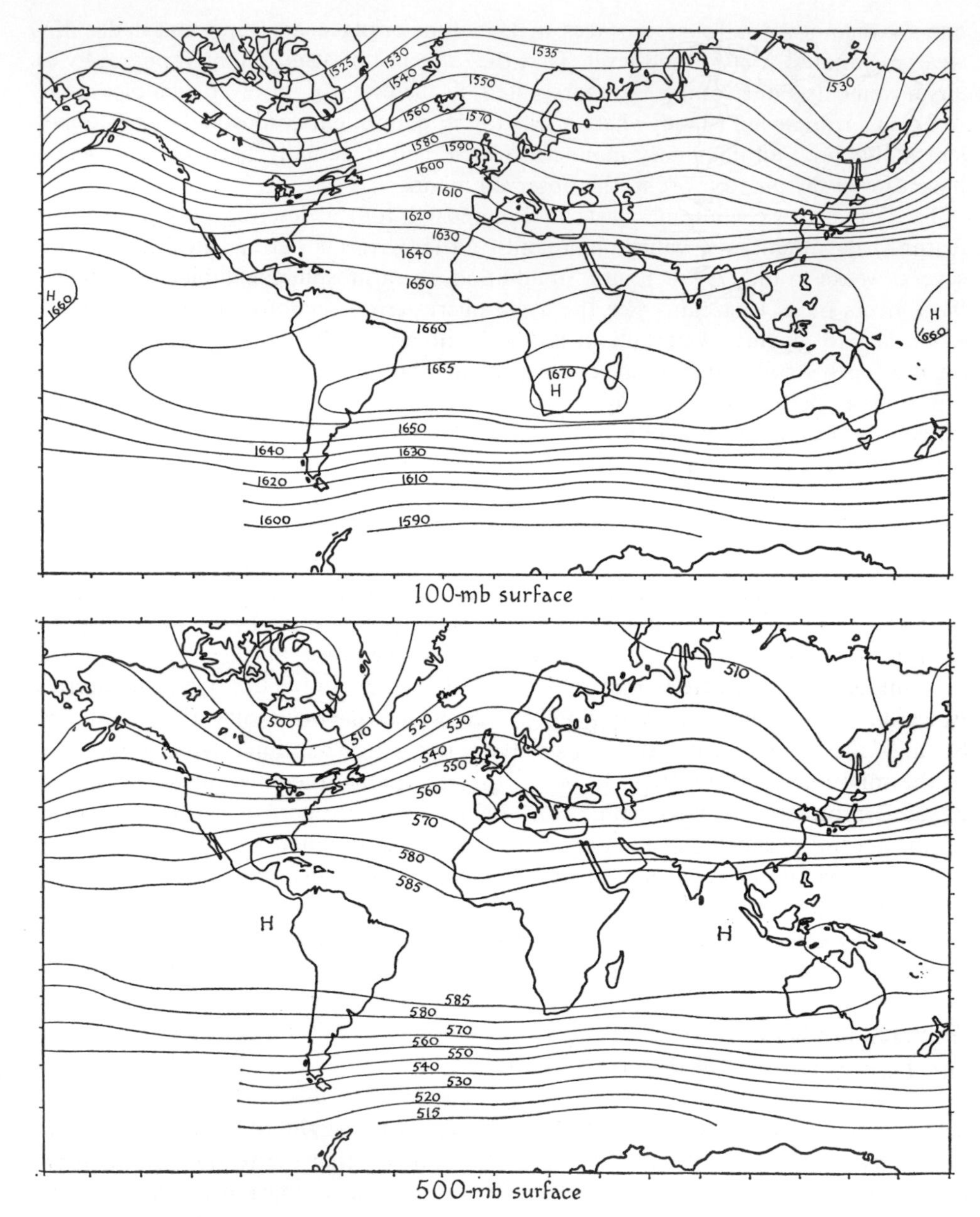

FIGURE 1.1.
Atmospheric contours for January.
Figures give heights of pressure
surfaces in geopotential decameters, 1949-53.
(After H. Heastie and P. M. Stephenson, *MOGM*, vol. 13, nos. 103 and 104, 1960.)
ICAO height equivalents: 500 mb = 18,298 feet (5,574m);
100 mb = 53,083 feet (16,180m); 700 mb = 9,882 feet (3,012m).

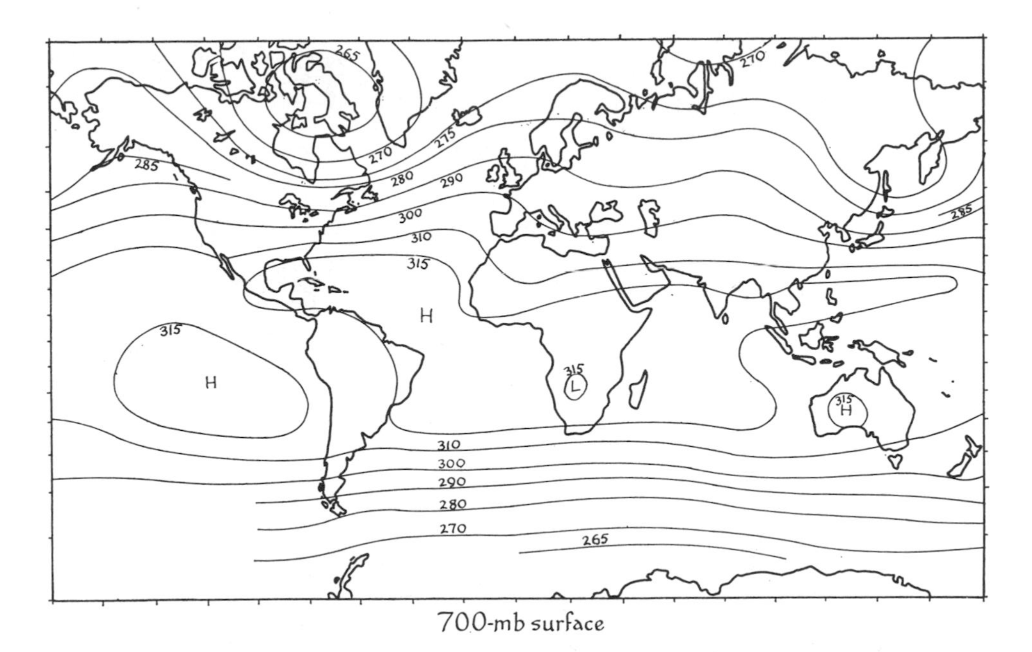

A number of investigations made on mountains have become important landmarks. Around 1750, Bouguer determined the height of the freezing level in the atmosphere in various latitudes, from observations made during a geodetic expedition to Peru. In the mid-nineteenth century, a number of mountain observatories were set up—Puy-de-Dôme, Mont Ventoux, Pic du Midi, Santis, Brocken, Sonneblick, and Jungfraujoch—all of which have become world-famous for meteorological work. Important discoveries have been based on their observations: for example, Hann in 1876 deduced the existence of the temperature inversion in anticyclones, and the subsidence above the inversion. But in general, mountain observatories could never have provided systematic data for three-dimensional climatology, for, apart from their uneven geographical distribution, the data they supply are for a single level only, and are much biased by the local topography. Other techniques had to be developed.

The first technique was the kite. Wilson and Melville, in Glasgow in 1749, made the first attempt at scientific exploration of the atmosphere, using kites to lift minimum thermometers to considerable heights. In 1752, Franklin at Philadelphia found that a kite became electrified when flown into thunderclouds, and was thus able to prove the electrical nature of lightning. Fisher and Parry made use of kites carrying self-registering thermometers on Arctic voyages in 1822, and Archibald in 1885 introduced steel piano wire as a means whereby kites could reach quite impressive heights. From then until the early twentieth century, kites were the chief aid to upper-air exploration.

Eddy sent up a barothermohygroanemograph, suspended beneath a kite, from Blue Hill Observatory (Boston, Mass.) in 1894, to an altitude of 436 m. The technique was improved by Rotch, and the U.S. Weather Bureau organized a network of 17 sounding stations. Kite soundings were also made regularly by Teisserenc de Bort at Trappes, by Rykatcheff at Pavlovsk, by Dines and Cave near London, and by Hergesell and Assmann in Germany. A Hargrave-type box kite (first introduced in 1893

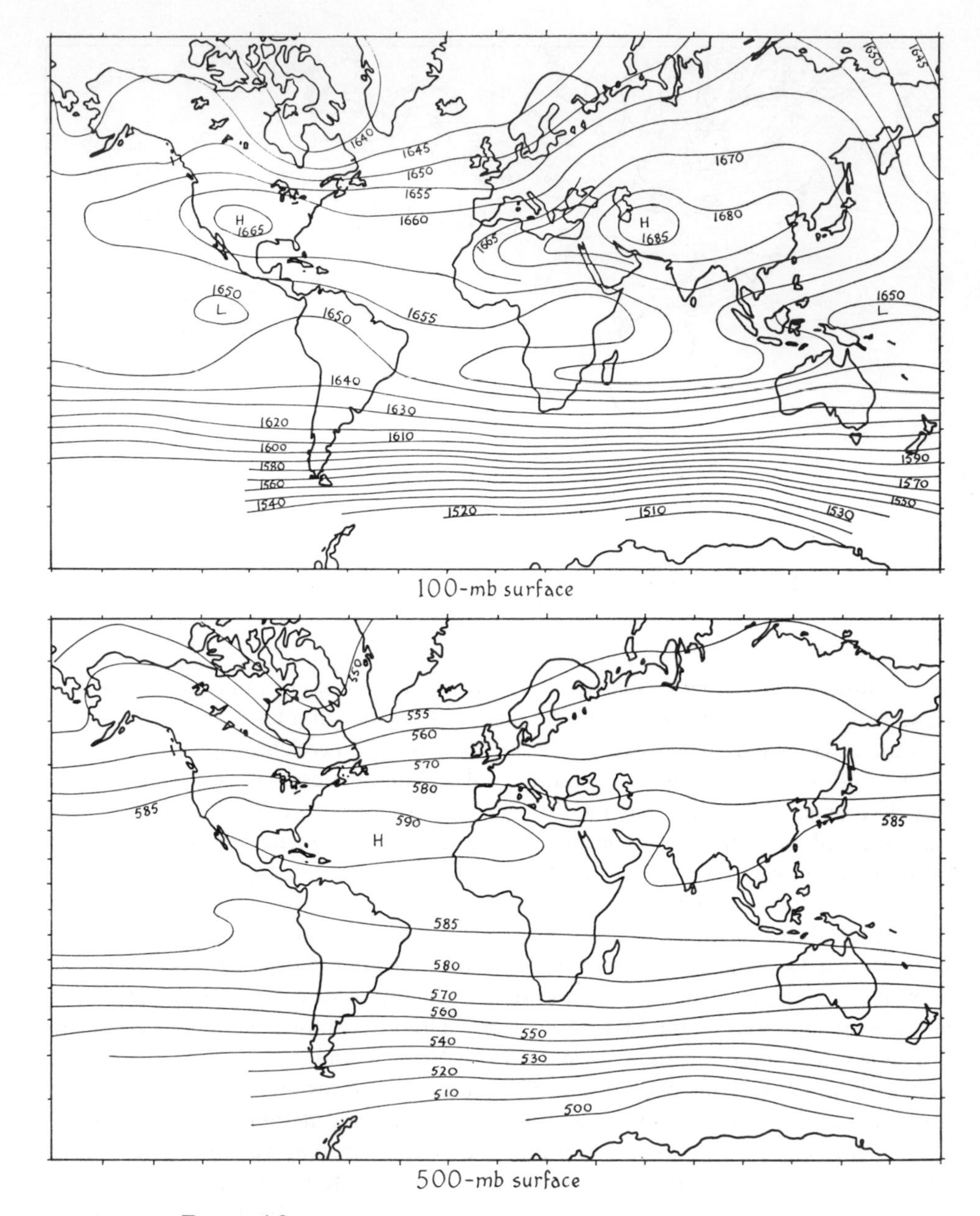

FIGURE 1.2.
Atmospheric contours for July.
Figures give heights of pressure
surfaces in geopotential decameters, 1949-53.
(After Heastie and Stephenson.)

ICAO height equivalents: 500 mb = 18,289 feet (5,574m);
100 mb = 53,083 feet (16,180m); 700 mb = 9,882 feet (3,012m).

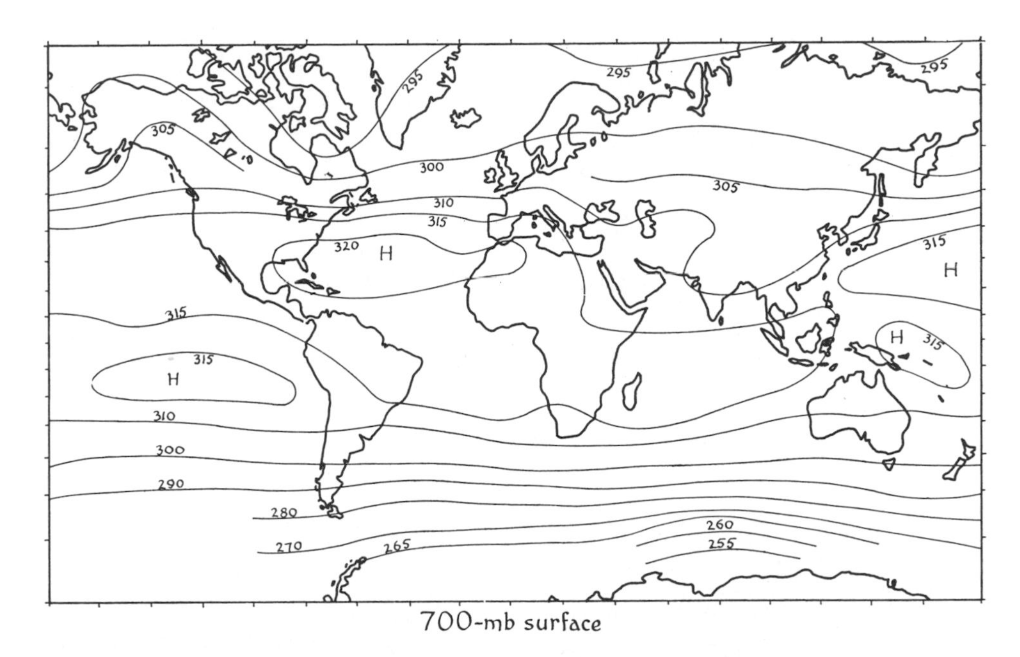

at Sydney) was normally used, because the tension on the wire with this type could exceed 100 kg with a strong winch. A serious limitation was that the kites could work successfully only in fairly high winds.

The kite technique was inexpensive and simple, but the resulting soundings were for only the lower layers of the atmosphere. Still, the first accurate information about wind distributions in the atmospheric friction layer came from kite soundings: they were used by Abbe to study sea breezes in the United States, and by Hergesell to make the first (1904-6) observations of the trade winds, from a yacht near the Canary Islands.

The second technique was the balloon. Hydrogen, the lightest gas, was discovered by Cavendish in 1781. The same year (1783) that the Montgolfier brothers made their first experiment in ballooning, the French physicist J. A. Charles—the inventor of the hydrogen balloon—made a balloon ascent from the Jardin des Tuileries in Paris, carrying a barometer and a mercury thermometer with him. This first free meterological balloon ascent is usually taken as marking the birth of the science of aerology. In 1784, the Van Bochaute process for making coal gas was perfected, and the first balloon filled with this gas was released near Louvain by Minkelers in 1784. It had envelopes made of varnished silk or goldbeater's skin. However, the hydrogen balloon had more meteorological advantages.

Lavoisier drew up a program for scientific upper-air measurements in 1784, for the Paris Academy of Sciences. In this year, too, the first manned-balloon ascent for purely meteorological purposes was made by Jeffries and Blanchard, who flew from London to Dartford carrying laboratory-tested instruments. In 1785, Jeffries and Blanchard crossed the Straits of Dover; during their flight the first angular measurements of a balloon from the ground were carried out, enabling the height of the balloon to be found.

During the early part of the nineteenth century, free ballooning techniques improved considerably, so that in 1804 Biot was able to reach 7,000 m. Gay-Lussac made the

FIGURE 1.3.
Atmospheric pressure-surface contours for January. Units are geopotential decameters. (After Heastie, Stephenson, and Hofmeyr.)

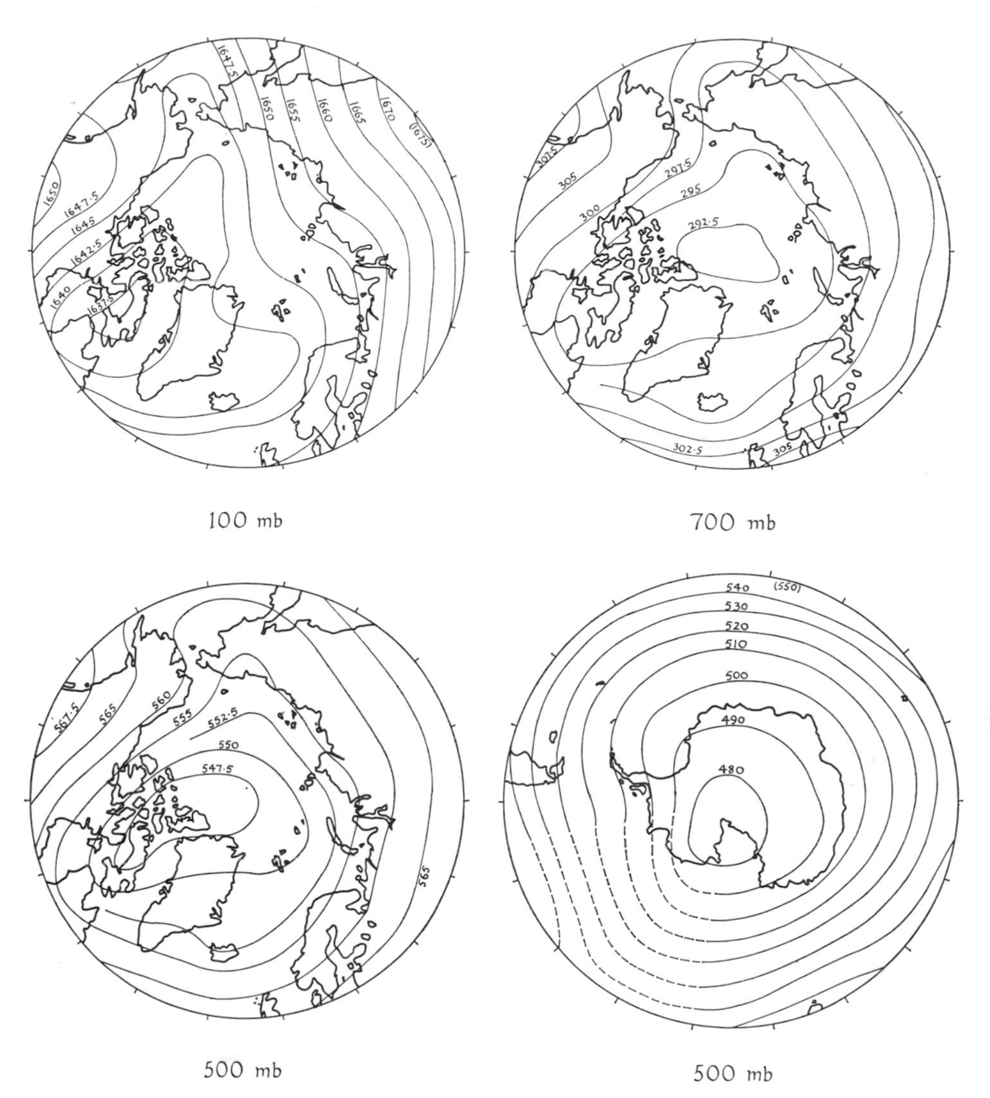

FIGURE 1.4.
Atmospheric pressure-surface contours for July. Units are geopotential decameters. (After Heastie, Stephenson, and Hofmeyr.)

first high-altitude ascent for meteorological purposes, measuring temperatures above the Paris region up to 23,000 feet, and discovering an average decrease of 1°F per 340 feet. In 1848, Barral and Bixio made the first meteorological ascent in cloud, and Welsh and Green measured pressure, temperature, and relative humidity over England from the surface up to 20,000 feet. In 1862 Glaisher and Coxwell made similar measurements over London to almost 30,000 feet.

Most balloon ascents carried a barometer and a thermometer in the nacelle. Arago proved in 1841 that the readings of these instruments did not yield true atmospheric pressures and temperatures. Welsh made four ascents near London in 1852, using a hand-aspirated psychrometer, which reduced the effect of solar radiation on

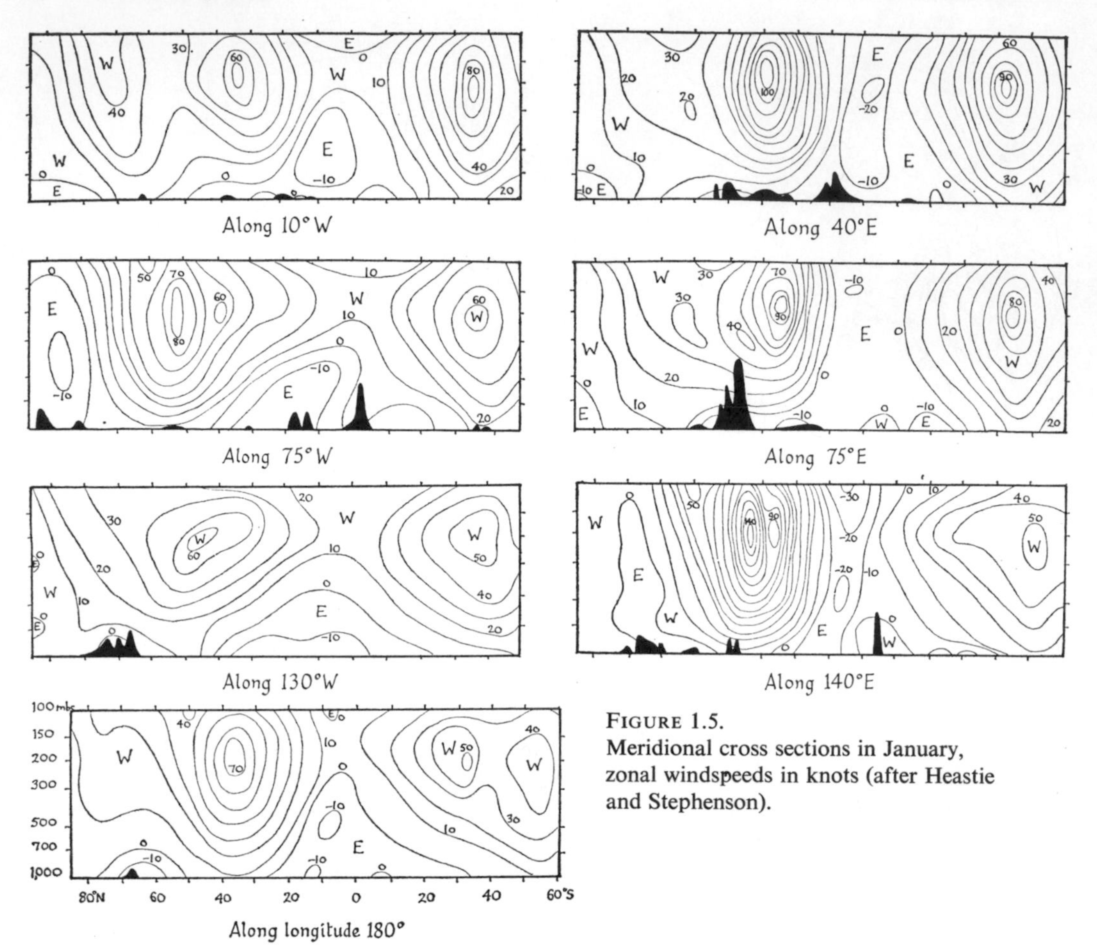

FIGURE 1.5.
Meridional cross sections in January, zonal windspeeds in knots (after Heastie and Stephenson).

the thermometers. The aneroid barometer was invented by Vidie in 1844, and Glaisher made a number of ascents using this instrument, achieving 11,300 m in 1862. Recording instruments were first introduced in 1881, when du Hauvel and Duté-Poitevin carried two barographs in their nacelle.

Balloon measurements in the nineteenth century covered a remarkable range. In addition to atmospheric pressure, temperature, and humidity, radiation was measured (using black-bulb thermometers), and observations were made of electrical (with an electrophore), chemical (by sampling), and magnetic (with compasses) properties of the atmosphere, as well as of optical and acoustical phenomena. Unexpected features were revealed; for example, clouds were found to develop much more rapidly than had been anticipated, the tops of cumuli rising more rapidly than the balloons; and abrupt changes were found in the upper-air winds.

The *Deutsche Verein zur Förderung der Luftschiffahrt*, set up in Berlin in 1881 to promote aerial navigation, had as one of its chief activities the scientific exploration of the free atmosphere by means of balloons. Under Assmann, Berson, and Süring, many advances in instrumentation were made by this organization, for example, the Assmann aspirated psychrometer (1891) and the Assmann ventilated barothermohygrograph (1893). The latter instrument was specifically designed for suspending from a balloon nacelle, and in effect was the first aerological sonde.

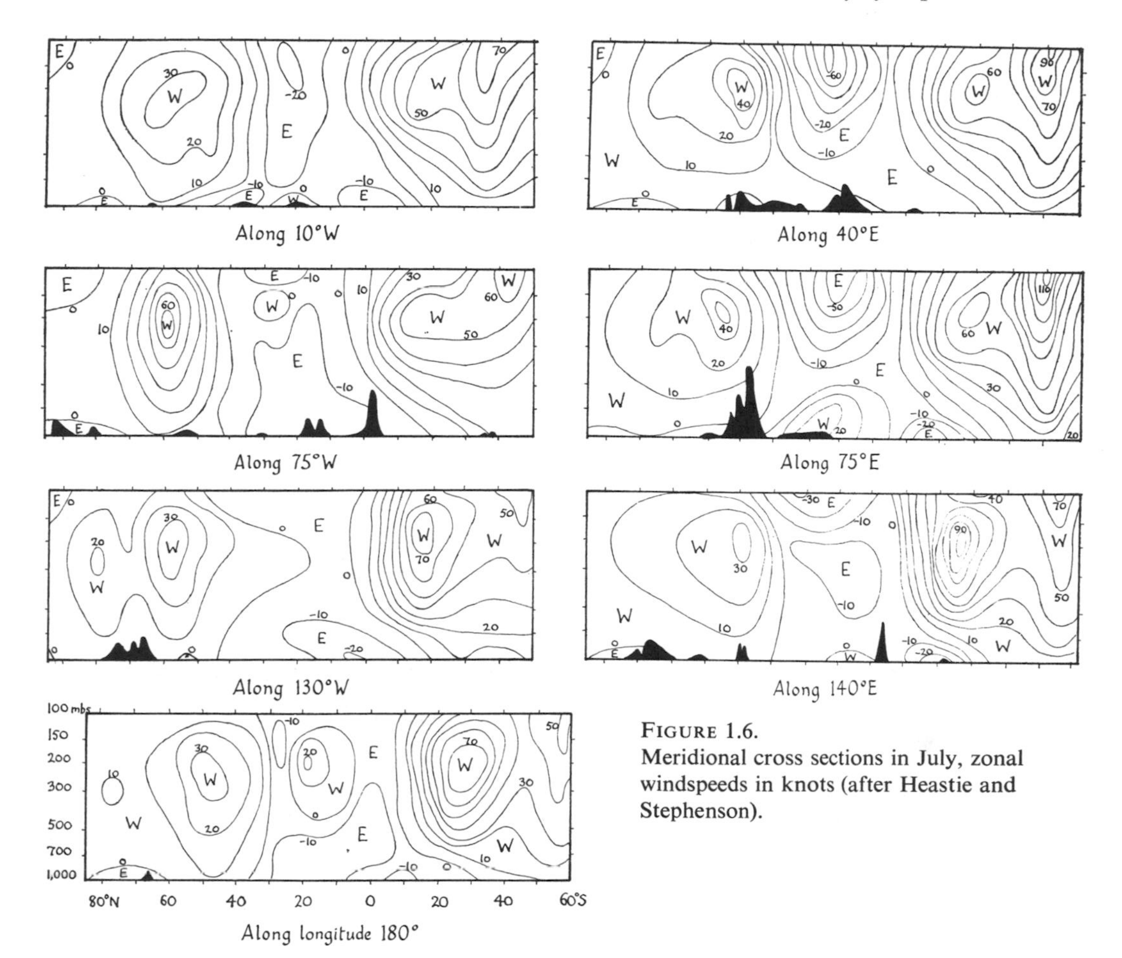

FIGURE 1.6.
Meridional cross sections in July, zonal windspeeds in knots (after Heastie and Stephenson).

A few figures will indicate the rapid progress of scientific balloon ascents in the twentieth century. In 1932 Picard reached 17 km altitude; in 1957 Kittinger attained 30 km. Also in 1957, Simons, a U.S. flight surgeon, reached 102,000 feet in 32 hours using a helium-filled plastic balloon. Stevens and Anderson in 1935 measured the vertical distribution of ozone and the variation in cosmic-ray intensity with latitude up to 22.5 km; since the position of their balloon was accurately determined by three different methods (terrain photography, angular measurement from the ground, and recording of air temperature and pressure), they were able to verify completely the hydrostatic assumption for the atmosphere for the first time.

Sounding balloons and pilot balloons, as distinct from man-carrying balloons, had their origin in the late eighteenth century in Paris and at Annonay, when small balloons were released and allowed to move freely: their subsequent paths obviously indicated something about air movements at whatever height the balloon was at. The first real wind-sounding was made by a group of astronomers who, in 1783, observed from the highest points in Paris the horizontal projection of the trajectory of a balloon launched from the Champ de Mars.

Forster in 1809 observed the drift of small free-floating balloons filled with inflammable gas, and Schreiber in 1885 employed two surveyors' theodolites, at the ends

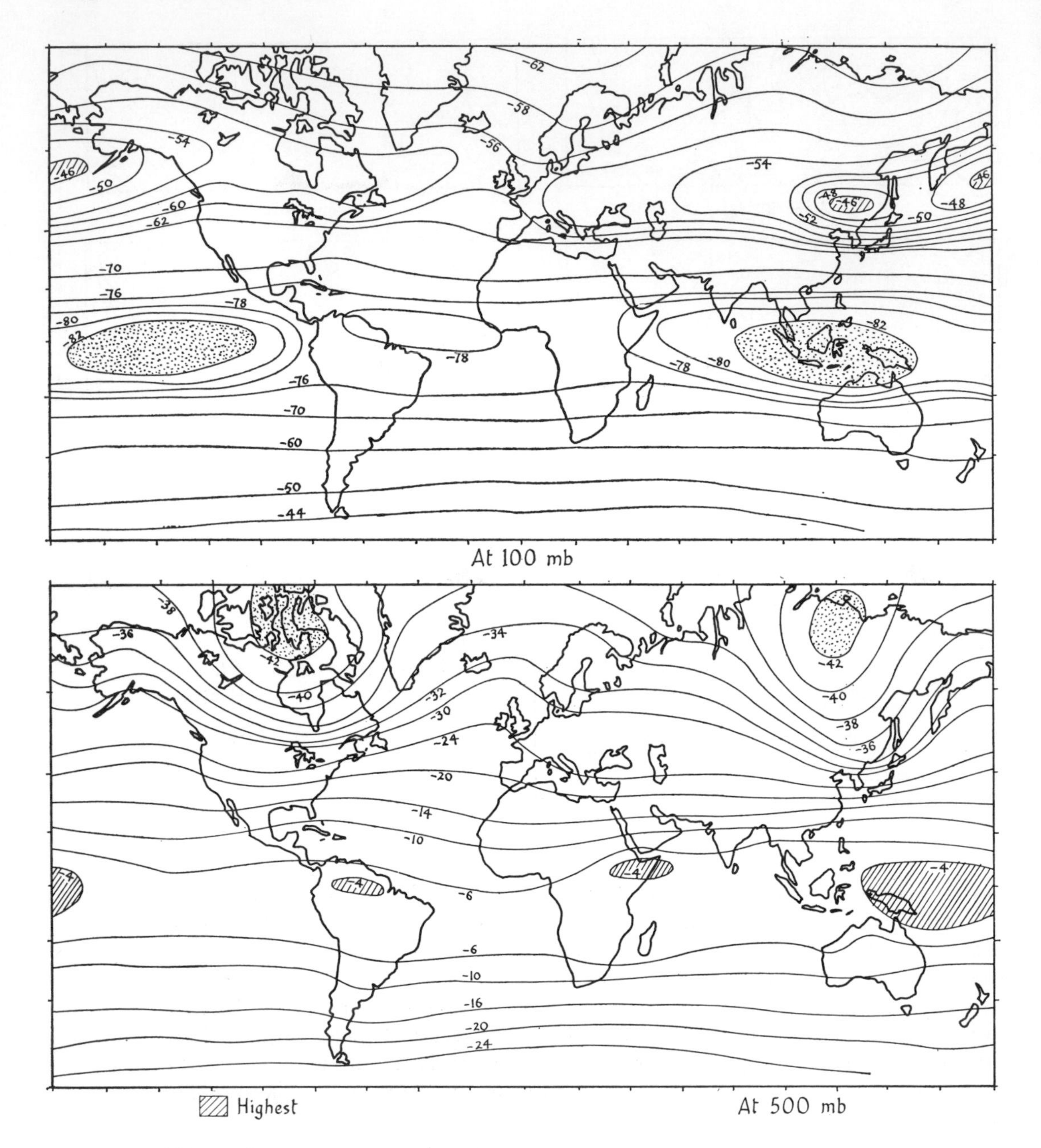

FIGURE 1.7.
Air temperatures in January in °C
(after N. Goldie, J. G. Moore, and E. E. Austin, *MOGM*, vol. 13, no. 101, 1958).

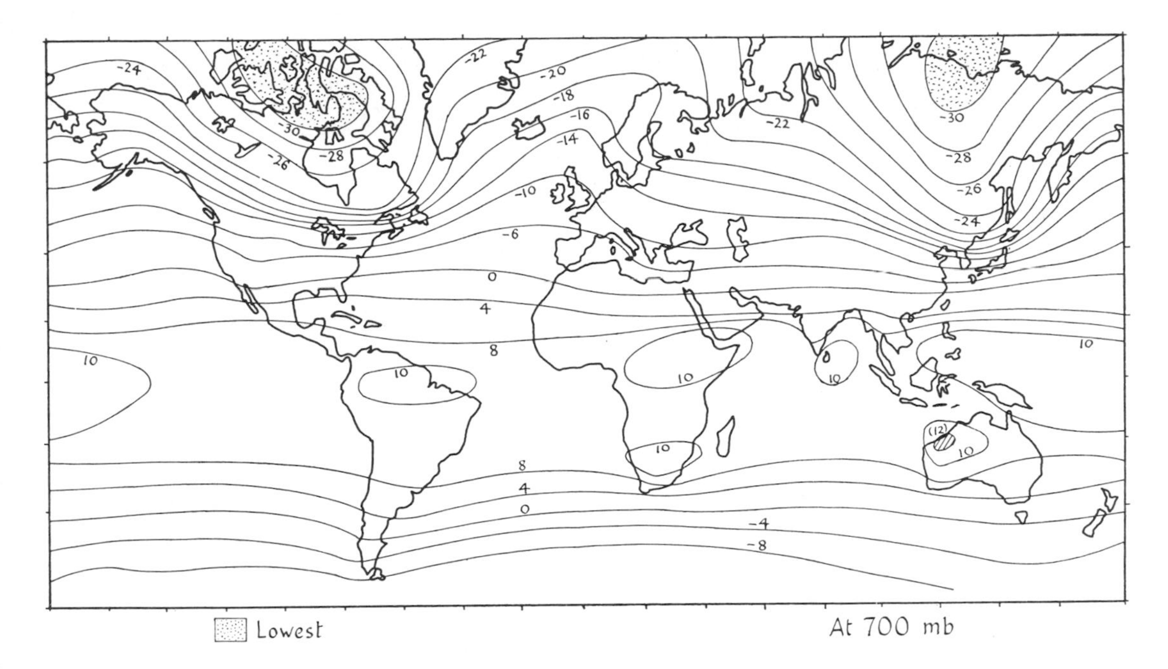

of a baseline 5 km long, to follow the movements of a large free balloon and thus determine its drift due to wind.

Hermite and Besançon produced in 1892 a 26-cubic-meter balloon of paraffined paper, weighing 20 grams, which carried suspended beneath it a maximum-minimum thermometer and a Vidie barometer. The balloon attained a height of 16 km, and is considered to be the first true sounding balloon. The following year, a barothermograph was added, both pressure and temperature being recorded on the same drum to reduce weight.

The International Meteorological Organization Paris Conference in 1896 set up the *Commission for Aerology*, whose main object was to be the organization of atmospheric exploration by means of sounding balloons of the type used by Hermite and Besançon. Members of the Berlin organization had experimented with the simultaneous launching of manned and sounding balloons, to check the accuracy of the automatic instruments carried by the sondes. The Commission for Aerology carried this idea further, arranging International Aerological Days, the first of which was November 14, 1896. These days were marked by simultaneous ascents of both free and manned balloons from Berlin, Munich, Paris, St. Petersburg, Strasbourg, and Warsaw. They continued, with expanded coverage, until April 1939 and were especially important in the International Polar Year of 1932–33.

One of the first discoveries made possible by the Commission for Aerology was that of the stratosphere, by Teisserenc de Bort in 1902, using 100-cubic-meter balloons of varnished paper. The discovery of the stratosphere was a great stimulus, and most national meteorological organizations set up aerological divisions, with sounding stations, as a direct result. The Commission also had improvements made in the sondes: Assmann introduced rubber balloons for the first time, and deduced the existence of the stratosphere simultaneously with Teisserenc de Bort's discovery of it. Assmann's balloons were able to penetrate to greater heights than those of de Bort, because a smaller rubber balloon than paper balloon can achieve a given height;

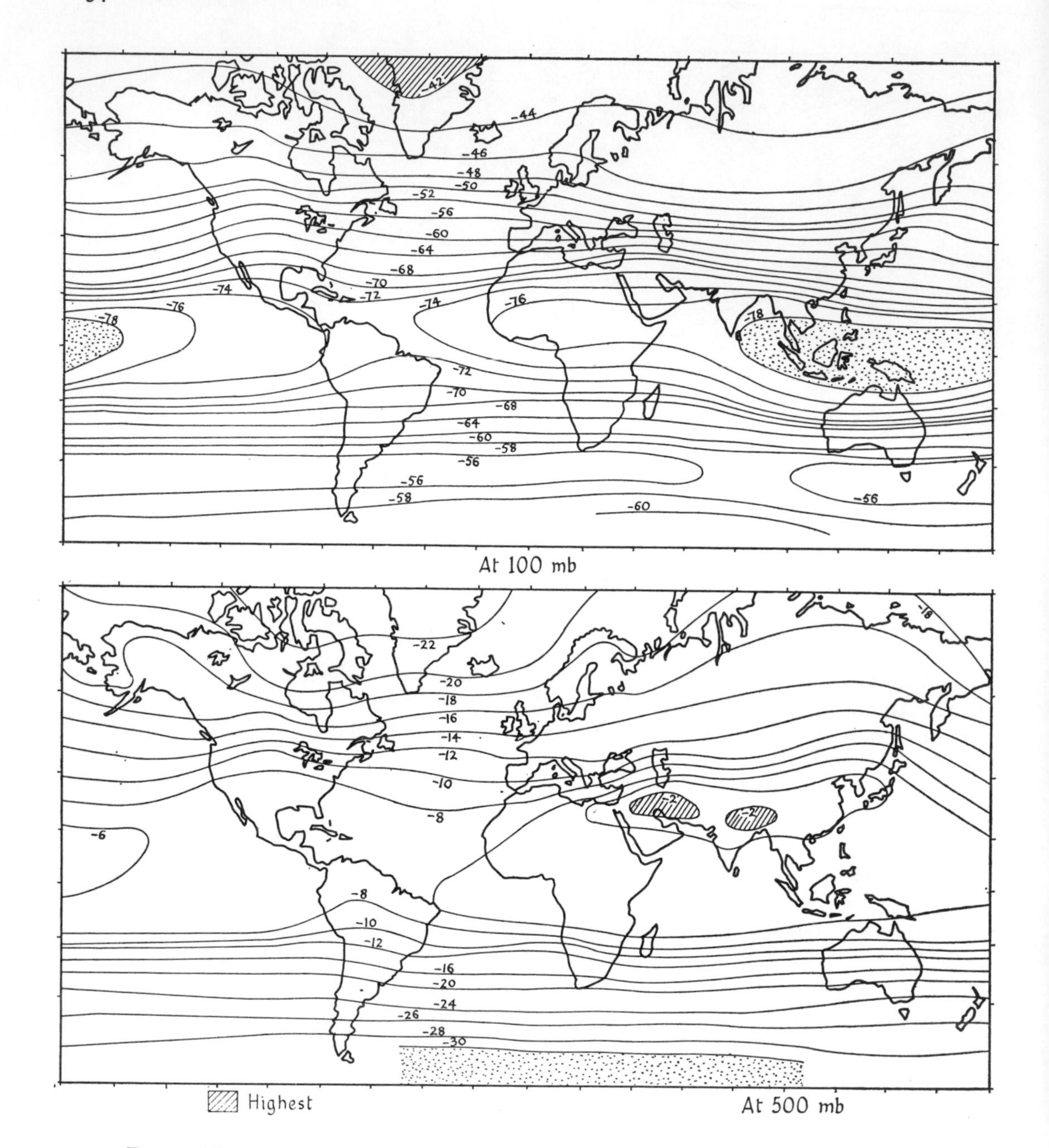

FIGURE 1.8.
Air temperature in July in °C. Note that the shading emphasizes: (a) very warm air at 100 mb; (b) very cold air at 500 and 700 mb; (c) the cold pole at 700 mb in the northern hemisphere. (After Goldie, Moore, and Austin.)

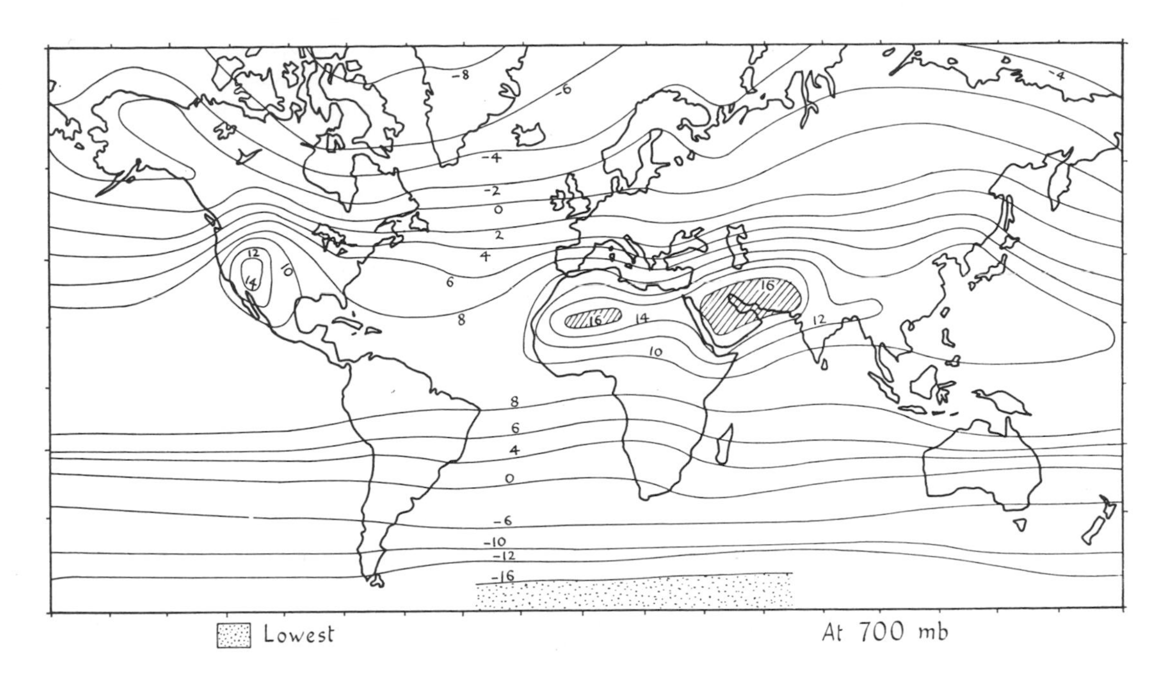

also, the constant rate of ascent of the rubber balloon permits better ventilation for the thermometers. Hergesell in 1904 introduced the practice of tandem balloons: after the first balloon bursts, the second acts as a parachute, allowing the spot where the sonde falls to be easily located. The first theodolite specially designed for balloon tracking was produced by de Quervain in 1905, and during the following ten years, Cave in the United Kingdom carried out the first systematic pilot-balloon observations, using this instrument.

Jaumotte in 1923 modified the Marvin-Bosch meteorograph normally used in sounding balloons by reducing its time-lag. The resulting fineness of trace enabled the structure of frontal surfaces to be clearly delineated during flights through frontal zones. Jaumotte also produced a very small version of this meteorograph, about 9 cm long and weighing only 30 g, which was well-suited to balloon soundings and permitted one launching per hour for two or three days, at critical times during the development of cyclones.

In 1937, Simpson and Scrase attached altielectrographs to balloons, providing the first soundings of thundercloud electricity. In the 1960's, radiosondes not only measure and transmit data concerning pressure, temperature, humidity, and hydrometeors, but also provide observations of airflow, ozone, air conductivity, the average vertical distribution of electrical potential gradient in the atmosphere, and the net flux of long-wave radiation.

The development of telegraphy was of enormous significance in the evolution of sounding-balloon techniques. The "universal telemeteorograph" constructed by Olland and von Baumhauer was exhibited at the World's Fair in Philadelphia in 1876. This instrument consisted of a transmitter and receiver linked by telegraph wires. An obvious meteorological application was to use the cable of a kite as the medium for transmission, but this was not made a practical proposition until 1917, by Herath and Robitzsch.

The first experiments that applied radio transmission to meteorological soundings were carried out in Germany in 1921, when Herath attached a spark transmitter to a balloon. Two years later, in the United States, Blair obtained radio signals from a balloon up to a height of 3.5 km.

Two notable achievements occurred in 1927. Bureau made the first true radiosonde, which could automatically transmit data back to earth by mechanical means. Later, the electromagnetic action of the sensitive surfaces of the instrument was employed to transmit data. Also in 1927, Idrac and Bureau at Trappes used a short-wave transmitter of low power, attached to a balloon, to make a radio link between Earth and stratosphere for the first time. Rapid progress, however, depended on the development of the more advanced forms of thermionic tubes for use in the airborne transmitters. Important advances were made in 1928 by Molchanov, who first successfully applied telechronometry to sounding balloons, and three years later, when Duckert synchronized transmitter and receiver. The first British radiosonde was produced at the National Physical Laboratory in 1938, and was based on the principle of audio-frequency modulation by means of a variable inductance. An instrument developed by Dymond at Kew, working on the same basis, became the standard Meteorological Office instrument.

The tracking of balloons by radio was a separate problem. Radio direction finding was first applied to locating thunderstorms, more precisely, to locating the atmospherics produced by lightning flashes. The technique was developed in the U.S.S.R. in 1895 by Popoff, who employed an untuned circuit to receive electromagnetic disturbances. Cave and Watson Watt, at Aldershot in 1915, used an early type of radiogoniometer to make the first directional observations of atmospherics, and in 1926, the now standard technique of the cathode-ray oscilloscope was first applied to locating them.

Radio direction finding was first applied to the tracking of pilot balloons by Blair and Lewis in 1928 in the United States. Simultaneous measurements at two stations allowed the azimuth of the balloon to be found up to distances of 15 km. In 1940, however, adaptation of the first American-produced radar (see Appendix 1.3) rendered use of two stations superfluous, and made possible the following of balloon transmitters for distances of up to 150 miles, by what was in effect a radiotheodolite.

In 1943, the British Meteorological Office adapted a standard Army primary radar, normally used for antiaircraft gun-laying (set A.A. no. 3, Mark 2, type GL3), to follow a balloon-borne reflector. Observation from a single station gave range, azimuth, and elevation of the balloon, and hence upper winds could be determined accurately.

Development of secondary radar in Germany during the Second World War resulted in the "*Fledermaus*" system, which involved automatic retransmission, instead of reflection, of the signal received by the balloon. An audio-frequency-modulated signal from the ground station, transmitted on 300 megacycles, was retransmitted on 30 mc by a transponder suspended beneath the balloon, the phase difference between the incoming and outgoing signals enabling the slant range of the balloon to be determined for considerable distances. After the war, Jones, Hooper, and Alder in Britain developed a radar-theodolite system using pulsed signals, the time delay between the pulses in the returning signal—which was on a centimeter wavelength—allowing very accurate calculation of the angular position of the balloon transmitter to be carried out.

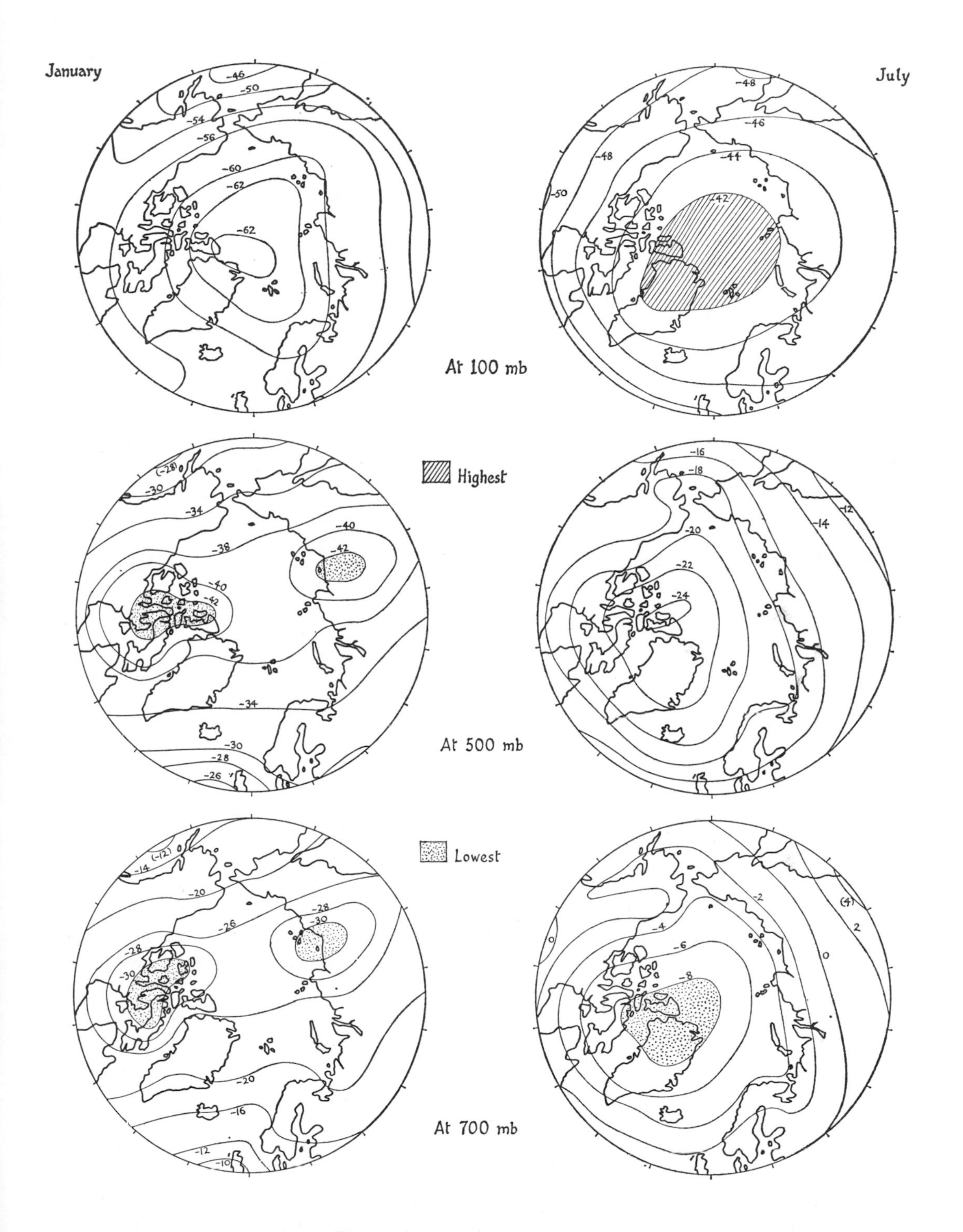

FIGURE 1.9.
Air temperatures in °C
(after Goldie, Moore, and Austin).

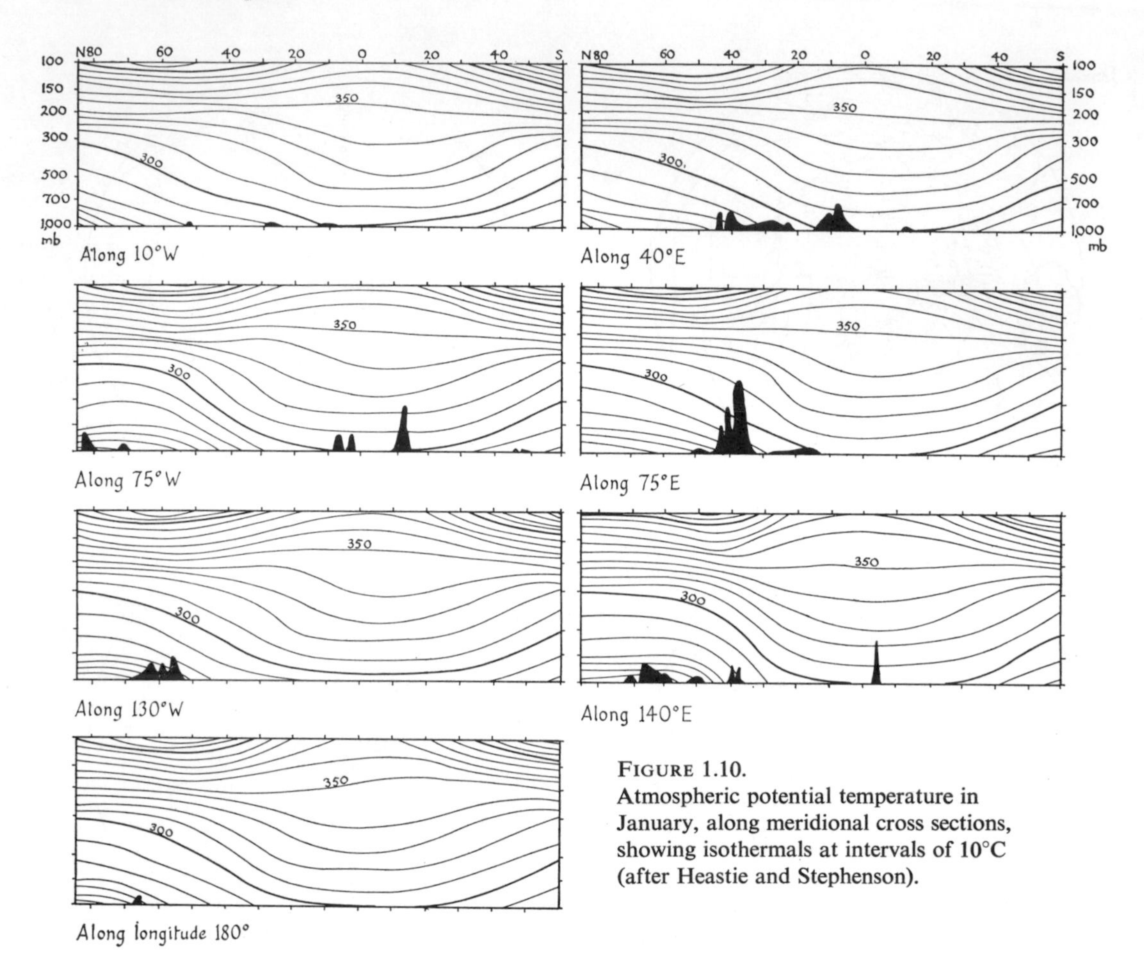

FIGURE 1.10.
Atmospheric potential temperature in January, along meridional cross sections, showing isothermals at intervals of 10°C (after Heastie and Stephenson).

The network of balloon-sounding stations now includes 500 in the northern hemisphere and 100 in the Southern Hemisphere. The oceans create particular difficulties that can be overcome to some extent by the use of transosondes (transoceanic sondes), which had their origin in the incendiary-carrying Japanese balloons that frequently crossed the Pacific during the war to reach the west coast of North America. They were essentially free balloons, held at a preset altitude by automatic means instead of by a navigator.

Tests were carried out in the United States between 1947 and 1950 with constant-volume polyethylene "Skyhook" balloons, and Orville proposed using these, on constant-level (isobaric) flights, for transoceanic soundings. The transosondes were to be equipped with radiosondes for automatic launching at regular intervals, so that soundings of the atmospheric layers beneath the balloon could be made. A radio attached to the transosonde was to collect data from the radiosondes and automatically retransmit it to coastal ground stations. Radio direction-finding stations were to precisely locate the transosonde every two hours or so, giving its trajectory, and thus allowing the pattern of atmospheric flow at the appropriate isobaric level to be determined. Transoceanic soundings carried out by the U.S. Naval Research Laboratory between 1953 and 1956 proved that transosondes not only allow soundings to

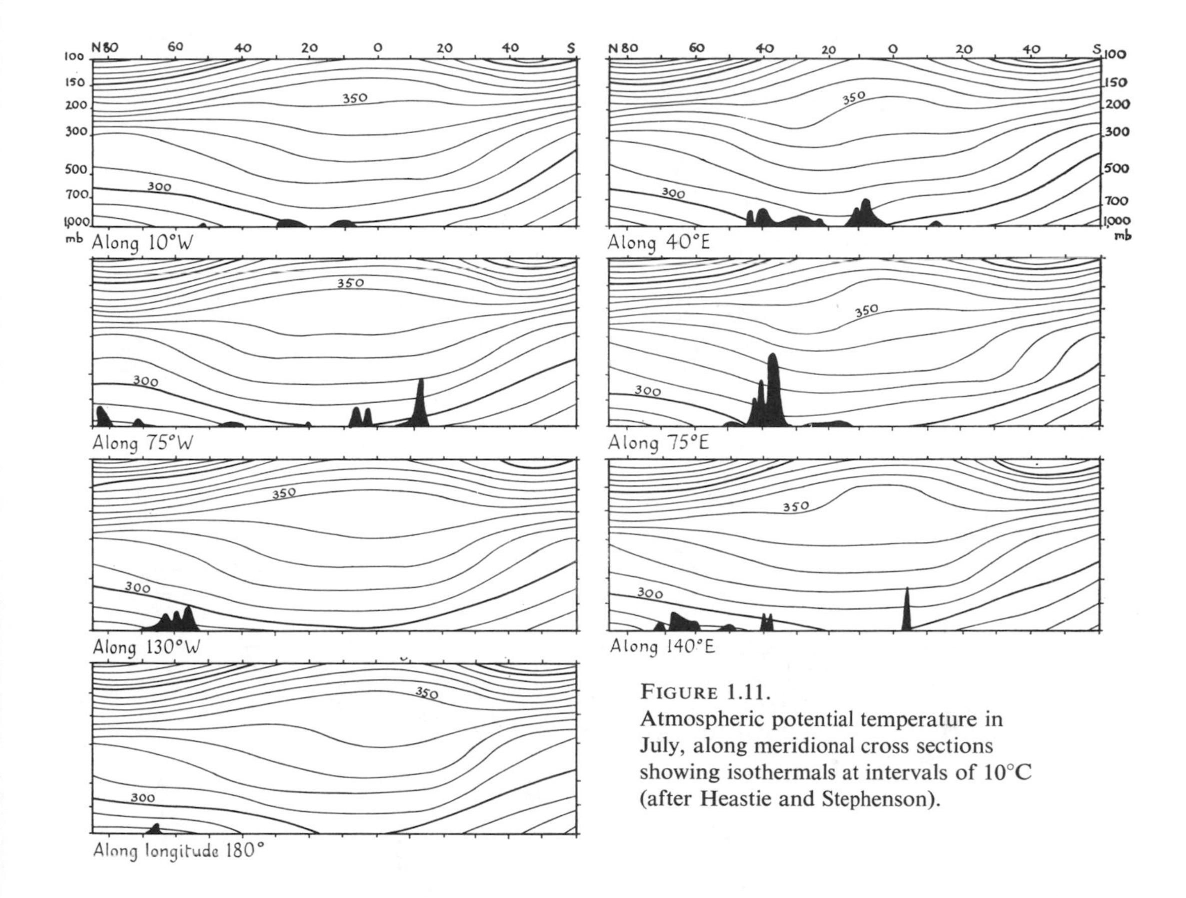

FIGURE 1.11.
Atmospheric potential temperature in July, along meridional cross sections showing isothermals at intervals of 10°C (after Heastie and Stephenson).

be carried out above extensive areas devoid of ground stations (see Figure 1.14), but also furnish data on the synoptic acceleration of the air, which the normal network of sounding stations cannot provide.

The third technique of atmospheric exploration uses aircraft. The first aerological observations from aircraft were made by von Hiddessen in Germany. Early work involved equipping the aircraft with a meteorograph, in addition to which the pilots made visual observations whenever possible. Dines produced a very light (50 gram) meteorograph in 1907, which eliminated the normal clockwork mechanism, and was both robust and easily calibrated. Instead of recording changes in terms of time, it recorded air temperature and relative humidity as functions of pressure. The low price of the Dines meteorograph made it also very suitable for attachment to balloons.

Classic observations of clouds and temperatures up to 3 km over France were made by Douglas in 1915. A psychrometer for use in aircraft was developed by the British Meteorological Office in 1918, the year in which Dobson produced an efficient aircraft barothermograph. A suitable frost-point hygrometer was not introduced until 1942, by Dobson.

During and since the Second World War, specially equipped aircraft—usually converted bombers—have been widely employed to gather data on, among other phenomena, tornadoes, thunderstorms, clouds, and jetstreams. More and more light aircraft are also being instrumented to study, for example, land and sea breezes,

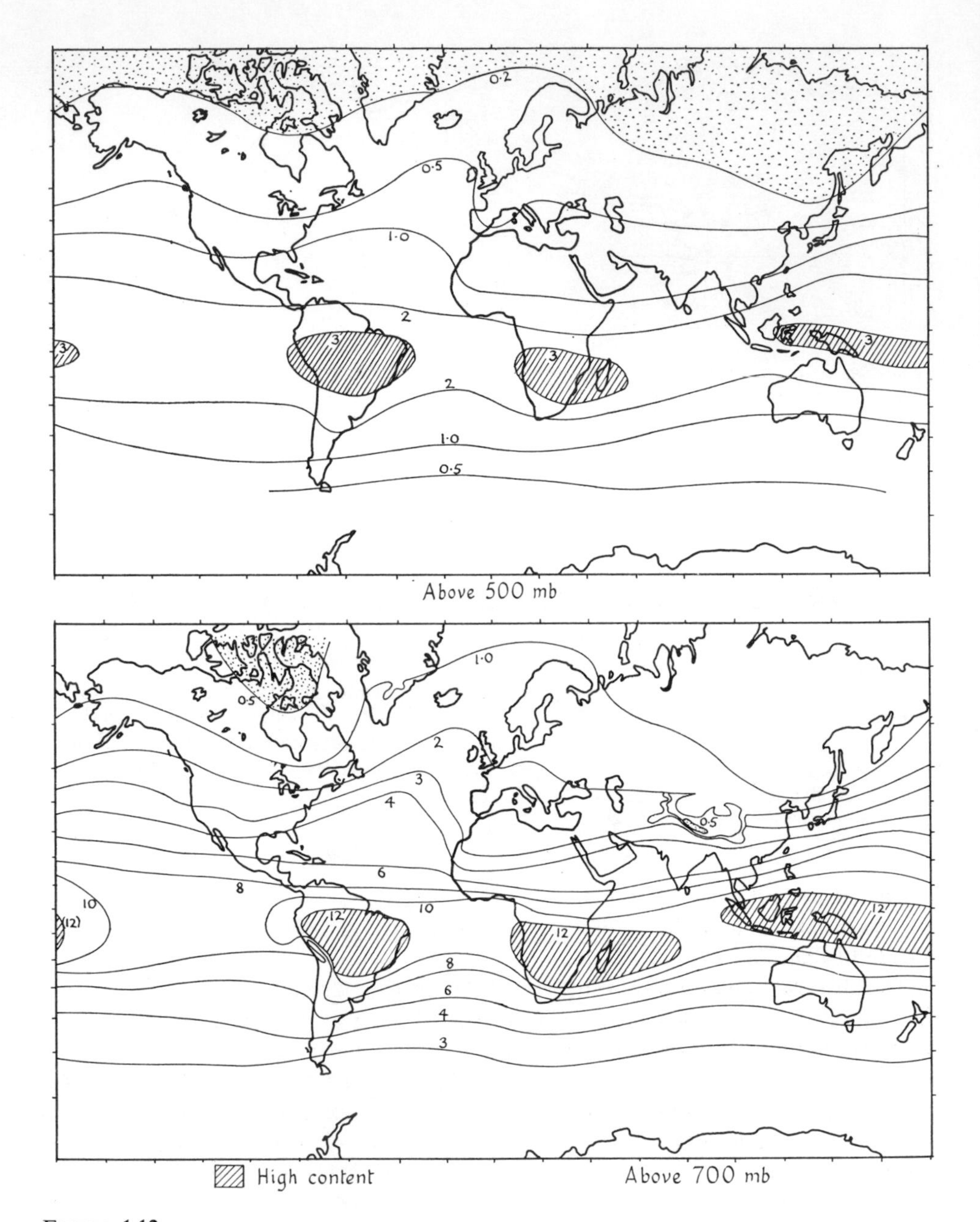

FIGURE 1.12.
Atmospheric water-vapor content in January, 1951-55. The units are decigrams of water vapor above one square centimeter. (After J. K. Bannon and L. P. Steele, *MOGM*, vol. 13, no. 102, 1960.)

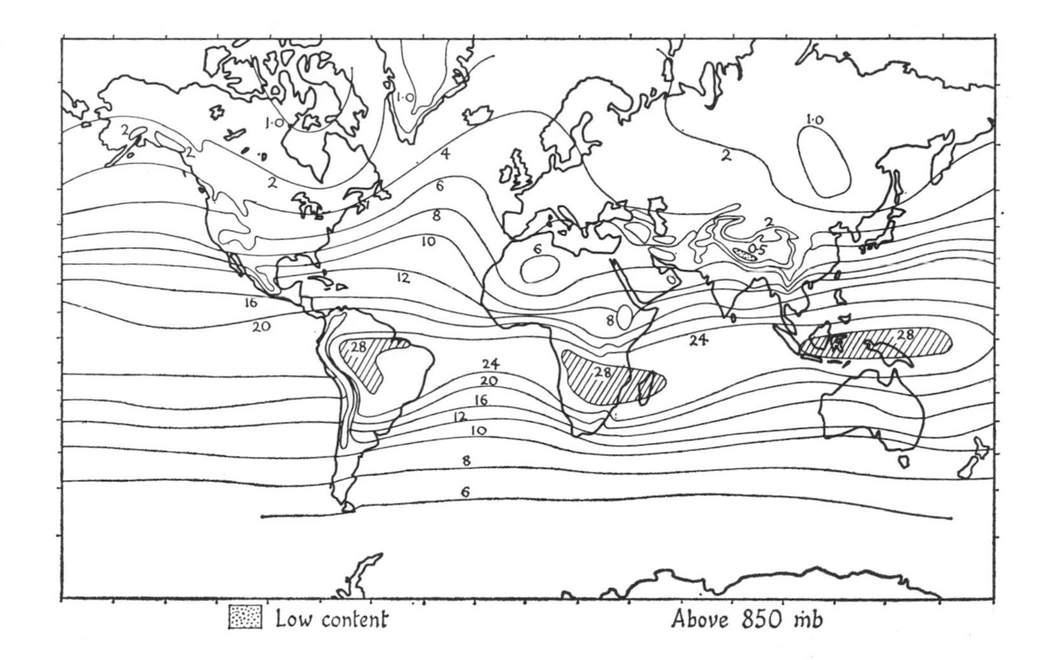

orographic waves and airflow characteristics generally, and atmospheric heat and radiation balances over different types of topographic surfaces.

Aircraft soundings can provide observations and measurements only up to 6 or 7 km above mean sea level. Theoretically, sounding balloons can reach up to 40 km, above which the buoyancy forces acting on the balloons are negligible, but in practice 30 km is the usual limit. Above this altitude, therefore, other techniques must be used. Radio-wave propagation, the abnormal propagation of sound, and observations of aurorae, solar particles and cosmic rays, the light of the night sky, and shooting stars all provide indirect evidence of conditions in the upper atmosphere. More importantly, rockets and guided missiles generally provide direct measurements.

From 1946 on, German V-2 rockets were adapted for meteorological purposes in the United States, and were found to attain 100 km easily. By July 1, 1954, 266 rockets had been fired by the Americans, reaching altitudes of 71 to 389 km. A number of standard meteorological rockets have been developed in America: the Aerobee, reaching 60 km; the Viking, 120 km; and the Deacon-Skyhook or Rockoon, a small, inexpensive rocket launched from a plastic balloon at heights up to 25 km, capable of reaching 100 km. Other countries also developed meteorological rockets, in particular, Britain, France, and the U.S.S.R. The British Skylark rocket reaches 300 km, and the French Veronique rocket 100 km.

Meteorological rockets can take air samples for chemical testing, measure the ion density of the upper atmosphere and its cosmic-ray intensity, and observe the ultraviolet spectrum. Data on the ordinary climatic elements must be obtained by techniques quite different from those used at the ground or in balloon soundings. Atmospheric pressure at great heights must be measured by vacuum techniques: the gauges used determine pressure as a function of either atmospheric thermal conductivity or ionization intensity, depending on the height. Air temperature is calculated from observations of atmospheric pressure and the specific and molar masses of the air,

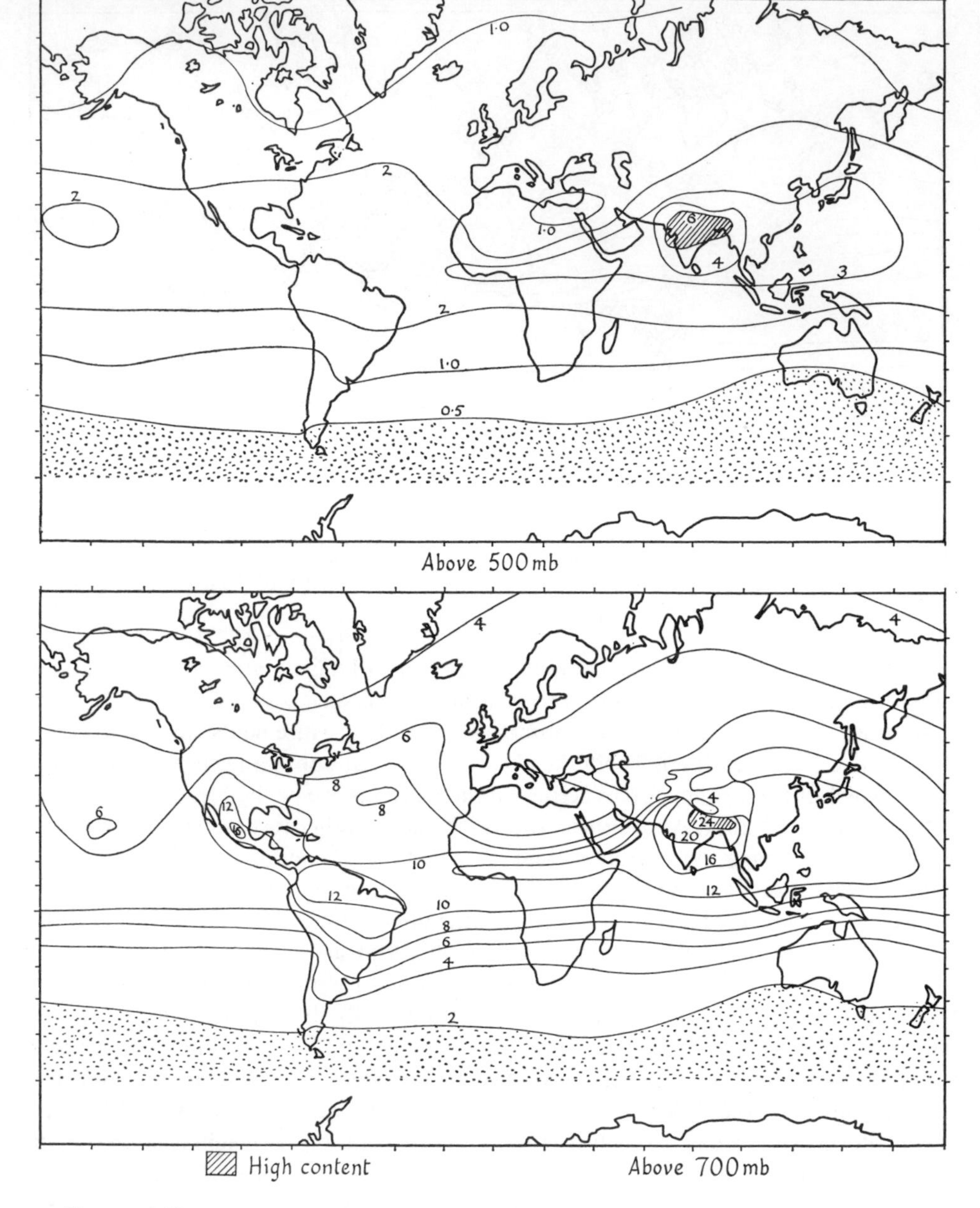

FIGURE 1.13.
Atmospheric water-vapor content in July, 1951-55. The units are decigrams of water vapor above one square centimeter. (After Bannon and Steele.)

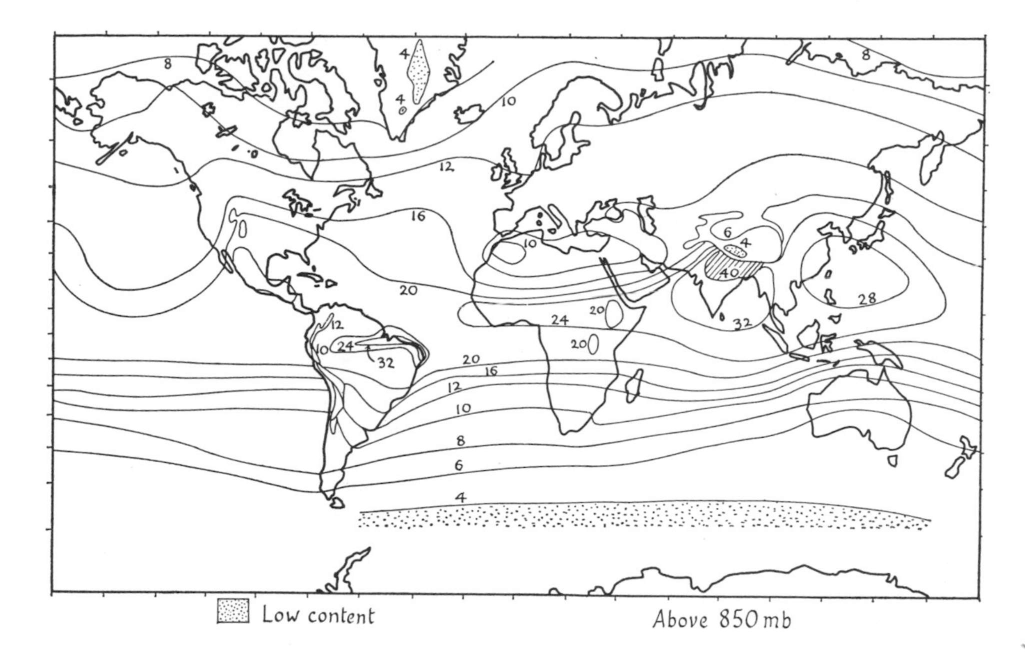

the latter being deduced from a law relating to the ultrasonic displacement of a cone.*

Winds above 30 km can be inferred from the trajectory of a guided missile. The trajectory is determined both optically and by radio direction finding, from the position angle of the axis of the missile (found by means of the gyroscopes mounted in the missile) and from the angle between this axis and the direction of the relative airspeed. This method yields instantaneous winds—not mean winds as with both radiosonde and radar-sonde determinations—since the rockets reach 60 km in 150 seconds.

Rockets are only capable of providing soundings along a single vertical for a very short time. If measurements are required over extensive areas of the Earth's surface or over longer periods of time, then artificial Earth satellites are the most suitable instrument carriers.

The main conclusion to be drawn from our brief survey of geographical exploration of the Earth's atmosphere is that today large-scale investigations are costly in both money and manpower, as well as being highly specialized. But the individual worker has a vast field to explore in the analysis of published data, since the organizations making the actual explorations are generally more concerned with the perfection of the new techniques than with the climatological analysis of the resulting data.

The climatologist needs a knowledge of past techniques of atmospheric exploration, not only because much of his work will be with historical weather data, whose limitations he must know, but also because he may have to use some of these methods himself if he wants to carry out original work. Very frequently, the climatologist interested in some local or regional problem finds that no published soundings are available for his area. In that case, rockets are clearly too expensive, and, particularly

* On the determination of atmospheric pressure from rocketsonde temperature data, see F. J. Schmidlin, *MWR*, 94 (1966), 529.

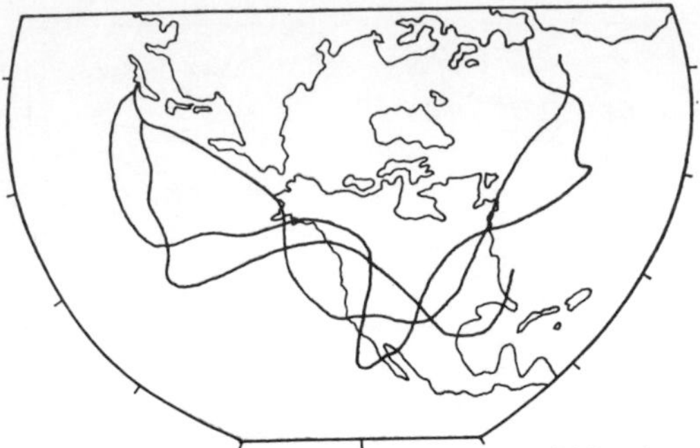

Released from Oppama, Japan at 300 mb, January-February 1956

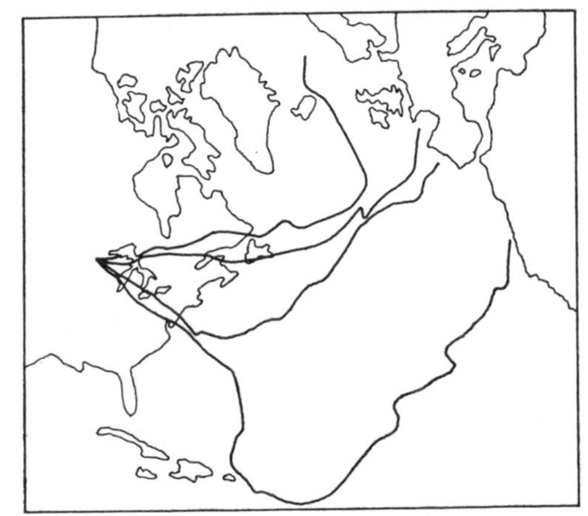

Released from Minneapolis at 300 mb, 1953

FIGURE 1.14
Transosonde trajectories (after Anderson and Mastenbrook).

in densely populated countries with numerous airways, it will be hazardous—if indeed not forbidden—for him to launch his own radiosondes. Therefore he must use pilot balloons, tethered sounding balloons, temporarily instrumented light aircraft, or some of the older methods. Unlike the pioneers, however, he has the benefit of plastics, printed circuits, and transistorized equipment to save him both money and weight. The classic literature of atmospheric exploration is therefore still significant for both climatologist and geographer.[6]

Standard Methods for Measuring Amospheric Pressure*

Atmospheric pressure is the only well-known meteorological quantity that is accurately and regularly measured but is not a weather element itself. It can be shown theoretically to have a close relationship to temperature, winds, and the other elements generally, and is therefore a very suitable basic quantity to which these elements may be referred.

Pressure is the force the atmosphere exerts because of its weight; this force acts

*This account is based primarily on *MOHSI*, chapter 2.

equally in all directions. In the usual definition, pressure equals force per unit area; for the atmosphere, a force of 10^6 dynes per square cm is defined as producing a pressure of one *bar*. This is a rather large unit for practical purposes, and so 10^3 dynes per square cm, the *millibar* (mb), is used as the working unit.

At sea level, atmospheric pressure is not far from 1,000 mb, but it drops very rapidly with height, falling to around 700 mb at 10,000 ft and 500 mb at 18,000 ft (see Appendix 1.4). The variation of pressure with height can be expressed very simply in mathematical form in the *hydrostatic equation*, which is derived from basic definitions. If this equation is applied to two points in the atmosphere, one vertically above the other, a second relation can be deduced, known as the *hypsometric equation*, which connects pressure and height with temperature (see Appendix 1.5). This is very useful in climatology, because it provides a means of determining the "true" temperature of the air, i.e., its temperature before it is affected by topography. Thus the effect of topography on an airstream may be deduced by comparing its "true" temperature with its observed temperature at ground level. The hypsometric equation also provides a way to reduce observed values of barometric pressure to a standard level, a necessary procedure if the values are to be used to study general, not purely local, pressure distributions.

Atmospheric pressure-contour charts depict features that cannot be perceived by the physical senses. The question of their reliability is therefore paramount, and it is necessary to consider in some detail just how accurate the measurement of atmospheric pressure can be.

A first distinction is between barometers, which measure the atmospheric pressure at the level of the instrument at the time of observation, and barographs, which provide a continous record. Barometers are more accurate than barographs, and the latter must frequently be checked against the former, which are the standard instruments for measuring atmospheric pressure.

There are three basic types of barometer. The familiar *Torricellian barometer*, which in effect balances the weight of the atmosphere against a column of mercury, is the most accurate instrument. Mercury is a very suitable liquid, because its vapor pressure is negligible at ordinary temperatures, and its large specific gravity makes the mercury column of convenient length. It does not wet the walls of the glass tube in which it is placed, hence its meniscus is convex upwards, permitting easy reading.

In the *elastic barometer*, an elastic membrane, secured at its edges, is deformed when the air pressure on one side exceeds that on the other. The *aneroid barometer* is of this type, and uses two membranes to form the walls of a chamber. It is small and easily portable, but is less accurate than the mercury barometer, and needs checking at regular intervals against the latter.

The third type of barometer, the *hypsometer*, is very suitable for expeditions or investigations where the readings of the working instrument cannot be compared with those of a standard mercury barometer. It depends on the fact that the boiling point of a liquid—pure distilled water, usually—is a function of atmospheric pressure. The boiling point of a liquid is the temperature at which the vapor pressure of the liquid equals the pressure of the air above its surface. Consequently, once the appropriate function has been determined, the temperature at which the liquid boils measures the atmospheric pressure. The instrument requires a fair length of time in which to make an observation, and a very accurate and sensitive thermometer is necessary. However, the sensitivity of the hypsometer increases with decreasing pressure, so that it is an extremely useful instrument for work in high mountains, or for radiosonde soundings.

The hypsometer is direct-reading, and requires no corrections to give true atmospheric pressure. With the other types of barometer, on the other hand, corrections must be applied. All barometers may be subject to two types of error, which are difficult to allow for routinely. First, *temperature* affects the readings of all barometers, and all parts of the instrument must be at the same temperature to render this effect negligible. Second, *wind* produces either a pressure or a suction on all surfaces exposed to it, and causes rapid and irregular oscillations known as "pumping" in the barometric readings. In moderate winds, barometers in ships are likely to be up to 1 mb in error; in strong winds an indoor barometer on land may read up to 4 mb differently from an outdoor instrument of the same type a short distance away. This effect may be minimized by sealing the indoor instrument in a case that is connected, via a tube, to a static head placed in the open air and well above local obstructions.

Meteorological *mercury barometers* are of two main types. In both types, a vertical glass tube contains the barometric column, and a cistern containing a reservoir of mercury seals the lower end of the tube. An attached thermometer measures the mean temperature of the instrument. When atmospheric pressure changes, the mercury levels in both tube and cistern change. In the *Fortin barometer*, the scale is fixed, but the cistern is raised or lowered to the zero of the scale before a reading is taken. In the *Kew barometer*, the zero of the scale is adjustable.

In practice, Fortin barometers are fixed, whereas Kew barometers hang freely in a gimbal mounting. *Marine barometers* are Kew barometers specially adapted for mounting aboard ships. All these instruments are subject to errors arising from capillarity, vacuum, temperature, and verticality effects, to index error, and to personal bias in sighting the top of the mercury meniscus. Marine barometers have additional errors due to the motion and vibration of the ship, which amount to not more than 1 mb usually. The *vacuum effect* causes readings to differ from the true pressure by as much as 3 mb; the other errors are much less, usually not more than 0.1 mb, but they must still be allowed for in accurate work. With marine barometers, the lag of the instrument may result in appreciable error if atmospheric pressure is changing at 1 mb per hour or more (see Appendix 1.6).

In addition to being aware of these sources of error, the observer using a mercury barometer must apply to his reading, in order to obtain a representative pressure value, corrections for gravity, for temperature, and for height above mean sea level (see Appendix 1.7). None of these corrections is negligible.

By international agreement, mercury barometric readings are reduced to certain datum levels, assuming certain atmospheric properties. Below an altitude of 500 meters (1,600 feet), the datum level is usually mean sea level. Above 500 m, the datum levels normally taken are 1,000, 1,500, 2,000, 2,500, 3,000, and so on, geopotential meters above mean sea level.

In the United Kingdom, an isothermal atmosphere is assumed, and barometric readings are reduced by using the air temperature at station level to calculate the correction, for observations at stations up to 1,000 feet above mean sea level. Above 1,000 feet, a hypothetical vertical air column is assumed to exist from the station down to mean sea level, and the observed pressure is reduced to mean sea level on the assumption that the temperature of this air column increases downward, from station-level air temperature at the top, at a rate of 0.5°C per 100 m.

In North America, barometric readings are reduced by employing a temperature based on the mean of the current air temperature at the station and the air temperature recorded 12 hours previously. This temperature is taken to be that at the top of a hypothetical air column extending vertically downward to the appropriate datum level. A humidity correction is also applied, based on the mean dewpoint over the previous 12 hours.

Meteorological aneroid barometers have one elastic membrane that is fixed at its center, another that is secured only at its periphery. Changes in atmospheric pressure cause changes in the thickness of the diaphragm chamber between the membranes; the movement of the center of the second membrane is magnified by a system of levers, which activate a pointer moving over a scale.

To minimize errors due to temperature effects, the air pressure inside the chamber must be very small, and a spring must be used to keep the chamber from collapsing; hence the actual quantity measured is the change in the deflection of the spring as the load (i.e., the atmospheric pressure) on the outer surface of the membranes changes. If the instrument is taken through a complete cycle of pressure changes, the spring develops a small residual deflection. Also, the deflection for a given pressure is different when pressure is rising and when it is falling. Thus aneroids have an error due to hysteresis, and must be checked frequently against a mercury barometer.

Aneroids exhibit a small, long-period error arising from secular changes in the internal structure of their constituent metals. More important errors result from changes in temperature. The elastic properties of the diaphragm, or of the spring, depend on the appropriate values of Young's modulus, or the rigidity modulus, for the material, which varies with temperature. The temperature coefficients of these moduli are negative for most alloys and metals, and therefore the pressure recorded by the aneroid apparently increases when the temperature of the instrument increases. This error can be compensated for by leaving a small volume of air inside the diaphragm chamber, or by introducing a bimetallic link in the system of magnifying levers.

Because of these errors, pressures derived from aneroid readings are only accurate to within about 1 mb, except for the precision aneroid, which is as accurate as a standard Kew barometer.[7] This instrument employs a two-stage beryllium-copper alloy capsule, fixed at one side, the other side moving as the air pressure changes and at the same time deflecting a bar pivoted in jeweled bearings. The bar is kept in contact with the aneroid capsule by a hairspring, which exerts much less pressure than the gears and levers of the conventional aneroid. The displacement of the free end of the bar is measured by a micrometer which displays its contact with the bar electrically on a cathode-ray tube. The micrometer is calibrated directly in millibars and tenths from 930 to 1,055 mb. Alternatively, the micrometer may be replaced by a visual counter mechanism. The whole instrument is effectively sealed off from ambient temperature changes by a metal case. Some of these instruments show a tendency to zero-shift, and therefore must be checked at regular intervals against a standard barometer. However, they are easier to check than mercury barometers, and represent a considerable improvement on the conventional aneroid.

The simplest form of barograph uses the aneroid principle, with a recording pen replacing the pointer. These *aneroid barographs* employ a bellows-type diaphragm chamber with internal springs. The bottom surface of the diaphragm is fixed to a brass base plate, and the vertical motion of the top surface of the diaphragm is transmitted to a pen arm via a magnifying lever system. The "control" of the instrument depends on the crossectional area of the aneroid chamber: the greater this area, the clearer and more detailed the record (see Appendix 1.8).

There are three standard types of aneroid barograph: open-scale, small-scale, and marine. With the open-scale instrument, the aneroid unit consists of two or three chambers; with the small-scale type, it comprises seven or eight chambers. Both barographs have weekly clocks normally, covering a pressure range of 950 to 1,050 mb. In the open-scale type 1 mb is represented by 1.8 mm on the chart, but in the small-scale instrument, 0.75 mm represents 1 mb, which is too compressed a scale to allow pressure tendencies to be estimated from the chart to within less than ± 1.5 mb.

Open-scale barographs are accurate to ± 0.2 mb usually, but their error varies with the ambient temperature and the altitude of the instrument. For stations up to 250 feet above mean sea level, these barographs are usually set to read mean sea-level pressure. Above 250 feet, they are set to read the pressure at station level, the mean station pressure being roughly at the center of the chart. An instrument set correctly at mean sea level will not give a true reading if the outside air temperature changes: at 250 feet above sea level, a barograph reading correctly at 55°F will be 0.4 mb in error when the temperature drops to 30° or increases to 80°F, and this error increases with altitude.

Marine barographs are open-scale instruments in which the aneroid bellows are immersed in a brass cylinder filled with oil. This immersion considerably reduces the lag coefficient of the instrument, since the bellows can only expand or contract by forcing the oil to flow through a narrow annular gap where the connection from the bellows to the lever system passes through a hole in the diaphragm (see Appendix 1.9).

Both small-scale and open-scale barographs are normally mounted on a rigid horizontal support, away from sunshine or artificial sources of heat, and on rubber pads to eliminate the effect of vibration. For marine barometers, a special antivibration mounting has to be employed. This consists of a metal tray, which can swing about an axis parallel to its longer edge, suspended by elastic cords from fixed brackets. The oscillation of the tray prevents the motion of the ship from throwing the pen off the chart.

Ordinary open-scale barographs cannot be carried on ships, because the resulting pen trace shows a broadening (up to 2 or 3 mb in rough weather) which bears no relation to the true barometric tendency. This broadening of the trace results from the rolling and pitching of the ship (causing movement of the pen arm), vibration from the ship's engines, wind gusts (causing short-lived pressure changes), and pressure oscillations due to the rise and fall of the ship. Makeshift arrangements, for example, a soft woolen blanket, folded eight times and placed beneath an ordinary open-scale barograph, allows traces to be obtained which show broadening only in heavy seas—wind force 5 to 6—because of the rolling and pitching of the ship.[8] With the oil-damped marine barometer, the broadening of trace does not exceed 0.3 mb.

Microbarographs record very small oscillations and changes in atmospheric pressure. The classic *Shaw-Dines microbarograph* records the rate of change of pressure, not the pressure itself. A large reservoir, lagged to keep its temperature constant, is connected to the outside atmosphere through a fine capillary "leak." The size of the leak decides the lag, which differs for each instrument. The difference in pressure between the air in the container and the outside atmosphere is measured by a float in a mercury bath, the movement of the float being transmitted to a pen arm. An outlet tube from the container passes through the mercury inside the float, and the leak can be bypassed by connecting the reservoir directly to the atmosphere by a valve. The recorded deflection of the instrument is proportional to the difference in pressure between the outside and the inside of the float. When atmospheric pressure is changing steadily, the deflection is constant, proportional to the rate of change of pressure multiplied by the time-constant of the instrument (see Appendix 1.10).

The Shaw-Dines microbarograph is not portable, since the mercury has to be removed before it can be transported. It therefore has never been a common standard instrument, although because of its very open scale—15 mm representing 1 mb with its usual magnification of 40—it is capable of a very detailed record.

Another classic instrument, more sensitive than any microbarograph, is the *Dines float barograph.* With this instrument, pressures can be recorded with an error of less than 0.1 mb, and read from the chart to the nearest 0.001 inch. Thus it provides a permanent record as detailed as one obtained by making a series of consecutive readings with a mercury barometer or a precision aneroid. Like the Shaw-Dines microbarograph, the mercury must be removed before the instrument can be transported.

Essentially, the Dines float barograph records the changes in level of one arm of a siphon-type mercury barometer. Two reservoirs are employed, the lower one being exposed to the atmosphere, the upper one being sealed by mercury in an adjacent U-tube. A glass float, resting on the mercury surface, is maintained coaxially with the reservoir by means of a ring of floating steel balls. Temperature changes are compensated for by the air beneath the float. A counterweighted steel tube, suspended from a pulley, fits over a glass stem on top of the float. A second pulley has a recording pen attached to it, and two fixed pens draw datum lines on the chart, enabling expansion or contraction of the latter due to temperature or humidity changes to be taken into account.

The most accurate standard instruments are thus the mercury barometer and the Dines float barograph. In the United Kingdom, Kew and Fortin barometers provide the working standards, and a siphon-type barometer maintained at the National Physical Laboratory forms the primary standard, accurate to − 0.01 mb. In practice, the index error of a portable "secondary standard" barometer is found by comparing its readings with those of the primary standard, and then the station barometers are checked against this secondary instrument.

From this survey of standard pressure-measuring instruments, it is evident that there is considerable variation in both accuracy and sensitivity. Therefore, before climatological deductions may be made from any pressure charts, the types of instrument used in obtaining the original data must be ascertained.*

Pressure Maps and Their Climatological Significance

One serious drawback in any work involving pressure maps is that it is often difficult, if not impossible, to obtain accurate continuity and comparison between surface and upper-air charts. One reason is that, unlike the surface chart, the upper pressure

* Occasionally, series of pressure data may be found that were derived from nonstandard instruments. For details of a brine barometer, see P. R. Corbyn, *W*, 22 (1967), 378.

surfaces undergo a systematic diurnal variation in height. This variation may be a real feature of the atmosphere, or it may represent an effect of solar radiation on the temperature elements in radiosondes.[9] The amount of variation increases with height. Over Britain, it increases from 5 feet at 700 mb to 170 feet at 100 mb in winter, and from 35 feet at 700 mb to 370 feet at 100 mb in summer. This diurnal height variation amounts to more than four times the height variation brought about by the passage of synoptic disturbances at 100 mb.

Another reason for the lack of continuity between the two types of maps is the relative inaccuracy of the sea-level pressure chart, which is based on extrapolations to mean sea level made by individual stations in terms of fictitious air columns. The user of the chart cannot always be certain how these fictitious columns were defined, and in any case the procedure means that the sea-level pressure chart is a function of both air temperature and air pressure, whereas the upper-air map is a function of pressure alone.

Particularly in high plateau regions, the hypothetical sea-level pressure patterns are much distorted by horizontal temperature gradients, whereas the actual surface pressure pattern could not be. The winter Asiatic anticyclone provides a good example of this. The mean sea level map shows a closed 1,030-mb isobar over the Siberian high plateau, yet the real surface pressure is around 700 mb. The real Asiatic high is not centered over Mongolia or the Irkutsk region at all, as it appears to be from atlas maps, but over the Peking area. In fact, the Siberian anticyclone is not a single gigantic high at all. Instead, one relatively small anticyclone actually occurs over North China, and a shallow one over the Baikal region. The pressure gradient between these two highs is very steep, and the outward flow of cold air from Siberia is mainly due to temperature effects, not to any squeezing action by the anticyclones.[10]

Three pressure parameters other than the normal pressure values offer certain advantages for use in charts.[11] These are the pressure altitude (z), the altimeter correction (D), and the altimeter setting (P), which are all based on the use of a *Standard Atmosphere*, i.e., an agreed-on set of measurements that most closely approximate the conditions in the real atmosphere and that at the same time conform perfectly to basic theory (see Appendix 1.11).

The parameter z is the height above any selected level in the atmosphere at which the actual pressure at the selected level would be in the Standard Atmosphere (see Appendix 1.12). It is not related linearly to barometric pressure as measured by the height of a mercury column, because of the compressibility of the air. The parameter D is the difference between the height h of the station above mean sea level and the value of z for that station; i.e., it is the vertical displacement of the Standard Atmosphere from its standard position that is required to obtain the observed pressure at the height of the station above mean sea level. The parameter P is the pressure which would occur at sea level if the Standard Atmosphere were between the station and sea level.

Maps of constant D are more accurate than the usual mean sea-level pressure maps: they precisely describe the actual pressure distribution, and more closely represent the true gradient wind flow. In addition, the observed pressure for any station can be readily found from a D map, whereas it cannot be found from a sea-level map without knowledge of the reduction technique used (see Appendix 1.13). Thus D maps often differ considerably from sea-level maps, as Figure 1.15 indicates.

Atlas maps of mean sea-level pressure tend to give the impression that the patterns exhibited represent stable "normal" conditions for the Earth. Actually, there are a good many "norms"; in the North Atlantic the location and period of maximum normal (i.e., 30-year average) pressure has moved southward from Iceland and Scandinavia in 1870 to Portugal and Madeira in 1919–21.[12]

Although the mean annual pressure at a given location tends to show a high degree of correlation with the mean annual pressure for the following year (see Appendix 1.14), 10-year averages of atmospheric pressure at any one place show considerable variations. For example, the averages for England and Wales show steady increases of up to 3.6 mb between 1922 and 1957 in spring, summer, and autumn. In winter, the 10-year averages over the period 1924 to 1943 had maxima of 1,016.4 and 1,015.6 mb at the beginning and end of the period, respectively, and a minimum of 1,012.8 mb in the years between 1934 and 1943.[13]

Deviations of atmospheric pressure from its mean value are of great significance for both weather and climate. Thus severe winters in Britain are due to easterly airflows, associated with a stationary anticyclone over northern Europe, that displace the normal westerly flow. This happens about once very ten years. February 1947 was a good example. The normal pressure gradient from 1,021 mb at Horta, in the Azores, to 999 mb at Reykjavik, Iceland, was reversed to 1,009 at Horta and 1,023 at Reykjavik, which created severely cold days in Britain, with heavy snowfall. February 1895 and January 1940 had above normal pressure in Britain, giving rise to intensely cold nights but no great snowfall. In January 1881 pressure was below normal over Europe, but the pressure gradient was not steep: very snowy conditions resulted, but not intense cold.[14]

October 1948 to September 1949 witnessed a long period of abnormally high pressure over Europe and the North Atlantic. The anomaly area extended from central Asia to the Mississippi, the greatest excess (more than 2.5 mb) being centered over Belgium. In March 1953, very high mean pressure again occurred over Britain, the

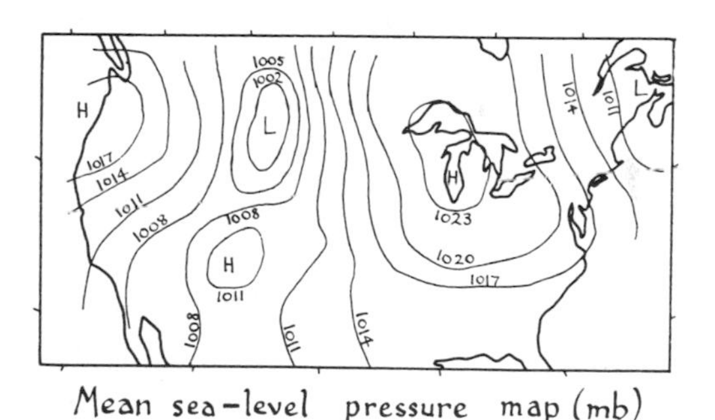

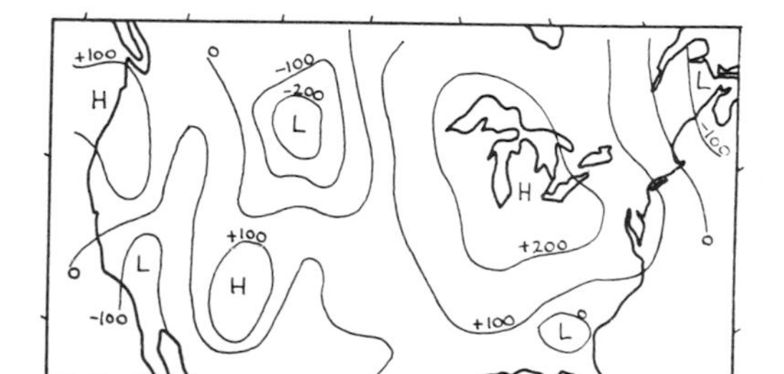

FIGURE 1.15.
Pressure maps. Both maps are for 4:00 A.M. G.M.T., August 25, 1944. (After Bellamy.)

monthly 9:00 A.M. G.M.T. means varying from 1,031 mb in the Thames valley to 1,023 mb in the Orkneys.[15]

Taking a world-view of the variability of mean sea-level pressure, two homogeneous zones occur in the northern hemisphere, one centered on 65°N and the other near 35°N, such that pressures in one zone correlates positively with pressures in the same zone and negatively with pressures in the other zone.[16] The mass shifts of air between these zones represent in effect the sum of two processes: (1) the normal seasonal mass transport of air due to insolation variations; and (2) an irregular mass transport of air across the equator. Process 1 causes large movements of air from tropical latitudes in one hemisphere to tropical latitudes in the other, whereas process 2 causes large mass shifts between the north and south polar regions.

Vast seasonal exchanges of air take place between the continents and between the northern and southern hemispheres (see Figure 1.16).[17] Air moves from continent to ocean from February to July, and from ocean to continent from August to January. The flow is most rapid at the equinoxes, and is much more pronounced over Eurasia than over North America. The summer decrease in pressure over the northern hemisphere continents results in runoff of huge masses of air, leading not only to enlargement of the oceanic subtropical anticyclones north of the equator, but also, since some of the air spills over into the southern hemisphere, to an increase in the size of the southern oceanic subtropical high-pressure cells.

Although solar control is clearly the primary cause of the seasonal movements of the atmosphere, the chain of cause and effect is very complex. Thus the seasonal movement of the doldrums and the subtropical anticyclones occurs two months after the sun is vertically overhead. The axes of the anticyclones move only half the distance the doldrums do.[18]

The areas with the greatest variability of daily pressure vary with height (see Figure 1.17). Thus at sea level, the greatest variability occurs over Novaya Zemlya and in the region of the Icelandic and Aleutian lows in the northern hemisphere. At 500 mb, the maximum variability is found over North America, but this continent exhibits the least variability found in the northern hemisphere at 700 mb.[19]

Correlation studies have demonstrated the existence of widespread oscillations of air quite distinct from any periodic movements. The most remarkable of these are the *Southern Oscillation* and, to a lesser extent, its northern counterpart, the *North Atlantic Oscillation* (see Appendix 1.15).

Blanford in 1887 discovered that droughts in India were associated with high surface pressures over Mauritius, Australia, and much of Asia; in 1897 Hildebrandsson found that pressures at Sydney and Buenos Aires showed opposite tendencies. The latter was in effect the discovery of the Southern Oscillation. In 1902 the Lockyers demonstrated that the surface pressures around the Indian Ocean and those in Argentina generally varied inversely. Walker found in 1924 that the mean surface pressures of Buenos Aires, Cordoba, and Santiago in April and May were related to Indian monsoon precipitation during the following June to September period. In 1932, he and Bliss confirmed that the effect noted by the Lockyers was a very large-scale one: it represented a tendency for air to be removed from the whole Pacific area, from Japan to South America, at the same time as air accumulated over the Indian Ocean, and vice versa. This tendency forms the Southern Oscillation: a mass pulsation of air which has no regular periodicity, but is closely correlated with itself six months ahead, and is related to, but not controlled by, sunspot variations.

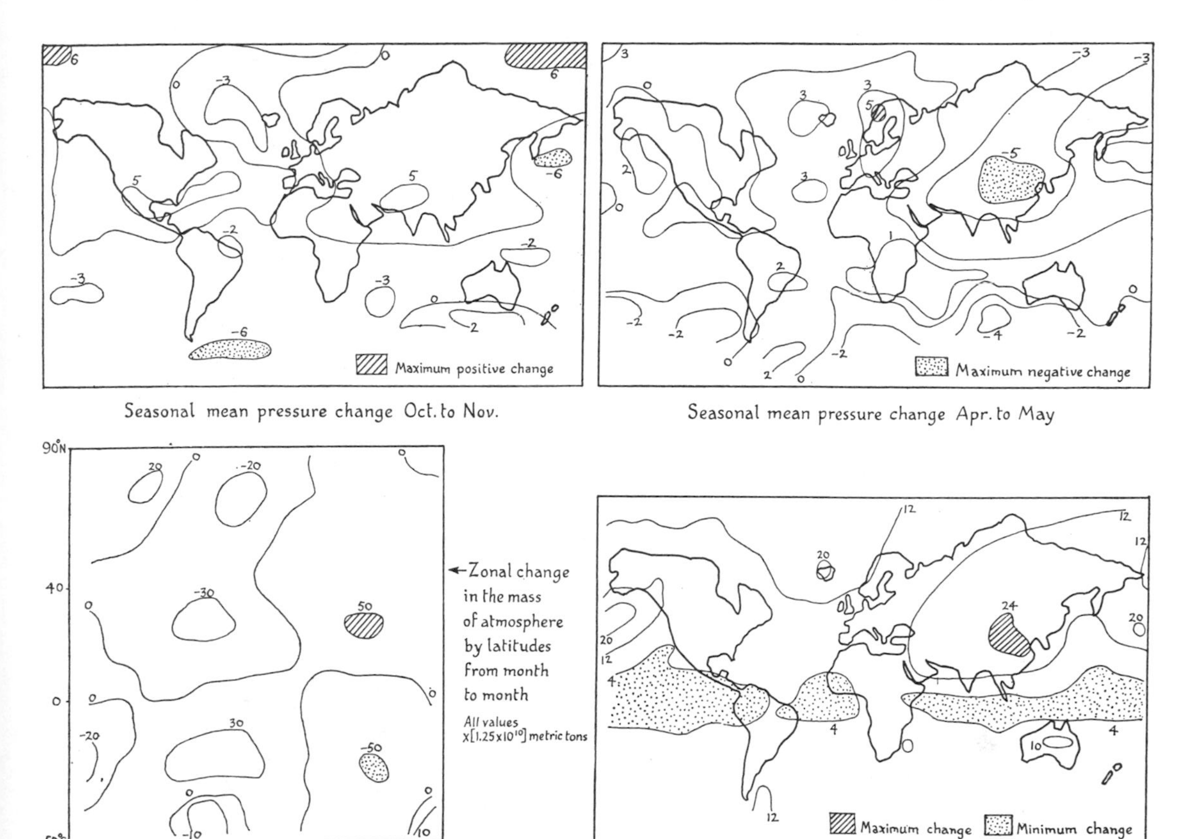

FIGURE 1.16.
Global pressure oscillations. All pressures in millibars. (After A. H. Gordon.)

It should be emphasized that the Southern Oscillation is a tendency only, and does not form the dominant part of the total observed variation of pressure at any one place. The same tendency appears in rainfall figures from Australia, India, and Java, and also in some temperature records. It is somewhat related to the pressure oscillations observed to take place between the Azores and Iceland and between comparable areas in the North Pacific.

Figures 1.18 and 1.19 indicate the extent of the Southern Oscillation and its counterpart in the North Atlantic. By regarding the Oscillation as a gigantic air-pressure balance whose scales are centered near Djakarta and near Easter Island, it is possible to map its extent from correlations of pressure between each of these two locations, taken in turn, and all other available stations in the world. The maps prove that the Southern Oscillation has worldwide ramifications: both Arctic and Antarctic areas seem to be linked with the Djakarta variations. The Djakarta scale is zonally developed, particularly along the equator, whereas the Easter Island scale is meridionally developed.

During the period 1949 to 1957, a winter-type circulation dominated both northern and southern hemispheres. Over the North Atlantic region, the amplitude of the

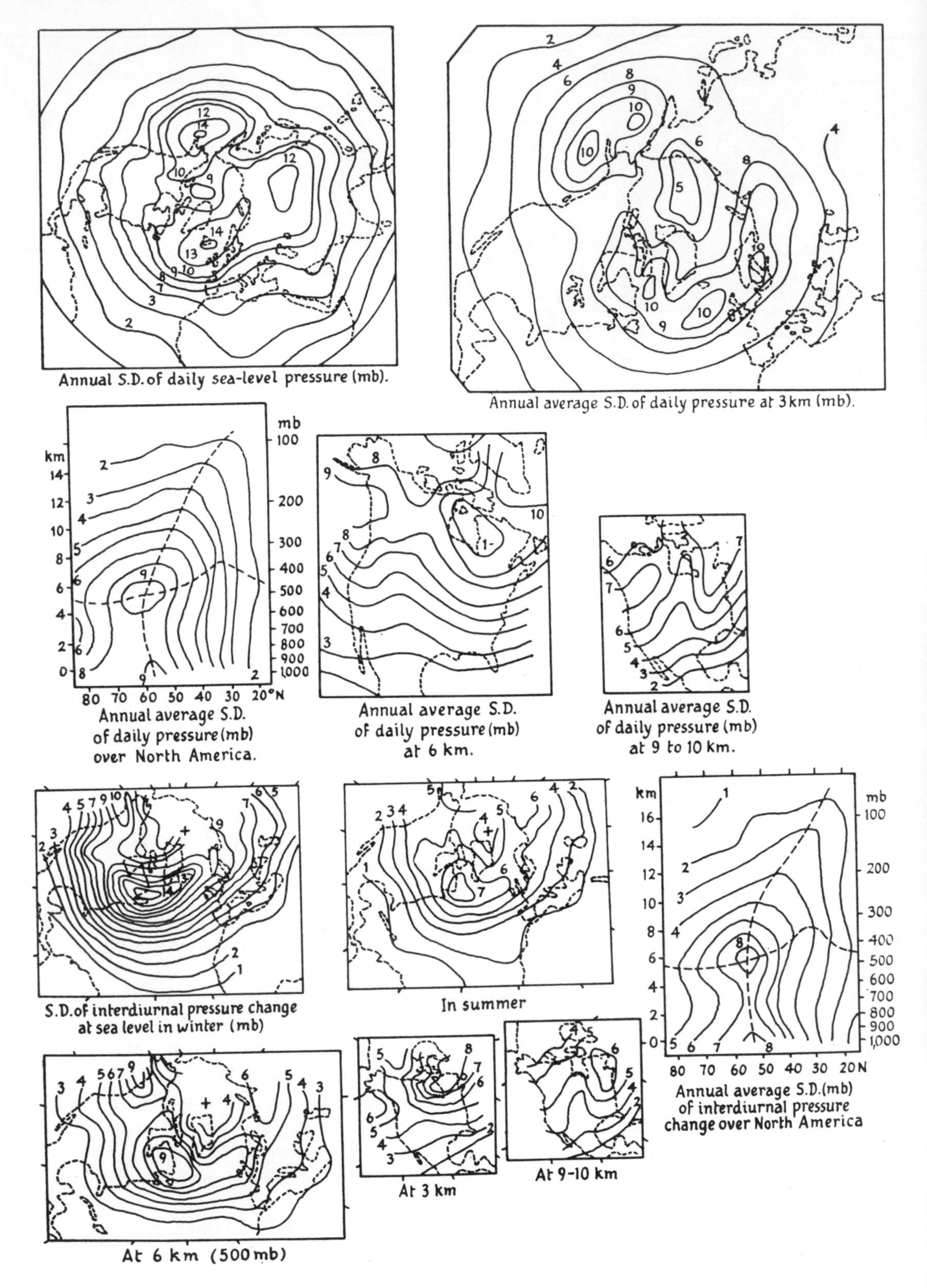
Annual S.D. of daily sea-level pressure (mb).
Annual average S.D. of daily pressure at 3 km (mb).
Annual average S.D. of daily pressure (mb) over North America.
Annual average S.D. of daily pressure (mb) at 6 km.
Annual average S.D. of daily pressure (mb) at 9 to 10 km.
S.D. of interdiurnal pressure change at sea level in winter (mb)
In summer
Annual average S.D. (mb) of interdiurnal pressure change over North America
At 6 km (500 mb)
At 3 km
At 9-10 km

Southern Oscillation exceeded that of the North Atlantic Oscillation by 24 per cent. During 1931 to 1939, however, a summer-type circulation prevailed, and the North Atlantic Oscillation was well-developed over the North Atlantic region, whereas the Southern Oscillation was not. During this period the average length of both oscillations was the same (20.8 months), but during 1949–57 the length of the Southern Oscillation was 29.8 months as against 20.8 months for the North Atlantic Oscillation. Clearly, therefore, these oscillations are not features with a simple annual variation.

The Southern Oscillation appears to act for the most part like a stationary wave, but occasionally it shows the characteristics of a progressive wave. After a long and excessive pressure difference between a high-pressure cell over Easter Island and a low-pressure area over Djakarta, there will be a sudden large increase in the amplitude and wavelength of the Southern Oscillation. Afterwards, both amplitude and wavelength gradually decrease, but the general circulations of both atmosphere and ocean in the South Pacific area are accelerated until the next relaxation.[20]

Recurrence tendencies in atmospheric pressure, brought to light by means of harmonic analysis and autocorrelation techniques, are quite marked in certain areas of the globe. Thus Batavia pressures tend to recur at intervals of 2.34, 3.36, 5.97, 7.32, 8.47, 11.12, and 15.87 years.[21] Cycles of 11 and 36 years, approximately, appear to be worldwide, originating in the tropics, then converging and losing intensity poleward. The 11-year pressure cycle varies according to the 36-year rhythm; there is also a 16-year cycle due to interaction between the 11-year and 36-year cycles.

A 20-day pressure recurrence is almost worldwide, with some remarkable exceptions: for example, it does not occur at Kew. Cycles of 72 days and factors thereof, quite common in many areas, are not present at Kew either. The reason seems to be that, in theory, the pattern of pressure oscillation on the Earth's surface at any time consists of a number of adjacent high-pressure and low-pressure cells, separated by lines of zero amplitude which may intersect one another. The amplitudes of annual pressure oscillation are greatest over the Atlantic near Iceland and (in opposite phase) over the continent of Europe. The coastline of northwest Europe has the smallest amplitude, suggesting that Kew may lie on an amphidrome or line of nodes.[22] A 36-day cycle is also well-marked in northwestern Europe, and a 30-day oscillation occurs in winter, with its maximum amplitude over the North Sea.

Pressure shows well-developed tendencies to persist. Thus surface pressures for the eastern North Atlantic and those for western Europe correlate positively with each other if the pressures in successive months are compared. The persistence is particularly marked in winter and at northerly stations, suggesting that variations in the extent of Arctic sea ice is the controlling factor.[23]

Significantly, the persistence patterns of surface pressure for the period 1899 to 1953 differ from those for 1932 to 1951. Two large areas of high positive correlation exist in the southeastern United States and in the east-central Pacific from December to March: these are displaced southward in January, returning northward in February. Two areas of negative correlation are in the Gulf of Alaska and in the region of the Icelandic low. From March to September, there is high positive correlation

FIGURE 1.17.
The geographical pattern of standard deviations of pressure (after Klein).

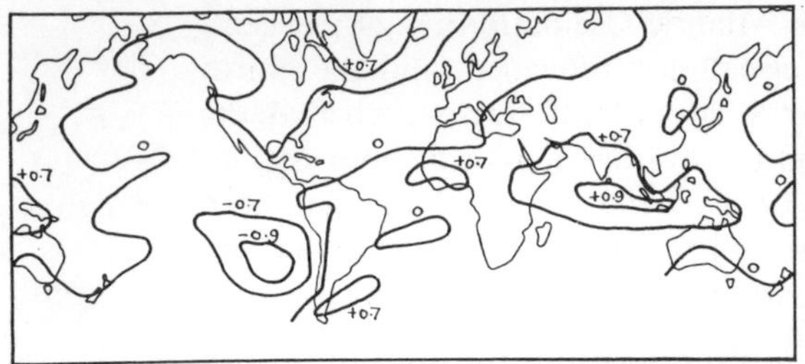

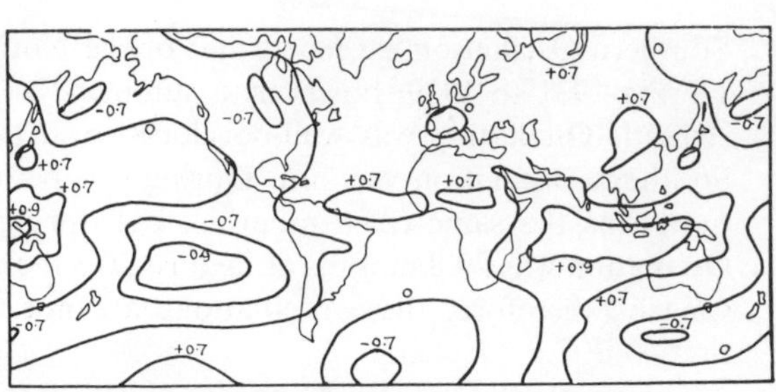

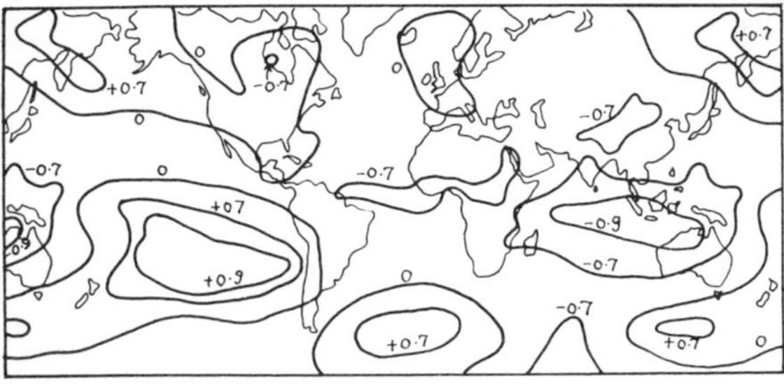

FIGURE 1.18.
The Southern Oscillation. Yearly pressure departures from normal for stations specified. (After Berlage and de Boer.)

over the South Atlantic, and a belt of minimum correlation (i.e., least persistence) moves northward over North America. The area of maximum persistence is usually found just north of the normal ice-snow limit in spring and early summer in the northern hemisphere. The period September to December shows much negative correlation of monthly pressures, October and November in particular being months of very transient pressure distribution.[24]

A comparison of the persistence of surface pressure distribution for most of North America with a comparable area centered on Greenwich indicates that the latter exhibits much greater persistence for periods of up to three days, possibly because more small-scale pressure features occur in North America than in Europe. The persistence correlations between the two areas are unrelated; both show high persistence in winter and summer and low persistence in spring and autumn, but with great differences. For example, the greatest persistence for North America is found in July, the greatest for Europe in January. The annual oscillations in persistence are out of phase if the curves for the two areas are compared, but the semiannual oscillations are in phase. In general, the surface persistence is closely related to the rate of change of windspeed with height in the zonal component of the geostrophic wind up to 500 mb, low persistence being associated with a high rate of change. Both North America and Europe behave very similarly in persistence, however, during the few days immediately following large geomagnetic disturbances.[25]

The tendency for a time-scale graph of the changes of pressure to present a mirror image of itself after certain days, termed *symmetry points* by Napier Shaw in 1906 and first discovered in pressure data by Weickmann, has given rise to much controversy over the physical reality of the phenomena.[26] Weickmann found a pressure periodicity of 140 days, by assuming that the time interval between successive symmetry points should be a measure of half the fundamental period of the pressure oscillation inducing them. Also on the basis of symmetry points, pressure waves of

24 and 36 days were discovered. Symmetry points may be merely persistence features associated with pressure surges. Using data for Melbourne, for example, they appear to be readily explainable as chance phenomena. In northwest Europe, in contrast, physically significant symmetry points occurred 36 days apart during the winter of 1932–33, the first one being near January 1.[27]

Atmospheric Tides

Day-to-day pressure variations as revealed by an ordinary barograph are very irregular, despite the fact that their primary cause, the effects of the sun on weather and climate, is rigidly periodic. To separate the irregularities from the periodic phenomena is a central problem of climatology, and weather forecasting suffers because this cannot yet be done with complete success.

Barograms from tropical and nontropical stations show considerable differences (see Figure 1.20). The nontropical stations indicate large irregular variations due to the passage of synoptic disturbances. The tropical stations show a regular variation of much smaller amplitude, representing tidal effects in the atmosphere; as the tidal motion heaps air up at or draws air away from a certain locality, so a barograph situated there will record an excess or a deficit of pressure.

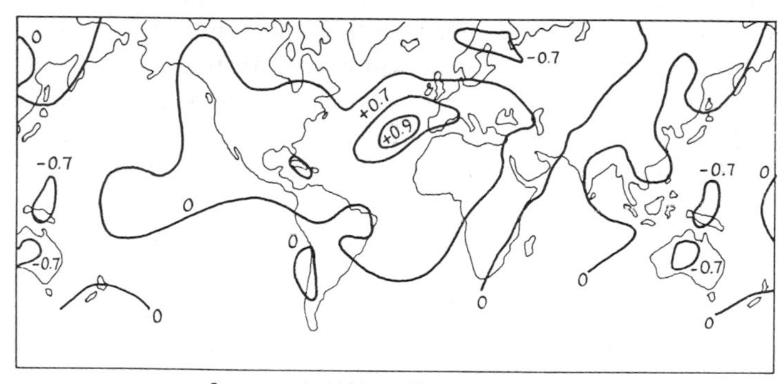

January 1, 1931, to January 1, 1939

July 1, 1949, to July 1, 1957

FIGURE 1.19.
The North Atlantic Oscillation. Annual pressure departures from normal at Ponta Delgada. (After Berlage and de Boer.)

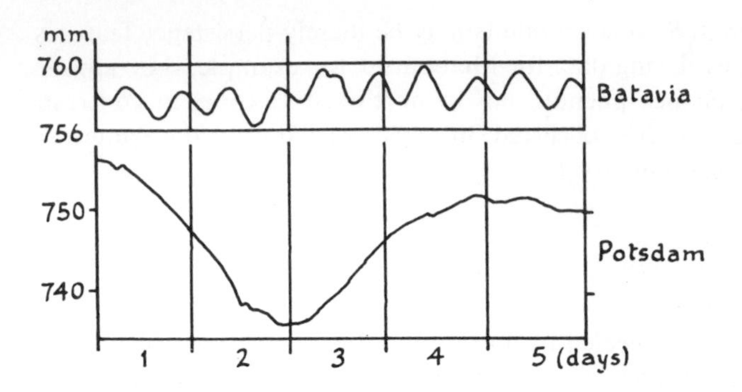

FIGURE 1.20.
The daily pressure wave (after Bartels).

Atmospheric tides are periodic worldwide motions in which the atmosphere oscillates as a single system.[28] They are excited by both the sun and the moon, as are the oceanic tides, but are modified by the solid Earth's surface much less than the latter. Their amplitudes are small near the ground, where they are swamped by the effects of cyclones and anticyclones, but can be quite large at 60 km and above, where the resonant periods of oscillation depend on the density and temperature of the air.

In the tropics, and in a few middle-latitude stations, the solar excitation is just large enough to be visible as a semidiurnal pressure variation on barograms for individual days. In extratropical regions, on the other hand, Fourier analysis must be employed to reveal the oscillations. By this means, the components S_n(n = 1, 2, 3, 4, . . .) of the daily pressure variation due to the sun can be abstracted from data representing averages for many days in a given season or month (see Appendix 1.16). The components may vary from place to place because of local phenomena such as land and sea breezes, but they show such over-all uniformity that they are obviously worldwide.

The moon's tidal effect is purely gravitational, the sun's effect both gravitational and thermal. The solar semidiurnal oscillation, S_2, is the predominant component of regular pressure oscillation all over the world. If the sun's action was mainly gravitational, this semidiurnal oscillation would still predominate, but the lunar semidiurnal oscillation, L_2, would be greater than the solar one. In fact, S_2 is sixteen times as great as L_2. This implies that the sun's thermal action has a greater influence on pressure than its gravitational attraction, yet the temperature variation observed at most places on the Earth's surface is diurnal, not semidiurnal. Atmospheric tides thus pose an interesting theoretical problem.

Newton recognized that the validity of a universal theory of gravitation meant that tides must occur in the atmosphere as well as in the oceans, but thought they would be inappreciable. Laplace took account of the Earth's rotation in his dynamic theory of tides, and showed that his theory was also applicable to the atmosphere, provided that it could be assumed to be isothermal and of constant composition and that vertical accelerations could be neglected.[29] Kelvin in 1882 suggested that the atmosphere might have a natural free period of oscillation nearer to 12 hours than to 24 hours, so that the semidiurnal thermal influence of the sun is magnified by resonance; i.e., the oscillation due to the daily variation of air temperature is regarded as a forced oscillation. Rayleigh, Margules, and Lamb also considered that the S_2 oscillation was a resonance effect, but in 1924 Chapman proved that it had both thermal and tidal components, of equal orders of magnitude, both being magnified 100 times by resonance. Pekeris in 1937 showed that, because of its stratification, the atmosphere could have two free periods, one being nearly 12 hours, which would result in a pressure variation both highly magnified and reversed in phase at high levels.

The main objection to the resonance theory is the improbability of its apparently fortuitous tuning to a 12-hour period, which, it has been shown, may be due to a slowing down of the Earth's rotation by tidal friction. It has also been shown that the S_2 oscillation may not be a resonance effect at all, but may result purely from daily temperature variation operating through the divergence of the isallobaric wind (see Appendix 1.17).

The search for the actual tides in pressure data illustrates the value of harmonic analysis in extracting small systematic variations of known period when they are apparently obscured by random variations. The S_2 component was fairly easy to isolate. It has a range of about 2 mm of mercury in the tropics, appearing as a regular twice-daily rise and fall, but, unlike the ocean tides, showing no perceptible changes in the times of high pressure throughout the month. High tide in the atmosphere recurs daily at the same local solar time, approximately 10:00 A.M. and 10:00 P.M. This may be explained by assuming the S_2 oscillation to be the result of the superposition of two sine waves, representing a gravitational component and a thermal component, both with 12-hour periods, but with different phases. The gravitational sine wave has its maxima at the times of solar transit, i.e., local noon and midnight, whereas the thermal component excites a pressure oscillation of the same order of magnitude but earlier in phase.

The amplitudes of the S_2 tide decrease from low to high latitudes, and since its phase depends on local time, it may be regarded as a sinusoidal wave of pressure that sweeps across an area every 12 hours. At the North Pole, the S_2 wave depends on universal, not local, time; it must therefore consist of two components, one of which rotates with the sun and has zero amplitude at the poles and greatest amplitude at the equator, the other remaining stationary with respect to the Earth's surface and being of appreciable amplitude only near the poles.

The solar diurnal tide S_1 is not worldwide, but is a thermally induced oscillation with a greater amplitude in summer than in winter. It is much affected by local variations in cloud and sunshine, and by topography; it is greatly magnified at the bottom of deep valleys. The S_3 and S_4 components of the daily variation in pressure are also thermal, not tidal, effects, and are due to the 8-hour and 6-hour components (i.e., T_3 and T_4), respectively, of the daily variation in air temperature. S_3 is worldwide but of irregular occurrence; S_4 is even more irregular. S_3 and T_3 have opposite phases in opposite hemispheres; the phases are reversed from summer to winter.

The lunar tides are much smaller than the solar ones. The L_2 oscillation is only 0.06 as large as the S_2 oscillation, although the gravitational tide potentials of the moon and the sun are in the ratio of 11 to 5, and is thus very difficult to determine. Laplace in 1823, using eight years' data for Paris, achieved a nonsignificant result. Lefroy in 1842, using 17 months' data for St. Helena, first determined the magnitude of L_2; it proved to have an amplitude of only 0.06 mm of mercury. Chapman in 1918 made the first satisfactory determination for nontropical regions, finding an amplitude of 0.01 mm from 40 years' observations at Greenwich. The L_2 tide decreases in amplitude polewards, from less than 90 microbars at the equator (see Appendix 1.18 and Figure 1.21), and shows a remarkably low value near the coast of northwestern North America. It is therefore not as uniform geographically as the S_2 tide, and undergoes some considerable annual changes. Maximum pressure is usually about one hour

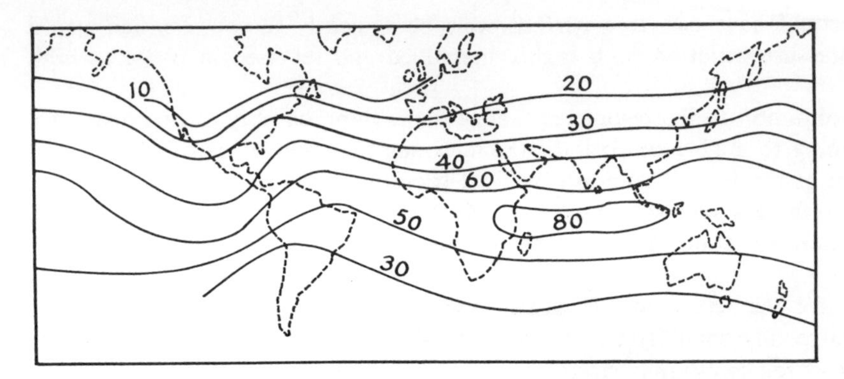

FIGURE 1.21.
The lunar semidiurnal tide. Lines of equal amplitude (microbars) of the annual mean lunar semidiurnal tide in barometric pressure. (After Chapman.)

before local transit of the moon. A lunar semidiurnal variation is also present in air temperature, but has a very small amplitude (0.01°C for Batavia). The lunar diurnal tide, L_1, is much more difficult to isolate. In theory, it should reverse itself fortnightly, when the moon crosses the equator.[30]

All the tides discussed so far have been sought in surface pressure data. There is also some evidence for stratospheric tides. Transosonde data indicate semidiurnal oscillations over the United States at 300 mb.[31] This tide has a vertical axis, and an antibaric flow, its maximum effect being on pressure near the equator. Its latitudinal component has a minimum near 10:00 A.M. local time, coinciding with a pressure maximum at the surface, and its meridional component has a minimum near 7:00 A.M. Winds at 100 mb over the British Isles show both diurnal and semidiurnal oscillations associated with solar tides.[32] The semidiurnal oscillation is of comparable magnitude to, and in phase with, the surface semidiurnal oscillation: the southerly and westerly components of their amplitudes average 0.9 knot at the surface and 0.6 knot in the stratosphere. Thus winds due to atmospheric tides appear to be fairly constant in Britain, at least up to 100 mb (16 km).

The annual temperature oscillation must induce an annual pressure oscillation. If the Earth's surface exhibited no irregularities, the temperature oscillation would be a standing wave, whose amplitude would depend only on latitude. Computing the pressure oscillation that would be caused by such a temperature oscillation shows that it would be higher in winter than in summer at all latitudes except at the equator, and would be 180 degrees out of phase with the temperature oscillation. Winter pressures due to this oscillation would reach a maximum at latitude 40°N and minima at the equator and poles. These theoretical results agree with the observed north-south pressure distribution at sea level, which agreement suggests that an annual meridional tide does exist in the atmosphere.[33]

Regular Variations in Atmospheric Pressure

The most marked regular periodicities shown by atmospheric pressure include the diurnal and semidiurnal cycles, and the annual and semiannual cycles. As with atmospheric tides, it is convenient to regard these cycles as due to waves of pressure that move across a point on the Earth's surface every 24 hours, 12 hours, 12 months, and 6 months, respectively.

The observed daily variation of pressure may be considered the algebraic sum of sine waves with periods of 24 hours, 12 hours, 8 hours, 6 hours, and so on. The 24-hour wave varies greatly in time and place, and is largely a local phenomenon due to the daily temperature cycle; its amplitude is usually greater in winter than in summer. The 12-hour and 8-hour waves are more regular and are worldwide. The former is due partly to tidal effects, partly to thermal effects, as already described. It consists of two components: an equatorial wave that reaches its maximum at the same local time at all places on the same latitude, and a polar wave that reaches its maximum at the same universal time at all places on the same latitude. The polar wave is very small (about 0.12 mb), but is predominant at high latitudes, since the equatorial wave decreases in amplitude as latitude increases.

The amplitude of the diurnal pressure wave is greatly affected by the terrain variations and local weather of the continents, but is not appreciably affected by changes in the temperature of the surface of the oceans. Thus over the oceans the diurnal pressure wave is clear-cut and not complicated by local influences as happens over the land masses. In general, the regular components of the daily variation in pressure can be deduced more easily from data for sea areas than from data for land areas. These regular components indicate that the month-to-month changes in the daily variation of mean sea-level pressure become both larger and more systematic as latitude decreases.[34]

Figure 1.22 gives some examples of the diurnal variation of pressure. Since the regular components of this variation have not been routinely determined at climatological stations, because the harmonic analysis is quite laborious, few general statements can be made about them. It is known that the amplitudes of the diurnal and semidiurnal waves are greater on days with a large temperature range, and that the maxima of the semidiurnal waves occur earlier on days with a small temperature range.

At Gibraltar (see Figure 1.23), the diurnal variation of pressure is almost negligible on the barograms, never being greater than 0.3 mb except in summer around 6:00 P.M. G.M.T., when it has a mean value of 0.6 mb. The first four harmonics of this diurnal variation, combined, indicate a variation in amplitude of 0.4 mb throughout the year, but the daily time for the maximum of the combined harmonics is constant at around 10:00 A.M. G.M.T. throughout the year.[35]

On the plateau of southern Rhodesia, the semidiurnal wave is very regular, but the diurnal wave fluctuates over a range of 0.5 mb, with a mean amplitude of 0.6 to 0.8 mb, and is sometimes undetectable. The semidiurnal wave is very steady in phase and extremely uniform in amplitude, which decreases from 1.52 mb in the west to 1.36 mb in the east of the country. On the plateau, changes of barometric pressure over three-hour periods generally reflect the diurnal wave rather than synoptic disturbances. In Greenland, the diurnal variation of pressure at Northice has a range of 0.3 mb, with maxima at 1:00 P.M. and 9:00 P.M. G.M.T. and minima at 4.30 A.M. and (though not as marked) 5:30 P.M. G.M.T.[36]

The mean annual amplitude of the semidiurnal pressure wave for the United States (see Figure 1.24) depends mainly on latitude, but with important departures from symmetry. The amplitudes decrease sharply from their latitudinal average above the east coast, then increase 100 to 200 miles inland, remain fairly uniform in the central regions, and reach their maximum along the continental divide. They decrease in the

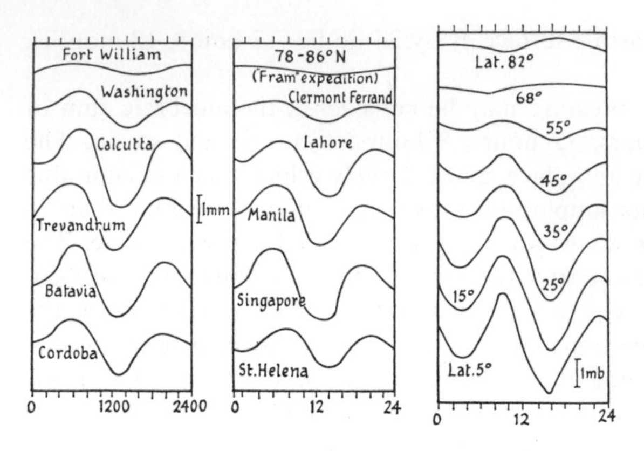

FIGURE 1.22.
Examples of daily pressure variation (after M. F. Harris, *JM*, 12, 1955, 394).

Great Basin, increase again along the Sierras, and reach their lowest values along the west coast. The mean annual phase of the semidiurnal oscillation is more complicated, and does not depend on latitude. Orographic influence is even more marked than with amplitude: early phases are typical of the Appalachians and the Sierras, late phases along the west coast and in the Great Plains. The earliest phase (9.2 hrs) is at Marquette, Mich., the latest (11.2 hrs) at Tatoosh Island, Wash. Both amplitude and phase show a seasonal variation. The variation in phase is 29 minutes along the west coast and 50 minutes on the east coast. Amplitudes reach maxima in spring and autumn (except along the east coast, where a January maximum is found) and minima in December and, especially, June.[37]

The amplitude of the semidiurnal oscillation at Singapore varied greatly during 1948–57; the variations suggest that for Southeast Asia the oscillations may be caused by the convective flux of heat in the layer between 1 km and the tropopause.[38] It is to be expected that the diurnal, semidiurnal, and shorter-period oscillations

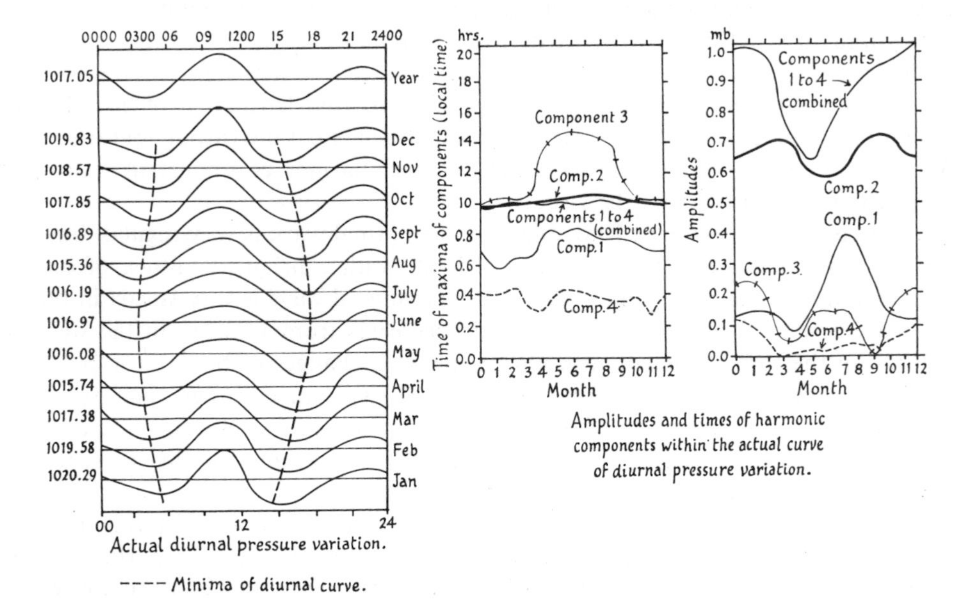

FIGURE 1.23.
Components of diurnal pressure variation at Gibraltar (after Hurst).

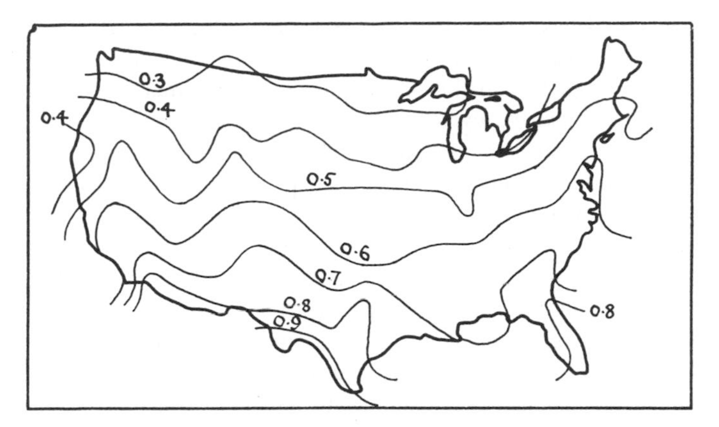

Mean annual amplitude in millibars

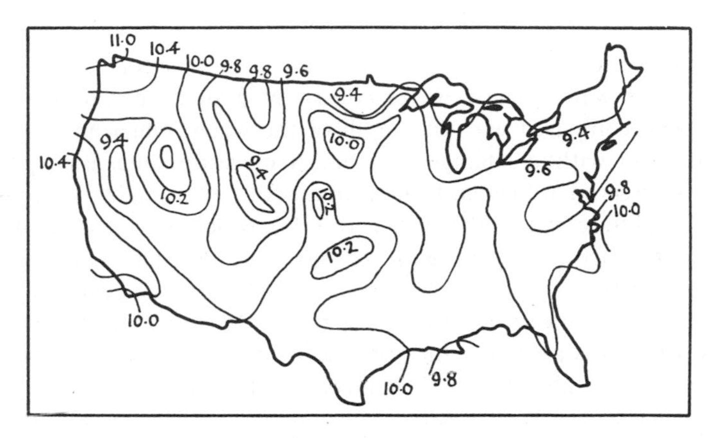

Mean annual phase-angle in local time

FIGURE 1.24.
The semi-diurnal pressure wave in the United States (after J. Spar).

recorded at the ground will reflect changes in the upper atmosphere; pressure variation and vertical motion are closely related, rising surface pressure generally accompanying descending air near the ground, and vice versa. At 8 km in middle latitudes, 1 to 2 km below the mean height of the tropopause, there is an isopycnic level, at which air density is nearly constant. Above this level, the fluctuations in the mass of the atmosphere are opposite in phase to those below it. The diurnal pressure wave has a primary maximum at the Earth's surface, a tropospheric minimum at 3 km, and a secondary maximum at the isopycnic level. In low latitudes, the amplitude of the pressure variations during the day is greater above 8 km than below. Besides this latitudinal effect, there is also a slight difference between normal continental areas and continental areas that are periodically subjected to marine influences in the rate at which the amplitude varies with height.[39]

The annual pressure wave is simply the pressure variation induced by the well-known annual oscillation of temperature. A theoretical treatment of the process shows that the Aleutian and Icelandic lows and the winter high-pressure area over Asia are thermally maintained pressure systems. In other words, because surface pressure undergoes an annual cycle, the climatological features we know as the Aleutian and Icelandic lows can be proved to be real phenomena, not merely the statistical consequences of frequent migrating cyclones.[40]

The semiannual pressure wave has an amplitude varying from less than 1 mb in the tropics to 5 mb in middle and polar latitudes. It is most easy to discern in the southern hemisphere, where, during the equinoctial months, its amplitude in latitude 40° reaches its maximum simultaneously with minimum amplitudes south of latitude 60°. The semiannual oscillation may originate thermally in the equatorial belt, the only zone where the annual variation of insolation shows the required two maxima. Alternatively, it may arise from meridional differences in incoming radiation; an increase in these differences intensifies latitudinal air currents, and causes opposing pressure variations in middle and polar latitudes. The zonal air currents exhibit the same semidiurnal oscillation as pressure.[41]

Irregular Variations in Atmospheric Pressure

It has been known since 1844 that "solitary" waves may be formed on the surface of shallow bodies of water. These waves consist of a single elevation of height, not necessarily small compared with the depth of the water, which may travel for considerable distances without change in form. A corresponding atmospheric solitary wave was first observed at Plevna, Kansas, in the form of a single rise in pressure, with a maximum amplitude of 3.4 mb (see Figure 1.25), which remained constant for more than three hours, during the passage of the wave from Plevna to Pittsburg. The wave was an elevated ridge of cold air 150 km long, and produced scattered showers.[42]

Solitary waves have the same effect as orographic barriers, forcing the air to rise up their slopes; the type of weather they produce depends on the stability of the upper air. If the air aloft is stable, the wave results in the formation of stratus, which moves with it. If the wave breaks, convective-type cumulus then forms by turbulence. With conditionally unstable air aloft, cumulonimbus and possibly thunderstorms result; if the instability is very great, the intense convective updrafts may even cause tornadoes to form.

Abrupt variations of surface pressure in the form of extensive pressure waves are very frequent, particularly at night, in the central midwestern states.[43] These usually move in coherent and continuous lines across the region, bringing about considerable changes of pressure (at least 0.03 inches and occasionally as much as 0.30 inches)

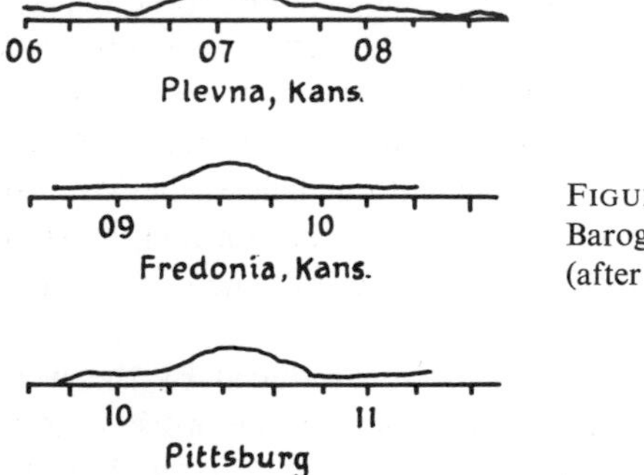

FIGURE 1.25.
Barograms of a solitary wave (after Abdullah).

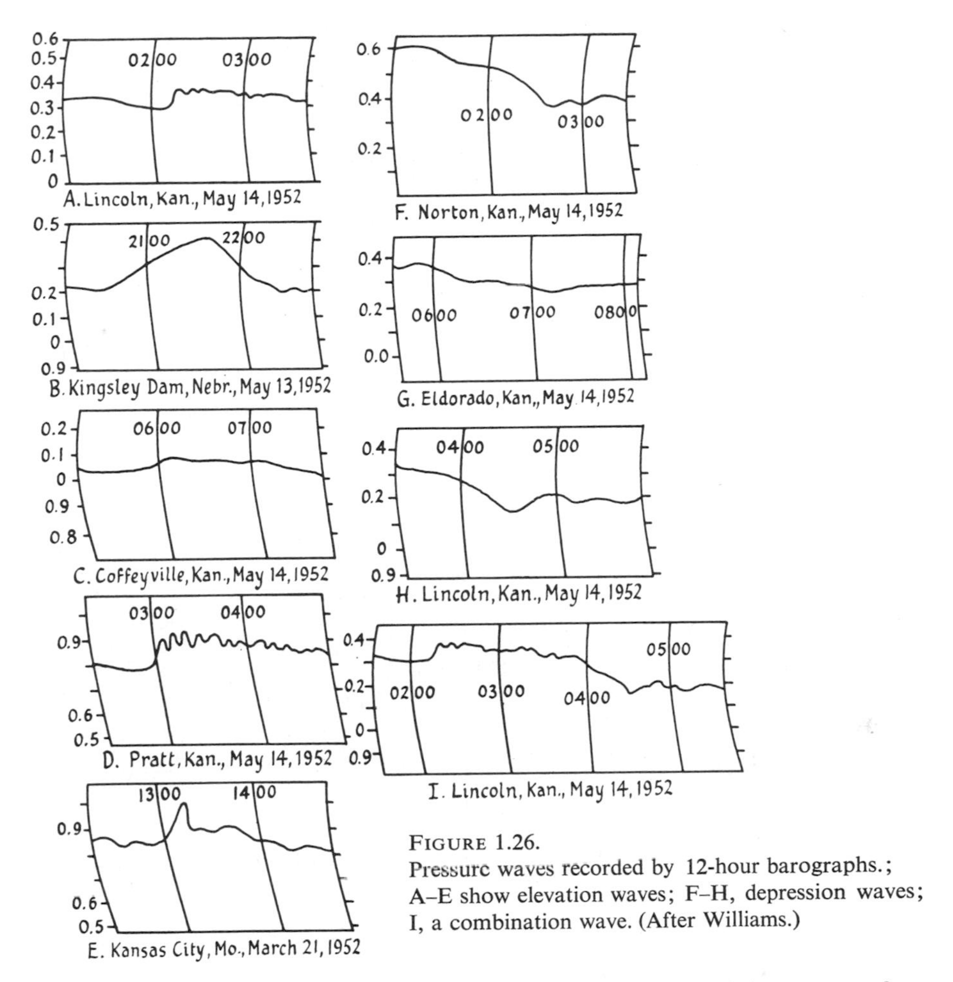

FIGURE 1.26.
Pressure waves recorded by 12-hour barographs.; A–E show elevation waves; F–H, depression waves; I, a combination wave. (After Williams.)

in a short period of time. Two types are found: *elevation waves*, which increase surface pressure; and *depression waves*, which decrease surface pressure (see Figures 1.26 and 1.27).

Elevation waves often coincide with synoptic features, in particular with surface and upper-air cold fronts and with squall lines. Thunderstorms, or at least heavy rain showers, cumuli congesti, and brief rapid variations in wind direction and velocity are usually found along their length. The waves move at right angles to their length, and the extremities of the wave line move most rapidly, so that after a time the line marking the pressure wave on the map becomes convex toward the direction of motion.

Depression waves do not usually coincide with synoptic features, and thunderstorms are rarely associated with them. Diurnal and topographic factors are very important in their formation. Often, a depression wave follows an elevation wave after an interval of one or more hours: in such a *combination wave*, thunderstorms normally develop abruptly at the onset of the pressure elevation and end abruptly at the trough in the depression wave. Although depression waves do not coincide with severe or even bad weather areas, their over-all importance is great. The pressure fall associated with them is 50 to 100 times greater than that normally found in troughs or in intense depressions, and they are significant in the genesis of tornadoes.[44]

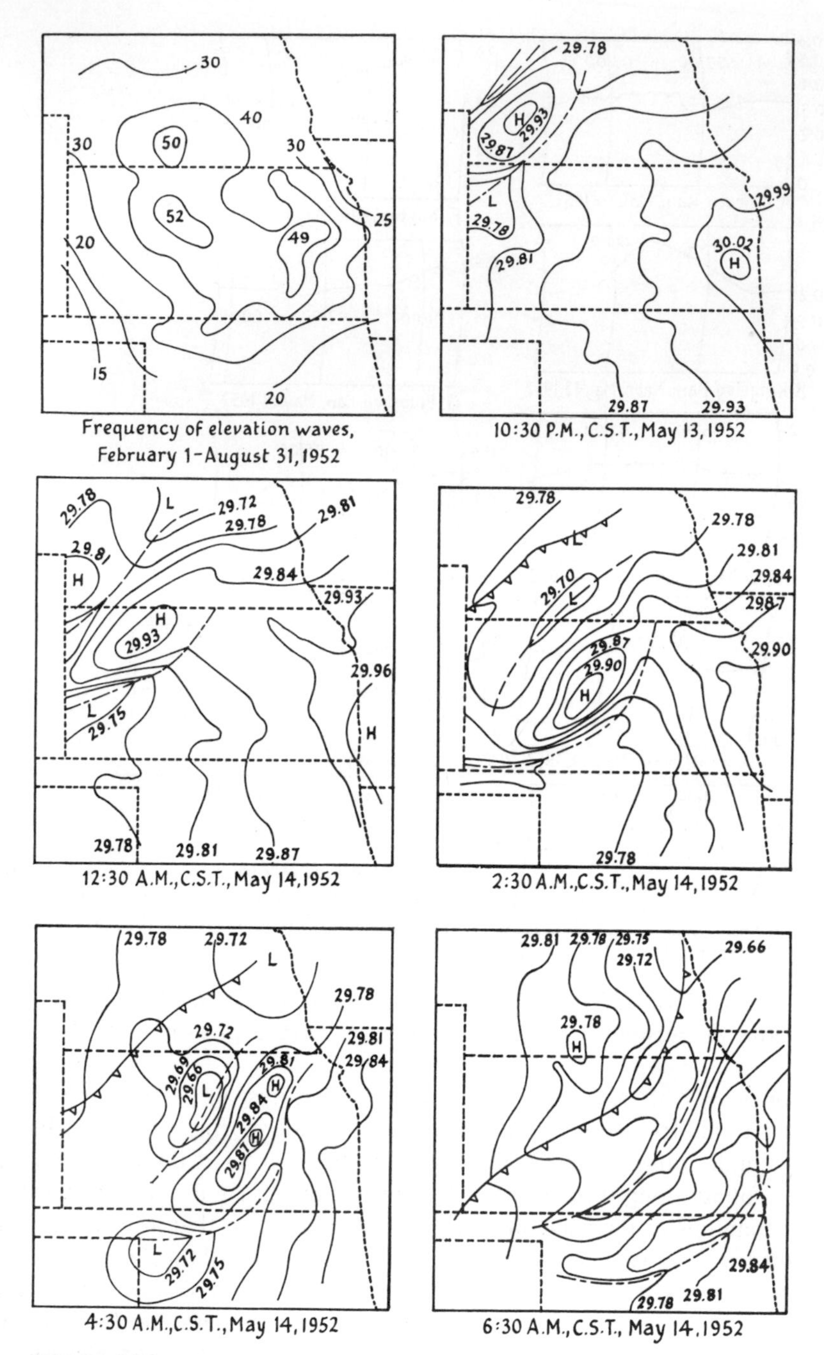

FIGURE 1.27.
Pressure waves in the central midwestern United States. Microsynoptic charts, indicating elevation-type (–·–·–) and depression-type (––––) waves. (After Williams.)

On December 6, 1949, a remarkable cloudbank some 300 to 500 feet thick, with a rear (westward) boundary that looked like the edge of the Antarctic ice barrier, passed over Washington, D.C., producing an unusual pressure trace. The pressure drop could be followed from south of Lake Michigan (at 8:00 A.M.) to the coast east of Washington (at 6:00 P.M.), its speed increasing from 43 miles per hour in Indiana to 77 miles per hour in Maryland. The leading edge of the cloudbank preceded the pressure drop over most of the area, but the latter overtook the cloudbank in Pennsylvania. The pressure was propagated below 3 km on a warm frontal surface, with a speed much greater than that of the winds in the lower atmospheric layers. It moved through the surface pressure pattern, not with it, and seems to have been produced by an expansion wave that moved east from the Midwest, propagated between two inversion surfaces. The cloudbank moved in this same layer, but was advected by the prevailing winds.[45]

Abrupt changes in pressure that can be traced as a linear zone across the map are often termed *pressure jumps.* On the usual scales of weather maps, the width of the line representing an isobar is too thick to show such a jump, which is commonly more than 2 mb within 5 miles, and which travels at 50 miles per hour or even faster. Hence it is often better to plot pressure jumps as lines of discontinuity on a map (see Figure 1.28), although they are not discontinuous at all.[46]

A pressure jump frequently precedes a squall line by some hours, but there is not necessarily any connection between the size of the jumps and the severity of the squalls. Jumps may trigger latent convection as they move through an instability zone, resulting in squalls or violent showers. Tornadoes often form in the zone of interaction between two jumps. For example, on March 18, 1954, 15 tornadoes appeared on a pressure-jump line in eastern Kansas; and on August 16, 1954, two jumps moving across Washington, D.C., were accompanied by tornado funnels aloft, hail, violent winds, a temperature drop of 20°F in 1.5 hours, and an increase in humidity of 40 per cent in 1.25 hours (see Figure 1.29).[47]

Experimental pressure jumps may be induced by operating a piston in an open channel that contains water flowing under gravity, the water surface simulating an inversion (see Figure 1.30). These experiments suggest that if a cold front accelerates behind an inversion, a pressure jump will be induced. If the front then decelerates,

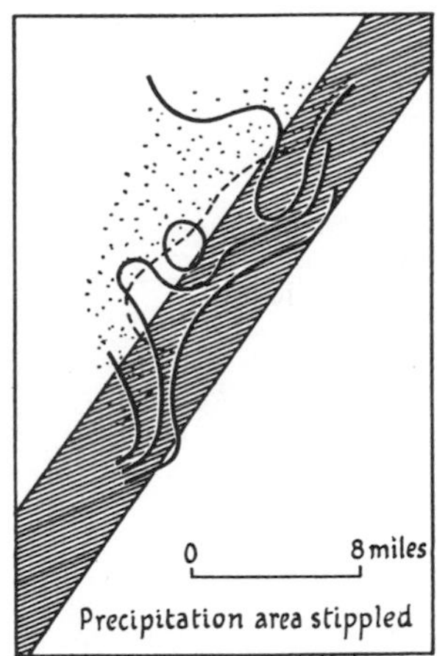

One-minute synoptic map
for 10:07 P.M., E.S.T., May 16, 1948
Shaded area shown as one isobar
on U.S.W.B. daily weather map.

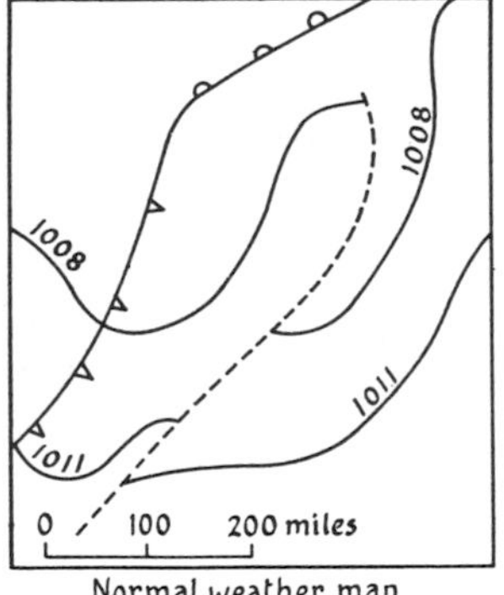

Normal weather map
for 10:30 P.M., E.S.T.

FIGURE 1.28.
Synoptic maps of pressure jumps (after Freeman and Tepper).

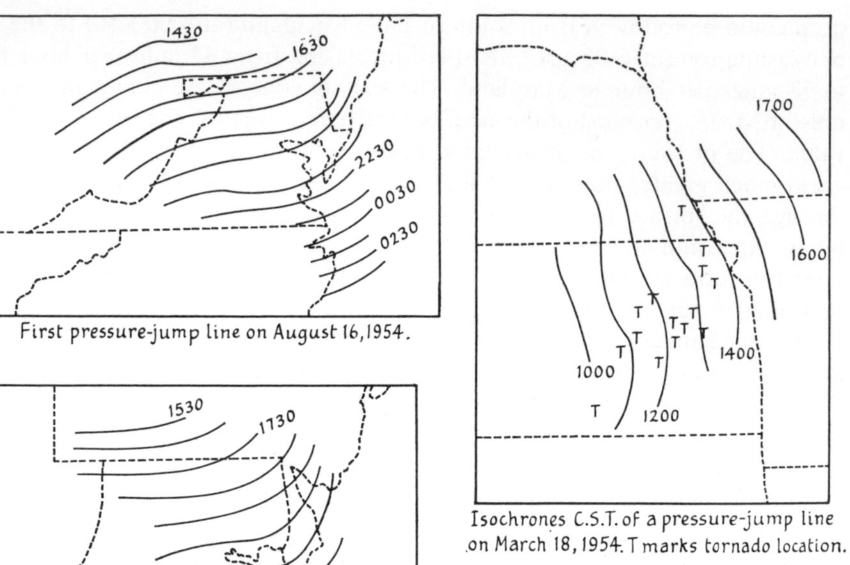

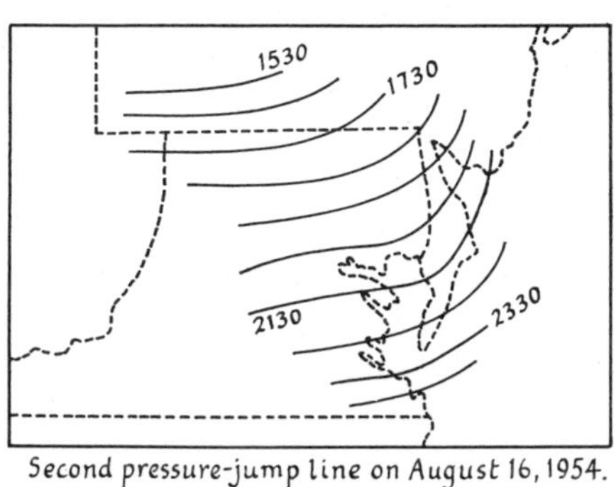

FIGURE 1.29.
Isochrones of pressure-jump lines (after Newstein).

Examples of barograph records of jumps

Jump induced by piston action on an inversion

Effect of a cold front

FIGURE 1.30.
Theory of pressure jumps (after Tepper).

the jump will move ahead, to be followed, and eventually overtaken, by a rarefaction wave. In other words, a pressure jump is a gravity wave in an inversion surface in the warm sector of a depression, and can generate squall lines and very violent weather.[48]

Sensitive microbarographs record not only the irregular pressure oscillations that are associated with pressure jumps, squall lines, cold fronts, hailstorms, and thunderstorms, but also some that are associated with low-level turbulence or convection caused by diurnal effects and cloud cover in air masses. The passage of a cold front may be marked by a train of long, low pressure waves that persist for several hours, especially near sunset, when five of more gravity waves often form on the frontal surface and set up pressure oscillations with a six-minute periodicity.[49]

Pressure pulsations are frequently a local effect of thunderstorms, and often have an amplitude of an inch of mercury. More rarely, there are even larger pulsations, such as one that brought strong easterly surface winds, with rainfall bursts and a squall line, to the eastern states (see Figure 1.31).[50]

Marked pressure oscillations with an amplitude of several mb per sec and a period of five to fifteen minutes often follow the reversal of land-to-sea breezes at La Jolla, Calif.; there are sometimes preceded by a pressure pulse. They seem to be gravity waves with a wavelength of 4 to 10 km, which remain coherent for at least four wavelengths, their crests being at right angles to the wind shear between the upper and lower winds.[51]

Some remarkable pressure oscillations are due to unusually developed major synoptic features. A cyclone that struck the east coast of Iceland on January 25, 1949, produced both a pressure rise of 33 mb and a fall of 26 mb within three hours

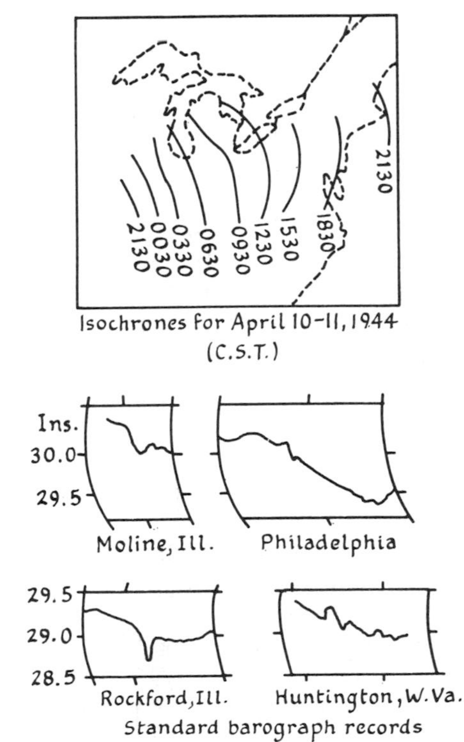

FIGURE 1.31. Pressure pulsations (after Brunk).

at Dalatangi, and a rise of 18 mb in one hour at Hòlar. In Alaska, a pressure rise of 18.5 mb in three hours and a rise of 11.1 mb in one hour on December 18, 1959, was caused by the rear of a sharp surface trough containing an occluded front. A surface-pressure oscillation of 9 mb with a period of 20 to 40 minutes was recorded at Okinawa during the passage of typhoon Emma in 1956.[52]

Periodic surface-pressure variations in the south and Midlands of England, caused by wave motion along the horizontal temperature discontinuities that are set up by the outflow of cold air from intense outbreaks of thunderstorms over the English Channel, have little apparent effect on the weather apart from causing pronounced oscillations in wind velocity. They are important, however, in transferring energy from small-scale weather systems over large distances, quite independently of wind-speed and wind direction. They can also cause rapid local formation or dispersal of layers of stratocumulus.[53]

Local pressure irregularities or "barograph antics" that are difficult to explain are occasionally reported. On September 30, 1950, oscillations resembling damped waves, with a period of 15 minutes and an initial amplitude of 0.8 mb, were reported at Khartoum. No pressure oscillations were reported elsewhere in the Sudan, and there was no change of air mass. The only observed effects were that the wind veered 60° and increased its speed 7 knots at the surface. The oscillations were probably associated with the intertropical front, located WSW to ENE near Khartoum at the time.[54]

Barograms for British stations often indicate the passage of small pressure troughs associated with large vertical wind shear in the lower troposphere and medium instability. On other occasions, the troughs have no obvious associations: for example, rapid oscillations of 2.5 mb at Birmingham on August 23, 1944, and a fall of 4.5 mb in half an hour at Yarmouth the following day. The former was accompanied by a brief squall, the latter by high-level thunderstorms.[55]

Barograph antics must have some consequences for weather, but they are not always easy to discover. Pressure waves recorded in Wales and the English Midlands on July 6, 1952, could be traced for several hundred miles, but had no apparent relation to the weather, which remained quiet during the passage of the waves, even though the strongest wave had an amplitude of 3.5 mb. In the Bristol Channel area, the same waves coincided with high winds and thunderstorms, even though their greatest amplitude was only 2.0 mb. The waves moved with the speed of the wind at 600 mb. They were probably caused by surface pressure increases, caused in turn by convergence, which forced the ascent of an inversion at 900 mb, the pressure increases being propagated horizontally as waves. Such pressure surges in quiet conditions in middle latitudes are the counterparts of tropical hurricanes. In both phenomena, a rapid change in pressure moves through the atmosphere like a wave, so that the pressure rise at the center of the feature is compensated for by a pressure fall outside.[56]

The rapid movement of an asymmetric pressure wave across northern France and northern Germany on August 13–14, 1952, caused pressure falls of 3 to 4 mb in 70 minutes over an area 60 miles in diameter, yet during this time surface winds remained light and variable, rarely reaching 15 knots. The pressure fluctuations here also moved with the speed of the 600-mb wind. A pressure drop of 5.3 mb in 30 minutes on July 4, 1952, in Holland, brought heavy gusts of wind on a day already rainy because of thunderstorms. During June 16–21, 1940, thunderstorms and showers from westerly winds were recorded in southern Rhodesia, although fine weather with upper easterlies almost always prevails at this season. The barograms exhibited the usual diurnal pressure wave of the tropics, but disturbed by very small oscillations, which appear to have been the critical feature in the unusual weather.[57]

Microoscillations of pressure are generally termed *short-period* if they have a period of less than one minute, and *long-period* if their period is one minute or more. The short-period oscillations have been recorded accurately, since 1936, on electromagnetic microbarographs. They are associated particularly with turbulence (not only at ground level but also aloft, since their amplitude increases as convection increases), and are propagated through the atmosphere in the same direction and with the same speed as the wind in the lower layers of the atmosphere (not necessarily the surface wind).[58]

The long-period oscillations can occasionally be recorded by a standard microbarograph, if a speeded-up drum is used. A microbarograph with a drum speed of 1.1 cm per hour recorded oscillations with a period of 6 to 8 minutes and an amplitude of 0.7 mb on February 1, 1951, in Massachusetts; nearby microbarographs with standard drum speeds of 0.25 cm per hour recorded nothing.[59]

Short-period microoscillations that were recorded in Alaska and Puerto Rico during the summer of 1946 and that showed a diurnal variation can be explained in terms of convective cells in the form of ring vortices.[60] A ring vortex has descending air around its outer periphery and ascending air at its inner periphery; deceleration of the descending current as it approaches the ground increases pressure at the surface. As a cell passes through a recording point, first a pressure increase will be recorded, then a decrease, followed by an increase associated with the ascending air, another decrease, and finally an increase due to the descending air current. There are thus two complete oscillations of pressure as each convective cell moves through.

Long-period microoscillations have been recorded both at ground level and in the high atmosphere. Classic ground-level examples, with periods between 2 and 50 minutes, were described by Johnson in 1929 and by Solberg in 1936. Such oscillations are frequent at Blue Hill, Mass., where during 1932, for example, their periods ranged from 5 to 50 minutes, with an average of 18.5 minutes. On February 15, 1948, four oscillations were recorded which remained well-defined for three hours. Each had a 38-minute period, was superimposed on a rising barometric trend, was out of phase with oscillations in both temperature and windspeed, and took place five hours after the passage of a cold front. They were in phase with variations in the height above Blue Hill of the frontal surface separating cold air below from warm air above: the ground pressure was lowest and the temperature highest when the frontal surface was nearest the ground.[61]

Constant-level balloon flights over the central United States have revealed pressure oscillations with periods of 4 to 10 minutes and amplitudes of a few mb, occurring between the 100-mb and 300-mb levels. These oscillations are of shorter period than those observed near the ground with standard microbarographs, probably because the upper layers of the atmosphere are more stable than the surface layer. They can be computed quite satisfactorily from air-parcel buoyancy theory, once the vertical rate of decrease in temperature is known.[62]

Surface microoscillations also occur at the center of such features as dust devils. At Phoenix, Ariz., for example, a pressure drop of 0.075 inches of mercury in 6 seconds was recorded.[63]

Topography can cause both long-period and short-period oscillations. Eddies can cause fluctuations in deep valleys walled by steep-sided mountains; the winds will fluctuate in sympathy with pressure, but temperature and humidity need not. Fluctuations caused in this way and on the order of 2 mb were recorded at Great Langdale, Westmorland, by an open-scale barograph with a drum rotating once in 6 hours, 23 minutes.[64]

Larger-scale effects have been discovered at Gibraltar. Pressures associated with easterly surface winds are 0.5 mb higher at Windmill Hill than at North Front, although the two sites are separated by only 2.5 miles horizontally and less than 400 feet of altitude. With westerly surface winds, pressure at Windmill Hill is 0.1 mb lower than at North Front.[65]

In general, every station has, because of its site and exposure, some *barometric error*, which may vary considerably from day to day, or even from hour to hour. Usually it is not more than 0.5 mb for any one observation, but it may reach 2 mb or more at well-exposed (particularly coastal) sites and at stations considerably above sea level and in irregular terrain.[66] The error can be measured by comparing the observed pressures with those on a synoptic chart, since the isobars on the latter are smoothed to avoid depicting purely local irregularities. The pressure variation which constitutes the error normally comes about because the airflow around the station is distorted by changes in topography or exposure: as windspeed increases, pressure drops, and vice versa. The error increases as the windspeed increases.

Volcanic eruptions and nuclear explosions produce another type of irregular pressure oscillation. Examples that are especially worth discussion are the effects of the Krakatoa eruption of August 27, 1883, the Siberian meteorite of June 30, 1908, the Kamchatka volcano of March 30, 1956, and the Soviet nuclear explosions of October 1961.[67]

The Krakatoa explosion repeatedly disturbed barometers all over the world; the pressure wave from it circled the Earth three times before becoming too weak to be recorded by an ordinary barograph. A typical disturbance occupied about one hour at a station: a sudden rise in pressure was often followed by repeated oscillations, a rapid drop, and a final rise. The calculated velocities of the pressure waves ranged from 674 to 726 miles per hour, both direct and reverse (antipodal) waves being recorded. The direct wave moved westward from Krakatoa to Europe, taking 36 hours 24 minutes, 36 hours 30 minutes, and 36 hours 50 minutes to circle the Earth on its three successive journeys; the reverse wave, moving eastward from Krakatoa to Europe, required 36 hours 46 minutes and 36 hours 4 minutes for its two recorded transits.

The pressure waves caused by the impact of the Great Siberian meteor (see Figure 1.32) had maximum amplitudes of 760 microbars for the direct wave and 170 microbars for the antipodal wave at Potsdam. The wave fronts had a velocity in Britain of 323 m per sec, and the troughs a speed of 318 m per sec, as compared with a mean velocity of 315 m per sec for the Krakatoa waves. The Bezyymyannaya Sopka volcano in Kamchatka produced pressure waves whose speed in Britain was measured as 329 m per sec at Lerwick and 326 m per sec at Kew and Eskdalemuir. The barometric disturbance it created lasted for 12 to 13 minutes, culminating in a rapid pressure fall of about 0.2 mb.

These figures are important because of what they can reveal about the nature of the waves. According to fundamental theory (to be discussed in Chapter 2), the atmosphere has a natural period of vertical oscillation. This natural period will cause any long-period pressure waves propagated through the atmosphere to travel at a speed

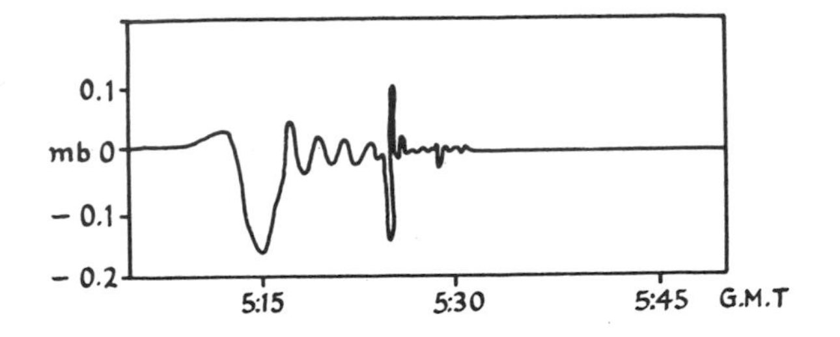

FIGURE 1.32.
The great Siberian meteor of 1908. Composite pressure trace for English stations, 3,500 miles from impact point.

between that of sound at the ground (330 m per sec) and that of sound at the tropopause (300 m per sec). The exact velocity and form of these waves depends on the temperature structure of the atmosphere; since the stratosphere prevents them from escaping, they will be reflected and rereflected between the solid surface and the tropopause. As a result, there will be several wave patterns, each moving through air at a slightly different temperature and so moving at a slightly different speed. The barometric disturbances will be complex, although the waves causing the disturbances are relatively simple.

The pressure waves from volcanic eruptions and nuclear explosions travel at about the velocities required for them to be waves of this type, and also have the required periods of several minutes. They seem in fact to be genuine acoustic-gravity waves, and to have the properties deduced from physical theory. The record of their passage commences with a dispersive wavetrain, in which the period decreases from a maximum of 9 minutes to a minimum of 0.5 minute. The dispersion of the waves is controlled by atmospheric structure: they move most readily along meridional tracks. Their passage may locally cancel the latitudinal winds of the general circulation.

It is well-known that acoustic waves in the atmosphere exhibit both audible and silent zones. Microbarographic records of the pressure waves due to hydrogen-bomb explosions show the same pattern. The Marshall Islands detonations in 1954 produced pressure waves that did not encircle the Earth. Both eastward-moving and westward-moving waves developed: the eastward-moving wave had a velocity 50 meters per second greater than that of the westward-moving wave. No barometric disturbances were recorded at stations to the west of, and comparatively close to, the Marshalls, but stations to the east recorded a pressure wave. Thus Midway Island, 2,570 km from Bikini Atoll, recorded the waves; but stations on Guam, Eniwetok, Saipan, and Ponape, between the Mariana Islands and Midway, Ponape being only 340 km from the latter, recorded nothing.

The exact forms of barometric disturbances due to nuclear explosions show considerable variety, although certain basic patterns can be discerned. The Marshall Islands explosions produced barograph traces in Japan of a different form from those of a U.S.S.R. explosion. The latter gave rise to noticeably different traces at three separate stations at distances between 85 and 155 km (because of the pattern of wave dispersion, not because of orographic or instrumental effects); the time required at the three stations for the passage from first trough to seventh crest was 7.2, 8.0, and 8.4 minutes.

The Russian nuclear explosion (of over 50 megatons) in the Novaya Zemlya region on October 30, 1961, produced a strong pressure wave that could be traced round the Earth several times (see Figure 1.33). The first antipodal wave was exceptionally strong and clear, probably because of unusual atmospheric conditions. During the passage

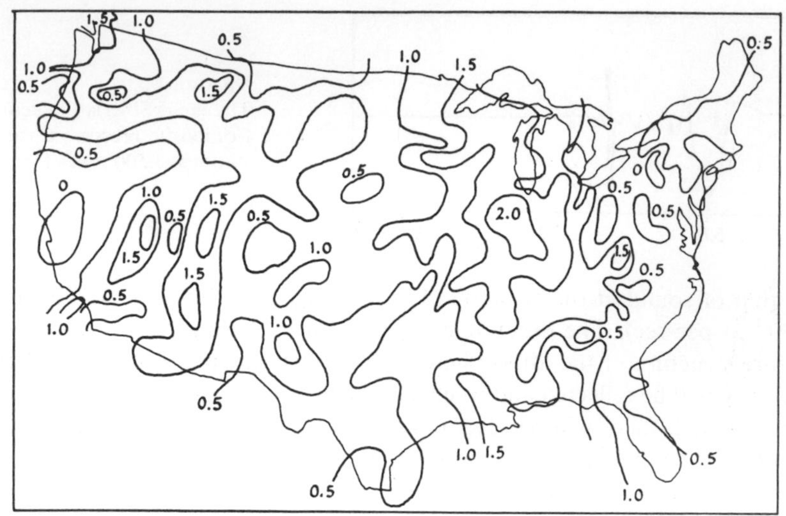

Maximum amplitude of primary wave (mb)

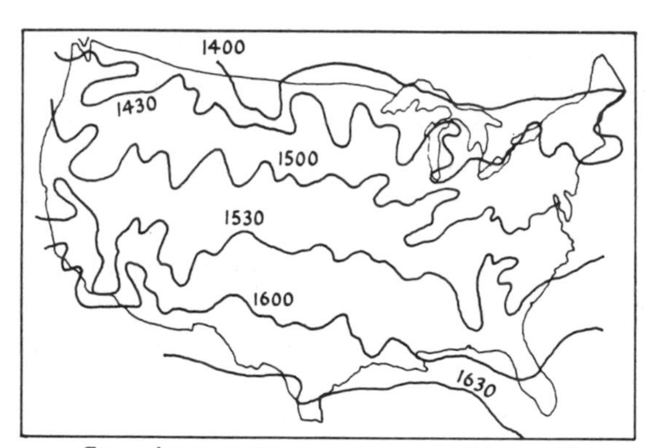

Time of arrival of primary wave on October 30

FIGURE 1.33.
Pressure waves from the Soviet 50-megaton explosion.

of the first direct wave (see Figure 1.34), the trace was very disturbed by local "noise" in the atmosphere. An initial pressure rise was followed by a sharp fall 4 minutes later (the maximum peak-trough amplitude was 1,000 microbars), which was succeeded by twenty or so waves that decreased in amplitude and in period, from 4 minutes to 2 minutes. The antipodal wave was clearer than the direct wave; its maximum amplitude of 580 microbars was followed by a train of 26 waves whose periods decreased from 9 minutes to 2 minutes. The direct wave had become much degraded by the time of its second passage at the recording point, its maximum amplitude only amounting to 220 microbars. Similarly, the indirect wave on its second passage was also considerably damped, with a maximum amplitude of 200 microbars. The direct wave was recorded on a third passage, with a maximum amplitude of 160 microbars. There was no obvious evidence of further passages of either wave, although a fifth passage of the direct wave may have been recorded at 9:40 A.M. G.M.T. on November 5.

The times for the complete transit around the Earth of the pressure waves from the

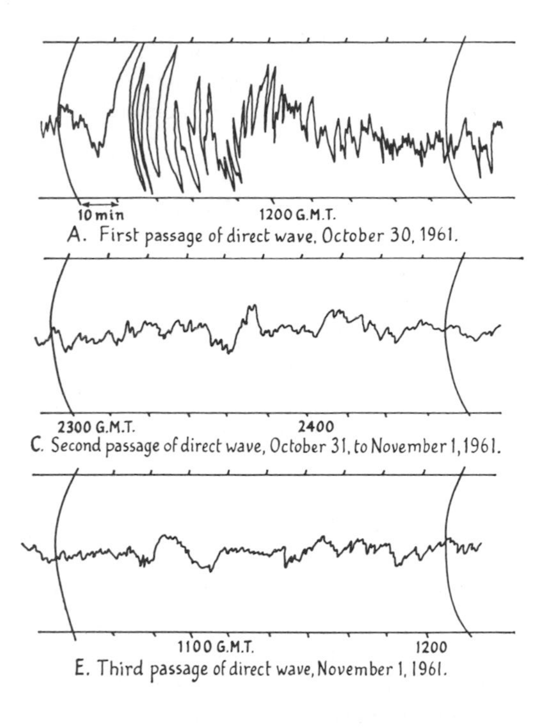

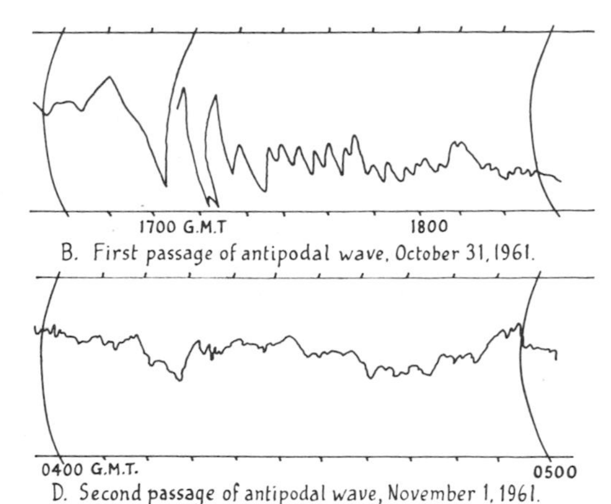

FIGURE 1.34.
Barometric disturbances produced by the Soviet explosion.
Barograph A: full-scale vertical deflection equals 400 microbars.
Barograph B: full-scale vertical deflection equals 400 μbars before 17:10 G.M.T., 800 μbars afterward.
Barograph C: full-scale deflection equals 800 μbars. (After R. V. Jones.)

Novaya Zemlya explosion were 36 hours, 17 minutes, and 35 hours, 20 minutes, for the first and second circuits of the direct wave, and 35 hours, 20 minutes, for the first transit of the antipodal wave. The third and fourth passages of the forward wave took 35 hours, 20 minutes, each. These waves were obviously on the same order of magnitude as the Krakatoa waves.

A nuclear explosion of 30 megatons near Novaya Zemlya on October 23, 1961, produced a pressure wave with an amplitude of 350 microbars at Aberdeen. The October 30 detonation, although it released only 60 to 70 per cent more energy than that of October 23, produced pressure waves with amplitudes three times greater than those due to the smaller explosion.

These explosions indicate how much energy is required to induce comparatively small barometric disturbances—disturbances that are not recorded at all on an ordinary barograph—since one gram of TNT produces an energy release of approximately 4×10^{10} ergs. Even with very powerful nuclear explosions, man appears to have little chance of altering the geography of the atmosphere on any appreciable scale.[68]

The Mechanics of Pressure Change

Since pressure is normally taken as the basis for studying the geography of the atmosphere, and since changes in this geography are the key to both description and explanation of climate and weather, we should examine pressure change in quantitative terms. It is difficult to do so, because pressure changes cannot be localized.[69] It is particularly difficult to do so by examining individual pressure centers, because pressure systems on the weather chart, and their associated wind and temperature patterns, result from complex series of events that follow the pressure changes themselves. The center of a migratory pressure system does not exist as an entity; it is merely the intermediate zone between a pressure rise and a pressure fall.

Pressure change is usually plotted on weather maps as the *pressure tendency*, i.e., the change in pressure recorded during the three hours immediately preceding each observation. Lines joining places with equal pressure tendencies form *isallobars*, and the isallobaric topography of an area several hundred miles across must be taken into account when one is studying the weather changes at any one locale. Observation indicates that a developing pressure system is frequently associated with the intensification of an upper trough and with a stratospheric temperature maximum, which is produced by air descending for many hundreds of miles upstream from the surface system. Thus pressure change at a point on the Earth's surface is accompanied by significant changes in other climatic elements over a wide area. These changes differ in the lower and higher atmospheres. Surface pressure systems move (more or less) with the speed of the horizontal surface wind, but systems in the middle troposphere and above move much slower than the horizontal wind blowing through them.

Theories of pressure change may be classed as either *thermal* or *dynamic*. The thermal theories originated with Ferrel, who proposed in 1889 that cyclones are formed by localized atmospheric heating at ground level, which causes mass outflow of air aloft and a fall in surface pressure. He attributed the local heating to release of latent heat by the condensation of water vapor. Although very great quantities of heat are produced by such condensation, they are now regarded as insufficient to produce cyclones.

Temperature measurements at mountain stations proved the existence of "dynamic" pressure systems, i.e., surface cyclones with cold air aloft (cold lows) and surface anticyclones with warm air aloft (warm highs), which obviously could not be explained by the thermal theory. Hann and others, from 1901 onward, began to explain pressure change as a dynamic consequence of wind variations, and regarded cyclones as the result of the damming and acceleration of upper-air currents.

Both types of theory encounter difficulties. The thermal theories must explain the local temperature changes that they postulate as the main causes of pressure change, and the dynamic theories must explain how wind variations can arise from the accelerations that accompany local pressure change.

THERMAL THEORIES OF PRESSURE CHANGE

Two relatively simple thermal models of pressure change, one based on nonadiabatic, and one on advective, temperature changes, were described by Austin in 1951. In the *nonadiabatic model* (see Figure 1.35), a pressure gradient will be created if an air column is heated uniformly from ground level to a height z (see Appendix 1.19). The pressure gradient accelerates air out of the column, with the outflow increasing as height increases, reaching maximum at height z. Pressure will fall at the ground, and upward motion will develop in the air column, into which air will flow near its base. Local heating therefore results in a pressure fall at the base of the column and a pressure rise in the surrounding air. Local cooling produces a rise in pressure at the base of the column and a pressure fall in the surrounding air.

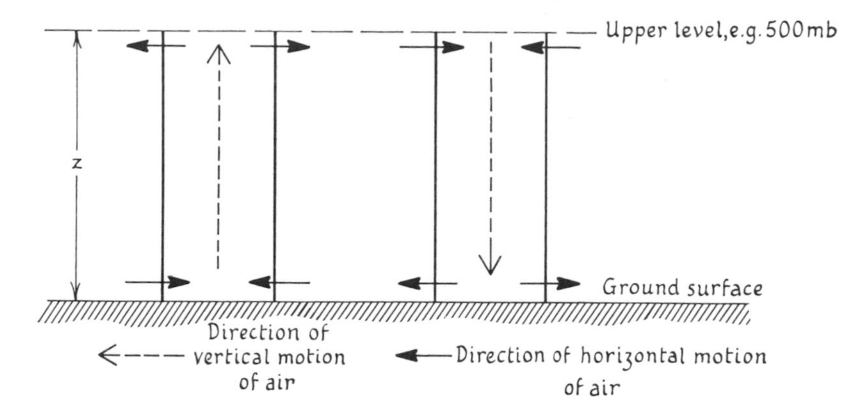

FIGURE 1.35.
A model of nonadiabatic pressure change.
Left: air heated uniformly from ground to level Z.
Right: air cooled uniformly from ground to level Z.

The most important nonadiabatic temperature changes are due to the rising and falling of the air as it flows over warmer and colder surfaces, the horizontal mixing of different air masses, heating or cooling of the air by falling precipitation, and local and regional differences in the vertical heat flux between the atmosphere and the Earth's surface (see Appendix 1.20).

In the *advective model*, the horizontal flow of cold or warm air causes local temperature differences, which in turn cause local pressure changes. The direct heating (or cooling) of the nonadiabatic model, in which the heat source (or heat sink) is stationary, is replaced by localized heating (or cooling) due to a moving heat source (or sink). The horizontal movement of warm air into a region of cold air leads to mass outflow aloft and a fall in pressure at ground level; a movement of cold air into a region of warm air results in mass inflow aloft and a pressure rise at the surface. Thus the advective model assumes that an overturning motion occurs between adjacent cold and warm masses of air, as a result of convection. Advection seems to control surface pressure change more in North America than in Europe. Although three-hour isallobaric centers do tend to occur in regions of strong advection in the lower troposphere, the advection is not necessarily the cause of the pressure changes.[70]

In 1920, von Ficker and other German workers proposed horizontal temperature advection in the stratosphere as the primary cause, and the thermal effects it induces in the troposphere as secondary causes, of sea-level pressure changes. Observations since then indicate that no single level in the atmosphere and no one process can be regarded as the sole cause of pressure changes at ground level. Surface pressure variation is the sum of mass changes at all levels of the atmosphere above the point of observation. It is rarely found that surface pressure rises as colder (denser) air advances or falls as warmer (lighter) air approaches, because the advection decides the net change in mass of the air above the observing point. The pressure tendency recorded at the latter need not necessarily have even the same sign as the net change in mass.

The algebraic signs of pressure variation change near the tropopause. In the lower atmosphere, mean maximum pressure fall coincides with mean maximum temperature rise (and vice versa), as would be expected intuitively, but at 8 km maximum pressure fall coincides with maximum temperature fall, and likewise for rises. In other words, at mean sea level, changes in pressure and temperature are 180° out of phase, whereas at 8 km they practically coincide. The maxima and minima of the pressure curve are displaced backward with height, those of the temperature curve are displaced forward, and the displacements are greater over North America than over Europe.

One would expect good correlations between pressure and temperature because of the diurnal waves of pressure and of temperature, and because of the properties of the long waves in the westerlies. In the long waves, the isobars and isotherms are more often in phase than out of phase, so that pressures and temperatures at a station should rise and fall together during the passage of cold troughs and warm ridges. In the shorter of the long waves, the isobars and isotherms may be up to 180° out of phase; pressure and temperature will be opposite in sign during the passage of warm troughs and cold ridges. At 3 km and above, the effect of these shorter waves is swamped by that of the longer waves, and there is usually a positive correlation between temperature and pressure changes. For the United States, this correlation leads to the rule of thumb (which works best for 12-hour changes) that a temperature change of 1°C is associated with a pressure change of the same sign and of 1 mb at the same level. At 700 mb, a 1°C change in temperature is associated with a change of 35 to 40 feet in the height of the pressure surface.[71]

Observations for North America show that pressure change can be separated into two components, each important to the thermal theories.[72] First, there is a *development component*, in which warming or cooling in the stratosphere leads to the intensification or weakening of a surface pressure system as a whole. Second, there is a *motion component*, in which warming or cooling in the lower troposphere leads to local falls (increases) in pressure. The motion component is responsible for the movement of a pressure system across the Earth's surface; the development component is responsible for the deepening or filling of cyclones.

When a low deepens at the Earth's surface, the column of air above its center obviously loses mass. The question is: which layer of the atmosphere loses the most mass? It has been found empirically that as a cyclonic storm deepens, the lower two thirds (by weight) of its central air column becomes thinner, i.e., colder and denser; the upper third becomes thicker and warmer, and loses enough mass both to counteract the increase of mass in the lower layer and to deepen the low.[73] Thus rapidly deepening cyclones are controlled by processes near the tropopause, rather than near the ground.

Solar radiation changes correlate somewhat with pressure changes. The average monthly north-south gradient of solar radiaton in the northern hemisphere is correlated with the average monthly sea-level pressure pattern two weeks later.[74] In late autumn and early winter, when the radiation gradient increases, air tends to accumulate in high latitudes and pressures consequently rise. However, later in the winter a decrease in the radiation gradient is associated with rising pressure.

Radiational cooling of the stratosphere in polar regions induces pressure change, by favoring the development of unstable pressure patterns. In these high latitudes, pressure changes at the 50-mb level and above are transmitted downward in damped form, being recorded at sea level as very minor effects. In low and middle latitudes, pressure changes in this part of the stratosphere are not due to radiational cooling, but instead appear to be the damped remnants of important pressure changes that develop at the Earth's surface and are then transmitted upward.[75]

DYNAMIC THEORIES OF PRESSURE CHANGE

A study of the dynamics of fluid flow shows that the movement of the atmosphere must both determine the pressure changes in it and depend on the instantaneous distribution of pressure. Thus pressure variability can be proved from dynamic theory to be controlled by the mutual interactions between air movement and the distribution of atmospheric pressure.[76]

Assuming a steady atmosphere with no disturbances, Jeffreys proved from dynamic theory (in a series of papers ending in 1938) that depressions and anticyclones must be essential parts of the general circulation, not merely irregularities superimposed on a generally smooth (average) flow. He also showed that deep streams of northward or southward moving air must develop to transport momentum meridionally, otherwise the east-west movements in the atmosphere, which form the predominant part of the general circulation, would be destroyed in a few days by friction with the Earth's surface. These various circulations are logical (mathematical) consequences of certain properties of the atmosphere—in particular of the hydrostatic relation, which is a consequence of the universal law of gravitation—and they in turn produce by dynamic means the observed distribution of mean atmospheric pressure.[77]

Priestley in 1948 carried this dynamic reasoning further, by proving that the breaking up of a latitudinal pressure belt into discrete closed systems is inherent in the dynamics and geometry of atmospheric flow, and is not a consequence of the geographical distribution of land and sea. He proved that the number of closed pressure systems in latitudinal zones must be greatest in middle latitudes, that the size of the systems must decrease as latitude increases, and that the upper limiting size of cold cyclones and warm anticyclones must be larger than that of warm cyclones and cold anticyclones. The geographical distribution of land and sea can cause only relatively minor modifications of the broad pattern of pressure determined by the laws of fluid dynamics.[78]

By means of the hydrostatic equation and the gas laws, one can determine theoretically the vertical motion that the air must undergo given specific values for pressure change, pressure gradient, windspeed, and rate of movement of the pressure system as a whole. The operative equation (see Appendix 1.21) shows that air tends to flow upward or downward toward the level where pressure is falling most rapidly (i.e., where cyclonic development is strongest), and to flow upward or downward away from the level of most rapid pressure rise. Let us consider a typical situation: the equation shows that air within a pressure system that is intensifying at the rate of 5 mb every 3 hours at a height of 1 km above mean sea level must develop a vertical velocity of 20 cm per sec at that height. The equation also indicates that air which is moving horizontally toward low pressure must also rise vertically at a speed of 5 cm per sec if its pressure decreases 5 mb every 3 hours in the horizontal direction.

Using basic concepts of fluid dynamics, the hydrostatic equation can be transformed to yield the *pressure tendency equation*, several forms of which have been derived (see Appendix 1.22). This equation gives the rate of change of pressure with time for any height in terms of the horizontal and vertical components of air movement at that height, the density of the atmosphere at that height and the gravitational acceleration, by means of the hydrodynamic concept of *divergence*. To use the tendency equation directly requires that the velocity of the air be accurately measurable to within 1 cm per sec, a very difficult task in practice. It is more usual to employ it as a basis for judging the validity of models of pressure change; for example, the Bjerknes–Holmboe model (see Appendix 1.23).

Empirically, certain aspects of pressure change have been discovered, that are not obvious from dynamic theory. Thus important lag relationships exist in the intensification or weakening of upper troughs and ridges, i.e., the pressure changes move downstream. This effect is well in evidence at 500 mb over North America and adjacent sea areas, but is less well-developed over the eastern Atlantic and western Pacific.

There are frequent weakening ridges or intensifying troughs over central North America, and frequent intensifying ridges or weakening troughs over the Pacific and the Atlantic between latitudes 40°N and 50°N. In winter, ridge development tends to prevail over oceanic areas in the northern hemisphere, while trough intensification predominates over the North America continent and off the coast of California, with maximum trough development at 90°W and 120°W. The mean rate of movement of both troughs and ridges also shows a marked variation with longitude, which one would not expect from simple dynamic theory.[79]

In general, atmospheric geography and its changes provide a sound foundation for a geographical study of weather and climate, but by no means completely explain the observed variations of the meteorological elements. Each climatic element must therefore be considered separately, especially since different techniques have been evolved for the study of different elements, and the causes of all variations must be determined.* Since the ultimate synthesis of all the elements cannot be achieved on geographical grounds alone, an alternative is to study the variations exhibited by each element in terms of the same mathematical or statistical model. Turbulence theory provides one such model.

*Chapters 1 and 4 through 7 of *Techniques* discuss climatic elements individually.

Notes to Chapter 1

1. K. Schneider-Carius, for example, combined a geographical training with considerable experience on meteorological research flights. The results of his investigations into the lower five or ten thousand feet of the atmosphere—the layer above that in which the microclimatologist works, but below that which mainly interests the modern meteorologist—are summarized in *Die Grundschicht der Troposphäre, Probleme der Kosmischen Physik*, vol. 26 (Leipzig, 1953). See also the review by F. Loewe in *GR*, 45 (1955), 146.

2. Articles describing personal observations of climatological features little-known to science may occasionally be found in such magazines as *Flying* (monthly, Chicago) and *Sailplane and Gliding* (bimonthly, London). See also reliable autobiographical accounts by experienced pilots, such as E. K. Gann, *Fate Is the Hunter*.

3. On a local scale, such writers as Gilbert White, *The Natural History and Antiquities of Selborne, in the County of Southampton* (London, 1789; this book

has had numerous later editions, e.g., Everyman, 1950) and W. H. Hudson regarded the study of climate and weather as a part of natural history. On a grander scale, Charles Darwin, whose *Voyage of the Beagle* was based on his voyage around the world between December 1831 and October 1836, and Alexander von Humboldt, *Voyage aux regions equinoxiales du nouveau continent, fait en 1799-1804*, 30 vols. (Paris, 1807 onward), and *Fragments de geologie et de climatologie asiatiques* (Paris, 2 vols., 1831) made the first attempts at scientific description of climatic conditions on the basis of extensive travels.

4. J. S. Malkus, *SP*, 43 (1955), 461.

5. Basic sources for the historical review are: J. Van Mieghem, *BWMO*, 7 (1958), 31 and 52; *MOHUI*, pp. 3-12.

6. For a classic account of the exploration of the free atmosphere, see N. Shaw, *Manual of Meteorology*, vol. I (Cambridge, 1926), especially pp. 207-33.

7. C. Hinkel, *MM*, 91 (1962), 154.

8. H. H. Lamb, *MM*, 78 (1949), 51.

9. D. H. Johnson, *MOPN*, no. 116 (1956).

10. E. Gherzi, *W*, 9 (1954), 386.

11. J. C. Bellamy, *JM*, 2 (1945), 1.

12. D. J. Schove, *CPRMS*, p. 217, and *GPA*, 49 (1961), 255.

13. J. Glasspoole, *MM*, 88 (1959), 137.

14. Anon., *MM*, 76 (1947), 94.

15. E. V. Newnham, *MM*, 79 (1950), 129; L. F. Lewis, *MM*, 82 (1953), 250.

16. E. N. Lorenz, *JM*, 8 (1951), 52; W. H. Klein, *JM*, 8 (1951), 332; B. N. Charles, *JM*, 12 (1955), 407.

17. A. H. Gordon, *BAMS*, 34 (1953), 357.

18. A. H. Gordon, *QJRMS*, 77 (1951), 302.

19. W. H. Klein, *JM*, 8 (1951), 332.

20. H. P. Berlage and H. J. de Boer, *GPA*, 46 (1960), 329.

21. H. J. de Boer, *MM*, 76 (1947), 212.

22. J. Wadsworth, *MM*, 83 (1954), 120; R. P. Waldo Lewis and D. H. McIntosh, *MM*, 82 (1953), 301.

23. C. E. P. Brooks, *BAMS*, 35 (1954), 314.

24. J. Namias, *BAMS*, 35 (1954), 112.

25. R. Shapiro, *JM*, 15 (1958) , 435, and *JM*, 16 (1959), 569.

26. G. T. Walker, *QJRMS*, 72 (1946), 263; U. Radok, *QJRMS*, 74 (1948), 196; J. Wadsworth, *QJRMS*, 74 (1948), 407.

27. On the effect of reducing observed pressure values to hypothetical mean sea-level values (particularly the effect of the Siberian anticyclone), see J. M. Walker, *W*, 22 (1967), 296. Walker, *W*, 23 (1968), 291, has also demonstrated that surface atmospheric pressure increases as "sea level" gets higher as one progresses up the St. Lawrence Seaway. I. Y. Ashwell, *MM*, 94 (1965), 52, discusses the problems of converting observed pressures in central Iceland to mean sea-level values. On the maximum surface pressure observed so far (above 1,065 mb, in the Canadian Arctic), see J. H. Bradley, *MWR*, 45 (1964), 708. On the correlation between monthly mean-pressure patterns and surface weather over central England, see R. F. M. Hay, *MM*, 96 (1967), 167, and *MM*, 97 (1968), 76. W. Schwerdtfeger, *JGR*, 72 (1967), 3543, discusses changes in the mass of the atmosphere over Antarctica. C. F. Palmer and W. D. Ohmstede, *T*, 8 (1956), 495, described barometric oscillations in equatorial regions that appear to indicate solar control of surface pressure. G. W. Brier, *T*, 4 (1952), 262, examined the correlation between sea-level pressure and sunspot activity. H. L. Stolov and J. Spar, *JAS*, 25 (1968), 126, find no evidence that surface pressure variation in North America is related to solar flares. For recent work on the Southern Oscillation, see A. J. Troup, *QJRMS*, 91 (1965), 490; see also D. J. Schove, *GPA*, 55 (1963), 249, and D. J. Schove and H. P. Berlage, *GPA*, 61 (1965), 219. G. I. Roden, *JAS*, 22 (1965), 280, discusses the results of a power-spectrum analysis of pressure variations along the Pacific coast of North America. Atmospheric pressures of at least 1,066 mb, recorded over and around Baffin Island on January 18, 1958, were not far below the value of 1,070 mb which is taken as an improbable extreme for the design of barometers; see A. Court, *BAMS*, 50 (1969), 248. For the climatology of pressure-contour height variability in the stratosphere, see K. W. Johnson and M. E. Gelman, *MWR*, 96 (1968), 371.

28. General reviews of the whole field will be found in: S. Chapman, *CM*, pp. 510-28; M. V. Wilkes, *QJRMS*, 78 (1952), 321.

29. P. S. Laplace, *Traité de mécanique céleste* (Paris, 1799), II, part 4, 294-98.

30. S. Chapman, *JM*, 8 (1951), 133; M. W. Jones and J. G. Jones, *JM*, 7 (1950), 14.

31. J. K. Angell, *JM*, 15 (1958), 566.

32. D. H. Johnson, *QJRMS*, 81 (1955), 1.

33. J. Spar, *JM*, 14 (1957), 83. For recent work on atmospheric tides in general, see: T. R. Visvanathan, *MWR*, 94 (1966), 307; G. W. Brier, *loc. cit.*, p. 311; and J. R. Probert-Jones, *QJRMS*, 93 (1967), 126. See: P. Thrane, *T*, 10 (1958), 415, on solar tides; B. Haurwitz and A. D. Cowley, *MWR*, 94 (1966), 303, on lunar tides; J. A. Fejer, *Reviews in Geophysics*, 2 (1964), 275, on tides and magnetic effects; B. Haurwitz and A. D. Cowley, *MWR*, 93 (1965), 505, on tidal effects observed above North and Central America; and B. Lettau, *JGR*, 70 (1965), 3255, on tides in the high atmosphere. On diurnal tides, see R. S. Lindzen, *QJRMS*, 93 (1967), 18. On details of lunar and solar barometric tides for Australia, see B. Haurwitz and A. D. Cowley, *MWR*, 96 (1968), 601, and for Copenhagen, see S. Chapman, *QJRMS*, 95 (1969), 381. For a theory of thermal tides in the atmosphere, see B. G. Hunt and S. Manabe, *MWR*, 96 (1968), 753. On stratospheric tides, see W. L.

Webb, *Rev. Geophysics*, 4 (1966), 363. For diurnal tides in the upper atmosphere, see C. O. Hines, *JGR*, 71 (1966), 1453. For diurnal tides inferred from rocket soundings, see: O. W. Thiele, *JAS*, 23 (1966), 424; R. J. Reed, D. J. McKenzie, and J. C. Vyverberg, *JAS*, 23 (1966), 247 and 416; N. J. Beyers, B. T. Miers, and R. J. Reed, *JAS*, 23 (1966), 325. On semi-diurnal tides, see M. F. Harris, F. G. Finger, and S. Teweles, *MWR*, 94 (1966), 427, and, for the 30-to-60-km layer, R. J. Reed, *JAS*, 24 (1967), 315. For an apparent correlation between diurnal and semi-diurnal tides, both solar and lunar, and precipitation variation, see G. W. Brier, *MWR*, 93 (1965), 93. For recent thoughts on the theory of atmospheric tides, see F. B. A. Giwa, *QJRMS*, 93 (1967), 242, on the choice of upper boundary conditions, and *QJRMS*, 94 (1968), 192, on the use of response curves; see also R. E. Dickinson and M. A. Geller, *JAS*, 5 (1968), 932, on the introduction of Newtonian cooling into the theory. R. Sawada, *JAS*, 22 (1965), 636, theorizes on the possible effects of the oceans on the atmospheric lunar tide. Theoretical discussions of the diurnal tide are given by R. S. Lindzen, *MWR*, 94 (1966), 295, and by S. Kato, *JGR*, 71 (1966), 3201 and 3211. G. A. Corby, *QJRMS*, 93 (1967), 368, and R. S. Lindzen, *GPA*, 62 (1965), 142, go back to Laplace's original tidal equations, Lindzen to discuss the asymmetry of the diurnal tide.

34. B. Haurwitz, *BAMS*, 36 (1955), 311; S. L. Rosenthall and W. A. Baum, *MWR*, 84 (1956), 379.

35. G. W. Hurst, *MM*, 87 (1958), 294, and *MM*, 89 (1960), 171.

36. N. P. Sellick, *QJRMS*, 74 (1948), 74, and 78-81; R. A. Hamilton, *QJRMS*, 85 (1959), 168.

37. J. Spar, *BAMS*, 33 (1952), 339 and 438-41.

38. R. Frost, *MM*, 90 (1961), 110.

39. T. F. Malone, *Mass. Inst. Tech. and Woods Hole Oceanog. Inst. Papers Phys. Oceanog. Meteorol.*, vol. 9, no. 4 (1945).

40. J. Spar, *JM*, 7 (1950), 167.

41. W. Schwerdtfeger and F. Prahaska, *JM*, 13 (1956), 217. For details of the solar semi-diurnal pressure wave over North America, see L. Avery and B. Haurwitz, *MWR*, 92 (1964), 79. For the evidence for continuous, regular, nocturnal variations in atmospheric pressure, see K. L. Shrestha, *QJRMS*, 93 (1967), 254, and W. L. Donn and N. K. Balachandran, *QJRMS*, 94 (1968), 208.

42. A. J. Abdullah, *BAMS*, 36 (1954), 511.

43. D. T. Williams, *MWR*, 81 (1953), 278.

44. D. T. Williams, *MWR*, 82 (1954), 289.

45. M. Tepper, *MWR*, 79 (1951), 61.

46. J. C. Freeman, *BAMS*, 31 (1950), 324.

47. H. Newstein, *MWR*, 82 (1954), 255; J. B. Holleyman and J. M. Hand, *MWR*, 82 (1954), 237.

48. M. Tepper, *JM*, 7 (1950), 21.

49. W. Donn, R. Rommer, F. Press, and M. Ewing, *BAMS*, 35 (1954), 301.

50. I. W. Brunk, *JM*, 6 (1949), 181.

51. E. Gossard and W. Munk, *JM*, 11 (1954), 259.

52. E. Tryggvason, *MWR*, 88 (1960), 256; M. A. Emerson, *MWR*, 88 (1960), 18; C. L. Jordan, *MWR*, 90 (1962), 191.

53. I. J. W. Pothecary, *QJRMS*, 80 (1954), 395.

54. L. S. Matthews, *W*, 6 (1951), 185.

55. C. K. M. Douglas, *MM*, 78 (1949), 309.

56. M. H. P. Hoddinot and S. E. Ashmore, *W*, 10 (1955), 16.

57. S. G. Crawford, *W*, 8(1953), 192; H. R. Reinders, *loc. cit.*, p. 31; N. P. Sellick, *loc. cit.*, p. 223.

58. W. H. Roschke, *BAMS*, 35 (1954), 20.

59. A. P. Crary and M. Ewing, *TAGU*, 4 (1952), 499.

60. R. D. M. Clark, *JM*, 7 (1950), 70.

61. J. M. Mitchell, *BAMS*, 30 (1949), 105.

62. G. Emmons, B. Haurwitz, and A. F. Spilhaus, *BAMS*, 31 (1950), 135.

63. H. L. Demastus, *BAMS*, 35 (1954), 497.

64. R. S. Scorer, *W*, 8 (1953), 343.

65. H. H. Aslett, *MM*, 78 (1949), 83.

66. A. W. Lee, *MM*, 77 (1948), 201.

67. R. Yamamoto, *W*, 10 (1955), 321, *W*, 11 (1956), 170, *BAMS*, 37 (1956), 406; K. H. Stewart, *MM*, 88 (1959), 1; R. V. Jones, *N*, 193 (1962), 229; R. Araskog, U. Ericsson, and H. Wägner, *N*, 193 (1962), 970; W. L. Donn and M. Ewing, *JGR*, 67 (1962), 1855; H. Wexler and W. A. Hass, *JGR*, 67 (1962), 3875.

68. R. S. Scorer, *Proc. Roy. Soc.*, 201, series A (1950), 137. For recent work on solitary waves, all involving theory rather than observation, see R. R. Long, *JAS*, 21 (1964), 197, and L. H. Larsen, *JAS*, 22 (1965), 222, on solitary waves in the westerlies, and M. C. Shen, *JAS*, 24 (1967), 260, on solitary waves in an atmosphere with a thermal inversion. For details of pressure waves traveling at sonic velocities from a squall line, see V. H. Goerke and M. W. Woodward, *MWR*, 94 (1966), 395; for surface pressure waves associated with the motion of a hurricane, see C. L. Jordan, *MWR*, 94 (1966), 454; and for pressure jumps at Malta, see T. H. Kirk, *MM*, 92 (1963), 51 and 89. A. J. Abdullah, *JGR*, 71 (1966), 1953, discusses the effects of a head-on collision between two pressure jumps. For a description of pressure waves generated by the Japanese earthquake of June 16, 1964, see G. G. Bowman and K. L. Shrestha, *QJRMS*, 91 (1965), 223. On pressure waves generated by nuclear explosions, see R. A. McCrory, *JAS*, 24 (1967), 443, and R. F. MacKinnon, *QJRMS*, 94 (1968), 156; for a purely theoretical study, see D. B. van Hulsteyn, *JGR*, 70 (1965), 257 and 271. For discussions of the acoustic-gravity waves involved in the propagation of these disturbances, see: D. G. Harkrider, *JGR*, 69 (1964), 5295; R. L. Pfeffer and J. Zarichny, *GPA*, 55 (1963), 175; R. H. Clarke, T, 15 (1963), 287; and R. F. Mackinnon, *QJRMS*, 93 (1967), 643. On the mathematical theory of the dispersion of large pressure disturbances, see M. K. Arora,

QJRMS, 91 (1965), 218. For a rapidly moving pressure rise over the northeastern United States, interpreted as a very long gravity wave, see N. K. Balachandran and W. L. Donn, *MWR*, 92 (1964), 423. For pressure-jump lines in the Persian Gulf area, see D. J. Clark, *MM*, 97 (1968), 368. For pressure waves traveling at sonic velocities from a squall-line storm in Colorada, see V. H. Goerke and M. W. Woodward, *MWR*, 94 (1966), 395. For background fluctuations in surface atmospheric pressure with a 30-to-90-minute period, caused by jetstream winds, see T. J. Herron and I. Tolstoy, *JAS*, 26 (1969), 266 and 270. For surface pressure variations along the Texas coast during the approach of hurricane Carla, see C. L. Jordan, *MWR*, 94 (1966), 454. On the use of nuclear explosives in exploring the atmosphere, see W. L. Donn and D. M. Shaw, *Rev. Geophysics*, 5 (1967), 53.

69. J. M. Austin, *CM*, pp. 630-38, is a basic source for discussion of pressure-change theories.

70. B. Haurwitz and E. Haurwitz, *HMS*, no. 3 (1939); J. M. Austin, *JM*, 6 (1949), 358.

71. S. L. Hess and H. Wagner, *BAMS*, 29 (1948), 219.

72. J. M. Austin and R. Shapiro, *JM*, 8 (1951), 191.

73. J. Vederman, *BAMS*, 30 (1949), 303.

74. M. F. Harris, *MWR*, 81 (1953), 193.

75. J. M. Austin and L. Krawitz, *JM*, 13 (1956), 152. For details of how to determine areas of rising or falling pressure in the free atmosphere, see S. Evjen, *T*, 2 (1950), 212.

76. The elements of fluid dynamics as applied to climatic phenomena are reviewed in Chapter 5. Classic accounts include N. Shaw, *Manual of Meteorology*, vol. IV (Cambridge, 2d ed., 1931), and *PDM*, pp. 160-312. For a classic treatment using vector methods, see H. Solberg, *Geofys. Publ.*, vol. 5 (1928).

77. H. Jeffreys, *QJRMS*: 48 (1922), 29; 50 (1924), 61; 51 (1925), 97; 52 (1926), 85; 53 (1927), 401; 64 (1938), 336.

78. C. H. B. Priestley, *QJRMS*, 73 (1947), 65, and 74 (1948), 67.

79. J. M. Austin *et al.*, *BAMS*, 34 (1953), 383.

CHAPTER 2

Atmospheric Properties and Processes

Climate is an abstraction from physical phenomena. Physics, the most advanced science dealing with such phenomena, is essentially concerned with the properties of matter and energy, and has evolved refined techniques for analyzing these properties. If the climatologist can express his problems in terms of physics, he can hope to solve them. This chapter outlines the extent to which climatic data reveal features of the properties of the Earth's atmosphere, and of the processes taking place within it, that can be expressed in the terms of the physicist.

Atmospheric Stratification

The most obvious physical property of the atmosphere as a whole is that it is stratified into "shells" that surround the Earth. The credit for correctly postulating the layered character of the atmosphere should probably go to Mairan, who in 1754 proposed a model atmosphere of three layers.[1] His lower layer went up to 10 or 12 km and was the layer of storms and hydrometeors; his middle layer, from 10 or 12 to 80 km, controlled barometric pressure; and his upper layer was the Auroral Zone, from 80 to over 1,000 km. Mairan's ideas were considered to be preposterous until proof of the existence of the tropopause and the stratosphere was announced in 1902.

Before 1902, it was believed that since atmospheric heat is derived from the Earth's surface rather than from direct absorption of solar energy, temperatures ought to decrease with height in the atmosphere. In 1902, on the basis of 236 free-balloon ascents from Trappes (Paris) during a three-year period, Teisserenc de Bort announced to the French Academy of Sciences that the actual rate of decrease of temperature with height became zero at approximately 11 km above Paris, and above that level was very weak or even reversed.[2] This discovery began the development of the "classic" picture of the atmosphere, which regarded it as consisting of two concentric shells: an inner shell or *troposphere*, which was essentially unstable and in which convective and mixing processes predominated; and an outer shell or *stratosphere*, which was stable and characterized by a lack of convection or mixing.

In the last fifty years, new atmospheric shells have been recognized. A recommendation of the WMO in June 1962 sets out the international conventional limits of these shells,[3] which are separated by boundaries or "pauses" as suggested by Napier Shaw many years ago. The *tropopause* is the upper boundary of the troposphere, and separates the latter, in which temperature decreases with height, from the stratosphere, where temperatures usually increase with height. The stratopause and the mesopause are now also recognized. The *stratopause* is found at the top of the stratosphere, usually around 50 to 55 km above the Earth's surface. The *mesopause* is at the top of the mesophere, usually at 80 to 85 km up. The *mesosphere* is thus the shell bounded by the stratopause and the mesopause; it is generally characterized by another decrease of temperature with height. The *thermosphere* is the shell above the mesopause; in it temperature usually increases with height, usually to the outer limits of the atmosphere.

Other shells or "spheres" may be found described in the literature.[4] The largest is the *magnetosphere*. The atmosphere, defined as the gaseous envelope which travels with the Earth in its orbit around the sun, extends outward to between 20,000 and 80,000 miles. The greater part of the atmosphere consists of a tenuous, ionized gas or plasma, whose behavior is directly controlled by the Earth's magnetic field, and which may therefore be conveniently called the magnetosphere; it extends downward to the bottom of the ionosphere.[5] Rainfall at the Earth's surface, the distribution of ozone, and the occurrence of noctilucent clouds are all directly influenced by processes within the magnetosphere. Charting of its features is an important task for the climatologist.

The *chemosphere* is the shell between 32 km and 80 km in which chemical reactions are important.[6] Ozone is produced here by solar radiation; water vapor is dissociated; and compounds of oxygen and nitrogen, and free ionized sodium, appear. It is the region in which temperature rises to a maximum (at 55 km), and then decreases to a minimum (at 80 km). Between 400 and 1,000 km, appreciable amounts of ionized matter are found as permanent constituents of the atmosphere. Above 1,000 km is the outer atmosphere or exosphere, from which particles can escape the gravitational and magnetic fields of the Earth.[7]

The *turbosphere* is that region of the lower atmosphere—up to 60 km, at least—in which turbulence is sufficient to mix the permanent constituents of the air, and to overcome the diffusive tendency of each to distribute itself independently according to its molecular weight. Below the *turbopause*, turbulent mixing is the dominant process; above the turbopause, diffusion is more important than mixing, and the composition of the atmosphere changes by diffusion rather than by the photochemical dissociations by sunlight that take place in the rest of the chemosphere generally. Another distinction sometimes made is between the homosphere and the heterosphere, separated by the *homopause* at about 100 km. The constituents of the upper layer (the *heterosphere*) are physically different from those of the lower layer (the *homosphere*).[8]

Four distinct strata can be recognized in the troposphere.[9] The ground layer, less than 2 m thick, is controlled by vegetation and soil conditions, buildings, and so on. Its properties decide the rate of exchange of heat, moisture, and momentum between the Earth's surface and the overlying air, and hence this layer is very important in causing local modifications in climatic conditions.

The *friction layer*, between 2 and 2,000 m, is the stratum in which friction between the atmosphere and the Earth's surface is important. Its upper boundary, defined as the level at which the actual wind equals the gradient wind, is determined by inversion conditions: from a height of 1,000 to 1,500 m over continental Europe and the United States, it drops to between 400 and 600 m over the poles, and is also low over the oceans. This layer is, in effect, the planetary boundary layer, and can itself be divided into two layers (except when convection is very intense).

The lower of the two layers, the bottom friction layer, varies in height from 2 to 100 m. Its characteristic feature is that the Austausch coefficient A (i.e., the coefficient of turbulent mass exchange) increases with height to a maximum that depends on latitude and surface roughness, among other things. Clouds are rare in it, but fog and haze occur if A is low and if humidity increases with height in it, as is quite common.

The upper friction layer is characterized by a decrease in A with height. The thicker the bottom friction layer, the thicker the upper friction layer; when convection increases, the two layers merge into one.

On hot afternoons, when convection is most intense, the friction layer can be observed visually to consist of two distinct strata, even when upper and lower friction regions cannot be distinguished. The boundary between these two strata, the *peplopause*, in cloudless conditions is visible as either a haze top, or as a sudden change in visibility, say from four or five miles near the ground to 30 or 40 miles immediately above the peplopause. On cloudy days, the peplopause is often found above a layer of turbulence clouds, fractocumulus, fractostratus, or stratocumulus. Generally, the peplopause acts as a blocking surface to descending air, similar to a subsidence inversion in an anticyclone.[10]

The *advection layer* is that portion of the troposphere in which horizontal motions exceed vertical motions by a factor of 100 to 1,000. Stratified clouds prevail here, and distinct cloud layers are common. The wind increases with height, and weather changes are due to variations in the lapse-rates brought about by descending or ascending air.

The stratification of the atmosphere is caused by both gravitational and thermal effects and is clearly important in the generation of climates. Two further points must be noted. First, atmospheric shells are not necessarily spherical or concentric, and should not be thought of as closed systems; there may, in fact, be considerable interchange of mass between them, as there is between troposphere and stratosphere, particularly in the vicinity of the tropopause discontinuity in middle latitudes. Eddy-mixing processes, too, are important above as well as below the tropopause, and possibly up to 100 km or more. The original distinction between troposphere and stratosphere in terms of presence of and absence of mixing, respectively, is now known to be unimportant.[11]

Second, it is unwise to define an absolute upper limit to the climatologist's interest in the atmosphere. The upper stratosphere and the mesophere are as much subject to "weather" variations as the troposphere, and the geographical distribution of temperature above 30 km is the reverse of that at sea level: temperatures increase poleward in summer and equatorward in winter. There are substantial day-to-day variations in upper stratospheric winds, and diurnal and semidiurnal tidal motions are present at 10 km, 80 km, and above, with amplitudes increasing from 0.5 to 40 m per sec.[12]

THE TROPOPAUSE

Of all the atmospheric discontinuities, the tropopause has been the most studied, and daily charts showing its height were published for some years in the British *Daily Aerological Record.* The WMO defines (1957) the "first tropopause" as the lowest level at which the lapse-rate decreases to 2°C per km or less, provided also that the average lapse-rate for the next 2 km nowhere exceeds 2°C per km. If, above the first tropopause, the average lapse-rate within any 1 km exceeds 3°C per km, then a "second tropopause" is defined by the same criterion; it may be either within or above this 1-km layer. The mean picture shows two clearly defined tropopauses: a high equatorial tropopause, and a low tropopause in high and middle latitudes, overlapping in the subtropics to form a region of double tropopauses. Actually, tropopauses are not continuous surfaces, but rather are made up of successive leaves, which rise toward the equator, the lowest leaf being over the cold pole of each hemisphere. The boundary of each leaf may be regarded as a tropopause break and can be outlined on a chart. These breaks prove to be very significant in surface-weather analysis.

The best level for tropopause chart analysis is at 50 mb. The criterion for locating the positions of the tropopause on a 50-mb chart is a difference in potential temperature of 15° to 20° between standard synoptic stations. Tropopause breaks may then be indicated by broken lines on the chart. These breaks, which may be several thousand miles long, are often boundaries of upper air masses (i.e., the intersections of upper fronts with the base of the stratosphere). They can also be rather self-contained phenomena, in which case the shape of the breakline has definite meaning for weather. There is heavy snowfall beneath the southern and eastern ends of a breakline, and the more curved the breakline, the heavier the snowfall. Curved breaklines move more easily in an east-west direction than straight ones, and can tap more moist air. Circular breaklines are important, because heavy snowfall is associated with a low, circular tropopause.[13]

Today the tropopause chart is a common feature of daily weather analysis (see Figure 2.1). These charts show that the tropopause exhibits folds as well as breaks; it dissolves and reforms by subsidence; it "funnels" into discontinuous centers of relatively low height above the Earth's surface; and sometimes tongues of high tropopause form and occlude. The tropopause moves with the air as a material surface, but transfer of air takes place across it, both upward and downward.[14]

Since changes in the height of the tropopause are brought about by large-scale physical processes, it must be possible to correlate them with subsequent weather developments at the ground. For example, in Japan, an increase in tropopause height and a decrease in tropopause temperature indicate that advection of tropical air can be expected, and cold fronts during the Bai-u season tend to change into normal warm fronts. Pacific typhoons tend to be located under a trough in the tropopause, moving into the region of interference between two tropopause troughs lying in the temperate westerlies and accompanying the jetstream. The typhoons then rapidly decay over

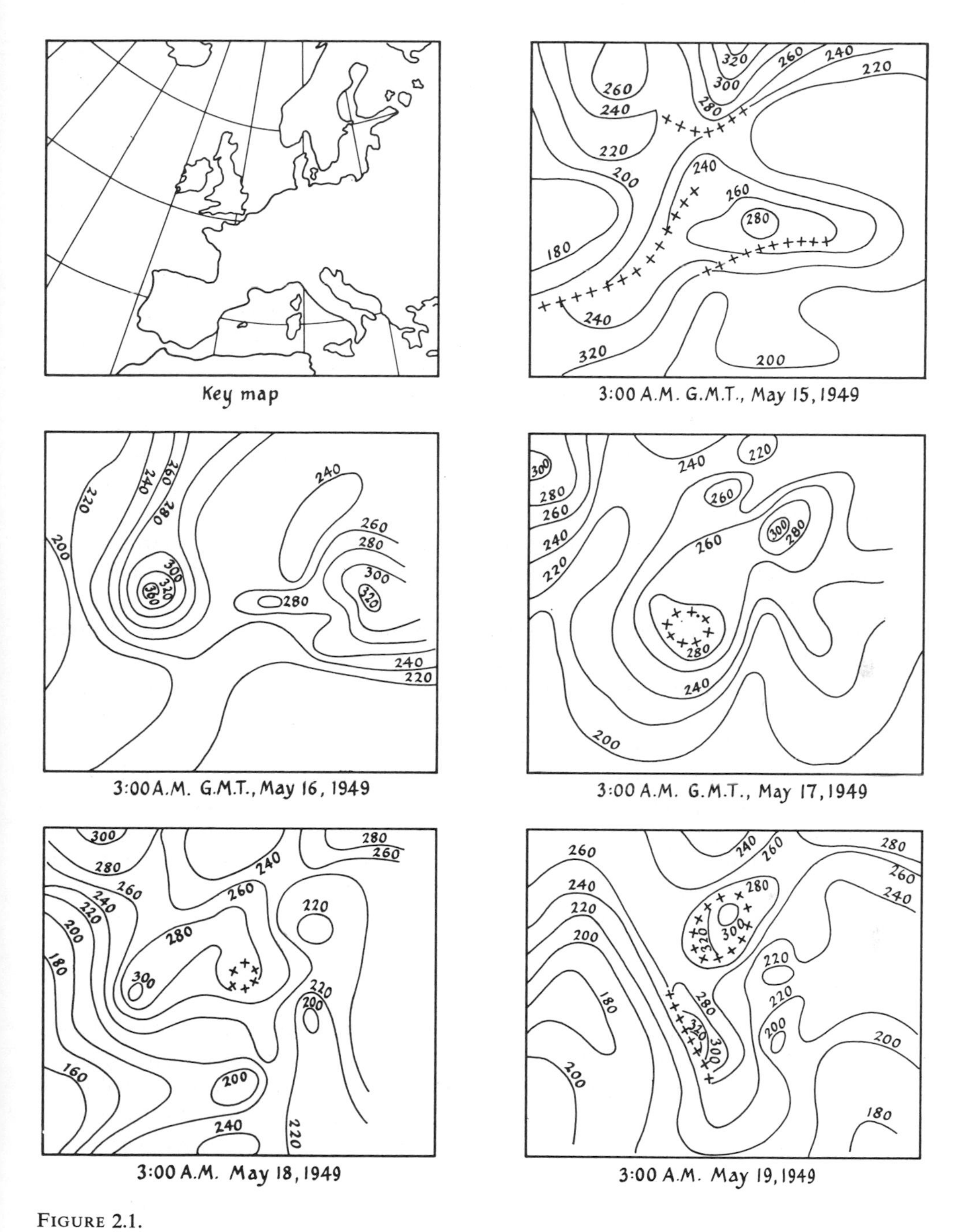

FIGURE 2.1.
The tropopause.
——isobar on the tropopause (mb).
+ +folded tropopause.
(After J. S. Sawyer.)

Japan. Deep cold waves in Europe are associated with the advection of cold air between an Atlantic anticyclone and a primary depression over northeast Europe, under a descending tropopause, which pattern is gradually replaced by vertical movement over a wide area and a rising tropopause. The long-wave patterns in the middle-latitude westerlies closely follow the contours of the tropopause, but the airflow is kept from being purely horizontal by dynamic developments; it is uncertain whether the tropopause heights control the long-wave patterns or vice versa.[15]

The relation of the tropopause to the general circulation of the atmosphere is of great importance for global weather patterns. The height of the tropopause varies with the temperature of air at the tropopause. High temperatures accompany a low tropopause, hence variation in the meridional movement of heat in the atmosphere should be associated with the variations in tropopause height.

The marked temperature and density discontinuity at the tropopause could be due to a slow flow of air toward the equator below it, and away from the equator above it. There is evidence that air in the lower stratosphere in middle and high latitudes has drifted there from the upper 4 km of the tropical troposphere. A mean net circulation seems to exist in the high troposphere and low stratosphere, with air entering the stratosphere in tropical latitudes, moving poleward, then sinking back into the troposphere in higher latitudes. The rate of transport would fully replace all stratospheric air within two years.[16]

The mechanism by which the tropopause is maintained differs in polar and tropical regions. The polar tropopause is probably caused by radiation from the layers of the lower atmosphere that contain water vapor, and then maintained by exchange of air with lower latitudes. The tropopause over Antarctica disappears on many days in winter, when there is an absence of such exchange, and the intensive radiative cooling of the lower air layers then ultimately leads to an increased lapse-rate in the lower stratosphere. In tropical regions, the layer at which maximum cooling of the atmosphere takes place is some 4 or 5 km below the tropopause, which in these regions therefore cannot be caused by radiation from the water-vapor layers. The tropical tropopause is probably the result of slowly ascending cold air being mixed with warmer air from higher latitudes.[17]

INVERSIONS

An inversion is defined as an increase of temperature (or moisture) with height above the Earth's surface for some minimum distance. Some inversions are caused by large-scale subsidence, as in the trade-wind inversions; others are very small-scale features due to local radiational cooling. All types of inversion are of direct significance for weather and climate. Although the term "inversion" usually refers to a thermal inversion, moisture inversions are also quite important.

Apart from standard radiosonde and micrometeorological techniques, some novel methods have been used for observing inversions. Smoke plumes (Figure 2.2) are useful for visual observation. Plumes will fan during inversion or near-isothermal conditions, loft during upper inversions, and fumigate when a nocturnal inversion is being dissipated by heat from the morning sun. Smoke plumes can travel long distances without appreciable dilution during inversions.[18]

At twilight, most of the light received at the ground has been scattered by molecular or larger dust particles. If there is a rapid increase in atmospheric dust content with

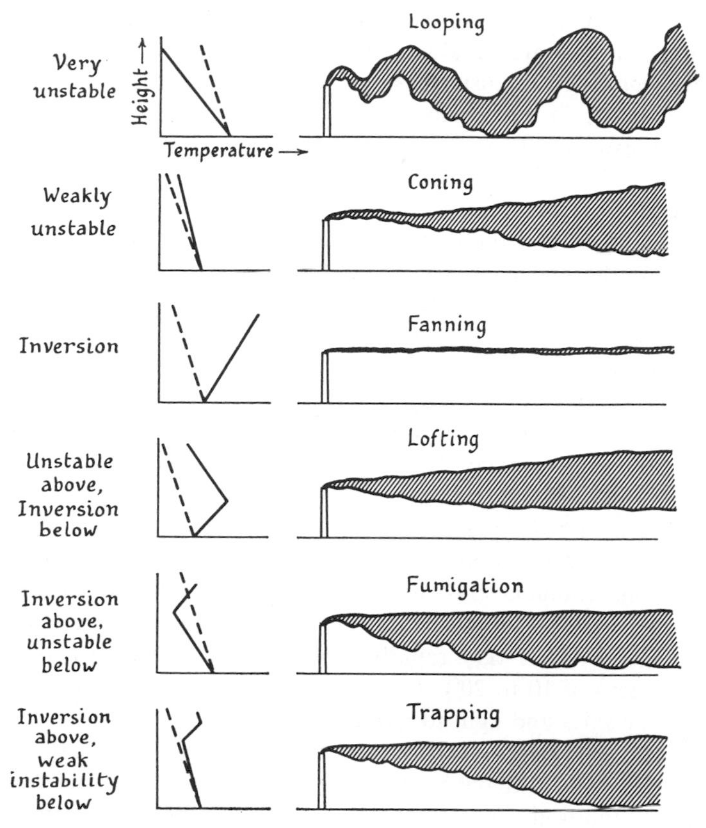

FIGURE 2.2.
Atmospheric stability and plume behavior. The broken line shows the dry-adiabatic lapse-rate, the solid line the actual lapse-rate. (After Bierly and Hewson.)

height, there will be a sudden increase in the amount of light reaching the ground when the Earth's shadow reaches the tropopause in the direction of observation. The increase in dust content is often associated with an inversion. The increase has been measured by gathering the light from a cone of one degree in angular width in a searchlight mirror. The light was interrupted at 250 cycles per sec by a rotating slotted disc, after which it fell on a photomultiplier tube, the output voltage of which measured the intensity of the light and hence of the dust concentration. This technique enabled temperature soundings to be plotted up to 100 km over Australia, as against 20 km for normal radiosonde soundings.[19]

Under convective (clear sky) conditions, the intensity of noise from an aircraft in flight is moderate, and varies in accordance with the inverse-square law. Noise from an aircraft at the same height and position but under an inversion is intense, and does not vary in conformity with the inverse-square law. When an aircraft is at the same height but above an inversion, its noise is almost undetectable, according to measurements made in California.[20]

Observations from aircraft over the North Pacific, the Gulf of California, the Red Sea, and the Persian Gulf show that when inversions occur at sunset, the red band in the sky just above the sea horizon is split into two layers. The lower layer has an angular height of half a degree, and is a much deeper red than the upper layer.[21]

A measure of inversion intensity has practical value, since low, relatively intense inversions often result in widespread smog. A double inversion, with both lower (radiation) and upper (subsidence) inversions, can cause extremely heavy pollution. Inversion intensity is a function of atmospheric stability, that is, of the gradient of potential temperature through the inversion layer, the potential temperature difference through the same layer, and the height of the inversion. A measure of this type for Los Angeles shows that mean inversion intensities there vary from 12.5 in November to 3.1 in May, with a maximum recorded intensity of 42.6.[22] *

Different geographical conditions give rise to different types of inversion. A basic distinction in cold regions is between maritime and nival inversions. Maritime inversions are formed by winds and turbulence in a mixing layer not in contact with the Earth's surface. Nival inversions are winter surface inversions produced by radiational cooling from snow or ice. Trade-wind inversions are maritime inversions in warm latitudes.[23]

Surface inversions in polar regions are mainly due to the low heat-conductivity of snow. When air temperatures above a snow surface are much less than 0°C, heat conduction from the ground layers beneath the snow cannot compensate for the energy lost by radiation from the snow surface. Hence the temperature of the air layer in contact with the snow falls very rapidly, and an inversion forms. At Maudheim, temperature increases of 10 to 20°C from the snow surface to 10 m above are quite common with clear skies and light to moderate winds.[24]

In Arctic regions generally, inversions of 5°C per 100 m are usual, with reversed lapse-rates of 1°C per m near the Earth's surface. Their depth varies from 200 m to 2 km. With calm conditions, inversions are usually located close to the ground, but with appreciable wind, they become elevated. In warmer latitudes the height of inversions depends on the synoptic conditions. At Lisbon, inversions form below 1,000 feet if there is warm air from the southwest or south, but develop at 7,000 to 10,000 if there are divergent anticyclonic airflows with a northerly component.[25]

Both large-scale and small-scale geographical conditions are significant for inversion development. The frequencies of inversion in the United States correspond well with the major geographical divisions: inversions are very frequent above the Appalachian Mountains and the Great Plains, moderately frequent above the Rockies, and infrequent above the Great Lakes and the Gulf-Atlantic coast. Inversions form at 300 to 500 m above the sandy deserts of the U.S.S.R. because the air is warmed by particles of dust and sand carried up during strong winds; the particles absorb solar radiation and also carry heat upward from the hot ground surface. On the microscale, a temporary inversion that may have been 350 feet deep was formed over the Cumberland coast of Britain because the air was cooled 40°F by hail.[26]

Surface heating varies enormously with geographical conditions, and has an important effect on an existing inversion. Surface heating causes convection currents, which rise up to the inversion, spread out laterally beneath it, and cause continuous entrainment of warmer upper air into the convection layer. Hence the inversion will

* E. Robinson, *BAMS*, 33 (1952), 247, employed the measure

$$I = \frac{(\Delta\theta)^2}{3 + z\Delta z}$$

where z is the height of the inversion and Δz the thickness of the inversion layer (both in units of 100 m) and $\Delta\theta$ is the potential temperature difference through the inversion.

rise and there will be a mean transfer of air downward through it. Material (such as pollutants and water vapor) beneath the inversion cannot penetrate it, but material above the inversion can easily sink downward through it.[27]

There is usually a strong correlation between winds and inversions. Nocturnal windspeeds often have a pronounced maximum below 3,000 feet, at the top of the inversion, forming a low-level jet that is important in squall-line formation. If there is a jet, the nocturnal inversion will be stable, but if there is not, the inversion will break down chaotically.[28]

In the lowest 300 m of the atmosphere over Honjo (Japan), the largest inversion gradient occurs just before midnight, and thereafter varies with windspeed. The inversion layer consists of a strong lower sublayer, an almost isothermal middle sublayer, and a weak upper sublayer. Cooling, due to a change in wind direction, commences in the middle sublayer and continues for one hour after surface temperatures begin to rise at dawn. After temperatures begin to rise in the middle sublayer, a cold zone sometimes develops in the lower sublayer at 30 to 100 m. The layer of maximum windspeed descends after sunset, to reach its lowest elevation, about 60 m in winter, just after midnight.[29]

Certain weather conditions, such as fog, are typically associated with inversions. During fogs, there is usually a very shallow inversion at the ground (50 feet deep at Cardington, 1951–54), with a nearly isothermal layer above it extending to the top of the fog, then a steep inversion layer above the fog. With high inversions several km deep, the extensive stratus developing above the polar basin in summer, and the deep haze forming above continents in winter and above seas in summer, are the large-scale counterparts of the local surface fogs common with low inversions.[30]

In Alaska, the extent of ice fog depends on the height and strength of the low-level inversion. During periods of intense radiational cooling, an unstable vertical temperature gradient may exist within the fog layer. As a result, ice particles are transported downwards, causing very low horizontal visibilities at ground level.[31]

The connection between inversions and cloud cover is very close. In California, the tops of summertime stratus coincide with the top of the strong inversion on the southeast side of the North Pacific subtropical anticyclone. Widespread stratiform clouds characterize the surface inversion layer in North Baffin Land, where the cold pole of the atmosphere in the northern hemisphere is situated at 100 mb. Cumulus clouds are common to the north of Puerto Rico; here the ocean surface is isothermal and no thermals can occur, but there is an inversion at 600 m, and clouds form where pressure oscillations, with a wavelength of about 2 km, enable the lower moist layers to break through the inversion.[32]

Temperatures at 1,000 to 2,000 feet, between a double inversion at Silver Hill, Md., dropped appreciably when the upper (subsidence) inversion lowered and the lower (nocturnal radiation) inversion began to break down because of development of a steep wind gradient. The temperature drop was enough that it would have caused thin low stratus if the air had been moist enough.[33]

Atmospheric Energy

Although the atmosphere receives its energy supply ultimately from the sun, most of this energy is first reflected or radiated from the Earth's surface or from various substances within the atmosphere, rather than absorbed directly. Consequently the reflecting power of the Earth's surface is of prime importance in deciding how much energy is locally available to the atmosphere.

ALBEDO

If a beam of short-wave solar radiation strikes a rough surface, it is diffusely reflected, that is, scattered in all directions. For an ideal, extensive rough surface, the amount of radiation reflected in any direction is, theoretically, independent of the direction of the incident beam, but for most natural surfaces, the amount of radiation reflected in a given direction will depend on the angle of incidence of the beam.

To define the total reflectivity or *albedo* of a surface, we must consider a small horizontal test surface a short distance above it, and assume that neither surface emits radiation of its own. The albedo of the original surface is then defined as r/i, where r is the intensity of the radiation reaching the test surface after being reflected by the lower surface, and i is the intensity of the radiation passing downward through the test surface. Albedo is usually given as a percentage.[34] It should be noted that the term albedo refers to the reflecting power of a surface for all wavelengths of energy, whereas the *reflectivity* of a surface is its response to one specific wavelength or range of wavelengths.

Albedo may be measured by means of a pyranometer, which is sensitive to short-wave radiation but not to radiation with wavelengths exceeding 4μ. The instrument is directed toward the sky and the ground in turn, and the readings compared. It is not rapid in response and is unsuitable for extensive surveys.

For rapid geographical surveys, the beam reflectometer is useful. This is a standard pyrheliometer mounted upright at the focus of a parabolic mirror, which is usually fixed to the underside of an aircraft wing. Observations using such an instrument over the Great Plains of North America show that albedo, and therefore surface heating, varies enough from day to day to determine where or not a high-level temperature inversion will break down. Similar observations over Wisconsin indicate that there albedo has two basic seasonal values, depending on whether there is snow or not. Albedo was between 10 and 20 per cent over agricultural land during the snow-free season, was 50 per cent over wooded hills during the snow season, and was 80 per cent over a frozen lake.[35]

On a world or continental scale, albedo is mainly determined by variations in large-scale physiographic features and glacierization, in vegetation cover, and in the duration of snow cover. For example, areas of dense boreal forest show only small seasonal differences in albedo, whereas northern forests transitional to tundra show large seasonal differences.[36]

The albedo of land surfaces changes considerably during the day, being relatively high in the early morning and evening, when the sun is low, and relatively low when the sun is high and its rays penetrate deeply into the vegetation cover and are absorbed. The albedo of a snow surface does not change much during the day, but the albedo of snow-free surfaces during the day depends both on the height of the sun and on cloud cover, increases in which reduce the dependence of albedo on solar elevation.[37]

The albedo of a water surface under direct radiation varies with the height of the sun, from very small values at the time of high sun to almost 100 per cent when the sun is near the horizon. For diffuse radiation, albedo depends much less on solar

Typical Albedo Values[a]

Water Surfaces

Sea, winter,	6 in latitude 0°,	21 in latitude 60°
Sea, summer,	6 in latitude 0°,	7 in latitude 60°
Inland water bodies, 7		

Bare Surfaces

Fresh snow	75–95	Dark soil	5–15
Old snow	40–70	Grey soil (dry)	20–35
Sea ice	30–40	Grey soil (wet)	10–20
Sand dune (dry)	35–45	Concrete (dry)	17–27
Sand dune (wet)	20–30	Pavement (wet)	9
Hard rock	12–15	Pavement (dry)	18
Chalk	2	Road (black)	5–10
Ploughland	13		

Vegetation-Covered Surfaces

Desert	24–30	Deciduous forest	10–20
Savanna (dry season)	25–30	Coniferous forest	5–15
Savanna (wet season)	15–20	Grass	14–37
Chaparral	15–20	Green meadow	10–20
Open woodland	10–18	Dry leaves	29
Closed woodland	4	Tundra	15–20
Green forest	3–10	Field crops	3–25

[a]All values refer to wavelengths less than 4μ (after Sellers and others).

elevations, and varies only from 8 to 10 per cent.[38] Fresnel's law of reflection applies to water surfaces if the angle of elevation of the sun exceeds 15°; albedo then varies with the direction of the incident beam. The albedo of a water surface is noticeably affected by waves or ripples, which increase net reflection of direct light when the sun is near its zenith, and decrease it when the sun is near the horizon (see Appendix 2.1). The albedo of wet ground is only half that of dry ground, mainly because of total internal reflection within the water films covering the ground.

On a global scale, the albedo of the planet Earth is about 35 per cent, and that of the average world cloud cover about 50 per cent. From photometric measurements of the relative intensity of light on the dark and bright sides of the moon, it is possible to calculate the visual albedo of the planet Earth; it comes to 39 per cent. Clearly, a considerable amount of the solar energy (both visible and invisible) reaching Earth does not ultimately find its way into the atmosphere, but is reflected back and lost into space.[39]

ENERGY PROCESSES

The portion of solar energy that reaches the planet Earth and is not reflected is absorbed by the Earth's surface, or by the atmosphere, and then provides the latter's internal energy supply. The energy of the atmosphere consists of four components: the kinetic energy of the major wind systems; the potential energy of the Earth's gravitational field; the latent energy of water vapor; and the air's internal energy, which depends only on temperature (for an ideal gas). The energy content of the atmosphere at any instant reflects the balance between these different components and their transformations. One measure of how this energy content varies from one moment to the next is provided by the atmospheric energy equation (see Appendix 2.2.).

In general, it may be stated that the rate at which atmospheric energy increases equals the rate at which the atmosphere gains energy from or loses energy to the Earth's surface, reduced by the rate at which the atmosphere does work. The term energy is used here in its usual physical sense, as the property of a system that decreases when the system does work, and increases when work is done on the system; the amount of increase or decrease of energy equals the work done. The atmosphere is not a closed system, because it gains or loses heat by contact with the Earth's surface. The value of energy concepts as applied to the atmosphere is that they provide a means for studying the temperature changes that are caused by air movements and vice versa.[40]

The amount of energy in the atmosphere may be estimated from measurements of temperature and pressure, which allow the relative contributions of potential, kinetic, and internal energy to the total energy to be separated (see Appendix 2.3). The importance of the conversion of internal and potential energies into kinetic energy was described in an early model by Margules, who demonstrated mathematically, from basic energy equations, how the subsidence of a cold mass of air and the associated lifting of an adjacent warm mass, as happens, for example, in an extratropical frontal depression, must of itself generate winds (see Appendix 2.4). A later model by Starr indicates that kinetic energy is generated in areas of divergence (e.g., anticyclones) and dissipated in areas of convergence (e.g., cyclones).[41] Thus kinetic energy in the atmosphere can be created or destroyed solely by hydrodynamic processes, quite apart from its dissipation by friction between moving air and the Earth's surface.

The chief climatological interest of atmospheric energy processes lies in evaluating to what extent either (a) the permanent and semipermanent features of the atmosphere, or (b) the deviations of these features from their average positions, are most effective as energy sources or sinks. The kinetic energy of the mean motion of the atmosphere increases from summer to winter, as the total of potential and internal energy decreases.[42]* The increase in kinetic energy is less than 20 per cent of the de-

---

* J. Spar, *JM*, *6* (1949), 411, showed that the kinetic energy per unit volume of air of the mean monthly circulation is given by $(\nabla\bar{p})^2/2\lambda^2\bar{\rho}$, where $\bar{\rho}$ refers to the mean monthly density of the atmosphere, $\nabla\bar{p}$ is the mean monthly horizontal pressure-gradient, and λ is the Coriolis parameter.

crease in potential and internal energy; in fact, kinetic energy is a relatively small part of the total energy of the atmosphere. If all the January kinetic energy were completely transformed into heat, the mean temperature of the atmosphere would increase by only 0.2°C.

The atmospheric energy cycle consists of net generation of potential and internal energy by nonadiabatic processes, conversion to kinetic energy by adiabatic processes, and net dissipation of this kinetic energy by viscosity, which converts kinetic energy back into potential and internal energy.[43] The rate at which kinetic energy is dissipated is about 2 per cent of the rate at which solar radiation is absorbed. This rate of conversion appears to be the maximum possible for our existing distribution of solar radiation. The general circulation is probably operating near its maximum possible intensity, even though, in a thermodynamic sense, the efficiency of the atmosphere is very low (see Appendix 2.5).

A useful practical measure of potential and internal energy is provided by the concept of *available potential energy*, which is the maximum amount of potential and internal energy available for conversion to kinetic energy by adiabatic processes (see Appendix 2.6). Considerable available potential energy is generated if the atmosphere is heated when its temperature is high or cooled when its temperature is low. Available potential energy is converted into kinetic energy when warm air rises and cold air sinks; kinetic energy is converted into available potential energy when warm air sinks and cold air rises. We can derive a cycle of available potential energy that will allow us to follow the development of global climatic differences.

For example, during a two-week period in late December 1958 and early January 1959, the energy processes over North America dominated the entire northern hemisphere. Available potential energy built up to a strong maximum, producing large-scale eddies that increased kinetic energy. The changes in these energy parameters were closely related to variations in the poleward transport of heat, large values of which correspond to a high rate of conversion from available potential energy to kinetic energy. In terms of weather, this energy conversion manifested itself as an extensive mass of cold air in the Canadian Arctic that subsequently penetrated southward to the United States and the northeast Atlantic.[44]

There is a very critical phase relationship between the heating and temperature fields in the deep layers of the atmosphere. This relationship determines whether local potential energy will be generated or destroyed. Its importance is demonstrated by the long waves in the westerlies, in which the magnitude of energy production increases as the wave amplitude increases.[45]

To maintain the wind systems of the world, the kinetic energy that is dissipated by friction between the atmosphere and the Earth's surface must be replaced by conversion of potential energy to kinetic energy in the atmosphere. A large part of this conversion takes place in the main meridional (Hadley-type) circulations rather than in disturbances of the general circulation. In the northern hemisphere, the rate of conversion is 35×10^{10} kilojoules per second during winter.[46] In high latitudes, however, mean meridional circulations are unimportant, and the conversion is mainly carried out by cyclones.

Within a limited area, the rate of conversion in developing cyclones can reach 20×10^{10} kj per sec, and such conversion is quite important in the westerly belt. Tropical cyclones are very effective converters of potential to kinetic energy. Hurricane "Hazel," for example, which was transforming itself into an extratropical cyclone over the North Atlantic on October 15, 1954, provided the surrounding atmosphere with large amounts of kinetic energy produced by the direct solenoidal circulations. Computations over an area of 370×10^{10} m^2 for the 1,000- to 200-mb layer show that, at 3:00 P.M. G.M.T., potential energy was being converted into kinetic energy at the rate of 19.7×10^{10} kj per sec. Kinetic energy was being dissipated through friction at the sea surface and in the free atmosphere at the rate of only 2×10^{10} kj per sec, and a net outflux of kinetic energy was in progress at the rate of 18.7×10^{10} kj per sec. There was thus a slight local decrease in kinetic energy in the hurricane for this specific time and air layer; the balance must have been corrected by the conversion of available potential energy into kinetic energy in the layers above 200 mb.[47]

Only three or four disturbances of the same size as "Hazel" would be required to produce enough kinetic energy export to maintain the circulation of the entire atmosphere north of latitude 30°N. It follows that we should expect the normal climatological picture to include a few centers of strong energy production (i.e., cyclones), surrounded by extensive quiet areas, in which part of the kinetic energy generated in these centers is converted back into potential energy by the rising of cold air and the sinking of warm air.

Even without the presence of a tropical cylone, large amounts of kinetic energy can be produced and exported by the latent energy released by condensation. For example, during September 1957, a warm anticylone of thermal origin was both produced and maintained over the Gulf of Mexico and the southeast United States by such energy. The energy of water vapor in the atmosphere is very great. For example, a typical tropical cyclone produces 0.3×10^{15} kilocalories of kinetic energy, yet it releases 12×10^{15}k cal of latent energy; the kinetic energy generated amounts to only one-fortieth of the latent energy of the moist air. The net upward flux of latent energy in the equatorial belt has been estimated at 123,000 cal per cm^2 per year. Water vapor is the dominant medium of heat transport in the atmosphere.[48]

For the whole atmosphere, it can be shown theoretically, and without considering any effects of friction at the Earth's surface, that sources and sinks of kinetic energy must always exist together. It can further be shown that kinetic energy is generated mechanically in divergent high-pressure areas, and dissipated in convergent low-pressure areas. The strong winds normally associated with cyclones result from work done by the pressure forces on the converging airflow at the cyclone boundaries and from the convergent transport of energy into the low, not from generation of kinetic energy.[49]

One can map the sources and sinks of kinetic energy in the atmosphere alone, as integral atmospheric properties that are completely independent of the underlying Earth's surface. For example, kinetic energy is generated by eddies in the upper troposphere in specific atmospheric regions, in particular to the north and east of the core of the Polar Front jetstream. The distance between the jet core and the center of the region of maximum kinetic energy depends on the strength of the jet, both being greatest in winter and least in summer.[50]

The generation of kinetic energy by atmospheric disturbances may be critical in determining the heat balance in certain atmospheric regions. The Antarctic atmosphere receives heat energy from the southward movement of disturbances and from the mean meridional atmospheric circulation, and loses heat by radiation to space. The relatively small difference between the rates of gain and loss is responsible for the observed seasonal changes of temperature. Energy processes within the Antarctic atmosphere are, in fact, quite decisive in its climatology. The entire Antarctic atmosphere, from 950 to 75 mb, radiates about 10^{22} calories per year into space, and sinks downward about 0.35 cm per sec throughout the year, yet its temperature increases in the spring. The spring warming of the Antarctic stratosphere is dynamic, due to subsidence and advection of air heated by the kinetic energy of southward-moving cyclones; it is not due to direct absorption of solar radiation by the atmosphere.[51]

Because atmospheric energy computations are very complex, it has become customary to divide the movements of the atmosphere into units whose energy properties can be conveniently visualized. Usually the upper-air circulation is split into its so-called Reynolds components, mean flow and eddy flow, the latter representing the deviations from the former. Fourier analysis is applied to the observed data to separate the components, and then the energy content of each component wave or eddy is studied separately. Some results of this procedure will now be described.

Daily data for January 1951, analyzed by finding the harmonic components of the heights of pressure surfaces around the entire circle of latitude 55°N, were found to contain 11 waves. Most of the kinetic-energy movement appeared to be in the very long (planetary) waves for west-east (zonal) motion, and between wave-numbers two and eight for north-south (meridional) motion. Most of the kinetic and sensible heat flux was centered in the long waves.[52]

Daily data on the 500-mb surface in the northern hemisphere for January 1949 indicated that disturbances of wave-number three (that is, the third harmonic component of the system, which makes three complete waves around the planet) appeared to be the main source of energy for the mean zonal flow. On only three out of the 31 days did these disturbances actually draw energy from the mean motion.[53]

Spectra of the wave components at 500 mb, around latitude 45°N for the months December 1949 to February 1950, showed a double maximum, indicating the existence of two distinct wave-number bands. There must be two distinct types of large-scale disturbance in the atmosphere: (a) long planetary waves, with maximum kinetic energy at wave-number four, and associated with the major orographic and oceanographic features of the Earth; (b) rapidly moving transient disturbances of about wave-number eight, which are probably the result of baroclinic instability. Data for 1951 show that the kinetic energy of the standing waves is only 10 per cent of the total kinetic energy of the atmosphere, and is less than 20 per cent of the energy in all the long waves, although that in the long standing waves is more than 85 per cent of all the kinetic energy in standing waves. Thus only a few of the large waves can be standing, but most of the standing waves must be large.[54]

The kinetic-energy spectrum of meridional flow in the northern hemisphere shows interesting latitudinal differences. Maximum kinetic energy at latitude 30°N is associated with wave-number seven, at 45°N with wave-numbers five and six, and at 60°N with wave-number four. The zonal flow has a different spectrum, with maximum energy at wave-number one, then a decrease in energy as the wave-number increases. The maxima change little from year to year, but there is considerable variation in all latitudes, and a given month shows great variation from year to year. In general, kinetic energy shifts seasonally, to increase the length of the wave of maximum energy in winter and reduce it in summer.[55]

During the period March 1, 1959, to July 12, 1960, the maximum energy concentrations at 500 mb were at higher wave-numbers at latitude 60°N, and between 5°N and 25°N, than at 40°N. Energy concentrations in general appear to be strongest in early summer and weaker in winter, and the wave-numbers shift with the seasons, so that the longer waves carry the most energy during summer, the shorter waves during winter.[56]

Detailed study of periods of stratospheric warming illustrates the complexity of atmospheric energetics. Between January 25 and February 9, 1957, a bipolar warming was centered at 50 mb. During the first (amplifying) phase, the meridional temperature gradient decreased poleward as usual at high latitudes in the northern hemisphere. Cyclonic disturbances drew kinetic energy from the mean zonal flow, the decrease in the kinetic energy of the zonal flow being greater than the decrease in available potential energy. The flow of energy—corresponding to baroclinic instability—was as follows: available potential energy in the mean zonal flow was converted into "eddies" of available potential energy; the eddy available potential energy was converted into eddies of kinetic energy; the eddy kinetic energy was transformed into zonal kinetic energy; finally, the latter was converted back into zonal available potential energy in the indirect mean meridional circulations, involving ascending motion in tropical and high latitudes and descending motion in middle latitudes. All these energy parameters were computed from wind, pressure, and temperature data.[57]

During the second phase, the energy of the cyclonic disturbances decreased and the potential energy of the mean zonal flow increased. In comparison with the amplifying phases, the energy flow was reversed, except that eddy kinetic energy continued to be converted into zonal kinetic energy, which, however, continued to decrease, probably because of energy transfer to the troposphere.

Explanation of these energy transactions depends on the fact that the atmosphere above 30 mb behaves differently in winter from the underlying stratosphere. In the lower stratosphere, energy lost by radiation is replaced by upward movement of long-wave energy from the troposphere. Above 30 mb, zonal available potential energy is generated by radiation. Sudden warmings of the stratosphere result from baroclinic amplification of long waves that are generated orographically in the lower stratosphere.

Another period of stratospheric warming during January 1958 was preceded by a pronounced expansion in the tropospheric circumpolar westerlies. The kinetic energy of wave-number one increased rapidly as energy was transferred to it from the zonal flow and from other waves. Wave-number two then increased in energy simultaneously with the transfer of kinetic energy from wave-number one back into the zonal flow.[58]

ATMOSPHERIC ANGULAR MOMENTUM

Force is rate of change of momentum. Consideration of the momentum of the atmosphere should tell us much about the winds and why they vary from place to place and time to time. The vast size and extent of the global wind systems, as well as the power behind the winds in, say, a depression, indicate that it would be of great use in explaining weather if we had current charts showing the distribution of momentum in the atmosphere, for such charts would show us where and when to expect changes in windspeed and wind direction, which changes would later affect the synoptic picture of the weather. If, for example, two airstreams with identical trajectories but different windspeeds exist side by side, then future movements of these airstreams and future weather developments in them generally will depend on the direction in which momentum is being transferred from one airstream to the other.

From the principle of conservation of angular momentum, if a body of air moves poleward across the Earth, it must acquire an apparent acceleration toward the east, since its angular velocity must increase as the radius of its rotation with the rotating Earth decreases. Likewise, if a body of air moves equatorward, its westerly component must increase. Hence, by reversing the process of reasoning, we can use measurements of the winds themselves to work out the transport of momentum across the face of the Earth.

The absolute angular momentum of the atmosphere is the sum of two components, (1) the momentum of the air relative to the Earth, and (2) the momentum of the rotation of the Earth (see Appendix 2.7). The angular momentum of the ring of air enclosed between two circles of latitude may be changed by the transport of momentum across its boundaries, by friction between the atmosphere and the Earth's surface, or by variations in the pressure exerted by large mountain ranges on the atmosphere.[59]

Angular momentum must be transported between different latitudinal rings if the mean zonal winds are to be maintained against frictional losses. Easterly winds prevail between latitudes 30°N and 30°S, hence in these regions a positive torque, and a gain in momentum, is transmitted to the atmosphere by its contact with the Earth's surface. This momentum gain must be lost in the middle-latitude westerlies, and therefore one would expect a net poleward transport of momentum. The transport of the momentum caused by the Earth's rotation involves the mass transport of air across latitude circles, and hence must average out to zero. The transport of the air's own momentum is a poleward flux, accomplished mainly by the mean meridional circulation in tropical latitudes, and by horizontal mixing processes in atmospheric disturbances north of 25°N. Two horizontal flow patterns give the required transport in the disturbances, which takes place mainly in the upper troposphere (see Figure 2.3, A).

If the airflow pattern is sinusoidal in the westerlies, then the winds on the eastern

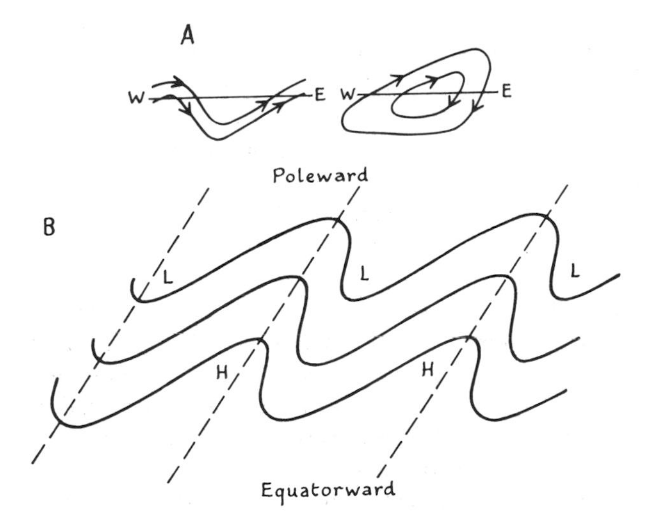

FIGURE 2.3.
Typical horizontal flow patterns that give rise to large-scale transport of angular momentum toward the pole (after Starr).

side of the troughs will be from the southwest and the winds on their western side from the northwest. In such a situation, no net momentum transport could occur. Yet there must be a net momentum transport from the equator toward the poles, otherwise friction between the atmosphere and the Earth's surface would bring the atmosphere to a standstill. Clearly then, cyclones and anticyclones must exist if momentum is to be carried from low to high latitudes. (The relatively few occasions when "tilted troughs" occur in the upper westerlies, with streamlines from the northeast to the west of the troughs, also allow a net northward transport of momentum (see Figure 2.3, B).

Friction between the Earth's surface and the atmosphere obviously tends to cancel the motion of the air. Friction has the effect of injecting westerly angular momentum into the surface easterlies, and extracting it from the surface westerlies. In order to compensate for the loss of angular momentum from the westerlies, westerly angular momentum must be transported from low latitudes to high latitudes as fast (on the average) as it is injected into the surface easterlies. Therefore the momentum source must be in the surface easterlies, and the sink in the westerlies. One would expect the momentum transport to reach its maximum value somewhere in the subtropical high-pressure belt between these two systems.[60]

The flow of the air across the Earth's surface produces a horizontal force, the boundary shearing stress, which represents the rate at which horizontal momentum is transferred to a unit area of the surface. The shearing stress can be calculated if the drag coefficient of the Earth's surface (which is a function of its roughness) at the required location is known.[61]

The surface drag of the Earth on the atmosphere is an index of the accumulation or depletion of momentum, and hence describes the likely changes in the atmospheric circulation. For example, the mean input of angular momentum into the surface easterlies is approximately 10^{34} g per cm^2 per sec, which corresponds to a mean zonal surface drag of about 0.2 dyne per cm^2. In the southern hemisphere, north of latitude 70°S, the mean zonal surface stress is everywhere less than 2 dynes per cm^2.[62]

Continuity of momentum within the atmosphere requires that the change in momentum in any volume of air equals the difference between inflow of momentum to and outflow of momentum from this volume, plus the difference between momentum sources and sinks within the volume. In addition, surface drag must be taken into account if observed features of the wind belts are to be explained. Frictional interaction between the Earth's surface and the trade winds produces a large eastward torque on the atmosphere, which requires a continuous flow of momentum from the Earth to the atmosphere to balance it. Since continued accumulation of this momentum within the atmosphere would destroy the trade winds, the Earth must exert an opposite torque on the atmosphere in the surface westerlies belt, to remove the excess momentum. The "flow" of torque required to bring about the required balance represents a stress of 50 to 100 dynes per cm^2 in the subtropics.

The computation of the actual flow of angular momentum in the atmosphere is only one aspect of the more general problems of energy and mass flow, which are crucially important in explaining variations in the general circulation.[63] The essential preliminary procedure in all these computations is to split the observed variations of the property studied—i.e., wind velocity for momentum, humidity for water transport, absolute temperature for total energy flow—into the mean flow and variations from the mean flow, with respect to both time and space (see Appendix

2.8). The contribution of each mechanism in the atmospheric circulation can then be assessed separately. Thus

$$\textit{total momentum transport} = A + B + C,$$

where A represents the contribution of the mean meridional circulation to the momentum flow, B the contribution of instantaneous meridional circulations or "standing eddies," and C the contribution of large-scale horizontally moving eddies (i.e., cyclones and anticyclones).

Actual computations for latitude 30°N show that C is the dominant mechanism: the poleward transport of momentum by cyclones and anticyclones is enough to replace the drain by surface torque in the westerlies. Thus the primary thermal drive of the atmosphere manifests itself in the generation of individual disturbances, rather than in the generation of mean meridional circulation as described in old climatology textbooks: mechanism A proves not to be strong enough to produce the required poleward flux of momentum to maintain the wind belts.

In more tropical latitudes, the picture is somewhat different: the mean meridional circulation contributes about half of the required poleward momentum transport, and is stronger in winter than in summer, as reflected in the seasonal variation in momentum flux. The middle-latitude westerlies also show a marked seasonal variation, which causes the momentum flow across latitude 30°N to be twice as great in winter as in summer.[64]

In general, the relative angular momentum of the atmosphere as a whole increases from July to January. There is substantial transport of momentum across the equator to provide for this maximum in the northern hemisphere winter. The relative angular momentum of the northern hemisphere increases from 0.1 units in July to 12.7 units in January, whereas that of the southern hemisphere decreases from 10.7 units in July to 6.9 units in January (the unit being 10^{32} g per cm^2 per sec).

In the southern hemisphere, mechanisms A and B also cannot maintain the balance of momentum, and the transient eddies (C) dominate as in northern hemisphere middle latitudes. However, the standing eddies (B) appear to be the main agents for interhemispheric exchange of momentum.[65]

Mechanism C, the eddy flux, is responsible for most of the momentum transport in middle and subtropical latitudes, which takes place mainly between the 200-mb and the 300-mb levels. The lower half of the atmosphere, despite the wind variability in it, makes no appreciable contribution to the net eddy flow of momentum in these latitudes. The northern and southern hemispheres differ in the transport of momentum by eddies: maximum flux takes place in winter in the northern hemisphere and in autumn in the southern hemisphere.[66]

Locally, angular momentum is conserved. That is, variations in angular momentum at any one place can be accounted for with enough accuracy for most climatological purposes in terms of the horizontal movement of momentum toward or away from that place. The flow of relative angular momentum in a unit time at a single station may be quite easily computed. Over Valentia, Ireland, for example, the average direction of momentum flow is northward from the surface up to 9 km, due to the mean flow, but is southward from 9 to 20 km up. The momentum flow due to eddies is northward when pressure is decreasing and southward when pressure is increasing, up to 16 km.[67]

Various correlations between momentum and other atmospheric properties have been demonstrated. The product of the mean zonal wind velocity and the rate of poleward momentum transport gives the rate of poleward movement of kinetic energy across a latitude circle. For the northern hemisphere, the computations indicate that intense northward movement of kinetic energy occurs in the layer between 100 mb and 300 mb, particularly in the vicinity of the core of the polar-front jet stream. Changes in the speed of the middle-latitude westerlies are related to variations in the meridional flow of angular momentum. Accumulation of relative angular momentum transported into latitudinal rings over periods of about five days causes changes in the intensity of (and latitudinal displacement of) the belts of maximum windspeed between 30°N and 60°N.[68]

ADVECTION OF HEAT AND WATER VAPOR

Movements of energy in the atmosphere are generally termed *advection* when the movement is horizontal, i.e., parallel to the Earth's surface, and *convection* when the movement is vertical. Convection is usually a local phenomenon, involving a distinctive process, and varying considerably with local geographical and atmospheric conditions, whereas advection is normally regarded as a global, or at least regional, phenomenon.

As with momentum transport, the advection of heat can best be studied by separating the mean flow from the eddy flow. Thus the total movement of heat in the atmosphere is separated into an eddy flux, due to migratory pressure systems; a standing-eddy flux, due to semipermanent features of the atmospheric circulation; and a toroidal flux, due to the mean meridional circulation. In addition, it is convenient to divide the total heat transport into a movement of sensible heat, and a movement of latent heat.

The general poleward movement of sensible heat in the northern hemisphere, in the layer between 850 mb and 500 mb, exhibits a marked annual cycle, with a rapid buildup from August to November, and a less rapid decline between February and June. The rate of poleward flux in December is quite variable, but other months show very similar patterns from year to year. North of latitude 45°N, three regions are especially important in this heat transport, in particular, the area in which cold air moves southward to the rear of the mean trough line along the east coast of Asia.[69]

An early computation of heat transport through the atmosphere over Larkhill, Wiltshire, indicated that here the eddy fluxes of sensible and latent heat are approximately equal, even though the latter derives more heat from the Earth's surface (see Appendix 2.9). For some levels of the atmosphere, the heat flow due to eddies is, unexpectedly, from regions of lower mean temperature to regions of higher mean temperature. The eddy flux of sensible heat over Larkhill has a winter maximum, and shows two reversals in sign in the vertical direction, one in the troposphere and the other near the tropopause. Over two-month periods, the eddy flux varies by 0.35×10^7 cal per cm per min, which increase would raise the temperature of the atmosphere an average of 7°C throughout a latitudinal ring 2,000 km wide, unless it were compensated for by an equivalent outflow.

Computations for North America show that there the eddy transport of sensible heat is poleward in the middle-latitude troposphere, but equatorward in the stratosphere. The eddy flux reaches a maximum in middle latitudes, that is quite variable seasonally, being much stronger in winter than in summer. The eddy flux of latent heat over North America is also poleward in middle and high latitudes, has its maximum at the ground, and has a slight maximum in middle latitudes, which shifts northward from winter to summer.[70]

For the northern hemisphere generally, the greatest poleward eddy flux of sensible heat occurs slightly north of 50°N, with its maximum in the lowest levels of the atmosphere and a secondary maximum at 200 mb. For radiation balance, the maximum flux should be closer to latitude 40°N. There is still a poleward eddy flux at 100 mb, where the meridional temperature gradient is reversed compared with that at the Earth's surface; this reversal indicates that, in these high levels of the atmosphere, eddies build up, rather than dissipate, north-south temperature contrasts.[71]

Detailed computations of the heat flux over India show that there the local eddy transport is too small to move enough energy poleward to correct the radiation imbalance. The maximum flux of sensible heat is directed equatorward in winter—the over-all global flux has a poleward maximum in winter—although the mean annual flux is directed northward at all levels. The vertical variation of flux over India indicates that maximum heat transfer is in the middle and upper troposphere, instead of in the lower troposphere as is general in the northern hemisphere.[72]

Computations for a single station, Valentia, prove that, even with a northerly airstream flowing, the total northward eddy flux of sensible heat may not be appreciably reduced. A marked diurnal variation in eddy flux is also apparent at this station: there is maximum heat movement in the morning and minimum in the evening at all levels up to 300 mb, but a morning minimum and evening maximum above 300 mb.[73]

Eddy transports of heat are very significant in the generation of weather systems. For example, the rate at which eddies transport sensible and latent heat from ocean to atmosphere is between 200 and 400 cal per cm^2 per day outside hurricane areas, but is between 2,000 and 3,000 kcal per cm^2 per day in the center of hurricane areas. Hurricanes tend to remain near, or to move along, specific areas, from which they draw their energy. When the eddy flux of energy to such areas weakens, the hurricanes weaken, and vice versa. Hence advection of eddy energy toward or away from hurricane areas is critical in determining whether a storm will be formed or dissipated.[74]

Determining the magnitude and direction of latent heat advection largely depends on assessing the horizontal flux of water vapor. Figure 2.4 presents the climatological pattern of water-vapor transport over North America for 1949, based on radiosonde observations of windspeed and humidity (see Appendix 2.10). The moisture transport patterns are closely related to the precipitation patterns.

At any level in the atmosphere, the total horizontal movement of water vapor consists of two parts: (a) the advective transfer due to the mean motion; (b) the eddy or residual transfer due to the passage of synoptic disturbances. Considerable variation is found in the effectiveness of these components, even over such a small area as England.

Unlike the transport of sensible heat, where the north-south movement is the major factor, both meridional and zonal transports are appreciable in water-vapor flux. For southern England, computations indicate that during June to August 1954, the average zonal transport was eastward in the layer from the surface to 350 mb, reaching a maximum of 400 to 500 g per cm per 100 mb per sec below 900 mb. Both northward and southward transfers also occurred, but their precise vertical distributions varied from station to station. The largest meridional transfer below 800 mb was about 100 g per cm per 100 mb per sec, and the average meridional flux was

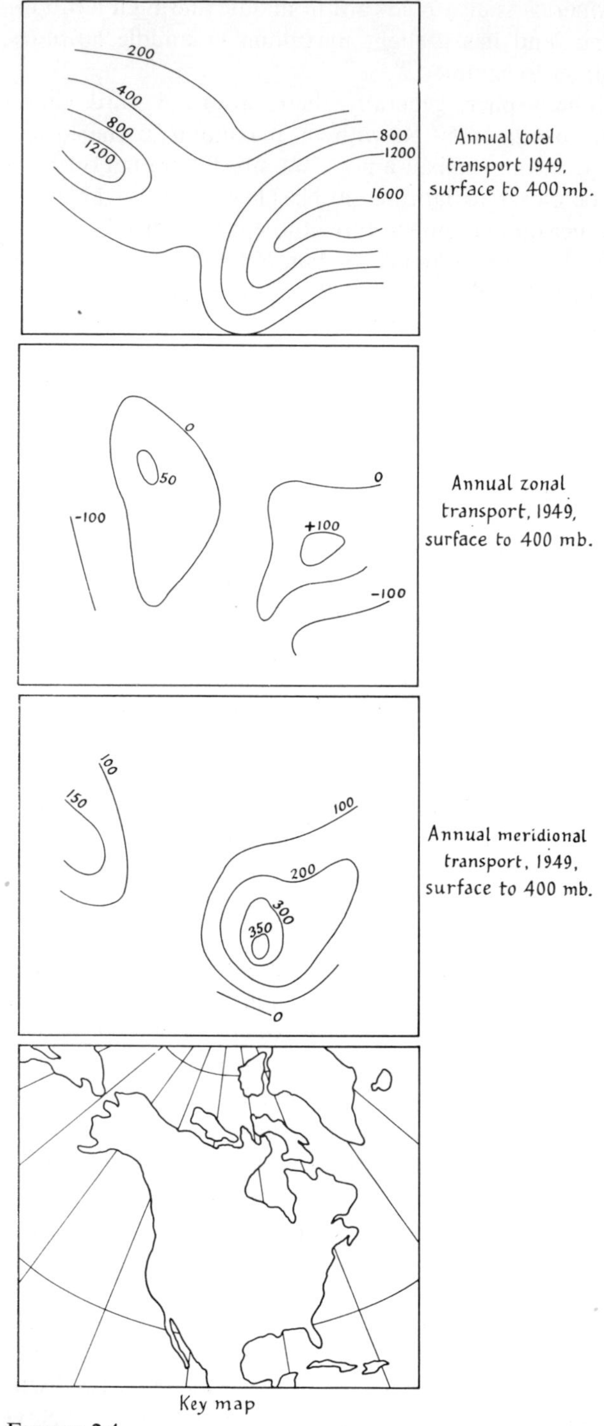

FIGURE 2.4.
Water-vapor transport in North America. Units are grams per cm per sec. (After Benton and Estoque.)

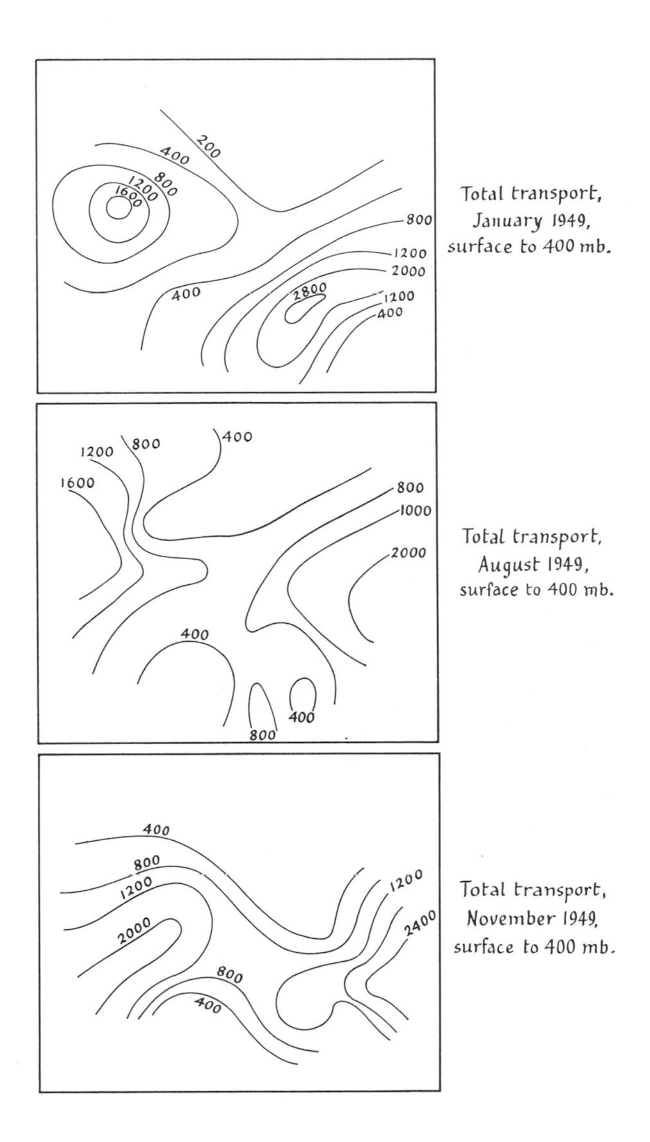

Total transport, January 1949, surface to 400 mb.

Total transport, August 1949, surface to 400 mb.

Total transport, November 1949, surface to 400 mb.

directed northward. Near 350 mb, both zonal and meridional transports were very small. (See Appendix 2.11).

During the three months June through August 1954, there was a net inflow of 31 g per cm^2 of water vapor at low levels over southern England, with a net outflow of 23 g per cm^2 at higher levels. This represented an increase in storage of 1 g per cm^2 during the period. There was a surplus of precipitation over evaporation of 7 g per cm^2. There was an average convergence from the surface to 850 mb, with divergence above, decreasing to zero, between 300 and 200 mb. The eddy transfer was less effective than the large-scale vertical motion in bringing about a net upward movement of moisture, but the fact that even the latter was not enough to balance the moisture transfer during the period indicates how important small-scale turbulence and local convective systems must be in bringing about the required balance.

Viewing the latent-heat transfer on a broader scale, Figure 2.5 presents the advective transfer patterns in 1956 over Australia, with their associated values for convergence and divergence, which correspond, respectively, to net inflow and net outflow of water vapor over the continent. On a still broader scale, Figure 2.6 gives the patterns of convergence of the water-vapor flux—i.e., indicating increasing moisture content of the air—for the northern hemisphere for January and July, based on computations over a five-year period. The advective transfer approximates fairly well to the total flux of moisture, but the divergence of this transfer does not agree so well with the divergence of the total flux.[75]

CONVECTION WITHIN THE ATMOSPHERE

Energy transfer in the atmosphere and between the Earth and the atmosphere is brought about by conduction, convection, radiation, and turbulent diffusion. At the Earth's surface, conduction and convection are important; we deal with turbulence in the next chapter, and with radiation in Chapter 4 of *Techniques*.

Convective transport of energy involves the movement of air, in particular, upward movement from the Earth's surface into the atmosphere. Motion in the atmosphere is mainly caused by the action of gravity on adjacent masses of air that have been heated differently. Large-scale motion is mainly that of pressure systems. Small-scale motion, mainly stirring due to turbulence, often stems from the degeneration of motion on large scales. On a medium scale, the motion involves an overturning process, in which vertical and horizontal velocities are of comparable magnitudes, and which is essentially convection. Convection is thus important both for heat transfer and for air movement.[76]

Two types of convection may initially be defined. *Free convection* is air motions produced by buoyancy forces, which are caused within a mass of air by imposed temperature differences. Since these motions do not depend on an external source of kinetic energy, their final effect is to reduce the potential energy of the system. The heat flow in free convection is always directed upward, whatever the direction of the mean temperature gradient. *Forced convection* involves air motions induced by external forces, for example, in an airstream moving over a topographic obstacle. Heat transfer in this case is always down the gradient of potential temperature—usually from regions of low temperature to regions of high temperature—and the final effect is to increase potential energy. An element of free convection is usually present in forced convection.[77] The instantaneous rate of convection at a point in the atmosphere can be found quite simply if simultaneous values for temperature, density, and vertical velocity are available (see Appendix 2.12).

Slow convection within the atmosphere manifests itself in distinctive cloud patterns. For example, cellular patterns, involving stationary clouds with steady air movement through them occur in altocumulus in the northeast trades; billows or rolls of cloud transverse to the general wind direction are quite common in most regions; and mammatus clouds, which occur widely, are caused by inverse convection.

Cellular convection is found in unstable fluids in which there is no general motion; the fluid is divided into a number of polygonal cells, termed *Bénard cells* by Brunt, each of which exhibits ascending motion in its center, outward motion at its top,

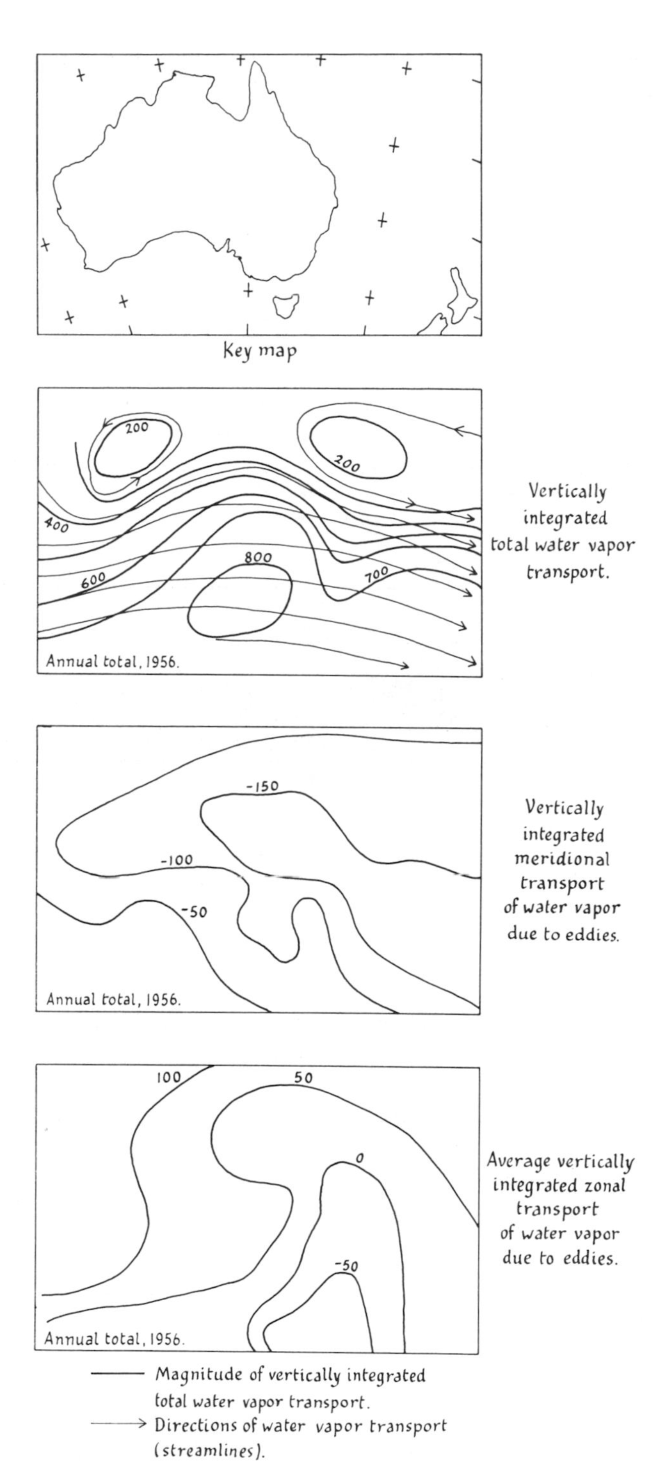

FIGURE 2.5.
Water-vapor transport in Australia. Units are grams per cm per sec. (After Hutchings.)

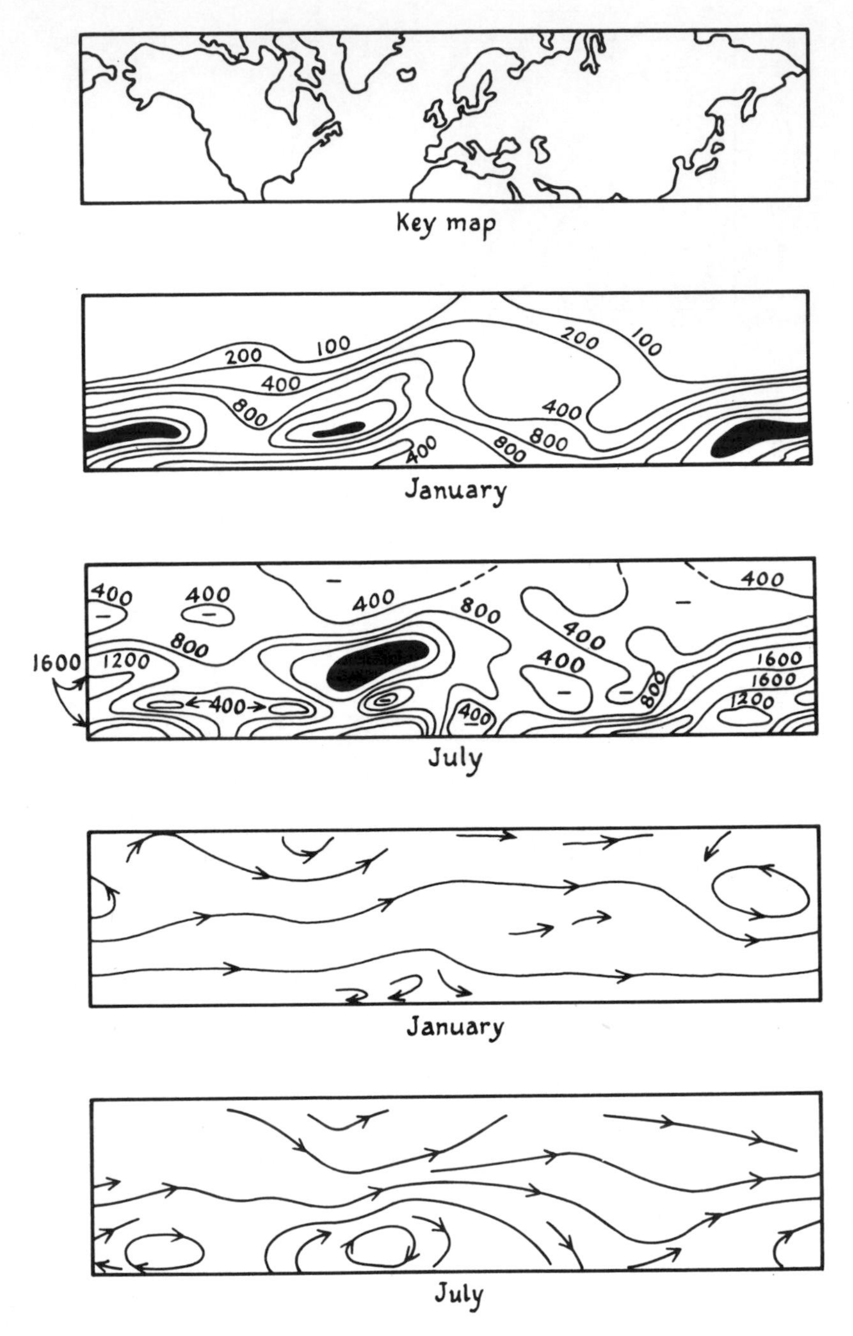

FIGURE 2.6.
Water-vapor flux due to the mean winds, 1951–1955. Units are grams per cm per sec. The minus sign indicates less than 400 units, the solid areas more than 2,000 units, and the arrows the direction of transport. (After Bannon, Matthewman, and Murray.)

downward motion at its outer margin, and inward motion at the bottom of the cell. If there is a large shear in the fluid motion, the fluid becomes divided into a series of parallel rolls, adjacent rolls rotating in opposite directions.[78]

Bénard cells can occur in the atmosphere only under certain conditions of wind structure and stability, and so are likely to be infrequent and ephemeral. They have, however, been observed by radar. Observations at Cambridge, Mass., revealed a fine-scale cellular structure in the lower levels of the atmosphere. The cell tops were at 5,000 feet; there was a marked inversion at 4,000 to 5,500 feet, and unstable air at the surface. Observations at East Hill, Dunstable, Bedfordshire, proved that cooling of the top of a higher stratocumulus layer by radiation after sunset produced Bénard cells that were 1,750 feet high and had an average diameter of 4,500 feet. Quite possibly, layer clouds frequently break up into cellular clouds at dusk, but rarely do large ice crystals form to act as radar targets.[79]

Other types of convection cell are more frequent. One type was observed to develop in nearly calm, cloudless conditions over moist, black, peaty soil. The cells resembled inverted funnels, or hyperbolic figures of revolution, being sharp-edged from the soil surface up to a height of 2 meters, but indistinct above this height because of mixing. Cells were present every few meters in all directions, their form made visible by the fog or steam moving through them, and the amount of heat convected upward within each cell could be estimated.[80]

Convection does not always follow the same pattern. The exact pattern will depend on the rate at which surface temperatures are changing in the locality, how much water vapor there is in the air, whether there is any wind shear, and the roughness, nature, and thermal capacity and conductivity of the ground surface. Large-scale convergence and divergence may also influence the local pattern.

The most usual type of convection over variegated countryside on sunny days is a form of penetrative convection, the *thermal bubble*. In cellular convection, the whole fluid circulates, whereas in penetrative convection, masses of fluid or air rise through an environment much larger than themselves. Glider pilots many years ago believed that thermals were shaped like bubbles, and some undoubtedly are. Observations of soaring birds led to the same conclusion. However, not *all* thermals are bubble-shaped.[81]

Several models of thermals have been proposed. In the Malkus model, thermals are regarded as "chimneys" in which air rises in a jet and mixes with its surroundings. This mixing causes the jet to move horizontally into surrounding air that has a different velocity. The model shows that air within a thermal in a shearing airstream moves at a different speed from its environment.[82]

In the bubble theory of Ludlam and Scorer, the top of a rising thermal bubble is assumed to be spherical, since pressure within it should be uniform because the air has negligible density, but its bottom is roughly horizontal and very disturbed by the extremely turbulent air below the bubble. The rising bubble steadily sheds its outer skin into its disturbed wake, until it either becomes completely exhausted or flattens out at a stable layer in the atmosphere. The air above the rising bubble is lifted and cooled, and then drains down the outside of the bubble and partially mixes with the wake. The wake is a favored place for the ascent of further bubbles. The wake of a clear bubble is bouyant, but that of a cloudy bubble may sink if it is cooled enough by dilution with its clear environment. Above the tectonic limit, rising bubbles penetrate previously undisturbed, usually stable, air, and waste away. When the tectonic limit reaches condensation level, cloud bases are uniform and flat; below this level, the cloud base is more ragged.[83]

Figure 2.7 shows the stages in the formation of a thermal bubble. Convection begins where the isotherms bulge upward over a thermal source, the upward motion initially being organized like a convection cell, whose depth decides the size of the bubble. The bubble begins to appear when potentially cooler air adjacent to it descends to the ground surface. The cold air then spreads out over the thermal source and the bubble becomes detached, forming a well-defined cap. Newly formed bubbles are sharp-edged, but after a bubble develops a turbulent wake, it begins to be eroded. The bubble theory is based on laboratory experiments, which enable the velocity of a rising thermal bubble to be determined from its size and the buoyancy force (see Appendix 2.13).

The critical feature in bubble formation is obviously the creation of a suitable isotherm field at the Earth's surface. A thermal can persist only if fresh bubbles are being continuously generated by a ground source. The occurrence of good thermal sources is very much a geographical problem, being involved with the distribution of certain orographic features, rock and soil types, vegetation areas, and building zones. However, thermal convection is also much influenced by the current synoptic situation, by the time of year and of day, and by geographical features.

Thermals are more frequent, and often stronger, over hills than over adjacent lowlands. This is partly a height effect—upland locations are usually warmer than nearby valley stations by mid-afternoon—and partly a slope effect. The latter acts both mechanically and thermally. Air forced to move up sloping ground helps to initiate upcurrents and to reduce stability. Sloping surfaces normally become warmer than horizontal areas because they are more perpendicular to the sun's rays—which are rarely overhead at a given place—and so the intensity of solar heating per unit area is greater than for horizontal surfaces. Thus a horizontal temperature gradient is automatically produced if there is both sloping and flat ground within a limited area.[84]

Mountains surrounded by flat plains or broad valleys are very effective thermal generators. Early in the day, the most intense thermals are produced over isolated hills, but later on, hill ranges, especially if they are fringed by extensive lowlands, become more effective. During midafternoon, southwest-facing slopes with pronounced valleys to the southwest are good sources, particularly in southwesterly air-streams, when orographic uplift intensifies the upcurrents and any clouds forming over the slopes will not drift between the slopes and the sun. Thermals over hills are often intensified by anabatic winds.

Other good thermal sources are where cool surface areas lie downwind from warm surface areas, as, for example, at the edge of a lake or wood. Here the sharp horizontal temperature gradient close to the ground causes the cool air to undercut the warm air.

Localities in which surface temperatures in sunlight rise higher than in surrounding areas—for example, patches of bare rock or sand in the midst of wet clay or green vegetation—are also effective generators of thermals, as are extensive man-made surfaces with a high thermal capacity (asphalt or macadam roads, concrete runways, buildings, etc.), which store up heat and so can warm the overlying air even when they are in shadow.

Spaces where warm air accumulates, between trees or within tall cereal crops, for example, can produce thermals when gusts of cold air displace the warm air. Although

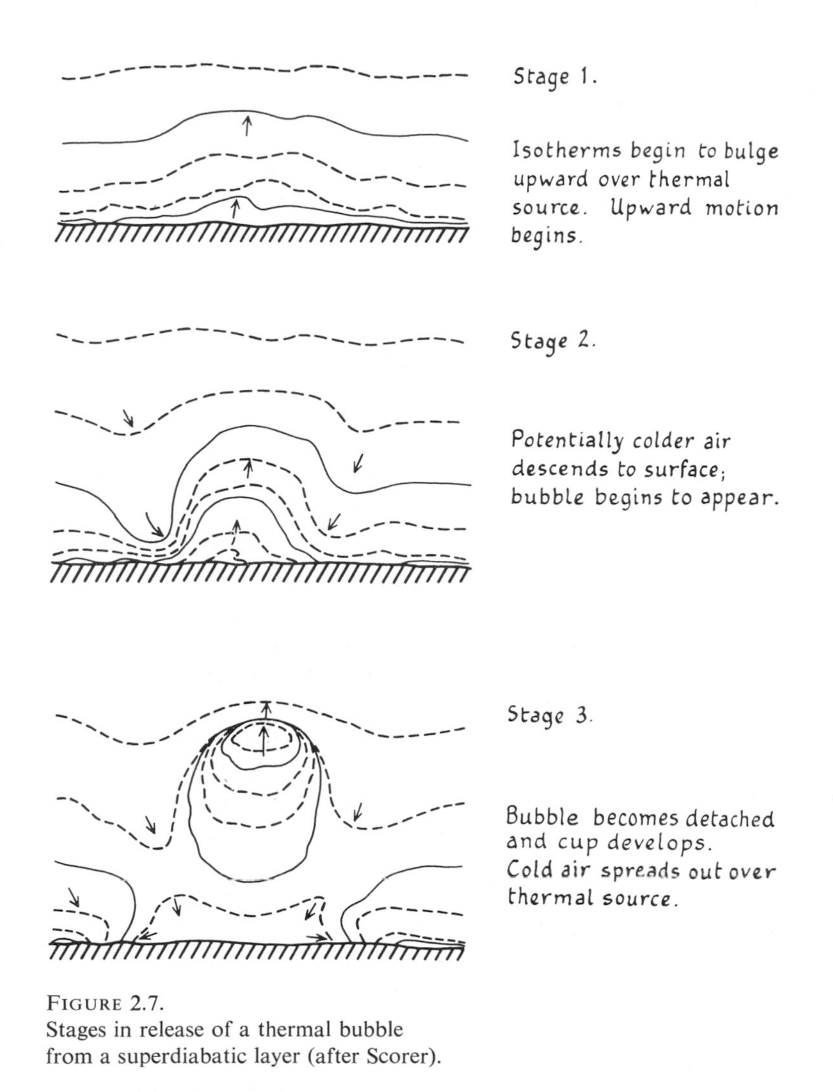

Stage 1.
Isotherms begin to bulge upward over thermal source. Upward motion begins.
Stage 2.
Potentially colder air descends to surface; bubble begins to appear.
Stage 3.
Bubble becomes detached and cup develops. Cold air spreads out over thermal source.

FIGURE 2.7.
Stages in release of a thermal bubble from a superdiabatic layer (after Scorer).

woods are cool spots during the day, because any warmth tending to accumulate must be used to transpire moisture taken in by the tree roots, air trapped between the tree trunks may provide thermals when surrounding open fields begin to cool during the evening.

Grass fires can both initiate and maintain thermals for up to one hour, as has been observed in the veldt of Northern Rhodesia, where columns of grey smoke were seen above the fires on still days, with white clouds forming at the tops of the columns. Some of these clouds drifted away and persisted like ordinary cumulus; others reached their maximum size in 10 to 15 minutes and then gradually disappeared.[85]

Above surfaces without good thermal sources, i.e., very uniform surfaces with a low thermal capacity such as the greater part of the Sahara desert, no large bubbles can be created and so intensive convection does not normally occur. Under such conditions, the lapse-rate increases to that appropriate to the transport of heat by small bubbles, becoming superadiabatic, up to 4,000 feet above the ground in Egypt. If the wind is then retarded by a surface obstruction—an embankment or the trees of an oasis, for example—the air may acquire a rotary motion about a vertical axis, so that a dust devil is generated.[86]

Not all thermals are tied to fixed ground sources; instead, some behave like plumes. Aircraft observations in Australia show that in windy fair-weather cumulus conditions over rough horizontal terrain, a region some tens of meters deep develops immediately above the ground, in which temperatures fluctuate randomly and no distinct zones of rising warm air exist. Immediately above is a shallow layer in which zones of steady uniform temperature are interspersed with zones in which temperatures fluctuate about some higher value, the sizes of both fluctuating and quiescent temperature zones varying greatly. Above this second layer, there occur thermals of a few hundred meters wide; the excess of their temperatures over that of their surroundings decreases steadily with altitude. Before the cloud base is reached, the pattern of thermal up-currents with intervening quiet areas becomes lost once more in a layer of small random temperature fluctuations. Over smoother terrain, a similar pattern of convection was also observed, but the bottom layer of randomly fluctuating temperatures was less than 10 m deep.[87]

Isolated thermals are normally associated with small local sources of buoyant heat, and plumes with large heat sources (e.g., an active volcano, as an extreme case). Large cumulus clouds are roughly midway between these extremes, consisting of a chain of thermals that have risen in rapid succession to form a "thermal plume." It seems probable that the isolated thermal is the final result of all convective processes in the atmosphere. This form is generally reached at heights varying from 200 to 2,000 m above the ground; in England, the thermal level is usually about 500 m.

Isolated thermals move rather like vortex rings in laboratory experiments. The center of the thermal rises at just over twice the rate of the bubble cap, and, relative to the observer, the buoyant air descends on the edge of the thermal. As an isolated thermal rises, it mixes with the surrounding air, more mixing taking place in front (to windward) of the thermal than behind, and traverses a cone of double angle 15° (see Figure 2.8). Wind-shear effects complicate the application of this vortex model to actual thermal motions.[88]

Since thermals must move considerable masses of air upward, continuity requires that compensating downcurrents be present. In general, downcurrents between adjacent thermals move less air downward than the thermals move upward, and the compensating downcurrents are far away from the thermals. A limiting distance between a thermal and its compensatory downcurrent is set by the maximum atmospheric signal velocity. A thermal of 20 minutes' duration can be compensated for by downcurrents up to 360 km away; one lasting several hours can be compensated for by downcurrents many times farther away.[89]

Whatever the form taken by thermal convection in the atmosphere, its main result is always to mix thoroughly the lowest layers of the air. The maps in Figure 2.9 illustrate some climatological aspects of the mixing process, i.e., the mean maximum depth of vigorous convective mixing over the United States. This mean mixing depth represents the maximum height above the ground of the layer within which the vertical structure of the lower atmosphere varies diurnally. It is less than 1,500 m for almost the entire United States (except southern Florida) from October to February; very shallow (often under 500 m) in December and January; but practically unlimited between March and September over the Rocky Mountain states and Florida. Where the mean mixing depth is topped by a conditionally unstable layer instead of

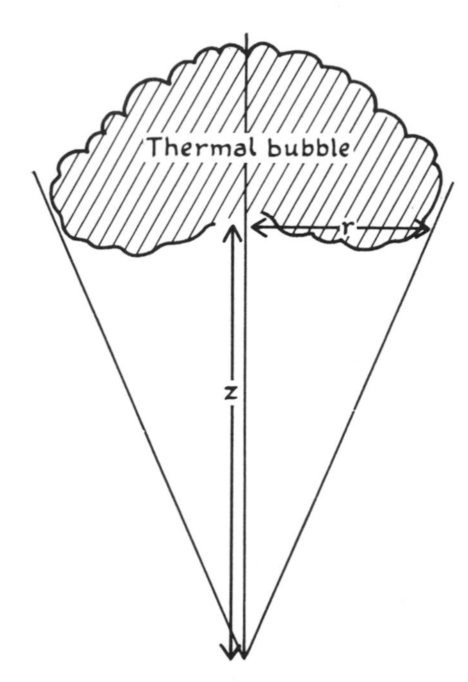

FIGURE 2.8. A thermal bubble. The ratio z/r is constant for a given thermal, but varies from one thermal to another. Its value can vary between 1 and 7, but is usually about 4.

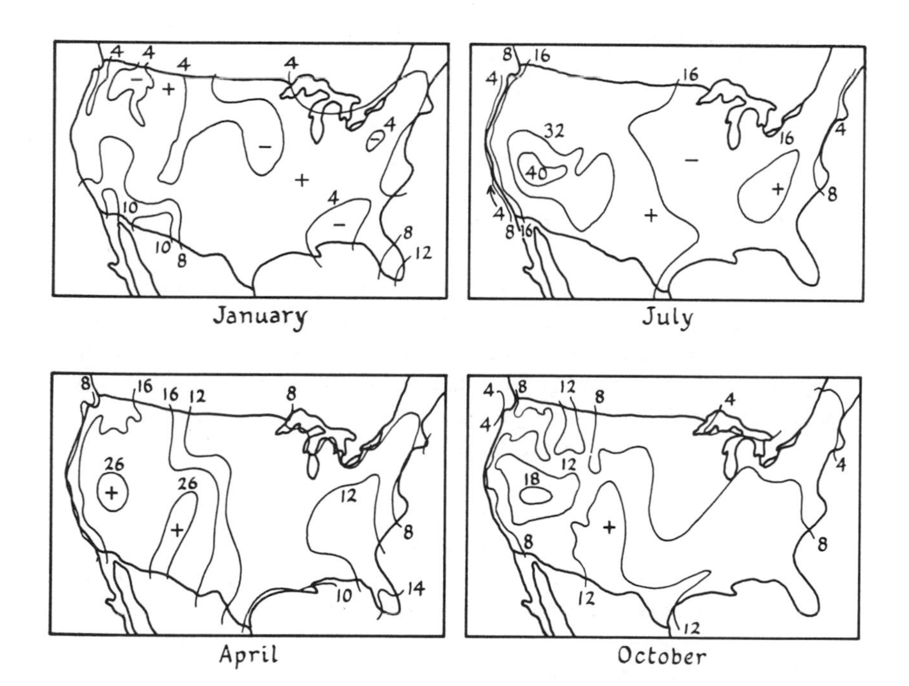

FIGURE 2.9.
Mean maximum depth of convective mixing layer, in hundreds of meters above ground surface (after Holzworth).

by the more usual stable layer, pentrative convection may rise locally well above these limits.[90]

In microclimatological studies of convection, direct measurement of the quantities involved is very intricate, and it is often more convenient to estimate them indirectly by means of theoretical considerations. Convection theory follows either one of two approaches: (a) the parcel model, in which convection is regarded as a heated air parcel rising through an environment in which temperature either is constant or slowly changes with height; (b) the cell model, in which convection is considered to occur in a cellular pattern.

The parcel model, which assumes that the horizontal movement of the parcel is very small and can be neglected, is applicable only to small-scale features. The cell model is unrestricted in scale, but assumes that all velocities involved are small. A further difficulty with the latter model is that when air is heated from below, its potential energy is increased, and the resulting generation of kinetic energy may be associated with patterns of convection other than a purely cellular one. The two models can be shown to be mathematically similar.[91]

Two types of parcel model have been recognized. In the closed-parcel model of Normand, no mixing takes place between the parcel and its environment. In the open-parcel model of Priestley, continuous mixing takes place between the parcel and its environment. Since entrainment of environmental air within the rising parcels is very common in such convective phenomena as thunderstorms, cumulus clouds, and even very small-scale turbulence, the closed-parcel model is not always very useful.[92]

The *open-parcel model* is particularly useful in microclimatology, because it describes quantitatively the connection between convective and turbulent phenomena. It deals most effectively with the lowest layer of the atmosphere, which is in contact with the Earth's surface, and within which the height variation of the eddy fluxes of energy can be neglected. Above this atmospheric surface layer, turbulence decreases with height until ultimately surface frictional effects are negligible and air motion is controlled by gravity and horizontal pressure gradients rather than by thermal effects. The depth of the atmospheric surface layer varies from place to place, and from hour to hour and from day to day at the same place, usually being between a few meters and several hundred meters thick.[93]

Quantitatively, the open-parcel model states that the parcels rise and lose heat adiabatically without disturbing their environment, accelerating under a buoyancy force that is proportional to the excess of their temperature over that of the environment, and finally oscillating about an equilibrium level where their temperature excess is zero. The model also specifies a theoretical upper limit of convection, which parcels will expend all their kinetic energy in attaining. Actual cumulus clouds very rarely rise to this upper limit.

The most important parameter deciding the intensity of convection in the open-parcel model is the dimensionless Richardson number, Ri. If there is a normal lapse-rate in the lower atmosphere, convection may be either free or forced, and the intermediate region (characterized by values of Ri between 0.02 and 0.05) is typical of the real atmosphere. The transition from free to forced convection (and vice versa) is rapid, and, if Ri equals -0.03 at 1.5 meters above the ground, both free and forced convection can occur. The usual situation is for a layer of forced convection to be overlaid by a layer of free convection, separated by a very shallow transitional layer that is lower the lower the windspeed and the greater the lapse-rate (see Appendix

2.14). Under conditions of very strong lapse-rate and small wind shear, however, convection appears to attain a constant intensity that is quite independent of *Ri*, and is characteristic of the free convection conditions in arid locales during daytime.[94]

By means of the open-parcel model, the characteristic features of the microclimatic vertical temperature profile in a given locality, under calm conditions, can be deduced from measurements of potential temperature at a single height, and the magnitude of the vertical flux of sensible heat from ground to air during the day can be deduced from measured vertical wind and temperature profiles (see Appendix 2.15). The theory involves a positive constant, *C*, which determines the rate of convection of heat away from the ground surface. Under extreme conditions, *C* equals 8, which is the fastest that residual heat needs to be convected away. With the sun high over a dry lake bed in California, this rate was found to be 0.7 cal per cm^2 per min, i.e., just over one-third of the value of the solar constant. Under average conditions, *C* equals 3 or 4, corresponding to a transfer of heat by convection of about 0.2 cal per cm^2 per min on clear summer days in latitudes up to 40°.

If the size of the parcel is assumed to remain constant during its ascent, then the open-parcel model predicts the mode of behavior of the rising elements, and hence the temperature and cloud structure, in that locality. Depending on the lapse-rate and a parameter *K*, several modes are possible (see Figure 2.10). In *oscillatory convection* (mode I), the parcel rises and then executes damped oscillations about an equilibrium level. In *asymptotic convection* (mode V), the parcel rises, then falls back to a new equilibrium level at an exponentially decreasing rate; this type of convection is quite frequent in the atmosphere, but the closed-parcel model does not predict its occurrence. In *absolutely buoyant convection* (mode IV), the rate of ascent of the parcel increases exponentially with time, so that it continues to rise indefinitely.

The open-parcel model also predicts that large upcurrents of hot air may develop in

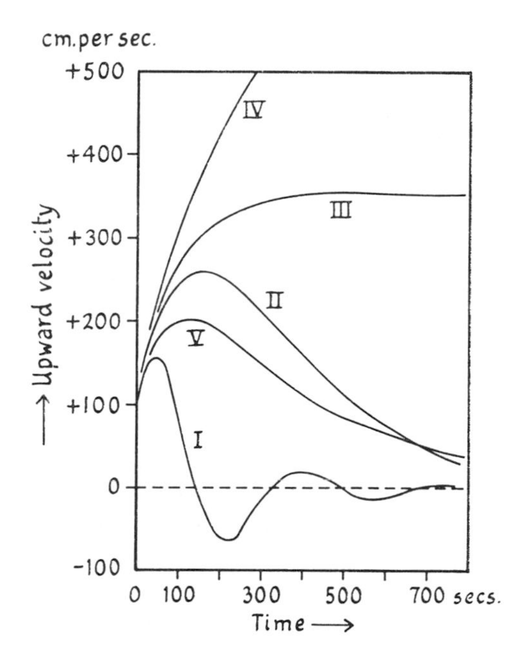

FIGURE 2.10.
Modes of convection for an open parcel:
I, oscillatory;
II, transitional;
III, transitional;
IV, absolutely buoyant;
V, asymptotic.
(After Priestley.)

stable layers in the free atmosphere when these layers are penetrated either by thermals from the ground or by cumulus tops rising from an unstable air layer below. The theory then allows an estimate to be made of the upper limit of these currents.[95]

In this consideration of atmospheric energy concepts, we have seen that existing physical concepts are quite adequate to explain the existence of many climatic phenomena, both large-scale and small-scale, that require energy to be maintained. These explanations are quantitative and based on fundamental principles. It follows that one could deduce the geographical distribution of these climatic phenomena from the fundamental principles, without recourse to actual observation, apart from very precise measurements to estimate certain parameters. This conclusion clearly establishes climatology as a physical science, but the fact that the parameters apply only under certain specified conditions means that geographical empiricism is still necessary in order to discover the existence of real climatic phenomena in real localities.

Vertical Motion in the Atmosphere

The general impression conveyed by most climatology textbooks is that almost all motion in the atmosphere is horizontal. Detailed maps showing mean wind patterns are familiar sights, and are used for explaining the global distribution of precipitation and other elements. Yet the truth is that average winds, i.e., mean horizontal motions, are not nearly as important as net vertical motions in explaining the geographical pattern of rainfall, cloudiness, and so on. The rest of this chapter will therefore deal with vertical motion in the atmosphere; horizontal motion will be dealt with in the next chapter, as part of the general circulation.

VERTICAL VELOCITIES AND THEIR MEASUREMENT

Vertical motion is the only component of atmospheric movement that is not generally recorded. It can be measured directly over small local areas, for example, in connection with local turbulence and thunderstorm research, provided it does not exceed 10 cm per sec. But the average vertical motion over large areas (10^{14} cm^2 or more), which are the vertical velocities most significant for global studies of precipitation, cannot be directly measured, and so must be estimated indirectly.[96]

Two main methods for estimating vertical velocities have been developed. The *kinematic method* employs observed horizontal winds and the equation of continuity to estimate vertical motion from the horizontal flow into, and out of, a given volume of air below a particular height (usually about 10,000 feet). The *adiabatic method*, which assumes that rising or sinking air will cool or warm at the known (adiabatic) rate, estimates the resulting vertical velocities from the observed temperature distribution, on the assumption that under these conditions temperature changes along a constant-pressure surface will depend only on the vertical motion.

The kinematic method requires a dense network of wind-reporting stations, and hence can rarely be used for motion above 3 km. It gives instantaneous values for vertical motion, averaged over an extensive areas. The adiabatic method does not require observed winds—geostrophic winds are sufficient—and yields values for vertical velocity that are averaged over considerable distances and times (usually 12 hours). Since it ignores the effects of nonadiabatic temperature changes, the adiabatic method may give misleading results for neutral or near-neutral stability, whereas the kinematic method may give misleading results in bad weather areas where the horizontal winds are not uniform. The adiabatic method may also give erroneous results if the vertical velocities so obtained are used to calculate energy conversions. The effects of latent heat may seriously influence the results obtained from either method. The immediate effect of the release of latent heat is on vertical motion: the rate of upward movement increases toward the centers and times of maximum condensation rate, with increased rates of downward motion some distance outside the condensation area.[97]

The computational equations, derived from the basic theoretical formulations, require, for the kinematic method, wind observations resolved into components directed towards the north and east, and, for the adiabatic method, graphs of observed local temperature at a given height as a function of time. The equations are not in general applicable to the lower layers of the atmosphere, or under conditions of very low stability (see Appendix 2.16).

A third method, more involved than the kinematic and adiabatic techniques, is based on vorticity. Since the change in vorticity along a three-dimensional air trajectory is a measure of the degree of hydrodynamic divergence, the vertical motion can be inferred from the latter, usually as 12-hour averages of vertical velocity. Alternatively, an equation derived by Fleagle gives the vertical velocity as an instantaneous value in terms of the geostrophic windspeed and vorticity, the static stability, and the baroclinicity of the local atmosphere. Fleagle's equation shows that, generally, upward motion accompanies positive vorticity and downward motion accompanies negative vorticity, and that, regardless of baroclinic effects, air currents moving poleward should converge and consequently move upward.[98]

Some approximate formulae for estimating vertical motion are available. For example, Thorn employed observed lapse-rates and horizontal differentiation of pressure to obtain the vertical velocity. Schaffer developed an expression which depends only on the hydrostatic equation and the gas laws to yield the vertical motion, and which requires data on the rate of movement of the prevailing pressure system at the point in question. Penner produced a synoptic method enabling maps of vertical motion to be constructed by combining charts of thickness and of vorticity. Kessler showed that, when no horizontal advection is in force, vertical velocities may be estimated from the distribution of water content in the air, and the resulting rate of precipitation then determined.[99]

The local geographical pattern of vertical motion is usually quite intricate. For example, Thornthwaite and his collaborators found considerable variations in vertical velocities over a single field. Within the air layer between the ground and 32 meters above, over a field of undulating, diversified farmland, both net upward and net downward motions were measured and found to correspond in general to the microconfiguration of the ground surface. For example, strong net upward motion occurred over a "hillcrest" only a few feet high. At no place was the net vertical motion found to equal zero, which it is often assumed to do in microclimatology.[100]

Observations using tetroons at Yucca Flat, Nevada, provide another illustration. These showed that during the day, when vertical velocities often exceeded 5 knots and caused the tetroons to oscillate vertically as much as 10,000 feet, helical circulation patterns developed with axes parallel to the north-south oriented valley floor, the circulations being in alternate senses on opposite sides of the valley. During the night, variations in vertical velocity were only about half a knot, caused tetroon oscillations of 100 feet or so, and were more closely tied to local topography than during the day.[101]

The geographical pattern of vertical velocities on the continental scale is more clear-cut than the local one. In the lower stratosphere above the United States in winter, for example, air moving through the long-wave pattern in the upper westerlies moves upward when going from a warm trough to a cold ridge. The vertical velocity pattern in the short-wave component of the circulation is more complex, but tends to be upward near ridges and downward near troughs. From data for September 18–27, 1957, one of the World Meteorological Intervals of the IGY, vertical velocities of 10 to 30 cm per sec were computed by the vorticity method for the middle troposphere. These velocities were upward in the cold air and downward in the warm air (i.e., opposite to the intuitive expectation), which gave good agreement with the observed distribution of humidity and precipitation.[102]

A few examples will illustrate the connection between vertical motions, weather, and climate. Vertical motion is the best single predictor of convective summer precipitation during the nighttime over the eastern United States, where large-scale vertical velocities are produced in the summer by diurnal variation in the momentum exchange between atmosphere and ground. These velocities are significant for clouds and precipitation in three ways: (a) they bring the air layers towards saturation more rapidly than would otherwise happen; (b) they release convective instability; (c) they lower the level of free convection. Because the computed velocities represent averages over large areas, values as low as 1 cm per sec are important for precipitation release; these may reflect actual vertical velocities of about 20 cm per sec along squall lines and other very active zones.[103]

Important variations in regional vertical motion can be found synoptically; vertical velocities may completely change in character within 200 miles, from very large positive to very large negative values. Despite this, over Europe and the North Atlantic the height profile of mean vertical velocities is parabolic, with the largest vertical motion values occurring in the midtroposphere. High values of vertical velocity have also been found near the point of an occlusion.[104]

Large fluctuations from day to day in upper-air temperatures are due to changes in vertical motion rather than in horizontal or longitudinal motion. Vertical velocities and meridional wind components are correlated positively below 100 mb and negatively above 100 mb, between latitudes 20°N and 50°N. This indicates that northward-moving air below 100 mb in these latitudes must gradually ascend, and southward-moving air gradually subside, whereas above 100 mb, northward-moving air must subside and southward-moving air must rise.[105]

The fact that large-scale vertical motion over areas the size of cyclones and anticyclones can be determined from fundamental equations (in particular, those dealing with vorticity and the first law of thermodynamics) implies that it should be possible to predict the actual, or future, geographical distribution of precipitation as a property of the atmosphere, quite independently of topographic or any other geographical considerations. In the model proposed by Smebye, the equations are first used to determine a mean value of vertical motion during the period in question (24 hours originally), and the assumption is then made that all the water vapor condensing during the period falls out of the atmosphere as rain. It is then possible to compute the precipitation that will result if no latent heat is released, and if no external heat sources are present. Corrections may be made later for the effect of latent-heat release, and also for orographically induced precipitation (see Appendix 2.17). The resulting maps for the United States agree quite well with the observed distributions, considering that the theoretical method does not take convective precipitation into account.

Combining the vorticity method of assessing vertical velocity with synoptic methods for determining the amount of precipitable water in the atmosphere above a specified region, and introducing allowances for the effects of initial unsaturation, orography, latent-heat release, friction, instability, and cloud-base height (all of which can be assessed quantitatively), enables precipitation distributions and amounts to be mapped quite precisely from atmospheric properties alone, as displayed on either current or prognostic charts (see Figure 2.11).[106]

OROGRAPHIC AND LEE WAVES

Although it has been obvious from the earliest days of meteorology that there must be "waves" or at least wave-like action in the atmosphere near mountains, since air cannot flow through a mountain barrier but must flow over it, only within the last twenty years or so has the real magnitude and meaning of this wave motion become apparent. The subject is extremely complex, but it is so important for explaining the geographical pattern of such weather phenomena as cloud development, rainfall distribution, and visibility variation, that it is worth considering at some length.[107]

Wave clouds, elongated patches of cloud in the shape of flattened lenses or almonds,

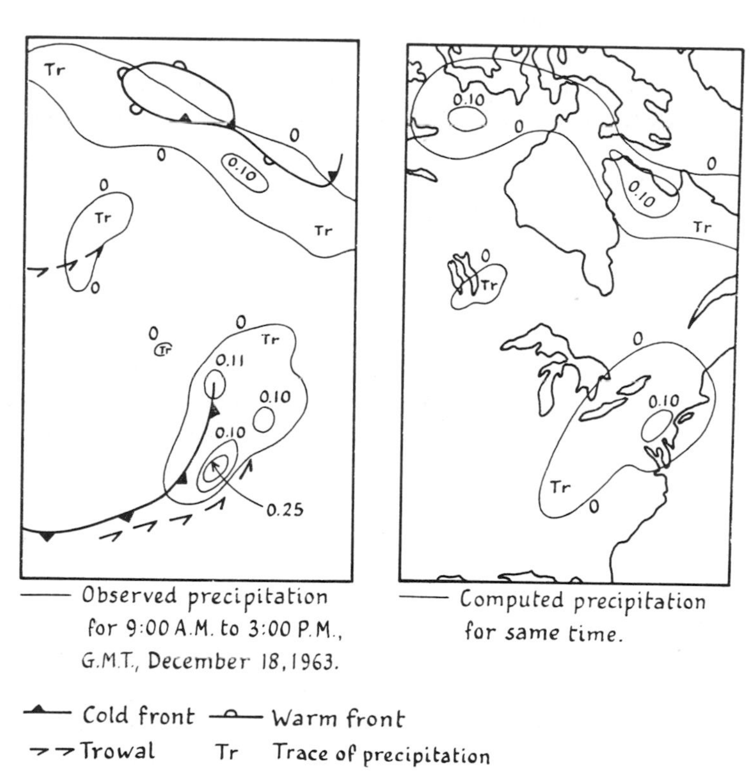

FIGURE 2.11.
Computation of precipitation from vertical motion, in inches (after Harley).

have long been commonplace in Greenland, Bohemia, Silesia, and Cumberland. Their connection with gigantic *standing waves* in the atmosphere, which move through a depth of air many times greater than the height of the mountain barrier producing them, was first demonstrated by glider pilots. Standing waves, with associated rolls of cloud, were discovered in the Alps (in association with the Moazagotl clouds), in Italy (the Contesso del Vento), in England (the Helm wind of Crossfell and the Longmynd waves in Shropshire), and in California, Africa, Japan, and New Zealand, among other places. The most spectacular example is the Bishop wave in California, in which gliders have soared to well over 40,000 feet. So strong are the downdrafts encountered in the Bishop wave (more than 3,000 feet per min) above and near the Sierra Nevada, that it is impossible for a small aircraft to cross the range when the wave is in operation. Even high-powered aircraft are affected. On March 5, 1950, a P38 aircraft encountered a vertical current above the range of 8,000 feet per min, and by feathering his propellers the pilot was able to use his aircraft as a sailplane.[108]

Observations of the Sierra Nevada waves made from sailplanes, which were tracked by radar and by cine-theodolites, allow us to build up a visual picture of the waves that is of fairly general application.[109] Three cloud types are typical. First, the *cap cloud*, or *föhnwall*, covers the ridge summit and flows down the leeward slopes (with associated downdrafts of as much as 5,000 feet per min) like a waterfall. Most of this cloud extends upwind on the windward side of the range; its leeward edge is stationary, being a wall of condensed vapor that is constantly dissipating and being replenished.

Second, the *rotor cloud* forms a line of cumulus or fractocumulus parallel to the crest of the range, and may be found anywhere from immediately in the lee of the crest to ten miles downwind from it; the updraft area is well-defined no matter where the cloud is located. Although the cloud is stationary, it is very turbulent, being characterized by a boiling motion both within it and below it. It is constantly being formed on its windward side and dissipating on its leeward side. Updrafts (to windward) and downdrafts (to leeward) may both be as much as 5,000 feet per min. The base of the rotor cloud is sometimes below the ridge summit, sometimes level with it. Its top may be twice as high as the highest point of the ridge, or may merge with the overlying lenticular clouds, so that a solid cloud mass extends up to the tropopause.

Third, the *lenticular clouds* form stationary, smooth rolls and layers in the laminar flow region of the atmosphere, sometimes reaching up to 40,000 feet. They are often stratified because of the layering of moisture content in the air. In addition, in polar regions, lenticular clouds can form at heights of about 80,000 feet under wave conditions, in the form of mother-of-pearl clouds.

The Sierra wave clouds can extend several hundred miles parallel to the crestline of the ranges, exhibiting well-defined leading edges. Their form shows that both the amplitude and the intensity of the waves decreases downstream. The strongest waves are when the wind component perpendicular to the Sierras is 25 knots or more. Although the actual wind direction during waves may vary as much as 50° from perpendicular, the greatest wave activity is when the general wind direction is perpendicular to them. (This correlation between wind direction and wave activity is not universal.) A very strong increase in windspeed with height inhibits the Sierra wave, and leaves stagnant air in the valley.

Observations of the Bishop wave in 1952 were made with two pilot-balloon theodolites some 1 to 3.5 km apart, standard 100-gram pilot balloons being observed every minute up to a height between 7 and 10 km. Small variations in the rate of ascent of the balloons was at first ascribed to microturbulence, but a graphic plot of the rate of ascent and of the ground profile as functions of the distance from the observing station along a line perpendicular to the front of the Sierra range, distinctly shows the presence of waves. During March 1951, a series of updrafts and downdrafts of between 300 and 1,000 feet per min, caused by waves, was observed, and during this period, a glider used the wave to soar to a height of more than 30,000 feet.[110]

Other pilot-balloon observations have demonstrated the existence of waves of lesser amplitude elsewhere. Figure 2.12 shows the results of observations made on March 12th at Aberporth. These are plotted differently from the Bishop data. Observations were made of the positions of the balloons every minute, as before, but each position is represented on the diagram by a short line, which is in effect a small section of a streamline. The midpoint of each short line represents the mid-height of the balloon above station height in any minute, and the length of each line gives the increase in distance from the station during each minute; i.e., the difference in height between the ends of a given line represents the difference in height between the observed and estimated rates of ascent of the balloon in still air. Both 150-inch and 90-inch balloons were used, the former rising at 760 feet per min and the latter at 540 feet per min. Assuming that any deviation of *observed* increase in height per minute from the *estimated* rate in still air is due solely to the existence of vertical currents, then the plotted streamline sections indicate that the maximum values of such vertical currents are 350 feet per min upward and 650 feet per min downward.[111]

The general conditions associated with waves have been well-established by observation. In the western United States, the intensity of mountain waves is found to be a function of (a) mountain height, (b) degree of slope of the lee side of the barrier, and (c) the strength of the airflow. Waves occur especially when a cold front approaches the mountains from the northwest and/or an upper trough approaches from the west, provided a stable layer or an inversion exists to windward of the range; preferably, the top of the stable layer should be just above the ridge summits, and certainly not above 600 mb.

The strongest Bishop waves are found when vertical wind shear is large, with an inversion (or at least a stable layer) near the mountain crests. The usual synoptic conditions associated with strong waves (see Figure 2.13) involve a low-pressure area at the surface over western Washington, with high-pressure centers over northwest Canada and (weaker) over the southwestern United States. The typical upper-air situation shows a deep, cold low, giving a strong westerly airflow at all levels above the Sierra Nevada, with a jetstream centered at 200 or 300 mb over northern California, where a strong temperature gradient is in force, particularly at 500 mb.

Mountain waves are widespread, and particularly frequent during autumn, winter, and early spring, in the lee of the Appalachian mountains. Most of them occur when a jetstream is within 200 miles of the mountains, but very few are discovered when the windspeed perpendicular to the crestline of the ranges at 850 mb exceeds 40 knots. The wave clouds normally form in lines parallel to the mountain ranges, but if a strong jet develops, it breaks up these lines into bands or streets of clouds perpendicular to the ranges.[112]

These North American observations suggest that any mountain or hill range whose crests project at least 300 feet above the surrounding area can effectively produce waves. Over low mountains, wave effects can extend to heights more than 25 times that

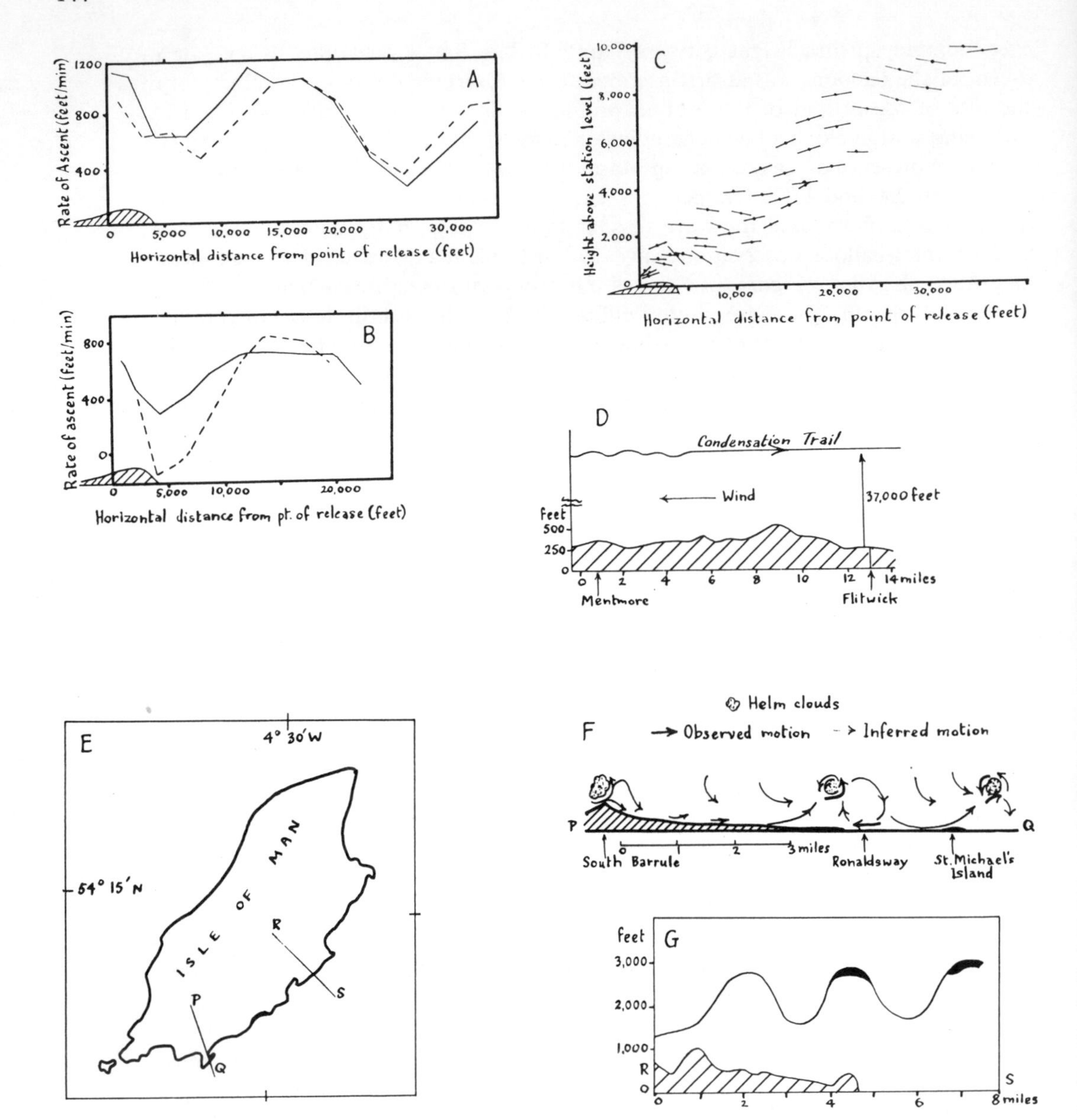

FIGURE 2.12.
The empirical study of orographic waves: I.
A–C, Standing waves at Aberporth, Cardiganshire:
A, as observed by 150-inch balloons, assumed rate of ascent 760 feet per minute;
B, as observed by 90-inch balloons, assumed rate of ascent 540 feet per minute;
C, streamline segments for paths of balloons.
D, Wave pattern in a condensation trail.
E,F,G, Standing waves in the Isle of Man.
(After W. G. Harper, W. H. Kavanagh, A. R. Laird, and F. W. Ward.)

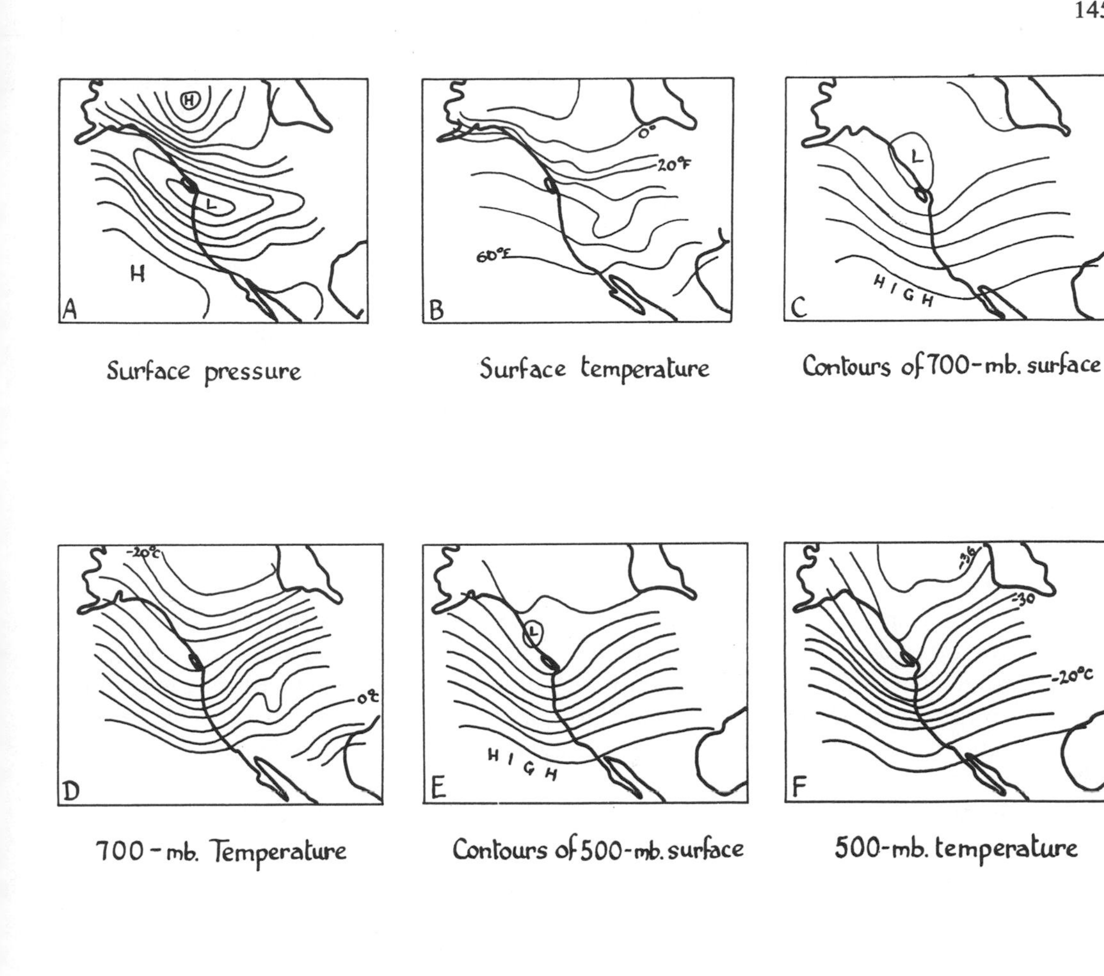

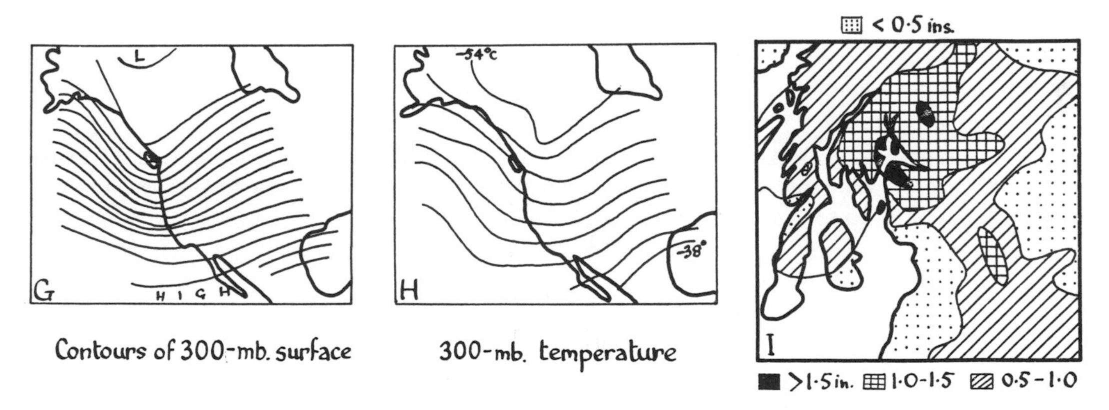

FIGURE 2.13.
The empirical study of orographic waves: II.
A–H, Mean patterns for the 18 strongest mountain-wave situations, 1949–1952.
I, Rainfall for 24-hour period ending 9:00 A.M. G.M.T., February 15, 1950. (After De V. Colson and J. S. Sawyer.)

of the range. Observations in the Ochils, Scotland, show that a small hill projecting only 800 feet above the surrounding ground can affect airflow to leeward up to at least 10,000 feet.[113]

European observations generally lead to empirical conclusions similar to those for North America, and also provide some additional evidence about waves. Thus in the French Alps, windspeeds must exceed 10 m per sec at the summit before the mountains can produce waves. If the ridge crest is concave toward the wind, wave effects are more pronounced than for a straight crestline; if the ridge crest is convex toward the wind, wave effects are less pronounced. The Alpine waves are strongest when the wind direction above the ridge is constant with height; if the wind direction above a certain height becomes parallel with the mountain range, then waves do not occur above that height.[114]

A classic case of lee waves generates the *helm wind* of Crossfell in northern England, a strong local wind blowing down the smooth western slope of the northern Pennines in stable easterly airstreams that are particularly associated with winter anticyclones. The rotor cloud here is recognized locally as the *helm bar*. Helm-bar clouds have also been observed with standing waves at Ronaldsway in the Isle of Man, where the airflow is rather more complex than in Crossfell. With a synoptic situation involving straight isobars from the direction 340° to 350°, the surface wind at Ronaldsway shows rapid changes because of a "helm wind" produced by the South Barrule hills in the southern part of the island. These hills, although only some 1,500 feet high, have an unbroken slope leading down from the summit, similar to that at Crossfell. Associated with the helm wind is the helm-bar cloud: three bars have been found to occur in the Ronaldsway district. For example, on June 28, 1952 (see Figure 2.12), four rolls of cloud were seen, each oriented southwest to northeast, the first roll being 6 to 8 miles long and occurring just above the crest of the hills. The second cloud, i.e., the first bar, was 4 miles to the south of the crest, the third was 2.5 miles to the south of the second, and the fourth was 2.5 miles south of the third. The clouds were stratocumulus or cumulus, with bases at 1,700 feet and tops at 3,000 feet, although the fourth cloud was thinner and less well-developed. No clouds or cloud fragments were present between the rolls. When the wind variations stopped, the cloud rolls changed to detached cumulus or broken stratocumulus; when convection ceased, the rolls broke down completely.[115]

Waves of a more regular pattern have also been noted in the Isle of Man. The Irish Sea area generally appears to be a particularly interesting area for waves. On one occasion, up to 100 consecutive waves were found across it.[116]

The Pennines are also very important in the initiation of waves. The necessary synoptic situation involves an anticyclone to the south but close enough to provide a subsidence inversion, with low pressure to the north giving a vigorous westerly airstream. Waves then result parallel to, and in the lee of, the Pennines. They also form immediately ahead of a warm front approaching from the west or northwest if there is an inversion at the frontal surface. A Pennine wave in the Vale of York on September 12, 1956, had updrafts of 1,000 feet per min on the upwind side of the cloud plume that formed at the wavecrest about 5,000 or 6,000 feet over Dishforth, and downdrafts of 800 feet per min on its leeward side.[117]

Information collected at Northholt from BEA pilots regularly flying over the British Isles showed that during October 1953 to April 1954, their aircraft encountered 66 lee waves between heights of 4,500 and 10,500 feet during the hours of 8 A.M. to 10 P.M.

Although the strongest vertical currents in the waves were usually found near mountains or very hilly regions, the waves showed no special geographical distribution. Waves must be related to specific orographic features, but the complicated topography of the British Isles means that there must be much damping or reinforcement of orographically induced waves. Some very extensive wave-trains were encountered, for example, from Northolt all the way to Scotland and Ireland.

Aircraft observations in general indicate that certain conditions favor widespread lee-wave activity. First, the lower layers of the atmosphere (but not necessarily all the way to the ground) must be quite stable, with comparatively low stability aloft. Second, windspeeds at the summit must be greater than 8 to 13 m per sec, depending on the profile and size of the ridge, and above the summit must either increase or at least remain constant with height up to the tropopause. Third, the general wind direction must be within 30° of being perpendicular to the ridge, and must not change substantially with height. The scale of the terrain, and the degree of irregularity of the topography, are also important.[118]

In specific regions, these conditions may be broken down in more detail. In the British Isles, a stratification of stability has been found to be very conducive to wave formation: (a) instability (i.e., a high lapse-rate) from the ground up to at least 2,000 feet above the surface; (b) a stable layer above this, between 800 and 10,000 feet deep; (c) a second unstable layer above the stable one. In addition, the windspeed at 950 mb should be at least 15 knots.

Wavelengths usually average 3 to 8 miles for the first 10,000 feet above the British Isles, and are longer above this height. Vertical windspeeds in the ascending or descending parts of the waves often reach 500 feet per min, and sometimes reach 900 or even 2,000 feet per min. The waves may or may not be turbulent; turbulence is usually marked if there is a wind shear, as when there is a jetstream. The waves show a distinct diurnal variation, being most pronounced in the morning and evening (fortunately for glider pilots), and very weak in the early afternoon, when thermals and convective activity are generally strongest. The wavelengths tend to increase as the morning progresses, and the usual early morning temperature inversion is transformed by insolation into a very low unstable layer. Wavelengths decrease in the late afternoon and evening, because of the radiative cooling of the ground surface and hence of the lowest air layer.

Although most correlations discovered between clouds and wave motions have emphasized either the rotating roll cloud of the helm-bar type, or the almond-shaped lenticular roll of altocumulus typical of wavecrests in which there is no rotary motion, other patterns have been found. Orographic waves over Malta have a distinct effect on sheets of broken altocumulus approaching from the northwest: the cloud is transferred across the wave bands from crest to crest, but entirely disappears in the intervening troughs.[119]

Another correlation is that waves are nearly always found above stratocumulus, especially above flat or even-topped stratocumulus sheets. One would expect intuitively that stratocumulus with an undulating or crested top would reflect wave, or at least turbulent, motion, yet such clouds rarely have marked lift or sink above them. With extensive flattened stratocumulus sheets covering much of the country, waves have been found above the clouds, for example, over the entire London-to-Belfast air route.[120] This indicates that even in a region of relatively slight topographic irregularity such as the British Isles, waves can persist for 60 to 80 miles leeward of the high ground producing them. An interesting connection between orographic waves and cumulus clouds is that convection sometimes takes place from the top of cloud in the wave crests, even when there is little or no convection at the ground.[121]

Lee waves over the Alleghany Mountains of central Pennsylvania tend to encourage convection in the lower layers of the atmosphere under certain conditions; under other conditions, showers moving into the lee-wave areas become less intense or even disappear, particularly in midmorning and late afternoon. Whether a wave encourages or inhibits the cumulus growth depends on the phasing relationships between the cloud and the wave and between the wave and the terrain. If the wave is interrupted at the mountain crest because it is out of phase with the ridge, then convection will continue, and the cumulus will grow larger, without downdrafts developing.[122]

Besides visual evidence of the existence of lee waves derived from cloud observations, instrumental evidence from aircraft participating in the vertical motion itself, and estimations from pilot-balloon ascents, other methods may be used to determine the incidence of waves in specific localities.

For example, trees in mountainous areas often provide indirect evidence of the effects of orographic waves, since they indicate that the wind does not blow uniformly over a given topographic barrier, and does not necessarily always blow parallel to the ground surface (even excluding gusts). Wind damage to trees on lee slopes is almost always worst than on windward slopes, especially opposite a gap or pass in the hill range; and the damage is worst on lee slopes of less than 8°. Lee slopes show a sheltered area just below the summit; windward slopes show the most damage on the lower part of the slope; isolated hills show the most damage on the lower part of their flanks. Trees that are blown down are seldom less than 30 feet tall, and are usually over 45 feet tall. Hence there must be a "skin layer" of air next to the ground surface, some 25 or 30 feet deep, in which the effect of wave motion is not very noticeable; that is, there may be aerodynamic separation of airflow in the lee of the ridge.[123]

Direct visual evidence of waves in the atmosphere is occasionally presented by condensation trails, if the wave motion is smooth and not turbulent. An example of such a contrail was seen just before sunset on March 10, 1955, from East Hill, Dunstable. It was made by an aircraft at 37,000 feet, and indicated a wavelength of 1.5 miles. The waves remained stationary, despite a wind of 70 knots, and were apparently caused by airflow over a spur of the Chilterns that projected some 300 feet above the general level of the countryside. The path of the aircraft was parallel to, and to the southeast of, the main ridge of the Chilterns.[124]

Careful examination of radiosonde soundings has shown that the presence of waves can be inferred from them. Features of the sounding (as plotted on the tephigram) that are believed to be temperature inversions may in fact be due to orographic gravity waves. British data for the period November 1953 to April 1954 showed 48 definite waves out of a total of 1,440 soundings.[125] Two methods were used to study the waves in detail.

First, the audio frequency of the note transmitted by the sonde—which is controlled by pressure—was plotted on a graph at intervals of 20 seconds. The mean rate of change of this frequency was then evaluated for each of 30 minutes, using a protractor and a conversion table, and the corresponding rate of change of pressure with frequency was found from the calibration curves. The mean rate at which the balloon crossed the isobaric surfaces could then be evaluated for each minute, and these values could be converted into vertical velocities by multiplying them by the rate of change of the isobaric surface height that was appropriate for the mean pressure and temperature during each minute. Vertical currents exceeding 120 feet per min can be detected by this method.

The second method involved plotting rate of ascent against height, and finding the mean rate of ascent for each case. A departure from the mean rate of ascent then indicated the magnitude of the vertical currents involved.

The data for this six-moth period showed a wide geographical distribution of waves over the British Isles. Leuchars had 27 waves, Liverpool 14, Stornoway 4, and Aldergrove 3. The soundings were taken at about 2:00 A.M. and 2:00 P.M. G.M.T. each day, so that only a very small part of the total time available was actually sampled. The wavelengths averaged between 6 and 7 miles, the smallest being 2.6 and the longest 14.4 miles. The maximum vertical currents were found between 5,000 and 10,000 feet; i.e., the lee-wave amplitudes appear to increase in magnitude up to some specific height in the lower troposphere, and then to decrease above it.

The waves were most common near a front, especially to the rear of a cold front, again indicating that a preferred condition for wave formation is stability in the lower troposphere, with an increase in windspeed with height. The wave regions tended to be advected with the front, and to be maintained with it during its passage across the British Isles. Other waves were found in the warm sectors of depressions, or in anticyclonic westerlies, with low-level stability and an increase in windspeed with height. The waves usually occurred downwind of high ground, but not always. The Liverpool waves were caused by the hills of Northern Ireland and the Isle of Man in northwesterly winds; the Stornoway waves in south-southwesterly airstreams were caused by the hills of southwest Ireland as well as the hills in Lewis itself; and the Aldergrove waves were all in northwesterly airstreams.

Wave motions over the Rocky Mountains were discovered indirectly by means of cross sections depicting potential-temperature distributions. Very complex temperature patterns appeared, in which the isentropes exhibit a general sinusoidal pattern. In one pattern there was a nodal surface at 17,000 to 20,000 feet in the lee of the mountains, with ascending motion above the surface and descending motion below, and with low stability. In another pattern there was a very low nodal surface, a second node at 30,000 feet, and high stability. The development of standing lee troughs in this region is linked with migratory pressure waves. When a Pacific low-pressure cell crosses the west coast, the airflow over the mountains is increased, with intensification of the lee troughs. As the low moves across the Rockies, there is an additional drop in pressure in the lee, further intensifying the trough.[126]

Two final empirical points about orographic waves must be mentioned: their periodicity and their upper limit. Although lee waves are "standing," they are not truly stationary, but periodically release both large and small eddies. The small eddies are air that frequently sweeps down lee slopes for several hours at a time, the exact motion consisting of a series of spurts separated by quiet intervals, with a periodicity of a minute or two. The large eddies give rise to wave clouds that sometimes shift downwind for a few minutes, moving more slowly than the wind, then suddenly jump back upwind to their original position, the whole process being repeated many times. Both phenomena may be explained by boundary-layer separation.[127]

The existence of *boundary-layer separation* in the atmosphere is shown by everyday observation. For example, a wind often blows up the lee slope of an orographic barrier, near the ground and against the prevailing wind; obviously, the flow at the surface and the general airflow at cloud level must have separated from each other. More generally, however disturbed (turbulent) the airflow is near the ground, particularly in valleys and between hills, it may simultaneously be smooth (laminar) higher up in the atmosphere; for example, ragged low clouds may have smooth lenticular clouds above them. Above the boundary-layer separation, the airflow follows steady streamlines—the lowest streamline may even follow the ground contour—but below it, standing or unsteady eddies between the topographic irregularities replace the streamline flow.

The boundary layer becomes separated from the main airflow when the air moves up the pressure gradient. When it is in operation, topographic obstacles may result in stagnant air, or reversed flow, at the ground (see Appendix 2.18). In general, separation lessens the effect a mountain has on the airflow, and hence reduces orographic wave activity, but it is itself inhibited by the existence of lee waves or katabatic winds.

The maximum height at which lee waves can be or ought to be present in the atmosphere is still uncertain. They have been encountered in well-developed form at 44,000 feet above mean sea level in the Sierra Nevada in California, and undoubtedly extend much higher. Mother-of-pearl clouds, first reported over the Norwegian mountains at a height of 26 km, and since seen over Scotland and many other areas, are very probably wave clouds. Orographic waves are, in fact, likely to be not uncommon in the stratosphere.

For a useful summary of the facts about lee waves, we may consider Förchtgott's series of empirical models, published between 1949 and 1952.[128] These models were based mainly on observations during glider flights in the Bohemian mountains.

According to Förchtgott, the deformation of an airstream by an orographic barrier causes three distinct types of waves. First a simple *downwind wave*, or a series of such lee waves, whose intensity decreases with altitude, may be produced by the lee slope, the wave characteristics depending in part on the geometry of the slope. Such waves result in quasistationary regions of strong turbulence. In downwind waves, the low, rotor-like roll clouds, and even the higher, more extensive bands of billow cloud, often change position, within short periods, moving slowly downwind. Then they make sudden backward leaps against the wind, because of the periodic release of eddies, with a nearly stable flow in between.

Second, an *obstacle wave*, a smooth deformation of the airflow above the orographic obstacle, is common when there is both a pronounced inversion and an increase in wind strength with height above the mountain crest. Such a wave is produced by the windward slope of the orographic barrier, and its characteristics depend on the geometry of this slope. Thermals are often continously released from the windward slope during obstacle waves.

Third, the *composite wave* combines the first two types; conditional instability in the upper air is necessary for such a wave to occur. Although composite waves can occur anywhere, they are more frequent in high mountain regions.

Förchtgott classified the airflow over mountains into four distinct types, and prepared empirical models for each type. He found that orographic waves, according to his models, only occur if (a) the air is stable, or at least not unstable, (b) the wind direction is roughly perpendicular to the mountain barrier, and (c) the wind direction varies little with height. He also found that mountain ridges with steep lee slopes were most favorable for lee waves, but the topography of the windward slope was unimportant.

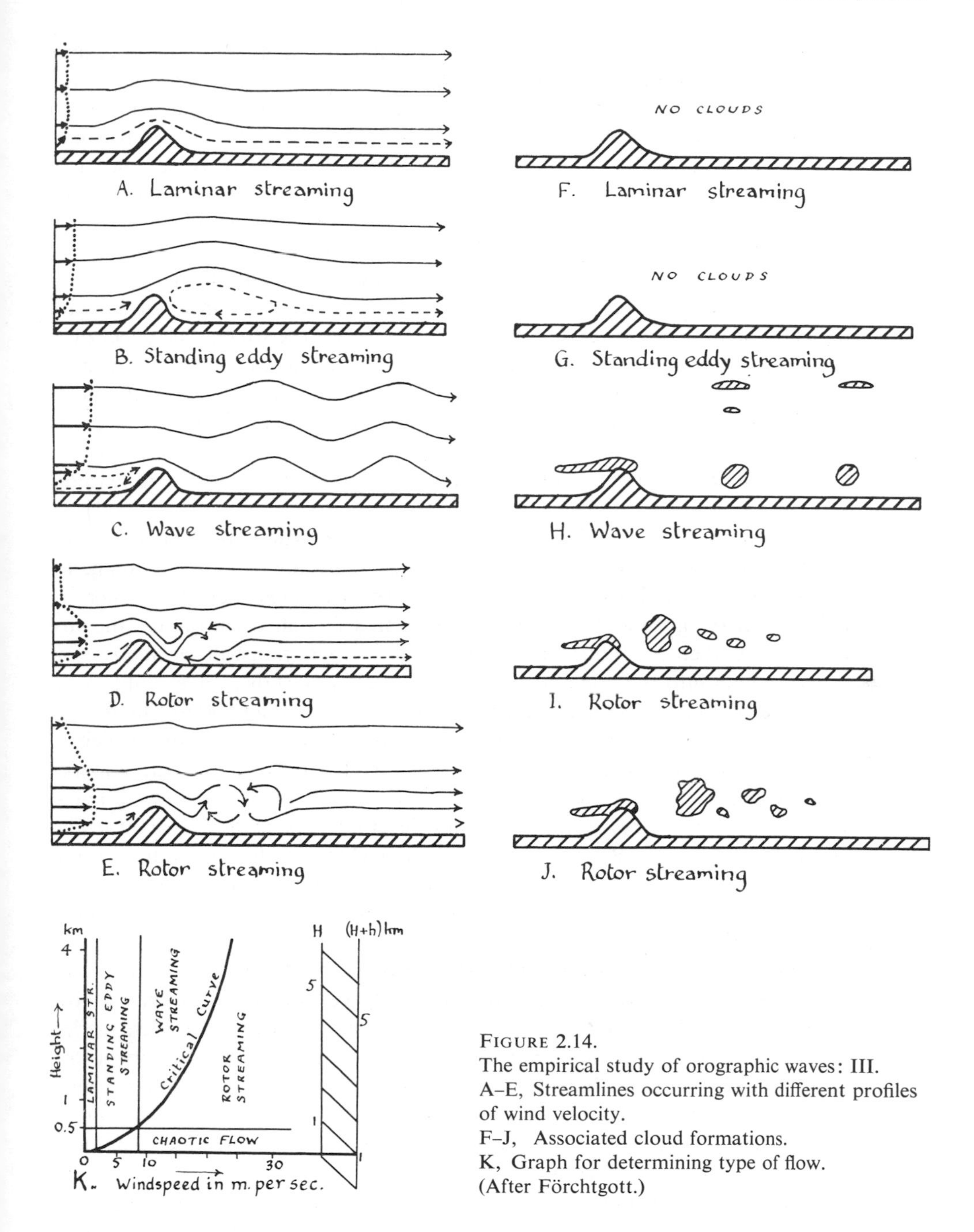

FIGURE 2.14.
The empirical study of orographic waves: III.
A–E, Streamlines occurring with different profiles of wind velocity.
F–J, Associated cloud formations.
K, Graph for determining type of flow.
(After Förchtgott.)

Förchtgott's models are illustrated in Figure 2.14, the distinguishing criteria being the variation of wind with height, and the ratio of windspeed to the height of the mountain ridge above the lee plain. It should perhaps be emphasized that these types are not supposed to represent *all* possible cases, and that the model only applies to an airstream in which the airflow is *barostromatic*, i.e., statically stable, so that the effect of gravity in the airstream is stratified. *Static stability* is defined as the difference between the actual (environmental) lapse rate and the dry adiabatic lapse-rate, and is very important for a study of waves.[129] If the air is statically stable, the effect of a given mountain on vertical motion in an airstream will penetrate much higher than if the airstream is statically unstable. In static instability, the airflow is not periodic or even steady, and hills or mountains are more important as sources of thermals than as rigid barriers for the production of waves. With static instability, the characteristic clouds are therefore cumulus, and the lens-shaped altocumulus or stratocumulus typical of the wave systems of a barostromatic airstream are absent.

When conditions are not favorable for waves, except in very light winds, the mountain barrier causes a *wake* in the airstream. The wake may consist of a single standing eddy, or of a row of eddies or rotors, as in a vortex street. The effect of a wake is to alter the lowest laminar streamline into such a shape that the airstream is not disturbed to any great height. For a wake to occur, the lowest streamline must intersect the ground at some point. This can most easily take place if the Earth's surface is aerodynamically rough in the area in question, for example, if there are irregular patches of woodland, built-up areas, or some sharp topographic feature, such as a cliff.

Thus a mountain does not always disturb the airflow over it to a great height, particularly if the flow is not barostromatic, because the wake can alter its shape to one that does not demand a disturbance of the airflow to great heights. If there is laminar flow near the ground, as when the airstream is statically neutral and the airflow aerodynamic, then the influence of a given hill or mountain will reach 20 or 30 times its own height into the upper atmosphere. In such an airflow, wave clouds have been seen at 30,000 feet in Hampshire, formed over hills less than 800 feet high.

Figure 2.14 shows that very light winds may cause *laminar streaming*, which is merely a vertical displacement of the streamlines directly above the orographic obstacle, the displacement dying out with height. However, even very small topographic irregularities can cause very prominent cloud developments at great heights. The Clent and Lickey Hills near Birmingham, both relatively insignificant features on the relief map of England, cause hill-wave cirrus at 30,000 feet because of laminar streaming, the clouds being visible at Dunstable some 80 miles away. This phenomena is usually termed *hill lift*.[130]

With slightly stronger winds, *standing-eddy streaming* results, in which the surface streamline separates from the ground, and a stationary lee eddy forms, so that airflow is reversed on the lee slope. With still stronger winds, *wave streaming* results; the lee eddy disappears, and a system of lee waves develops, associated with stationary vortex systems in which there is turbulence. Finally, with very strong winds that extend through a vertical depth much less than the height of the mountain, *rotor streaming* results. Here there are no significant waves, but a double vortex develops, as does heavy turbulence, especially close to the lee slope.

The critical values of windspeed that determine the type of flow depend on the height of the mountain ridge above the lee ground. For example, with a ridge 900 m high, a 12 m per sec wind is necessary for wave streaming. For an isolated hill, much higher windspeeds are necessary to produce a given type of flow than for a long ridge of equal height. Förchtgott provides an empirical graph (see Figure 2.14) from which the flow type appropriate to specific sets of conditions can be found.

Such empirical models are very laborious, and perhaps dangerous (if pilots rely on them, for example), to construct, and so a question arises: To what extent can theoretical considerations help indicate the type of wave motion likely to be present given certain specific conditions? In other words, to what extent can it be shown from purely theoretical physical concepts that wave motion is an essential property of specific types of airstream? Before discussing this point, it may be necessary to indicate the importance of orographic wave phenomena in practical as opposed to theoretical climatology.

The significance of waves lies in their effects on certain weather elements. Rainfall distribution, for example, is definitely controlled to some extent by orographic waves. The usual 24-hour or 12-hour readings of nonrecording rain gauges are very misleading, in that they provide a poor sample of the true precipitation, and often filter out the really interesting falls. Radar, on other hand, usually indicates only the instantaneous amount of precipitation that is actually falling at the time of observation. The rainfall due to orographic waves will normally last for periods in between these two observing intervals, so that one would not expect to find direct evidence of such rainfall very often in the published data. However, lee waves do explain some apparently anomalous precipitation distributions. For example, in the Scottish Highlands, during the 24-hour period ending at 9:00 A.M. G.M.T. on February 15, 1950 (see Figure 2.131), maximum rainfall was over low ground in the lee of an orographic barrier, presumably under the ascending part of the first lee wave, and not over the highest ground or the windward slope of the highlands.[131]

There is evidence that airstreams in which there are orographic waves produce very different amounts of rain from other airstreams, even if they have identical trajectories and similar life histories. For example, the rainfall at high-level stations south of the path of a given low in the British Isles is nearly three times greater than that at neighboring low-level stations, yet the records at high-level and low-level stations north of the path show very little difference in precipitation. The main reason is that easterly winds north of the path of a low usually decrease with height, producing an air current in which the effect of orographic uplift is confined to a much shallower depth of the atmosphere than in the westerly airstream to the south of the path.[132]

Waves profoundly influence the concentration of suspended particles in the atmosphere—dust particles have been carried thousands of feet upward in the Bishop wave, for example—and this affects the climatology of radiation balance at the ground. Although reports of very strong and very localized surface winds are apparently sporadic, they may fit into a logical pattern when they are related to waves. (Furthermore, such a pattern might be very relevant to collapses of cooling towers and other structures.)[133]

THEORY OF OROGRAPHIC WAVES

The fundamental principle behind the theory of lee waves in the atmosphere is that an airstream, if disturbed vertically by some obstacle, for example, an orographic barrier that forces the streamlines upward, will begin to oscillate vertically because of gravity and its own inertia. The atmosphere, in effect, has a natural period of oscillation, so that if an air particle in a stable environment is displaced vertically from its equilibrium level, buoyancy forces will cause the particle to oscillate vertically about this level, with a period of oscillation that depends primarily on the vertical variation of temperature within the airstream (see Appendix 2.19). The greater the stability, the shorter the period of oscillation. In a wind blowing parallel to the ground, such vertical oscillation becomes a wave-like oscillation, whose wavelength is given by the product of the period of oscillation and the horizontal windspeed.[134]

A real airstream will contain several of these natural wavelengths; for example, it might consist of a slow-moving lower layer of great stability, which would have natural wavelengths of, say, 2 miles, and a fast-moving upper layer of lesser stability, which would have larger natural wavelengths, of about 15 miles. On crossing a mountain barrier, the two layers would not, of course, oscillate independently; the airstream as a whole would oscillate, with a period that is some function of the distribution of the natural wavelengths with height. A primary task of the theoreticians was to devise a method for finding this function in terms of easily measured quantities, such as windspeed, temperature, and lapse-rate.

Queney in 1936 devised a wave equation that involved the use of three parameters, k, l_s, and l_f.[135] The first of these, k, is simply the wave-number of the disturbance in the direction of the wind. Thus the first lee wave encountered moving downwind from the orographic obstacle would have the wave-number 1, the next would have the number 2, and so on. Thus it was assumed that the wave motion is harmonic in the direction of the wind, with a wavelength of $2\pi/k$. The two other parameters refer to the effect of the stability of the airstream (l_s), and the effect of the Earth's rotation (l_f) on the wave motion (see Appendix 2.20).

Queney provided a series of models (see Figure 2.15) for the interpretation of his equation, the different models corresponding to different values of k, l_s, and l_f, and hence to different airflow types. The first three models apply to wave motions produced by a ground surface consisting of simple sinusoidal undulations of infinite transverse extent. If k is greater than l_s, small-scale waves result, of length less than 2 km, the wave being in phase with the undulations of the ground surface. If k is greater than l_f, and l_s in turn is greater than k, then medium-scale waves result, of length between 2 and 200 km, the wavecrests being displaced in position with height. And if k is less than l_f, then large-scale waves result, again in phase with the surface undulations, and with lengths greater than 200 km.

The second set of models applies to wave motions produced by a single obstacle. A small ridge up to 100 m wide results in a simple symmetric wave over the ridge, with no train of lee waves, since l_s and l_f are both negligible. For an obstacle several kilometers wide, l_s is appreciable, but l_f is still negligible; a single wave of more complicated form will appear above the hill, but still no train of lee waves. However, with an obstacle several hundred kilometers wide, both l_s and l_f are important, and a train of lee waves results, involving horizontal as well as vertical displacements of the streamlines.

Queney's pioneer work undoubtedly proved that orographic waves were an essential property of the atmosphere, but his model atmosphere did not agree particularly well with the real atmosphere. Lyra in 1940 applied the equations of motion to the flow of a stably stratified airstream over a mountain ridge of rectangular cross section. He was able to show that the effect of a single obstacle was to make the air much higher up in the atmosphere execute more than one oscillation (see Figure 2.16). The oscillations increased in amplitude with height, and decreased downstream, with most of the disturbance downstream from the mountain. More complicated ridges could be fairly effectively simulated by superimposing several rectangles until the true cross section was approximated, but a serious difficulty was that if the angular mountain was replaced with one of smooth outline, the series of lee waves almost vanished. Another

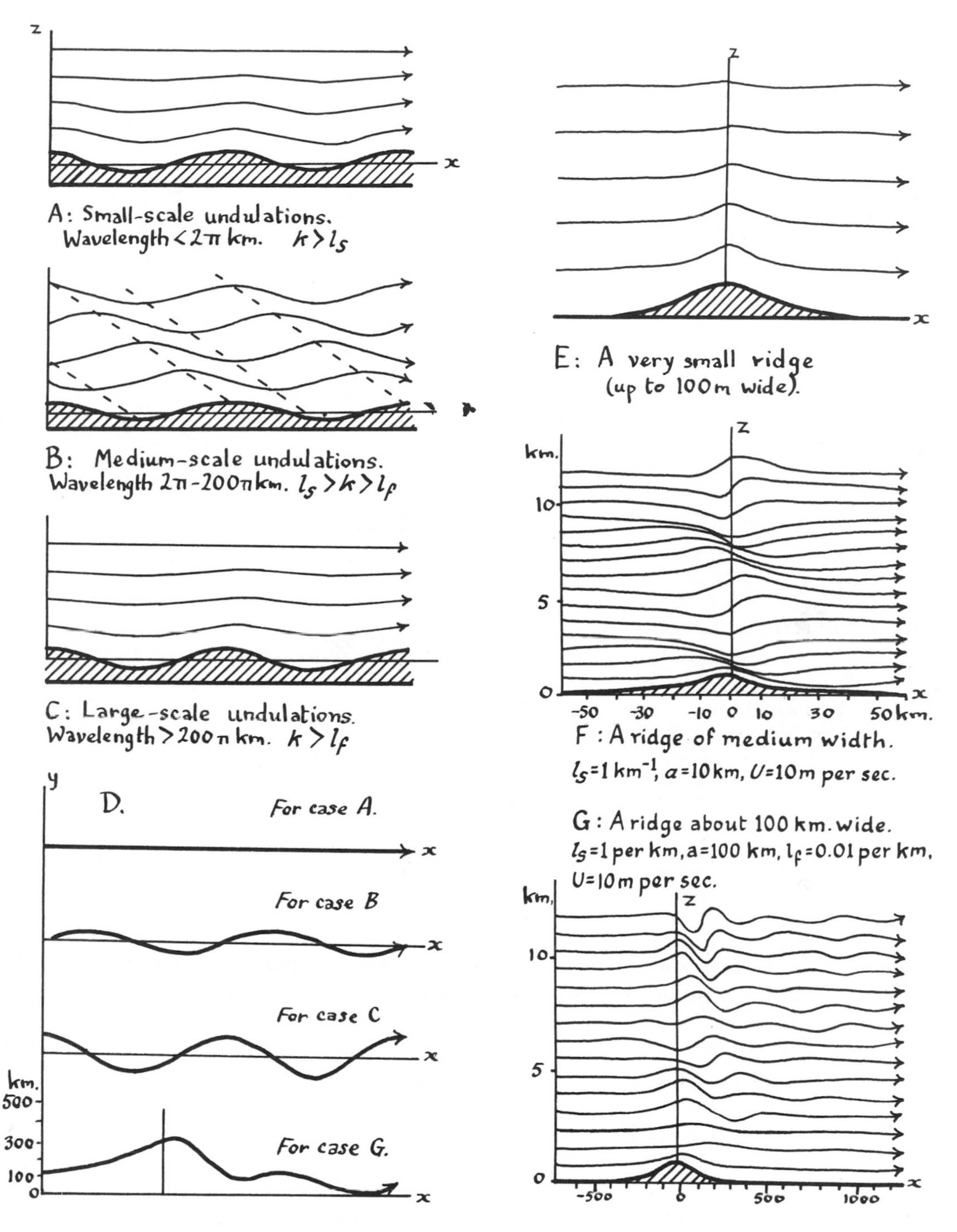

FIGURE 2.15.
The theoretical study of orographic waves: I. Queney's models.
A–C, Streamlines produced by an infinite sinusoidal ground profile.
D, Horizontal displacements for A,B,C,G.
E–G, Streamlines for flow of a uniform airstream over a ridge. (After G. E. Corby.)

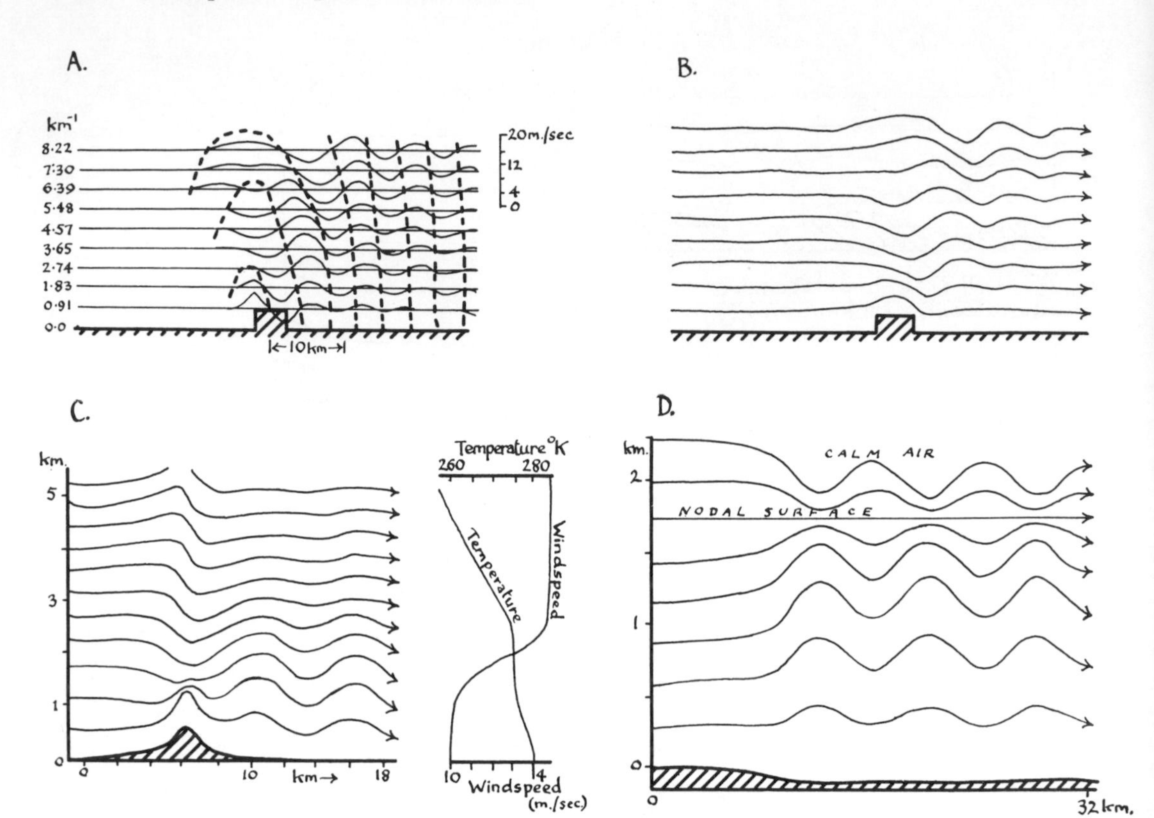

FIGURE 2.16.
The theoretical study of orographic waves: II.
A, Lyra's model, showing vertical component over mountain of rectangular cross section.
B, Streamlines for the above (wind 15m per sec).
C, Scorer's model for airflow over a ridge in conditions suitable for lee waves.
D, Scorer's model, a special case.
E–G, Queney's model for: (E) statically neutral airstream (aerodynamic flow);
(F) and (G) barostromatic flow. F assumes SALR with wind of 13m per sec or DALR with wind of 8m per sec.

objection to Lyra's theoretical model was that it indicated no upper limit to the disturbance, so that even a tiny hill ought to result in wave oscillations well into the stratosphere.[136]

Thus Queney and Lyra explained the observations that small hills could produce disturbances in the upper troposphere vertically above their position, but their models were not so successful at explaining lee wavetrains. A more realistic series of models for this latter purpose has been provided by Scorer.

In 1949 Scorer first explained the train of waves in the lee of a ridge by supposing the airstream to consist of more than one layer, and his model took into account the observed fact that the amplitude of lee waves usually increases for some distance upward before it begins to decrease. He designed a parameter that describes the wind and stability characteristics of an airstream fairly realistically (see Appendix 2.21). For a wave to remain stationary relative to the mountain producing it (and most lee waves are, of course, *standing waves*), Scorer found that the main criterion was that the square

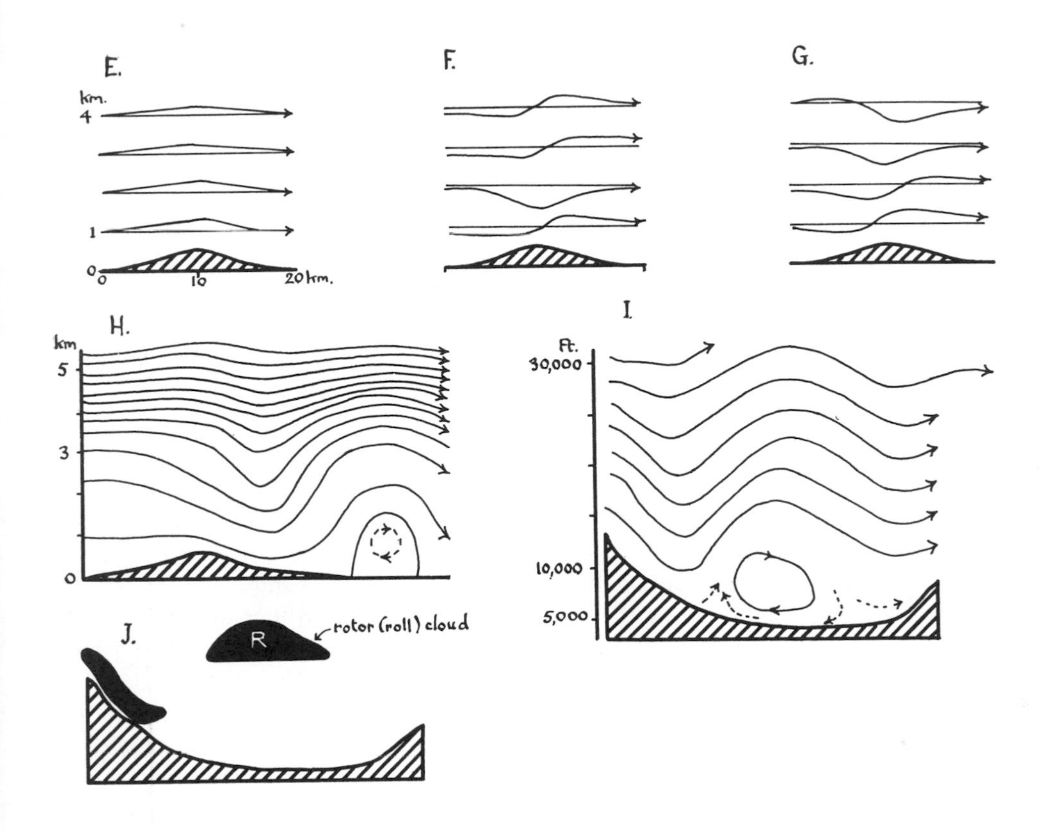

F, Parameter l is 1.5 per km.
G, Parameter l is 1 per km, with a wind 1.5 times that of F.
H–J, Large amplitude mountain waves:
H, theoretical streamlines;
I, actual streamlines February 16, 1952, Owen's Valley;
J, cloud pattern for I.
(After Corby, Scorer, and Klieforth.)

of his parameter *l* should decrease with height. He then produced a series of models with different numbers and types of air layers (see Figure 2.16). The three-layer model proves to be the best approximation to the real atmosphere, which agrees with the empirical evidence that most waves are associated with a three-layer troposphere consisting of a stable layer sandwiched between two layers of lesser stability. The model also indicates that the length of the waves should increase in the morning (i.e., increase as the lapse-rate in the lower layers of the air decreases) and decrease in the later afternoon and evening, again agreeing with the observational evidence.[137]

Sawyer extended thc theory to show that, for ridges less than 200 or 300 m high, lee waves may arise from variations of *l* with height additional to those postulated by Scorer. He also showed theoretically that (a) downslope flow on the lee side of an orographic ridge ought to be stronger than upslope flow on the windward side, from atmospheric properties alone, and the crest of the streamlines crossing the ridge must be tilted upstream with height; (b) upward displacement of an airstream crossing a ridge can reverse with height at quite low levels; (c) if two or more lee waves are present, the shortest wave dominates in the lower troposphere and the longest dominates in higher levels of the atmosphere.[138]

To predict the occurrence of lee waves by Scorer's model, it is of course necessary to determine the value of the parameter l. This can conveniently be done by using Wallington's lee wave scale, which consists of a piece of celluloid on which has been inscribed a right-angled triangle (see Figure 2.17).[139] The scale is placed over a tephigram on which has been plotted the upper-air sounding nearest in time and space to the area under consideration, such that the point x is placed directly over the point on the ascent curve at which it is required to determine l. The line PH must be parallel to the dry adiabatics on the tephigram. A straight line is then drawn along the environment lapse-rate at x to meet PK at z, giving a value for the static windspeed at the height in question. Next l is found for several points on the sounding; if l decreases with height, waves are likely to occur.

The best conditions for the production of waves involve a layer at least 200 mb thick, near the ground, and in which l^{-1} is considerably less than in the next layer higher up, which must be at least 300 mb thick. If l^{-1} is very large near the ground, then a thicker layer above it of small l^{-1} will be necessary for waves to occur, with another layer of large l^{-1} above that. The amplitude of the lee waves is at a maximum at the top of the layer where l^{-1} is small, i.e., at the level where l^{-1} begins to increase substantially with height. The wavelengths will be less than $2\pi l^{-1}$ in the upper layers, and more than $2\pi l^{-1}$ in the lower layers. With a narrow, well-defined ridge, the first lee wave downwind from the ridge will be three-quarters of a wavelength from the ridge crest.

All these deductions apply to Scorer's three-layer atmosphere, not necessarily to the real atmosphere. But conditions approximating those in the model are quite frequent, and airstreams suitable for lee waves are very common in the British Isles, especially in winter.[140] The larger-amplitude waves are more likely to occur in airstreams that contain a shallow layer of great stability, than in a thick layer of slight stability throughout. Near the ground, the adiabatically mixed lower atmosphere reduces the wave amplitude and increases the wavelengths. Wave amplitude is also reduced if the mountain barrier becomes broader or narrower than the optimum width for the prevailing airstream.

Although the parameter l can be used to determine whether or not waves are likely to occur in a particular airstream, it will not of course indicate the precise structure of the wave. For this purpose it is necessary to compute the actual streamlines. For an ideal two-dimensional ridge of infinite length, the vertical displacement of an individual streamline from its undisturbed level at a particular height above the ground can be computed (see Appendix 2.22). In general, the higher the ridge, the greater the wave amplitude.

Two other consequences result from the wave equation. If an airstream flows over several ridges all the same height, the biggest waves will be in the lee of the ridge whose width parameter, b, is equal to $W/2\pi$, where W is the length of the lee wave. Narrower or broader ridges will produce waves of lesser amplitude, as already noted. In addition, there will be a resonance effect between the airstream and the mountain width that may swamp the effect of the mountain height; i.e., even a large mountain will not set off waves if the mountain is too wide for resonance with the natural wavelength of the airstream. In general, because of resonance, large-amplitude waves occur only if the natural wavelength of the airstream corresponds to the size of the

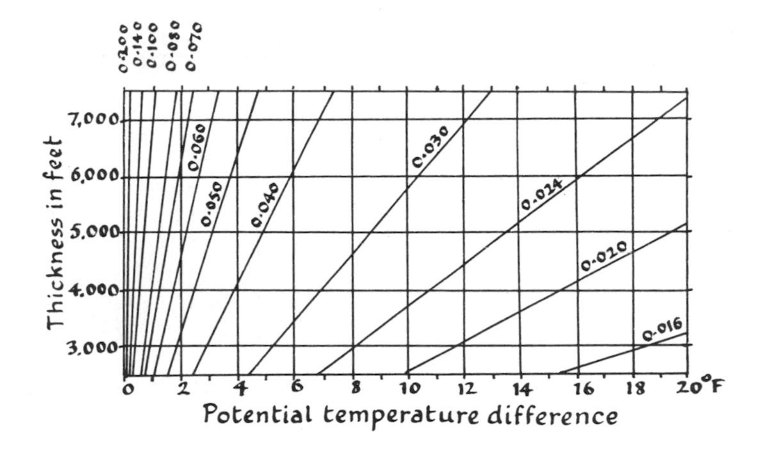

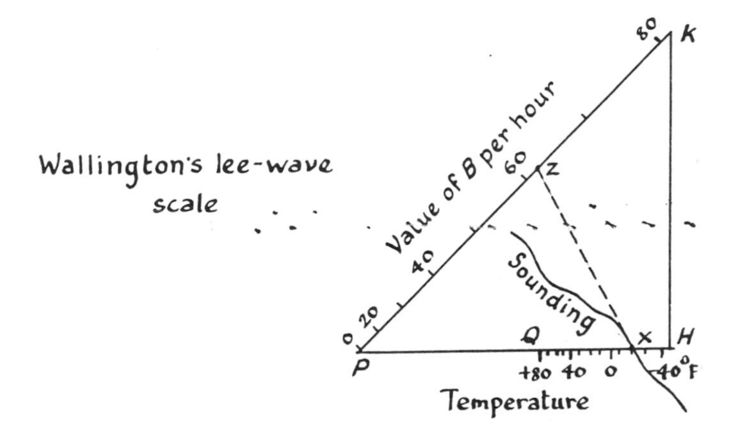

FIGURE 2.17.
Graphic aids for studying orographic waves. In the nomograph, the sloping lines give values of $1/Ul$, to give values of $1/l$ in miles (U in knots). To use Wallington's scale, it should be drawn on celluloid, then placed over the current tephigram, such that PH is parallel to the dry adiabatics. The appropriate temperature on scale HQ must coincide with the point on the sounding at which it is required to determine B, the scale of which indicates static stability.

ridge. Since the wavelength increases with windspeed, large mountains require stronger winds than small mountains if they are to produce large-amplitude waves.

Actual topographic barriers are, of course, not ideal simple ridges, and the pattern of vertical motion they cause is not simple. For airflow over a broad mountain range, the highlands of Wales, Wallington has shown that the air does not undergo simple

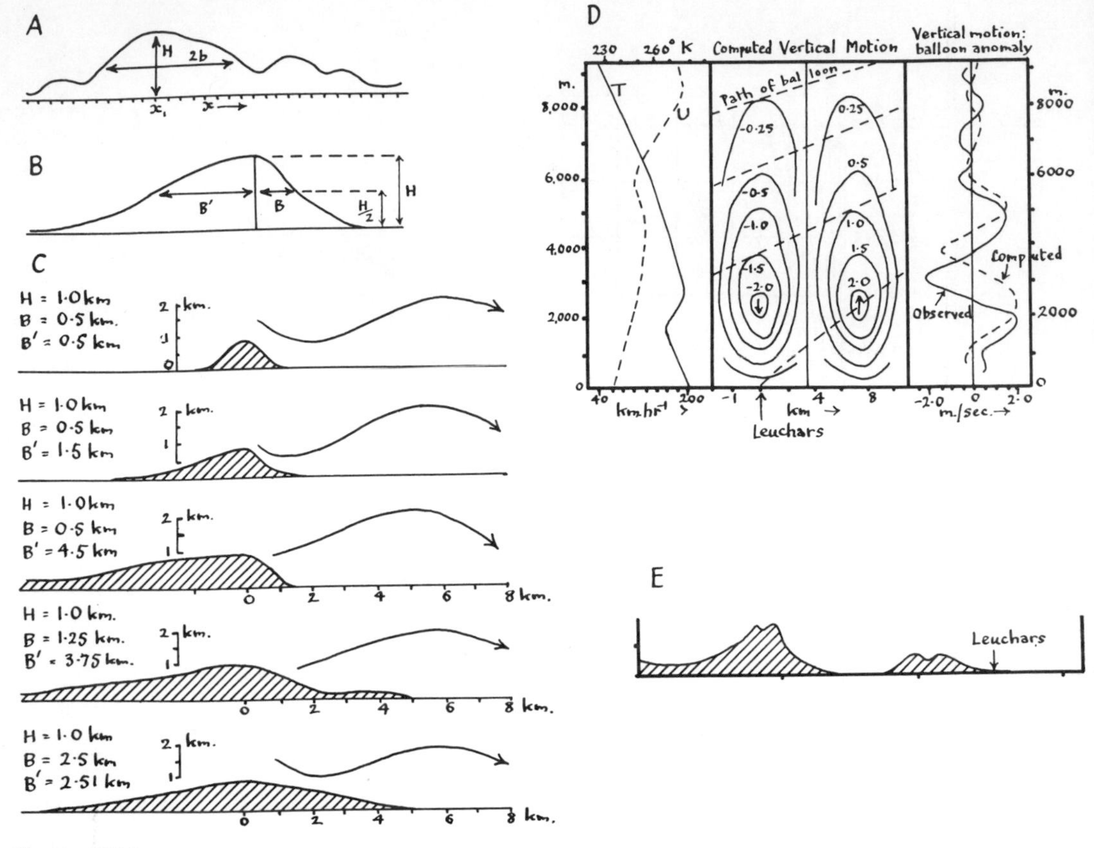

FIGURE 2.18.
The theoretical study of orographic waves: III.
A, Mountain profile specified by 80 heights at equal horizontal intervals.
B, Specification of an asymmetric ridge.
C, Vertical displacement of streamlines for different types of ridge:
H, height parameter of ridge; B,B′, width parameters; airstream amplitude factor, 2 per km; Wavelength, 8 km.

lifting as it crosses the barrier, and descends over some parts of the range.[141] It is possible, however, to predict theoretically the airflow over complex real barriers by algebraically synthesizing several ideal ridges (see Appendix 2.23). For example, a simple asymmetric ridge (see Figure 2.18,B) can be represented by the upwind and downwind halves of two elementary ridges of different size. More complicated orographies need many ideal ridges: Figure 2.18,F, shows that the topography in the direction 290° from Leuchars can be represented quite well by the synthesis of six simple ridges.

A different equation must be used to compute the displacement of the streamlines if several ridges are used instead of one ideal ridge. The new equation involves computing a "mountain factor", M, and a "phase displacement," X, for the various wavelengths. For example, suppose that 80 values of height at equal horizontal intervals along its cross section are necessary to specify the height profile of a ridge (see Figure 2.18,A). Fifteen elementary ridges are necessary to describe this particular ridge. Taking the maximum height of the ridge as H, its abscissa (i.e., the horizontal distance of H

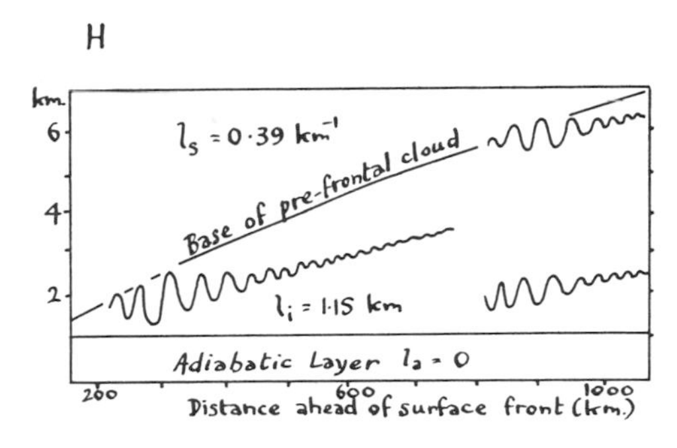

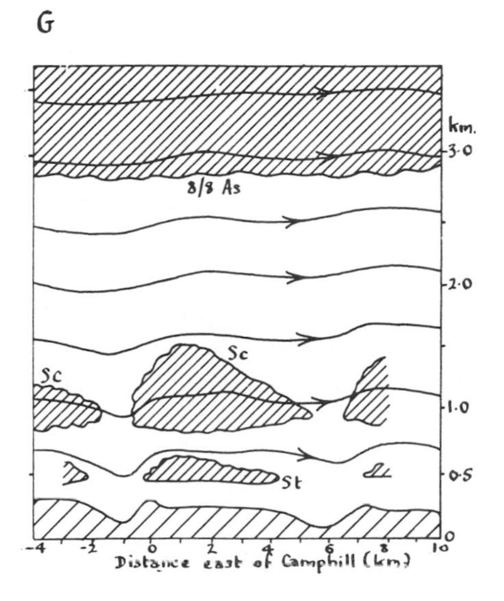

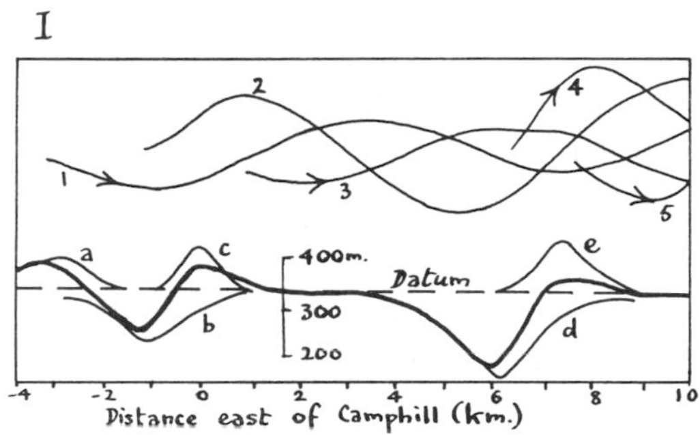

E, Profile of high ground 290° from Leuchars.
F, Six ideal ridges.
G,H,I, Lee waves ahead of a warm front, Camphill, Derbyshire.
G, Computed streamlines and observed clouds, 10.00 A.M. G.M.T., July 16, 1954.
I, Topographic section through Camphill: a–e, idealized ridges; 1–5, lee-waves due to a–e.
(After C. E. Wallington.)

from some arbitrary origin) as x_1, and half the distance between the two nearest points on either side of the summit (whose height is $\frac{1}{2}H_1$) as b_1, these values of H_1, b_1, and x_1 are used as parameters of the height, width, and position of a first elementary ridge. This elementary ridge is then subtracted from the actual mountain profile by reducing all the 80 heights along the profile by the height values of the ideal ridge at these 80 points. The maximum numerical value of the reduced (i.e. residual) profile is then taken to be H_2; x_2 and b_2 are found; the profile of the secondary elementary ridge is computed and in turn subtracted from the reduced profile; and the process is continued until the residual profile shows no significant height variation.

This method has been found to be particularly useful for waves having lengths between 10 and 40 km. It shows (see Figure 2.18,C) that for mountain profiles of similar height and similar over-all width, asymmetric ridges are better generators of large-amplitude waves than symmetric ridges. Also, the gentler the upwind slope of an asymmetric ridge, the closer the first lee wave trough is to the ridge summit.

Waves caused by an isolated hill of irregular shape are more difficult to study than those caused by ridges, but by superimposing several ridges that are oriented at various angles on a plane but that all intersect at the same point, it is possible to represent the topography by an integral equation. Using a two-layer model airstream, with a discontinuity in the parameter l at the interface between the two layers, in each of which l is independent of height, it can be shown that an isolated circular hill produces a lee wave pattern similar to the pattern of waves made by a ship moving across deep water. The waves are confined to a wedge-shaped region, the area of which depends on the characteristics of the airstream: 15° is a typical value for the half-angle of the wedge. Unlike what happens with a simple ridge, the amplitude of these waves decreases downstream, even if there is no frictional drag.[142]

An oval isolated hill produces appreciable lee waves, but the amplitude of the waves is greatly reduced if the major axis of the hill is turned 30° or more from being perpendicular to the airstream. Small, steep-sided oval hills produce almost as much displacement as larger, circular hills with less steep sides. If the hill is obliquely inclined to the airstream, the wave-pattern is very asymmetric, the displacement being greatest over the end of the hill that is farthest downwind. Isolated hills lying along the wind direction should produce only very small waves (see Figure 2.19).[143]

A good example of a lee wave computation, and how it can explain observed features of the weather, are the lee waves found ahead of a warm front in the Peak district of England on July 16, 1954. Generally, lee waves occur in two zones ahead of a warm front: low-level waves are present 150 to 250 miles ahead of the front, and both high- and low-level waves 400 to 600 miles ahead. In the example in question, the two sets of waves were as follows (see Figure 2.18,H). Most distant from the front were waves of length 5.4 km and very small amplitude, the wavelengths increasing to 10 km ahead of the front, the maximum wavelengths (associated with negligible amplitudes) of 16 km occurring 810 km ahead of the front. This first set of waves had maximum amplitudes at two levels, one just above the adiabatic layer and one just below the frontal surface, with a nodal surface of no amplitude midway between the two. Closer to the front was a second set of waves, which exhibited a single level of maximum amplitude, their lengths increasing from 5 to 15 km toward the front.

The model used to compute the streamlines (see Figure 2.18,G) consisted of three layers: an adiabatic layer, from the ground to 855 mb, in which $l = 0$; a stable layer between 855 and 690 mb in which $l = 1.15$/km; and a layer above 690 mb in which $l = 0.39$/km. Three simple ridges and two simple valleys were superimposed on a datum level at 340 m above mean sea level to approximate the topographic profile. The observed cloud patterns agree well with those one would expect from the model streamlines.[144]

More complex models derived from Scorer's original model show that when the amplitude of the lee waves becomes very large, regions containing an airflow in the opposite direction to the mainstream (rotors) appear. The helm wind of Crossfell has been known for many years to exhibit rotor flow in the helm-bar clouds, when no reversed flow exists at the ground. In the Owens Valley of California, a similar type of flow occurs on a vastly greater scale. Here rotors develop under the crests of lee waves in airstreams flowing from west to east across the Sierra Nevada; they are usually invisible, unlike the helm-bar rotors, but occasionally are shown by dust or smoke. On April 25, 1955, a sailplane investigating the wave was broken up in severe turbulence, and the pilot was carried in his parachute very rapidly downwind in the main airstream, i.e., eastward across the valley, below the roll cloud. He then encountered calm air 1,300 m above ground level, below which he drifted *westward* in a 25-knot wind to land on the western side of the Owens Valley, below the leading edge of the roll cloud.[145]

The above discussion has been based almost entirely on the theoretical approach first developed by Scorer. There have been other, quite independent approaches. Colson produced a model in 1950 that mathematically described a westerly airstream of infinite width moving over a north-south mountain range. He showed that if a quasistationary wave exists to the west of the range, then the variation with latitude of the Coriolis parameter must result in a deepening and abrupt eastward displacement of the leeward trough as the wave moves across the mountain.[146]

Airflow around orographic barriers is as important for climatology as airflow over them. A good example is the occurrence of very strong (supergradient, nonkatabatic) winds over Barter Island, a small flat island off the northern coast of Alaska. These winds result from the airflows being forced around a knob of the Brooks Range, which extends from east to west some 50 or 60 miles away. A mathematical model of the phenomenon was produced by supposing a circular cylinder to be set over the knob, so that the latter is represented by circular arcs, and then computing the distribution of air velocities produced by such a barrier in a steady, horizontal, frictionless, irrotational flow of an incompressible fluid. The velocities so determined match the observations quite well, indicating zero vertical movement at both windward and leeward slopes of the barrier, and maximum velocities at the sides. The phenomenon is an example of "corner effect," in which a horizontal airflow is forced to move around the side of a topographic barrier; in this example the effect is intensified by the stable Arctic air. In this region, the influence of the orographic barrier extends almost 100 miles to the north of the Brooks Range during periods of strong winds, and twice as far during severe storms.[147]

Although it may seem fairly obvious that air coming up against a mountain barrier must move over the barrier, why shouldn't the air instead flow around the barrier? Putting the question in another way: when will an airstream surmount an orographic barrier, and when will it merely flow round it? Profound differences in local weather and climate depend on whether airstreams normally flow over or around particular mountains or hills.[148]

Air always has difficulty in rising over a mountain, and prefers to flow around it. The first air particles arriving at the mountain may not possess enough kinetic energy to rise to its top. Energy will therefore be expended in tilting the isentropic surfaces upward over the mountain, so that later air particles will have to depart less from the existing isentropic surfaces to rise over the mountain. As soon as one air particle succeeds in riding over the crest of the mountain, an "isentropic path" will have been laid over the barrier. Since air particles can move along isentropic surfaces without using extra energy, the air may then easily follow this path.

Application of the equation of continuity and of Bernoulli's equation to the problem shows that we do not yet have a complete explanation. The equation of continuity implies that if the air does rise over the summit of the mountain, then its mean velocity at the summit must be greater than its velocity would be at the same height above mean sea level upwind. Bernoulli's equation conflicts with this implication, because it indicates that the air must slow down as it rises up the mountain against gravity (see Appendix 2.24).

This contradiction can be resolved by taking wave theories into account. First, the theory of lee waves shows that a given airstream may not be compatible with a given mountain. If the length of the wave that could be induced by the mountain does

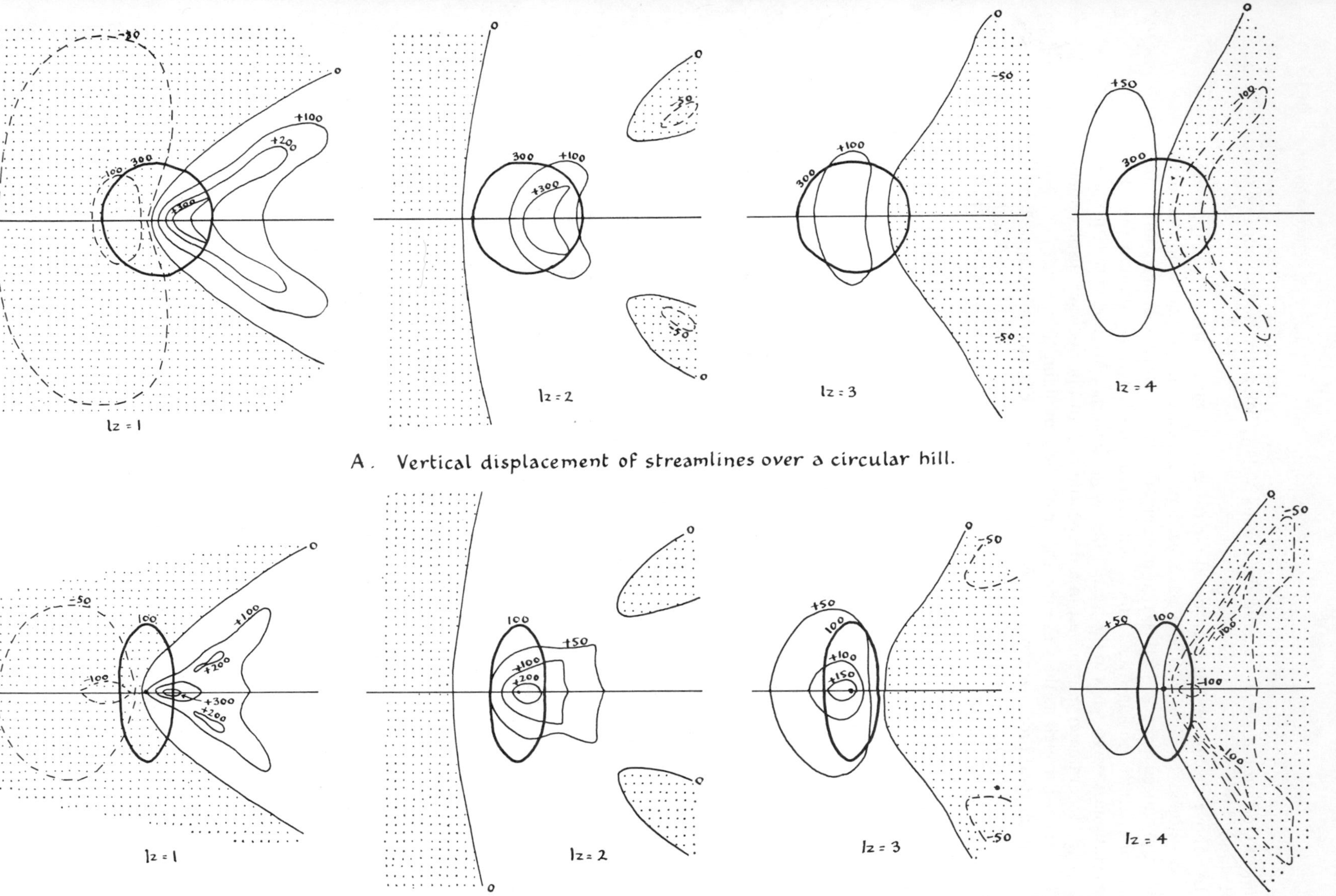

A. Vertical displacement of streamlines over a circular hill.

B. Vertical displacement of streamlines over an oval hill lying normal to the wind direction.

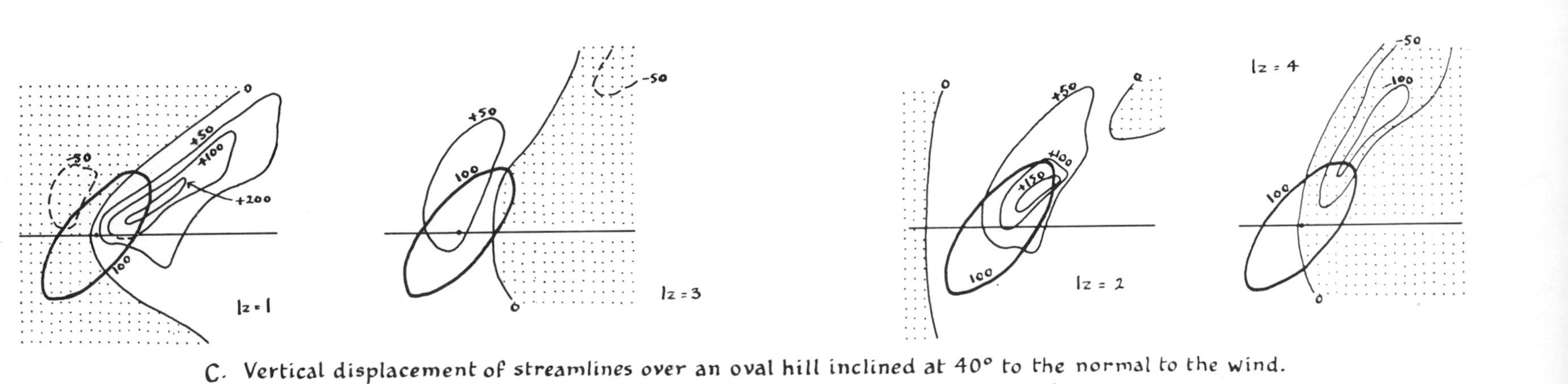

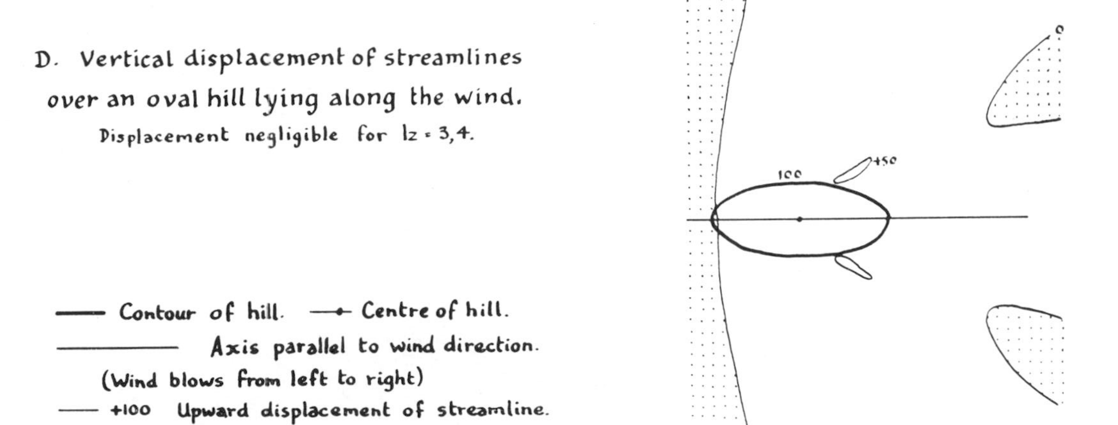

FIGURE 2.19.
The theoretical study of orographic waves:
IV. $1z = 1,2,3,4$ approximately correspond
to heights of 1,2,3,4 km. Stippled areas: areas of downward displacement. Units are arbitrary.
(After R. S. Scorer.)

not coincide with the natural wavelength of the airstream, then no lee waves can develop. Air arriving at the mountain that fails to flow up and over it will therefore flow toward low pressure, against the geostrophic wind and thereby gaining in kinetic energy. This deflection across the isobars may begin anything from a few to 1,000 miles upwind of a large mountain barrier; in such a distance the air can easily acquire sufficient kinetic energy to enable it to rise over the mountain. This additional energy may appear as potential energy on the lee side of the barrier, in the form of a cross-isobar flow of air towards high pressure, or as kinetic energy in the form of lee waves superimposed on the airstream. The kinetic energy goes not necessarily appear only in the air layers below the mountain top.

Second, wave motion at high levels in the atmosphere produces horizontal hydrostatic pressure gradients in the direction of motion. If the waves are not in phase at all heights above a given point on the ground, then an upper wave-trough—in the windward part of which air will be moving downward and thus be adiabatically heated—will warm the air, resulting in a pressure minimum near the ground. Therefore low-level air moving toward the given point will be speeded up.

From Scorer's lee wave theory, a given mountain may cause large positive or negative displacements of air at any level, or may cause no displacement at all, depending on the value of the parameter l. If the airflow is not laminar near the ground, the air may only be displaced a small fraction of what it would be if the flow were laminar. This illustrates the extreme complexity of the problem of the geographical distribution of short-period rainfall; since mean rainfalls are determined by the daily values, average rainfall distribution must also be a complicated problem. Every case of orographic rainfall, given our present stage of knowledge, must be studied separately, on its own merits. No general rules for the effect of a particular mountain on rainfall can be given, because the nature of the vertical displacement of the airstream, and hence the rainfall, is very largely a property of the airstream in question, and airstreams can vary considerably in their physical properties.

Paradoxically, the effect of orographic waves on vertical motion may be important even where there are no mountains, not even distant ones. For example, in deducing climatic distributions from first principles, one of the simplest cases is what happens when an airstream, of known physical properties, flows across a small, flat, heated island. By means of a mathematical model devised by Malkus, the resulting vertical air motions may be split into two components: a heat-diffusion component, only important near the island; and an airflow that behaves as if it were moving over an imaginary mountain. The shape and size of this *equivalent mountain* may be determined from the temperature profile along the ground surface of the island, the speed and stability of the basic airflow, and the coefficient of eddy-heat exchange in the atmosphere directly above the island. The equivalent mountain grows and decays during a single day, and the mountain corresponding to a given island may vary by more than one order of magnitude from one sunny day to another, because its height is inversely proportional to the windspeed. Once the size of the mountain has been determined, using the temperature criteria just mentioned, then the airflow pattern and the distribution of vertical velocities over the island may be computed by applying Scorer's theory of lee waves (see Appendix 2.25).

Vertical wave motion in the atmosphere provides a fascinating field for geographical, as well as meteorological, research. Not only are its consequences profound geographically, but it also provides some observable features, the characteristic almond-shaped stationary wave clouds. Wave-like rolls of cloud are not necessarily due to lee waves, however, because parallel rolls of stratocumulus frequently lie along the direction of wind shear, whereas wave clouds normally lie across it. Roll clouds should only be ascribed to waves if they can be definitely associated with some specific feature of the ground.[149]

There is plenty of scope for the correlation of wave clouds and wave motions with ground features. For example, plateau edges cause waves to form; if the windward edge of the plateau is the main causative factor, then the waves usually form above the plateau. Coastlines with a sea breeze blowing in under an offshore airstream often cause waves to form. Circular hills often give rise to only one lee wave. Steep-sided ridges, if they are long enough, and especially if they are relatively undissected and are backed by large expanses of flat ground in the lee, may give rise to an almost unlimited number of waves.

Although the orographic waves described in this chapter have been large-scale, or at least medium-scale phenomena, there is evidence that they occur on small scales too. Small-scale lee standing waves have been recorded in New Jersey, within the lowest 100 feet of the atmosphere, and set up by very modest relief features.[150]

Even with the largest-scale wave phenomena, where the relationship between the trend of the major orographic barrier and the direction of the wind in the wave-bearing airstream seems to be obvious, new effects may be discovered by careful observation. For example, in eastern Colorado, the main mountain ranges trend north-south, but orographic waves are strongest and most frequent with northwesterly winds, not with westerly winds as one would expect. The theories of lee waves we have now simply assume that airstreams normal to a mountain range will be most significant in generating waves, yet wind direction is apparently very significant. The Colorado waves can probably be explained in terms of the change in the Coriolis parameter as one moves longitudinally along the mountain range, which change causes pressure to decrease more on the eastern (lee) side of a north-south mountain range when the airflow is from the northwest than when it is from any other direction.[151]

Notes to Chapter 2

1. J.-J. D. de Mairan, *Traité physique et historique de l'aurore boreale* (Paris, 2d ed., 1754).
2. G. Ohring, *BAMS*, 45 (1964), 12.
3. J. S. Sawyer, *QJRMS*, 89 (1963), 156.
4. G. W. Wares, *BAMS*, 34 (1953), 221.
5. F. D. Stacey, *MM*, 93 (1964), 25.
6. N. C. Gerson, *BAMS*, 32 (1951), 34.
7. H. Flohn and R. Penndorf, *BAMS*, 31 (1950), 71 and 126.
8. S. Chapman, *BAMS*, 31 (1950), 288 and 387.
9. Flohn and Penndorf, *op. cit.*
10. See, in particular, K. Schneider-Carius, *Die Grundschicht der Troposphäre*, *Probleme der Kosmischen Physik*, vol. 26 (Leipzig, 1953).

11. R. E. Newell, *QJRMS*, 89 (1963), 167.

12. P. A. Sheppard, *SPIAM*, p. 509. The term *triptosphere* has been suggested by W. B. Johnson, *BAMS*, 48 (1967), 896, as a name for the region of the lower troposphere where frictional effects are important, but T. E. A. van Hylckama, *BAMS*, 49 (1968), 388, considers the term *peplosphere* to be more correct, and F. Loewe, *BAMS*, 49 (1968), 1060, suggests that "peplostratum" is a better term than "triptosphere." For exchanges between troposphere and stratosphere, see E. F. Danielson, *JAS*, 25 (1968), 502. For measurements of atmospheric stratification from observation of twilight scattering, see E. K. Bigg, *T*, 16 (1964), 76.

13. W. M. Culkowski, *BAMS*, 37 (1956), 199.

14. J. S. Sawyer, *MOGM*, no. 92 (1954).

15. H. Yamashita, *J. Meteorol. Res.*, 4 (1952), 179. K. Uwai, *TGPM*, 23 (1952), 309; *TJMR*, 4 (1952), 58 and 118; *TJMR*, 5 (1953), 1. W. Hesse, *Z. Meteorol.*, 3 (1949), 129. J. S. Sawyer, *QJRMS*, 77 (1951), 480.

16. A. H. R. Goldie, *M. O. Meteorol. Res. Papers*, no. 734 (1952), and *MM*, 82 (1953), 195. P. B. Storebø, *JM*, 17 (1960), 547.

17. M. J. Rubin, *JM*, 10 (1953), 127. P. B. Storebø, *op. cit.* On the tropopause over Canada, see A. J. Kantor, *JAM*, 6 (1967), 593, and on the penetration of the tropopause over Texas by tall thunderstorms, see M. J. Long, *JAM*, 5 (1966), 851. For details of the mass flow of air from troposphere to stratosphere across the tropopause in North America, see R. M. Smith, *T*, 20 (1968), 76.

18. For examples, see *MAE*.

19. E. K. Bigg, *JM*, 13 (1956), 262.

20. R. L. Ives, *BAMS*, 40 (1959), 149.

21. *Ibid.*, p. 311.

22. E. Robinson, *BAMS*, 33 (1952), 247.

23. A. D. Belmont, *PAS*, p. 215.

24. G. H. Liljequist, *PAS*, p. 167.

25. S. Petterssen, W. C. Jacobs, and B. C. Haynes, *Encyclopedia Arctica*, vol. VII (1951). A. E. Cole, *Servico Meteorologico Nacional, Memorias*, no. 2 (1954).

26. C. R. Hosler, *MWR*, 89 (1961), 319. V. A. Bugaev, V. A. Dzhordzhio, and V. R. Dubentsov, *Akademia Nauk SSSR*, *Izvestia*, *Geog. ser.*, no. 3 (1952), p. 44. B. H. Imrie, *MM*, 80 (1951), 231.

27. R. Jalu, *La Météorologie*, no. 39 (1955), p. 247.

28. J. R. Gerhardt, *JAS*, 19 (1962), 116.

29. I. Akita, *TJMR*, 7 (1955), 490.

30. K. H. Stewart, *M.O. Meteorol. Res. Papers*, no. 912 (1955). K. Schneider Carius, *AMGB*, 2 (1950), 97.

31. G. B. Bell, Jr., and E. Robinson, *Stanford Research Institute Sci. Report*, no. 19 (1954) and no. 21 (1955).

32. M. Neiburger, *IUGG Assoc. Meteorol.*, *Oslo 1948*, *Trans.*, 2 (1950), 134. H. Flohn, *Polarforschung*, 21 (1951), 58. P. A. Langwell, *JM*, 10 (1953), 187.

33. For the theory of the control of inversions by surface temperature changes, see F. K. Ball, *QJRMS*, 86 (1960), 483, and S. Fritz, *MWR*, 86 (1958), 129. On the function of radiative cooling in the maintenance of middle-tropospheric inversions over Arizona, see D. O. Staley, *QJRMS*, 91 (1965), 282. For details of the climatology of temperature inversions, see: H. C. Shellard and R. F. M. Hay, *MOSP*, no. 10 (1961), on the northeastern Atlantic; P. Y. Haraguchi, *MWR*, 96 (1968), 177, on subsidence inversions over the Pacific; E. N. Lawrence, *MM*, 97 (1968), 270, on low-level inversions at midday over Crawley, Sussex; and H. W. Baynton, J. M. Bidwell, and D. W. Beran, *JAM*, 4 (1965), 509, on low-level inversions at Point Arguello, Calif. For an example of both subsidence and nocturnal inversions in southern England, see R. F. Zobel, *QJRMS*, 92 (1966), 196. For a micrometeorological analysis of a temperature inversion in Red Butte Canyon, Utah, see A. H. Thompson, *JAM*, 6 (1967), 287. For the effects of temperature inversions, see E. R. Stephens, *WW*, 18 (1965), 172, on air-pollutant trapping, and P. Kruger, *T*, 18 (1966), 576, on movement of radioactive aerosols.

34. See *M*, p. 165, for the basic physics of albedo.

35. P. M. Kuhn and V. E. Suomi, *JM*, 15 (1958), 172. K. G. Bauer and J. A. Dutton, *JGR*, 67 (1962), 2367.

36. P. Larsson and S. Orvig, *McGill Univ. Publications Meteorol.*, no. 45 (1961).

37. For details, see *HBES*, p. 38.

38. *Ibid.*, p. 39.

39. S. Fritz, *JM*, 6 (1949), 277. For the climatological significance of albedo in central Canada, see J. D. McFadden and R. A. Ragotzkie, *JGR*, 72 (1967), 1135. For the albedo of the ground surface in particular areas, see: K. J. Hanson and H. J. Viebrock, *MWR*, 92 (1964), 223, on the northeastern United States; Lin-sien Chia, *QJRMS*, 93 (1967), 116, on Barbados; R. G. Barry and R. E. Chambers, *W*, 21 (1966), 60, on Hampshire and Dorset, England; and R. G. Barry and R. E. Chambers, *QJRMS*, 92 (1966), 543, for a map of summer albedo for England and Wales. For the technique of airborne albedo measurements over Antarctica, see M. C. Predoehl and A. F. Spano, *MWR*, 93 (1965), 687, and for results of this work, see A. F. Spano, *MWR*, 93 (1965), 697. For albedo determination from TIROS photographs, see J. H. Conover, *JAM*, 4 (1965), 378; and for Earth albedo as observed by TIROS VII, see A. Arking and J. S. Levine, *JAS*, 24 (1967), 721. On the control of albedo by solar elevation, see F. Volz, *GPA*, 60, no. 1 (1965), 197. For the effect of albedo on evapotranspiration rates and on energy budgets in O'Neill, Nebraska, see I. Seginer, *Agric. Meteorol.*, 6 (1949), 5. On the relation between albedo and snow cover in central Japan, see *JPC*, 3 (1966), 88. For aircraft measurements of the albedo of stratus clouds and of the Earth's surface, see M. Griggs, *JAM*, 7 (1968), 1012.

40. J. E. Miller gives an account of energy equations in *CM*, p. 483.

41. V. P. Starr, *CM*, p. 568.

42. J. Spar, *JM*, 6 (1949), 411.

43. E. N. Lorenz, *DC*, p. 86. See also A. H. Oort, *MWR*, 92 (1964), 483.

44. J. S. Winston and A. F. Krueger, *MWR*, 89 (1961), 307.

45. See P. F. Clapp and F. J. Winninghoff, *MWR*, 91 (1963), 494, for the mathematics.

46. E. Palmén, *GPA*, 49, no. 2 (1961), 167.

47. E. Palmén, *ASM*, p. 212.

48. S.-K. Kao, *JM*, 11 (1954), 352. H. Riehl, *ASM*, p. 381. See also M. B. Danard, *JAM*, 3 (1964), 27.

49. F. Burdecki, *JM*, 14 (1957), 206. V. P. Starr, *CM*, p. 568.

50. S.-K. Kao and W. P. Hurley, *JGR*, 67 (1962), 4233.

51. M. J. Rubin and W. S. Weynant, *MWR*, 91 (1963), 487.

52. R. M. Henry and S. L. Hess, *JM*, 15 (1958), 397.

53. B. Saltzman, *JM*, 15 (1958), 259.

54. R. M. White and D. S. Cooley, *JM*, 13 (1956), 67. B. Saltzman and A. Fleisher, *JAS*, 19 (1962), 195. See also A. Wiin-Nielsen, J. A. Brown, and M. Drake, *T*, 16 (1964), 168, and A. H. Oort, *T*, 16 (1964), 309.

55. R. Shapiro and F. Ward, *JAS*, 20 (1963), 353.

56. B. P. Harper, *JM*, 18 (1961), 487.

57. For the mathematical details, see R. J. Reed, J. L. Wolfe, and H. Nishimoto, *JAS*, 20 (1963), 256.

58. S. Teweles, *MWR*, 91 (1963), 505. The literature on the climatology of atmospheric energy is now extensive. On the climatology of energy exchange between eddies and the zonal flow, see A. Wiin-Nielsen, J. A. Brown, and M. Drake, *T*, 16 (1964), 168. For the climatology of kinetic energy in the atmosphere, see: E. C. Kung, *MWR*, 94 (1966), 67 and 627, and 95 (1967), 593, on North America; S. L. Hasthenrath, *GPA*, 78, no. 3 (1967), 173, on the Caribbean and Gulf of Mexico; E. O. Holopainen, *T*, 15 (1963), 26, on Britain; S.-K. Kao and V. R. Taylor, *JGR*, 69 (1964), 1037; B. Saltzman and S. Teweles, *T*, 16 (1964), 432, for the northern hemisphere in general; and E. Palmén, *T*, 18 (1966), 838, for northern extratropical latitudes. For kinetic energy spectra, see S.-K. Kao and E. S. Sands, *JGR*, 71 (1966), 5213, and for kinetic energy, jetstreams, and cyclone tracks, see S.-K. Kao and G. R. Farr, *JGR*, 71 (1966), 4289. On kinetic-energy conversion, see: A. Wiin-Nielsen and M. Drake, *MWR*, 94 (1966), 221; J. A. Brown, Jr., *T*, 19 (1967), 14; A. Wiin-Nielsen, J. A. Brown, and M. Drake, *T*, 15 (1963), 261. On available potential energy, see: R. A. Anthes and D. R. Johnson, *MWR*, 96 (1968), 291, on Hurricane Hilda; J. A. Brown, Jr., *T*, 16 (1964), 371, on tropospheric diabatic heating correlation; J. L. Corcoran and L. H. Horn, *JGR*, 70 (1965), 452, on infrared radiation correlation; M. D. B. Danard, *JAM*, 5 (1966), 81, on latent-heat release; A. Wiin-Nielsen, *MWR*, 92 (1964), 161, and A. Eddy, *JAM*, 4 (1965), 569, on conversion into kinetic energy, and D. R. Johnson, *T*, 19 (1967), 517, on terrestrial radiation correlation. On stratospheric warming energetics, see A. H. Oort, *T*, 16 (1964), 309, on the northern hemisphere I.G.Y.; J. S. Perry, *JAS*, 24 (1967), 539; M. A. Lateef, *JGR*, 69 (1964), 1481; R. E. Newell, *GPA*, 58, no. 2 (1964), 145. On atmospheric energetics during the winter 1963 stratospheric warming, see P. R. Julian and K. B. Labitzke, *JAS*, 22 (1965), 597. For the energy budget of the atmosphere in the Caribbean–Gulf of Mexico area, see S. L. Hasthenrath, *JAS*, 23 (1966), 694, and for energy exchange at the air-sea interface, see G. D. Robinson, *QJRMS*, 92 (1966), 45. On the annual variation and spectra of atmospheric energy, see A. Wiin-Nielsen, *T*, 19 (1967), 540; and on atmospheric energy balance in relation to climatic change, see W. D. Sellers, *JAM*, 3 (1964), 337. According to V. P. Starr and J. M. Wallace, *GPA*, 58, no. 2 (1964), 138, large-scale eddy energy processes in tropical regions of the middle troposphere operate in the inverse to a normal heat engine; i.e., colder air rises and warmer air sinks. E. R. Reiter, *MWR*, 97 (1969), 200, introduces a new symbolism for distinguishing mean and eddy motions. For the effects of sea-surface conditions in the equatorial Atlantic on the energy of hurricanes, see I. Perlroth, *T*, 21 (1969), 230. On the theory of kinetic energy in the troposphere in relation to the quasi-biennial oscillation, see A. J. Miller, *MWR*, 97 (1969), 142. For atmospheric energetics, see P. J. Smith, *T*, 21 (1969), 193, on North America, March 1963; see also A. F. Krueger, J. S. Winston, and D. A. Haines, *MWR*, 93 (1965), 227, on the climatology of the northern hemisphere. For a climatology of the energy budget of the stratosphere during the IQSY, see R. E. Newell and M. E. Richards, *QJRMS*, 95 (1969), 310.

59. For a discussion of these orographic effects, see B. Bolin, *AGP*, 1 (1952), 95.

60. This description follows P. A. Sheppard, *SP*, 40 (1952), 89.

61. For details, see J. E. Vehrencamp, *EAFM*, p. 99, and P. A. Sheppard, *AGP*, 9 (1962), 77.

62. *Ibid.* G. O. P. Obasi, *JAS*, 20 (1963), 516.

63. C. H. B. Priestley, *QJRMS*, 77 (1951), 200. V. P. Starr and R. M. White, *QJRMS*, 78 (1952), 62.

64. Y. P. Rao, *QJRMS*, 88 (1962), 96.

65. G. O. P. Obasi, *op. cit.*

66. C. H. B. Priestley and A. J. Troup, *TP*, no. 1 (1954).

67. R. L. Pfeffer and B. Saltzman, *JM*, 12 (1955), 500. M. Doporto, *SPIAM*, p. 444.

68. E. Palmén, *ASM*, p. 212. E. N. Lorenz, *JM*, 9 (1952), 152. W. G. Leight, *JM*, 11 (1954), 10. For the climatology of angular momentum, see: G. B. Tucker, *T*, 12 (1960), 134, on the northern hemisphere, and

QJRMS, 91 (1965), 356, on the divergence of momentum in the lower equatorial stratosphere; F. A. Berson and A. J. Troup, *T*, 13 (1961), 66, on the equatorial trough; A. J. Miller, S. Teweles, and H. M. Woolf, *MWR*, 95 (1967), 427, on momentum at 500 mb in the northern hemisphere; and Wan-cheng Chui and H. L. Crutcher, *JGR*, 71 (1966), 1017, on spectra for North and Central America. On meridional transport of momentum, see R. Murray, A. E. Parker, and P. Collinson, *QJRMS*, 95 (1969), 92; and see E. O. Holopainen, *T*, 19 (1967), 1, on the mean meridional circulation. See H. F. Hawkins and D. T. Rubsam, *MWR*, 96 (1968), 617, for a momentum budget of Hurricane Hilda; P. A. Gillman, *GPA*, 57, no. 1 (1964), 161, for a theory of vertical transport of momentum; and E. C. Kung, *MWR*, 96 (1968), 337, for a theory of momentum exchange between the atmosphere and the Earth's surface. For a theory of momentum transport by internal gravity waves, illustrated from lee waves over North Wales, see F. P. Bretherton, *QJRMS*, 95 (1969), 213.

69. D. A. Haines and J. S. Winston, *MWR*, 91 (1963), 319.

70. R. M. White, *QJRMS*, 77 (1951), 188.

71. V. P. Starr and R. M. White, *AFGP*, no. 35 (1954).

72. M. S. Rao, *JAS*, 19 (1962), 468. M. S. Rao and R. Ramanadham, *JAS*, 20 (1963), 350.

73. N. S. Montgomery, *SPIAM*, p. 453.

74. E. L. Fisher, *JM*, 15 (1958), 164.

75. J. W. Hutchings, *JM*, 18 (1961), 615. J. K. Bannon, A. G. Matthewman, and R. Murray, *QJRMS*, 87 (1961), 502. On the theory of transfer of atmospheric properties in the lower stratosphere, see R. J. Murgatroyd, *QJRMS*, 91 (1965), 421. For the climatology of atmospheric water vapor advection, see: E. M. Rasmusson, *MWR*, 95 (1967), 403, on North America; R. G. Barry, *QJRMS*, 93 (1967), 535, on northeastern North America; and S. L. Hasthenrath, *JAM*, 5 (1966), 778, on the Caribbean and Gulf of Mexico. For the water balance of the atmosphere in the northern hemisphere during the I.G.Y., see V. P. Starr, J. E. Peixoto, and A. R. Crisi, *T*, 17 (1965), 463. For the diurnal variation of atmospheric water-vapor flux over the Caribbean and Gulf of Mexico, see S. L. Hasthenrath, *JGR*, 72 (1967), 4119. For the daily meridional transfer of atmospheric water vapor between the equator and 40°N, see H. M. E. van de Boogaard, *T*, 16 (1964), 43. For the exchange of water between sea and atmosphere in Hurricane Betsy, see H. G. Östland, *T*, 20 (1968), 577. For monthly evaporation in Southern Sweden computed from atmospheric water balance, see A. Nyberg, *T*, 17 (1965), 473.

76. See F. H. Ludlam and R. S. Scorer, *QJRMS*, 79 (1953), 317, for a general review.

77. G. D. Robinson, *QJRMS*, 77 (1951), 61.

78. For details, see *PDM*, p. 219, and *CM*, p. 1255.

79. H. Foster, *JM*, 9 (1952), 437. R. F. Jones, *MM*, 81 (1952), 152.

80. R. A. Bryson, *TAGU*, 36 (1955), 209.

81. R. S. Scorer, *MM*, 83 (1954), 202.

82. J. S. Malkus, *TAGU*, 30 (1949), 19.

83. Ludlam and Scorer, *op. cit.* R. Scorer, *MM*, 83 (1954), 202.

84. P. B. MacCready, Jr., *W*, 10 (1955), 35.

85. E. E. Foster, *BAMS*, 34 (1953), 134.

86. Ludlam and Scorer, *op. cit.*

87. J. Warner and J. W. Telford, *JAS*, 20 (1963), 313.

88. B. Woodward, *QJRMS*, 85 (1959), 144. For details, see W. S. Hall, *QJRMS*, 88 (1962), 394.

89. P. B. MacCready, Jr., *op. cit.* Ludlam and Scorer, *op. cit.*

90. G. C. Holzworth, *MWR*, 92 (1964), 235.

91. For the proof, see S. B. Kraus and C. H. B. Priestley, *GPA*, 51 (1962), 199.

92. Sir C. Normand, *QJRMS*, 72 (1946), 145. C. H. B. Priestley, *MM*, 83 (1954), 107.

93. K. H. Jehn and J. R. Gerhardt, *JM*, 10 (1953), 10.

94. C. H. B. Priestley, *QJRMS*, 81 (1955), 139. R. J. Taylor, *QJRMS*, 82 (1956), 89.

95. See C. H. B. Priestley, *MM*, 83 (1954), 107, for details. On the prediction of the depth of afternoon convection (i.e., the surface mixing layer) in the United States, see M. E. Miller, *MWR*, 95 (1967), 35. For aircraft measurements of convection in England, see D. R. Grant, *QJRMS*, 91 (1965), 268. On convection over burning cornfields on Salisbury Plain, England, see D. J. Ride, *W*, 20 (1965), 238. For observations of thermals at Cardington, Beds., see G. W. Paltridge, *MM*, 97 (1968), 56. For conditions in clear-air thermals (i.e., convection below cloud base) in Australia, see J. Warner and J. W. Telford, *JAS*, 24 (1967), 374. For a mathematical experiment in the prediction of dry and moist convection, including the resulting rain, in terms of thermal bubbles, see G. Árnason, R. S. Greenfield, and E. A. Newburg, *JAS*, 25 (1968), 404. For a theory involving two superposed layers with different properties, heated from below, see D. Ray, *GPA*, 65, no. 3 (1966), 168: the resulting convection gives rise to cellular cloud patterns as observed from satellites. For a mathematical model of convection in a conditionally unstable, stagnant air mass, see S. M. A. Haque, *T*, 16 (1964), 147, and for a theory of convection in an atmosphere that is stable for dry adiabatic descending motion, but unstable for saturated ascending motion, see H. L. Kuo, *T*, 17 (1965), 413. On a dynamic theory of convection, see Y. Sasaki, *T*, 19 (1967), 45.

96. For general reviews, see A. Miller and H. A. Panofsky, *BAMS*, 39 (1958), 8, on the connection between large-scale vertical motion and weather, and H. A. Panofsky, *JM*, 3 (1946), 45, and *CM*, p. 639, for the mathematics involved.

97. A. Wiin-Nielsen, *MWR*, 92 (1964), 161. E. J. Aubert, *JM*, 14 (1957), 527.

98. R. G. Fleagle, *JM*, 15 (1958), 249.

99. W. A. Thorn, *JM*, 4 (1947), 73. W. Schaffer, *JM*, 6 (1949), 212. C. M. Penner, *JAM*, 2 (1963), 235. E. Kessler III, *JM*, 16 (1959), 630.

100. C. W. Thornthwaite, W. J. Superior, J. R. Mather, and F. K. Hare, *The Measurement of Vertical Winds and Momentum Flux*, *PC*, 14 (1961), 5. Thornthwaite, Superior, and Mather, *Vertical Winds Near the Ground at Centerton, N.J.*, *loc. cit.*, p. 95.

101. J. K. Angell and D. H. Pack, *MWR*, 89 (1961), 273.

102. E. S. Epstein, *MWR*, 87 (1959), 91. L. Oredsson, *T*, 16 (1964), 411.

103. R. C. Curtis and H. A. Panofsky, *BAMS*, 39 (1958), 521.

104. E. Knighting, *QJRMS*, 86 (1960), 318.

105. *MM*, 79 (1950), 314. A. C. Molla, Jr., and C. J. Loisel, *GPA*, 52, no. 2 (1962), 166.

106. W. S. Harley, *JAM*, 4 (1965), 305. For vertical motion over the Caribbean, see S. L. Hasthenrath, *T*, 20 (1968), 163, on its climatology, and D. P. Baumhefner, *MWR*, 96 (1968), 218, for a case study. For the climatology of the variability of vertical velocities in the lower stratosphere, see R. E. Newell and A. J. Miller, *JAM*, 7 (1968), 516. On the magnitude of large-scale vertical motions in the upper stratosphere, see M. Kays and R. A. Craig, *JGR*, 70 (1965), 4453; and on vertical motions at 100 mb over the Caribbean, see M. A. Lateef, *MWR*, 96 (1968), 286. On the theory of terrain-induced vertical motion, see L. Berkofsky, *JAM*, 3 (1964), 410. For synoptic studies of subsidence, see C. J. Boyden, *MM*, 93 (1964), 138 and 180. On the computation of precipitation from vertical motion, see T. H. R. O'Neill, *JAM*, 5 (1966), 595, and N. Besleagă, R. Stoian, and A. Doneaud, *T*, 17 (1965), 111.

107. For general surveys of orographic wave effects, see R. S. Scorer, *SP*, 40 (1952), 466, and G. A. Corby, *QJRMS*, 80 (1954), 491.

108. DeVer Colson, *BAMS*, 35 (1954), 363. See also *W*, 20 (1965), 162.

109. C. F. Jenkins, *AFSG*, no. 15 (1952).

110. DeVer Colson, *BAMS*, 33 (1952), 107.

111. A. R. Laird, *MM*, 81 (1952), 337.

112. C. V. Lindsay, *MWR*, 90 (1962), 271.

113. A. J. Fyfe, *W*, 7 (1952), 137.

114. N. Gerber and M. Berenger, *QJRMS*, 87 (1961), 13.

115. G. Manley, *QJRMS*, 71 (1945), 197. F. W. Ward, *MM*, 82 (1953), 234.

116. W. H. Kavanagh, *MM*, 86 (1957), 46. R. S. Scorer, A. E. Slater, A. A. J. Sanders, *Gliding*, 4 (1953), 24, 26, 27.

117. F. P. U. Croker, *MM*, 86 (1957), 146.

118. *BWMO*, 7 (1958), 83. See also "M.O. discussion on orographic waves," *MM*, 87 (1958), 80.

119. R. K. Pilsbury, *MM*, 88 (1959), 17

120. R. K. Pilsbury, *MM*, 84 (1955), 313.

121. *MM*, 87 (1958), 86.

122. D. R. Booker, *AMM*, 5, no. 27 (1963), 129.

123. *QJRMS*, 81 (1955), 488. R. S. Scorer, *QJRMS*, 81 (1955), 340.

124. W. G. Harper, *MM*, 85 (1956), 38.

125. G. A. Corby, *QJRMS*, 83 (1957), 49.

126. S. L. Hess and H. Wagner, *JM*, 5 (1948), 1.

127. R. S. Scorer, *QJRMS*, 81 (1955), 340.

128. J. Förchtgott, *Bull. Meteorol. Tchécosl.* (Prague), 3 (1949), 49. See also *QJRMS*, 81 (1955), 488.

129. See P. M. Breistein and H. D. Parry, *MWR*, 82 (1954), 355.

130. F. H. Ludlam, *W*, 7 (1952), 300, and *QJRMS*, 78 (1952), 554.

131. J. S. Sawyer, *QJRMS*, 81 (1955), 489.

132. *Ibid.*

133. See DeVer Colson, *BAMS*, 33 (1952), 107, on dust particles in the Bishop wave. For observations of occurrences of lee waves, see: R. H. T. Collis, F. G. Fernald, and J. E. Alder, *JAM*, 7 (1968), 227, on lidar in the Sierra Nevada; J. P. Kuettner and D. K. Lilly, *WW*, 21 (1968), 180, on the Colorado Rocky Mts.; D. W. Beran, *JAM*, 6 (1967), 865, on the Rocky Mts., associated with chinook winds; R. D. Reynolds, R. L. Lamberth, and M. G. Wurtele, *JAM*, 7 (1968), 353, on complex waves in New Mexico; P. Williams, Jr., and J. D. Quinn, *WW*, 18 (1965), 128, on the Wasatch Mts., Utah; G. A. Corby, *W*, 21 (1966), 440, on ESSA II satellite data on Britain; and A. Gray and W. J. Stewart, *MM*, 94 (1965), 8, on Acklington, Northumberland. For an explanation of the destructive gale of February 16, 1962, in the West Riding of Yorkshire in terms of lee waves, see C. J. M. Aanensen and J. S. Sawyer, *N*, 197 (1963), 654. On the significance of mountain lee waves as seen from satellite photos, see S. Fritz, *JAM*, 4 (1965), 31. For details of a dust-devil induced by the rotor of a mountain wave in Nevada, see J. Hallett, *W*, 24 (1969), 133.

134. For details, see G. A. Corby, *QJRMS*, 80 (1954), 491.

135. For details, see P. Queney, *Univ. of Chicago Dept. of Meteorol. Misc. Report*, no. 23 (1947).

136. G. Lyra, *Beitr. Phys. Freien Atmos.*, 26 (1940), 197.

137. R. S. Scorer, *QJRMS*, 75 (1949), 41; *QJRMS*, 80 (1953), 70; *MM*, 82 (1953), 99 and 232.

138. J. S. Sawyer, *QJRMS*, 86 (1960), 326.

139. C. E. Wallington, *QJRMS*, 79 (1953), 545.

140. G. A. Corby and C. E. Wallington, *QJRMS*, 82 (1956), 266.

141. C. E. Wallington, *MM*, 90 (1961), 213.

142. R. S. Scorer and M. Wilkinson, *QJRMS*, 82 (1956), 419.

143. R. S. Scorer, *QJRMS*, 82 (1956), 75.

144. C. E. Wallington, *QJRMS*, 81 (1955), 251.

145. R. S. Scorer and H. Klieforth, *QJRMS*, 85 (1959), 131.

146. DeVer Colson, *JM*, 7 (1950), 279.

147. W. W. Dickey, *JM*, 18 (1961), 790. See also *DMWF*, p. 607.

148. See P. A. Sheppard, *QJRMS*, 82 (1956), 528, and R. S. Scorer, *QJRMS*, 83 (1957), 271.

149. See R. S. Scorer, *MM*, 80 (1951), 99, for the reason.

150. See Thornthwaite *et al.*, *op. cit.* (1961) and Thornthwaite, Superior, and Mather, *op. cit.* (1961).

151. N. J. Macdonald and H. T. Harrison, *BAMS*, 41 (1960), 627. For a theory of two-dimensional mountain waves, see T. N. Krishnamurti, *MWR*, 92 (1964), 147, and *Rev. Geophysics*, 2 (1964), 593. For a three-layer model involving a highly stable middle layer, see R. P. Pearce and P. W. White, *QJRMS*, 93 (1967), 155. For a mathematical model of a stably stratified airflow that indicates lee waves produced by a barrier may be strengthened or weakened by the presence of another barrier downstream, see Y.-H. Pao, *QJRMS*, 95 (1969), 104. On the calculation of maximum vertical velocities in mountain lee waves, see S. A. Caswell, *MM*, 95 (1966), 68. For a comparison of actual with predicted positions of lee waves in the Middle East, see A. Cohen and E. Doron, *JAM*, 6 (1967), 669. For the theory of airflow over mountains with gentle slopes (i.e., large mountain barriers), see Z. Petkovšek and M. Ribaric, *T*, 17 (1965), 443. For a mathematical model of orographic rainfall over the Western Ghats of India, based on orographic wave theory, see R. P. Sarker, *MWR*, 94 (1966), 555.

CHAPTER 3

Atmospheric Turbulence and Diffusion

The study of meteorology and climatology encompasses a host of phenomena of different types and magnitudes, for example: depressions with a dimension of hundreds of square miles and a life of several days; anticyclones with dimensions of thousands of square miles and several weeks' duration; microfluctuations of the weather elements with a period of a few minutes or even a few seconds; and macrofluctuations with a period of several months or even years. The general circulation can be regarded as the smooth "normal" flow of the atmosphere on which have been superimposed irregular "perturbations" due to the existence of depressions, waves, and other pressure features. Similarly, the progress of weather and climate with respect to time can be viewed as consisting of a smooth "normal" trend in the elements, due to the influence of the sun (i.e., the planetary climate), on which is superimposed the day-to-day variation in the elements that we term the weather.

All these phenomena are linked together by the theory of atmospheric turbulence, which is an outcome of the general theory of turbulence first demonstrated experimentally by Osborne Reynolds, and later developed mathematically by G. I. Taylor. Turbulence theory is important for the study of day-to-day weather changes; it is important in the study of the general circulation of the atmosphere, for microclimatology, and for long-period climatology. To follow it in its entirety demands the use of advanced mathematical ideas; only an outline of the theory will be given here.[1]

Osborne Reynolds, in his classic experiments on the flow of liquids in pipes, which led to the distinction between laminar and streamline flow, regarded a turbulent flow as being composed of a mean motion, which remains steady or varies slowly with time, on which is superimposed a rapidly fluctuating flow whose mean value is zero. He showed that the mean motion itself is partly determined by the fluctuating flow, and hence the properties of the mean flow could be examined by regarding it as the only flow present, provided that additional apparent (frictional) forces—later termed the "Reynolds stresses"—were introduced, which could be derived from measurements of the fluctuations

Turbulence is hence a technique, not a term for specific phenomena, which is used to describe relatively small-scale motions: all motions that are smaller than the motion being investigated are regarded as turbulent (irregular) fluctuations superimposed on the latter. The smaller motions are usually agents of momentum transport, determining the nature of the main motion, and agents of diffusion, if the physical properties of the fluid or the medium considered differ from point to point.

The simplest illustration of atmospheric turbulence is given in Figure 3.1, which shows different scales of wind turbulence. The most rapid fluctuations in windspeed, measured by a hot-wire anemometer, have a frequency of 20 oscillations or more per second. Next come the oscillations recorded by a Dines anemometer, with a frequency of one every three or so seconds. Finally, the variations in daily windspeed, with a frequency of one oscillation every two or three days. Each of these can be regarded as turbulence of different scales. Other illustrations from strictly meteorological phenomena are the turbulent flow of the air over the rough ground surface, free-air turbulence in the upper atmosphere, turbulent motion in clouds, and stellar scintillation.

Random fluctuations in the refractive index of the atmosphere produced by turbulence cause the twinkling of stars and the scattering of radio waves. Collisions between water droplets within clouds result from the turbulent air currents present there. Regions of large vertical shear in the free atmosphere, for example, near jetstreams and inversions, frequently develop severe forms of clear-air turbulence that can be very unpleasant, even dangerous, to fly through. In some circumstances, the air flowing beneath an invertion can undergo a transition in its motion analogous to a hydraulic jump, which may result in considerable turbulence.

The turbulence of airflow near the Earth's surface, which is obviously brought about by the obstructions to air movement presented by buildings, trees, hedges, and so on, is largely responsible for the upward transport of heat and water vapor from the ground to the atmosphere. The general properties of this turbulent airflow near the surface are also very important in determining the variation of wind with height and with time of day.

The turbulent diffusion of heat and water vapor from the ground to the atmosphere, which is much affected by variations in the vertical stability of the air due to changes in the temperature lapse-rate, and a similar transfer from one level of the atmosphere to another, is responsible for the formation and modification of air masses. The vertical transport of momentum at the Earth's surface, which manifests as surface drag on the movement of the air, is responsible for the degradation of the kinetic energy of the wind systems. And finally, the quasihorizontal transfer of heat, water vapor, and momentum from low to high latitudes, which is accomplished mainly by lateral turbulent diffusion, is responsible for the creation of the Earth's climatic belts.

Early workers on atmospheric turbulence, for example, Taylor in 1915 and Schmidt in 1917, considered that atmospheric eddies acted as molecules do in the kinetic theory of gases. In other words, they believed that the "particles" of atmospheric motion, i.e.,

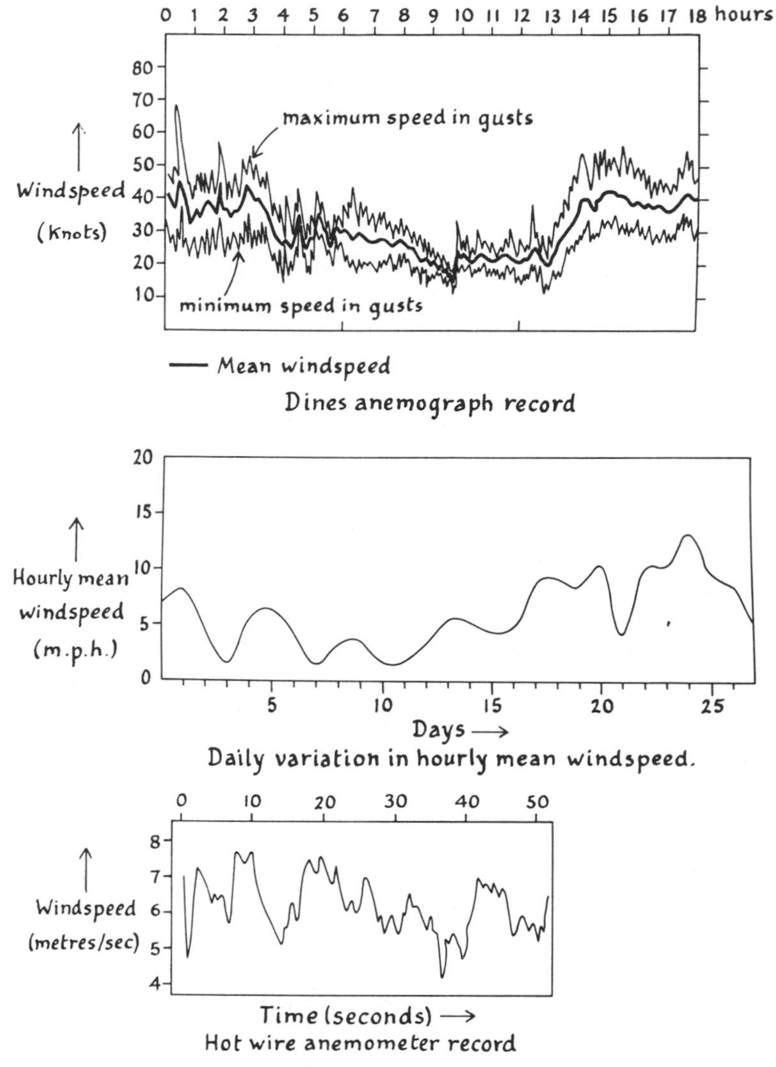

FIGURE 3.1.
Examples of turbulent fluctuations
in windspeed.

the eddies, conserved their physical properties (potential temperature and specific humidity, for example) until the particle became completely mixed with its surroundings, after which the simple laws of mixtures were obeyed. Analogous to the "mean free path" of gas molecules in the kinetic theory was the *mixing length* in atmospheric turbulence. The mixing length is the average distance an eddy, of a specific size and with certain specific physical properties, travels, before becoming completely mixed with its surroundings. The mixing length in the atmosphere is much longer (relative to scale) than the molecular free path, and much more persistent.

The picture of an eddy can be conceived as follows. If we had a large number of equally sensitive anemometers spaced small distances apart both horizontally and vertically, by examining the traces of these instruments we ought to be able to draw a line (strictly speaking, we ought to be able to trace out a three-dimensional closed surface) around the area in which, at a given time, the wind was above (or below) its average value. These lines would enclose *eddies*, or elements of turbulence in the atmosphere.*

Eddies grow, move, decay, appear, and reappear very rapidly. All the transport of latent and sensible heat, moisture, and momentum in the lower levels of the atmosphere is accomplished by the movement and mixing with new environments of these eddies, and this transport is termed the *eddy flux*. The picture of turbulence suggested by the eddy concept is thus of a process in which small bodies of air become detached from their original positions, and move into other regions or atmospheric strata, where they become absorbed by the surrounding air, having transferred some physical properties of their mother locality or layer to the new environment. Their essential function is thus in the *diffusion* of elements of weather and climate from one part of the Earth's surface, or from one stratum of the atmosphere, to another. The general result of turbulence would be to produce a general uniform vertical distribution of temperature, humidity, etc., but this is prevented by variations in pressure and solar heating.

No one has ever seen an eddy. The exact structure and shape of eddies is not known, apart from the fact that they have both horizontal and vertical axes. Eddies vary greatly in size, intensity, and duration, and "eddies within eddies" occur, since, occasionally, gusts of 15- or 20-minute duration are found superimposed on rapid wind pulsations. Very little is known about the properties and characteristics of eddies in relation to different air masses, the effect of variations in topography on the formation and dissipation of eddies, and whether or not the infrequency and persistence varies, for comparable eddies in different regions and latitudes. Since these are very important geographical questions, such a gap in our knowledge may appear strange.

There are several difficulties in trying to study turbulence by following the life-history of an individual eddy or group of eddies. Apart from the obvious difficulty in keeping track of a specific eddy, there are other, more basic drawbacks. The mixing-length hypothesis assumes that the properties of an eddy remain conservative during its life-history, and this may not always apply. Then there is the question of how to define an eddy. It is usually defined as a fluid element whose motion differs from that of the mean flow. But how is the mean flow to be defined; over what period of time should the average be calculated?

Despite these difficulties, the concept has resulted in certain very definite additions to knowledge. Lateral diffusion in the atmosphere, i.e., the horizontal transport by eddies of heat, water vapor, and momentum from one region to another, has been shown to have a coefficient one thousand times greater than that for vertical diffusion, which is the process involved in the formation and modification of air masses. Without the great importance that must therefore be ascribed to horizontal turbulence, the climatic zones as we know them would not exist, and the tropical and equatorial latitudes would become progressively hotter, the polar latitudes colder.

Turbulence investigations have shown that the standard method of measuring evaporation, by means of the net water loss from large open pans or small exposed dishes (*evaporimeters*), is of doubtful meteorological value. The data obtained from

---

* On the measurement of eddies by anemometers at fixed points at Porton Down, Wiltshire, Eng., and their display on a cathode-ray tube screen, see J. I. P. Jones, *JAM*, 5(1966), 816.

such measurements are rarely consistent, and turbulence theory explains why. The theory furthermore was used to develop the Thornthwaite-Holzman formula for evaporation, which enables the rate of evaporation to be determined indirectly from measurements of wind velocity and specific humidity (both susceptible to high accuracy of measurement) at different heights, one above the other, and so does away with the necessity for evaporation pans and their attendant difficulties.

The concepts of turbulent diffusion have been used to develop formulae that enable the patterns of pollution from one or several sources to be predicted. Some of these will be considered in more detail later. They are valuable geographically, since, by comparing actual pollution patterns with those predicted theoretically, it is possible to discover the effect of topography, in its broadest sense, on the distribution of pollution particles. On a larger scale, turbulence concepts enable the approach of the wind in the friction layer of the atmosphere near the ground to the geostrophic wind to be explained quantitatively.

It was recognized by Defant in 1921 that depressions, anticyclones, and other synoptic disturbances could be regarded as eddies forming in the general airflow. Lettau, adopting the eddy concept, found that the mixing length for the meridional motion of these synoptic "eddies" was on the order of several degrees of longitude.

From another point of view, not involving the eddy concept, it has been possible to compute the meridional flux of heat, water vapor, and momentum from upper-air data alone. Such computation has shown that the amount of heat transported by turbulent eddies is of the required magnitude to equalize heat losses and gains in the various climatic zones. Far more sensible heat, and latent heat in the form of water vapor, is transported zonally than is transported meridionally, and surface friction does not reduce the zonal transport to any extent. In other words, the zonal stresses that result from turbulence in the deep meridional currents are able to maintain the wide zonal circulations against the effects of surface friction.[2]

The maximum poleward flow of sensible heat and momentum, due to eddies, is in winter, associated with the increased pressure activity during that period. A maximum eddy flux of moisture in winter is prevented by the lower moisture content of air during that season. The vertical eddy flux shows changes of sign at certain levels. For example, over southern England, reversals occur both at the tropopause and within the troposphere. The eddy flux is northward in the lowest layers of the atmosphere at all seasons, southward in the upper troposphere, and northward again in the stratosphere. The flow of heat in latent form is insignificant above 400 mb, however, so that all appreciable flow of heat due to eddies in the atmosphere above southern England is northward.

All the above considerations illustrate the point that turbulent diffusion is a concept, rather than a phenomenon, which may be applied to a wide variety of entities in climatology that exist on very different scales. It is thus an important unifying concept. Some kind of universal law of diffusion must apply to all scales of atmospheric diffusion: the "scatter" of the particles whose distribution throughout the atmosphere is controlled by diffusion is proportional to the pth power of the distance through which they travel. This law applies to the world-wide movements of volcanic dust as readily as to the movements of smoke particles released by a garden fire. The parameter p has a numerical value varying from $\frac{7}{8}$ to 1, depending on the scale of the diffusion. The turbulent entities involved in this universal concept of diffusion are not necessarily the same as the familiar eddies of the atmospheric friction layer, and the name *turbulon* has been suggested for them.*

The exact definition of the physical counterpart of the concept of turbulence has given rise to controversy.[3] For example, not all fluctuating motions of a fluid are turbulent. An essential feature of turbulent motion is that it includes some type of stirring action that causes diffusion of physical properties. Another essential is that the fluctuating motions must be random; i.e., individual fluctuations must not be determined by the mean flow, although the average of the individual fluctuations is so determined. For this second feature to obtain, a turbulent fluid must have a chaotic vorticity. Thus a fluid is physically turbulent if it contains a stirring motion that has a chaotic distribution of vorticity. Diffusion may be shown mathematically to be a property of such a turbulent flow. An important implication of the chaotic vorticity of the fluid is that it is impossible to draw climatic mean charts that describe the spatial variations of the physical properties of the fluid with any physical validity. An essential concomitant of a turbulent fluid is that it possesses a specific direction of energy flow; this fact may be incorporated into a physical definition of turbulence, thus:

A turbulent fluid is one in which (1) *each component of vorticity is distributed irregularly and aperiodically in time and space;* (2) *energy is characteristically transferred from the larger to the smaller scales of motion; and* (3) *the mean separation of neighboring particles composing the fluid tends to increase with time.*[4]

The incorporation of the chaotic or random element in the definition has a practical as well as a theoretical importance. For example, an atmospheric phenomenon of great significance for aviation is clear-air turbulence. Not all clear-air turbulence proves to be turbulent in the physical sense: some of the "bumpiness" encountered by aircraft flying through cloudless air proves to be produced by atmospheric standing waves rather than by turbulence diffusion. The term *undulance* has been proposed for such nondiffusive turbulence. The following distinction may be made:

Turbulence describes the condition of a fluid composed of particles in a state of random motion that may be described only in probabilistic terms. Undulance describes the condition of a fluid composed of particles that are oscillating in a more or less regular manner. Turbulence is essentially diffusive, with a strong stochastic component; undulance is essentially nondiffusive, with a weak stochastic component.[5]

A simpler distinction is that turbulent motion is essentially irregular, whereas undulance consists of regular, wave-like motions. It is the association of *diffusion* with turbulent motions that is important for climatology.

Theories of Turbulence

The classical theory of turbulence is based on Reynolds' view of the physical nature of a turbulent fluid. Turbulence is defined as the difference between the actual motion and the mean motion; i.e., the orthogonal components of turbulent flow are described by

$$u' = u - \bar{u} \text{ in the east-west direction,}$$

* See the remarks by O. G. Sutton in *AFGP*, no. 19 (1952), pp. 25–27, and E. Inone, *loc. cit.*, p. 398.

$$v' = v - \bar{v} \text{ in the north-south direction,}$$

$$w' = w - \bar{w} \text{ in the vertical direction,}$$

where u, v, w, represent instantaneous velocities, $\bar{u}$, $\bar{v}$, $\bar{w}$, represent mean velocities, and u', v', w', represent deviations from the mean. These components may be incorporated into the equations of motion by introducing the effects of viscosity into the latter.*

The Eulerian equations of motion apply to an inviscid and incompressible fluid in a state of laminar flow. The effects of internal resistance to flow (i.e., viscosity) in such a fluid may be introduced by postulating the existence of a system of forces that would account for the observed effects. These forces are the orthogonal components of the *shearing stress*. In Figure 3.2, a fluid is contained between two extensive boundaries, the upper boundary moving steadily to the right, relative to the lower boundary. Observation shows that in such a situation the fluid does not remain at rest or flow uniformly, but moves in layers: the fluid particles in each layer glide over the particles of the underlying layer. Thus a shear develops within the body of the fluid, while at the boundaries the fluid particles cling to the boundary surfaces, so that there is no relative motion. A velocity gradient V/Z will be produced within the fluid, and, following Newtonian mechanics, a force (proportional to the velocity gradient per unit area of boundary surface) must exist if the fluid is to be maintained in a state of steady motion. This force is the shearing stress, described by the expression

$$\tau = \mu \frac{dv}{dz}$$

where dv/dz is the velocity gradient, μ is the coefficient of viscosity (see Appendix 3.1), and τ is the horizontal shearing stress. The latter may be defined as the force exerted (per unit area) by any given layer of particles in the fluid on the layer immediately below it. In a cube within the moving fluid, nine surface shearing stresses exist on the faces of the cube (see Figure 3.3). For example, for the face described by x = constant (shown shaded), there are three tangential stresses: τ_{xx} normal to the plane, and τ_{xy}, τ_{xz} tangential to the plane, using the notation that τ_{xx} represents the stress acting on face x = constant in the x direction, etc. In the absence of body forces, the nine stresses reduce to six independent components, defined as follows.

Three tangential stresses:

$$\tau_{xy} = \mu(\partial v/\partial x + \partial u/\partial y) = \tau_{yx},$$

$$\tau_{yz} = \mu(\partial w/\partial x + \partial v/\partial z) = \tau_{zy},$$

$$\tau_{zx} = \mu(\partial u/\partial z + \partial w/\partial x) = \tau_{xz}.$$

Three normal stresses:

$$\tau_{xx} = 2\mu(\partial u/\partial z),$$

$$\tau_{yy} = 2\mu(\partial v/\partial y),$$

$$\tau_{zz} = 2\mu(\partial w/\partial z).$$

These components of viscous stress may then be incorporated into the equations of

* See Chapter 4 for details of the Eulerian and Lagrangian equations of motion.

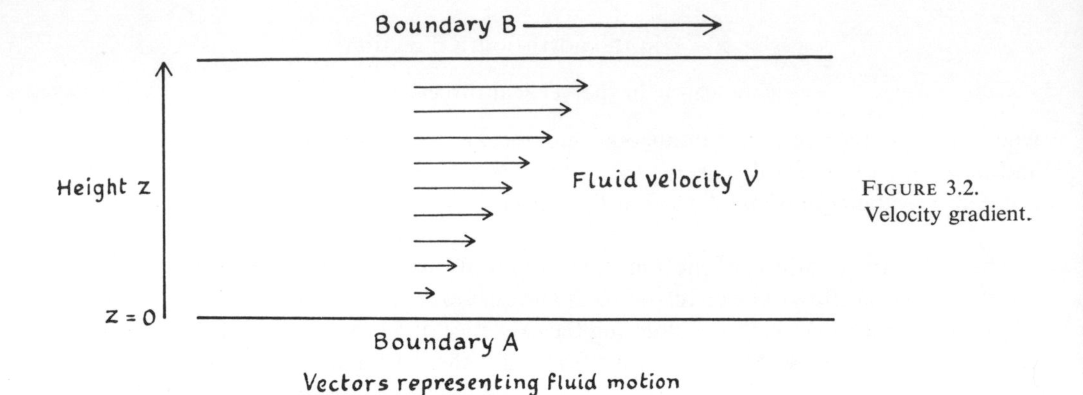

FIGURE 3.2.
Velocity gradient.

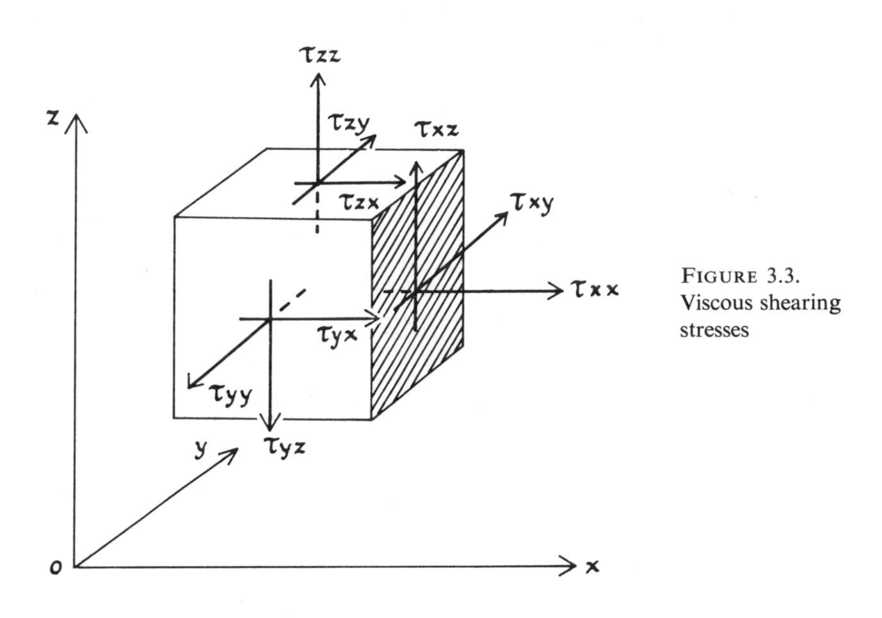

FIGURE 3.3.
Viscous shearing stresses

motion to provide the *Navier-Stokes equations of motion*. The vector form of these equations is represented by

$$\mathbf{a} = -\frac{1}{\rho}\nabla p + \mathbf{F} + \frac{\mu}{\rho}\nabla^2 v,$$

where $\mathbf{a}$ is the acceleration vector, $\mathbf{F}$ is the body force (per unit mass) due to gravity, ρ is the density and p the pressure of the fluid, v is its velocity, and ∇, ∇^2, represent the del and Laplacian operators, respectively. The complete form of the equations may be written, for the OX direction, in the form

$$\rho\frac{\partial u}{\partial t} = \frac{\partial}{\partial x}(\tau_{xx} - \rho u^2) + \frac{\partial}{\partial y}(\tau_{xy} - \rho uv) + \frac{\partial}{\partial z}(\tau_{xz} - \rho uw), \quad [3.1]$$

with analogous equations for the OY and OZ directions. These equations still refer to laminar flow. If they are made applicable to turbulent flow by replacing u, v, and w by $\bar{u} + u'$, $\bar{v} + v'$ and $\bar{w} + w'$, respectively, three new equations of motion result. The new equation for the OX direction may be written as

$$\rho \frac{\partial u}{\partial t} = \frac{\partial}{\partial x}(\bar{\tau}_{xx} - \rho \bar{u}^2 - \rho \overline{u'^2}) + \frac{\partial}{\partial y}(\bar{\tau}_{xy} - \rho \bar{u}\bar{v} - \rho \overline{u'v'}) + \frac{\partial}{\partial z}(\bar{\tau}_{xz} - \rho \bar{u}\bar{w} - \rho \overline{u'w'}).$$

Comparison between this equation and Equation 3.1 reveals that, in order to take into account turbulent motion, the viscous stress τ_{xx} of the laminar case is replaced by $\bar{\tau}_{xx} - \rho \overline{u'^2}$, and so on. In other words, the effect of turbulence is to introduce new terms in the equations of motion in addition to the viscous terms. These terms ($-\rho \overline{u'^2}$, $-\rho \overline{u'v'}$, $-\rho \overline{u'w'}$) are called the *Reynolds' stresses*; they depend on the magnitude of the velocity fluctuations from the mean, and not on the actual velocities. Since the Reynolds' stresses, for atmospheric motions, prove to be many thousands of times as large as the laminar stresses τ_{xx}, etc., the latter may usually be neglected.[6]

It should be noted that the Navier-Stokes equations as adapted for turbulent flow are essentially climatological descriptions of the balance of forces in a turbulent air current, because they are expressed in terms of mean velocities and deviations from the means. The equations cannot be used directly, because they contain six new unknown quantities that must be expressed in terms of the mean velocity components if the equations are to be employed to solve actual problems. Nevertheless, they are very useful as indicators that velocity fluctuations—like molecular agitations in the general theory of gases—are responsible for transporting momentum across a surface in the fluid. Consequently, the Navier-Stokes equations for turbulent flow are important tools for the study of momentum flow in the atmosphere and across the air/Earth interface. They show, for example, that vertical turbulence is responsible for the downward transfer of horizontal motion in the atmosphere.[7]

By assuming that the effects produced by turbulence, as described by Reynolds' modification of the Navier-Stokes equations, may be ascribed to the movement of discrete masses of fluid termed *eddies*, the equations of motion reduce to the single equation

$$\frac{\partial \tau}{\partial z} = \frac{\partial \bar{p}}{\partial x},$$

where $\bar{p}$ is the mean atmospheric pressure. This theory, devised by Prandtl, assumes that momentum is conserved during the motion of an eddy, and that the rate of transfer of momentum is not affected by pressure fluctuations. A parameter is then defined, the *mixing length*, which is a unique length characterizing the local intensity of turbulence at any level in the fluid. Unlike the "mean free path" in the kinetic theory of gases, the mixing length may be a function of position in the fluid and of the latter's mean velocity. By means of the mixing length, the basic parameters of the eddy theory (i.e., the *eddy shearing stress* and the *eddy viscosity*) may be expressed in terms of mean velocities (see Appendix 3.2). The mixing length is a convenient and useful parameter, but it has no precise physical significance and may be influenced by nonturbulent effects—for example, under steady-state diabatic conditions, it is affected by convective energy changes.[8]

An alternative to the mixing-length theory, also designed to overcome the mathematical difficulties of the description of turbulence provided by the equations of motion, is that of *exchange coefficients* (also called *Austausch* coefficients). It is also based on analogy with the kinetic theory of gases, and introduces the concepts of *virtual coefficients* of viscosity, conductivity, and diffusivity. The exchange-coefficient hypothesis states that the amount of any transferable conservative physical entity (e.g., heat, momentum, or water) that is transported in a unit time through a unit cross section of a plane parallel to $z = 0$ can be expressed as the product of a virtual coefficient of diffusion and the gradient of the mean entity (see Appendix 3.3). The advantage of this hypothesis is that it readily lends itself to verification by observation, although very precise measurements are required. Examples of its application may be derived as follows. The amount of any entity transported vertically in a unit time through a unit area of the horizontal plane is $\rho\overline{w's'}$, where ρ represents density, w represents vertical velocity, and s represents the amount of the entity present in a unit mass of air. Primes denote departures from the mean, and the bar denotes a mean quantity. The expression $\rho\overline{w's'}$ represents the flux of the quantity through a unit horizontal area in a unit time. Therefore:

(a) Momentum flux $= -\rho\overline{w'u'}$, the minus sign denoting a downward transport of momentum. This momentum flux is the same as the Reynolds stress, from the definition of the latter. According to the exchange-coefficient hypothesis,

$$\tau = \rho K_M \frac{\partial U}{\partial z},$$

where τ is the Reynolds stress, $\frac{\partial U}{\partial z}$ is the vertical gradient of velocity U, and K_M is the virtual coefficient of viscosity, renamed the *eddy viscosity* by Taylor.

(b) Heat flux $= c_p\rho\overline{w'T'}$, where c_p is the specific heat of the air at constant pressure, and T represents the absolute temperature of the air. According to the exchange-coefficient hypothesis,

$$H = -c_p\rho K_H \frac{\partial \theta}{\partial z},$$

where H is the heat flux, $\frac{\partial \theta}{\partial z}$ is the vertical gradient of potential temperature θ, and K_H is the virtual coefficient of conductivity, i.e., the *eddy conductivity*.

(c) Water-vapor flux $= \rho\overline{w'q'}$, where q represents the specific humidity of the air. According to the exchange-coefficient hypothesis,

$$E = -\rho K_E \frac{\partial q}{\partial z},$$

where E is the water-vapor flux (i.e., the evaporation), $\frac{\partial q}{\partial z}$ is the vertical gradient of specific humidity, and K_E is the virtual coefficient of diffusivity, i.e., the *eddy diffusivity*.

The mixing-length and exchange-coefficient hypotheses may be combined, but for the combination to be realistic, stability parameters must be introduced. One of these stability parameters, the Richardson number *Ri*, proves to be very useful in describing a criterion that is necessary if natural rather than wind-tunnel turbulence is to be described accurately. This criterion states that turbulence will increase or die away depending on whether *Ri* is less or greater than the ratio of K_M to K_H (see Appendix 3.4). In some cases—for example, momentum transfer in the lower layers of the atmosphere—K_M and K_H may be assumed to be equal, in which case the critical value of *Ri* is obviously unity. In other cases—for example, strongly convective situations in which K_H increases rapidly with height—it may not be assumed that $K_M = K_H$.

In general, air above a surface whose hottests parts do not coincide with its dampest parts will be characterized by inequality of K_H and K_W, and buoyancy forces will develop if K_H is greater than K_W. If the hottest and the wettest portions of the surface coincide, then $K_H = K_W$ and heat and water vapor will be transported upward together. These principles have many geographical applications at all scales.

The combination of mixing-length and exchange-coefficient hypotheses was largely due to Taylor's classic investigations aboard the whaler *Scotia* in the Grand Banks region off the coast of Newfoundland. Measurements made by means of kites flying a few hundred feet above the sea surface showed that the air temperature was usually higher at that level than at sea level in summer, because of the rapid cooling of warm westerly winds from the North American continent as they passed over the iceberg-ridden sea. A wave of cooling must therefore be set up at the air-sea interface, and the observed rate of penetration of this wave into the atmosphere proved to be explainable by a transfer process of the same nature as molecular conductivity, but much more vigorous. Taylor therefore introduced in 1913 the concept of a *mixture length*, i.e., a ratio of diffusivity to vertical turbulent velocity, which determined a coefficient of turbulent diffusion for heat, and which could be measured. Prandtl independently introduced in 1920 the idea of the mixing length, and assumed that turbulent fluctuations of wind velocity at any point depend only on the mean rate of wind shear and the mixing length. For the Prandtl theory to be valid, it was necessary also to assume that the mass of fluid, i.e., the *austausch*, which carries the physical property being transferred, is completely unaffected by molecular diffusivity or fluctuations in pressure gradient during the process. The difficulty with the mixing-length theory was in explaining the process by which the austausch transferred its property or properties to the surrounding fluid at the end of the mixing length. Taylor circumvented this difficulty by introducing a Lagrangian model of turbulence in which the diffusive properties of turbulence are described by determining the correlation between the velocity of a particle of diffusing matter at one instant and its velocity at a later time. The model involves a particle of diffusing material that is assumed to retain its identity as it moves through a turbulent environment. By adopting this model, Taylor was able to show that the pattern of diffusion may be described from a knowledge of (a) the Lagrangian time-correlation function, and (b) the mean-square velocity of the turbulent fluctuations. The time-correlation gradually decreases as time increases, and ultimately vanishes; diffusion for longer periods may be described in terms of a constant virtual coefficient, i.e., in terms of one of the eddy coefficients already mentioned. Thus in the Taylor hypothesis, the mixture length may be defined as a length that is proportional to the average distance a diffusing particle moves before its velocity becomes uncorrelated with its initial velocity.[9]

Taylor's hypothesis provides the greater part of the working concepts in contemporary turbulence theory, but before we describe it further, we must note an alternative to the mixing-length and exchange-coefficient hypotheses that has considerable practical application, the *Fickian theory of diffusion*, which is an application of Fourier's law of heat conduction to molecular diffusion (see Appendix 3.5). By analogy, von Fick[10] showed the process of molecular diffusion may be represented by a model described by the equation

$$\nabla^2 \chi = \frac{1}{d}\frac{\partial \chi}{\partial t},$$

in which χ represents the concentration of diffusing material (in say, grams per cubic centimeter of air), d represents the molecular diffusivity, and ∇^2 is the Laplacian operator. Similar types of diffusion equation were derived by Taylor and by Schmidt, who showed that

$$\frac{d\chi}{dt} = \frac{\partial}{\partial x}\left(K_x \frac{\partial \chi}{\partial x}\right) + \frac{\partial}{\partial y}\left(K_y \frac{\partial \chi}{\partial y}\right) + \frac{\partial}{\partial z}\left(K_z \frac{\partial \chi}{\partial z}\right),$$

in which $\frac{d\chi}{dt}$ represents the total instantaneous rate of change of concentration χ with time, $\frac{\partial}{\partial x}$, $\frac{\partial}{\partial y}$ and $\frac{\partial}{\partial z}$ represent partial rates of change in the OX, OY, and OZ directions, respectively; and K_x, K_y, and K_z represent the coefficients of diffusion in the OX, OY, and OZ directions, respectively. The Fickian theory (usually called the K theory) was applied to meteorology by Roberts, Calder, and Deacon. One great difficulty in the meteorological case, pointed out by Richardson in 1926, was that the coefficient of diffusion, K, depends on the scale of the phenomena being investigated. For example, K has the value 0.2 cm^2 per sec for molecular diffusion, and the value 10^{11} cm^2 per sec for large-scale diffusion due to anticyclones and cyclones.[11]

The combination of the K theory with turbulence concepts was largely carried out by Sutton, who developed the ideas presented by Taylor in his *statistical theory* of turbulence. The statistical theory is essentially a continuous-variation theory of turbulent diffusion, in which the velocity of the air particles is regarded as varying continuously with time along the path of each particle. The sizes of the eddies may then be expressed in terms of the differences in velocity between one point and another in the atmosphere. Space-correlation coefficients are employed to define a parameter termed the *scale* of the turbulence. The scale parameter is a measure of the average size of the eddies producing the turbulent flow; it is analogous to the mixing length, but is independent of any model. Since application of the theory shows that the scale parameter has no upper limit in the atmosphere, all eddies, whatever their size, must be taken into account if a complete description of the restless atmosphere is to be provided. The *microscale* of turbulence is a length parameter that describes the average size of the small eddies that bring about most of the energy dissipation effected by turbulence (see Appendix 3.6).[12]

The statistical theory also makes use of time-correlation coefficients, i.e., *autocorrelation coefficients*, defined as the coefficients of correlation between eddy velocities at a fixed point at regular intervals of time (see Appendix 3.7). Assuming that the pattern of turbulence does not change with time, the diffusion is then completely specified by the autocorrelation coefficient and the mean energy of the eddying motion. Taylor introduced an important relationship between the autocorrelation coefficient and the *spectrum of turbulence*, i.e., the total range of all possible variations in velocity at a fixed point in the atmosphere. This relationship states in its simplest form that the autocorrelation function (i.e., the curve describing the variation in autocorrelation coefficients at a point for all eddy sizes) and the spectrum function (giving the fraction of the total energy associated with eddies of one size) are Fourier transforms of one another. Therefore, if the autocorrelation function is known, the spectrum function can be found, and vice versa (see Appendix 3.8). In practice, observations of wind-velocity fluctuations at a point are usually used to compute the autocorrelation function, which is then transformed to give the spectrum function. A graph is plotted of spectrum function against eddy frequency, resulting in a power-spectrum curve that indicates which range of eddy sizes contain the most energy, i.e., which components of the observed wind fluctuation are the most significant for turbulent diffusion.

As with power-spectrum analysis of time-series of climatic data, the processing of observations to yield a spectrum curve is quite complex, but it is necessary if the full range of eddy sizes at the point of observation is to be determined, since the actual observations present not a complete picture of the velocity fluctuations but a distorted one; i.e., the real fluctuations are seen through a *spectral window*. Each individual observation, however fine the resolving power of the measuring instrument, is of necessity an average during some period of time, and the lag or inertia of the instrument is always an important influence.

The application of the Taylor theory enables statistics to be compiled that describe the degree of turbulence associated with various scales of motion at fixed locations. If these statistical properties prove to be unaffected by changes in the axes of reference, i.e., if $u'^2 = v'^2 = w'^2$, the turbulence is said to be *isotropic*. In the actual atmosphere, isotropy is usually local; i.e., isotropic turbulence only occurs in small-scale eddies. This state of affairs is described in *Kolmogoroff's similarity theory*,[13] which says in essence that all turbulent motions, whatever their mode of origin and the form of their mean flow, possess local isotropy. The theory in effect describes a model that may be represented by a graph of eddy energy against frequency of wave-number. In this model, energy is transferred down the scale of eddies, i.e., from larger to smaller eddies, never the reverse. According to the similarity theory, during this process of energy transfer the eddies successively lose their directional preference, so that ultimately a stage is reached beyond which the motion of smaller eddies will be isotropic. The stage occurs in the *equilibrium range* in Figure 3.4. The properties of

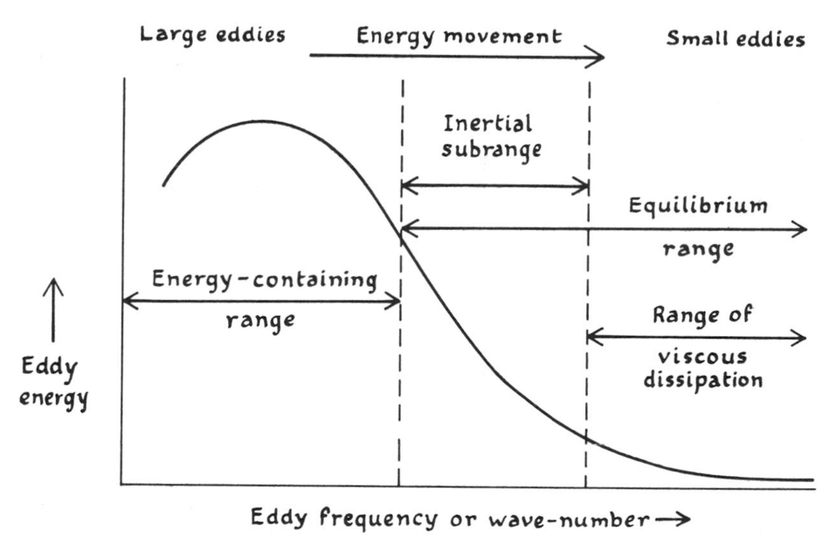

FIGURE 3.4.
Kolmogoroff's similarity theory: energy spectrum of turbulence.

eddies within this range depend on the energy they receive from the mean flow (which is ultimately dissipated by viscous action) and on the viscosity of the air. The Kolmogoroff theory states that the average properties of the eddies within the equilibrium range depend only on the viscosity and the rate of viscous dissipation per unit mass of air. Furthermore, it states that the average properties of the larger eddies within the equilibrium range, i.e., the eddies within the *inertial subrange*, are determined only by the rate of viscous dissipation of energy. Assuming equivalence in the time- and space-intensities of the turbulence, the theory enables the energy spectrum of eddies within the inertial subrange to be readily found.[14]

The statistical view of turbulent diffusion is very important, both theoretically and practically. It shows that the study of the spread of a cluster of particles involves more than the time-variations in velocity at a fixed point (i.e., the Eulerian approach) or the following of the motion of one of the particles (i.e., the Lagrangian approach). Instead, a new system of covariances must be used, expressed in the form of covariance spectra that relate the fluctuations undergone by two particles both at different times *and* simultaneously.

Sutton's development of the statistical theory of turbulence has resulted in the introduction of various formulae for the distribution of particles that are released from a certain type of source. These formulae may be applied to either isotropic or nonisotropic turbulence, and they describe the degree of dispersion of the particles after a given time in terms of a virtual coefficient of diffusion C and a stability parameter n (see Appendix 3.9). Different formulae must be used for an instantaneous point-source of particles (e.g., a sudden explosion at a point on or above the Earth's surface), a continuous point-source (e.g., the smoke coming from a smokestack), a continuous line-source (e.g., a line of smokestacks), an instantaneous volume-source (e.g., a large-scale explosion due to an accident at a nuclear-power station), an elevated line-source, or an elevated point-source (see Appendix 3.10).[15] Observation shows that the parameter C may take on a wide variety of values, and also indicates that the Sutton theory does not work so well near the surface of the sea—probably because of interface effects—or in wind-tunnel experiments.[16]

An alternative to the Sutton theory for the dispersion of large particles in the atmosphere is that devised by Rounds. Important divergences from the Sutton theory occur if it is applied indiscriminately to both particulate diffusions and gaseous diffusion. In the diffusion of heavy particles in the atmosphere, the relatively rapid travel of the particles across the trajectories of the eddies may result in appreciable reductions in the rate of dispersal of the particles.[17]

There is an extensive literature dealing with the experimental testing of the above-mentioned theories of turbulence. The investigations confirm the validity of the theories, and also indicate that, given the wide range of atmospheric phenomena grouped under the general term "eddy," it is advisable to recognize at least two fields of study: macroturbulence and microturbulence. They show that Eulerian variations in wind velocity are much easier to measure than Lagrangian variations, although a simple relationship between the two does seem to hold. Although observations show that there is an equivalence of space-variations (i.e., velocity fluctuations at a fixed time) and time-variations (i.e., velocity fluctuations at a fixed point) through the simple transformation $x = \bar{u}t$ in wind-tunnel measurements, it is very difficult to ascertain whether or not it applies in the atmosphere.[18]

Macroturbulence in Theory and Practice

Macroturbulence investigations are simpler than microturbulence studies, in the sense that they usually do not require special instrumentation but may make direct use of published synoptic data. Nevertheless, some observational programs have proved especially useful for macroturbulence determinations, transosonde balloon flights, for one. Statistics derived from the latter show that macroscale eddies with a period of about two days are prominent at 300 mb, corresponding to the average transit time of the transosonde through the average pattern of long-waves in the westerlies.[19]

Area-means appear to be more useful than time-means in the study of macroturbulence, because they give a better measure of the deviation of an atmospheric variable from the mean in terms of the mean itself. The "scale" of macroturbulent motion may be defined by performing a Fourier analysis on the planetary-size wind systems, from which equations may be derived for each scale of motion that governs their behaviour. The equations enable the rate of change of kinetic and available-potential energy in a disturbance of a given wave-number to be determined. The large-scale eddies in the atmosphere are shown to produce the kinetic energy of the mean zonal motion (i.e., they do not derive their energy from the latter). This contrasts with the old idea of the general circulation, in which the mean zonal motions were believed to be maintained by the mean meridional circulations. The production of kinetic energy by the large-scale disturbances of the global circulation is sufficient to completely regenerate the latter in about two weeks; at the macroscale, the rate of kinetic-energy generation is greater than the amount dissipated by turbulence.[20]

Large-scale synoptic disturbances (i.e., cyclones and anticyclones) in the upper troposphere appear to behave like isotropic "eddies" for harmonics with wavelengths between 60° and 20° in longitude, according to data for latitudes 20°N and 70°N. The large-scale motions of the atmosphere in general do not satisfy the conditions for homogeneous isotropic turbulence (see Appendix 3.11), and the departure from homogeneity is most marked for vertical turbulent fluctuations, but for many climatological purposes one can assume that synoptic-scale disturbances are two-dimensional isotropic turbulence. The spectra of this turbulence have received a great deal of attention. At 300 mb, for example, the spectrum of the mean seasonal circulation pattern dominates at low wave-numbers, but at higher wave-numbers the mean spectrum of the perturbations of the mean flow is more important. Power spectra for the disturbances at 200 mb resemble those at 300 mb in general, but the discontinuity introduced by the tropopause results in the cospectra of meridional winds with zonal winds and dry enthalpy being very different at the two levels. Detailed computations of the spectra at 500 mb for January 1949 show that disturbances of wave-number three were very significant as sources of the energy of the mean zonal circulation during this period. On only three out of the 31 days in this month did the disturbances of wave-number three draw energy from the mean motion; on the remaining days, the momentum carried by these disturbances was injected into those parts of the zonal flow where wind velocities were greatest. A spectrum analysis of the variation in daily hemispheric meridional circulation index at 500 mb, during March 1, 1959, to July 12, 1960, indicates that the maximum energy concentrations are at higher frequencies at 60°N than at 40°N, and at higher frequencies again between

5°N and 25°N (see Appendix 3.12). The concentrations of energy in the 500 mb eddies are greatest in early summer and relatively weak in winter, and the frequencies associated with the highest energy levels shift with the seasons: high energy content is associated with the low-frequency fluctuations in summer, but with the shorter wavelengths in winter. At 850 mb over the southeastern United States, most of the large-scale turbulent transfer is effected by disturbances with periods of about four days.[21]

In general, spectral analyses of large-scale disturbances reveal large spatial and temporal variations in the spectra. For example, a case study based on data for each day of January 1951 found that in an 11-wave circulation pattern around latitude 55°N, there was no evidence of a simple transfer of kinetic energy from low to high wave-number (i.e., in support of the Kolmogoroff theory) or in the reverse direction; instead the processes of energy transfer were much more complex. Although zonal motion was relatively simple, with the maximum kinetic-energy concentrations in the very long planetary waves, the meridional motion revealed one or more kinetic energy maxima between wave-numbers two and eight, and a decrease in energy at wave-number one and at those above eight.[22]

Spectral studies of large-scale distrubances are very relevant to the explanation of regional differences in the general atmospheric circulation, and hence to the explanation of the genesis of climatic types on the Earth. Implicit in the classic idea of the general circulation is the concept of a mean climatological state, which varies geographically and seasonally, and on which are superimposed the "eddies" of synoptic-scale turbulence. This concept implies that the observed differences in weather at some one place are "random" in the turbulence sense; i.e., they represent sampling errors introduced by temporal differences in the phasing and intensity of the large-scale disturbances passing over that place. However, this is an incomplete picture, because it does not take into account possible changes in the mean climatological state, i.e., climatic fluctuation.[23] Despite the incompleteness of the picture, the turbulence view of the general circulation proves to be very useful in the interpretation of situations that are themselves significant in climatic fluctuation. Blocking patterns are good examples of such situations, and variations in the extent of blocking action may be related to seasonal variations in lateral mixing and in the rate of northward heat transport across the westerlies. Blocking, lateral mixing, and northward heat transport all appear to be greater during the winters of years with very strong north-south temperature gradients than during those of years with more normal gradients. By regarding a period of abnormal northward or southward flow as a "surge" or turbulent eddy moving in a meridional direction, a coefficient of *lateral eddy viscosity* may be defined (see Appendix 3.13). The mean mixing length for these lateral eddies is on the order of 8.2×10^7 cm; their over-all mean period is about six days; and their most frequent period is about three days, the latter being the average interval between the passage of the major troughs in the upper westerlies, each of which contains a family of cyclones. The basic eddy period shows some geographical pattern, varying from 2.5 days in high latitudes, and 3 days in the heart of the westerlies, to 3.5 days in low latitudes.

The theoretical explanation of the general circulation as a macroturbulence concept relies mainly on austausch and mixing-length concepts. Lettau introduced coefficients of zonal and meridional exchange that enable the usual austausch coefficients to be modified to take into account such synoptic variables as mean air temperature, mean air pressure, latitude, and longitude.[24] In contrast to characteristic values of about 10^2 cm and $1-10^2$ grams per cm per sec for mixing lengths and exchange coefficients of microturbulence, the characteristic values in macroturbulence are about 10^8 cm and 10^7–10^8 grams per cm per sec.

The key to the turbulence concept of the general circulation is the movement of momentum.[25] As is explained in Chapter 2, friction between the winds and the surface of the Earth extracts westerly momentum from the surface westerlies in middle and high latitudes, and injects westerly momentum into the surface easterlies. Medium-scale convective turbulence is responsible for the downward movement of westerly momentum from the upper westerlies to replace that lost to the Earth's surface by friction; consequently, there is no need for an appreciable flow of air across the isobars in the lower troposphere in middle and high latitudes. However, a cross-isobaric flow toward the equator is necessary in tropical latitudes, because the downward flow of westerly momentum brought about by turbulence in the upper westerlies that overlie the trade winds must cease at the level of maximum speed of the latter; below this level, an upward transfer of westerly momentum must occur, to explain the presence of the surface easterlies. Consequently, westerly momentum must accumulate in the lower levels of the trades, which requires an equatorward (i.e., cross-isobar) flow to maintain the easterly momentum of the latter. Large-scale turbulence (i.e., the movement of long waves in the westerlies and upper cold lows) is responsible for transferring westerly momentum from the tropical upper westerlies to the upper troposphere of middle latitudes to replace that lost to the lower troposphere and the Earth's surface in those latitudes by friction.

Despite the prominence of large-scale and medium-scale turbulence in the momentum-transfer concept of the general circulation just described, small-scale turbulence is very important. In fact, the ultimate source of the energy for all the motions included in the general circulation of the atmosphere is the heat that has been pumped into the air from the Earth's surface by small-scale turbulence. The amount of energy provided in this way decreases with increasing latitude. Therefore, according to the turbulence view of the circulation, a meridional temperature gradient must be created. The planetary wind systems must arrange their internal dynamics in accordance with this gradient, i.e., through an increase in the strength of westerly (or a decrease in the strength of easterly) winds with height up to the tropopause. This interpretation of the circulation in terms of turbulence does not apply to air movements above the tropopause; although the total kinetic energy dissipated by turbulence within the mesosphere is on the same order as that in the troposphere, the restriction is justified in that about 90 per cent of the turbulent energy in the atmosphere dissipates in its lowest kilometer.[26]

The problem of global-scale diffusion is one in macroturbulence, and must be approached with other than austausch or mixing-length concepts. By means of probability concepts, it is possible to produce maps (see Figure 3.5) that indicate the most likely area to find particles that have been released into the general atmospheric circulation at a certain time and place.[27]

Microturbulence in Theory and Practice

Experimental verification of turbulence theories for microscale diffusion requires very precise measurements, usually necessitating special instrumentation, especially when vertical profiles of wind, temperature, or humidity are being observed. In such investigations it is usual to focus attention on the *differences* between the values

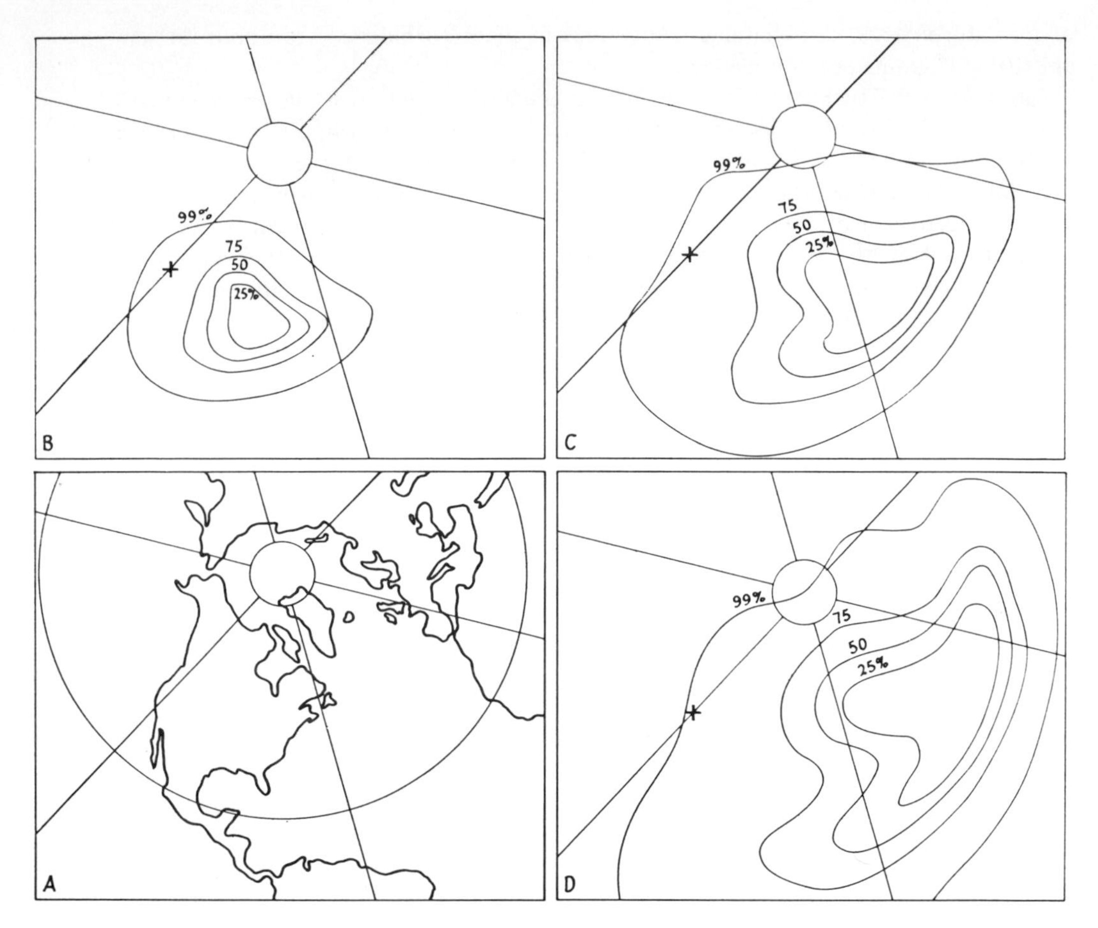

FIGURE 3.5.
Macrodiffusion probability fields (after S. B. Solot and E. M. Darling, Jr.).
A, Key map. Maps B, C, D, depict probability that a particle released from the point marked by a cross will be within a given area: B, after one day; C, after two days; D, after three days.
Maps B, C, D, refer to the 50,000-foot level, in January.

measured at successive heights rather than on the values themselves. The precise requirements of microdiffusion studies have resulted in refined types of sampling grid being employed for observational programs, and the difficulty of measuring microscale and mesoscale turbulence spectra in the free atmosphere has led to an ingenious radar-anemometer principle. Climatic data derived from measurements with a conventional anemometer must be corrected to take into account the fact that the anemometer measures velocities projected on to a horizontal plane, independently of their direction in that plane, if the data are to be of any use to the microclimatologist investigating frequency distributions of air motion in the turbulent lower levels of the atmosphere. Any automatic system for computing wind-direction statistics that may show evidence of microturbulent fluctuations must take into account the mathematical discontinuity that exists at true north in the 360° wind-direction notation.[28]

Spectra of microfluctuations of wind are produced by making power-spectrum analyses of very precise series of observations. The investigations are complicated by the wide range of eddy sizes at the microscale, and also by the thermal stratification of the lower layers of the atmosphere, which gives rise to buoyancy forces that may augment or suppress the turbulent vertical movements it is desired to measure. Various experimental techniques have been used to determine the Eulerian spectra (i.e., the spectrum of microfluctuations at a fixed point), for example, bidirectional wind vanes, tethered balloons, and hot-wire anemometers. The Lagrangian spectra (i.e., the spectra obtained from measurements made while following the air motion) are much more difficult to determine, but have been produced from records of the motion of soap bubbles released from a generator carried aloft by a captive balloon, and from fixes taken every ten seconds or so on balloons inflated to approximately neutral state and then released from a tower.[29]

Classic observations made by Taylor in 1927 at a height of 130 feet above the ground suggested that microfluctuations of wind not too close to the ground represent isotropic turbulence. However, observations made over gently undulating wooded country on Long Island, New York, at Brookhaven at a height of 91 m showed that the horizontal components of the turbulence were substantially stronger than the vertical component in the lower-frequency eddies. In general, at a point near the ground, in unstable or neutral conditions an isotropic limit may be recognized, beyond which the turbulence becomes nonisotropic. This limit occurs at a wavelength of approximately the same magnitude as the height of the observing point above the ground. The Brookhaven observations also indicated a "gap" in the spectrum near periods of about one hour, which has been taken to represent the natural division of microscale from mesoscale turbulence. However, other investigators did not find evidence for this gap in the spectrum of surface winds.[30]

Scrase found in 1930 that the eddy velocities in both horizontal and vertical directions in the lowest 20 m of the atmosphere over open downland are closely proportional to the mean horizontal velocity of the wind at speeds of more than 3 m per sec under negligible buoyancy conditions. The eddy velocity components were unequal at all heights, but the inequality decreased as the ground was approached. Observations by Giblett in 1932 at Cardington proved that, depending on stability conditions, four types of microturbulence could be recognized at a height of 50 feet above ground level. The amplitudes of these fluctuations increased as the air became more unstable, and their frequencies decreased. The fluctuations did, in fact, represent a portion of the microscale and mesoscale spectrum, the end members of which are, respectively, very small-scale mechanical fluctuations of wind near the ground under nocturnal cooling conditions, and the large-scale convective gusts associated with cumulonimbus clouds. Best found that, in general, and fairly close to the ground, the influence of thermal stability on microturbulence decreases as windspeed increases and as height decreases. Stability tends to decrease, and instability tends to increase, the degree of intensity of the turbulence, particularly in light winds.[31]

According to observations made by Smith at Cardington at 300 m, a decrease in the scale of microturbulence is associated with an increase in the vertical component of intensity of turbulence, although Webb showed that, over flat grassland in Victoria, Australia, scale first increased with a decreasing value of the Richardson number, and then decreased for values of $Ri < -0.05$. The type of convection, i.e., predominantly free or predominantly forced, must therefore have an effect on the size of eddies developing in the lower layers of the atmosphere. Observations from Australia, Brookhaven, and Nebraska, ranging over heights from 1.5 to 91 m above the ground, show that in general the vertical component of the intensity of turbulence is proportional to mean horizontal windspeed near the ground in both stable and unstable conditions, equally by day or by night. By contrast, the Cardington observations (over the range 150–1,000 m) indicate that the vertical component of the turbulence intensity is not dependent on the mean horizontal windspeed when there is vigorous convection.[32]

As a general approximation to conditions over a level site, the vertical component of turbulence intensity is independent of height above the ground in the lowest 100 m of the atmosphere, provided the stability is neutral. As stability decreases, the intensity of the vertical component progressively increases with height; and as stability increases, the intensity decreases with height within the same height range. The intensity of both lateral and longitudinal components of turbulence rapidly decreases with height under stable conditions, but in convective situations their intensities are independent of height in the lowest 30 m or more of the atmosphere. The lateral components differ at different places, corresponding to differences in the aerodynamic roughness of the topography.[33]

There is some evidence that eddy sizes are determined by the degree of temperature fluctuation. For example, Nebraska data show that, assuming a logarithmic vertical distribution of air temperature, two-thirds of the eddies observed at a height of 2 m above the ground had vertical extents of about 0.5 to 7.5 m on the average, but that they rapidly decreased in size as stability increased. Furthermore, the average size of the eddies increased with height under stable conditions. In inversion situations the increase in eddy size with height was more rapid than the increase in the log of the height; under lapse-rate conditions, the increase in eddy size with height was less rapid than the increase in the log of height with increasing distance above the ground. In general, thermal effects appear to result in many spectra for vertical wind fluctuations that show two peaks, which represent mechanical and thermal inducement of turbulence.[34]

The complete physical description and explanation of the spectra revealed by microscale atmospheric fluctuations is very mathematical, and involves the details of boundary-layer theory as applied to the air layers near the ground.[35] For the climatologist, the important point is that the physical state of the atmospheric boundary layer in any locality may be described in terms of four groups of nondimensional (i.e., universal) parameters. These parameters include: (a) numerical or physical constants, such as the Prandtl number; (b) ratios of physical quantities or forces, such as the Richardson number and the drag coefficient; (c) exponents or logarithmic derivatives of height, such as the β parameter in the Deacon wind profile; and (d) autocorrelation coefficients and other statistical parameters defined in the statistical theory of turbulence (see Appendix 3.14). Consequently, no microclimatic description of a locality can be complete unless it indicates the usual range of values taken there by these parameters. Use of the parameters achieves considerable economy in description, and lays a foundation for physical understanding. If certain assumptions are made about the form of the variation with time of a microclimatic element at some point, it is possible, by introducing the appropriate local values of the boundary-layer parameters, to predict theoretically the thermal or other microclimate in a given area.[36]

Boundary-layer theory is very useful in the prediction of theoretical vertical profiles of wind, temperature, and humidity, which enable physically sound interpolations and extrapolations to be based on precise but sparse measurements. When a turbulent stream flows over a smooth surface, boundary-layer theory recognizes that three distinct layers may exist within the fluid: (a) a *laminar sublayer* immediately adjacent to the surface, in which the shearing stress is mainly due to viscosity; (b) a *turbulent boundary layer* immediately above the laminar layer, in which the Reynolds stress is at least as strong as the stress due to viscosity; and (c) the main body of the fluid above the boundary layer, in which viscosity is negligible.[37] If the surface is too irregular to permit the existence of the laminar sublayer, as most parts of the Earth's surface are, then it is said to be an *aerodynamically rough* surface. The distribution of velocity with height within a fluid moving over such a surface is described by

$$\frac{\bar{u}}{u_*} = \frac{1}{k} \ln \frac{z}{z_0}, \qquad [3.2]$$

in which $\bar{u}$ is the mean velocity at a height z above the surface; u_* is a reference velocity, the *friction velocity* (see Appendix 3.15); k is von Kármán's constant; and z_0 is a constant for the surface, called the *roughness length*. The roughness length specifies the effect of the irregularities of the surface on the mean flow of the fluid. Equation 3.2 is valid only for a shallow layer of the fluid, i.e., for a layer in which the effect of the irregularities is felt equally at all distances above the surface. For deeper layers, different expressions must be used (see Appendix 3.16). The main difficulty in the atmospheric case is that dynamic and thermal factors, in addition to the purely geometric ones associated with a rough surface, must be taken into account.[38]

The friction velocity of the lower layer of the atmosphere is related to the density of the air and to the surface drag exerted by the underlying topography.[39] In turn, the surface may be described in terms of the drag coefficient (see Appendix 3.17), which ranges in value from 0.0028 to 0.008 for relatively smooth topography, and from 0.003 to 0.006 for rolling diversified topography. Over the sea, the drag coefficient depends on the windspeed, changing from 0.001 to 0.0026 at a windspeed of 6 m per sec. The roughness length of the topography naturally depends on its degree of unevenness. It varies from values of about 0.1 cm over a sea surface, through a range of about 0.25 to 0.3 cm over short grass, to 10 cm over full-grown root crops, and to much greater values over wooded country.[40]

Under conditions of neutral stability in the atmosphere, i.e., situations in which potential temperature is constant with height, the wind profile in the lowest few meters of the turbulent boundary layer does appear to conform to the logarithmic law described by Equation 3.2. When a vertical temperature gradient exists, however, buoyancy forces come into operation and the simple logarithmic law no longer applies. Observation shows that horizontal wind velocities measured increase more rapidly than the log of the height at which they are measured under stable conditions, and increase less rapidly than log z in unstable conditions. The vertical profile of horizontal windspeed in unstable conditions is consequently determined by a stability parameter, by a buoyancy parameter given by g/θ_m, where g is the local value of gravity and θ_m is the mean potential temperature of the profile, and by the vertical heat flux, as well as by z and the friction velocity (see Appendix 3.18). The atmospheric variables involved in nonneutral stability may be combined to form a characteristic length, the *Monin-Obukhov parameter*, which is a convenient measure of the stability of the airflow (see Appendix 3.19). Application of the Monin-Obukhov parameter to specific wind profiles in the atmosphere involves implicitly assuming the validity of the *similarity theory*, in which the vertical profiles of wind and temperature are regarded as having similar essential features. By assuming similarity, one can obtain a wind profile from a temperature profile, and vice versa. The theory then indicates that the vertical profile of wind velocity can take only two basic forms, depending on whether the atmosphere is stable or unstable, and the logarithmic profile represents their limiting form within small ranges of height.

Depending on the arrangement of the physical parameters, different forms of theoretical wind profile have been defined, each of which is valid over certain ranges of height and under certain stability conditions. An important point is that a height of 30 feet above the ground appears to be critical for wind-distribution theory.[41] Different techniques must be used (i.e., different types of theoretical profile) according to whether one must interpolate between ground level and 30 feet above it, or extrapolate upward from the 30-foot level. It is very significant indeed that the standard height for wind measurements for both climatological and synoptic (surface) purposes corresponds approximately to this critical level.

For adiabatic conditions below the 30-foot level, the *Prandtl profile* (i.e., the logarithmic profile already noted) is valid, and the friction velocity of the classical boundary-layer equation, 3.2, is replaced by air density and surface stress.[42] This profile is completely satisfactory for neutral conditions, and once the roughness parameter is known for the locality, a single observation of wind at any height above the site enables the surface stress to be determined, and hence the horizontal wind velocity at any required height within the lowest 30 feet may be predicted.

For diabatic conditions below the 30-foot level, the *Deacon profile* is suitable for nonneutral conditions, and the *Monin-Obukhov profile* is valid for both neutral and unstable stratifications. For mean wind profiles in the lowest 500 feet or so of the atmosphere, the *power-law profile* is valid above the 30-foot level. The *Swinbank (exponential) profile* applies in all types of stability at all heights below which the shearing stress and the vertical heat flux due to turbulence remain constant (see Appendix 3.20). The virtue of the Swinbank profile is that it is claimed to be independent of empirical constants. A somewhat unusual series of observations of the atmospheric friction layer in stable conditions constitute the *Leipzig profile*.[43]

Once the vertical profile of wind at a site is known, it is possible to predict the vertical flux of heat and moisture due to turbulence. This was realized to be a practical possibility after Thornthwaite and Kaser's empirical proof of the curvature of actual wind profiles was followed by a similar demonstration by Deacon that the potential temperature profile is also curved vertically. However, under certain conditions, one must also measure horizontal gradients in order to predict vertical energy transfer with precision, because shear stress is not necessarily constant with height for a short distance above the ground. Wherever shear stress is constant with height, the profile of vertical heat and water transfer is considerably simplified, because the logarithmic law of wind then applies, but wherever shear stress is not constant with height, distortions are produced in the vertical profiles of wind and temperature that are functions of distance from the upwind discontinuity in surface roughness that has given rise to the shear stress inconstancy. The irregularity in shear stress has been found to extend downwind for a distance more than ten times the height of the roughness elements causing it, and this distance increases with increasing height above the ground. The theory of turbulent transfer assumes that the eddy fluxes of momentum, heat, and water are constant both vertically and horizontally, but this assumption is not necessarily valid in a given location. For example, in a small irrigated area at

Davis, California, horizontal uniformity was not present, and the vertical fluxes of heat and water vapor were not constant with height even for fetches as long as 50 to 180 m. In terms of the eddy theory of turbulence, uniformity of eddy diffusion in the vertical and horizontal direction does not necessarily apply to microturbulence. Observations of microvariations in temperature at a height of 1 meter above the surface in cold air moving from over the sea to a strongly heated island in the Baltic showed that eddy conductivity increased both upward and horizontally: eddy conductivity values at a given height proved to be three or four times greater at a distance of 0.5 km from the shoreline of the island than over the sea.[44]

All wind-profile predictions involve considerable computational work, but the determination of the appropriate parameters is simplified by vector methods. The determination generally involves the precise measurement of wind velocities at one or more levels, for example, by means of a Sheppard anemometer, a very sensitive type of cup anemometer that enables measurements to be made with an error of not more than 1 per cent (possibly not exceeding 1 cm per sec). Alternatively, wind profiles up 30-feet over open, level ground may be estimated from the standard wind observations at 33 feet, provided the vertical temperature difference between two heights within the range 1 to 15 feet above the ground cover is measured. It is also possible in principle to determine wind and temperature profiles indirectly by measuring the temperatures of a large number of indicators suspended at different heights above the ground. These temperatures are controlled by energy exchange via convective and radiative processes between the indicators and their atmospheric environments, so that theoretical considerations should enable the vertical profiles of wind, heat, and water controlling these processes to be determined.[45]

Verification of theoretically computed profiles of temperature and water vapor requires the use of resistance thermometers, thermocouples or thermistors for the former case, and dry- and wet-bulb thermocouples for the latter.[46]

An interesting question relates to the changeover from the turbulent boundary layer of the atmosphere, in which micrometeorological influences are the dominant control, to the general planetary-scale motions, in which dynamic processes are strongest. Specifically, how does the microclimatic wind, as described by the various theoretical profiles, transform itself into the gradient wind? Or, to take another example, how does the microclimatic temperature profile change into the vertical temperature gradient of about 3°F per 1,000 feet that is often taken to be typical of mean conditions in the lower troposphere? The question is not easy to answer, because it involves a region of the atmosphere—the layer above the 300 to 400 feet associated with instruments permanently installed on masts, and below the lowest level for which routine radiosonde observations are available—that is very sparsely represented in the climatological record.

Observations made with a captive balloon at Cardington, Beds., show that the figure of 3°F per 1,000 feet is not necessarily universally valid. For example, the mean vertical temperature gradient within the lowest 2,000 feet of the atmosphere at Cardington varies from 3.1°F per 1,000 feet in winter, through 3.9 in autumn and 4.1 in spring, to 4.6°F per 1,000 feet in summer (all midday values). The value of the gradient decreases with increasing low cloud, but varies little with variations in the state of the sky or in the general windspeed. Mean temperatures derived from observations at the top of Blackpool Tower, Lancs., 418 feet high, show that mean temperatures there decrease with height at the rate of 2.3°F per 400 feet in the season of maximum lapse-rate; i.e., the summer. Temperature inversions are common during the period November to January inclusive, and similar results were noted for towers at Leafield (288 feet) and Rye (350 feet). Actual, rather than mean, lapse-rates at Blackpool were greatest around midday and in the afternoon, and the greatest values do not usually exceed 1°F per 100 feet.[47]

Measurements made of the vertical temperature and humidity gradients in the lower atmosphere over the Caribbean indicate that a homogeneous turbulent air layer is very characteristic of this region, immediately above the water. The turbulent motions within this layer are not due to the vertical temperature gradient, i.e., to convection, but to the flow of the air over a wind-roughened water surface. Typically, the homogeneous layer has a dry-adiabatic lapse-rate and a nearly constant humidity-mixing ratio. The height of the layer does not build up to the height one would expect (for a given horizontal windspeed) from turbulence theory, because its upward development is limited by the low level of lifting condensation.[48]

The modifying influence on the lower atmosphere of relatively cool ocean waters, the North Atlantic between New Jersey and Nantucket, was observed from 32 aircraft soundings in June 1945 to spread upward comparatively rapidly, within the lowest 1,500 feet of the atmosphere, during the first 50 miles of trajectory. For an initial air-water temperature difference of between 10 and 20°F, the depth of the modified air layer remained more or less constant at 600 feet during the next 250 miles of trajectory, provided surface winds were less than 15 miles per hour. If the windspeed increased considerably, and provided the lowest air layer was initially less than 5°F warmer than the water surface, the depth of the modified layer rapidly increased. A noteworthy discovery was that consecutive soundings (every 15 minutes or so) differed considerably in terms of temperature and humidity, associated with vertical motions induced by wave-like oscillations of the upper surface of the modified layer. These oscillations represented shearing waves, which form on discontinuity surfaces within a fluid. They had an amplitude of at least 150 feet, and were most marked in situations with a temperature inversion or a very strong vertical gradient of moisture (i.e., a large *hydrolapse*). Their occurrence indicates that dynamic influences must be considered, as well as purely turbulent processes, when the center of interest shifts to above the lowest few hundred feet of the atmosphere.[49]

In general, the magnitudes of the vertical temperature lapse-rate and the hydrolapse within the lowest layers of the atmosphere over the sea enable the prevailing type of air mass to be identified. For example, polar maritime air in the North Atlantic, in which the air is cooler than the sea surface, is characterized by a superadiabatic lapse-rate involving a temperature decrease of 1°F for each 50-foot increase in height. By contrast, returning polar maritime air (i.e., polar maritime air initially moving in a curved path toward the equator, so that it returns northward with some tropical maritime characteristics), with both surface air and sea surface at the same temperature, has an isothermal lapse-rate, and tropical maritime air, in which the air is warmer than the sea, typically has an inversion extending 60 feet above the sea surface. Over land surfaces, in the lowest 100 m of the atmosphere, lapse-rate conditions of temperature and humidity generally prevail during the day, with the temperature lapse-rate decreasing upward. During the night, inversion conditions generally occur. The changeover from lapse-rate to inversion conditions and vice versa usually occurs shortly before sunset and shortly after sunrise, respectively. The magnitude of the midday lapse-rate is usually greatest in summer and least in winter, and the vertical temperature and humidity gradients at some one instant in time differ considerably, depending on whether they are measured in rural, urban, industrial, or coastal areas.[50]

The vertical transfer of heat, momentum, and matter (water vapor, ozone, carbon dioxide, radon, etc.) across the Earth's surface and into the overlying atmosphere takes place at a rate that can be assumed constant (at the surface value of the flux) throughout the lowest few decameters of the atmosphere, because the variation in the total flux is only a few per cent of the surface flux. Assuming fully turbulent airflow, the vertical fluxes of heat, water vapor, and momentum may be evaluated quantitatively. The evaluations require a precise knowledge of the eddy viscosity, which in turn presupposes a precise knowledge of the local value of surface stress. Since surface stress may vary considerably horizontally, it is obvious that the vertical fluxes of matter and energy will be influenced by horizontally distributed geographical influences. Most evaluations of vertical fluxes made for a specific site by meteorologists interested in agriculture, for example, assume that the site forms part of a flat uniform terrain so that conditions are uniform horizontally and the net turbulent fluxes will be vertical only. A further simplification is that steady-state conditions are assumed, so that the mean properties of the atmospheric boundary layer do not change with time.[51]

If conditions are not steady, then departures from the simple situation arise that may involve nocturnal cooling (to which the nocturnal-cooling equations of Brunt, Frost, and Knighting apply*), leading-edge effects on the concentration of flux at the surface, and effects due to changes in surface roughness. Both leading-edge and surface-roughing effects may be allowed for by means of theoretical models. For the former, a two-dimensional diffusion equation is solved by assuming horizontal uniformity of flux across the wind direction, and negligible diffusion along the wind (see Appendix 3.21). The solution indicates that a long fetch is required to obtain complete adjustment to a new surface condition. For heights up to a few meters, the required ratios of horizontal fetch to height are 2:3 if 90 per cent adjustment is desired, and 4:5 if only 10 per cent adjustment is required. The required fetch at a given height varies inversely with the magnitude of the kinetic energy of the local wind system. The figures quoted apply to conditions over a surface of grass 15 to 20 cm high; if the Earth's surface at the site in question is rougher or smoother, then the local kinetic energy will be correspondingly larger or smaller, and the required fetch-height ratios will be smaller or larger, respectively, than those given here. For changes in surface roughness, the theoretical model assumes a progressive change in the value of the surface shearing stress downwind from the discontinuity in surface roughness, and the existence of an "internal boundary layer" (i.e., a new boundary layer induced by the new value of surface roughness) within the normal turbulent boundary layer of the lower atmosphere. The shear stress is assumed to vary continuously with height from the ground up to the top of this internal boundary layer. The model shows that the fetch-height ratio describing the minimum distance required before the new wind regime (within the internal boundary layer) is adjusted to the new surface roughness is about 70; this compares with a value of 80 determined empirically over a short-grass surface.[52]

There are numerous geographical sites and meteorological situations in which the models just described do not apply. For momentum transfer by turbulence, it is

* These equations are described in Chapter 4 of *Techniques*.

possible to estimate local wind variations caused by nonuniform geographical conditions by means of double-theodolite pilot balloons and other simple techniques. Observations made in valleys in Vermont, for example, show that the zones of greatest turbulence in summertime are located above the lee slopes of mountain ridges, associated with a sharp transition zone between the "wind-shadow" regime immediately above the lee slopes and the general regional airflow. The transition zone varies greatly in height and occurs intermittently in time, and is usually marked by persistent vertical gusts of air for 3 km downwind of the ridge crest line. The zones of greatest turbulence tend to be found at the height of the ridge crests under stable conditions; in inversion situations a zone of maximum turbulence may still be identified, although much diminished in intensity, at a height just above that of the ridge crests. Over the center of the valleys, the zone of maximum turbulence occurs at height above the valley floor approximately halfway between ridge crest and valley bottom.[53]

The question of the approach to the geostrophic or gradient wind may be regarded as an example of the application of the concepts of microturbulence to the atmospheric boundary layer as a whole, i.e., to a range of phenomena midway between local microeffects and planetary-scale macroeffects. The geostrophic wind equations for horizontal motion may be written for the atmospheric friction layer as

$$\frac{\partial \bar{u}}{\partial t} - 2\Omega\bar{v}\sin\phi = -\frac{1}{\rho}\frac{\partial p}{\partial x} + \frac{1}{\rho}\frac{\partial \tau_{xz}}{\partial z}, \qquad [3.3]$$

$$\frac{\partial \bar{v}}{\partial t} + 2\Omega\bar{u}\sin\phi = -\frac{1}{\rho}\frac{\partial p}{\partial y} + \frac{1}{\rho}\frac{\partial \tau_{yz}}{\partial z}, \qquad [3.4]$$

in which $\bar{u}$ and $\bar{v}$ are the mean air velocities in the east-west and north-south directions, respectively; Ω is the angular velocity of the Earth at the equator; ϕ is latitude; ρ is density; and p is pressure. The effect of friction in slowing down the wind from its geostrophic value is described by introducing virtual stresses, τ_{xz} and τ_{yz}, which vanish at the top of the friction layer.[54]

Provided the virtual (i.e., eddy) viscosity is constant throughout the friction layer, that the air motion is steady and unaccelerated, and that the pressure gradient is invariant with height, then the *Ekman spiral* provides a good approximation to the variation of wind with height between about 300 feet above ground level and the gradient-wind level (see Figure 3.6). The Ekman spiral approach to the problem is really an oversimplification, because it assumes that atmospheric mixing by turbulence is a large-scale analogy of molecular diffusion as described in Einstein's law for Brownian motion, which uses a generalized coefficient of diffusion (see Appendix 3.22), whereas in the atmosphere the value of the coefficient of diffusion varies with both horizontal and vertical position, and also with the scale of the phenomena and the depth of the layer being considered. There are occasions when the Ekman spiral provides a very useful method for interpolating winds at levels for which no data are available, but whenever a horizontal variation in temperature exists, the Ekman spiral is greatly distorted, because the assumption of a constant pressure gradient does not then apply. The assumption of a constant eddy viscosity throughout the friction layer is often invalid too; the wind varies more rapidly with height than it should according to the Ekman spiral where the eddy viscosity is small, and less rapidly with height where it is large.[55]

The low-level jet is essentially a friction-layer phenomenon, and is related to the

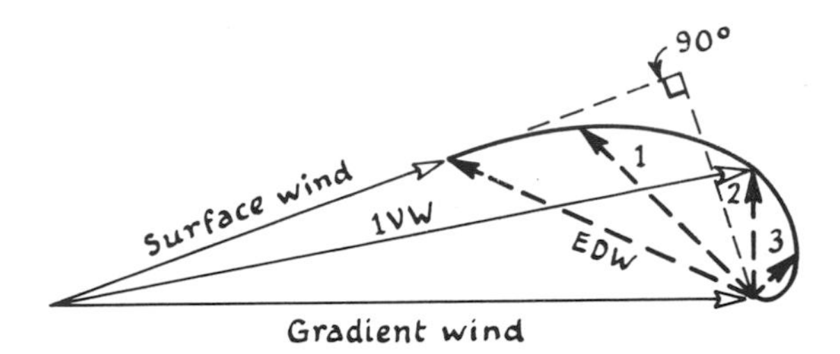

FIGURE 3.6.
Ekman spiral.
EDW, Ekman drift wind. IVW, Wind vector for wind approximately midway between surface wind and gradient wind. 1, 2, 3, Drift winds for layers shallower than the Ekman layer.

sudden relaxation of frictional influence that occurs around sunset. Convection during daylight hours brings about a downward transport of momentum, which is not entirely compensated for by a replenishing momentum flow from above the 2,000-foot level. Consequently, the air in the vicinity of the 2,000-foot level has insufficient momentum to achieve its equilibrium velocity as described by Equations 3.3 and 3.4. When convection dies down in the hours approaching sunset, the frictional retardation also decreases, and the air accelerates under the combined influence of the Coriolis and pressure-gradient forces. The final speeds attained within the 1,000- to 3,000-foot layer are often supergeostrophic; i.e., they exceed the equilibrium velocity described in Equations 3.3 and 3.4 by appreciable amounts.[56]

Applications of Theories of Turbulent Diffusion

Turbulence theory has many applications in climatology, which, provided geographical considerations are given due weight, point the way to the solution of numerous practical problems in such fields as agriculture, engineering, and transport. An obvious application concerns the diffusion of matter in the atmosphere; the matter may be man-made (e.g., factory smoke or nuclear fallout) or natural (e.g., water-vapor droplets, pollen, or the smoke from forest fires), but the underlying principles are the same for both. The only distinction to be made is that large particles behave differently from small particles.[57] Although particles of matter of all sizes tend to move away from each other, the smaller particles are more difficult to deal with than the larger particles, because the smaller particles are subject to random bombardment in an environment of fast-moving molecules, and thus acquire a component of motion (the *Brownian* component) that is additional to their turbulent motion. By contrast, the larger particles behave as if they are surrounded by a gaseous environment that forms a physical continuum. Small-scale turbulent motions result in a spreading out of a group of particles in the atmosphere as time passes, and the effect of this process can be described statistically. If large-scale motions are superimposed on the small-scale turbulence, or if the latter takes place in an atmosphere whose physical properties vary horizontally or vertically, then a deformation of the group of particles develops that cannot be described in simple statistical terms. With both small- and large-scale motions, the average distance between adjacent particles increases until the diffusion process affects the entire atmosphere, after which the mean distances between pairs of particles tend to be permanently stabilized.

The previous paragraph describes a model of atmospheric diffusion that obeys the laws of turbulence theory. For global-scale diffusion, this model appears to work quite well, but for local-scale diffusion, many modifications must be introduced because of atmospheric peculiarities induced by geographical effects. The latter problem may be simplified: if the direction and speed of the mean surface wind are known for the geographical site in question, the diffusion of matter from a given point within the site may be estimated by means of ordinary synoptic parameters.[58]

Figure 3.7 illustrates the effect of lapse-rate variations on the diffusion of matter released from a source such as a factory chimney. For each type, the ground-level concentration of matter at a certain distance from the chimney after a given interval of time may be computed from the equations of small-scale diffusion (see Appendix 3.23). It will be seen from Figure 3.7 that the pollution pattern known as *fumigation* may be associated with one of three diffusion types. Type I results from a temporal change in the turbulence régime. Type II results from the heating of the atmosphere from below as it passes over a man-created heat source, such as a city or an industrial area. The heat injected into the lower atmospheric layers maintains an unstable lapse-rate up to two or three times the general height of the buildings, but air moving over the city from adjacent rural areas in the evening will be stable as a result of cooling due to terrestrial radiation if the sky is clear. Consequently, a shallow unstable layer of air is capped by a deeper stable layer, and the effluent cannot rise through the latter. The factory effluent is therefore carried closer to the ground with increasing distance from the chimney. Type III represents a situation somewhat similar to type II, but in which the heat is injected into the lower atmosphere from a natural source. Examples are a shoreline across which onshore winds blow during daytime in spring and summer, or offshore winds during clear nights in autumn and winter. *Aerodynamic downwash* conditions result from one or both of the following processes: (a) downward movement of air in the lee of buildings because of eddying; (b) the formation of a Kármán vortex street in the lee of the chimney. Both processes depend on there being moderate to strong winds, since there is a critical windspeed that depends on wind direction, the position of the chimney or stack relative to the building, and the velocity of the effluent as it leaves the chimney.

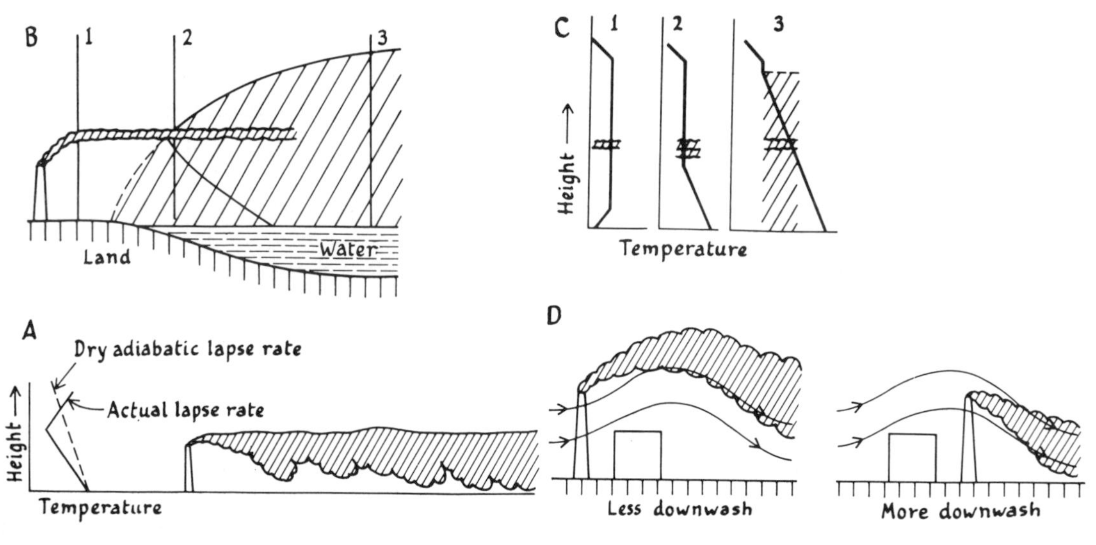

FIGURE 3.7.
Atmospheric structure and fumigation-type pollution (after E. W. Hewson and E. W. Bierly).
A, Type I fumigation. B, Type III fumigation. C, Lapse-rates associated with type III fumigation for different trajectories. D, Variation of aerodynamic downwash with position of stack relative to building.

The verification of the equations of small-scale diffusion in actual situations often indicates the presence of geographical effects that would otherwise remain undetected. For example, observations made in a coastal area covered with desert scrub in the United States reveal strong departures from the idealized model described in the Sutton equations, in which the logarithm of the concentration of diffused matter integrated in the crosswind direction varies linearly with the logarithm of downwind distance from the source. The departures proved to be due to the presence of quasistationary waves, in the form of a helix, with a wavelength of about 4,000 yards. The waves were probably induced by a steep seaward slope of the terrain two miles upwind of the observing point. In another investigation, the value of Sutton's diffusion coefficient C in the crosswind direction was found to be greater when there were marked variations in turbulence in time and area than when there was a uniform site. Transitional states could be identified when this effect was very prominent over the western extremity of Lake Erie: such a transitional state resulted, for example, from the advection of warm air aloft.[59]

Microdiffusion studies usually involve the use of tracer substances that are released into the atmosphere and then tracked at regular intervals of time at points arranged in the form of a sampling grid. For distances up to 10 km downwind from the release point, uranine dye is an effective tracer; for greater distances fluorescent pigments are useful. Nonhomogeneity of the wind field may introduce considerable problems in the interpolation of statistics derived from tracer studies. One investigation, concerned with diffusion under nocturnal inversions near Richland, Washington, involved the setting up of a diffusion model for shear flow in which the observed concentration of tracer particles at a point downwind from the source is related to the lateral and vertical coordinates of the point and the correlation coefficient between the lateral and vertical coordinates of the tracer particles.[60]

If the diffused matter consists of water vapor, then the microturbulence techniques constitute a way of determining the instantaneous rate of evaporation. The procedure was first applied in detail in classic work by Thornthwaite and Holzman in New Jersey, which involved adopting the mixing-length concept and developing very sensitive anemometers. The resulting equation has proved to give an accurate assessment of the rate of evaporation during periods of 10 to 60 minutes in neutral or near-neutral conditions. Very precise observations of windspeeds and temperatures at two heights, one vertically above the other, are required. For nonneutral conditions, an expression devised by Deacon and Swinbank may be used, which necessitates making observation of windspeed at three heights. For unstable conditions, Priestley's equation may be used, which assumes that the coefficients of eddy diffusivity and eddy conductivity are equal (see Appendix 3.24).[61]

All these profile methods for the determination of the rate of evaporation require an adequate fetch, so that the profiles measured are fully representative, thus ensuring that the flux of physical properties is constant with height. To obtain a 90 per cent adjustment to the difference between the surface at the observing site and the surface further upwind, a fetch-height ratio of 200:1 is necessary over grass 15 to 20 cm high. A larger ratio is necessary over a smoother surface, and a smaller ratio will suffice over a rougher surface. The lowest instrument must be mounted at least three times the height of the vegetation or other surface cover, and the other instrument(s) should be located at intervals of doubled height above the zero plane. Observations should be made under steady-state radiation conditions: i.e., there should not be excessive, increasing, or decreasing amounts of incoming or outgoing radiation. Measurements should not be taken near sunrise or sunset, or in situations with bright periods and scattered showers or with patchy cover of low cloud. The Thornthwaite-Holzman method should be applied when there are strong winds with a heavy overcast, preferably.

In the applications discussed so far, turbulence theory is used as a model to interpret data on a diffusion phenomenon that is not necessarily "turbulent" in the everyday sense. Two further applications must be mentioned, concerned with what is "turbulence" even to the layman.

The first application concerns wind gustiness in the everyday sense. Gustiness may be estimated empirically from photographs of smoke plumes, and it may be approximated by means of climatological data. However, neither of these procedures are of much use in engineering applications of climatology, for example, in the determination of the wind-load that is likely to affect a tower or some other structure to be erected at a site. The general principle is well-understood: the wind velocity impinging on the tower is transformed into a pressure in accordance with Bernouilli's law. The resulting force to which the structure is subjected depends also on its geometric properties. Provided that the wind is steady, i.e., that the properties of the airflow are invariant in time and in the horizontal, then the wind force may be regarded as static. Consequently, a "safe" structure may be ensured by designing the tower so that its natural frequency of oscillation does not coincide with the frequency of the large-amplitude fluctuations in wind velocity characteristic of the site as revealed by, for example, a Dines anemometer located at the site.[62]

However, such structures do occasionally fail because of wind effects, and turbulence theory explains why. The natural airflow within the atmospheric friction layer is frequently not steady, but consists of gusts due to eddies of a wide range of sizes, varying from a few centimeters or less in diameter to hundreds of kilometers. Consequently, it is not the "typical" size of gust as determined directly from the anemograph trace that must be used as a basis for the design of the tower, but rather the "spectrum" of gust sizes that may be derived from the anemogram by means of power-spectrum analysis.[63] If the frequency of any one of the gust "components" revealed by the analysis approaches the natural frequency of the structure, then resonance effects will develop that may cause the tower to be damaged or even to collapse. To prevent this from happening, the actual wind loads must be determined by power-spectrum analyses of wind frequencies for each direction separately, so that the ultimate loads to which the collapse strength of the structure may be equated can be established. For each direction, the relative contributions of the different sizes of gust to the total variance of the wind observations must be assessed. Thus the power-spectrum analysis will reveal all the sizes of gust (and their relative importance) that could give rise to the observed fluctuations in wind velocity, and the tower must be designed with this in mind. This is, of course, quite separate from the problem of designing for the maximum wind velocity likely to be experienced in, say, a 150-year period. The latter involves a problem in climatological prediction that is distinct from the problem in turbulence we are concerned with in this chapter.

The advantage of approaching the problem of wind loads via turbulence theory is that, by making use of power spectra, there is no necessity to define the form of a typical single gust. The *shape* of the power-spectral-density function of the atmospheric turbulence at the site in question must be established, and then the dynamic response of the structure may be calculated if its frequency-response function is known.[64]

The phenomenon known as *clear-air turbulence* provides a second example of the concurrence of the layman's idea and the scientist's idea of turbulence. Clear-air turbulence (CAT) is the name given to very violent air movements in areas devoid of clouds, which severely buffet aircraft and often create the impression of riding over cobblestones. CAT may cause severe damage or even structural failure in an aircraft, and it therefore constitutes a serious handicap for high-altitude jet operation. In general, CAT has been observed at all levels of the troposphere and lower stratosphere but single cases are usually restricted to volumes of air, whose dimensions vary from a few kilometers to tens of kilometers in the horizontal, and from a few tens of meters to a few hundreds of meters in the vertical. Its onset is almost always sudden and unexpected, and its duration rarely exceeds a few minutes. Because of its importance to aviation, CAT has given rise to a voluminous literature.[65]

Air layers in which CAT is reported are usually shallow, but sometimes may be as much as 10,000 feet deep. Within these layers, the CAT is always patchy and spasmodic; sometimes the turbulence is in an irregular pattern, rather like convection currents near the ground, at other times (particularly if it is of the "cobblestone" variety), it involves sharply defined, regular repetitions of the turbulent elements. Normally, the turbulence at a given point or over a given area is very temporary, but one patch of CAT formed to the east of the Isle of Wight, occupying a horizontal area about the same size as the island, and persisted all morning.[66]

Interesting geographical variations in the incidence of CAT have been reported. For example, on the main western European air routes radiating out from London, CAT is more than twice as frequent over the land, and four times as frequent over the coastal areas, than over the sea. Of the cases reported over land or over coasts, only one-third were associated with fronts, whereas three-quarters of the reported cases over the sea were associated with fronts, particularly warm fronts. The greatest concentration of CAT was reported to be over Troyes. Analysis of more than 1,800 cases of CAT reported over Colorado, Florida, Maine, and North Carolina between 3,000 and 18,300 m, indicated that the likelihood of CAT increased with increasing altitude from 3,000 to 7,400 m, decreased slightly between 7,400 and 14,400 m, and then rapidly decreased from 14,400 to 18,300 m. Most of the CAT layers were less than 400 m deep, with an average of 239 m. CAT maxima were at 7,400 m, 11,000 m, and 14,400 m (the first being due to fronts, the other two to jetstreams), and minima at 6,200, 9,500, and 12,800 m. The horizontal dimensions of the CAT patches were at least 10 to 20 miles. CAT occurrences reported over Agra, by contrast, revealed four maxima within the lowest 20 km of the atmosphere, with average heights of 1–2 km, 5–6 km, 12 km, and 17–18 km. The first and third of these maxima correspond to levels with a high frequency of inversions or isothermal layers; the second and fourth generally represent inversion levels in all seasons of the year; the second coincides with the transition zone between the lower and upper troposphere; and the fourth is very close to the tropopause.[67]

Apart from reports of CAT from Europe and North America, numerous instances have been recorded from Australia, the Middle East, and southern Asia. In all reports, certain synoptic and orographic situations appear to be conducive to CAT. The vicinity of jetstreams is a favored location for CAT, which seems to be induced by marked horizontal and vertical wind shear at the level of a jet axis on its anticyclonic side and at tropopause level on its cyclonic side. The strongly baroclinic layer beneath a jet axis in middle latitudes is also a favored one for CAT. The association of CAT with jetstream maxima may be inferred from a physical model that shows how, at a certain critical combination of vertical wind shear and static stability conditions, a steady airflow should break down into a series of helical vortices and streaky motions. The model suggests that CAT is observed to be associated with jet maxima simply because the required critical conditions are frequently met in such situations. Whether or not this suggestion is true, the association of CAT with jets is well-established. For example, many incidences of severe CAT over the British Isles occur in the zone of strong wind shear below, and on the low-pressure side of, the jet axis. Others occur in the vicinity of pronounced upper troughs, and near the tropopause. Such upper troughs usually incorporate a marked wind shear in the horizontal, and the most severe CAT is usually found in a zone running parallel with the axis of the trough.[68]

Other synoptic associations of CAT include deep thermal troughs between 300 and 200 mb, and thermal winds. For example, over the British Isles and western Europe, severe CAT is associated with a combination of a 500- to 300-mb thermal wind exceeding 35 knots and a horizontal wind shear exceeding 0.2 per hour.[69] But the importance of synoptics for CAT may be overemphasized. Although CAT over Britain and western Europe is likely to be severe on the low-pressure side of a jetstream axis, and to a lesser extent above the axis on its anticyclonic side, CAT may occur up to 50,000 feet (and also well above the tropopause) in almost any synoptic situation. The main difficulty in studying it synoptically is that the usual parameters employed in synoptic work are essentially macroscale ones, whereas CAT is probably the result of mesoscale processes.

CAT is often associated with orographic waves: high-level CAT, in particular, and standing waves are frequently related. Static stability is an important controlling factor in both, but because a higher degree of stability is necessary for wave formation than for CAT development, there will obviously be cases of CAT not associated with standing waves. A 30-mile long patch of CAT at 30,000 feet was reported over the southwestern portion of a gap in the Australian Alps, where the ground surface descends 4,000 feet fairly rapidly. The CAT was observed in the transition zone between the orographically induced flow (which extended to a considerable height) and the flow (less affected by orographic-wave effects) that was moving through the mountain gap. The main connection between CAT and orography is, in general, probably via standing waves, and an inference from the observations is that CAT is most likely above areas where the general level of the Earth's surface is changing rapidly with horizontal distance.[70]

The atmospheric parameters that appear to be of direct significance in deciding the occurrence of CAT include, apart from static stability, the degree of curvature of the vertical wind profile, the rates of horizontal and vertical variation with distance of windspeed (i.e., the wind shear), lapse-rates, and differential advection. The most useful single criterion appears to be the Richardson number, and there are also correlations with wind direction, temperature, and humidity.

In general, the intensity of CAT depends on the magnitude and sign of (a) the product of horizontal windspeed and the vertical gradient of horizontal windspeed, and (b) the product of horizontal windspeed and the vertical gradient of wind shear. Product (a) is a measure of the vertical gradient of horizontally disposed kinetic energy per unit mass of air. CAT is much more likely to be associated with negative wind shear, i.e., with situations in which windspeed decreases with height, than with positive wind shear. The abrupt wind shear associated with the layers immediately above an inversion is conducive to CAT development, and pronounced temperature inversions accompanied by very marked vertical wind shear are favored zones for the occurrence of CAT.[71]

Small values of the Richardson number are usually associated with severe CAT: *Ri* of less than 0.5 is a typical value, and CAT is almost certain if *Ri* is less than 0.01. A theory has been proposed that CAT is produced mainly in shallow layers in those zones of the atmosphere in which the air in a jetstream is being slowed down and its kinetic energy is being converted into potential energy. In such zones, the value of *Ri* is considerably reduced because the static stability is diminished while the vertical wind shear is increased. Changes in static stability and a reduction in the value of *Ri* may be brought about by differential advection. The vertical wind shear is assumed to be constant during the advection; the effect of the latter is to reduce the static stability, and hence to reduce *Ri* until it falls below the critical threshold required for the onset of CAT. After the release of the turbulence, the vertical wind shear rapidly increases. Differential advection appears to be important for CAT only when the initial lapse-rate is not too small; warm advection below a certain level in the atmosphere, or cold advection above it, can be shown to be insufficient to modify effectively the vertical structure of an atmosphere with a very stable thermal stratification.[72]

Quite apart from turbulence theory's use in the study of phenomena that are genuinely the products of turbulent diffusion, its concepts have value in the interpretation of climatic elements or entities that are not, at first sight, of a turbulent nature. For example, the initiation of thunderstorms in a calm atmosphere over land on hot afternoons has been interpreted—for cases in India in the premonsoon months—in terms of turbulence theory. The assumption is made that Bénard cells form, as a result of Rayleigh's limiting condition for the breakdown of a fluid heated from below into convective elements, and the coefficients of eddy viscosity and eddy conductivity are regarded as being approximately equal; the theory then shows that the moisture content of the air plays a minor part in the initiation of a thunderstorm under such conditions (see Appendix 3.25). This corresponds with synoptic experience in India, where thunderstorms occur when little or even no latent instability is indicated by the tephigram, i.e., when the tephigram has no positive area. The theory is essentially concerned with the prediction of the breakdown of the equilibrium of the air in terms of the mixing-length concept. If the lower layers of the atmosphere have a high water-vapor content, then the result of a breakdown in equilibrium will be the formation of large cumulonimbus clouds. If the lower layers of the air are fairly dry, a breakdown will result in the formation of CAT rather than thunderclouds.[73]

Notes to Chapter 3

1. O. Reynolds, *Phil. Trans. Roy. Soc. London*, series A, 174 (1883), 935, 186 (1895), 123. G. I. Taylor, *Phil. Trans. Roy. Soc. London*, series A, 215 (1915), 1, *Proc. Roy. Soc. London*, series A, 92 (1916), 196, and 94 (1918), 137. For the general accounts of turbulence theory followed in this chapter, see: P. A. Sheppard, *W*, 16 (1951), 42, and *QJRMS*, 84 (1958), 205; H. Charnock, *SP*, 46 (1958), 470; T. H. Ellison, *SP*, 49 (1959); 495; and pp. 400ff in G. K. Batchelor and R. M. Davies, eds., *Surveys in Mechanics* (Cambridge,

Eng., 1956). See also H. R. Byers, *General Meteorology* (New York). Standard accounts are by O. G. Sutton, *Atmospheric Turbulence* (New York, 1949), and by C. H. B. Priestley and P. A. Sheppard, *QJRMS*, 78 (1952), 448; see also *CM*, p. 492.

2. C. H. B. Priestley, *QJRMS*, 75 (1949), 28, and 77 (1951), 200.

3. E.g., see R. S. Scorer, *W*, 15 (1960), 110 and 287; G. B. Tucker, *loc. cit*, 145.

4. R. H. Steward, *JGR*, 64 (1959), 2112.

5. A. Court, *QJRMS*, 91 (1965), 234. For a review of problems of turbulent transfer in the lower stratosphere, see R. J. Murgatroyd, *QJRMS*, 91 (1965), 421.

6. For the derivation of these equations, see *HM*, pp. 448-52. See also O. G. Sutton (1949), *op. cit.*, p. 11. According to K. L. Calder, *QJRMS*, 92 (1966), 141, the equations of atmospheric diffusion in the standard form are not valid mathematically. According to A. B. Bernstein, *loc. cit.*, p. 533, Reynolds' turbulence equations are valid only when the bar operator represents an ensemble average, yet in meteorology the bar is usually taken as a time average. According to J. A, Businger and M. Miyake, *QJRMS*, 94 (1968), 206, the third moment (ie., $\overline{\rho' w' T'}$) in the steady-state expression for turbulent heat flux near the ground is negligible. See E. L. Deacon, *BAMS*, 49 (1968), 836, for details of errors in the measurement of the Reynolds' stresses. On the distinction between "total" heat flux and eddy heat flux, see J. A. Businger and J. W. Deardorff, *JAS*, 25 (1968), 521.

7. See *DMWF*, pp. 420-24, for a discussion. Byers, *op. cit.* (1944), pp. 591-603.

8. L. Prandtl, *Z. Angnew Math. u. Mech.*, 5 (1925), 136. J. A. Businger, *JM*, 16 (1959), 516. For a generalization of the mixing-length theory, see G. T. Csanady, *JAS*, 23 (1966), 667.

9. G. I. Taylor, *AGP*, 6 (1959), 101, and *op. cit.* (1915). For a vorticity-transfer theory of turbulence, see H. Lettau, *JAS*, 21 (1964), 453, and *JAS*, 23 (1966), 151. See also J. L. Lumley and R. W. Stewart, *JAS*, 22 (1965), 592.

10. For details of von Fick's theory, see *MAE*, p. 38.

11. G. I. Taylor, *Proc. London Math. Soc.*, 20 (1920), 196. W. Schmidt, *Der Massen Austausch in Freien Luft* (Hamburg, 1925). O. F. T. Roberts, *Proc. Roy. Soc. London*, series A, 104 (1923), 640. K. L. Calder, *Q. J. Mech. Appl. Math.*, 2 (1949), 153. E. L. Deacon, *QJRMS*, 75 (1949), 89. For an application of the *K* theory of *Fickian diffusion* to the horizontal diffusion of particles in the atmosphere over periods of one hour or more, see J. L. Heffter, *JAM*, 4 (1965), 153. For vertical diffusion at Cincinnati, Ohio, see R. A. McCormick and K. R. Kurfis, *QJRMS*, 92 (1966), 392. For a three-layer diffusion model for diffusion in unstable conditions at O'Neill, Nebraska, see S. Jaffe, *JAM*, 6 (1967), 297.

12. O. G. Sutton, *Proc. Roy. Soc. London*, series A, 135 (1932), 143, and *QJRMS*, 73 (1947), 257. F. Pasquill, *Atmospheric Diffusion* (New York, 1962), gives a convenient review of the statistical theory. For applications of the *statistical theory* of turbulence, see D. A. Haugen, *JAM*, 5 (1966), 646, on Lagrangian properties of atmospheric turbulence as deduced from diffusion experiments at O'Neill, Nebraska, and S. P. Zimmerman, *JAM*, 4 (1965), 279, on the estimation of diffusion parameters from observations of smoke puffs and contaminant deposition. See also S. P. Zimmerman, *JGR*, 71 (1966), 2439, and *JGR*, 72 (1967), 5153, on the parameters of turbulent atmospheres; and W. Klug, *QJRMS*, 91 (1965), 215, for mean values of dimensionless turbulence parameters near the ground at O'Neill, Nebraska, illustrating the influence of diabatic conditions on wind fluctuations. On the determination of the *scale* of turbulence, see F. Pasquill and H. E. Butler, *QJRMS*, 90 (1964), 79.

13. A. N. Kolmogoroff, *Compt. Red. Acad. Sci. U.S.S.R.*, 30 (1941), 301, and 32 (1941), 16. See also P. B. MacCready, *JGR*, 67 (1962), 1051.

14. For confirmatory observations that Eulerian space- and time-spectra are equivalent at wavelengths at least as long as 600 feet for a mean windspeed of 20 feet per sec and 900 feet for a mean windspeed of 30 feet per sec, see E. W. Bierley and G. C. Gill, *JAM*, 2 (1963), 145. See Pasquill, *op. cit.* (1962), for details on finding the energy spectra. On *Kolmogoroff's similarity theory*, see: K. Klug, *QJRMS*, 94 (1968), 555, for a test of it at O'Neill, Nebraska; A. E. Gill, *loc. cit.*, p. 586, on geostrophic adjustment; and L. O. Myrup, *JAS*, 25 (1968), 1160, on the *buoyant subrange* of the energy spectrum, defined as that part of the equilibrium range in which turbulence is suppressed by buoyancy forces associated with a stable environment, using data from the El Mirage dry lake, in the Mojave Desert of southern California. See also K. L. Calder, *QJRMS*, 92 (1966), 141, on the Monin-Obukhov similarity theory.

15. See F. V. Brock and E. W. Hewson, *JAM*, 2 (1963), 129, and F. V. Brock, *JAM*, 1 (1962), 444, for analogue computing techniques for solving the equations for a continuous point-source and a continuous line-source. See *MAE*, p. 48, for details of instantaneous volume-source formulae devised by Frenkiel and Kellog (both in 1952), Holland (in 1953), and Gifford. See A. W. Waldron, Jr., *JAM*, 2 (1963), 740, for an example of equations for an elevated line-source. See P. J. Meade, *Int. J. Air Pollution*, 2 (1960), 303, for details of a nondimensional form of the equations for an elevated point-source.

16. D. A. Haugen, M. L. Barad, and P. Antanaitis, *JM*, 18 (1961), 368. R. E. Munn, *JAM*, 2 (1963), 440. For applications of *Sutton's development of the statistical theory*, see the following.

On evidence for anistropic turbulence, see R. Bolgiano, Jr., *JGR*, 68 (1963), 4873, on tropospheric microturbulence; see also S.-K. Kao and B. L. Niemann, *JAS*, 22 (1965), 85, for a test of Lin's theory of anisotropic diffusion (C. C. Lin, *Proc. Nat. Acad. Sci.*, 46, 1960, 1147), which deals with the effect of small-scale motions on diffusion from microturbulence at 500 feet above the Mojave Desert. For point-source diffusion models, see M. E. Smith and J. A. Singer, *JAM*, 5 (1966), 631. On continuous diffusion in a vertically, nonuniformly stratified atmosphere, see F. K. Wippermann, *JAM*, 5 (1966), 640. For observations of diffusion from continuous point-sources, see J. J. Fuquay, C. L. Simpson, and W. T. Hinds, *JAM*, 3 (1964), 761, on Richland, Washington, and K. R. Peterson, *JAM*, 7 (1968), 217, who found that a radioactive plume emitted from the Brookhaven nuclear reactor, Long Island, N.Y., at a stack height of 355 feet was still well-organized at a distance of 160 nautical miles downwind of the stack, with a nearly uniform vertical distribution for at least 2,000 feet. For observations of diffusion from elevated continuous point sources, see: E. R. Walker, *JAM*, 4 (1965), 614, for Alberta, Can.; M. E. Smith, A. P. Hull, and C. M. Nagle, *JAM*, 6 (1967), 303, for Long Island, N.Y.; and R. E. Stewart, *JAM*, 7 (1968), 425, for the shore of Lake Huron. For a model of multiple point-source diffusion for Nashville, Tenn., see D. B. Turner, *JAM*, 3 (1964), 83. For short-range diffusion in southern Alberta, Can., from both continuous and instantaneous sources, see E. R. Walker, *QJRMS*, 92 (1966), 411. For short-range vertical diffusion under stable conditions at Porton Down, Wiltshire, see N. Thompson, *QJRMS*, 91 (1965), 175. For diffusion from low-level point-sources in St. Louis, see J. L. McElroy, *JAM*, 8 (1969), 19. For a test of Round's equations, see K. D. Hage, *JM*, 18 (1961), 534.

17. G. T. Csanady, *JAS*, 20 (1963), 201. See also F. Pasquill, *op. cit.*, p. 621. On the use of tracers in diffusion studies, see G. S. Raynor *et al.*, *JAM*, 5 (1966), 728, on dyed pollen grains and spores; see also J. D. Morton, *JAM*, 6 (1967), 725.

18. S. Corrsin, *JAS*, 20 (1963), 115. For details of the $x = \bar{u}t$ transformation, see Pasquill, *op. cit.*, Chapter 1.

19. J. K. Angell, *MWR*, 87 (1959), 427, and 92 (1964), 465. See also J. K. Angell and P. H. Pack, *MWR*, 88 (1960), 235, and 89 (1961), 273.

20. P. D. Thompson, *T*, 6 (1954), 150. B. Saltzman, *JM*, 14 (1957), 513. V. P. Starr, *T*, 5 (1953), 494.

21. Y. Ogura, *JM*, 15 (1958), 375. C. E. Buell, *JM*, 13 (1956), 413, and 14 (1957), 471. G. S. Benton and A. B. Kahn, *JM*, 15 (1958), 404. A. B. Kahn, *JAS*, 19 (1962), 1950. B. Saltzman, *JM*, 15 (1958), 259. M. A. Estoque, *T*, 7 (1955), 177.

22. R. M. Henry and S. L. Hess, *JM*, 15 (1958), 397.

23. R. C. Sutcliffe, *W*, 14 (1959), 163.

24. R. D. Elliott and T. B. Smith, *JM*, 6 (1949), 67. For a discussion of the mathematics involved, see Tschu Kang-kun, *Academia Sinica, Coll. Sci. P. Meteorol., 1919–1949* (China, 1954), p. 535.

25. P. A. Sheppard, *W*, 7 (1952), 382, gives a review of this topic.

26. E. M. Wilkins, *JAS*, 20 (1963), 473.

27. For full details on such maps, see S. B. Solot and E. M. Darling, Jr., *AFGP*, vol. 1, no. 58.

For observations of macroturbulence, see: S.-K. Kao and W. S. Bullock, *QJRMS*, 90 (1964) 166, on Lagrangian and Eulerian energy spectra of geostrophic wind velocities; R. E. Newell, J. M. Wallace, and J. R. Mahoney, *T*, 18 (1966), 363, on the climatology of large-scale mixing properties, between 15 and 75 km, in relation to the general circulation of the atmosphere; R. J. Reed and K. E. German, *MWR*, 93 (1965), 313, on an application of mixing-length concepts to large-scale eddy fluxes in the stratosphere; and R. J. Murgatroyd, *QJRMS*, 95 (1969), 40, on horizontal diffusivities at northern hemisphere latitudes and stations. See also G. P. Williams and D. R. Davies, *QJRMS*, 91 (1965), 471, for a mathematical model, based on the austausch concept, of a quasi-stationary meanflow picture of the general circulation; and R. C. Smith, *QJRMS*, 90 (1964), 338, for the use of an eddy viscosity varying with latitude for deriving a meridional profile of wind velocity in the lower troposphere from the zonal profile of velocity.

On the theory of macroturbulence, see: S.-K. Kao, *JAS*, 25 (1968), 32, for the governing equations and spectra, and *QJRMS*, 91 (1965), 10, for Eulerian-Lagrangian correlations and spectra; S. Panchev, *JAS*, 25 (1968), 933, for the *K*-coefficient of horizontal macroturbulent diffusion; S.-K. Kao and A. A. Al-Gain, *JAS*, 25 (1968), 214, on dispersion of clusters of particles between 850 and 200 mb; D. Djurić, *QJRMS*, 92 (1966), 231, on the role of deformation of large-scale horizontal airflow in the northern hemisphere on large-scale dispersion of clusters of air particles, in the absence of microturbulence; and H. Yoshihara, *JAS*, 25 (1968), 729, for a simplified mathematical model of large-scale air-sea interaction, which incorporates local energy balance at the air-sea interface.

28. E. W. Bierly and G. C. Gill, *JAM*, 2 (1963), 145, give full details on the sampling grids. A. Fleischer, *JM*, 16 (1959), 209, gives a detailed account of the radar anemometer. For the procedure for the corrections, see F. N. Frenkiel, *JM*, 8 (1951), 316. D. A. Haugen, *JAM*, 2 (1963), 306, provides a solution for the discontinuity. On difficulties in measuring the Reynolds' stresses in Kansas, and their effects, see J. C. Kaimal and D. A. Haugen, *JAM*, 8 (1969), 460.

29. For full details of the computations for power-spectrum analyses, see: H. A. Panofsky and R. A. McCormick, *AFGP*, no. 19 (1952), p. 219, and *QJRMS*, 80 (1954), 546; and especially H. E. Cramer, *AMM*, 4, no. 22 (1960), 12. For a review of the experiments on Eulerian spectra, see Pasquill, *op. cit.*

(1962). On the Lagrangian spectra, see J. G. Edinger, *AFGP*, no. 19 (1952), p. 241, and F. Gifford, *MWR*, 81 (1953), 179.

On instrumental techniques, see: J. C. Kaimal, J. C. Wyngaard, and D. A. Haugen, *JAM*, 7 (1968), 827, on the derivation of power spectra by means of a sonic anemometer whose horizontal axes are 120° apart, thus accommodating the range of wind directions normally encountered during any one-hour observation period under steady synoptic conditions; J. A. Turner, *loc. cit.*, p. 714, for the direct estimation of standard deviations of wind fluctuations from a wind vane; P. B. MacCready, Jr., *JAM*, 5 (1966), 219, on the theory of errors in mean windspeed determinations; F. R. Payne and J. L. Lumley, *QJRMS*, 92 (1966), 397, for one-dimensional wind velocity spectra derived from hot-wire anemometer measurements in aircraft of wind fluctuations in the flight-path direction; J. C. Kaimal, D. A. Haugen, and J. T. Newman, *JAM*, 5 (1966), 411, for a computer-controlled mobile micrometeorological station that measures, processes, and records turbulence statistics; J. C. Wyngaard and J. L. Lumley, *JAM*, 6 (1967), 952, for a spectral filter system that provides the first seven derivatives of fluctuations of windspeed from a hot-wire anemometer; B. A. Silverman, *JAM*, 7 (1968), 168, on attenuation of eddy diameters due to base-line effects when using sonic anemometers, infrared hygrometers, and transmissometers to measure microturbulence; and P. B. MacCready, Jr., and H. R. Jex, *QJRMS*, 90 (1964), 198, on wind-vane measurements of turbulent energy.

30. G. I. Taylor, *QJRMS*, 53 (1927), 201. H. A. Panofsky, *QJRMS*, 79 (1953), 150. R. H. Bushnell and P. O. Huss, *JM*, 15 (1958), 180. For observations of fluctuations in the vertical component of wind, see: J. K. Angell, *QJRMS*, 90 (1964), 307, for some between 600 and 2,600 feet at Cardington, Eng.; J. C. Kaimal, and D. A. Haugen, *QJRMS*, 93 (1967), 305, for tower measurements at Cedar Hill, Texas; and J. C. Kaimal *et al.*, *QJRMS*, 90 (1964), 467, for a comparison of sonic anemometer and bivane techniques.

31. F. J. Scrase, *MOGM*, no. 52 (1930). M. Giblett *et al.*, *MOGM*, no. 54 (1932). A. C. Best, *MOGM*, vol. 7, no. 65 (1935).

32. F. B. Smith, *QJRMS*, 87 (1961), 87. E. K. Webb, *CTP*, no. 5 (1955). H. A. Panofsky and R. A. McCormick, *QJRMS*, 86 (1960), 495.

33. R. J. Deland and H. A. Panofsky, *Sci. Rep.*, no. 2, USAF Project 19 (604) 1027, AFCRC-TN-57-262 (1957).

34. R. W. Longley, *JM*, 16 (1959), 140. N. Thompson, *QJRMS*, 88 (1962), 328. For statistics of microturbulence, see: I. R. Graham, *JAM*, 7 (1968), 90, on Fort Wayne, Indiana; G. H. Fichtl, *loc. cit.*, p. 838, on data from a 150-m tower at Cape Kennedy, Fla.; R. N. Sachdev and P. B. Rawlani, *loc. cit.*, p. 981, on data from an 8-foot mast on a flat roof in Trombay, India; and R. N. Swanson and H. E. Cramer, *JAM*, 4 (1965), 409, on lateral and longitudinal intensities of turbulence from a 62-m tower at White Sands, New Mexico.

For microturbulence spectra, see: S. Berman, *QJRMS*, 91 (1965), 302, on some near the ground, along the mean-wind direction; N. E. Busch and H. A. Panofsky, *QJRMS*, 94 (1968), 132, for a review of observations; L. O. Myrup, *T*, 21 (1969), 341, for those at 1,000 feet above El Mirage dry lake in the Mojave Desert); and H. A. Panofsky and I. A. Singer, *QJRMS*, 91 (1965), 339, on cross-spectra between wind components at various heights at Upton, N.Y.

For an energy budget of microturbulence near the ground at Upton, N.Y., see G. D. Hess and H. A. Panofsky, *QJRMS*, 92 (1966), 277. For observations of dissipation of microturbulent energy at Victoria, Australia, see P. Frenzen, *QJRMS*, 91 (1965), 28; see also F. A. Record and H. E. Cramer, *QJRMS*, 92 (1966), 579, for some above a nonhomogeneous surface at Round Hill, Mass. For a comparison of Eulerian and Lagrangian properties of microturbulence at the 2,500-foot level at Cardington, Eng., see J. K. Angell, *QJRMS*, 90 (1964), 57.

35. For a detailed review of the mathematical theory of the planetary boundary layer, see H. H. Lettau, *EAFM*, pp. 337–72.

36. H. H. Lettau, *TAGU*, 32 (1951), 189. For observations of microturbulence in the planetary boundary layer of the atmosphere, see: C. R. Dickson and J. K. Angell, *JAM*, 7 (1968), 986, on eddy velocities up to 2 km at Idaho Falls; G. A. McBean, *loc. cit.*, p. 410, on some within a forest canopy at Banff, Can.; D. H. Lenschow and W. B. Johnson, Jr., *loc. cit.*, p. 79, on some above a forest in northeastern Wisconsin; S. Pond *et al.*, *JAS*, 23 (1966), 376, on the sea surface off Vancouver Island, Brit. Columbia; H. S. Weiler and R. W. Burling, *JAS*, 24 (1967), 653, on direct measurement of surface stress, drag, and turbulence spectra at Vancouver Island, B.C.; and S. Pond, *BAMS*, 49 (1968), 832, on air-sea fluxes at Lough Neagh, Northern Ireland.

37. See *M*, p. 14, for a discussion of these layers.

38. See H. H. Lettau, *AFGP*, no. 19 (1952), p. 49, for a discussion. On the theory and measurement of vertical eddy fluxes of atmospheric boundary-layer properties, see: A. J. Dyer, *JAM*, 7 (1968), 845, on data from a 16-m mast at Hay, New South Wales; J. W. Telford and J. Warner, *JAS*, 21 (1964), 539, and *JAS*, 22 (1965), 463, for aircraft observations to 1,500 m; S. R. Hanna, *JAS*, 25 (1968), 1026, on using vertical velocities in North America; J. C. Kaimal, *QJRMS*, 94 (1968), 149, on the effect of vertical line averaging on heat flux spectra. On flux-gradients near the ground in unstable air, see H. Charnock, *QJRMS*, 93 (1967), 97, for Victoria, Austr. See A. B. Bernstein, *QJRMS*, 92 (1966), 560, and K. L. Calder, *QJRMS*, 93 (1967), 544, for a

dimensional argument applied to the flux-gradient equations.

39. For an account of this relationship, see C. H. B. Priestley and P. A. Sheppard, *QJRMS*, 78 (1952), 448.

40. T. H. Ellison, *op. cit.* (1956); see note 1.

41. A. K. Blackadar, *AMM*, 4, no. 22 (1960), 3, gives a survey of wind characteristics below 1,500 feet that clarifies this point.

42. L. Prandtl, *Beitr. Phys. für Atmos.*, 19 (1932), 188.

43. See H. H. Lettau, *T*, 2 (1950), 125 for details.

44. For the procedure for predicting vertical flux, see H. H. Lettau, *EAFM*, p. 305, and E. K. Webb, *AMM*, 6, no. 28 (1965), 27–38. C. W. Thornthwaite and P. Kaser, *TAGU*, 24 (1943), 166. E. L. Deacon, *MOGM*, no. 91 (1953). For details of the estimation of heat and moisture flow profiles based on the logarithmic law, see M. H. Halstead, *AFGP*, no. 19 (1952), p. 97. F. A. Brooks, *JM*, 18 (1961), 589. A. J. Dyer and W. O. Pruitt, *JAM*, 1 (1962), 471. A. Nyberg and L. Raab, *T*, 8 (1956), 472.

For heat and horizontal momentum flux determinations, see: W. C. Swinbank, *QJRMS*, 94 (1968), 460, on unstable conditions at Kerang and Hay, Australia; H. A. Panofsky, *loc. cit.*, p. 581, on cospectra for heat flux and wind stress; and M. Miyake and F. I. Badgley, *JAM*, 6 (1967), 186, on a hot thermocouple anemometer for measurements near the surface.

For heat flux determinations, see: J. W. Deardorff, *JAM*, 6 (1947), 631, on eddy conductivity of sensible heat in polar air on a trajectory from New England to Bermuda; A. Perry, *MM*, 97 (1968), 246, on the climatology of turbulent heat flux in the North Atlantic; and A. J. Dyer, *QJRMS*, 91 (1965), 151, on measurements at Victoria, Australia. According to J. W. Deardorff, *JAS*, 23 (1966), 503, measurements indicate that the eddy flux of sensible heat may be directed upward, in the lower atmosphere, when the vertical gradient of potential temperature is zero or the air slightly stable, which is not predicted by simple mixing-length theory. For heat and water-vapour flux determinations, see A. J. Dyer, *QJRMS*, 93 (1967), 501, on unstable air at Victoria, Australia.

For direct determination techniques for eddy fluxes, see: A. J. Dyer, B. B. Hicks, and K. M. King, *JAM*, 6 (1967), 408, on the Fluxatron; A, J. Dyer and F. J. Maher, *JAM*, 4 (1965), 622, on the Evapotron; and J. A. Businger *et al.*, *JAM*, 6 (1967), 1025, for a comparison of Fluxatron, Evapotron, and sonic anemometer at Hay, New South Wales.

45. S. M. Robinson, *JAS*, 19 (1962), 189. P. A. Sheppard, *Journal of Scientific Instruments*, 17 (1940), 218. D. N. Keast and F. M. Wiener, *TAGU*, 39 (1958), 858. G. S. Benton and W. Covey, *TAGU*, 33 (1952), 673.

46. E.g., P. A. Sheppard and M. K. Elner, *QJRMS*, 77 (1951), 450, made use of an electrical analogue of the wet-bulb equation to develop an apparatus to measure turbulent fluctuations of the mixing ratio in the lower atmosphere.

47. J. B. Shaw, *MM*, 84 (1955), 233. R. E. Booth, *MM*, 88 (1959), 110.

48. A. F. Bunker *et al.*, *PPOM*, vol. 11, no. 1 (1949–50).

49. G. Emmons, *PPOM*, vol. 10, no. 3 (1947).

50. A. H. Gordon, *MM*, 79 (1950), 141. "M.O. discussion," *MM*, 80 (1951), 109.

51. See P. A. Sheppard, *QJRMS*, 84 (1958), 205, for a review of vertical transfer processes. For an example of the last simplification, see E. K. Webb, *AMM*, 6, no. 28 (1965), 45.

For vertical profiles of temperature near the Earth's surface, see H. W. Baynton *et al.*, *QJRMS*, 91 (1965), 225, on one in and above a tropical forest in northwestern Colombia. For the critical value of *Ri* for the breakdown of nighttime inversions by turbulence, see R. Lyons, H. A. Panofsky, and S. Wollaston, *JAM*, 3 (1964), 136. For a theoretical profile of daytime temperature, related to buoyant-plume convection, see W. P. Elliott, *JAS*, 23 (1966), 678.

For vertical profiles of wind and temperature near the ground, see: G. E. McVehil, *QJRMS*, 90 (1964), 136, on stable air at O'Neill, Nebraska, and Antarctica; R. H. Thuillier and U. O. Lappe, *JAM*, 3 (1964), 299, on data from a 1,400-m tower at Cedar Hill, Texas; and C. A. Ratcliffe and E. M. Sheen, *loc. cit.*, p. 807, on automatic data collection from an 80-foot portable tower. See also J. P. Pandolfo, *JAS*, 23 (1966), 495, and R. K. Kapoor and A. Sundararajan, *JAS*, 25 (1968), 522, for theoretical profiles of wind and temperature for constant-flux boundary layers in which the eddy conductivity to eddy viscosity ratio varies with height.

For vertical profiles of horizontal windspeed near the surface, see: A. B. Bernstein, *JAM*, 6 (1967), 280, for measurement by means of cup anemometers; D. H. Slade, *JAM*, 8 (1969), 293, for measurements from tall tower in rough terrain, Philadelphia; W. F. Dabberdt, *JAM*, 7 (1968), 367, on the effect of a vertical cylinder on the wind profile above the frozen surface of Lake Mendota, Wisc.; T. Nishizawa and M. Namba, *TJC*, 3 (1966), 69, on one over a river surface in central Japan. For theoretical profiles, see: W. Cover, *JAM*, 3 (1964), 812, on the Swinbank profile; J. J. O'Brien, *JGR*, 70 (1965), 2277, on the diabatic profile at Dallas, Texas; A. B. Bernstein, *JAM*, 5 (1966), 217; H. Charnock, *JAM*, 6 (1967), 211, for a comparison of profiles; O. Essenwanger and N. S. Billions, *GPA*, 60, no. 1 (1965), 160, for a technique to enable the stationary (i.e., smooth) theoretical profile to be isolated from the nonstationary part of the observed wind profile; and W. Klug, *QJRMS*, 93 (1967), 101, for the determination of the eddy fluxes of heat and momentum from wind profiles at Victoria, Austr. See also: L. H. Allen,

Jr., *JAM*, 7 (1968), 73, for mean profiles of horizontal windspeed within a larch plantation at Ithaca, N.Y.; H. Lettau and J. Zabransky, *JAS*, 25 (1968), 718, for a semiempirical model of unrelated changes in wind-profile structure and *Ri* in air flowing from land over inland lakes, at Lake Hefner; and M. A. Estoque, *T*, 19 (1967), 560, for a theory of boundary-layer wind profiles.

For vertical profiles of wind to relatively high levels, see M. Armendariz and H. Rachele, *JGR*, 72 (1967), 2997, to 14,000 feet from the surface. See also L. J. Rider and M. Armendariz, *JAM*, 7 (1968), 293, for a comparison of simultaneous wind profiles derived from smooth (ROSE) and roughened (Jimsphere) 100-gm spherical balloons at White Sands, New Mexico, which indicates that the smooth balloon gives greater computed windshears.

For water-vapor fluxes and vertical profiles, see: U. Högström, *T*, 19 (1967), 230, on an instrument for direct measurement of flux; O. T. Denmead, *JAM*, 3 (1964), 383, on profiles within a forest canopy; and A. C. Dilley, *JAM*, 7 (1968), 717, on computer calculation of humidity gradients from psychrometric data for flux estimations. See also T. V. Crawford, *QJRMS*, 91 (1965), 18, for a theory of moisture transfer in free and forced convection.

52. H. A. Panofsky and A. A. Townsend, *QJRMS*, 90 (1964), 147. On surface roughness and its effects on microturbulence, see: H. A. Panofsky and A. A. Townsend, *QJRMS*, 90 (1964), 147, for a theory of how changes in terrain roughness affect wind profiles; E. F. Bradley, *QJRMS*, 94 (1968), 361, on how changes in roughness affect velocity profiles and surface drag on the Jervis Bay peninsula, New South Wales; P. A. Taylor, *QJRMS*, 95 (1969), 77, for a theory of wind and shear-stress profiles above a change in surface roughness; E. C. Nickerson, *JAS*, 25 (1968), 207, for a theory of how a neutrally-stratified boundary layer adjusts to a sudden change in surface roughness; E. W. Peterson, *QJRMS*, 95 (1969), 561, for a theory of how the lower part of a turbulent boundary layer, with neutral stability, adjusts after a sudden change in surface roughness; and J. Blom and L. Wartena, *JAS*, 26 (1969), 255, for a theory of the development of a turbulent boundary layer (neutral atmosphere) downwind of an abrupt change (and two subsequent changes) in surface roughness. See also N. Untersteiner and F. J. Badgley, *JGR*, 70 (1965), 4573, for the roughness parameters of sea ice.

53. B. Davidson, *JAM*, 2 (1963), 463. For additional references on microturbulence theory, see: J. B. Tyldesley and C. E. Wallington, *QJRMS*, 91 (1965), 158, on the effect of wind shear and vertical diffusion on horizontal dispersion of particles; F. Pasquill, *QJRMS*, 92 (1966), 185, on vertical diffusion from a ground-level source, on the basis of Lagrangian similarity theory; J. H. Gee, *loc. cit.*, p. 301, on an approximate method for the effect of thermal stability on turbulent diffusion; D. R. Dickson and H. G. Allbee, *JAS*, 24 (1967), 18, on the effect of wind shear and lapse-rate variations on friction velocity at the surface; J. P. Pandolfo, *JGR*, 68 (1963), 3249, on the relation between the nondimensional parameters of surface-layer turbulence under diabatic conditions; and J. R. Philip, *JAM*, 3 (1964), 390, for a mathematical model of turbulent transfer processes in air layers occupied by vegetation.

54. This follows O. G. Sutton, *CM*, pp. 500ff.

55. See A. K. Blackadar, *AMM*, 4, no. 2 (1960), 3, on the Ekman spiral, and *M*, p. 89, for a discussion of Brownian motion. On the approach to the geostrophic/gradient wind, see J. K. S. Ching and J. A. Businger, *JAS*, 25 (1968), 1021, for a theory of the response of the planetary boundary layer to a pressure-gradient force varying with time. On the Ekman spiral, see G. T. Csanady, *JAS*, 24 (1967), 467, on the physical characteristics of turbulent Ekman layers, and *JAS*, 26 (1969), 414, on the theory of diffusion of large cloud of particles in an Ekman layer.

56. Windspeeds of 60 to 70 knots are frequent in the 1,000- to 3,000-foot layer. E.g., on March 18, 1918, at Drexel, Nebraska, the mean windspeed increased from 5 to 70 knots in a vertical distance of 780 feet. See Blackadar, *op. cit.* (1960), for details. On phenomena midway between micro- and macroturbulence, see P. D. Tyson, *JAS*, 25 (1968), 381, for a mountain wind in valleys at Pietermaritzburg, South Africa, in which the maximum turbulent energy was found to be generated by waves of about 10 km in length and one hour in period, thus filling the spectral gap between microturbulence and macroturbulence; see also J. Neumann, *JAM*, 6 (1967), 587, for wind variability over periods of 1 to 100 hours at Beersheba, Israel. On mesoscale diffusion, see: N. Thompson, *QJRMS*, 92 (1966), 270, on vertical diffusion for distances of several km at Porton Down, Wiltshire; U. Högström, *T*, 16 (1964), 205, on smoke-puff photography of diffusion at ground level in Sweden; J. F. Pohle, A. K. Blackadar, and H. A. Panofsky, *JAS*, 22 (1965), 219, on quasihorizontal isentropic mesoeddies, causing lateral spread of contrails, and lateral meandering of smoke plumes in stable air; J. H. Gee, *QJRMS*, 93 (1965), 237, on the effect of wind shear on mesoscale diffusion; and S.-K. Kao and H. D. Woods, *JAS*, 21 (1964), 513, on the energy spectra of mesoscale turbulence along and across jet streams. See also, on turbulent convection in the atmospheric boundary layer: J. W. Deardorff, *JAS*, 22 (1965), 419, for a mathematical model for the 50- to 500-m layer, under nearly neutral stability; J. W. Telford, *JAS*, 23 (1966), 652, for a mathematical model of the rise of an isolated, turbulent convective plume in still, clear air, on which see also B. R. Morton, *JAS*, 25 (1968), 135; and T. Green III, *JAS*, 26 (1969), 441, on the effect of vertical variation in the coefficients of eddy viscosity and eddy conductivity on the circulation in Bénard cells.

57. J. G. Edinger, *BAMS*, 36 (1955), 211.

58. For details of a field program in diffusion in a relatively simple geographical site, see M. L. Barad, ed., *Project Prairie Grass*, *AFGP*, no. 59 (1958), 3 vols. See W. G. Tank, *BAMS*, 38 (1957), 6, for details of estimating with synoptic parameters. On micro-diffusion theory, see: E. J. Kotz, *JAS*, 23 (1966), 159, on heavy, settling particles with a sluggish response; G. T. Csanady, *JAS*, 24 (1967), 21, on prediction of fluctuations in the concentration of particles due to turbulent diffusion; and E. E. O'Brien, *JAS*, 23 (1966), 387, on prediction of the rate of growth of a small cloud of particles in a turbulent boundary layer.

59. N. L. Halanger *et al.*, *JAS*, 19 (1962), 99. E. W. Bierly and E. W. Hewson, *JAM*, 2 (1963), 390. On air pollution, see: E. W. Barrett and O. Ben-Dov, *JAM*, 6 (1967), 500, on vertical profiles of pollution particle concentrations in the atmosphere, from lidar observations; W. B. Johnson, Jr., *JAM*, 8 (1969), 443, on lidar observations of the diffusion and rise of a stack plume in west Pennsylvania; J. K. Angell and D. H. Pack, *JAM*, 4 (1965), 418, on tetroon estimations of lateral diffusion from a continuous point-source in Los Angeles; H. A. Panofsky and B. Prasad, *JAM*, 6 (1967), 493, on prediction of pollution from a large number of low-level sources in a narrow valley at Johnstown, Penn.; E. S. Mason, *JAM*, 7 (1968), 512, on an inexpensive constant-density balloon tracking technique for pollution trajectory determination; and V. J. Schaefer, *BAMS*, 50 (1969), 199, on inadvertent modification of the atmosphere by air pollution in the United States. On air-pollution potential, see: E. N. Lawrence, *MM*, 95 (1966), 241, on surface wind directions favoring pollution at Crawley, Sussex; P. Williams, Jr., *JAM*, 3 (1964), 92, on the relation of pollution to stability and wind-speed in the Salt Lake Valley, Utah; and G. C. Holzworth, *JAM*, 6 (1967), 1039, on the climatology of pollution in the United States.

60. R. K. Dumbauld, *JAM*, 1 (1962), 437. For details of the sampling grid, see M. L. Barad and J. J. Fuquay, *JAM*, 1 (1962), 257.

61. C. W. Thornthwaite and B. Holzman, *USDA Tech. Bull.*, no. 817 (1942). For a discussion of how to make the observations, see E. K. Webb, *AMM*, 6, no. 28 (1965), 45. On evaporation, see: P. S. Drysden, *JAM*, 6 (1967), 858, on direct measurement of evaporation from a bare soil by fluxometer; W. Brutsaert, *JGR*, 70 (1965), 5017, on a mathematical model for evaporation into a turbulent atmosphere as a molecular diffusion process; L. Machta, *T*, 21 (1969), 404, on the rate of evaporation in Hurricane Betsy, estimated from the turbulent diffusion of titrium; and R. G. Fleagle, F. I. Badgley, and Y. Hsueh, *JAS*, 24 (1967), 356, on estimation of turbulent flux of water vapor over the Indian Ocean, from aircraft and dropsonde measurements.

62. See *AMM*, vol. 1, no. 4 (1951). For details of the mathematics, see E. Cohen, *AMM*, 4, no. 22, (1960), 25.

63. H. E. Cramer, *AMM*, 4, no. 22 (1960), 12–18, gives full details.

64. See *ibid.* for the mathematics. On gustiness, see F. K. Davis and H. Newstein, *JAM*, 7 (1968), 372, on data from a 1,000-foot tower in Philadelphia; G. C. Gill, *JAM*, 8 (1969), 167; F. O. Okulaja, *JAM*, 7 (1968), 379, on an application of Gumbel's extreme-value theory to wind gusts recorded by a standard anemograph in Lagos, Nigeria; H. Arakawa and K. Tsutsumi, *JAM*, 6 (1967), 848, on the strongest gusts in the lowest 250 m at Tokyo; G. J. Jefferson, *MM*, 95 (1966), 279, on severe, low-level turbulence in sea areas near Cyprus; I. A. Singer, *JAM*, 6 (1967), 1033, on steadiness of the wind at Long Island, N.Y.; J. Förchtgott, *W*, 24 (1969), 255, on mountain-size lee gusts in the Little Carpathians and Low Tatra Mountains of Czechoslovakia; and A. Watts, *W*, 21 (1966), 188, and *W*, 22 (1967), 23, on gust cells associated with convective clouds. See also: R. M. Endlich and J. W. Davies, *JAM*, 6 (1967), 43, on turbulence measurements in the free atmosphere from rising balloons tracked by radar; P. B. MacCready, Jr., *JAM*, 3 (1964), 439, on the standardization of gustiness measurements from aircraft; E. Kessler, J. T. Lee, and K. E. Wilk, *BAMS*, 46 (1965), 443, on associations between aircraft and radar measurements of gustiness; and R. R. Brook and K. T. Spillane, *JAM*, 7 (1968), 567, on the effect of averaging time and sample duration on estimation of maximum wind gusts.

65. Very severe CAT at 30,000 to 40,000 feet over Edinburgh on April 14, 1954, rolled a Canberra aircraft on its back, as reported by D. C. E. Jones, *MM*, 84 (1955), 107). For a general account of CAT literature, see J. Van Mieghem, *BWMO*, 10 (1961), 18.

66. "M.O. discussion," *MM*, 88 (1959), 50.

67. H. S. Turner, *MM*, 88 (1959), 33. A. D. Anderson, *JM*, 14 (1957), 477. K. L. Sinha, *W*, 9 (1954), 3.

68. For details of the model, see L. H. Clem, *BAMS*, 36 (1955), 53. J. Briggs, *MM*, 90 (1961), 245.

69. Y. P. R. Bhalotra, *W*, 10 (1955), 329. D. C. E. Jones, *MM*, 83 (1954), 166.

70. H. S. Turner, *W*, 10 (1955), 294. U. Radok, *MM*, 83 (1954), 48. H. S. Turner, *MM*, 88 (1959), 33.

71. H. Lake, *AFGP*, no. 47. H. Arakawa, *JM*, 14 (1957), 188. R. Wilson, *MM*, 91 (1962), 131.

72. I. J. W. Pothecary, *MM*, 82 (1953), 175, and J. Briggs, *MM*, 90 (1961), 234. R. S. Scorer, *W*, 12 (1957), 275. For the theory, see E. L. Keitz, *JM*, 16 (1959), 57; see also R. J. Reed, *JM*, 17 (1960), 476. W. Schwerdtfeger and U. Radok, *JM*, 16 (1959), 588.

On CAT, see: R. M. Endlich and G. S. McLean, *JAM*, 4 (1965), 222, on gust intensity in CAT; N. Z. Pinns *et al.*, T, 19 (1967), 206, on power spectra of CAT data; D. Trout and H. A. Panofsky, *T*, 21

(1969), 355, on CAT spectra and energy dissipation rates for the 25,000- to 40,000-foot layer; R. L. Mancuso and R. M. Endlich, *MWR*, 94 (1966), 581, on CAT frequency as a function of wind shear and deformation; D. Atlas, K. R. Hardy, and K. Naito, *JAM*, 5 (1966), 450, on a theory of radar detection of CAT, on which see also J. J. Stephens and E. R. Reiter, *JAM*, 6 (1967), 911; and G. W. Kronebach, *JAM*, 3 (1964), 119, on an automated procedure for identifying CAT regions from upper-air data. On CAT and jet-streams, see E. R. Reiter and A. Nania, *loc. cit.*, p. 247; see also S.-K. Rao and A. H. Sizoo, *JGR*, 71 (1966), 3799. For CAT in the United States, see: R. M. Endlich and R. L. Mancuso, *MWR*, 93 (1965), 47; D. Colson and H. A. Panofsky, *QJRMS*, 91 (1965), 507, on a CAT intensity index; R. M. Endlich, *JAM*, 3 (1964), 261, on the mesoscale structure of CAT regions in the southeastern United States; and H. A. Panofsky *et al.*, *JAM*, 7 (1968), 384, for CAT case studies in the Rocky Mts. and Midwest. For CAT occurrences elsewhere, see: W. T. Roach, *MM*, 98 (1969), 65, for synoptics of CAT in the North Atlantic and western Europe; J. C. Lennie, *MM*, 98 (1969), 9, for its synoptics in the North Atlantic; and A. A. Binding, *MM*, 94 (1965), 11, on its association with the 300-mb circulation above the North Atlantic. For U.S. Air Force work, see J. A. Dutton, *BAMS*, 48 (1967), 813, on Project ALLCAT. According to E. R. Reiter and A. Burns, *JAS*, 23 (1966), 206, aircraft measurements over southern Australia during Project TOPCAT show that a wavelength region exists between 1,000 and 4,000 feet in which the atmosphere *receives* turbulent energy, because of gravitational shearing waves that break up into eddies below a critical wavelength; see also E. R. Reiter and H. P. Foltz, *JAM*, 6 (1967), 549, for Australian data indicating that CAT over mountainous terrain is "fed" by orographic waves. According to E. R. Reiter and P. F. Lester, *AMGB*, 17, series A (1968), 1, *Ri* in the free atmosphere depends on the scale of the layer over which it is measured.

73. N. Ramalingam, *N*, 185 (1960), 900.

CHAPTER 4

The General Circulation of the Atmosphere

The general circulation comprises the movements of the atmosphere on a worldwide scale. Since it is usually studied by means of data averaged over several days, so that minor, local, or day-to-day irregularities are smoothed out, any model of the general circulation must be generalized, and cannot include very many short-lived features of importance for local weather. The general circulation is the over-all pattern that must obviously affect local weather at some time or another, directly or indirectly, and is in this sense the greatest single terrestrial cause of climate and weather.[1]

Technically, the general circulation may be defined as the mean three-dimensional pattern of the meteorological elements, plus the "turbulence," the oscillations in or perturbations of the mean pattern, provided by changing, day-to-day synoptic weather patterns. Its basic features may be described in terms of global, seasonal vector mean winds as a function of height, or they may be derived by applying the geostrophic wind relation to mean pressure-contour charts (such as Figures 1.1 to 1.4).

Figures 4.1 and 4.2 indicate the main features of the general circulation as actually observed: its three-dimensional aspects must be particularly emphasized. For comparative purposes, it is convenient to separate the zonal (east-west) and meridional (north-south) components of the mean motion. Figure 4.2 gives these components for the northern hemisphere. The mean meridional circulation is about a meter per second in the lower and middle latitudes throughout a substantial depth of the atmosphere: this is a much weaker circulation than the zonal one, but can nevertheless create or destroy momentum at the rate of 10 m per sec.[2]

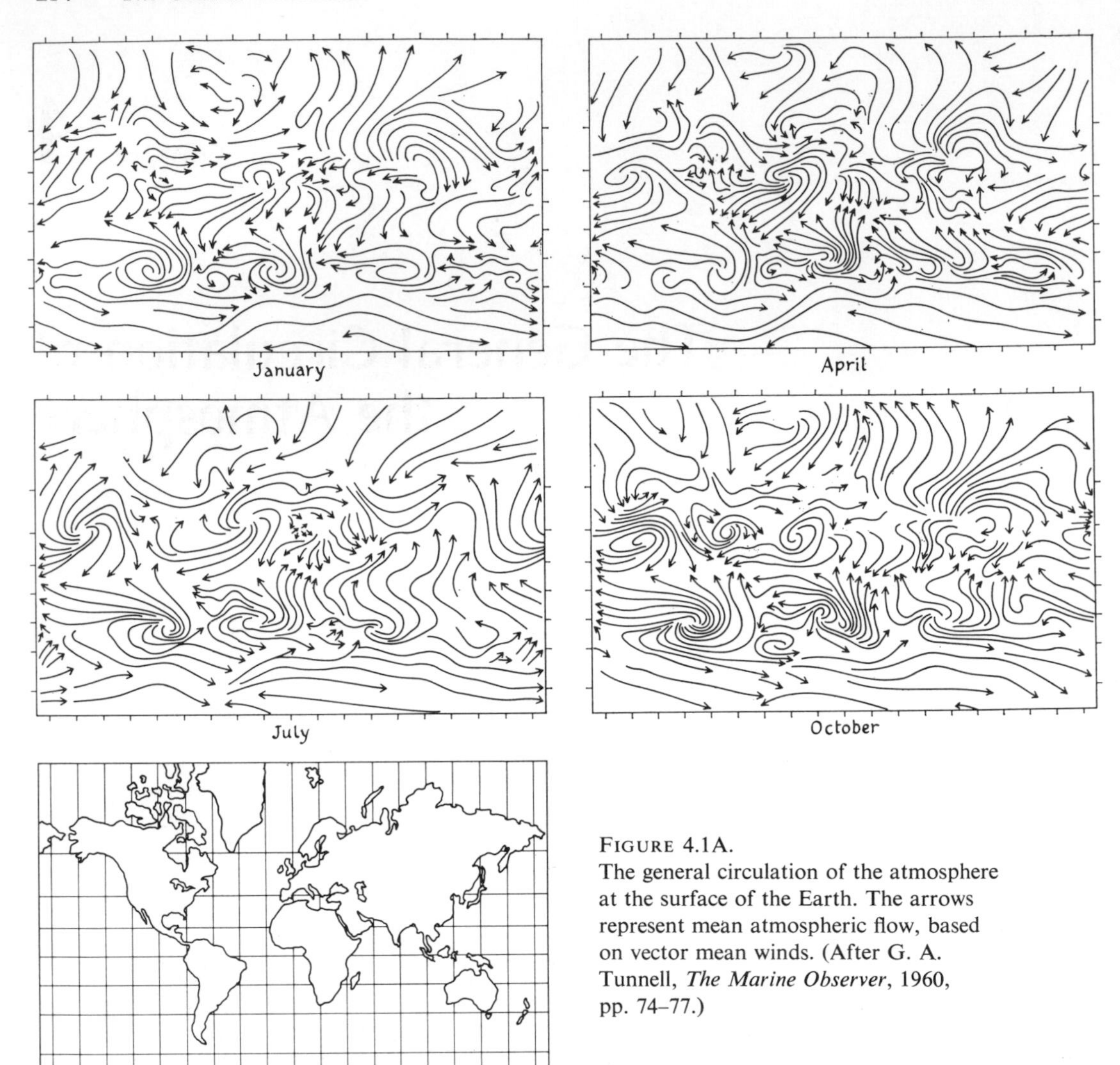

FIGURE 4.1A.
The general circulation of the atmosphere at the surface of the Earth. The arrows represent mean atmospheric flow, based on vector mean winds. (After G. A. Tunnell, *The Marine Observer*, 1960, pp. 74–77.)

The simplest model that incorporates the main features of the observed mean meridional circulation is given in Figure 4.3A. The three kinds of cells are in the troposphere in each hemisphere: the Hadley cells in the tropics, the Ferrel cells in middle latitudes, and the weak subpolar cells beyond these. Angular momentum is injected into the Ferrel cells as indicated by the arrows, and is later carried downward by small convective eddies. Convection is most intense in low latitudes, and thus for equilibrium to occur the Hadley cells must rotate faster than the Ferrel cells.[3]

The model devised by Palmén (Figure 4.3B) takes into account the existence of jetstreams, which are the dominant features of the actual circulation. Observation shows that the Hadley cells, which are directly driven by heat, are the most important single elements of the mean tropospheric circulation, but the Ferrel cells, driven by friction with the Hadley cells, prove to be more significant than Palmén envisaged.[4]

An independent circulation is generated by heating and cooling in the stratosphere, down to about 10 mb. Stratospheric winds reveal remarkable reversals in direction.

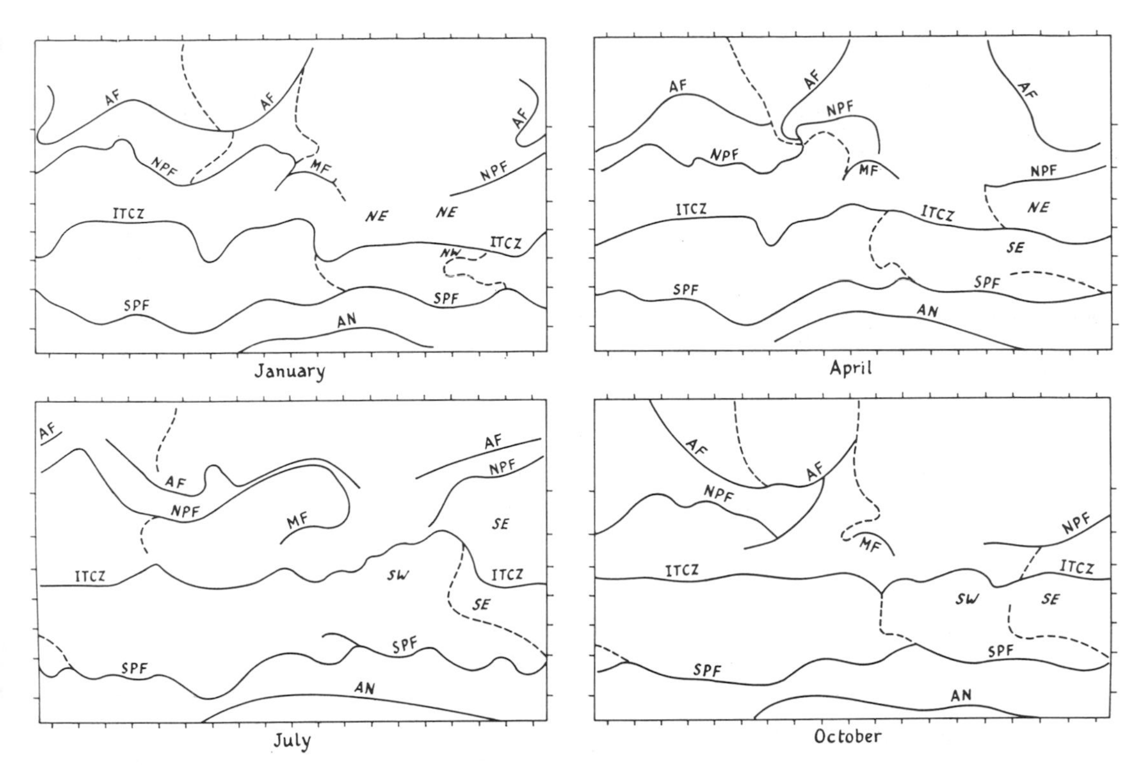

FIGURE 4.1B.
Approximate position of mean frontal zones at the surface. These charts use the key map of Figure 4.1A. *NE*, northeast monsoon. *NW*, northwest monsoon. *SW*, southwest monsoon. *SE*, southeast monsoon. *AF*, Arctic front. *AN*, Antarctic front. *NPF*, Polar Front of northern hemisphere. *SPF*, Polar Front of southern hemisphere. *MF*, Mediterranean front. *ITCZ*, Intertropical convergence zone. Meridional fronts are represented by broken lines. (After Tunnell.)

A stratospheric monsoon occurs in the northern hemisphere: westerlies change to easterlies in April above 10 mb, the reversal proceeding downward and southward from the polar regions, reaching 100 mb in late May. Easterlies prevail above 100 mb from May to August. In late August and early September, these easterlies revert back to westerlies.[5]

Over North America in April, stratospheric polar easterlies are separated by the middle-latitude westerlies from the tropical easterlies of the lower atmosphere, which move northward in that month. By July, easterlies prevail down to at least 15 km in low latitudes, to 20 km in middle latitudes, and to 15–17 km in polar latitudes. By September, the polar and tropical easterlies begin retreating to their minima, in November and December, respectively. In general, the picture at 10 mb is of a slow, steady transition from summer easterlies to winter westerlies, but during January and February 1958 this simple pattern broke down in a very complex manner.

General models of the upper atmospheric circulation have been produced by Kellogg and Schilling (1951), Murgatroyd (1957), and Batten (1961).[6] According to Batten's model, the major center of westerly winds is in the winter hemisphere, although these winds also cross the equator into the summer hemisphere. Easterlies occur in spring in the lower ionosphere, building down from the mesosphere as westerlies develop

A. Mean annual zonal velocity component at mean sea level, westerlies positive, in m per sec.

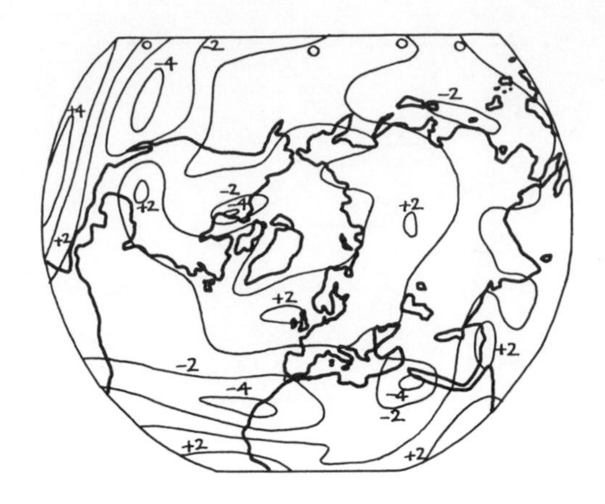
B. Mean annual meridional component at mean sea level, southerlies positive in m per sec.

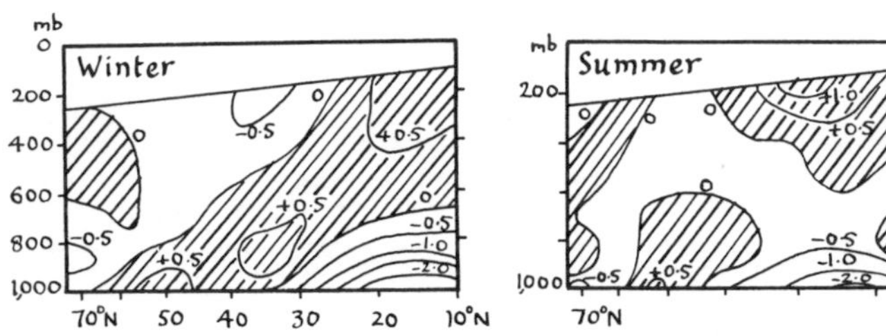

C. Mean meridional velocities in m per sec., southerlies shaded.

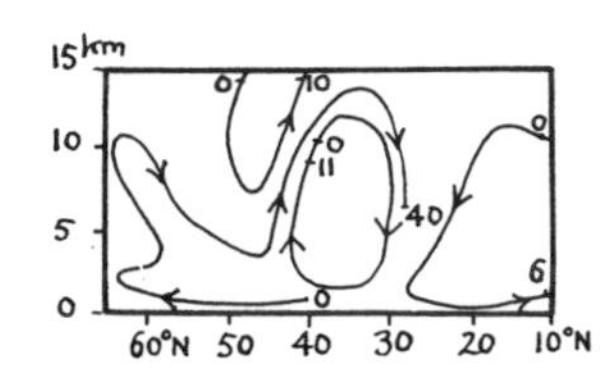

D. Mean winter trajectories, in periods of 10 days since particles were at initial positions.

FIGURE 4.2.
Components of the general circulation. (After Tucker).

aloft; in turn, the westerlies then build down as easterlies develop aloft. Small easterly centers occur in the lower mesosphere in late winter and spring. The stratospheric winds above the Pacific equatorial region are extremely variable.

Important elements of the stratospheric circulation include the Berson westerlies and the Krakatoa easterlies. The former, first discovered at 50 to 60 mb over central Africa, but now known to occur anywhere up to 10 mb, form a continuous ribbon around the equator; the westerlies and easterlies alternate, one half-cycle being 12 to 15 months. The Krakatoa easterlies occur at 25 mb; their existence was first inferred from the movement of volcanic dust after the Krakatoa eruption. Radar wind observations now indicate that Krakatoa westerlies also occur.[7]

The geographical importance of these stratospheric winds is that they make any simple, intuitive model of the atmospheric circulation untenable. In such a model, the rotation of the Earth from west to east is assumed to drag the lower part of the atmosphere with it, imparting a westerly motion to these layers. Thus slight variations in the momentum of the west-east rotating atmosphere would be interpreted at the Earth's surface as indicating winds apparently coming from different directions. A local excess of momentum in the atmosphere, causing the latter to move more rapidly from west to east than the Earth's surface, would be described as a west wind. A local deficit of momentum, causing the atmosphere to move less rapidly than the Earth's surface, would give rise to an east wind.

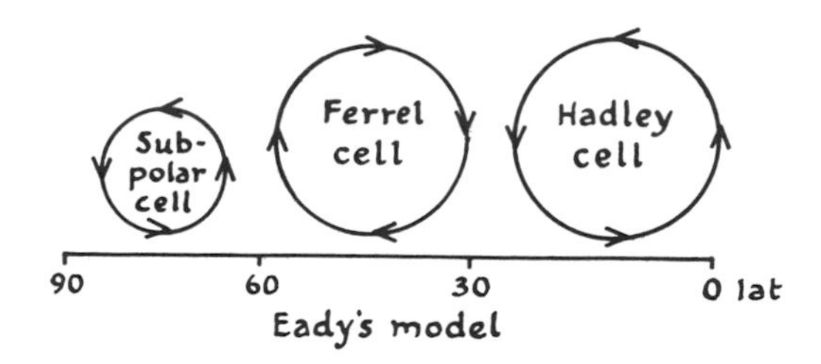

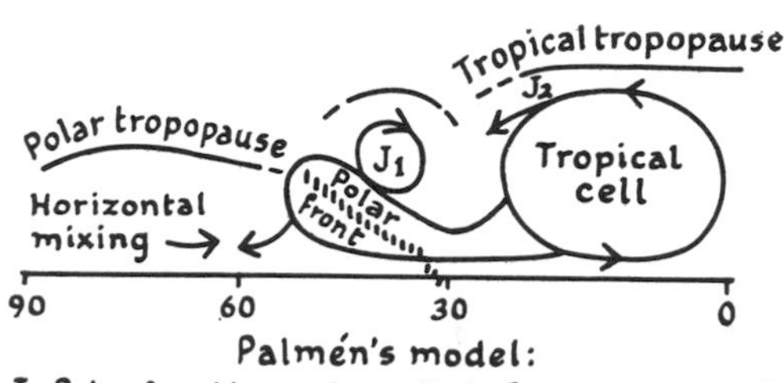

FIGURE 4.3.
Meridional circulation models.

Recent observations indicate that the high atmosphere contains many circulation features that cannot be explained by a simple intuitive model. Thus changes in the rotation of Sputnik 3 can be explained by the existence of a strong westerly wind, accompanying the Earth's rotation, but well above any region of possible frictional drag with its surface. Slow oscillations, representing a balance between inertia and Coriolis forces and static stability, have been measured by radarsonde theodolites at Crawley. They have a period of 12 hours or so, and are due to disturbances with vertical and horizontal dimensions of 1 km and several 100 km, respectively. The complexity of data for the high atmosphere has made it necessary to split the circulations observed into mathematical components.[8]

The circulation models discussed so far have been based on wind observations. Models have also been based on diffusion. In Brewer's model, based on the diffusion of ozone and water vapor (see Figure 4.4A), air rises through the equatorial tropopause, which acts as a cold trap owing to its low temperature (around 80°C). The cold, dry air then moves horizontally poleward, finally sinking in middle and high latitudes. According to Dobson's model, ozone-enriched air arriving via Brewer's meridional circulation is stored in the region of the stratospheric polar-night jet, i.e., in the cold pool over the winter pole. From here, it gradually sinks into the lower stratosphere at temperate latitudes in late winter and spring.[9]

In Spar's model (Figure 4.4B), which is based on the diffusion of radioactive debris, the main exit for air from the stratosphere is through the gap in the tropopause, in which turbulent mixing takes place. More mixing takes place in the polar stratosphere (particularly in winter) than elsewhere, and much less mixing in the equatorial stratosphere than in the Brewer-Dobson model, which describes only one part of the whole circulation. The highest parts of the Brewer-Dobson circulation reach 80,000 feet; above this the atmosphere was envisaged by Brewer and Dobson as stagnant to large-scale meridional and vertical motion. This stagnant region is moist, and meridional transfer is effected by small-scale turbulent diffusion. The height of the transition from meridional-circulating to meridional-stagnant air varies both in time and in latitude.[10]

In the Goldsmith-Brown model (Figure 4.4C), rising air at the equator does not

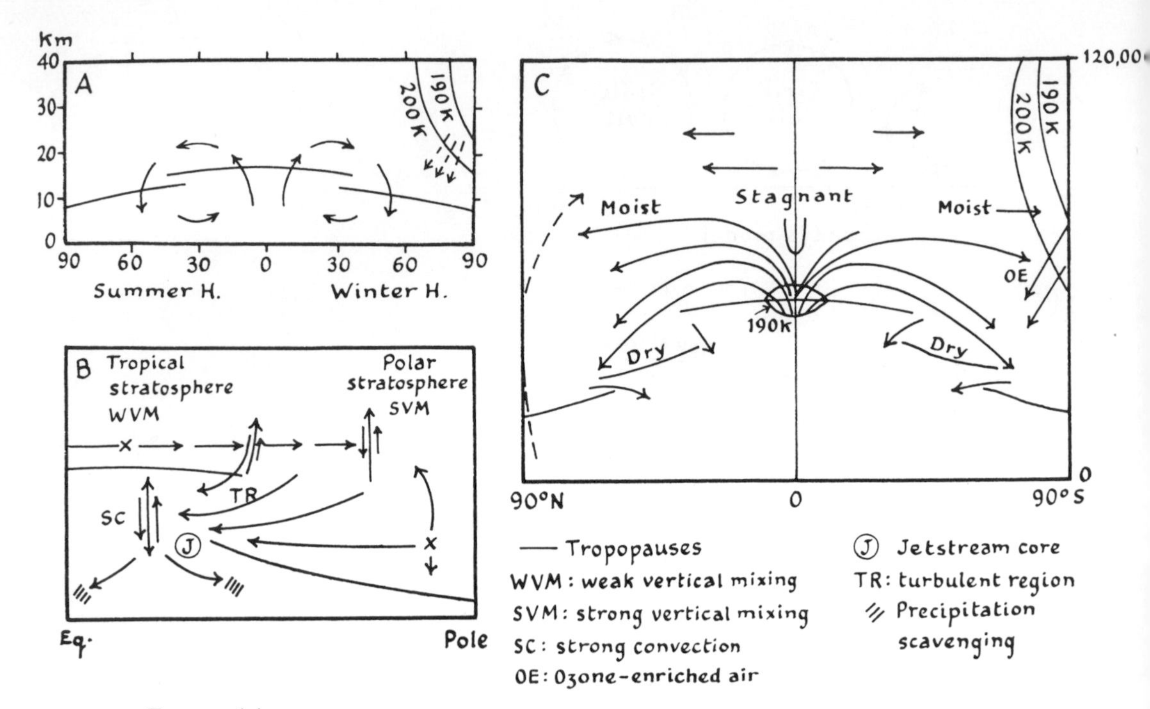

FIGURE 4.4.
Stratospheric-tropospheric circulation models (A, after Brewer and Dobson; B, after Spar; C, after Goldsmith and Brown).

reach great heights, but turns poleward almost immediately above the tropopause. The meridional circulation is rapid just above the tropopause, the air taking slightly more than two months to reach temperate zones. The upper flow is much slower, so that air remains in the ozone-producing layers for about a year. From there, ozone-enriched air is fed slowly into the polar-night jet, where it becomes available for transfer into the middle-latitude lower stratosphere by the Dobson mechanism.[11]

These circulation models for the high atmosphere are significant for weather at the Earth's surface because they demonstrate that air is definitely exchanged between troposphere and stratosphere. Surface weather systems, especially if described in terms of air masses and fronts, cannot be regarded as closed systems, although the error in so regarding them may not be serious in daily weather analysis. A further complication for climatological work is that the atmospheric and oceanic circulations must be considered a complexly integrated single system, if a complete study is to be made of the general circulation. If the stratospheric and oceanographic complications are ignored, the general circulation can be regarded as a consequence of part of the solar radiation received by the Earth being transformed into kinetic energy.[12]

Measurement of the General Circulation

The zonal and meridional components of the general circulation are usually measured by means of indices, which indicate the strength of the mean current and the large-scale weather type. The *zonal index* is found by averaging surface pressure values at equal intervals along two latitudinal circles, and finding the difference between average pressures for the two latitudes. This difference may be converted into a measurement of wind by using the geostrophic relation. Zonal indices are usually computed for the middle-latitude westerlies (using latitudes 35 and 55°N), the polar easterlies (55 and 70°N), and the subtropical easterlies (20 and 35°N).

The *meridional index* measures the total flow of air across a selected parallel of latitude. The *solenoidal index* is computed as for the zonal index, but using thickness isopleths instead of isobars; it is thus in effect a thermal-wind index. *Transport indices* measure the flux of any particular physical property of the atmosphere across a latitude circle: for example, momentum (equal to meridional velocity × zonal velocity) and heat content (meridional velocity × temperature and water-vapor content).[13]

These indices provide indicators of weather type under certain, especially extreme, conditions. At intermediate values, however, a given index may be associated with several weather types. The daily values (see Figure 4.5) exhibit large, irregular fluctuations, which often show no definite periodicities or correlations with weather features. On other occasions, and at certain places, however, distinct correlations do occur.

At Walla Walla, Washington, three-day running means of temperature and precipitation for January (1899–1939) show good correlation with similar means for high and low index values. Temperature singularities occur as follows (the index singularity being in parentheses): warm, January 4 (high index), 17 (low), 26 (low); cold, January 7 (low index), 19 (low), 20 (high). The first precipitation maximum is a high-index singularity, while the second is a low-index one.[14]

A very close correlation exists between the index cycle and the February rainfall minimum in Hawaii. Figure 4.6 gives rainfall graphs for five stations whose annual rainfalls vary from 28.6 to 162.3 inches, but each of which shows a marked decrease in rainfall in February. The explanation is that the northern circumpolar vortex begins to expand in late January or early February, and the ensuing southward migration of the westerlies weakens the trades, which form important components in Hawaiian rainfall. The rainfall increases again when the trades return in March. The southward migration of the westerlies makes possible a longer stationary wavelength between the quasistationary Asiatic coastal trough and the next trough downstream. The position of the latter is critical for rainfall, bringing frequent Kona storms in January, but very few in February.[15]

Over the eastern United States, the lower the velocity of the zonal wind in the layer from 2,000 to 20,000 feet, the greater the possibility of precipitation. Strong zonal winds are associated with clear skies. Two types of mean zonal motion at 300 mb are distinctly related to precipitation distribution. In the first type, the contours are

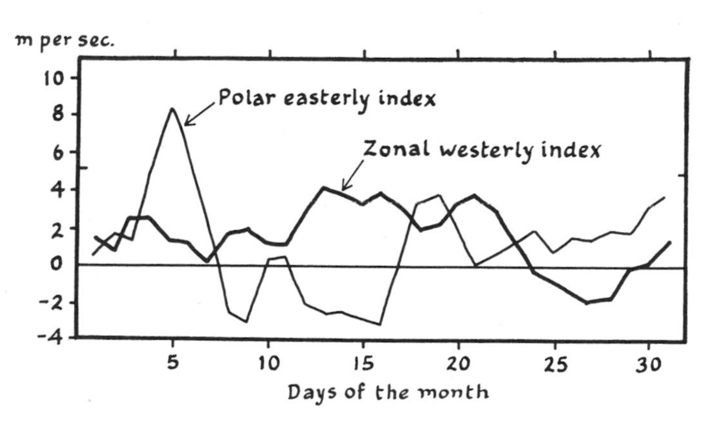

FIGURE 4.5.
Daily values of zonal indices in the half-hemisphere from the Rockies to the Urals, March 1950 (after Forsdyke).

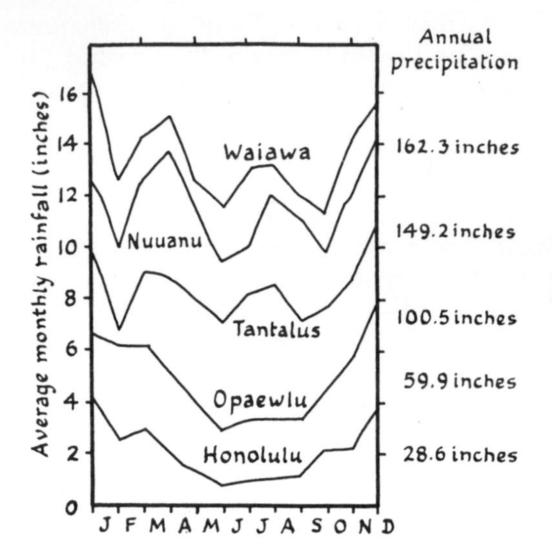

FIGURE 4.6.
A rainfall singularity in Hawaii (after Namias and Mordy).

closely packed in the trough, and there is a weak gradient in the ridges. These troughs tend to move anticlockwise, and most precipitation occurs 5° north of the jet stream and 10° ahead of the trough. In the second type, the troughs tend to move clockwise, with two precipitation maxima, one 3° south of the jet and 8° east of the trough, the other 3° north of the jet and 17° west of the trough.[16]

Since the mean meridional circulation is much less spectacular than the zonal motion, the use of zonal indices has tended to overshadow meridional studies. However, despite its comparative weakness, the mean meridional circulation transports sufficient heat poleward from the tropics to balance the net heat loss in higher latitudes. The heat flux across latitude 15°N comes to 1×10^{15} cal per sec, while the average vertical heat content of the air in that region is 78.5 cal per g. The sensible heat flux across the same region by eddies is only 0.03×10^{15} cal per sec, which is negligible in comparison with that due to the mean meridional motion. The poleward movement of moisture by eddies, however, is much more significant, amounting to 0.2×10^{15} cal per sec between 1,000 and 500 mb.

During the northern hemisphere winter, there is an equatorward current in the lower troposphere, averaging 1 to 3 m per sec near latitudes 10 to 15°N, and a poleward upper current centered at 200 mb and with the same speed. Between the two, from 700 to 400 mb, is an inactive layer with no net measurement. Thus, despite the apparently large poleward and equatorward components in the trades and the westerlies, respectively, the net north-south movement is very slight.[17]

Characteristic Features of the General Circulation

So-called "normal" maps of the general circulation are available from many sources, and are identical with the mean pressure charts discussed in Chapter 1, the link with the circulation being the geostrophic wind relation. If secular changes occur in the circulation, however, the term "normal" perhaps should not be applied to long-period

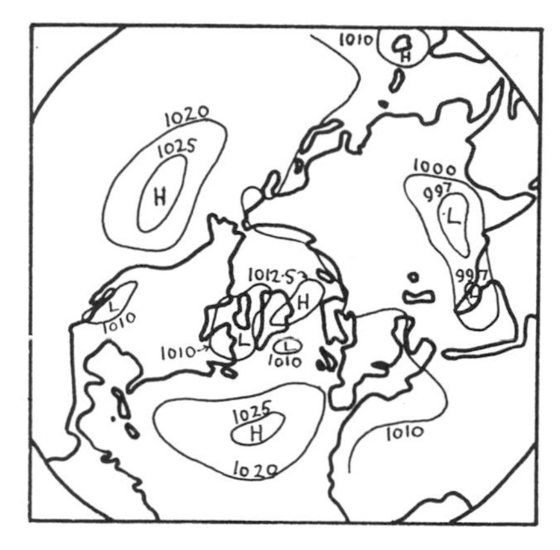

July mean sea-level pattern.

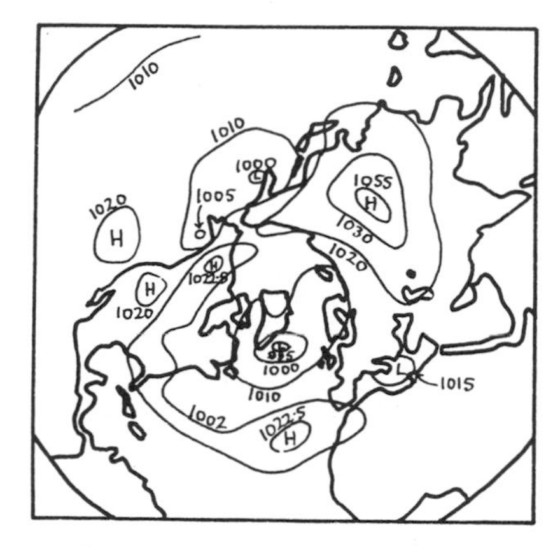

January mean sea-level pattern.

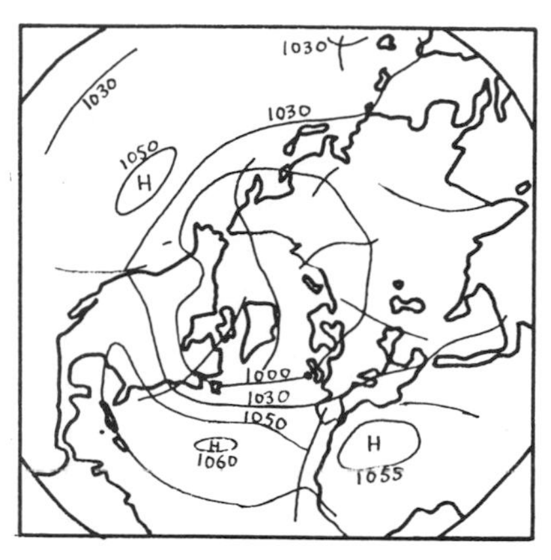

July mean 700-mb pattern.
——— Trough line

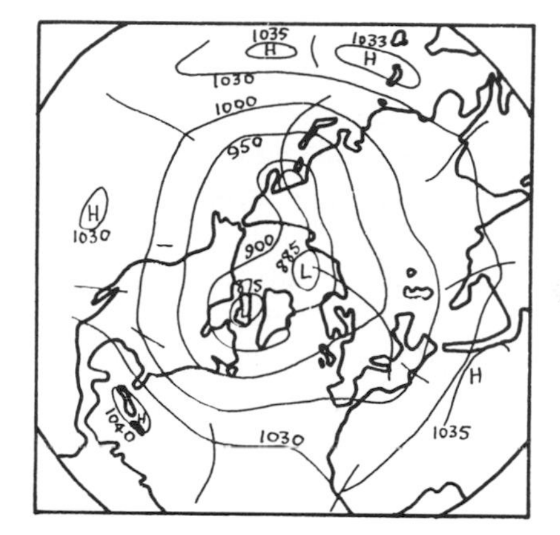

January mean 700-mb pattern.

FIGURE 4.7.
The general circulation, 1947–1958 (after O'Connor).

means of 30 to 50 years. Figure 4.7 therefore describes the atmospheric circulation in terms of 12-year (1947–58) means. The maps in Figure 4.7 differ in many details from the "normal" charts; these differences reflect the secular changes in the circulation, rather than errors in the normal charts, which are based on earlier periods.[18]

A number of features that appear on the 12-year charts are absent from the normal charts; these include the secondary lows in the Gulf of Alaska in January and February, the low centers over the Adriatic (April), Quebec (May), and southwestern Iceland (June), a weak high pressure center over the Philippines in July, an inverted trough over Cape Hatteras in September, and a low pressure center over the Bering Sea in November. Features that appear on the normal sea-level charts but not on the 12-year charts include high pressure areas in the Arctic during January, July, and August, low centers over South China (August) and the Davis Strait (November), and a low pressure area over Colorado in December and January.

At 700 mb, three polar vortex cells occur in both normal and 12-year charts, but those in the latter are not as deep. The cells over Baffin Island and the Siberian Arctic are further south in the normals, while a cell over Kamchatka occurs over the land in the normals, but mainly over the Bering Sea in the 1947–58 maps. A high occurs over the North Pacific in both charts, but in the 12-year maps it is tricellular, with an additional cell over the Philippines dominating during late winter. Troughs near Spitzbergen, Alaska, the Philippine Sea, and the Bay of Bengal, and highs over northeastern Siberia and the Caspian Sea appear only on the 1947–58 maps. These comparisons indicate that the general circulation probably varies with time: it is, in fact, a synoptic problem. The simple world pattern of subtropical anticyclones, travelling middle-latitude depressions, and variable polar easterlies, does not always appear.[19]

During February 20–27, 1947, an "inverted" general circulation pattern prevailed, with a large anticyclone in high latitudes and low pressure in middle latitudes. July 5–14, 1948, illustrated a "summer monsoon" type of circulation, with relatively high pressure over the oceans and low pressure over the continents, characterized by a meridional upper flow. March 8–11, 1949, presented a "double structure" type of circulation, with a series of anticyclones displaced to middle latitudes and a series of lows to the south; even at 500 mb, the circumpolar westerlies broke down, with cyclonic cells in low latitudes. March 13–18, 1949, exhibited a "meridional" type of circulation, with winter monsoon characteristics; large persistent continental anticyclones existed in the same latitudes as large (mainly oceanic) cyclonic cells, and the upper flow was grossly distorted.

Although mean synoptic maps, not normal charts, must be used to study the general circulation, certain phenomena can be recognized on even ultrashort-period mean charts. These include the circumpolar vortices, long waves, jetstreams, and blocking patterns.

THE CIRCUMPOLAR VORTICES

The northern and southern circumpolar vortices are the great whirls of westerly winds in high latitudes. The ratio of momentum in the southern vortex to that in the northern vortex varies from 1.5 to 1, for the year as a whole, to 4 (or more) to 1, for the northern summer-southern winter season. Therefore the southern circumpolar vortex can be regarded as the flywheel of the general circulation; changes in its size or intensity result in worldwide adjustments.[20]

The "roaring forties," the westerly belt that forms the equatorward edge of the southern vortex, are well-known for their strength. The momentum and energy of these southern westerlies are responsible for maintaining the position of the meteorological equator to the north (on the average) of the geographical equator, at least over the oceans. During the "Little Ice Age" in the northern hemisphere, 150 or so years ago, the southern westerlies were further south and weaker than today. As a result, the meteorological equator was nearer the geographical equator, and the equatorial low-pressure belt was wider.

The mean 500-mb patterns in the circumpolar vortices show a series of troughs and ridges, those in the northern vortex being much more pronounced than those in the south. The northern troughs and ridges appear to be real features of the circulation, anchored in position by the Rocky Mountains barrier. Those in the southern vortex are statistical features, representing successions of mobile features that reach their greatest amplitudes near certain preferred longitudes. The southern features are held

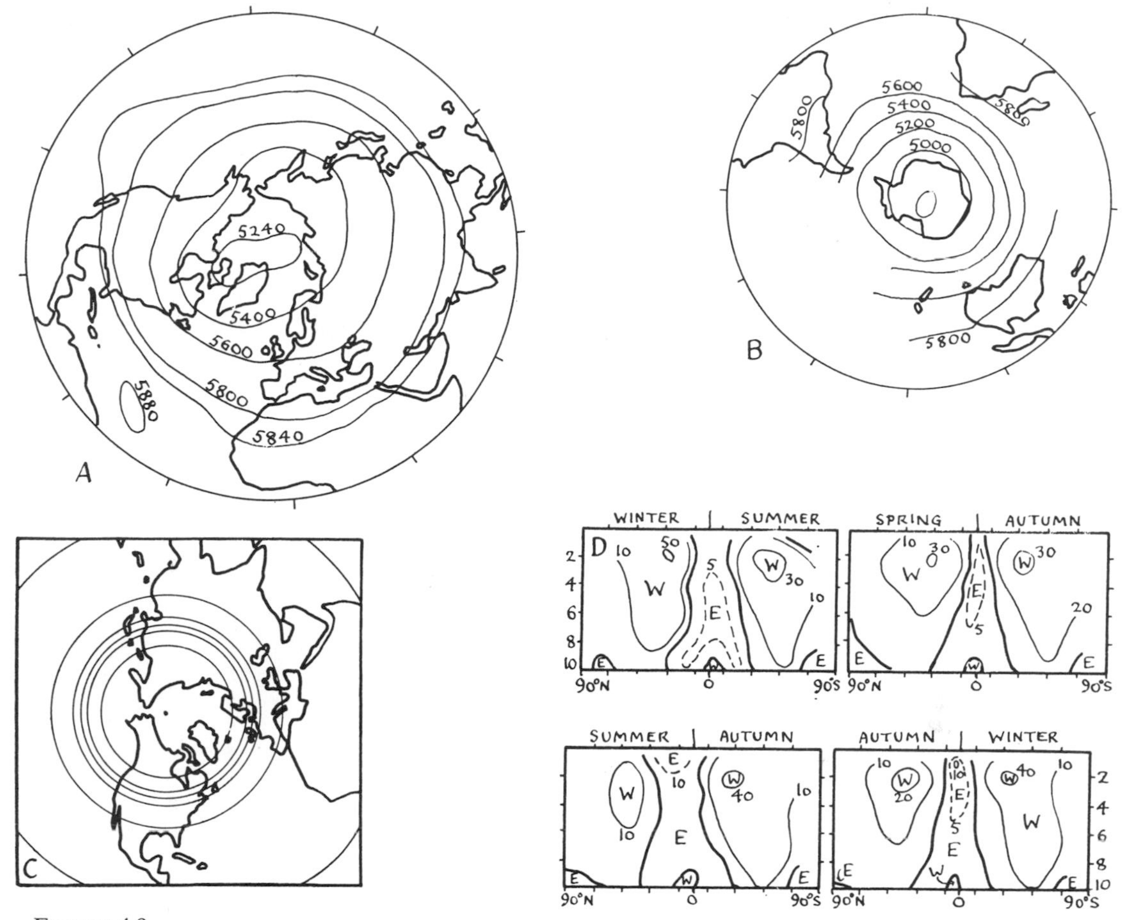

FIGURE 4.8.
Some features of the general circulation.
A,B, Comparison of northern and southern circumpolar vortices, giving contours in meters (after Lamb).
C, Model illustrating eccentricity of northern vortex (after La Seur).
D, Zonal cross section of mean westerly component in m per sec. Vertical scale in hundreds of mb. Sections are averaged over all longitudes in the northern hemisphere, and over 70°W–170°E in the southern hemisphere.

in position by thermal effects, particularly the cooling produced by the Antarctic continent. Figure 4.8 presents a comparison of the mean features of the two vortices.

Both vortices are frequently asymmetric, so that their centers of rotation do not coincide with the geographical poles. The asymmetry of the northern circumpolar vortex is occasionally so large (up to 10 degress of latitude) that the zonal index no longer accurately describes the true state of the westerly current. Periods of large asymmetry usually occur during the high-index stage of a poleward index trend, especially in winter, when the circumpolar current is strongest and concentrated in a narrow zone. These periods last about one week, usually. They result in a concentrated westerly current flowing at much lower latitudes over the Pacific than over Europe. Temperatures are above normal for the whole of the United States, and precipitation is heavy in the Pacific Northwest, absent in the southwestern United States, and moderate

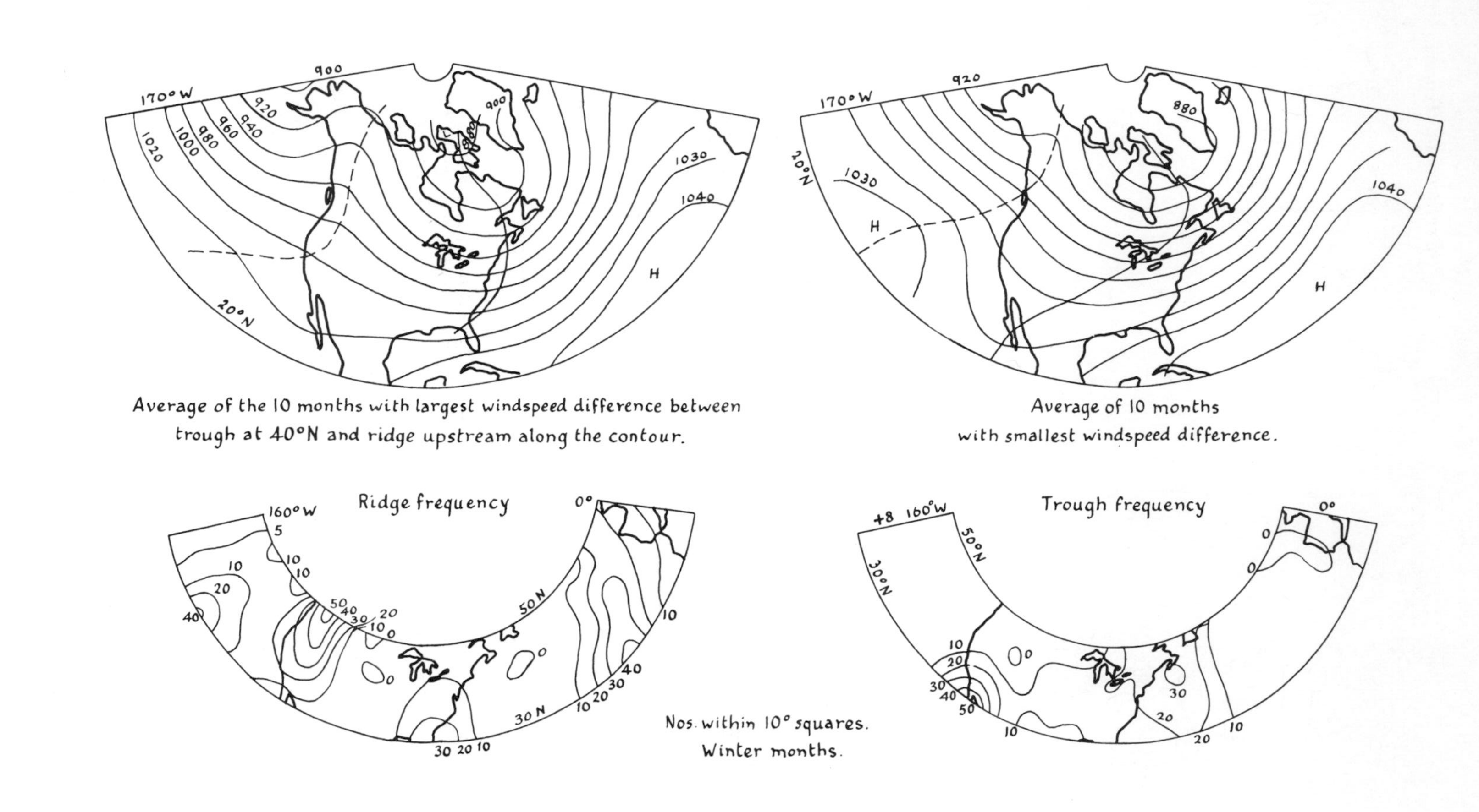

FIGURE 4.9.
Composite 30-day mean 700-mb charts during winter,
December 1932 to March 1951. A is at bottom left, B at top left (after Klein).

elsewhere. The asymmetries are somehow produced by surface friction and large topographic barriers, but the cause of the vortices themselves is uncertain. The vortices appear to be properties of the airflow, and the presence of an Antarctic continent does not seem to be a cause of the southern vortex.

The seasonal march in the strength of the southern vortex consists of a simple annual variation in the stratosphere, with a maximum in late winter, and a semiannual variation in the troposphere, with maxima at the equinoxes, when the mean temperature gradient is greatest. The semiannual variation of the tropospheric vortex is an integral part of the planetary circulation. The stratospheric vortex, however, disappears when the polar stratosphere becomes heated by incoming solar radiation or ozone concentration, and reforms when the incoming radiation diminishes.[21]

LONG WAVES

Daily aerological charts clearly indicate the existence in the atmosphere of wave-like isobaric patterns. When averaged over periods of days, these patterns resolve themselves into simple sinusoidal waves with a length of several thousand miles, the Rossby waves; usually three to six waves occur in a circuit of the globe. Figures 4.9 and 4.10 present some observed characteristics of these long waves. Figure 4.9,A, is of waves of relatively large dimensions, Figure 4.9,B, of waves of small amplitude and length, on which the effect of the Rocky Mountains is very great.[22]

Long-wave patterns often differ with height above the same region. If waves at two different heights are opposite in phase, a blocking pattern may result, in which the usual eastward progression of long waves is arrested; the disturbance of the zonal circulation then gradually spreads itself westward.

In a true zonal circulation, the long waves are in phase at all heights in the atmosphere. The long waves in both pressure contours and thickness patterns are also in

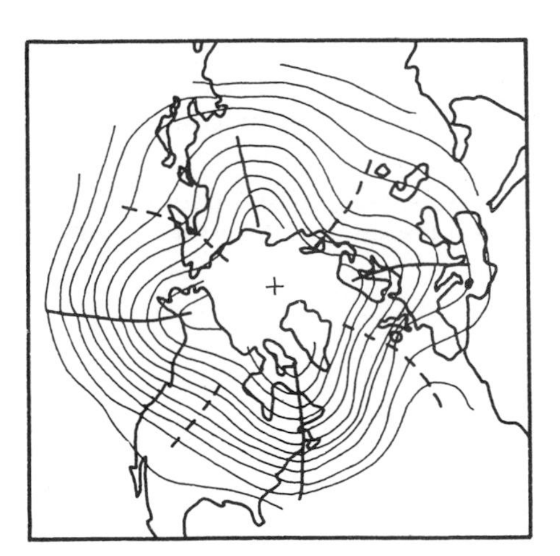

Area-mean contours.

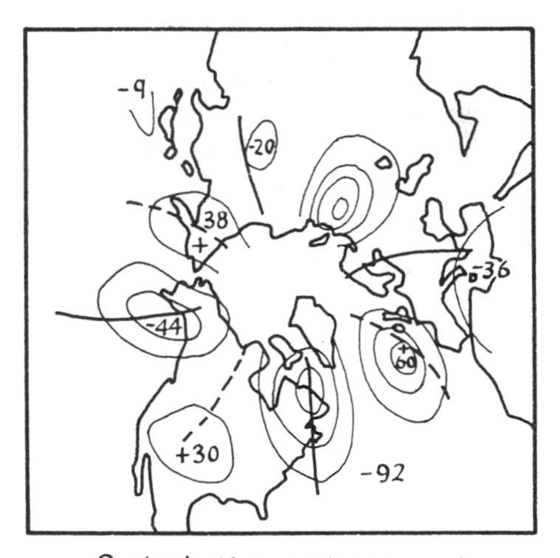

Perturbation centers based on area-mean heights.

FIGURE 4.10.
Area-mean synoptic charts for 3:00 P.M. G.M.T., April 13, 1954. Troughs are marked by solid lines, ridges by dotted lines. (After Wolff.)

phase; the wave forms increase in strength right up to the tropopause, then gradually disappear with increasing height in the stratosphere.

Long waves are clearly dynamic features, but may be much modified by geographical contrasts. In North America, the dynamic effect of the Rocky Mountains is to produce a major trough in the westerly current in their lee. Inertia processes then give rise to a dependent wavetrain downstream of the orographically anchored trough.

Daily aerological charts reveal the frequent presence of short waves, which appear as ripples on the pressure contours. The short waves may be stable or unstable. The unstable waves move in the same direction as the general wind through the long-wave pattern; they rapidly increase in size, and often reach the ground as important weather systems. The stable waves usually remain mere ripples on the long waves. They move much faster than the long-wave pattern, and in the lowest part of the atmosphere may cause the pressure contour and thermal patterns to become opposite in phase. Long waves sometimes form from short, unstable waves, and vice versa.

Rapid increases in the amplitude of long waves may have great repercussions on surface weather. During November 13–18, 1958, a major upheaval in the zonal index over the United States was followed by a rapid change to near-normal conditions. The resulting weather embraced very low temperatures in the southwestern United States, very high temperatures in the southeast, heavy snowfall over the Rockies and the north and west central plains, and very rapid cyclogenesis over the central United States, with tornadoes and windstorms in the area from Texas to Illinois.[23]

Long waves are not always suitable for everyday weather analysis, because they are not always progressive, and sometimes show poor day-to-day continuity. By averaging aerological charts in space instead of in time, for example, by using a 20-degree square- or diamond-shaped grid, the long-wave part of the airflow may be filtered out. The space means are subtracted from the total observed (24-hour) height change of the pressure surfaces. The residuals indicate variations in the mass of the atmosphere due to short baroclinic waves. These waves form small, symmetric, closed centers (see Figure 4.10), which maintain good continuity and never exhibit retrograde motion. They show a close relation to surface developments and to jet-stream locations. Space-meaning a long-wave chart around a latitude circle, and subtracting the means from the respective pressure-surface heights, eliminates the mean zonal flow, and reveals new types of closed cyclonic and anticyclonic centers, the *perturbation centers*, which show good continuity and do not suffer from the disadvantages of long waves when day-by-day analysis is required.[24]

JETSTREAMS

Pure long waves are essentially climatological phenomena, in that their characteristics are not immediately obvious from a synoptic chart, but must be determined by statistical analysis of data averaged over definite periods of time or space. In other words, it is not possible to infer the existence of a pure long wave, at a particular time in a given area, by actual observation.

Jetstreams, in contrast, are very much real features of the atmosphere, although the major climatological jet shown in Figures 1.5 and 1.6 does not necessarily have the depicted form at all times. These figures make it clear that jetstream phenomena are the dominant features of the general circulation as studied in the vertical plane.

A glance at almost any upper-air chart covering an extensive area will reveal that on most days ribbons or zones of relatively high-velocity winds exist over many parts of the Earth's surface. The question then arises: which of these are jetstreams? The convention adopted by the WMO regards all airstreams with velocities exceeding 60 knots (30 m per sec) as jetstreams. The definition states:

> "A jetstream is a strong narrow current, concentrated along a quasihorizontal axis in the upper atmosphere or in the stratosphere, characterized by strong vertical and lateral wind shears, and featuring one or more velocity maxima. . . . Normally a jetstream is thousands of kilometers in length, hundreds of kilometers in width, and some kilometers in depth. The vertical shear of wind is of the order of 5-10 meters per second per kilometer and the lateral shear of the order of 5 m per sec per 100 km. An arbitrary lower limit of 30 m per sec is assigned to the speed of the wind along the axis of a jetstream."[25]

Applying this criterion to synoptic upper-air charts shows that the main jetstreams are as follows: (a) the main circumpolar, or Polar Front, jets, most marked at around 300 mb between latitudes 40° and 60°; (b) the westerly subtropical jets, near 200 mb in both hemispheres, and around lattitude 30°; (c) the equatorial jets, easterly, above 200 mb, and of up to 100 knots; (d) the stratospheric subpolar jets, strongest at 30 km or higher, which vary from strong westerly jetstreams in winter to moderate easterly jetstreams in summer; and finally (e) local or regional jetstreams, which are isolated strong winds in the middle and upper troposphere, affecting areas the size of, for example, the English Midland counties.[26]

The Polar Front jets are highly variable in position, and do not appear clearly on the mean charts, but do produce a general "jetstream zone" in middle latitudes that shows up very prominently on average zonal cross sections. The strongest Polar Front jets in winter are found where troughs in the middle-latitude westerlies dip into the latitudes of the northern subtropical jet, especially near the east coasts of Asia and North America and over the Middle East. Middle-latitude depressions are often related to Polar Front jets, their movement corresponding in a general fashion with longitudinal variation in the speed of, and movement of waves along, the jet. Polar Front jets often break, and are connected to surface fronts in a complex manner; pronounced fronts extending all the way up to the tropopause are nearly always associated with jetstreams (the latter tend to be found above the 500-mb location of the Polar Front), but jets not accompanied by distinct fronts often occur.

Subtropical jetstreams are not normally associated with fronts, and typically reach velocities of between 100 and 200 knots. They usually are very steady in winter, but weaken and shift poleward in summer; the northern hemisphere subtropical jet in particular is very weak in summer.

Certain wind variations, both horizontal and vertical, are typical of different parts of a given jetstream, the variations being in wind shear, rather than in wind direction. On the anticyclonic side of a jet, horizontal variations in wind shear are limited by the condition that the absolute vorticity of the current must not fall below zero. This means that in a straight airflow in a middle-latitude jet, the anticyclonic wind shear does not exceed 10 m per sec in 100 km, approximately the value of the Coriolis parameter, but the total observed shear often approaches this value. On the cyclonic side of a jet, horizontal wind shears are often two or three times as great, and may be five or six times as great without being unusual.

The strongest vertical winds are found 1 km below the tropopause, on the average, although very often the axis of the jet lies in the tropopause break region. In middle latitudes, the rate of vertical wind shear changes little with height, although windspeeds decrease to approximately one-half the maximum value at 5 km above and 5 km below the maximum wind. With the subtropical jets, strong winds are confined to a shallow layer 100 mb above and 100 mb below the maximum wind. When a jet is associated with a front, the rate of wind shear through the front in the vertical direction is often 15 to 20 m per sec per km.[27]

The Polar Front Jetstreams. The existence of the northern hemisphere Polar Front jet was demonstrated in spectacular fashion during the Second World War, when USAF bombers over Japan found themselves stationary above their targets, while their airspeed indicators recorded 250 miles per hour. Recent transosonde observations show that probably the strongest tropospheric winds in the world (or at least in the northern hemisphere) are found above and to the east of Japan in winter: these may ultimately be due to either (a) the strong thermal winds resulting from the intense contrast in surface temperatures between land and sea on the east coast of Asia, or (b) the airflow around the Himalayas, or both, but the direct cause is the existence of the Polar Front jet in this region. Before the Second World War, computed rates of movement of cirrus clouds, and incidents such as the destruction of zeppelins at 20,000 feet by winds and the remarkable flight of a pilot balloon from Calshot to near Leipzig in 1923 at an average speed of 143 miles per hour, had suggested that very high-velocity winds must exist in the middle and upper atmosphere. It was not until the late 1940's and early 1950's that the real nature of these high-velocity ribbons became apparent. Now the Polar Front jets are recognized as possibly the most important constituents of the whole atmospheric general circulation.[28]

The possible existence of a Polar Front jet was postulated by Bjerknes in 1933: he suggested that the kinetic energy of the upper westerlies might not be evenly distributed with latitude, as the energy of the lower westerlies is, but might instead be concentrated in a relatively narrow latitudinal ribbon, possibly above the Polar Front. We know today that a well-defined major front, the Polar Front, is often present below the jet core, and intersects the ground to the south of the jet in the northern hermisphere. It is very distinct at 500 mb, where it forms a more or less continuous baroclinic zone around the Earth, but is very diffuse near the ground and in the upper troposphere.

Early studies concentrated on the synoptic relationship between temperature distributions in the upper westerlies and the Polar Front jet. For example, occasional cross sections revealed that the jetstream, the wind field, and temperatures are interconnected; the zone of lowest temperature on most cross sections is 4° of latitude south of the zone of strongest wind, and the zone of highest temperature 7° of latitude north of the latter; the contrast in atmospheric temperatures between high and low latitudes is mainly concentrated in the zone of strongest upper westerlies, and migrates with it.[29]

Other studies emphasized the layer of maximum windspeed in the atmosphere. Synoptic charts giving the height of this layer (see Figure 4.11) could be constructed because raw soundings near jetstreams produce a graph of windspeed against height that usually shows a sharp maximum or peak; the locus of these peaks in space defines a more or less continuous surface, the layer of maximum windspeed. Statistical analyses proved that the maximum windspeeds above Europe are normally just below

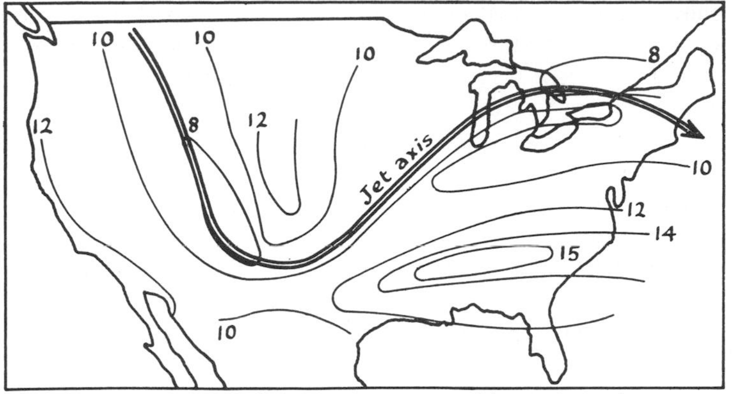

Mean height of the maximum-wind layer (in km), 3:00 P.M., G.M.T., February 27, 1954.

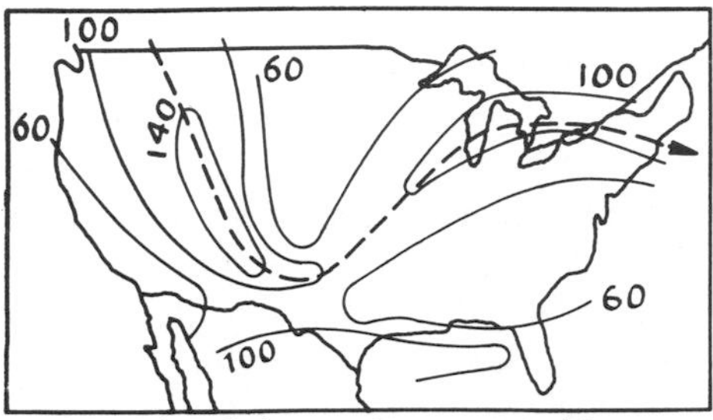

Isotachs of the maximum-wind layer (in knots), 3:00 P.M., G.M.T., February 27, 1954.

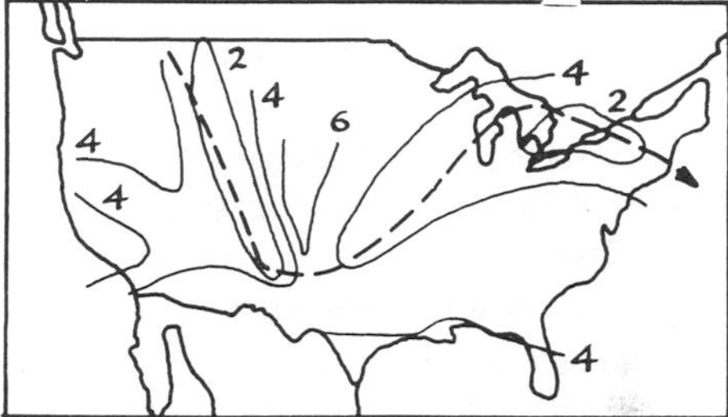

Thickness of the maximum-wind layer (in km), 3:00 P.M., G.M.T., February 27, 1954.

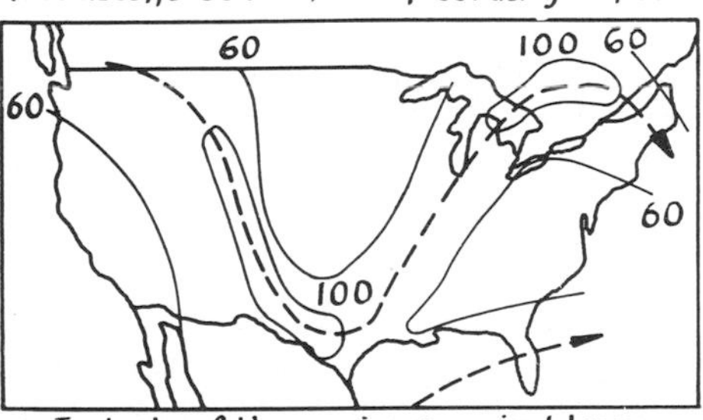

Isotachs of the maximum-wind layer (in knots), 3:00 A.M., G.M.T., February 28, 1954.

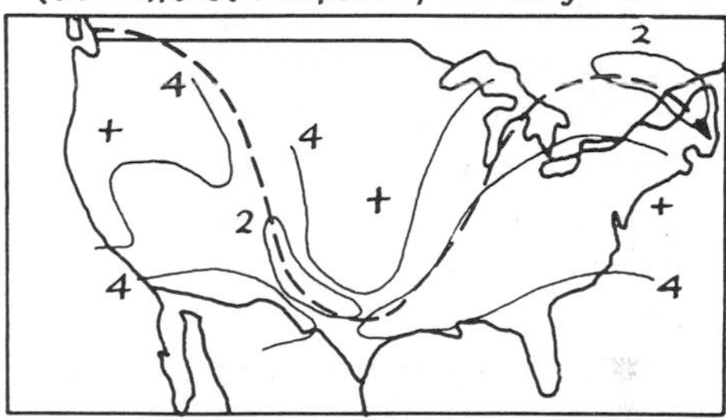

Thickness of the maximum-wind layer (in km), 3:00 A.M., G.M.T., February 28, 1954.

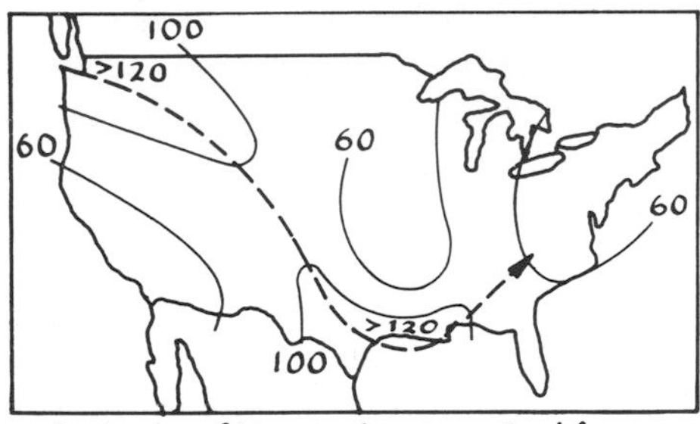

Isotachs of the maximum-wind layer (in knots), 3:00 P.M., G.M.T., February 28, 1954.

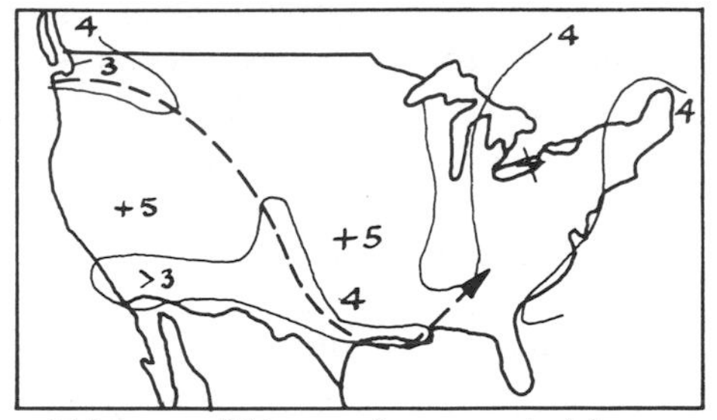

Thickness of the maximum-wind layer (in km), 3:00 P.M., G.M.T., February 28, 1954.

FIGURE 4.11. Mean height of the maximum wind-layer. Broken-line arrows indicates jet axis. (After Reiter.)

the tropopause. Above the British Isles, the maximum wind is most frequent at 20 to 40 mb below the tropopause, but there are local deviations; for example, the maximum winds are often found above the tropopause at Lerwick, but less often at Larkhill. At Gibraltar, the relation between maximum wind and the height of the tropopause is less good than for the British Isles. The monthly variations of tropopause height and maximum-wind height show parallel features, but are separated in time; the correlation between them is closer in winter than in summer.[30]

Statistically the thermal winds must reverse in direction below the tropopause. Other considerations also indicate the importance of the thin layer between the layer of maximum wind and the tropopause. It is here, in the uppermost layer of the troposphere, that mean vertical motion is approximately zero, and this fact led German meteorologists to term it the *nullschicht* instead of the maximum-wind layer. Above central Europe, the nullschicht is at an average height of 10 km. Below it, vertical motion is upward in cyclones and downward in anticyclones; therefore the former are usually colder than the latter at all heights up to the nullschicht, and the resulting thermal wind must produce maximum windspeeds at, or near, the nullschicht.[31]

The nullschicht forms the level toward which all net vertical motions in both troposphere and stratosphere are directed in deep depressions, and it marks the level of vertical divergence in deep anticyclones. At the nullschicht, maximum windspeeds are above gradient-wind strength, and there is an average ageostrophic flow toward regions of higher pressure. In contrast, when the nullschicht is ill-defined, there is an average ageostrophic flow from regions of higher pressure toward regions of lower pressures. These facts have important consequences in the production of ageostrophic flow at the beginning, and end, of jetstreams in which the isobars run east-west in a straight line.

The existence of a nullschicht layer is really a necessary complement to the existence of an atmospheric friction layer near the ground. In the latter, windspeeds are usually much below gradient-wind strength, especially close to the Earth's surface. The constant drain of air because of the continual flow from high to low pressure near the ground clearly implies that windspeeds that are well above gradient strength must exist in the upper troposphere, in order that geographical differences in sea-level pressures may be maintained. The necessity for very high wind velocities in those layers of the atmosphere that are well above the friction layer is thus established, but the reason why they take the form of ribbon-like jets is another question.

The Polar Front jet in the northern hemisphere does not always form a continuous ribbon of high-velocity winds completely circling the Earth. Even data averaged over a decade show that it splits over the north Pacific, with a secondary branch passing north of Alaska. On some days it may be very disorganized; the upper westerlies will break up into separate vertical circulations, and there will be no clear-cut concentration of velocity. At other times, the jet is more organized, and consists of a succession of velocity maxima, alternating with areas of very weak winds along the "jet" axis. Individual jets usually have a life cycle of a week or more; sometimes there is no distinct jet, but wind velocities in middle latitudes are fairly uniform, with several narrow bands of higher velocity winds, termed *jet fingers*. Also, because the atmospheric circulation pole does not coincide with the geographical pole, normally being displaced towards the central Pacific by up to 10° of latitude, the latitudinal positions of the Polar Front jet in its Pacific and European-Atlantic sectors may differ by as much as 20° in latitude.[32]

Precipitation distribution at the Earth's surface is closely related to the positions of the Polar Front jets, both globally and regionally. According to the synoptic models of the Bergen school, precipitation should be organized in long narrow bands that are parallel with fronts. Actual precipitation tends to be much more localized, and in fact rarely occurs in bands the size of frontal areas, so that the real distribution of rainfall is very different from that shown in the classic models.

Dynamic theory shows that areas of general precipitation should be associated with regions of vorticity change in the atmosphere. Clouds and precipitation ought to appear where relative vorticity in the upper troposphere decreases downstream, and clear or fine weather where the vorticity increases downstream. The frequency of *virga* indicates that precipitation is often falling in the upper air, although none is received at the ground; precipitation areas on relative vorticity charts are therefore larger than the areas of rainfall received at the Earth's surface. One would therefore expect to find rapid latitudinal variations in precipitation in the neighborhood of a jetstream, because absolute vorticity increases rapidly north of the jet axis (to a value much higher than the extremely high values found in polar regions), and decreases south of the axis, to near-zero vorticity some 200 km from the jet core.[33]

It is possible to design an empirical model (see Figure 4.12A) that describes the rainfall distribution to be expected at the ground from a "typical" jet. Much more rain (in 24-hour totals) falls immediately to the north of a jet axis over the British Isles than to the south of it. There is twice as much rain in the right entrance region of the jet than in the left, and twice as much rain in the left exit region than in the right. There are equal amounts of rain both left and right of the axis in the central portion of the jet. Thus the most rain falls to the right of the jet's confluence and to the left of its diffluence. This observed distribution of precipitation may be explained in terms of an acceleration of the vertical circulation of the atmosphere at jet-stream level: there must be upward currents to the right of the jet axis in its entrance region, and downward currents to the left (see Figure 4.12B). This is a direct circulation, since it is in the same direction as a natural overturning due to convection. An indirect circulation, in the opposite direction, must be present in the exit region to explain the precipitation distribution there. Therefore a cross-axis flow of air may be presumed to exist at jet level; observations suggest that this flow averages 10 knots above the United Kingdom.[34]

Local, daily movements of the Polar Front jet are certainly followed by changes in the distribution of precipitation, but it is very difficult to summarize these changes in a simple empirical model. For example, between September 19 and 21, 1958, heavy precipitation that was falling over the western Gulf states began spreading northeastward through the Ohio Valley without any surface cyclogenesis taking place. There was a time-lag between the maximum winds at jetstream height and maximum winds at the ground, which suggests that maximum winds along the jet axis can be used as precipitation criteria.[35]

During the period February 4–9, 1957, persistent rains falling in the southeastern United States were associated with a weak wave on a quasistationary front, with no significant cyclone development. The precipitation was very intermittent and patchy, moving from west to east. This spotty pattern was caused by features of the associated jetstreams. As individual wind maxima moved along the northernmost jet, the ascending motion north of the southernmost jet was alternately reinforced and opposed, the precipitation being during the periods of reinforcement, i.e., associated

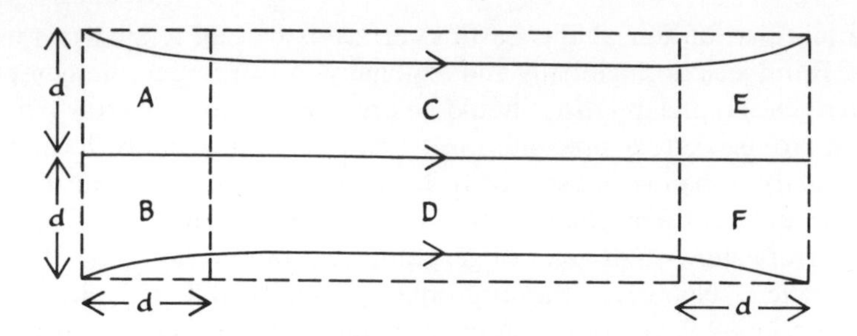

A. Sectors of a jetstream

→ 300 mb contours. d is approx. 200 miles
A: left center
B: right entrance
C: left center
D: right center
E: left exit
F: right exit

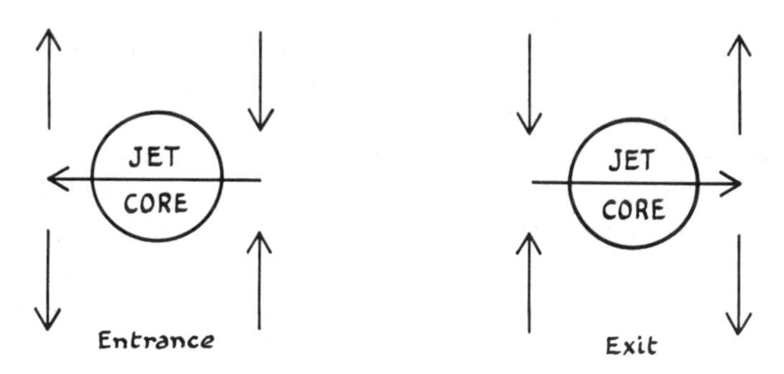

B. Cross sections of a jetstream, looking downstream, showing probable circulation.

FIGURE 4.12.
A jetstream model (after Murray and Daniels).

with the intensified upward vertical motion between the jets. The maps show a strong jet passing south of the Great Lakes at the beginning of the period, which weakened as a higher, subsidiary jet became more pronounced over the southern states. The northernmost jet intensified as the southernmost one weakened, and the greater part of the precipitation was found to the north of the higher (southernmost) jet.[36]

The Polar Front jet can be used as an index of synoptic development; its geographical position can be related to daily precipitation changes at points on the Earth's surface that may be several thousand miles away. For example, there is a strong connection between the position, velocity, and maximum shear of the jet at 700 mb over the western North Pacific and rainfall in the Hawaiian Islands. The critical location for the jet stream in this context is latitude 41°N, i.e., some 19° of latitude away from Hawaii. During winter, rainfall is much below average in Hawaii when the jet is south of latitude 41°N, and much above average when it is north of this line. In summer, the variation in rainfall as the jet changes position is less pronounced and regular, partly because the general circulation of the atmosphere is much weaker in summer, but also because small convection cells are set up above the islands during the summertime, and they affect precipitation enough to overshadow locally the more general influence of the jet.[37]

The general rainfall over the Hawaiian Islands increases with increasing horizontal wind shear across the jetstream. Rainfall is below average when the wind shear is less than 4° of latitude per day, and above average when the shear is more. When the maximum velocity of the jet exceeds 13° of latitude per day, rainfall is below average; when it is less, rainfall is above average. Hawaiian precipitation is thus least when a fast-moving jet is present in low latitudes, and greatest when there is a weak jet in high latitudes. The rainfall also increases as the anticyclonic wind shear increases south of the westerlies.

The rainfall of Hawaii thus becomes heavier as the zone of direct influence on the southern edge of the westerlies becomes narrower, and its distance from the islands increases. The explanation for this correlation is that when the jetstream is weak and far from the islands (i.e., when precipitation is heaviest), low-index circulation patterns prevail. At such a time, there are upper troughs and wedges of great amplitude, which bring about very great poleward and equatorward displacements of air, so that the energy of the atmosphere becomes concentrated within these displacements instead of in the basic westerly zonal flows. Much of the winter precipitation of Hawaii was known to be due to waves in the easterlies, before the jet correlation was discovered; the heaviest rainfalls occurred when the easterly waves came under the influence of the upper troughs, i.e., during low-index periods.

The relation of the Polar Front jet to surface weather is not simple, but the connection between the two is indubitable. A slow-moving jet brings changeable weather, because wave depressions that develop on the Polar Front (and are usually found underneath the jet), as well as the high-pressure ridges between them, tend to move in the same direction as the air in the jet. Depressions traveling along the jet usually deepen and stagnate at its exit. The movement of disturbances and that of surface temperatures are related to the jet by the thickness pattern: for jetstreams above the United Kingdom in winter, the position of the jet axes at 300 mb is related to the position of the maximum thermal winds between 1,000 and 500 mb, the mean position of the maximum thermal wind being located on the warm side of the jet axis and 60 nautical miles from it. During the course of its day-to-day movements, the jet remains tied to the line of strongest thermal wind, not to a particular thickness isopleth.[38]

Polar Front jetstreams in general are very closely related to fronts, so that, even if a jet has no associated fronts, one has a good idea where future fronts will appear. The jets usually follow the line of active cold or warm fronts, 400 to 800 miles ahead of the warm front, 200 to 400 miles behind the cold front, and cut perpendicularly across newly formed occlusions. Old occlusions are usually located parallel to a jet, and tend to have the thermal characteristics of simple cold or simple warm fronts. Strong fronts at the ground in middle and high latitudes are nearly always accompanied by jetstreams between 500 and 200 mb, but jets are by no means always accompanied by surface fronts. In the subtropics, for example, there is a prominent jet but no surface front, although a marked temperature gradient does exist across the jet in the upper air.[39]

For the stronger jets over Britain, there is a model that illustrates the relationship between surface fronts and the position of the jet at 300 mb. Most strong, warm-front jets are 120 nautical miles beyond the surface ridge ahead of the warm front, that is, between 300 and 450 nautical miles from the latter. If the axis of a jet is at an angle of more than 20° to a surface geostrophic wind of 30 knots or more through a warm front, the jet is unlikely to be strong. Most strong, cold-front jets over Britain are accompanied by surface geostrophic winds of at least 30 knots with little change in wind direction up to 300 mb. Most are found at a distance from the cold front that is proportional to the distance of the jetstreams from the tip of the warm sector of the depression; the separation between jet axis (at 300 mb) and front increases by an average of 140 nautical miles for each 10° of latitude.[40]

Distinct cloud patterns and cloud types are associated with Polar Front jets; the connection obviously depends mainly on peculiarities in the distribution of humidity. A strong humidity gradient exists across the jet, with moister air on its high-pressure side (i.e., usually to the south of the jet axis in the northern hemisphere), and a patch of relatively dry air usually found below the axis in the vicinity of the 500-mb frontal zone. High and medium cloud of layer type occurs on the high-pressure side of the jet axis, but not higher than the axis. Cloud is very rare more than 100 nautical miles from the jet axis on its low-pressure side. Generally, two cloud maxima are associated with the jet core. One occurs in the upper troposphere, 4° of latitude north of the core; the other is found just above the jet front, some 5° of latitude south of the core. As an average figure, the presence of the jet increases cloudiness in an extensive area about 5 per cent over-all.[41]

Jets are quite common in clear air, but when they are accompanied by clouds, cirrus is the dominant type. The cirrus is usually complex: ice-crystal streamers with long tufted streaks and shear features complicate the general structure of the bands, which lie along the jet, are often very long (one band was traced for at least 1,700 miles across the Atlantic, for example), and exhibit crossstriations. At lower altitudes, the jet is often marked by lines of altocumulus along the direction of the wind, which form billows or waves at right angles to the latter.[42]

Cirrostratus is found in the region between the jet front and the jet axis, and cirrocumulus in or slightly above the front (see Appendix 4.1). Cirrus and cumulus occur in bands on the equatorial side of the jet, usually parallel with it. There are no clouds at the jet axis or on its poleward flank, except possibly for mare's tail cirrus near the core of the jet. The cirrus bands usually lie within 200 nautical miles of the jet core, on its warm side, and in the entrance region to the jet. Sheets of cirrocumulus slope upward toward the jet core in the region of the jet front (see Figure 4.13), and cirrostratus sheets reach their maximum development above the front, some 5° of latitude equatorward from the jet core. Cirrus is most frequent in the elongated region that slopes equatorward from the jet core, usually on the warm side of the core, and in the vicinity of the polar tropopause; it often takes the form of a sheet with a sharp poleward edge 100 km equatorward from the jet core.[43] If the jet is weak, however, or if two jets are close together in latitude, cirrus may be present on both equatorward and poleward sides of the jet. Sharp discontinuities in the cirrus sheet near the jet core usually mark the location of the strongest winds, above the height of which there are normally no clouds.

The long bands of cirrus and cirrocumulus on the equatorward flank of the jet often show evidence of helical vortices, which give rise to very turbulent conditions. The

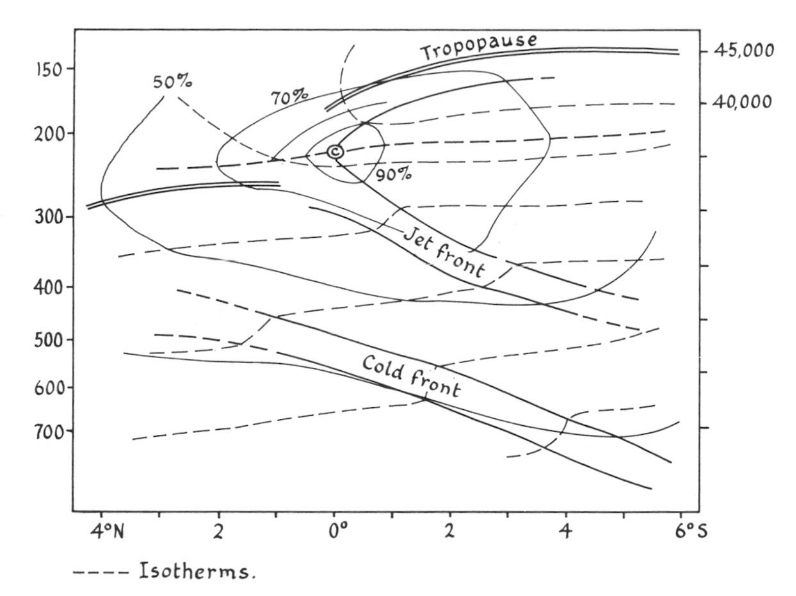

FIGURE 4.13.
Idealized model of Polar-Front jetstream in cross section perpendicular to the airflow (after English and McLean).

cirrus bands are created between parallel, horizontal vortices, which rotate in opposite directions; clouds form where the combined motion is upward. The small-scale structure of the cirrus elements indicates the nature of the lapse-rates and wind shear through the jet.[44]

Smoke drifts and aircraft condensation trails show that the horizontal wind field at the heights of the maximum winds in the jet is very streaky, and that pronounced vertical wind shears must be present. In other words, at or near the jet maximum there are critical combinations of stability and vertical wind shear at which the airflow breaks down from its normal, relatively smooth flow, and vortices and streaks of different velocity appear. Hence clear-air turbulence is very often associated with jetstreams, and is almost entirely on the cyclonic flank of the jet, especially in regions of pronounced vertical stability.

The movements of the Polar Front jet across the Earth have profound consequences for climate as well as for daily weather, and are themselves much affected by orographic barriers.[45] The upper westerlies can move over the Rocky Mountains and the Andes, but have to move around the great width of the Himalayas; the mean jet thus becomes split in two over Asia. In summer, the jet is entirely north of the Himalayas. The southernmost branch of the jet collapses at the onset of the monsoon, while the northward shift of the northernmost branch of the jet in late spring results in a great circulation change over the Indian subcontinent, and coincides with the onset of the monsoon. The jet maintains a fairly constant position over eastern Asia in winter, and the heaviest precipitation occurs beneath it, with a very rapid decrease in precipitation away from the jet axis. To the south of the jet, there are widespread descents of air over southeast Asia.

The effect of the Rockies and the Andes on the upper westerlies is to produce a pressure trough east of their crests, with a pressure ridge over the mountains, in conformity with the principal of conservation of absolute vorticity. The trough due to the Rockies occurs over the High Plains area; that due to the Andes occurs over the South Atlantic, the same distance downstream from the orographic barrier as in North America. In North America maximum precipitation is along or just north of the jet axis; over the west coast, the heaviest precipitation is directly under the jet. When the jet moves southward across the central United States, cyclogenesis increases in Texas and the Gulf states. Multiple jets are frequent in the lee of the Rockies, and in spring the anticyclone centered on Hudson Bay usually separates the jets above northern Canada from those above the central plains of the United States, the very dry weather of the anticyclone contrasting with the heavy snowfalls of the former area and the flood rainfalls of the latter. In North America generally, the fluctuations of seasonal precipitation from average correspond to fluctuations in the frequency of the jet in the vicinity.

In Europe, the building up of a winter continental anticyclone, and the consequent blocking of the general eastward progression of depressions and wedges, will cause the jet to split into two separate branches. One branch then lies from northeast to southwest between Iceland and northern Scandinavia; the southernmost branch extends west-northwest to east-southeast through the Azores and over southern Spain or North Africa. Such a situation causes temperatures to be well above average over northwestern Europe and the North Atlantic, and below normal temperatures over central and southwestern Europe, and precipitation to be generally well below average over western Europe.

Turning to the question of the importance of the Polar Front jetstream in the general circulation, a first consideration must be, what causes the jet? Its origin is not completely understood, but model experiments indicate that a jet is a necessary feature in any fluid circulating over a rotating globe similar to the Earth. Hence, the jet is a *dynamic* feature of the atmospheric circulation, a property of the airflow, not a consequence of variations in the Earth's surface. Observations confirm this deduction. The jet cannot be due to permanent, large-scale thermal contrasts at the Earth's surface, because jets are embedded in the westerlies in both the northern and southern hemisphere despite their great geographical contrasts, and the climatological (mean) jets intensify from west to east over the continents both winter and summer, irrespective of the great change in intensity of the temperature gradients from weak (summer) to strong (winter). Indirect evidence of the dynamic nature of the jet is also provided by temperatures at 200 mb in middle latitudes. These cannot be explained by advection of warm stratospheric air from the arctic and cold tropospheric air from equatorial regions, but only by dynamic processes.

The earliest coherent theory of the general circulation, that of Hadley in 1735, only referred to latitudes 30°N to 30°S, and essentially postulated a simple, steady meridional circulation to carry poleward the excess heat caused at the equatorial belt by radiation, in order to maintain a constant average temperature in the amosphere (see Appendix 4.2). This single-cell theory was later enlarged to give the three-cell theory, which is commonly found in elementary textbooks. In the latter model, the circulation consists of three meridional cells (see Figure 4.3A), the equatorward (i.e., trade wind) and poleward (i.e., Polar Front) cells being direct circulations driven by atmospheric heat and cold sources, and the middle-latitude cell forming an indirect circulation driven by frictional interaction with the adjacent cells. This model is consistent with the idea, current before the discovery of the jetstream, that the principle of conservation of angular momentum contained the key to the atmospheric circulation. The model required that in middle latitudes the prevailing westerlies near the surface should decrease in intensity with height, finally becoming easterlies (i.e., trade winds) in the upper atmosphere. Observations since the late 1940's clearly indicate that instead the middle-latitude westerlies actually *increase* in intensity with height, and in particular tend to be concentrated in a tube or ribbon of high velocity near the tropopause, the core of the jet often coinciding with a break in the tropopause. Hence the presence of the Polar Front jet in the actual atmosphere indicates that angular momentum conservation cannot be the leading dynamic principle behind the general circulation, and hence the simple circulation model showing three meridional cells cannot accurately describe the real atmosphere. What dynamic principle and what circulation model does in fact apply?

One explanation might be in terms of *energy* in the atmosphere, but a valuable analogy with the Gulf Stream disposes of this theory. In the North Atlantic Ocean, the horizontal gradient of density is concentrated into a very narrow zone (see Figure 4.14), with which the velocity current must coincide. In other words, three-dimensional study shows that the Gulf Stream is also a jet, which meanders and at times breaks up into multiple currents just like the atmospheric jetstream; it shows a similar large velocity shear across it, and also develops cut-off centers. The lateral dimensions of the Gulf Stream are only 1/25th those of the Polar Front jetstream; in the vertical, the dimensions of the two systems are in the ratio 1:10, as are the meander sizes. The shapes of the current profiles are almost identical in both, with equivalent lateral shears. Both show comparable thermal structures, with similar frontal zones; both exhibit marked variations in speed along their axes; and ageostrophic motions are important in each. The ocean does not receive latent heat from condensation in large quantities, whereas the atmosphere does. Consequently, heat energy cannot be a factor in the formation of the oceanic jet, and it is reasonable to assume that this deduction also applies to the atmospheric jet. A further factor possible in jet formation, that of fluid compressibility variations and its repercussions, is ruled out by the

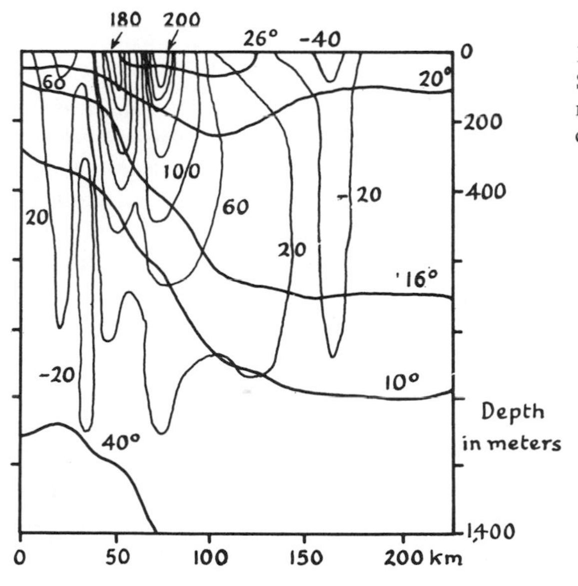

FIGURE 4.14.
Section through Gulf Stream SSE of Cape Cod near 38°N, 69°W, showing isotherms in °C, and current speeds in cm per sec.

fact that the atmosphere is compressible, the ocean is not, yet both contain jet streams.[46]

Rossby therefore proposed a *lateral-mixing theory* to explain jet formation, in which the mean zonal motion of the atmosphere is shown to be a consequence of horizontal mixing by large-scale atmospheric eddies. Provided two conditions are satisfied, mixing of this nature leads to the production, in the upper westerlies, of a narrow zone of very high windspeeds, and the meridional temperature gradient becomes concentrated at the equatorward boundary of the zone of mixing. The two necessary conditions are that absolute vorticity and potential temperature be conserved. If potential temperature remains constant for about a day, then if temperatures decrease from the equator toward the poles, southerly winds will transport more heat northward in the northern hemisphere than northerly winds transport southward. If the mixing brought about by the migrating anticyclones and depressions of middle latitudes ceases abruptly south of a given latitude, then the whole of the poleward temperature gradient will be concentrated near the equatorward boundary of the mixing zone. Conservation of potential temperature during periods of 24 hours does in fact obtain in winter in high latitudes, in the middle and upper troposphere.[47]

The principle of conservation of absolute vorticity is known to be valid from practical experience with the Rossby waves. Although the total amount of vorticity in the atmosphere must remain constant, it can be redistributed by lateral mixing. In high latitudes, cyclonic vorticity is absorbed by the Earth from the atmosphere; in low latitudes, cyclonic vorticity is transferred from the Earth to the atmosphere. Hence there is a flow of vorticity in the upper atmosphere from the subtropics to the poles, with a vorticity source in the subtropical high-pressure cells (where friction destroys anticyclonic vorticity) and a vorticity sink in the subpolar low-pressure cells (where friction destroys cyclonic vorticity). This picture obviously refers to mean conditions; the circulation on a given day would not necessarily show such a distribution of vorticity sources and sinks.

If absolute vorticity is continually being transported from low to high latitudes, as it undoubtedly is in the mean sense, this will eventually lead to values of anticyclonic wind shear south of the jet axis that reach the critical value for dynamic instability. (Dynamic instability in this sense is when an air particle, displaced north or south from its equilibrium position, continues to move in the direction of displacement, and shows no tendency to return to its original position.) Rossby showed from theory that in a zonal current, dynamic instability will occur when the absolute vorticity becomes anticyclonic, or where the anticyclonic shear increases to a certain value. In other words, the profile of constant vorticity, when followed poleward from the equator, should break down, because of the development of dynamic instability, at latitude 34°, and a zone with zero vorticity should occur, with a width of 5.5° of latitude, near latitude 30°. The observations show that this prediction is true.

When discussing the characteristics of the Polar Front jet, we pointed out that a transverse flow of air seems to take place across it. Such a transverse flow would be equatorward in the critical latitudes, and would intensify when the state of dynamic instability was reached, since no opposition would be exerted against lateral displacements in the atmosphere. Thus part of the air in the latitudes of the jet would be flung southward, with a compensatory northward flow at the surface because of the resultant fall in surface pressure beneath the jet and rise in surface pressure further south. Hence a meridional circulation cell, a reverse cell, results. In this cell, cold air is raised and warm air is lowered; the air sinks equatorward away from the jet, giving fine dry weather, and rises below the jet maximum, giving cloud and precipitation.

Therefore, according to the theory of lateral mixing, both mixing due to large-scale eddies, and also the reverse circulation due to a meridional circulation cell, must have a hand in the formation and maintenance of the Polar Front jet. In the real atmosphere, however, two distinct jet phenomena can be recognized. The first is the "synoptic" jet that appears on the daily upper-air charts, and consists of meandering segments, each of which have a life history of a week or so; the second does not necessarily show up on daily charts, but is a "mean" or climatological jet, obtained by averaging the meridional cross sections. A mean jet occurs in subtropical latitudes as well as in middle latitudes, and is maintained by the mean meridional circulation of the trades; it fluctuates far less than the synoptic jet of middle latitudes. The mean jet of the southern hemisphere is almost exactly similar to the northern one. Both exist at 12 kilometers, in latitude 25° in winter and between latitudes 35° and 45° in summer, and both have average speeds of 80 to 100 knots in winter and 40 knots in summer. These facts clearly establish that the mean jets must be dynamic features, not thermal effects due to geographical differences.

The two jets are clearly indicated in Palmén's model (see Figure 4.3B) of the atmospheric circulation in winter. This is the simplest model in current use that accurately describes the conditions now observed in everyday synoptic work. However, even this model does not indicate the regional or "local" jets that are frequent in various parts of the world, and that cannot be explained by the lateral-mixing theory.[48]

Rossby found that these local jets have three important characteristics. First, they are always associated with a very strong concentration of solenoids and intense vertical wind shears, and, therefore, of momentum. Second, westerly airstreams entering the United States in summer usually show no pronounced speed maxima at 200 mb, where the jet normally occurs, but after they cross the Rockies, a prominent local jet appears just above 300 mb. This suggests that the local jet is caused by the loss of momentum suffered by the airstream in crossing the Rockies, because of ground friction and the general retarding action of the mountain barrier. Third, the analogy with the oceanic Gulf Stream is again apt, for it also shows streaks of velocity concentration that may be regarded as local jets. The California current, in contrast, is more diffuse, and does not exhibit local jets; since it is driven by local wind systems, which add momentum to the current, it seems unlikely that local additions of momentum to the atmosphere could give rise to local jet streams. Rossby therefore proposed that local velocity concentrations must be inherent properties of stratified currents in both the atmosphere and the ocean.

According to Rossby's theory of local jets, stratified currents, which lose momentum to their environment, must gradually take on a distribution of velocity with height that will transfer the least momentum per unit time across a vertical section normal to the current. This presupposes that the air or water cannot escape sideways from the current, so that the latter must transport a fixed amount of fluid per unit time. When the currents achieve a state of minimum momentum transfer, they shrink in depth and increased velocities result at their center. This velocity increase must be followed by a pressure change, i.e., by a fall in pressure in the direction of the motion, and hence there must be distinct limits, both geographically and in time, to the existence of local jetstreams.[49]

Local jets may be either offshoots of the main circumpolar jet, intensifications of former parts of the main jet that become separated from it and lose their vigor, or completely distinct entities, originating in a pattern of confluent streamlines (see Figure 4.15). In such a pattern, a concentration of isotherms is built up along the line *AB*, and then, in order for the situation to be relatively stable, i.e., for geostrophic balance to be maintained, either a zone of strong winds must develop near the ground opposite in direction to the jet winds, or a strong vertical wind shear must arise in the upper troposphere, with a narrow belt of high velocity flow (i.e., a jetstream). A jet does, in fact, usually result.

The importance of confluent flow in the origin of jetstreams has been shown by theorists other than Rossby. Members of the University of Chicago group proved that a jet develops from confluence on the eastern side of an upper ridge or trough. Namias and Clapp found that thickness isopleths and contours become concentrated where cold air from the north follows such a trajectory that it comes to flow alongside warm air from the south, and in this concentration a jet develops; the confluent pattern results from a train of upper long waves in one latitude overtaking a wave train in another latitude. According to Boyden, however, middle-latitude jetstreams develop more often from the deformation of a single baroclinic flow, particularly on the eastern side of an upper ridge or trough, than from the confluence of two airstreams. This involves, dynamically, the downward propagation through the contour pattern of an amplification and intensification of the flow, which can take place at speeds greater than that of the zonal wind velocity.[50]

The main circumpolar jet appears to be formed not by combination of local jetstreams, but more probably by lateral mixing, as already described. The lateral-mixing theory requires the Coriolis parameter to vary with latitude, i.e., it requires a spherical Earth. However, laboratory experiments indicate that both the Polar Front jet and local jets can be produced in a *flat*, rotating, saucepan-shaped vessel. Therefore, although lateral mixing undoubtedly plays its part in the formation or maintenance of the circumpolar jet, it does not appear to be essential for jetstream formation. All

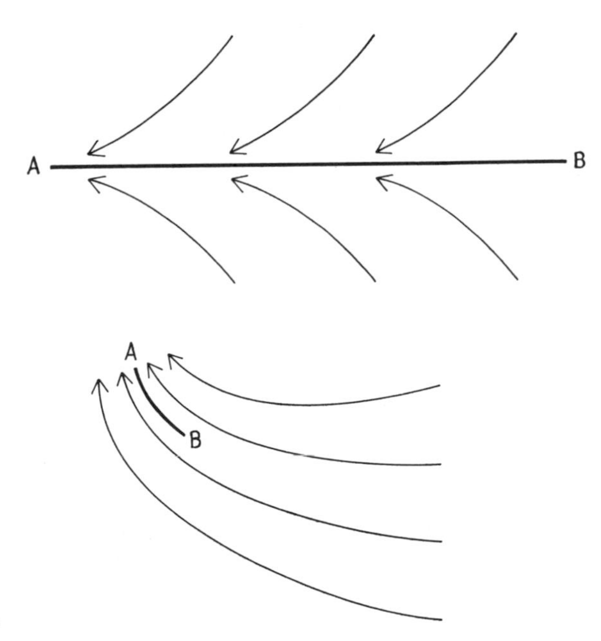

FIGURE 4.15.
Confluent streamline patterns.

that seems to be necessary for the development of a Polar Front jet is that the Earth should rotate at a certain critical angular velocity, and exert a certain critical frictional force against the bottom layer of the atmosphere. The investigations that led to this simple conclusion have involved complicated mathematics, but their importance lies in the fact that they have proved the Polar Front jets to be *very* basic features of the atmosphere, their *existence* not dependent on varying geographical conditions at the Earth's surface in any way. This is not to deny the important influence of the Rockies, the Andes, and the Himalayas on the *movements* of the Polar Front jet.[51]

The subtropical jetstreams are probably the result of a slow drift of air in the middle and upper troposphere from equatorial regions towards latitude 30°. In accordance with the principle of conservation of angular momentum, since the radius of the Earth in latitude 30° is less than the equatorial radius, the velocity of the air relative to the Earth's surface must increase to 200 knots or more in the latitudes of the subtropical jets. Two jetstreams reaching speeds of over 200 knots often occur over North Africa and the central Mediterranean regions; the subtropical jet proper is the higher of the two. Occasionally the two merge, as do the two found over the subtropical Pacific in winter.[52]

The stratospheric jetstreams appear to be due to different processes from the other jets, being maintained by differential heating and cooling by radiation in the ozone layer of the atmosphere. They are seasonal phenomena; for example, there is a stratospheric jet with its center between 25 and 50 mb over the Canadian Arctic in winter.[53]

We have said very little so far about the "microgeography" of the Polar Front jetstream, i.e., the detailed pattern of windspeeds and other elements within it. Aircraft observations show that the pattern is much more complicated than one would expect. One set indicates that "fingers" of very high wind velocities occur at distances of 10 to 20 miles apart, and cause temperature variations of up to 3°C in a distance of eight or nine nautical miles.[54]

Empirical models of jetstream structure date from the classic model compiled in 1948 by Palmén and Newton. They constructed it by means of isotachs computed from geostrophic winds derived from averaging 12 vertical cross sections along longitude 80°W for December 1946, at a time of strong Polar Front conditions. Later models are based on aircraft observations, for example, those due to: Riehl, Berry, and Maynard, based on five flights at a single level in the fair-weather regions of jetstreams; Endlich and McLean, based on flights made during the United States' Project Jetstream in 1956 and 1957, each involving two or more tracks at different altitudes; and Brundidge and Goldman, based on results from 14 Project Jetstream flights, each involving tracks at three or more levels, 2,000 feet apart, both above and below jet cores, the aircraft flying normal to the wind, to produce a series of different jet models that fit into different sectors of a wave-cyclone system.[55]

Figure 4.13 gives a generalized model across the Polar Front jet, and Figure 4.16 illustrates typical average windspeed profiles through the latter associated with different synoptic types. Aircraft observations made during Project Jetstream showed the structure of the jetstream front, the deep and stable baroclinic zone that gives rise to very strong horizontal and vertical wind shears near the jet core. The front was found to be better defined above the maximum wind than below it, and sometimes exhibited a double structure: the upper core strengthened while the lower core weakened, and

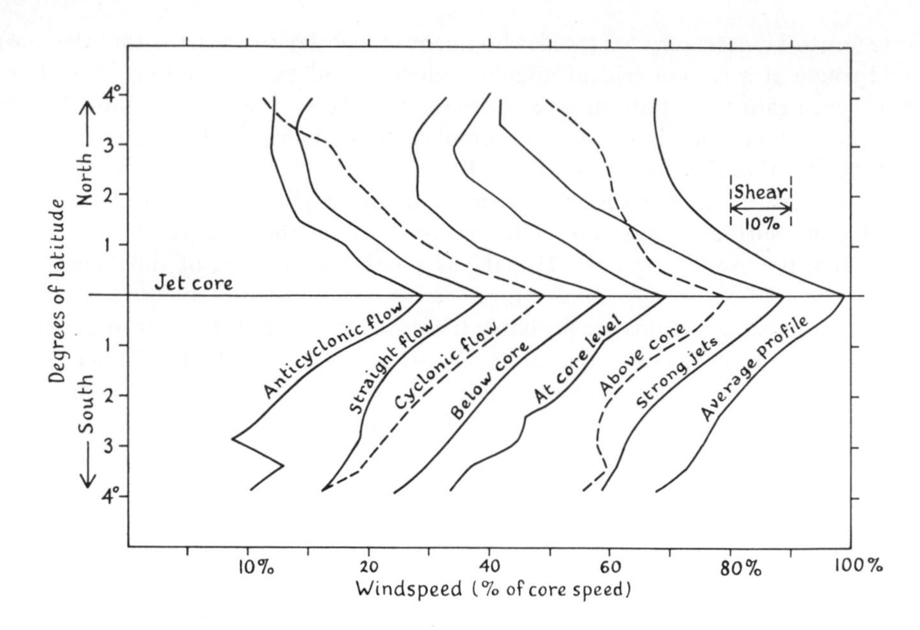

FIGURE 4.16.
Average horizontal windspeed profiles
(after Endlich and McLean).

at the same time the cyclonic shear to the left of the lower core increased. Apart from the jetstream front, the observations also revealed the existence of stable baroclinic zones on the anticyclonic side of the jet, and an upward bulge of the isentropic surfaces south of the jet.[56]

British meteorological flights show that a tongue of stratospheric air sometimes protudes into the baroclinic zone below and on the low-pressure side of the jet core, within well-marked frontal boundaries. Tropopause folding is present, with the formation of a frontal zone within which statospheric air may descend to midtroposheric levels. The main mechanism that causes mixing of stratospheric and tropospheric air is clear-air turbulence, which is very much associated with jetstreams. Turbulence is present about 5 per cent of the time in the atmosphere over land areas, only 0.2 per cent of the time over water areas, but 12 per cent in jets generally, and as much as 40 per cent of the time in some sectors near the jet core.[57]

The complicated relationships between Polar Front jets, local jets, and other jetstreams, the tropopause, and the stratospheric-tropospheric interchange of air, probably are best illustrated by the concept of a *jetstream complex*, in which two main middle-latitude jets are recognized (see Figure 4.17). The latter form the Polar Front jet when combined, but when they are separated, an interpolar jet also exists. On occasion, all the jets in the complex may merge to form one broad belt of high-velocity winds.[58]

Low-level Jetstreams. The movement of the air in the lowest 5,000 feet or so of the atmosphere is far from uniform. Imposed on the mean motion are organized systems, of about 500 feet to a mile or more in horizontal dimension, which move in apparently random paths, sometimes quite close to the Earth's surface. The relative movements

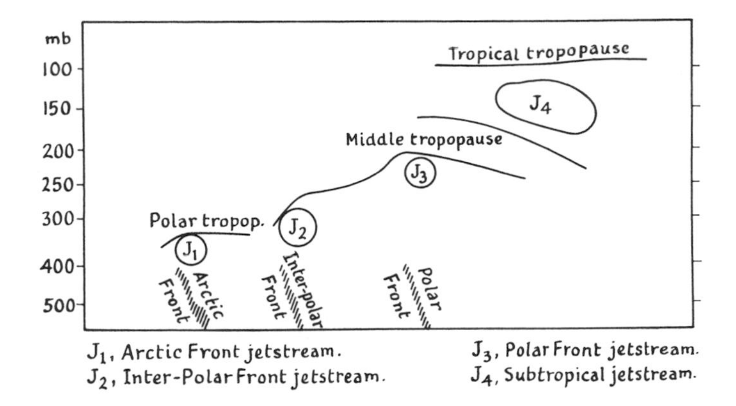

FIGURE 4.17.
The jetstream complex, in cross section
(after Serebreny, Wiegman, and Hadfield)

of adjacent systems may cause two different airstreams to come close together, producing considerable wind gradients and low-level jetstreams. These jets are much smaller-scale features than the main midtropospheric and stratospheric jetstreams already described, and they prove to have very intimate connections with local surface weather.

The low-level jet was first clearly recognized as a boundary-layer phenomenon during the Great Plains Turbulence Field Program in O'Neill, Nebraska, in 1953. Very high windspeeds were found at heights of 400 to 800 m above the ground (see Figure 4.18), with extreme values of positive and negative wind shear below and above the maximum wind. Nighttime observations indicated a gradual build-up of a westerly wind

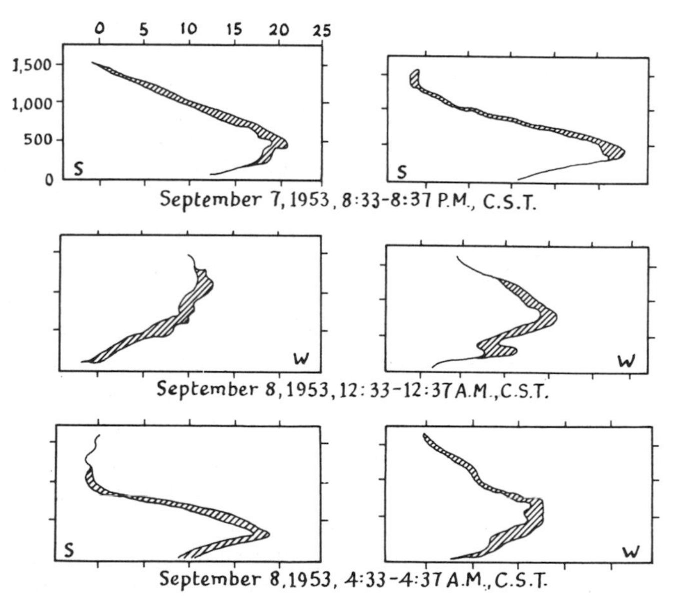

FIGURE 4.18.
Pibal swarm ascents, envelopes of overlapping 24-second displacements (after Chamberlin *et al.*).
S: southerly components (on left). W: westerly (on right).

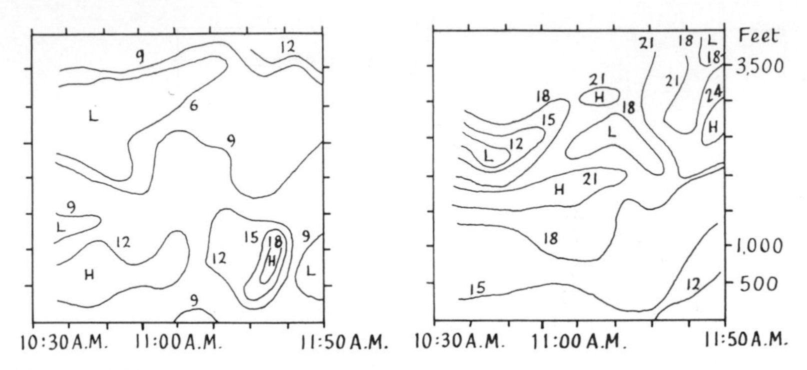

FIGURE 4.19.
Wind-velocity components at Suffield on April 7, 1955. Southerly on left, westerly on right; horizontal scale gives G.M.T. (After Longley.)

maximum at about 400 m, at the expense of initial westerly momentum at 700 m. The phenomenon proved to be related to boundary-layer stability: windspeeds reached a maximum intensity somewhere between 1,000 and 2,000 feet around sunrise, but this maximum rapidly dissipated as insolation increased at the ground and adiabatic conditions were reestablished.[59]

Observations of pilot balloons released over rolling prairie in Alberta during an 80-minute period in 1955 revealed the existence of low-level cores of stronger winds that were reversed to the normal positive gradient of wind velocity with height (see Figure 4.19). Possible causes of wind variations of this type include the development of surface inversions, during which the air begins to execute inertial oscillations; funneling of air flow by irregular topography; differential heating of the surface of the ground; and convergence behind a rising mass of warm air. The Alberta example proved to be unrelated to nocturnal inversions, and was not an inertial oscillation.[60]

Over the Midwest United States, low-level jets cause nocturnal thunderstorms. Production of low-level convergence by the jet is important, and when the jet axis runs north-south or northeast-southwest (indicating the presence of moist air, usually), thunderstorm incidence is high. Squall lines are also caused by the jets.[61]

The development of a low-level jet shows an orderly progression (see Figure 4.20). Thus on the night of February 22–23, 1961, observations of a low-level jet at Cedar Hill, Texas, showed that vertical mixing commenced at 300 feet in the early evening, and proceeded upward, slowly at first but more rapidly later, until it influenced the temperatures and windspeeds at 1,200 feet. Observations made on March 15, 1961, at the same site and also on May 16, 1961, show the importance of low-level jets in breaking down a low-level inversion. The latter process, and also the dissipation of the jet itself by turbulent mixing due to friction at the ground, proceed upward very rapidly. The development of a strong wind shear across a jet is an important destabilizing factor in the breakdown of the steep wind gradient. The temperature profile during the breakdown of an inversion by a low-level jet is usually as follows: (a) initial warming in the upper part of the boundary layer before sunrise, because of subsidence; (b) increased warming in this region immediately after sunrise; (c) cooling in the middle part of the layer after sunrise; (d) subsequent lifting of the elevated inversion.[62]

The general conditions that favor low-level jets vary from place to place, but observations made in an area of the western plains of North America, centered on

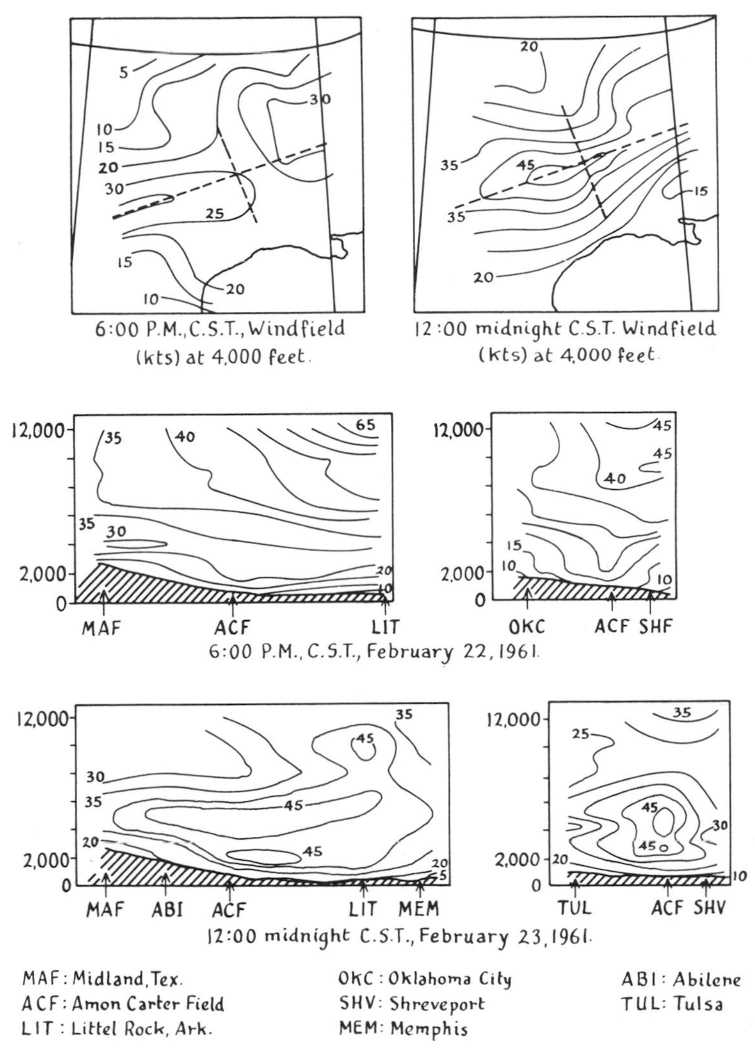

FIGURE 4.20.
The development of a low-level jet.

Oklahoma City, suggest that the optimal conditions for jet development involve (1) a steady southerly airstream, with no fronts in the vicinity, (2) a warm area of low pressure to the west, (3) a fair east-to-west sea-level pressure gradient, (4) strong insolation during the afternoon for air moving into the locality of future jet development, and (5) a cloudless night. Once a jet has developed, the factors that cause it to be sharply defined, and very low, involve an opposing thermal wind, a nocturnal inversion following a sunny afternoon, and Richardson numbers, in the shear layer below the jet nose, of less than 0.25, which is the critical value for the onset of mild turbulence in the jet region.[63]

Low-level jets may exist on micro, meso, and synoptic scales. For example, the thunderstorm or squall line caused by the convergence involved in a normal low-level jet on the mesoscale may generate a secondary low-level wind maximum, because of subsidence in a precipitation-cooled cold dome (i.e, a mesohigh) intermediate in scale between synoptic and local-scale features. This happened on May 8, 1961, at Dallas, Texas, when synoptic-scale jets developed in a southwesterly airstream in advance of a squall line; the mesoscale jet was located in the mesohigh between the squall line and a cold front, and the microscale jet was a characteristic feature of the vertical structure of the cold outflow from the latter.[64]

By convention, the term "low-level jet" is normally restricted to wind maxima in the lowest 1,500 meters of the atmosphere, in which (1) the windspeed generally increases with height, and (2) the windspeed of the maximum wind exceeds the windspeed of the next higher minimum wind by at least 5 knots.[65]

Some Special Problems of the General Circulation

Jetstreams and similar phenomena tend to give the impression that the transformations of the general circulation take place in an orderly manner and are predominantly zonal in character. On many occasions, neither of these impressions is justified, and in fact many quite distinct problems arise.

THE INDEX CYCLE

It has been well-established for many years that the atmospheric general circulation is driven by the temperature contrast between equatorial and polar regions. This contrast in turn is maintained by the balance of radiative energy exchange between the Earth's surface, the atmosphere, and outer space. In the middle troposphere, the general circulation oscillates between a predominantly zonal type of motion (high-index stage) and predominantly meridional type of motion (low-index stage), a full oscillation from one type to the other and back again forming the index cycle, which takes three to six weeks, on the average, to complete itself.

At times of high index (see Figure 4.21), the midtropospheric westerlies form a broad and well-developed current in middle latitudes, with very poorly defined and insignificant long waves superimposed on them; temperature deficits occur in high latitudes, and temperature surpluses in low latitudes, the isotherms running approximately east-west. At times of low index, the upper westerly belt becomes compressed into a narrow jet or jets, and is displaced southward in the northern hemisphere. Well-defined long waves are superimposed on the westerlies, with large meridional amplitudes, so that strong positive and negative temperature anomalies exist adjacent to each other in the same latitude, with high temperatures in relatively high latitudes and low temperatures in relatively low latitudes. Finally, the meandering westerly belt of the high-index stage breaks down into closed cyclonic and anticyclonic cells, leaving cold pools in low latitudes and warm pools in high latitudes; simultaneously a new westerly belt begins to form in poleward regions.

The concept of the index cycle dates from 1939, when Rossby introduced the zonal index as a measure of the strength of the middle-latitude (35–55°N) westerlies in the northern hemisphere, and found the index was related to the longitudinal positions of the centers of action in the atmosphere. High values of the zonal index indicated strong upper westerlies in middle-latitudes, low values of the index indicated weak westerlies. Willett in 1948 emphasized that variations in the zonal index were associated with both latitudinal and longitudinal shifts in the belt of maximum upper westerlies, and that the concept of a hemisphere-wide low- or high-index stage was false. For example, a low-index pattern in middle latitudes is contemporaneous with a high-index pattern in subtropical latitudes.[66]

Some winters are characterized by low-index patterns, others by high index, but from week to week in every winter the zonal index will vary widely about its seasonal average value. In particular, each winter has four to six weeks during which the zonal westerlies decrease to very low strength and then recover; i.e., there is a gradual decline from high to low indexes, followed by a similar rise. The intensity and duration of this phenomenon vary from year to year, but it is almost always a feature of late February (see Figure 4.21). It has been explained by Namias in his "containment" theory, as follows.

Maintenance of the heat balance of the atmosphere requires that equatorial-polar interchanges of energy operate at a certain intensity, which could in fact be provided by the observed short-period and week-to-week oscillations of the zonal index. However, the presence on occasion of blocking anticyclones means that these oscillations cannot always operate, so that there is not enough variation in zonal index during the year for heat-balance requirements. In the "containment" theory, the cold air in polar and subpolar regions is contained within its boundaries by the strong midtropospheric westerlies created by large-scale convergence. Cold air is thus both produced and stored in polar latitudes, the atmosphere acting as a gigantic condenser that is discharged by cold-air outbreaks, which are due to certain cellular-types of blocking action above the Atlantic that are effective only when an abundant supply of very cold air is available, normally in late February. How much "containment" of cold polar air there is before the year's primary index cycle begins determines the duration and intensity of the subsequent low-index phase. Thus in 1947 and 1958, each extreme southward displacement of the middle-latitude upper westerlies in the northern hemisphere was preceded by a long period during which the westerlies were stronger and farther north than usual.[67]

Well-defined index stages do not always occur. Sometimes, for example, the zonal index does not change its character simultaneously in all parts of the northern hemisphere, but changes successively in different areas, the transformations all originating from a single point.[68] On high-index days, well-defined meridional mean temperature profiles exist; on low-index days, these profiles are less clear or even absent. Transformations from low to high index are associated with pronounced cooling in polar regions and warming in middle latitudes, and with the gradual disappearance of meridional circulations. Transformations from high to low index are associated with the development of numerous branches of the meridional circulation; there is descending air in the western parts of developing upper troughs, with ascending air in the eastern parts and the western portions of the ridges; at the same time, the cold dome over the north pole breaks up and moves south over northeastern Siberia, the Greenland-Iceland region, and northeastern Canada.

The main problem posed by the index cycle is that its variations indicate that certain highly abnormal types of circulation—for example, a contracted circumpolar vortex in summer and an expanded vortex in winter, which give rise to very cool summers and very cold winters, respectively, in middle latitudes—are, dynamically, inherently stable phenomena. It thus appears that the general circulation has several stable modes in which it can operate, not just one.[69]

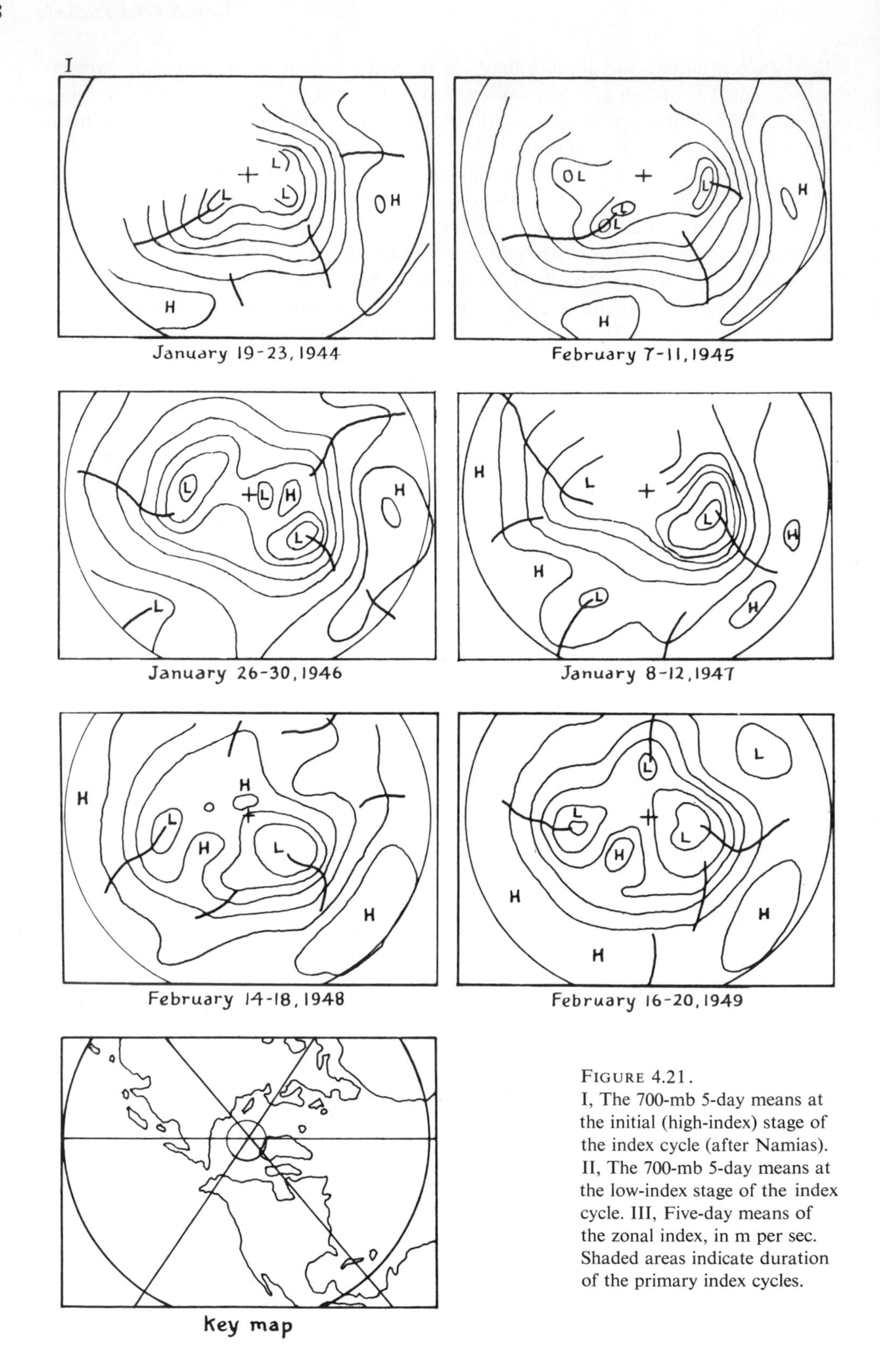

FIGURE 4.21.
I, The 700-mb 5-day means at the initial (high-index) stage of the index cycle (after Namias). II, The 700-mb 5-day means at the low-index stage of the index cycle. III, Five-day means of the zonal index, in m per sec. Shaded areas indicate duration of the primary index cycles.

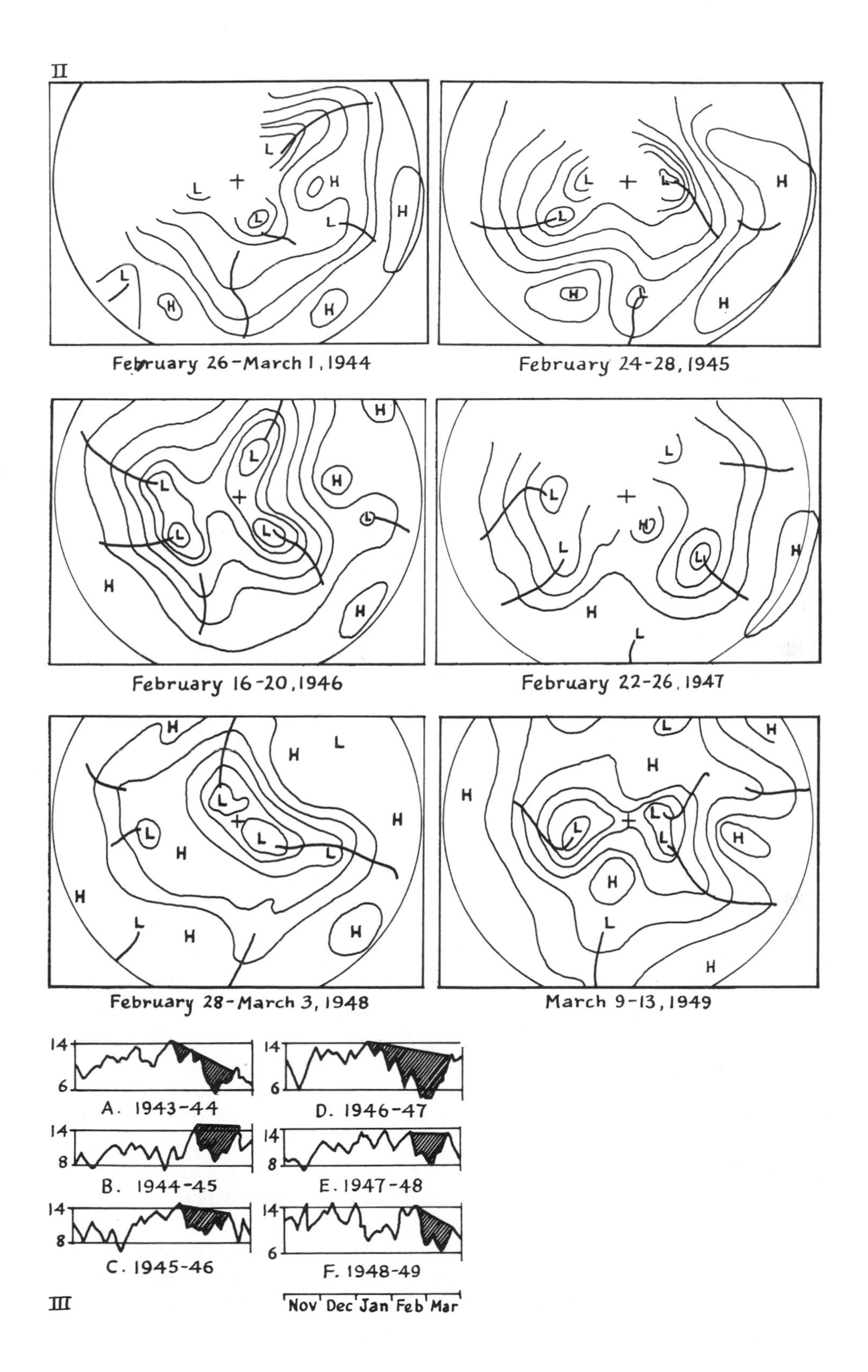
II
February 26-March 1, 1944
February 24-28, 1945
February 16-20, 1946
February 22-26, 1947
February 28-March 3, 1948
March 9-13, 1949
A. 1943-44
B. 1944-45
C. 1945-46
D. 1946-47
E. 1947-48
F. 1948-49
Nov Dec Jan Feb Mar
III

BLOCKING ACTION

Blocking action is a diminution or retardation of the zonal circulation in a limited sector of a hemisphere at all levels in the troposphere, the retardation tending to spread westward (i.e., upstream). It is a slow process, lasting from three to thirty days, and affects large areas—much of the northern hemisphere, for example—in which extreme low-index patterns prevail, with a warm anticyclone in high latitudes and a cold cyclone in lower latitudes, both of which usually move upstream, i.e., retrograde. Within the blocked sector, the zonal index may either increase or decrease as the blocking continues, depending on the orientation of the axis of the block to the latitudinal band for which it is computed. For a wide latitudinal band, there may be little longitudinal change in the strength of the mean zonal flow with the onset of blocking action.[70]

Alternatively, a blocking pattern may be defined as a state of the general circulation in which the normal zonal flow is interrupted in a sector or sectors of the hemisphere by strong and persistent meridional flow. The pattern is characterized by a persistent anticyclone in middle and high latitudes that obstructs the normal eastward movement of cyclones and migratory anticyclones, and by abnormally deep depressions in lower latitudes upstream or downstream of the block. Pressures and temperatures within the troposphere are above normal in the blocking high; pressures are above normal and temperatures below normal in the stratospheric part of the anticyclone. The usual latitudinal distribution of surface pressure is completely inverted in a blocking pattern, with cyclones in the subtropics and anticyclones in middle and high latitudes.[71]

The normal migratory cyclones and anticyclones sometimes cannot transport enough heat between equator and poles to maintain the heat balance of the atmosphere, especially at high levels of the atmosphere, and larger-scale "eddies," in the form of blocking highs, become necessary from time to time. In other words, blocking anticyclones are caused by the accumulation of heat in low latitudes and the consequent necessity for the general circulation to readjust itself so that this heat may be carried toward the poles. Several waves of blocking highs may follow each other in rapid succession; for example, on May 28, 1955, a blocking anticyclone over the British Isles recovered in less than 24 hours in the face of an invasion of Atlantic air; the amount of heat redistributed is therefore very considerable.[72] The blocking anticyclone is so large that it produces a semipermanent deformation of the westerlies; i.e., the latter are split by the block, which in effect lengthens the belt of westerly winds around the hemisphere and thus increases the heat transport out of tropical latitudes into polar regions by means of disturbances. In the region where the westerlies bulge northward on the poleward side of the block, the higher (subtropical) tropopause moves poleward with them, so that increased heat transport is possible in the troposphere in these latitudes. Individual disturbances along the periphery of the bulge carry much heat away from its center, which therefore decreases. In the trough on either side of the bulge, heat transport by migratory cyclones and anticyclones is directed inward, toward the trough line, so that heat is at first gained there.

Blocking patterns are associated with easterly polar outbreaks over Europe in winter and with warm anticyclones in the Azores region. For the Atlantic-European sector of the northern hemisphere, blocking was present on more than half the days during the period January 1949 to December 1952, inclusive. During the period 1949–

1956, in the same sector, 115 blocking spells were recorded, the length of individual spells varying from 3 to 54 days, with a mode at 7 days and a median at 12 days. Most blocking patterns occurred in spring, least in summer, with a secondary maximum in autumn. February blocks were the longest-lived; April blocks were the shortest; August blocks tended to be very persistent. During March and May, blocks tended to move from east to west, and during July and September from west to east; the blocking patterns also tended to shift from Europe to the Atlantic in spring and back again in late summer and autumn. The percentage of blocks moving west to east (progressing) decreased with increasing latitude, particularly over the Atlantic, while the percentage of blocks moving east to west (retrogressing) increased with increasing latitude, particularly over Europe.[73]

Data for 1899–1938 reveal an annual oscillation in the frequency and distribution of blocking anticyclones in the Atlantic-European sector. During January, blocking highs tended to occupy a band extending northeastward from the Azores towards Finland; by May, the main incidence of blocking shifted away from the European continent and lay in a belt running northward from the Azores through the Norwegian Sea; from July to September inclusive it was concentrated north of the Azores; during October, blocking was more frequent over the land than over the sea areas; and in December the southwest-to-northeast pattern became established again.[74]

Blocking patterns were especially frequent and persistent over northern Europe in all seasons from late 1958 to late 1960 inclusive. The increased blocking brought about a more or less steady decline in precipitation, particularly in autumn. The autumn rains on the west coast of Norway failed to develop in 1959–60 as the westerlies decreased, and in autumn of 1960 Bergen recorded a total of only 88 mm in comparison with its normal 199 mm, making this the driest autumn for at least 66 years. The physical cause of the high incidence of blocking was in the complex feedback mechanisms between the atmosphere, the sea surface of the North Atlantic, and the surface of the Scandinavian peninsula, in particular its lack of winter snow cover. Long periods of blocking over northern Europe such as these seem to be initiated by winters in which the prevailing north-south airstream between southern Greenland and southern Scandinavia is much stronger than normal.[75]

Two distinct types of blocking pattern can be recognized over the Atlantic-European sector of the northern hemisphere (see Figure 4.22). The *diffluent block* is the most common type; this is formed by the disruption of a long-wave pattern in the upper westerlies, usually by the high-latitude part of an upper ridge "toppling" eastward over the trough immediately downwind. The *meridional block* is due to instability developments when a small, mobile, east-moving upper disturbance, associated with a developing surface anticyclone or ridge, progressively increases in size and slows down. A diffluent block often changes into a meridional block within the same spell of blocking.[76]

Over the North American sector of the northern hemisphere, some form of blocking "wave" often proceeds upstream against the westerly current. Large areas of above-normal pressures at 10,000 feet were more frequent over the Bering and Davis Straits than over northern Europe during 1948–1957 inclusive, the positive anomalies at 700 mb coming a week or two later than the appearance of strong blocking highs over Europe. Thus "blocking surges" may move westward across the Atlantic.[77]

Jetstream developments are often incorporated in blocking patterns. In the case of the diffluent block, the Polar Front jet splits into two distinct branches, one going

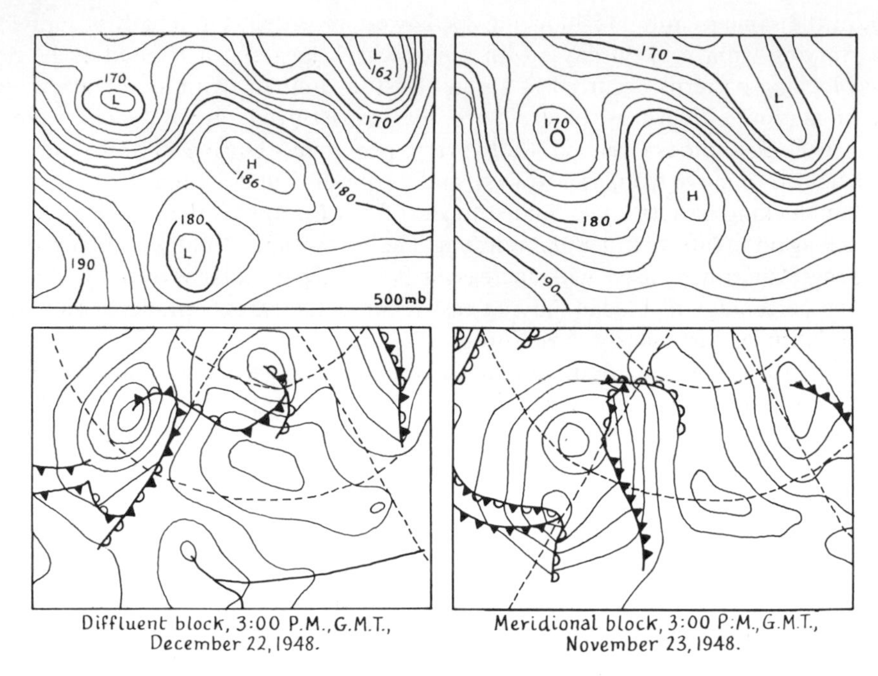

FIGURE 4.22.
Blocking patterns.

northeastward around the warm anticyclone, and the other southeastward around the cold low. Berggren, Bolin, and Rossby explain blocking in terms of instabilities in the zonal circulation, in which upper long waves in the westerlies grow abnormally, with partial cutting-off of warm (anticyclonic) and cold (cyclonic) vortices being followed by splitting of the jetstream in this manner. Rex introduced a model blocking pattern in which a clear-cut jet at 500 mb abruptly splits into two branches. Regions of mesoturbulence are present downwind of the bifurcation, the characteristic feature of which is a quasistationary upper-level warm ridge or anticyclone that may move slightly eastward or (more slowly) westward, in company with a surface high-pressure area that Rex regarded as the definitive blocking anticyclone.[78]

Persistence, the tendency for a weather phenomenon or climatic feature to maintain its existing position for some time, introduces problems in the study of both the index cycle and blocking patterns. For example, during the colder part of the year, low-index periods are more persistent than high-index ones, suggesting that there are feedback processes connected with variations in sea surface temperatures, snow cover, and soil moisture, all of which exert a conservative influence on the general circulation during periods of very weak airflow, which tends to maintain existing patterns as a consequence. Atmospheric events themselves have a bearing on persistence too. Thus when the height of the 700-mb surface is very variable and its changes pronounced, persistence is usually very weak; but when the height changes are small and roughly similar along a given latitude circle (as at midsummer and midwinter), then persistence is very marked.[79]

If the persistence of mean monthly patterns is examined by using correlation coefficients, there is evidence that the persistence itself undergoes a climatic fluctuation. Data for 1932–1951 analyzed in this manner show that persistence has a marked geographical pattern, but a different pattern would probably be given by data for other years, in view of this fluctuation. During December and January a high degree of persistence is found over the east-central Pacific and over the southeastern United States; these areas of high positive correlation vanish or are displaced southward in January and early February, but reappear in late February and early March. During the period March to September, the geographical pattern of persistence is different each month, except for a belt of high correlation over the South Atlantic. Gradual but regular progression of a belt of minimum correlation (i.e., very poor persistence) is found in North America: this is situated over the southern United States in March-April, moves northward as far as latitude 65° by July-August, then retreats southward in August-September.[80]

A comparison between North America and the East Atlantic-Europe area shows that the latter exhibits much more persistence than the former for up to three days. However, the averaging process in itself can generate spurious persistence.[81]

The areas of greatest persistence of mean monthly weather patterns are linked with the subtropical high-level anticyclones, which act through reservoirs of anomalous heating or cooling of the atmosphere provided by the underlying surface. For example, abnormally dry or abnormally moist soils exert a feedback effect. Thus in the spring of 1957, exceptionally moist soils in Texas (due to severe storms and floods that brought to an abrupt end several years of drought) served as a cooling reservoir for the vaporization of some of the heat normally used for the spring-to-summer buildup of the upper anticyclone in this area. The result was that pressures were much further below average in Texas than elsewhere. The severe droughts of summer 1957 over much of the eastern United States resulted in temperatures being well above normal, despite the fact that the general circulation—specifically, the mean monthly 700mb pattern—indicated that subnormal temperature should have resulted from the deep upper cold trough on the east coast. The dry soils created a reservoir of heat, like a desert; the heat thus released to the atmosphere caused higher daytime maximum temperatures than would have been thought possible in view of the prevailing circulation pattern.[82]

Dynamic persistence is a specially defined form of persistence that involves the linking of different and often widely scattered areas. It is a concept that essentially transfers the effect of local persistence from one area to another via a synoptic link. Dynamic persistence is said to exist when the general circulation satisfies two conditions: (a) there must be local persistence in a certain area from season N to the following season $N + 1$; and (b) there must be a contemporaneous synoptic connection between the weather in the area of local persistence and the weather of some other (not necessarily adjacent) area during season $N + 1$. For example, the weather (particularly precipitation) in the months December to February inclusive for the North Pacific and North America is connected via dynamic persistence with pressures in the general area of the Philippines and northern Australia during the previous September to October.[83]

STRATOSPHERIC CIRCULATION

For many decades, the movements of air in the stratosphere could not be measured directly, but had to be estimated by indirect means: smoke drifts, sound propagation, meteors, noctilucent clouds, and so on. The development of radiosondes able to reach

100,000 feet and above, and in particular the introduction of rockets, enabled more precise data to be accumulated. The first problem in studying the stratospheric circulation was thus to ensure that the data employed to analyze it really referred to the circulation.[84]

The first idea of the distinction between the stratospheric and tropospheric circulations arose late in the last century. After the Krakatoa eruption of 1883, the fact that the resulting dust traveled from east to west at a height above 25 km at least twice around the world meant that there must be easterlies in the stratosphere and in those latitudes. The *Krakatoa easterlies* were thus recognized: a belt of high-speed winds (about 30 m per sec), confined to the tropics, in particular between latitude 15°N and 15°S, and extending completely around the world. They were regarded as being present within the layer from 30 km to at least 40 km, their upper limit varying from month to month.

In 1908 and 1909, von Berson in Central Africa and van Bemmelen in Batavia independently discovered the presence of westerlies above 20 km, embedded in the Krakatoa easterlies. These *Berson westerlies* were a narrow belt of westerly winds, forming a thread rarely more than 7° of latitude in width and centered on 2°N, with a base near 20 km and, like the Krakatoa easterlies, circling the world. Both Berson westerlies and Krakatoa easterlies were confirmed in the Pacific stratosphere in the late 1940's and early 1950's, during observations connected with the atomic tests. Mean cross sections show the Berson westerlies in the lower stratosphere, with the Krakatoa easterlies above. The problem was to explain the presence of two circumglobal rings of oppositely moving airstreams, one above the other and quite close together latitudinally.[85]

The problem was soon complicated by the discovery that, in extratropical latitudes, the stratospheric circulation sometimes reverses. Thus radiosonde observations over Belmont (New Jersey) in 1948–1949 showed that, although below 60,000 feet the winds were predominantly westerly, with maximum speeds at 40,000 feet, between 60,000 and 120,000 feet the summer easterlies became winter westerlies. The easterly flow commenced three weeks after the vernal equinox and ended three weeks before the autumnal equinox. The observed data were consistent with the existence of a stratospheric circumpolar vortex that is cyclonic in winter and anticyclonic in summer. Later observations confirmed that, particularly in middle and high latitudes in the northern hemisphere, the stratospheric airflow between 25 and 60 km is westerly in winter and easterly in summer, reversing very rapidly twice a year near the equinoxes. The breakdown of the stratospheric circumpolar vortex in spring is accompanied by subsidence, which leads to a great increase in the ozone concentration in the atmosphere.[86]

Tentative theoretical solutions of the problem were put forward by Goldie and by Palmer. Goldie used Bjerknes' circulation theorem to postulate the existence of two independent stratospheric circulations frictionally driven by the tropospheric meridional circulation. Palmer extended this hypothesis, with special emphasis on the stratosphere in the tropics, by introducing the possible effects of ozone and vertical motions. In Palmer's model (see Figure 4.23) of the circulation between 10 and 60 km over the central Pacific, three distinct meridional cells are recognized in the stratosphere. Two of these are directly driven cells; the third—rotating in the opposite direction to the other two—is an indirect cell, driven by turbulent exchange with the direct cells. The indirect cell is very weak, with an average maximum component of

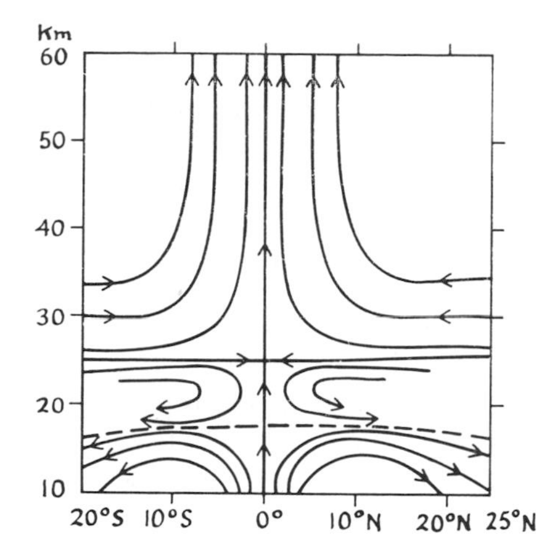

FIGURE 4.23.
Palmer's model of the meridional circulation between 10 km and 60 km at the equator in the Central Pacific. Broken line indicates the tropopause.

only 2 m per sec. One important implication of the model is that the comparatively high velocities of the Krakatoa easterlies are secondary features, arising from peculiarities in the temperature distribution (i.e., a thermal-wind effect), which are due in turn to the combined effects of vertical motion and the radiative properties of ozone.[87]

Data for 1955–59 for the northern hemisphere, collected in atlas form, show that the mean circulation patterns of the stratosphere in some months are very similar from one year to another, but others may be very dissimilar. Thus the patterns for May and June are much the same each year, and also those for January to some extent, but the mean patterns for February, March, and April vary considerably from year to year, particularly in high latitudes. This means that average circulation maps constructed from long-period means for the months February to April inclusive may be very misleading as estimates of monthly, let alone daily, conditions. During the period January to June, the general trend is for stratospheric temperatures to rise in high and middle latitudes, with a change from westerly to easterly winds in the layer between 25 and 50 mb, but deviations do occur—for example, in February-March 1957 and 1958, when temperatures decreased in high latitudes and the westerlies increased. A marked feature in the 25-to-50-mb layer is a reversal from easterly winds one year to westerlies the next, followed by a return to easterlies during the third year.[88]

Westerly stratospheric flow covers the entire northern hemisphere twice each year for about forty days, as the westerlies commence in the winter hemisphere and the zonal westerlies of the opposite hemisphere die down to reverse to summer easterlies. The summer easterly circulation is a steady flow that dominates the summer hemisphere and penetrates into the equatorial regions of the winter hemisphere, thereby limiting the expansion of the winter westerlies. The maximum rate of stratospheric wind shear is in latitudes 15° to 20° of the winter hemisphere, from which zone disturbances are propagated towards the winter pole. In middle latitudes, despite the steadiness of the summer easterlies, the stratosphere contains numerous disturbances that are so much smaller than those normally visible on tropospheric (particularly surface) charts that they usually are not visible on stratospheric synoptic maps. The summer easterlies reveal a relatively large variance at intermediate scales, which is not due to solar or lunar tides and therefore must result from disturbances. In addition, the meridional component of the stratospheric wind has a diurnal variation of about 1 m per sec in latitudes below 20°N in the northern hemisphere, and this variation may possibly exist in middle latitudes too.[89]

The layer between 30 and 70 km is one of considerable unrest in winter, although the incidence of disturbances may vary considerably from year to year. Thus winter 1960 over North America experienced the northward movement of tropical easterlies for a period of 10 days, during which winds from the base of the stratosphere to at least 50 km were affected. Winter 1961, however, experienced two easterly periods: one over western North America was associated with a stratospheric anticyclone centered over the Aleutians, whilst the other over the southern United States was due to a contraction of the stratospheric circumpolar vortex.[90]

Large disturbances undoubtedly do exist from time to time in the stratosphere. For example, stratospheric shearlines have been observed over the Caribbean. In January 1960, after a period of strong but variable easterlies in the layer between 100,000 and 120,000 feet, a large disturbance in the form of an equatorial shearline—similar structurally to those observed in the lower troposphere of the Caribbean in the hurricane season—moved in from the equatorial zone. Such a phenomenon implies there is a transition zone between the stratospheric easterlies of the summer hemisphere and the stratospheric polar-night jetstream of the winter hemisphere. The function of the shearline appears to be to transfer heat and absolute angular momentum from equatorial to tropical regions.[91]

According to Palmer's model (see Figure 4.23), although steady wind systems prevail in the lower stratosphere, the layer between 90 and 200 mb (18 and 12 km) is relatively turbulent, and rarely in a settled state. Although the disturbances that develop within this layer are "turbulent" in relation to the steady stratospheric flow, they form much more slowly than surface disturbances, and may remain stationary for as long as two weeks, in which case they will show up on mean monthly maps.

The tropical stratosphere provides perhaps the most interesting problem. The circulation in this part of the atmosphere may be divided into three components: the long-term mean, the annual cycle, and the 26-month cycle usually termed the *tropical* (or equatorial) *stratospheric wind oscillation*. During the period 1954–60, the mean winds in the tropical stratosphere were everywhere easterly—except possibly for very weak westerlies near the equatorial tropopause—and increased with height; they were strongest between 10° and 15°N. The annual cycle during this period had an amplitude varying from 10 m per sec at 25°N to zero at the equator; it essentially involves a build-up of the easterlies to a peak in late July or early August. The 26-month cycle, involving alternate easterly and westerly bands of winds that circle the Earth, has its greatest vertical amplitude at 25 mb—where windspeeds reach a maximum of 20 m per sec—and has disappeared before it reaches the tropopause; horizontally, it is most marked above the equator, but is still just evident at 30°N. The phase of the cycle varies with height: each band of easterly or westerly winds appears first at the highest levels, then gradually propagates downward, with a lag of just over 1 km per month. The fluctuation at a given height is in phase at all places in the same latitude.[92]

The facts of the tropical stratospheric wind oscillation may be expressed in a way slightly different from a 26-month cycle. Thus mean monthly zonal winds within 20° of latitude of the equator may be regarded as incorporating (a) an almost biennial oscillation in the middle and upper stratosphere and (b) a relative wave in the lower stratosphere, 180° out of phase with the biennial oscillation. Both waves progress upward in height from one biennium to the next. They merge near the equator, are most distinct from each other between 9°N and 15°N, and are still evident as separate occurrences (although much less intense) from 15°N to 20°N. According to this theory, in sum, the observed double cycle of winds is caused by the combination of two fundamental waves, one of which is annual, the other biennial, both being propagated upward with different phases at different heights.[93]

The 26-month oscillation involves only the zonal component of the stratospheric wind: the meridional component is unaffected. Since the oscillation apparently existed in 1908–1917, it may be surmised that: (1) if Krakatoa had erupted in 1882 or 1884 instead of 1883, the dust would have circled the Earth from west to east instead of in the opposite direction, and the "Krakatoa westerlies" would have been the first stratospheric winds to be recognized; (2) if von Berson had made his observations in 1907 or 1909 instead of 1908, he would not have discovered his westerlies.[94]

The 26-month oscillation was first proved to exist from northern hemisphere observations. The search for it south of the equator has brought to light some interesting features. For example, the oscillation is very pronounced at 70,000 feet over Hobart in latitude 42° 53′S—a very high latitude for an "equatorial" oscillation—and is present with increased amplitude at 100,000 feet. In fact, the oscillation occurs from the equator to at least 50°S and up to 90,000 feet.[95]

The 26-month oscillation extends its influence to middle and polar latitudes of both hemispheres in terms of induced oscillations in temperature, total ozone content of the atmosphere, and tropopause height. The biennial maxima of temperature and ozone tend to drift poleward at the rate of 0.2 m per sec, but the drift is not detectable poleward of latitude 40°.[96]

It is still uncertain whether the 26-month oscillation is due to solar influences, or is a dynamic feature of the atmosphere. Much of the circulation of the lower stratosphere is in part driven by the tropospheric circulation, and the oscillation may therefore to some extent reflect changes in the latter. Whatever its origin, its importance as a climatological phenomenon at variance with the normal annual progression of events is well-established.[97]

The effect on the stratospheric circulation of "explosive" increases in temperature is worthy of mention. For example, the abrupt warming of the stratosphere over Europe that began in June 1958, and spread through the entire northern hemisphere within 13 days, was accompanied by a distinct wind-shift in the stratosphere. During January and February of the previous year, an explosive stratospheric warming was associated with the sudden disappearance of the circumpolar vortex in the stratosphere, and its replacement by light winds. In midwinter of 1963, the abrupt warming initially appearing over southeast Canada at 10 mb coincided with a change in the stratospheric circulation in northern latitudes from one dominated by a circumpolar vortex to a nearly symmetrical bipolar one. General conclusions appear to be that (a) major abrupt warmings are associated with retrogression of circulation features in the middle stratosphere, and (b) the entire atmospheric circulation between 25 and 55 km reacts as one regime to the "explosion." The initial warming usually develops on the eastern side of a bipolar stratospheric trough that extends into middle latitudes. The warm-air center then moves north or northwest. With the arrival of the warm air in the stratosphere, in most northern latitudes the circumpolar vortex splits into two southward-moving portions, one or both of which fill rapidly. The circulation change is at its maximum intensity at 45 km, where it originates, then spreads both downward and outward through the atmosphere, eventually affecting the entire stratosphere and the lower mesosphere.[98]

WATER BALANCE AND THE GENERAL CIRCULATION

Some regions of the Earth show an excess of precipitation over evaporation, others an excess of evaporation over precipitation: very rarely does precipitation exactly balance evaporation. Regional excesses or deficits of moisture must be balanced by water transported by the general circulation of the atmosphere.[99]

Two types of water-balance equation may be written. The hydrological equation,

$$P = E + R_L + \Delta S_L,$$

applies to a land surface. In this equation, P represents precipitation, E represents evaporation, R_L is the runoff from the land, and ΔS_L (which may be positive or negative) is the increment in land storage. In the aerological equation,

$$PE = + R_A + \Delta S_A,$$

R_A represents the advective convergence of water vapor in the atmosphere, and ΔS_A the storage or depletion of moisture in the atmosphere above the region in question; R_A and ΔS_A may readily be computed from meteorological data, and thus the aerological equation for water balance is easier to solve than the hydrological one.

One can thus estimate from meteorological data alone which parts of the Earth will have excesses or deficits of water. This may be done using average figures on an empirical basis,[100] or it may be done on a more fundamental basis employing actual figures. Water cannot be created or destroyed within the atmosphere, and there is no significant net flow of water into or out of the atmosphere as a whole, except over a period measured in geological terms, as when water is added to the atmosphere by the release of "fossil water" from the combustion of coal and oil. It is therefore possible to derive a fundamental equation of water balance in the atmosphere from first principles (see Appendix 4.3). The map derived from this equation shows large centers with an excess of atmospheric moisture in the northwestern United States and western Canada, in northern South America, in East Africa, in the Soviet Union, and in eastern India and Pakistan, Burma, and the Indo-China region. Areas deficient in atmospheric water are the subtropics, the Gulf of Mexico, the central North Atlantic and central North Pacific, the east coast of Asia centered on the Sea of Japan, and the great desert areas of the northern hemisphere. On a more detailed scale, Africa north of the equator shows a deficiency in the west and an excess in the east.

In general, 2.5 cm of precipitable water (as measured by a rain gauge) are stored in the atmosphere. The whole atmosphere contains 0.5 cm more moisture during the northern summer than during the northern winter, hence the oceans should be shallower during the former. However, the storage of water on the lands in the form of ice may have a greater effect in the northern winter. Atmospheric water storage in the northern hemisphere averages 1.9 cm in January and 3.4 cm in July; that in the southern hemisphere averages 2.5 cm and 2.0 cm in January and July respectively. The total mass of water in the oceans varies from month to month, with a yearly amplitude of 3 g per cm^2, one-third of which is taken up in winter by snow on the ground, particularly in Siberia. The average depth of the oceans decreases by 3 cm in the northern spring and autumn, but the actual monthly decreases vary from one or two cm in the tropics to 1 m in the Bay of Bengal. One-half the fluctuation in sea level is due to the liquid and solid water in the atmosphere, the other half to the variations in water storage in the land masses. The atmosphere contains only ten days' supply of average precipitation; since this atmospheric water represents latent energy, the circulation of the atmosphere would "run down" in ten days if its supply of solar energy was cut off.[101]

The tropical atmosphere as a whole is much moister than that in middle and high latitudes (see Figure 4.24), and therefore some moisture must be carried poleward by the mean meridional circulation. Computations show that this is true in low latitudes, where the poleward moisture transport is due to the steady motion of the Hadley cells, and the eddy flux of water vapor may even be equatorward at times. In middle latitudes, however, the poleward moisture flux is largely due to large-scale eddies (anticyclones and cyclones), i.e., to unsteady motions, rather than to the meridional circulation.[102]

The total meridional flux of water vapor across a parallel of latitude may be expressed as the sum of the transport by the mean meridional circulation, by transient eddies, and by standing eddies. Each of these three transports can be evaluated quantitatively. From the speed of the wind in the mean meridional circulation, the total mass of air involved in the transport of moisture in low latitudes can be determined, and from this the meridional flux of water vapor can be obtained (see Appendix 4.4). The divergence (positive or negative) of the moisture flux gives the excess of evaporation over precipitation in the atmosphere (see Figure 4.25). The mean divergence of water vapor between latitudes 20°N and 40°N, i.e., the mean loss of water from the atmosphere during its movement from low to middle latitudes across the subtropical high-pressure belt, comes to 4.5 grams of water per cm^2 of air, which corresponds to an increase in the mean difference of evaporation minus precipitation of 4.5 mm per month.

A high rate of moisture transport across an area does not necessarily mean that the area will experience a high precipitation. To a greater or lesser extent, a given area may maintain its own balance between evaporation and precipitation. For example, precipitation anomalies, such as excessively wet or excessively dry areas, must be

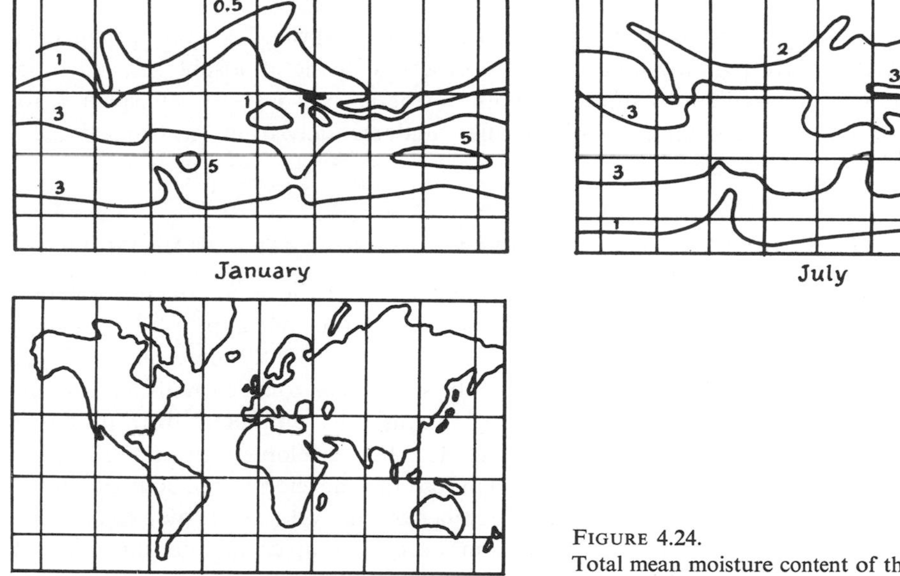

FIGURE 4.24.
Total mean moisture content of the atmosphere, in cm of precipitable water (after Bannon).

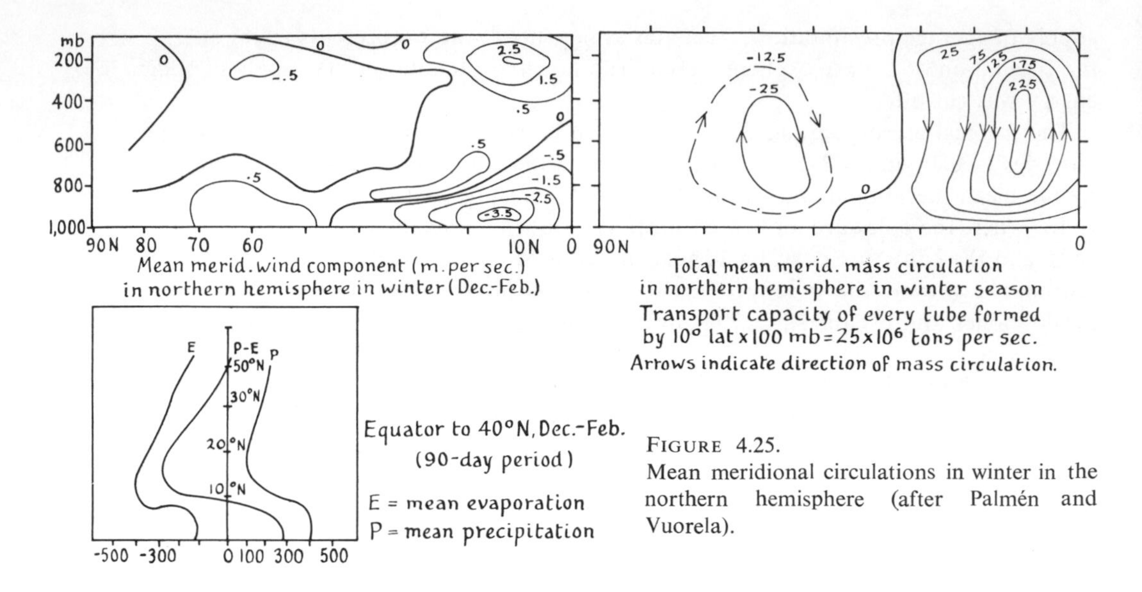

FIGURE 4.25.
Mean meridional circulations in winter in the northern hemisphere (after Palmén and Vuorela).

self-maintaining for the most part, especially land areas in summer. In some regions, therefore, water must be moving through a hydrological cycle that is almost closed locally, involving a water movement from the ground to the atmosphere and back again in an almost continuous process. In other areas, advection of moisture will be very important. Whether the water balance of a given area is controlled primarily by advection or by local processes is decided by the nature of the general atmospheric circulation over the area.

Some Regional Problems of the General Circulation

In view of the great regional differences in the Earth's surface, it is very pertinent to ask whether corresponding differences are observed in the atmospheric circulation. Are the general circulations similar in both northern and southern hemispheres, for example, and do the polar circulations differ fundamentally from the tropical circulations?

COMPARISON BETWEEN THE NORTHERN AND SOUTHERN HEMISPHERES

Observations show that the mean surface circulation of the southern hemisphere is stronger than that of the northern hemisphere. This does not appear to be due to the southern hemisphere circulation as a whole being inherently more powerful than that of the northern hemisphere; both have the same basic zonal currents,with similar rates of generation of disturbances. Rather, because of the different distributions of land and sea in the two hemispheres, cyclones that develop north of the equator cannot drift with the westerlies as much as those that develop in the southern hemisphere. Therefore a comparatively large number of anticyclones are able to enter the northern westerlies. Even though the cyclones are of comparable intensities in both hemispheres, the greater number of intense anticyclones north of the equator means that average surface windspeeds in the northern hemisphere westerlies are lower than those in the southern westerlies.[103]

The average zonal index of the southern hemisphere westerlies is 3 m per sec higher than that of the northern westerlies for the same reason, but this does not necessarily imply that the southern hemisphere circulation is a prevailing zonal flow. The mean value of 5.1 m per sec for the southern zonal index is made up of very varied daily values, and the summer range from 9.5 m per sec for the highest weekly average of these to 1.4 m per sec for the lowest weekly average over a two-year period, far exceeds the comparable northern hemisphere range. Nevertheless, the southern hemisphere westerlies are very strong, and usually prevent Antarctic air from reaching middle latitudes in very intense outbursts. Occasionally, polar anticyclones developing in or near Antarctica are strong enough for the outflowing currents from them to break through the zonal circulation and ultimately reinforce the subtropical high-pressure ridge.[104]

It is not correct to assume that a purely zonal circulation exists south of the equator even at high altitudes, although the upper circulation does seem comparatively simple by northern hemisphere standards. Considerable deviations from zonal flow have been discovered, for example, over Australia and New Zealand at 200 mb. Only a few of these deviations—a closed anticyclone that appears over Australia in summer being the main instance—are linked with definite seasons or with distinct regions, so that upper troughs and ridges are likely at any time of year. An abrupt transition from a winter flow pattern to a summer one is characteristic at 200 mb, and sometimes coincides with a transition in the reverse direction in the mean zonal flow of the northern hemisphere. Splits in the southern hemisphere zonal circulation, involving the development of a confluence-diffluence pattern (see Figure 4.26), tend to occur just south of Australia or near New Zealand.[105]

The circumpolar trough of low pressure in the southern hemisphere is the counterpart of the Icelandic and Aleutian lows in the northern hemisphere. Unlike the latter, which weaken in summer (the Aleutian low almost disappears), the circumpolar trough varies little in intensity or in position from season to season. The subtropical high-pressure belt forms the southern counterpart of the subtropical anticyclones of the northern hemisphere, and although it appears to be a continuous belt on mean maps, in fact it consists of permanent high pressure only in the eastern part of the South Pacific, where the airflow is both impeded and diverted by the Andes. Elsewhere, the subtropical "ridge" is the statistical result of a succession of migratory anticyclones, contrasting with the quasipermanent nature of the northern hemisphere cells. A further difference is that whereas the southern high-pressure belt decreases in intensity from winter to summer, the northern subtropical anticyclones decrease from summer to winter, that is, at the same time.

Unlike the northern hemisphere westerlies, the southern westerly belt always maintains its continuity, and is less subject to meridional phases: thus the exact form taken by the index cycle differs in the two hemispheres. High-index patterns in the southern hemisphere are characterized by a broad belt of strong westerlies covering the southern oceans, with fronts oriented west-east, and waves developing on and moving rapidly along the fronts. A true low-index pattern as found in the northern hemisphere is not normally observed south of the equator. The transition from high to low index involves a shift in the zonal westerly belt and the development of a pronounced trough-and-ridge pattern in the upper westerlies, instead of a complete breakdown into

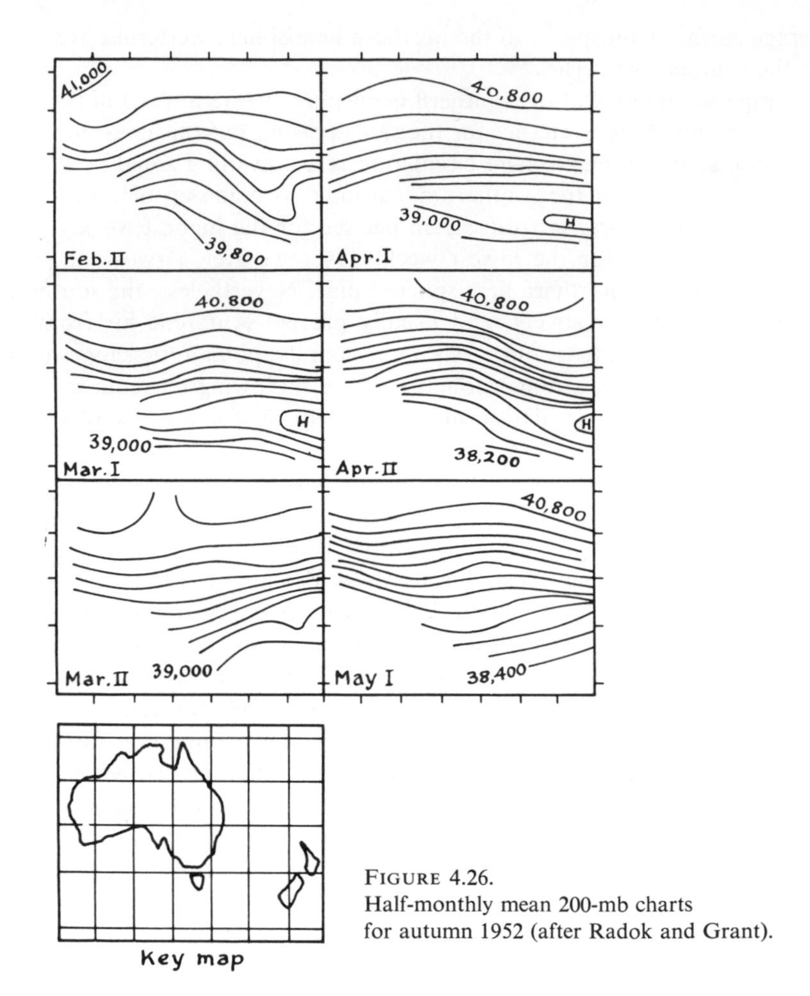

FIGURE 4.26.
Half-monthly mean 200-mb charts
for autumn 1952 (after Radok and Grant).

cells and meridional flow as in the northern hemisphere. High values of the zonal index generally tend to dominate the southern hemisphere in winter and spring, and comparatively low values in summer and autumn.[106]

Circulations north and south of the equator differ from each other in ways less obvious than the broad patterns just described. For example, a mean poleward drift of air in the lowest layers of the surface westerlies, due to the frictional drag between the Earth's surface and the atmosphere, exists all year in the southern hemisphere, but only in summer in the northern hemisphere. As another example, the eddy flux of momentum exhibits a seasonal variation over a large part of the southern hemisphere that involves maximum values of eddy flux in summer and minimum values in winter; in the northern hemisphere, the maximum momentum flux due to eddies is found in winter and the minimum in summer. The eddy flux of enthalpy reveals no difference between the hemispheres: both show maxima in winter and minima in summer.[107]

Exchanges of both air and energy take place across the equator each year, so that interhemispheric circulations must exist. Thus the Hadley cell of the winter hemisphere extends across the equator for at least 5° of latitude into the summer hemisphere, so that a mean meridional circulation must be in operation in transequatorial regions. Even here, differences between the hemispheres are apparent: the Hadley cell of the southern hemisphere, during the southern winter, does not extend quite as far north as the northern Hadley cell extends south during the northern winter. The mass of air in the northern hemisphere is at a maximum in midwinter and at a minimum in midsummer. The maximum monthly rate at which mass has to be carried into the southern hemisphere amounts to 1.86×10^{12} metric tons and takes place in May and June. In terms of a mean meridional flow across the equator, however, this mass transfer is equivalent to a circulation with a strength of only 0.003 knot.[108]

On some occasions, standing waves may extend over both sides of the equator, so that the circulation of one hemisphere influences the weather of the other. There is some evidence that large-amplitude flow patterns existing in the midtroposphere of the northern hemisphere during winter tend to induce responsive features at the same longitudes in the southern hemisphere.[109]

A marked seasonal flow of heat obviously must take place across the equator, representing the difference between the net heat gain and the net heat storage in each hemisphere. This heat transport is southward from May to October inclusive, and northward across the equator during other months. The rate of heat flow is about 2.9×10^7 cal per cm per min in January, 0.4 cal per cm per min in April, -2.2 cal per cm per min in July, and -0.8 cal per cm per min in October (the negative sign indicating a southward flow). The magnitude of this interhemispheric heat exchange is thus comparable with that of the heat transport across latitude 40°N, i.e., between tropical and middle latitudes, which averages 2.1×10^7 cal per cm per min for the year with a maximum of 3.8×10^7 cal per cm per min in January. Since the heat transport by eddies across the equator is negligible, both transequatorial meridional circulations and ocean currents must be the chief media. The ocean currents, although weak, can transport appreciable quantities of energy, since the heat capacity of a layer of sea water 2.5 m thick is the same as that of the entire vertical extent of the atmosphere.[110]

Both heat and vorticity appear to be transferred across the equator in the western Pacific by means of irregular disturbances of the upper westerlies. In this region, the 100-to-400-mb layer is particularly important for interhemispheric exchanges, as would be expected from the fact that the transequatorial winds at 40,000 feet are twice as strong as those in the lower troposphere.[111]

A strong flow of momentum takes place across the equator, directed towards the summer hemisphere, because of a standing-eddy flux associated with high-level monsoonal circulations over the Indian Ocean–East Pacific area. All three types of momentum transport (i.e., the toroidal, the standing-eddy, and the transient-eddy fluxes) are in the same direction and reach their greatest intensity above 250 mb. In all seasons except spring, the momentum transport across the equator is more than 10 per cent of the total momentum produced in the northern hemisphere: clearly, dynamic models designed for one hemisphere alone are of doubtful utility.[112]

CONTRASTS BETWEEN THE POLAR CIRCULATIONS

Since the north and south polar regions differ greatly in their distribution of land, sea, and ice-cover, it is a reasonable first assumption that their atmospheric circulations might also differ greatly. The climatic conditions over the Antarctic continent ought logically to be different from those over the Arctic Ocean; in fact, even the marine climates of the two polar regions are very different. Thus mean temperatures are symmetrically concentric around Antarctica throughout the year (see Figure 4.27,B), but are asymmetric in the north polar region. The average latitude in which the 32°F

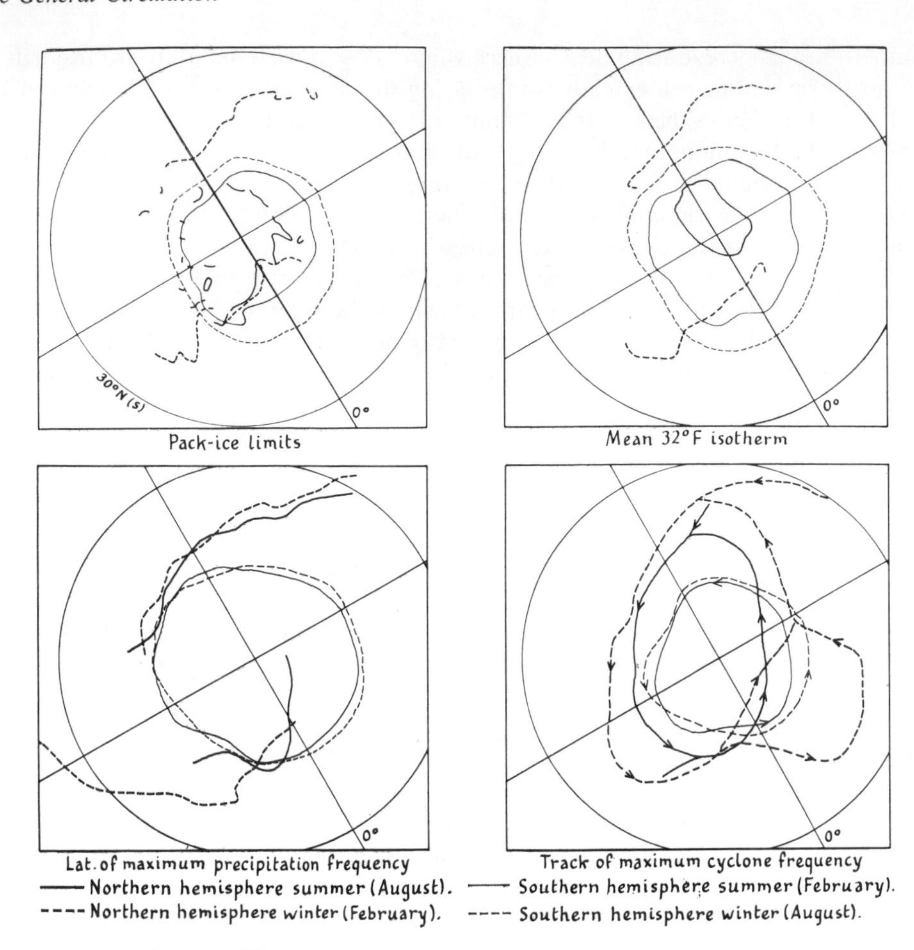

FIGURE 4.27.
Comparison of conditions in Arctic and Antarctic marine climates (after Holcombe).

isotherm is found shifts only 7° of latitude with the seasons in the southern hemisphere, from 64°S in summer to 57°S in winter, but shifts 22°, from 78°N in summer to 56°N in winter, in the northern hemisphere. As a consequence, Antarctic summers are colder than Arctic ones. Similar concentric symmetry in Antarctica holds for cyclone and precipitation frequencies (see Figure 4.27, C and D). Cyclone tracks are generally more scattered in the northern hemisphere than they are south of the equator. The average latitude of the zone of maximum cyclone frequency varies from 62°N in summer to 57°N in winter, compared with a southern hemisphere variation of only 2°, from 66°S in summer to 64°S in winter. The mean latitude of maximum gale frequency in the northern hemisphere changes from 62°N in winter to 46°N in summer, compared with a variation south of the equator from 56°S to 54°S.[113]

Average mean-sea-level pressures south of latitude 45°S are much lower than those of corresponding northern latitudes, so that there is an over-all deficit of mass in the southern hemisphere. The southern upper westerlies are stronger and more symmetrically disposed around the pole than are those of the northern hemisphere, and shift less with the seasons than do the latter. The southern hemisphere 500-mb circulation exhibits one mean trough, in longitudes 100–110°E over the Indian Ocean, compared with the two major troughs of the northern hemisphere over northeast Canada and eastern Asia. Unlike the orographically induced northern troughs, the southern trough is not a dynamically controlled feature, but a thermal effect associated with the quasipermanent outflow of cold air from Antarctica between longitudes 50°E and 150°E.

The subtropical high-pressure maximum is some 5° of latitude nearer the equator both in summer and winter in the southern hemisphere than in the northern; that is, the southern climatic zones are displaced equatorward. Thus a weakening of the southern westerlies, or a southward displacement of them, would lead to a southerly shift in all the climatic belts of the world. Since the early nineteenth century climatic fluctuations south of latitude 40°S have been opposite in phase to those of the rest of the Earth, and the question naturally arises whether the Antarctic continent might be a cause of this phenomenon.

Great differences are found in the elements of the atmospheric circulations on the fringes of the polar regions. Blocking patterns are very rare in the southern oceans, and any that appear are soon obliterated by the renewed strength of the westerlies upstream from the block. There are no southern hemisphere counterparts of the blocking anticyclones of latitudes 50–70°N, or of the persistent ridge that in winter often connects the Azores and Siberian anticyclones and the high-pressure system over the Arctic Sea ice north of Alaska. Instead, the southern hemisphere often reveals a pattern in which a high-pressure belt links the polar anticyclone, usually found over eastern Antarctica near 80°S, with the southern subtropical high pressures.

The frequencies of cyclones and anticyclones indicate distinct differences between the two hemispheres, as is confirmed by daily figures of central pressures for disturbances. Sub-Antarctic depressions are 10 to 20 mb deeper than those of comparable latitudes in the northern hemisphere; Antarctic anticyclones are 20 to 40 mb *less* intense than those of the north polar region. In general, sub-Antarctic cyclones are both deep and very large (in the area they cover), especially in winter, compared with their northern relatives. Contrasting with this, anticyclonic activity in Antarctica is not very marked compared with the situation in the north. Indeed, on half the days in most years there is no room for the existence of any anticyclones south of the southern Polar Front and its associated jetstreams, whereas almost every day there is a polar anticyclone somewhere over the snow-covered parts of the Arctic. But there are days on which relatively low pressure prevails over the north polar basin with high pressure over the surrounding continents.[114]

The polar stratospheres of both hemispheres show both similarities and contrasts. Both are characterized by a westerly vortex that develops in winter in the stratosphere and mesosphere, giving rise (see Figure 4.28) to the *polar-night westerlies.* The polar-night westerlies of the southern hemisphere are symmetric around the pole; those of the northern hemisphere are eccentric, with a circulation center to the north of Siberia and a persistent warm ridge over Alaska and the Bering Sea. These vortices are associated with the long periods of intense radiative cooling during both the northern and southern winters, and since the poleward gradient of temperature in the stratosphere is usually concentrated into a narrow belt in high latitudes, they have a jetstream structure.[115]

The polar-night westerlies are quite distinct from the normal middle-latitude tropospheric westerlies—usually termed the *Ferrel westerlies*, after William Ferrel, who first described them in 1856—and are separated from them by a wide zone of very

The polar-night westerlies
12:00 midnight, G.M.T., January 8, 1958

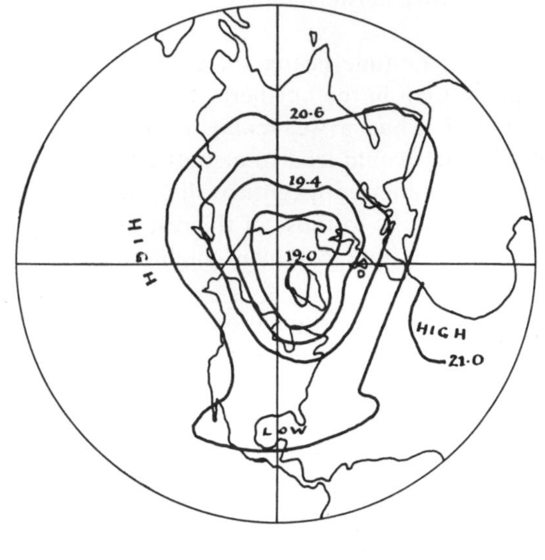

Contours of 50-mb surface in geopotential km.

The outer limit of
the Ferrel westerlies

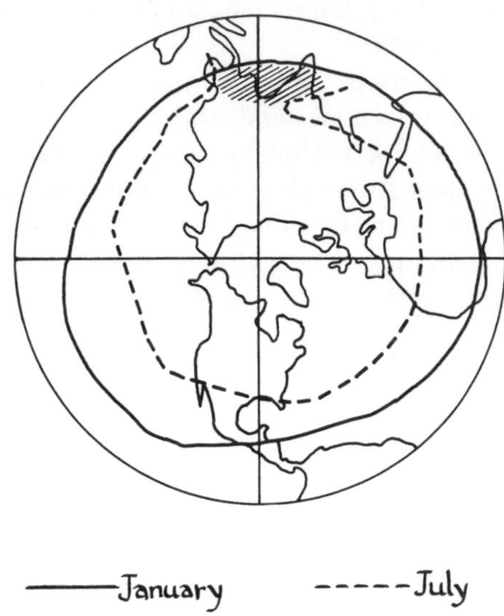

——— January - - - - - July

Indian monsoon westerlies

Typical cross sections through the westerlies

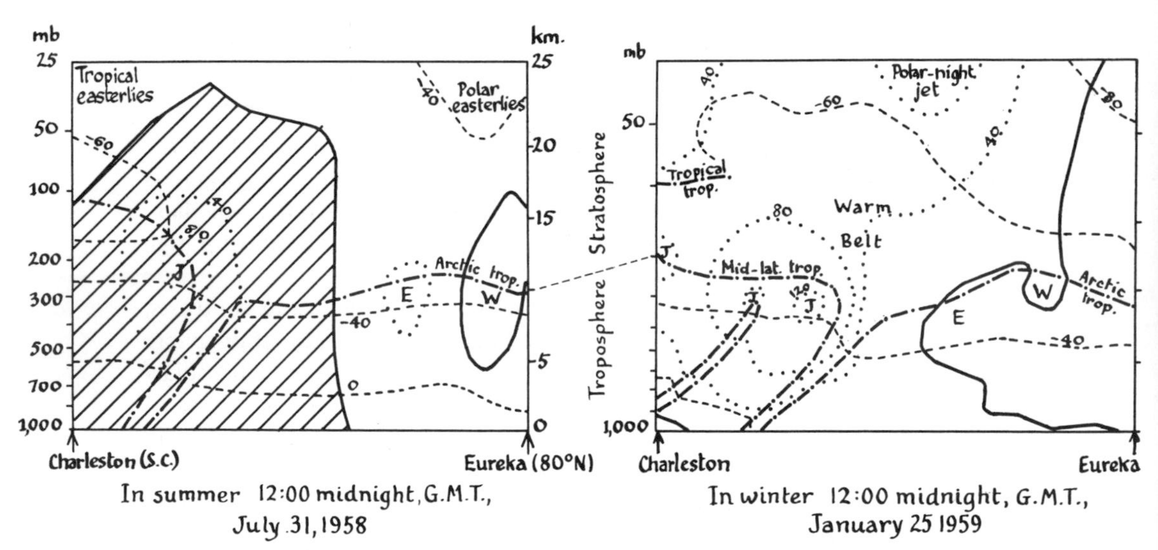

In summer 12:00 midnight, G.M.T., July 31, 1958

In winter 12:00 midnight, G.M.T., January 25 1959

J: jetstream cores —— zero isotach separating westerly from easterly components ···· isotachs (kts.)

-·-·- tropopauses and fronts - - - - isotherms (°C) E (W): easterly (westerly) wind maxima

FIGURE 4.28.
The westerlies (after Hare).

indifferent air movement. In general, the Ferrel westerlies occur south of latitude 55°N in the northern hemisphere and up to a height of 20 km in the lower stratosphere, whereas the polar-night westerlies are found north of latitude 60°N.[116]

THE ARCTIC CIRCULATION

Two groups of problems are posed by the general circulation of Arctic regions. First, two geographical problems exist: (a) what is the effect of the snow and ice cover and its variations on the atmosphere, especially on cooling and consequent formation of an anticyclone, and (b) what effect does Greenland have on the circulation? Second, the Arctic stratosphere also provides both geographical and dynamic problems.[117]

The atmospheric circulation pole in the north polar region is not over the geographical North Pole; it is displaced from the latter toward the Pacific in the troposphere and lower stratosphere, and toward the Atlantic in the middle stratosphere. A bipolar pattern of both pressure and temperature during the cold half of the year is caused by thermal differences between land and sea, albedo effects, and dynamic processes. The Gulf Stream provides a source of heat that causes the Icelandic low-pressure center to usually be located further north than the Aleutian low, and so allows more frequent extension of meridional ridges into the polar regions in the Atlantic sector. Important consequences follow for middle-latitude weather.

Asymmetric shifting of cold air to one side of the north polar region is associated with the development of blocking highs in middle latitudes in general. Large increases in pressure in midtroposphere arising in this manner in the Davis Strait–Greenland area in winter are related to decreases in pressure in northern and central Europe. Variations in the area of Arctic ice in the Siberian Arctic in winter are reflected by variations in average zonal wind components at 500 mb over Europe about one week later.[118]

Mean monthly sea-level pressure charts show a feeble surface anticyclone in July over the pack-ice area northwest of the Canadian archipelago and northeastern Greenland. In this season, thermal contrasts between polar and middle latitudes are at a minimum, so that the polar circulation is relatively weak and the circumpolar low-pressure belt is furthest north; cyclonic activity, however, is very marked along the Siberian coast and often extends to the North Pole itself. The Arctic surface anticyclone is smaller in October, and is centered either over northern Greenland or over the western part of the Canadian archipelago. Zonal windspeeds increase in November, and by January frequent cyclones cross the Kara Sea. Finally, in April, although warming is well-advanced in the middle-latitude continents, it is still cold in the Arctic, with a predominantly anticyclonic circulation.

An anticyclonic circulation thus prevails in the vicinity of the North Pole in the lower atmosphere during the winter half-year, although it is very small (see Figure 4.29). The anticyclone was formerly assumed to be maintained against frictional dissipation by radiational cooling, but in fact it can be maintained dynamically provided warm air accumulates in adjacent regions. The cooling of the lowest air layers by the cold surface causes the circulation pattern to deviate from being purely zonal, and the diverted airflow is warmed by advection in the surrounding areas, thus dynamically maintaining the low-level anticyclone.[119]

Despite the surface anticyclone, a cyclonic circulation prevails at 700 mb over the

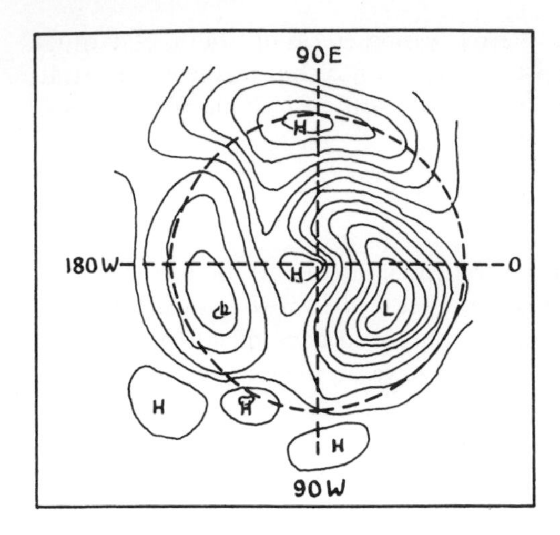

FIGURE 4.29.
Normal surface pressures
in November (after Fjørtoft).

American and Siberian Arctic in July, with frequent interruptions by a weak anticyclonic flow in the northeastern Greenland–Fridtjof Nansen Land area but rarely elsewhere. In January, two cyclonic cells are found at 700 mb over northeastern Canada and northeastern Siberia, thus elongating the tropospheric circumpolar vortex. In the layer from 700 to 200 mb, one low center remains north of Hudson Bay, and the other (composite) low remains over northern Siberia, but at 100 mb the two lows become merged to form a single elongated center over the North Pole.

The general picture of the Arctic circulation shows a mainly westerly flow throughout the north polar troposphere in January, with considerable asymmetry about the pole; three main troughs occur within the 700-to-200-mb layer, at 55°E, 135°E, and 80°W. The westerly flow increases with height because the coldness of the Arctic stratosphere results in the thermal pattern reinforcing the normal westerly pattern. As a consequence, the flow pattern becomes almost symmetric about the pole, in a cyclonic circulation, at 100 mb. In July, the tropospheric flow is also westerly, but much weaker than in January, and becomes more asymmetric with height. An anticyclonic circulation is centered on the pole at 200 to 100 mb. Above 25 mb, it also becomes symmetric about the pole. In April and October, westerly circulations are found at all levels in the troposphere, with major troughs at longitudes 60–80°W and 160–180°W, plus some minor troughs.[120]

The current understanding of the Arctic stratospheric circulation as greatly disturbed contrasts sharply with the old idea of the stratosphere as a dry and very stable part of the atmosphere.[121]

The polar-night westerlies—which first appear over the Arctic in September, when the summer easterlies reverse as the polar cap cools—are widest in December, after which they contract and intensify in strength. The vortex collapses sometime between late January and early April, usually within a period of from three to six weeks; that this collapse is before the return of the sun suggests that the vortex is dynamic in origin. The collapse of the polar-night vortex is accompanied usually by explosive warmings in the Arctic stratosphere, and by downward motions that result in large rises in temperature and increases in atmospheric ozone, both of which are important influences for large-scale meridional motion.

A stratospheric westerly jet, the Arctic jet, exists in winter at 60°N, with a center near 60 km; it is replaced in summer by an easterly jet in more southerly latitudes, centered at 50 km. Baroclinic waves that form within the polar-night vortex during winter include types resembling both the Rossby waves of the midtroposphere, with wave-numbers of four or thereabouts, and the shorter cyclone waves of the Ferrel westerlies. These waves reveal themselves quite well on the 25-mb chart; they are independent of the unstable baroclinic waves of the troposphere, from which they are separated by the broad barotropic warm belt within the lower stratosphere. They move from west to east, resulting in temperature variations of as much as 30°C in 36 hours, as a trough replaces a ridge or vice versa above some particular place. Their function appears to be to convert available potential energy into kinetic energy at stratospheric levels.[122]

The winter stratospheric circulation of the Arctic presents two facets: either (a) a warm, anticyclonic, barotropic flow associated with a strengthening and extension of a warm ridge over Alaska, involving a moderately warm (−50°C) and relatively weak anticyclone over the Canadian Arctic and a cool (−60°C), relatively weak cyclone near northern Scandinavia; or (b) the development of a large and intense (−75°C) cold low between Greenland and the North Pole, with a smaller, less intense, warm (−40°C) anticyclone just south of the Aleutians, together with a tendency for cold troughs to develop over eastern Canada. The outer portions of the cold lows are characterized by polar-night jetstreams and traveling baroclinic waves, the thermal effects of which may be felt over the entire stratosphere. Either type (a) or type (b) may dominate for long periods, so that a given winter month may be very different in successive years: for example, the polar-night jet may be almost absent one year, very strong the next.[123]

The warm anticyclonic current of type (a) may at times be very extensive. The dominant feature of the stratospheric circulation during winter 1958–59 was an intense warm anticyclone centered over the Aleutian-Alaskan area, which developed in October and persisted throughout the winter (see Figure 4.30). The anomalous warmth

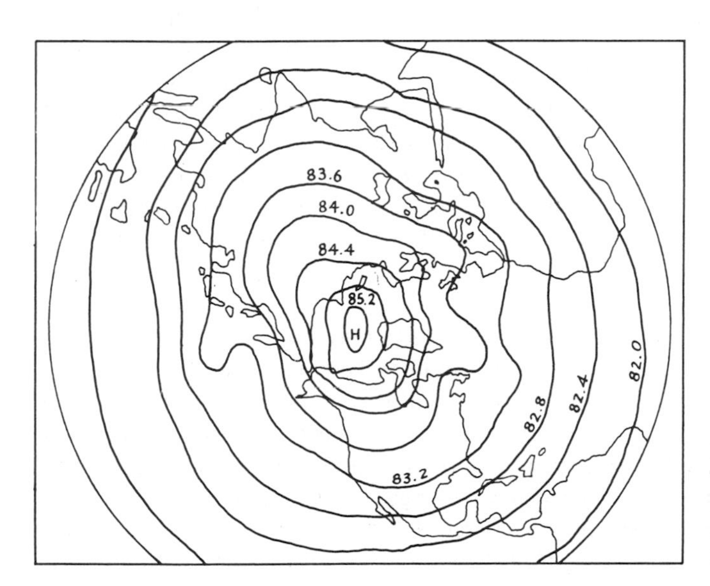

FIGURE 4.30.
The 25-mb chart for June 24, 1958. Contours in thousands of geopotential feet. (After Boville.)

resulted in the appearance of a strong baroclinic region over Alaska and eastern Siberia, through which moved strong waves in the polar-night westerlies. Changes in the position and intensity of the Aleutian stratospheric anticyclone are probably connected with interactions between winter disturbances of both the troposphere and the middle stratosphere.[124]

During winter, the Arctic stratosphere at 100 mb tends to return to either one or the other of two highly persistent synoptic patterns after a disturbance. The first pattern is an asymmetric single-wave one, involving a warm ridge over the Alaska–eastern Siberia area and an elliptical cold-core vortex over Eurasia. The second pattern is an eccentric bipolar one, i.e., a two-wave circumpolar vortex with a troughline extending from central Canada to central Siberia. A quasistationary trough and ridge are found over Eurasia and the Aleutians, respectively, in both patterns, but great differences between them are found over the Canadian and North Atlantic Arctic. Both patterns are asymmetric, their centers of gravity being displaced away from the pole toward the cold cyclones, a feature not consistent with either Rossby's theoretical or Fultz's experimental explanation of the fluctuations of the general circulation as being a consequence of its internal dynamics; instead, an external forcing agent is indicated. The arrangement of warm and cold areas in the patterns further suggests that both terrestrial and tropospheric influences are apparent in the Arctic stratosphere in winter. Other links also suggest that the tropospheric circulation affects that of the stratosphere: thus blocking highs in the troposphere over northwestern Europe at longitude 10°E in winter and spring have effects on the stratosphere, and in spring and autumn the cold polar vortex of the lower stratosphere over the Atlantic and northwestern Europe is affected by the cold tropical troposphere.[125]

During summer in the Arctic, the Ferrel westerlies and their wavetrains decrease with increasing height, until they reach zero at a surface sloping upward from 15 km at the North Pole to 20 km in latitude 50°N. At 50 mb and above, an almost undisturbed easterly anticyclonic flow prevails in the middle stratosphere because of a warm, flat-centered anticyclone centered near the pole. By August, cooling commences in the Arctic stratosphere, and a barotropic circumpolar westerly vortex, continuous with the Ferrel vortex of the troposphere, becomes established by September. In October and November, this becomes baroclinic and distinct from the Ferrel westerlies, so that the polar-night vortex then exists as a separate entity. The lower stratosphere below 20 km is, in summer, subject to vertical disturbances associated with waves in the Ferrel westerlies, or with cold lows within the latter. Closed cyclonic systems at 25 mb (25 km) are confined to the layer above the Ferrel westerlies.[126]

The Greenland ice cap is the only remaining large ice sheet in the northern hemisphere. Its crest reaches between 8,000 and 10,000 feet; it extends some 1,200 miles north and south. It would appear reasonable to suppose that it might have some effect on the general circulation.

A problem that diverted the study of polar meteorology for many years was that of the "Greenland glacial anticyclone." According to the geologist W. H. Hobbs, air resting on the ice cap is subject to intense radiational cooling, which leads to the accumulation of cold air in the central part of the ice sheet; atmospheric pressure should therefore increase and an anticyclone develop. The accumulating mass of cold air finally breaks out to the coast in a downslope surge or "stroph" of the glacial anticyclone. Hobbs suggested that strophs ultimately led to the generation of the main frontal zones of the North Atlantic, so that the Greenland ice cap acts as the "North

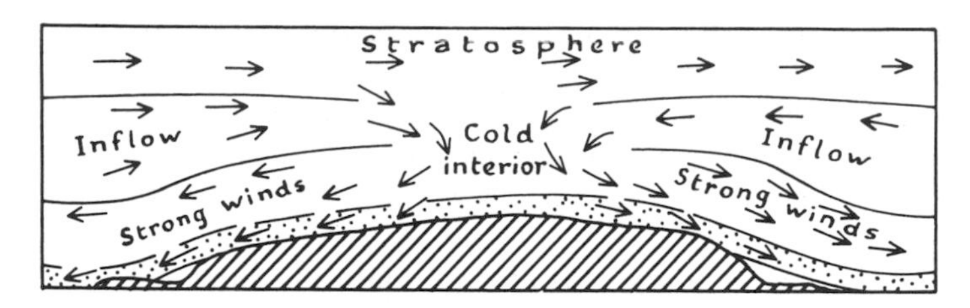

FIGURE 4.31.
Hobb's model of the Greenland glacial airflow. Stippled area above glacier surface indicates heavy air.

Pole of the winds," a unique all-season wind pole above which upper-air currents from high latitudes descend and then, via the outblowing anticyclonic circulation, return to lower latitudes as surface winds. Similar glacial anticyclones were postulated for Antarctica, and also for the continental glaciers covering parts of middle latitudes during the Pleistocene period.[127]

Hobbs proposed an empirical model (see Figure 4.31) involving a sharply defined cold core above the center of the ice cap. Within this core, air descended from heights of above 4,000 m, and subsequently fed the outblowing surface winds, which formed a zone around the central core. The strength of the outblowing winds varied widely, from near-calm to surges of 100 miles per hour or more, but their direction remained constant: for every point on the surface of the ice cap there should be a wind direction that deviates from the downslope direction only as much as required by the Earth's rotation. The importance of the outblowing surface winds in the scheme reflected Hobbs' reliance on observations made during South Polar and Greenland expeditions, which he thought neglected by meteorologists. Finally, the model suggested that eastward-moving cyclones should be deflected northward along the west coast of Greenland when they encountered the glacial anticyclone, but during the short summer period depressions might invade the western periphery of Greenland, i.e., the periglacial fringe, for a short distance.

Hobbs' hypothesis possessed three drawbacks. First, it stated that the glacial anticyclone prevented cyclones from crossing the ice cap, yet synoptic analyses indicate that they do, quite frequently. It is probably more correct to say that it is the upper perturbations, overlying the surface cyclones, that cross the ice plateau, rather than the depressions that appear on the surface synoptic chart; nevertheless, cyclonic activity is common above all parts of the Greenland ice cap. Second, major outbreaks of cold air from the center of the ice cap are not due to the "strophs of the glacial anticyclone," but result simply from certain favorable states of the general circulation. Excessive strophs require both a katabatic surface wind and a regional airflow in the same direction, so that the surface wind produced by the general circulation is reinforced by cold-air drainage. Furthermore, the actual regime of the surface winds is much more complex than that indicated by Hobbs' radial wind pattern.[128]

Third, there is the question of the alimentation of the ice cap. If no cyclones can penetrate the central part of Greenland, the source of the snow or ice that replaces the annual ablation loss of the ice sheet needs to be explained. Hobbs believed that the ice cap was nourished by hoar frost from cirrus clouds, which are common over the surface of the ice. However, although rime may condense on the lower slopes of the ice cap and so form an important alimentation source there, hoar frost cannot be significant, because any subsiding air over central Greenland will be very dry and in any case will be warmed adiabatically. Fronts and cyclones move across the ice cap in all seasons (but especially in autumn and early winter), and it is these, plus orographic effects, that provide the heavy snowfall necessary to maintain the ice sheet.[129]

The Greenland glacial anticyclone is thus of doubtful validity. In general, synoptic studies show that the Arctic does not have a distinct meteorology or climatology of its own, and that the Arctic circulation is intimately related to that of middle latitudes. Nevertheless, the peculiar geographical conditions of Arctic regions do have some effect on their atmospheric circulation. Whatever the state of the general circulation in the north polar regions, a temperature inversion nearly always exists at the surface or in the lowest 100 to 200 meters of the atmosphere over the pack-ice areas. Strong winds are very rare in these areas: an average of only one gale per year in the vicinity of the North Pole.[130] In Greenland, katabatic surface winds cause frequent gales that exceed 100 knots where the airflow is channeled into fjords. High mountains separate the Polar Basin from the Pacific, hence cyclones rarely penetrate the north polar regions from the latter. The area between Greenland and Norway forms the main channel both for the advection of warm air and warm water into the Arctic, and for the exit of drift ice from the polar seas. Consequently, depressions frequently enter the Polar Basin from the Atlantic, giving rise to more unsettled and warmer weather than is found on the Pacific side of the basin, which is characterized by relatively undisturbed weather, with a high frequency of anticyclones and more extensive areas of permanent sea ice.

In winter, the Arctic surface circulation is subject to numerous disturbances, with cyclones outnumbering anticyclones by two to one. In fact, the central Arctic basin has as many cyclones as any other part of the northern hemisphere. The main scenes of anticyclonic activity are Greenland, eastern Siberia, and eastern Alaska. Most of the anticyclones are colder than the cyclones. The Greenland anticyclones in particular are small, weak, and usually warm-core (whereas they should be large, intense, and cold-core according to Hobbs), although their centers are usually capped by oceanic air, which is slightly warmer than the air on their peripheries. Most Arctic cyclones are warm-core, and originate as old Atlantic depressions or secondaries, which either stagnate and fill at the head of Baffin Bay or weaken and soon fill during the course of a northeasterly movement from Iceland towards the Barents Sea. Cold-core cyclones occasionally form in the Arctic, usually developing in northern or northwestern Canada as weak surface expressions of strong vortices at 500 mb. A few warm-core cyclones penetrate into the Polar Basin from the Pacific, but these normally fill rapidly over Alaska.[131]

In summer, a secondary baroclinic zone—quite distinct from the main polar Front—develops along the northern shores of Siberia and North America. Cyclones that develop within this zone, and some that form on the main Polar Front to the south, quite often move into the central Arctic and stagnate near the North Pole, thus giving rise to the weak low-pressure center appearing there on mean monthly charts. In general, the summer disturbances of Arctic regions are very similar to middle-latitude depressions. There is no synoptic evidence for a semipermanent anticyclone over either Greenland or the North Pole, although small migratory anticyclones fairly frequently cross central Greenland and the Beaufort Sea area.[132]

If there is one circulation feature that can be taken alone as a circulation parameter for Arctic weather, it is probably the location of the cold-air pole of the troposphere. Individual cold-air poles (defined by 500-mb temperatures less than −30°C in summer and less than −40°C in other seasons) develop because of lifting processes in the center of Arctic cyclones, and their movement controls the height of the 500-mb surface over the whole of the north polar regions. The movement of the main cold pole may be followed from day to day, and the ultimate effects in terms of surface weather are widespread.[133]

THE ANTARCTIC CIRCULATION

The continent of Antarctica has an average height of about 2,500 m and is approximately seven times the size of Greenland: it also might be expected to have a marked influence on the atmospheric circulation. In particular, if the glacial anticyclone theory has any validity, it ought to be applicable in the Antarctic. No glacial anticyclone seems to develop over the south polar regions, and depressions and other systems frequently cross the center of the continent. Normal frontal clouds (cirrostratus, altostratus, nimbostratus, and embedded cumulonimbus) occur in Antarctica just as in middle latitudes. Apart from the fact that its radiation-balance conditions are very distinctive, the Antarctic continent does not appear to induce a meteorology and climatology of its own. There are, nevertheless, some interesting localized features. For example, the seasonal variation of atmospheric pressure over the Ross Sea sector is antimonsoonal, being greatest in summer and least in winter.[134]

In general, the circulation over Antarctica may be described as sometimes cyclonic, more often relatively anticyclonic, and occasionally characterized by warm anticyclones extending to great heights.[135] The south polar anticyclones, however, are not quasipermanent or constant, or even particularly intense. Eastern Antarctica (especially in the vicinity of 80°S, 60°E) seems to be the preferred area for anticyclones, which are found there on roughly 60 to 85 per cent of the days in each year. Cyclonic situations prevail over the whole continent about 5 to 15 per cent of the time. A characteristic feature is the tendency for long spells of one synoptic type to occur. Four types of circulation are especially persistent:

1. Anticyclonic circulations extending to a great height. The anticyclones may be relatively intense, but the central pressures are usually not high by northern hemisphere standards; 1,000 to 1,010 mb is a typical central pressure over coastal regions, and 1,025 to 1,030 mb is the highest pressure normally expected, usually on the outer fringes of Antarctica during blocking spells. The anticyclones sometimes are present over the continent for several months, but are often pushed out to sea by the movement of cyclones, or almost literally crowded out by the latter. Antarctic anticyclones usually weaken or retreat as the southern hemisphere subtropical high-pressure belt intensifies and extends southward. Cores of descending air at their centers are heated by subsidence, but the warmth is not evident at the surface, because there is a surface layer of very cold air immediately above the radiation-cooled surface of the ice sheet. The latter layer is usually separated from the subsiding air by a marked inversion. Thermal contrasts on either side of the edge of the sea ice in winter (around latitude 60°S) are very great; consequently, very deep and extensive depressions move along this

zone, and at times their influence may completely overwhelm the anticyclogenetic effect of radiational cooling on the ice-cap surface.

2. Cold-core depressions extending to a great height. Although the surface winds produced by these cyclonic circulations are relatively weak, the upper-level winds are very strong, reflecting the intense cold in the upper parts of the cyclones.

3. Migratory depressions. The centers of these usually are located some distance north of the Antarctic coast, normally associated with an upper ridge involving a sector of relatively warm maritime air aloft.

4. Mixed situations. These show no particular synoptic pattern in force at the surface, but a cyclonic circulation in the upper part of the troposphere.

The existence of an Antarctic glacial anticyclone seemed probable on the evidence of repeated pressure oscillations observed moving from east to west across the Ross Sea. These were attributed to the surges of a permanent glacial anticyclone radiating from a point situated at 80°S, 120°W. Later analysis suggests that the observed oscillations were produced by "dumbbell" rotations near the fringe of the continent, in which old, decaying cyclones and their associated occlusions move westward in the circulation around much larger and active cyclones over the ocean to the north.[136]

Observations up to 1951 indicated that, very broadly, the Antarctic circulation could be described as anticyclonic at the surface and cyclonic aloft; opinions differed on the areal and vertical extent of the anticyclonic circulation and on whether it is permanent or not. Later research indicates that the surface anticyclonic circulation is certainly not permanent, and may be quite shallow.[137]

A general model of the Antarctic circulation introduced in 1960 proposes the existence of a basic westerly current in all latitudes south of 40°S, except for a quasi-permanent anticyclone at 700 mb over eastern Antarctica and low (i.e., below 700 mb) easterlies near the coast of Antarctica. This zonal current is dynamically unstable, so that cyclonic disturbances are constantly created along the rim of the continent. These disturbances may be quasistationary—associated with instability in the form of waxing and waning troughs and ridges—instead of resembling the zonally migrating depressions of middle latitudes. Thermal contrasts between land and sea, and especially those caused by the distribution of pack ice, induce both low-level and (especially during the polar night) high-level baroclinicity, providing a source of kinetic energy for the migrating cyclones.[138]

Major cyclones enter the zonal current from the north, and then drift southward or southeastward with it until they enter the circumpolar low-pressure trough encircling Antarctica. They then either become quasistationary and ultimately merge with the trough, thus losing their identity, or move slowly westward, or move over the Antarctic continent.

Although the southern hemisphere upper westerlies do not exhibit the large-amplitude, semipermanent troughs and ridges of their northern hemisphere counterparts, they do reveal characteristic wave patterns. In particular, a type of blocking occurs when an intense ridge extends southward from the subtropical high-pressure belt and interferes with the basic zonal current. The development of cyclones by long-wave instability is especially favored by the presence of low-level easterlies, a situation also found in the polar easterlies of the northern hemisphere. Occasional anticyclonic cells within the high-latitude part of the basic zonal current move northward or northeastward associated with cold outbreaks from Antarctica.

The Antarctic continent affects the atmospheric general circulation mainly by means of outbreaks of polar air, which either set up temporary blocks in the middle-latitude westerlies of the southern hemisphere, or inject polar anticyclones into the zonal current that then drift with it. The idea of waves or surges of cold Antarctic air was first suggested by Simpson in 1919. The waves had a mean amplitude of the order of 20 mb and periods increasing from 3 to 7 days with decreasing latitude; the surges had periods of around one month, and amplitudes decreasing with latitude. These waves and surges of cold south polar air would obviously be expected to influence the weather of the southern continents.[139]

A characteristic feature of Antarctic polar outbreaks is that the western part of the cold air becomes bounded by a frontal zone; frontal zones of this type are very important for bringing unsettled weather to Tasmania. Cold Antarctic outbreaks are very frequent from November to February inclusive (i.e., they exhibit a summer maximum), and produce spells of fine, dry weather over inland districts of western Victoria and South Australia. Air reaching South Australia as a cold outbreak is not necessarily of recent Antarctic continental origin; it may be due instead to rapid cyclogenesis in the belt between 40°S and 45°S. The 500-mb circulation over the Australasian-Antarctic sector of the southern hemisphere frequently involves meridional flow prior to a cold outbreak; the meridional pattern is produced by a quasi-stationary anticyclone situated in middle latitudes at a distance of half a wavelength of the upper westerlies upstream from the developing cyclone that is giving rise to the cold oubreak.[140]

As with the Arctic stratosphere, the Antarctic stratospheric circulation changes from a cyclonic vortex in winter to an anticyclonic one in summer. This is of course a monsoon effect, but although the mass inflow-outflow associated with the change is of great magnitude, it appears to contribute little to the total seasonal shift of atmospheric mass between the southern and northern hemispheres. The total mass of air over the southern hemisphere undergoes an annual oscillation with a mean amplitude of about 10^{19} grams; this oscillation is very pronounced in the latitude belt between 10°S and 40°S, but is much smaller to the south. That is, the effect of the stratospheric circulation change over Antarctica does not appear to extend equatorward of 40°S.[141]

The changeover from a winter cyclonic vortex to a summer anticyclonic one—both are stable states—normally takes place around November, during an unstable phase of the circulation, but the process does not appear to take place regularly at the same time each year. In 1957 the winter vortex began to weaken in early October, and a closed anticyclonic circulation covered the continent by November 21. In 1958 the cyclonic vortex broke down gradually, so that an iregular pattern of cyclonic and anticyclonic disturbances characterized the Antarctic stratosphere at the end of November instead of a single anticyclonic vortex.[142]

Unlike the Arctic stratosphere, in which dynamic processes (especially those connected with advection and vertical motion) are of major importance, the Antarctic stratosphere is largely controlled by radiative processes. Notably, the rapid transformation of the stratospheric circulation in November is related to the spring-time maximum of the ozone content of the Antarctic stratosphere.[143]

The stratospheres of the north and south polar regions differ considerably in their thermal regime. In particular, seasonal temperature changes are much greater in the Antarctic stratosphere, but much higher temperatures in general are found in the Arctic stratosphere. The Antarctic stratosphere has a much greater temperature lapse-rate in winter than the Arctic stratosphere; the Antarctic tropopause virtually disappears during winter. Although both stratospheres are dominated by extensive cold cyclones during winter, the southern vortex is much more steady than the northern one. The Arctic vortex may move as far southward as latitude 60°N in midwinter, in which case warm anticyclones are able to reach the North Pole. The Antarctic vortex shows no comparable northward expansion.[144]

The dynamic result of radiative warming and cooling in the Antarctic stratosphere is felt in its degree of baroclinicity, which in turn affects the circulation. Intense cooling of the Antarctic stratosphere, which begins in March, results in the formation by April of the cold cyclonic vortex, which is associated with weak gradients in both temperature and pressure, and hence there is comparatively little energy available for the maintenance of disturbances. During June, thermal and pressure gradients tighten, and the Antarctic stratosphere is in a state of rather stable baroclinicity. Gradients weaken still more in August, then strengthen again in October. The center of the cyclonic vortex shifts from month to month, but is usually well to the north of the pole. The associated low-pressure area begins to fill in September, but during September and October both contour and temperature gradients become very tight, so that, despite the weakening of the low-pressure center the winds of the vortex increase and reach their maximum values (150 to 200 knots at 50 mb) during the springtime warming phase.

In early November, a rapid transformation takes place with the return of the sun, with both isotherms and pressure contours intersecting at much greater angles, indicating increased baroclinicity. Dynamic instability develops and the cyclonic vortex finally breaks down. During the remainder of the month, the remnant of the cold low of the Antarctic stratosphere moves over the Palmer peninsula and weakens. Warm air then covers the continent, with a warm pool over the South Pole. By the end of November, an extensive stratospheric anticyclone usually covers the entire Antarctic.

THE ATMOSPHERIC CIRCULATION OVER TROPICAL AND EQUATORIAL REGIONS

The atmospheric circulations in both polar and middle latitudes may be studied together as a single dynamic system. Chapter 5 of this book sets out to explain the circulation of the westerlies in terms of elementary dynamics, and Chapter 6 describes the synoptic picture of weather phenomena within the westerly belt; both may be regarded as basic to the description and explanation of circulation problems in all latitudes poleward of the subtropical high-pressures. Atmospheric circulations in tropical latitudes, however, constitute entirely different problems, and must be approached in a fundamentally different manner.

Although tropical meteorology is probably the most ancient branch of the science, modern climatology has been developed very largely by middle-latitude meteorologists, and the basic concepts and principles were first tested in temperate latitudes. The greater part of this book is, in fact, such middle-latitude climatology, and consequently, many of the ideas described herein must be considerably modified to be applied to tropical weather and climate. We have seen that the atmospheric circulations and weather phenomena of the polar regions do not differ radically from those of temperate latitudes, and that middle-latitude synoptic techniques on the whole work quite well in the Arctic and Antarctic. But tropical and equatorial regions have a distinct meteorology and climatology of their own, and very different synoptic models from those of the Bergen school are required if the tropical circulation is to be studied in a meaningful manner.[145]

The tropics cover a large part of the Earth, in particular, that part in which the main source of atmospheric energy is located. It would appear reasonable to expect considerable meteorological activity to be concentrated in tropical latitudes. Unfortunately, since observing stations are much more scattered in the tropics than in middle latitudes, the basic data is often lacking. Study of tropical meteorology on the basis of data from surface stations is complicated by the great influence of diurnal variations of pressure and temperature in tropical regions. Surface disturbances of the general circulation are largely controlled in the tropics by upper-air developments, observations of which have been particularly sparse.

Analysis of the atmospheric general circulation in middle and high latitudes has advanced largely because of the growth of the synoptic observing network and the evolution of synoptic techniques soundly founded on the application of hydrodynamic concepts to isobaric patterns. In the tropics, however, isobaric patterns are very ill-defined (except for tropical cyclones, which exhibit almost perfect isobaric symmetry, unlike middle-latitude depressions), and few distinctive patterns have been identified. The diurnal variation of pressure is usually much greater than the pressure variation due to the passage of tropical disturbances, and the reduction of observed pressures to mean sea level is often uncertain, particularly for tropical Africa. The geostrophic wind relation does not hold in tropical latitudes, because the Coriolis force is very small. Thus there is no readily applicable relation between pressure and wind in the tropics, and winds cannot be interpolated on the basis of pressure distribution. Although horizontal differences in atmospheric properties are evident in tropical and equatorial regions—for example, there is a great contrast between dry Harmattan air in West Africa and moist air from over the Gulf of Guinea—fronts and air masses in the middle latitude sense do not exist. Cloud and precipitation distributions, or the lack of them, must be explained largely in terms of hydrodynamic convergence and divergence, respectively. Orographic influences are especially marked in the tropics, as are diurnal rhythms in convection and other processes. Therefore local weather in these regions may often be dependent more on these than on the general circulation.

Certain tropical weather phenomena cannot be explained by the standard meteorological techniques evolved in higher latitudes. For example, not only are there extensive cloud systems (particularly cirrus spissatus, altocumulus lenticularis iridiscens, and altocumulus translucidus) that are not associated with convection, but a steady rainfall of 10 inches or more can take place without convection. Convective processes are probably responsible for most of the precipitation recorded in tropical latitudes, but the highest rainfall amounts are usually due to nonconvective processes. Most convective precipitation in the tropics falls in shower activity that is not random, but results from synoptic phenomena not known in extratropical regions. Orographic clouds in the tropics are often multilayered, and far more complex than those of higher latitudes. Tropical thunderstorms differ from temperate ones in that they usually produce very little precipitation but are characterized by excessive electrical activity, particularly lightning. Tropical sea breezes can often surmount mountain ranges 3,000 feet or more in height, and wind shears in the tropics may be extremely abrupt and not conform to the gradient-wind equation. Surface disturbances in tropical latitudes sometimes move from east to west at speeds completely different from those of the winds at any height in the overlying atmosphere. Some helical circulations in the tropical atmosphere seem to have no counterpart in extratropical regions; these become visible as cloud streets when they occur in a moist layer, and otherwise would have escaped detection.[146]

Even the "wet season" and "dry season" division of the year in many tropical countries is not as simple as it seems, for dry spells are found during the rainy season and wet spells within the dry season. Although tropical regions receive some of the highest incoming radiation totals of the world, the radiation balance of many tropical areas is controlled more by the dust content of the atmosphere than by the net radiation. Obviously the explanations of these and other problems must involve many aspects of meteorology, but the prevailing atmospheric circulation must have some greater or lesser effect in every case.

Three distinct schools of tropical meteorology have developed within the last thirty years. The first of these, the so-called *climatological school*, strove for simplicity. Its proponents accepted the statistically defined entities of doldrums, monsoons, and trades as real phenomena, and regarded the tropical circulation as consisting essentially of persistent trade-wind systems, which waxed and waned, and which were thermally driven, mainly by intense convection in the doldrum belt of light variable winds in the vicinity of the equator. They considered monsoon currents to be seasonal perturbations of the trades resulting from the effect of orographic factors and thermal contrasts between land and sea surfaces, and regarded diurnal variations in the climatic elements as arising from the same effects on a local scale. They therefore envisaged tropical weather as differing little from the tropical climate defined by annual and monthly means—both decided by a combination of astronomical factors and the different abilities of land and sea to absorb solar radiation—but they took no account of the development, in tropical oceanic regions, of the most violent type of storms on Earth, i.e., hurricanes and typhoons.[147]

The climatological school ignored the facts that (a) the steady trade-wind systems occupy only a relatively small part of the tropical oceans, in the eastern equatorial parts of the subtropical anticyclones; and (b) the anti-trades, which would be necessary to produce direct thermal drive of the climatological circulation, are only present in very restricted areas. Even for such apparently stable "phenomena" as monthly mean circulation patterns, the climatological concept of the tropical circulation leads to difficulties. In parts of the tropics, there are two complete reversals of circulation during the year at 200 mb, in May-June and in October. These reversals are completed within a few weeks, and coincide with the advance (May) and retreat (October) of the southwest monsoon over southern Asia. Thus one type of 200-mb flow pattern dominates the tropics and subtropics from Africa to the western Pacific during the period from November to April, and a completely different type from June to September. Not only is it impossible to make precise estimates of average wind conditions during intermediate periods from the mean monthly patterns, but the real tropical circulation cannot be represented by an average flow based on monthly means.[148]

The *air-mass school* of tropical meteorology originated through attempts to apply the Norwegian concepts of frontal and air-mass analysis to the tropics, on the assumption that tropical and higher latitudes differ only insofar as the former are characterized by easterly winds and high temperatures, rather than in any fundamental sense. Mean wind charts showed northeast and southeast trades, in the northern and southern hemispheres, respectively, converging along a line regarded as the "Equatorial

Front." However, even when such a front could be followed from day to day on synoptic charts (which was by no means common), its future movements could not be predicted by Bjerknes' dynamic theory, which works so well in middle and high latitudes. An alternative approach, in which organized lines of cumulonimbus were considered "tropical fronts" when they were accompanied by wind changes, proved equally useless for circulation studies. These tropical fronts were purely empirical, without any real dynamic basis, and rarely represented density discontinuities in the sense of the classic Norwegian fronts.[149]

The *perturbation school* represents the most popular group of ideas today. These were developed in Puerto Rico during the Second World War by University of Chicago meteorologists, who were concerned with applying Rossby's theories of the general circulation to the tropical atmosphere. Making use of a fairly realistic model devised by Dunn, who eliminated the semidiurnal pressure oscillation from Caribbean observations by means of 24-hour isallobars and thus was able to demonstrate the existence of westward-moving waves superimposed on easterly current, they found they could explain daily weather patterns in the West Indies and adjacent areas in terms of perturbations of a basic zonal current. Some of these perturbations were found to increase in amplitude and ultimately give rise to hurricanes. The perturbations take the form of small-amplitude pressure waves, which are associated with "highs" or "lows" in the isallobaric contour field; the latter follow the average hurricane tracks for the appropriate months. The trade-wind inversion is relatively low ahead of a moving pressure wave, bringing fine weather, and small cumulus; behind the wave the inversion is high or even absent, and numerous shower clouds characterize this region.[150]

Three circulation phenomena of equatorial and tropical regions are particularly important as climatic controls. The perturbation school in particular has achieved some success in explaining them. These are the trade winds, the Intertropical Convergence Zone (I.T.C.Z.), and the equatorial westerlies. In addition, the tropical counterpart of the index cycle, and the upper-air circulation in tropical latitudes, have important climatic consequences.

The perturbation school regards the *trade winds* as those portions of the circumpolar vortex that move more slowly than the Earth's surface in low latitudes, whereas in higher latitudes and in the upper atmosphere over tropical regions, the vortex moves faster than the Earth's surface. The trades appear to be very fundamental parts of the general circulation; the northeast Pacific trades, at least, can be shown to be a driving force in the worldwide circulation.[151]

The northeast trade of the Pacific is a source of both latent and sensible heat for the general circulation. It both gains and exports energy in its passage over the warm ocean surface, and the local heat it gains is an important source of the energy that maintains the trade current. A characteristic feature of this northeast trade is the existence within it of an inversion layer that rises 1 kilometer from the inflow end of the current (see Figure 4.32), despite continuous divergence and shrinking; i.e., although in going from 32°N, 136°W, to 21°N, 158°W, (along the trade current), the base of the inversion rises, the individual air particles *descend*, so that during their southwestward movement the air columns constituting the trade shrink vertically and spread out horizontally. The inversion rises because of the latent heat the trade collects as it passes over the sea surface.

The trade-wind inversion is an important regulating feature of the global general

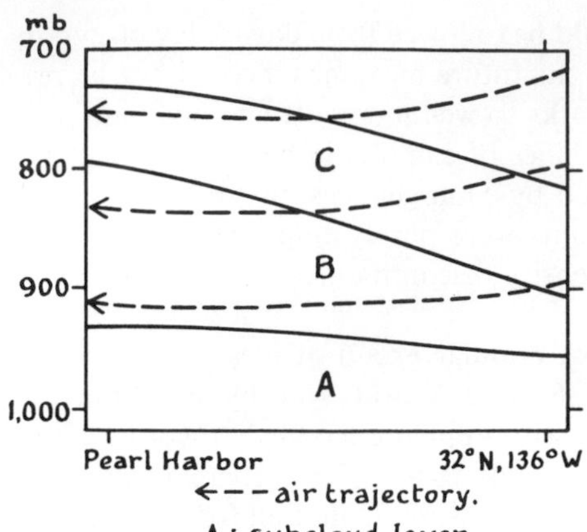

FIGURE 4.32.
The four layers of the trade-wind zone of the North Pacific. Arrow indicates air trajectory (after Riehl, Yeh, Matteus, and La Seur).

circulation and of climate in general. Part of the inversion layer is arid, and this aridity extends down to sea level in regions where there are cold ocean currents. Early observations by Piazzi-Smyth proved that in the Canary Islands the inversion is in the middle of the trade, at 1,750 to 2,500 feet, and therefore could not be a boundary between two airstreams of different direction, as was originally believed. Later work has shown that downward transfer of mass takes place through the trade-wind inversion, which therefore cannot be a discontinuity surface separating upper (dry) and lower (moist) airstreams. The inversion in general results from large-scale subsidence of air moving from high latitudes, particularly in the eastern parts of the subtropical anticyclones, where, because of dynamic effects of the Earth's rotation, the isobaric surfaces are higher than in the western portions. The exceptions to this are found near the coasts of Africa and America, where cold upwelling ocean currents cool the lower layers of the atmosphere. The cooling is transferred upward by turbulence, so that the inversion is strengthened by cooling from below as well as by subsidence from above.[152]

In the Atlantic Ocean, the trade-wind inversion layer is 400 feet thick on the average, but the thickness varies from a few meters to more than a kilometer. The inversion is strongest, both on the average and on specific days, where its base is lowest, and weakest where the base is highest. In the North Pacific, four distinct layers can be recognized within the trade wind (see Figure 4.32). Of these, the cloud layer—defined as between the base and top of the tropical cumuli—is of major importance. Large numbers of cumuli carry moisture upward from the sea surface to the base of the inversion every day. The tops of some of these penetrate into the inversion layer, break off and rapidly evaporate in the dry air, thus adding moisture to the lower part of the inversion layer, which gradually transforms itself into an upward extension of the cloud layer.

The global trade winds exhibit interesting contrasts with monsoon currents. Trades are essentially dry, tropical easterly winds; monsoon currents are essentially wet, tropical westerly winds. The difference is due to dynamic factors, not to thermal differences between air and water or land and sea; whereas trade winds consist of stable layers with extensive inversions, monsoon currents comprise deep layers of moist unstable air with intense shower activity.[153]

During times of high index in middle and high latitude, the trade winds are deeper and wider than usual. During times of low index—although real low-index patterns are very rare in the tropics and subtropics—the subtropical high-pressure ridge is weak or becomes broken in many places, so that westerly polar troughs and cut-off cold lows extend through the gaps from middle latitudes into tropical regions.

The narrow zone of apparently convergent winds between the trades of the northern and southern hemispheres that appears on mean charts (the I.T.C.Z.) is of considerable interest. Great confusion about the nature of this zone resulted from the work of the climatological school, when attempts were made to locate this mean convergence on daily synoptic maps. Actually, the equatorial rain belt (i.e., the so-called doldrum convection belt), is located *between* the northernmost position of the I.T.C.Z., in August and September, and the geographical equator, while dry parts of the trades are north and south of the belt. The precipitation is not due to convection in hot moist air; it is dynamic in origin, determined by the effects of vertical motion and divergence of the wind field on thermal stability.[154]

The equatorial rain belt is between latitudes 0° and 10°N over the eastern and central Pacific and the Atlantic, and also over equatorial parts of West Africa and the Sudan. Vertical motion in the equatorial zone is asymmetric, reaching a maximum of 1,300 m per day at a distance of 500 or 600 km south of the I.T.C.Z., i.e., between the geographical and meteorological equators, at a height of 4 to 5 km. To the north of the I.T.C.Z., surface convergence changes to divergence between 1.3 and 1.8 km up, but to the south of the I.T.C.Z., the convergence increases with height and does not change to divergence until a height of between 3 and 5 km. Therefore, although the inversion within the northeast trades is only destroyed very slowly by lifting as the air moves southwestward, the inversion within the southeast trades is soon destroyed by rapid lifting (hence decreasing stability) when the trades cross the equator to the south of the I.T.C.Z.

The *equatoral westerlies* are the zonal band of surface westerly winds found in the vicinity of the equator between the trades; they are completely unconnected with the middle-latitude surface westerlies. In general, their boundaries are characterized by increased cloudiness and poor weather. Several ideas on their origin have been advanced. Palmer regarded them merely as statistical entities, i.e., as the result of a series or train of vortices following one another from west to east, not as the result of the perturbation of a westerly circulation, because the trades do not meet the equatorial westerlies in a discontinuity surface that becomes perturbed and produces eddies. Against this must be set the fact that in some areas, e.g., the western North Pacific, the equatorial westerlies are very deep, and appear to be a genuine part of the general circulation. They have also been explained as due to a combination of seasonal variation in the latitude of maximum insolation plus radiation from cloud tops within the I.T.C.Z., locally reversing the equatorial solenoidal field to produce two new lines of convergence, one on each side of the equator; i.e., a cold source is generated within the I.T.C.Z., by radiation and evaporation, and the subsequent formation of two convergence lines results in the observed westerly airflow.[155]

The equatorial westerlies cannot be due mainly to differential heating of land and sea, because they prevail all year over the Gulf of Guinea between Cape Palmas and Cameroon Mountain, and also over the triangle formed by the west coast of Africa between latitudes 4°N and 6°S, and a point near 2°N, 27°E. They appear to form essential parts of tropical monsoons, but cannot be understood as caused by what is usually regarded as the monsoon effect. Thus, even on a completely land-covered globe, seasonally shifting tropical monsoon winds should be present.

According to Flohn, equatorial westerlies may be observed within an area that extends in a north-south direction from the heat lows of the subtropical parts of the continents to (and including) the I.T.C.Z., and that extends around the Earth in an east-west direction. At any one time, a single system of equatorial westerlies in one hemisphere may extend across the equator into the opposite hemisphere for a distance of 2 to 3° latitude. The equatorial westerlies form a belt that is widest in the summer hemisphere, but narrows during the transitional seasons and splits into separate segments near the equator. Occasionally, the equatorial westerlies may be merely the result of shallow cyclonic disturbances moving from west to east with deep tropical easterlies in the middle and upper troposphere. Equatorial westerlies are much more variable in both time and space than tropical easterlies: their persistence averages 60 per cent over the oceans (and is much less over land), compared with 90 per cent for the trades. In most regions, they exhibit a surface wind component directed toward low pressure, giving rise to southwesterly winds during the northern summer and northwesterly winds during the southern summer, both of which produce monsoons in some areas. Flohn restricts the term "monsoon" to tropical regions with a seasonal change in circulation from equatorial westerlies to trades and vice versa. He showed, in fact, that the most intense monsoons—i.e., those with the greatest seasonal variation of winds—should occur on a completely land-covered Earth. The equatorial westerlies over Africa are a natural experiment, representing what would be average conditions on a land-covered Earth, because they are only slightly deformed by the existence of the Gulf of Guinea, and somewhat more deformed by the plateaus and mountains of East and Central Africa.[156]

Frequent perturbations within the equatorial westerlies disturb their wind and pressure fields, and are responsible—via their accompanying convergence areas—for precipitation variations during the rainy season in most tropical countries. Convection appears to be a very minor cause of precipitation in these areas.

The three phenomena just discussed—trades, I.T.C.Z., and equatorial westerlies—differ so greatly from the circulation features found in middle and high latitudes that completely different models are necessary to relate them to the general circulation. The following account of the tropical circulation models largely accepts the concepts of the perturbation school, which regards the tropical circulation as a basic easterly zonal current that is deformed by various systems. Not all of the systems apply to the whole of the tropical atmosphere. In fact, there are parts of the tropical atmosphere where perturbation concepts may not be appropriate because of overriding geographical effects. For the western Pacific and southern Asia, the upper-air circulation is remarkably stable during the cool season, with the actual weather closely resembling the mean climatic conditions on most days. This may be brought about by the Indian peninsula and the Himalayan massif "anchoring" the general circulation by modifying the meridional circulation of the northern hemisphere so that the central cell (in the three-cell model) is of limited extent (with a maximum latitudinal coverage of only 5° in India-Pakistan and China, and even less in 90°E) while the intensity and rise of the equatorial cell is increased.[157]

The perturbation school divides the basic zonal current of the tropics into two parts: (a) a lower, nearly barotropic current, whose axis coincides with the equatorial trough in the lower troposphere; and (b) an upper current, which becomes increasingly baroclinic with increasing height and with increasing distance from the equator. Thus the model states that tropical easterlies decrease upward and often become westerlies in the upper troposphere of the subtropics. No distinct boundary is postulated between (a) and (b); instead, a transition zone intersects the Earth's surface at 15°N and 15°S, and then slopes upward to form a dome 20,000 to 30,000 feet high over the equatorial trough. This dome may be as high as 40,000 feet over the meteorological equator during periods of very deep tropical easterlies.

Both windspeed and pressure reach minima along the axis of the barotropic part of the basic current, windspeeds attaining their maxima on the outer boundaries of the barotropic current, usually below 10,000 feet; i.e., the speed of the low-level easterlies is at a minimum in the equatorial trough (in fact, average westerlies are present on some occasions) and increases poleward to reach maxima in the oceanic trades. Thus when geographical and meteorological equators coincide, the barotropic current is characterized by cyclonic shear throughout its breadth. An asymptote of convergence (representing the equatorial front, the intertropical front, or the I.T.C.Z.) marks the location of the meteorological equator in the current, indicating where horizontal convergence of velocity (a characteristic feature of the entire barotropic system) reaches its maximum. This asymptote is not necessarily present on streamline maps for any one time.

The upper baroclinic portion of the basic current is characterized by easterlies decreasing in speed both with height and poleward, becoming weak easterlies (or even westerlies) in the summer hemisphere and strong westerlies in the winter hemisphere, above the surface position of the subtropical anticyclones. A return current (somewhat similar to the antitrades) occurs above 30,000 feet, close to the meteorological equator in all longitudes. The mean meridional airflow is thus toward the equatorial trough in the lower troposphere, with a return poleward-directed flow in the upper troposphere. The lower convergent flow is more extensive and steady than the upper return current, the latter being sporadic in both position and time. The simplicity of the direct thermal (i.e., Hadley) cell of the tropics is usually obscured by the presence of long waves in the baroclinic current. The subtropical high-pressure cells alternate in the two hemispheres; i.e., the centers of the southern hemisphere cells lie in the same longitudes as the cols between the northern hemisphere subtropical anticyclones, particularly for the stationary elements of the circulation. Consequently, the airflow from one hemisphere to the other is superimposed on the Hadley cell, alternately working in one direction and then in the other, so that a breakthrough type of meridional circulation is typical of the layer of the tropical atmosphere beween 20,000 and 35,000 feet.

Westward-moving equatorial waves are found in the axial region of both barotropic and baroclinic parts of the basic current; these reach their greatest amplitude at the meteorological equator, and become rapidly damped to north or south of the latter. Developing instability in these waves causes westward-moving warm-core vortices to form in the lower layers of the atmosphere. These vortices may later extend upward into the baroclinic current and ultimately reach the tropopause. The more intense ones may develop into tropical cyclones. The degree of intensity of the Hadley cell decides the areas of the tropical oceans in which the waves become unstable. The transition from a weak vortex to a tropical cyclone is determined to some extent by the location and intensity of the upper long waves in the baroclinic current, which exert a selective influence on the existing vortices.

The outer and upper portions of the baroclinic current are favored regions for the formation of short waves, which result in cold-core cyclones whose intensity increases with altitude. These waves usually extend higher than 30,000 feet, and move in an irregular manner, unlike the equatorial waves. They may remain stationary for long periods, and may move out of the tropics entirely into middle latitudes. In the Hawaiian region, they form *kona storms*, bringing very heavy precipitation and violent squalls to the islands, although on the synoptic chart they appear to be very weak surface depressions. During intensification, the warm-core equatorial vortices within the barotropic current extend upward from the surface of the Earth, deriving their energy from the latent heat of vaporization. By contrast, the cold-core cyclones within the baroclinic current extend downward toward the surface as they intensify.

Disturbances of the basic zonal current of the tropics may be divided into two types: (a) broad systems, such as tropical cyclones and monsoons, that are common to many regions; (b) small-scale systems limited in area, for example, disturbance lines and harmattans in West Africa, and sumatras in Malaya. The disturbances may also be subdivided according to whether they involve vortices or wave-like perturbations.[158]

Probably the most important broad-system circulation feature of the tropics is the *equatorial trough*. This feature is the boundary between the trade winds of the northern and southern hemispheres, and was originally identified as the "intertropical front." Although confluence between the northeast and southeast trades does take place in a trough of low pressure, it does not always lead to convergence. The equatorial trough usually forms a continuous belt around the Earth, approximately coinciding with the equator, except for a gap in summer over northern India and Pakistan; it is far from uniform, and active sectors alternate with inactive ones. The active sectors move along the trough, and usually involve convergence. They indicate the preferred areas for bad weather, and do not move continuously by advection from one place to another, but jump in an erratic fashion. This type of motion emphasizes the essentially geographically bound nature of tropical weather, which is largely a consequence of the local distribution of divergence and convergence. Convergence and divergence appear and disappear in situ, unlike vorticity (the main control of weather in middle and high latitudes), which is advected with the wind field.

The tropical vortices include both cyclones and anticyclones; the tropical cyclones comprise both warm-core and cold-core systems. The warm-core cyclones usually originate from the intensification of surface disturbances in the equatorial trough. They derive their energy mainly from the release of latent heat of condensation during the intense convection, and develop upward from the Earth's surface, sometimes ultimately becoming hurricanes or typhoons. The cold-core cyclones are essentially upper-level features (above 20,000 feet), whose intensity is greatest in the high troposphere and decreases downward. Occasionally a cold-core cyclone develops downward, resulting in a large increase in convection within the lower troposphere, in which case it becomes transformed into a warm-core cyclone. Some cold-core cyclones originate in the upper portion of the tropical troposphere, probably maintained by available potential energy associated with the baroclinic part of the basic zonal current. Others form in middle latitudes as a result of differential subsidence of cold air from high latitudes. Both types move in an irregular, sluggish manner, usually from east to west in the summer hemisphere and from west to east in the winter hemisphere, but there are numerous examples of the reverse type of movement. In general, the cold-core cyclones tend to move toward the poles more often than toward the equator, but they may remain quasistationary for several days. Unlike warm-core cyclones they normally do not develop into hurricanes or typhoons: their surface expression is usually restricted to a weak, wave-like perturbation of the basic easterly current, but above the surface they cause extensive areas of medium and high cloud.

Tropical anticyclones are usually small, in both size and intensity, compared with middle-latitude anticyclones, and the large subtropical anticyclones are best regarded as parts of the basic zonal current of the tropics. Both low-level and high-level anticyclones are found in the tropics. The low-level anticyclones occur below 20,000 feet, mainly in regions and seasons in which the equatorial trough is developed more than 10° of latitude from the equator, for example, in the western portion of the North Pacific in winter, West Africa, and tropical North Africa. They usually move westward and produce above-average convection and cloudiness, particularly on the poleward margins of the tropics. The high-level anticyclones are found above 15,000 feet, and have a warm-core structure due either to subsidence or to the release of latent heat. The subsidence variety are more or less cloud-free, whereas the latent-heat variety are associated with fairly intense convection and extensive development of medium and high cloud.

The *clockwise eddy* is the low-latitude counterpart of the front of middle and high latitudes, and develops particularly in the area north of New Guinea in winter, when easterlies are temporarily in force south of the equator, and over Borneo and the Celebes Sea. A clockwise eddy usually forms when a branch of the South Pacific trades is diverted toward and across the equator, and then curves anticyclonically toward the east or northeast. Divergence is very marked in the weak flow of the northeastern sector of these eddies. They are very sporadic, and their persistence varies from one to three days in early summer to as many as five days in September. They rarely reach as high as 5,000 feet, and are usually most pronounced at 2,000 feet.[159]

The most well-known tropical disturbance, apart from the tropical cyclone, is the feature which is termed the *easterly wave* in the Caribbean and the *equatorial wave* in the Pacific. Both of these wave-like perturbations of the basic zonal current are manifestations of the same fundamental kind of disturbance, but they take place in different latitudes and disturb flows with different characteristics. Easterly waves are embedded in trades that are divergent in the mean, whereas equatorial waves are embedded in the easterlies of the equatorial trough, which are convergent in the mean. The equatorial wave consists of, essentially, isolated areas of divergence surrounded by areas of convergence; the easterly wave consists of isolated areas of convergence surrounded by areas of divergence. Consequently, the associated weather patterns may be expected to differ. Since the perturbation of the pressure field within the equatorial trough caused by an equatorial wave is usually much less than that caused by an easterly wave, it is very difficult to detect the presence of an equatorial wave from a study of pressure alone.

The basic work on the features that were later recognized as easterly waves was carried out in the Pacific by Visher (1925), in Africa by Regula (1936), and in the Caribbean by Dunn (1940), but it was the analysis of Second World War observations in the Caribbean by Riehl and others at the University of Chicago Institute of Tropical Meteorology in Puerto Rico that made a definitive study of easterly waves possible.[160] Their analysis revealed that easterly waves are similar in size to the wave-like perturbations of the middle-latitude westerlies, but are less well-marked on the surface chart

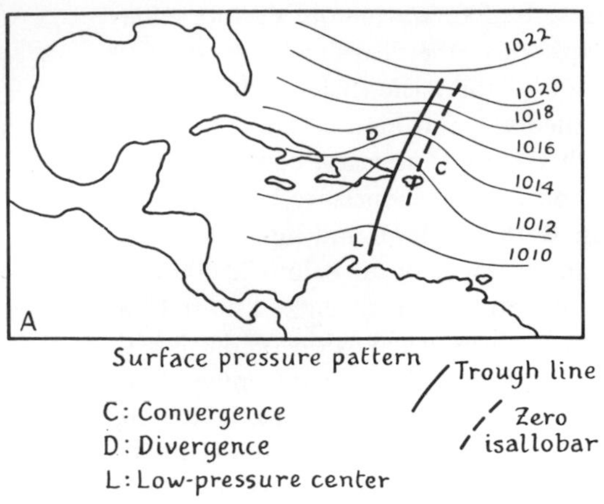

FIGURE 4.33.
Model easterly waves
(after Riehl).

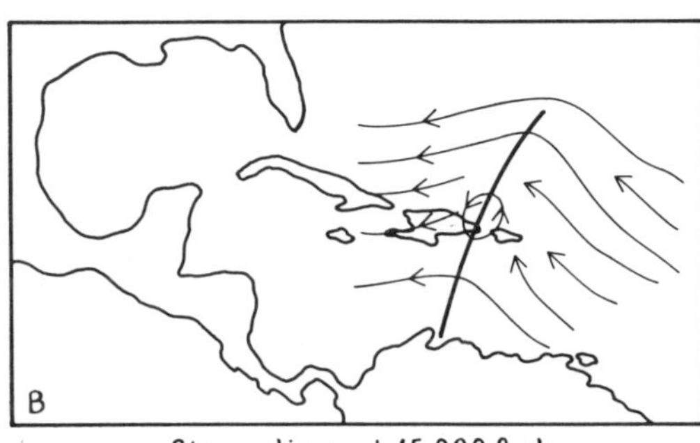

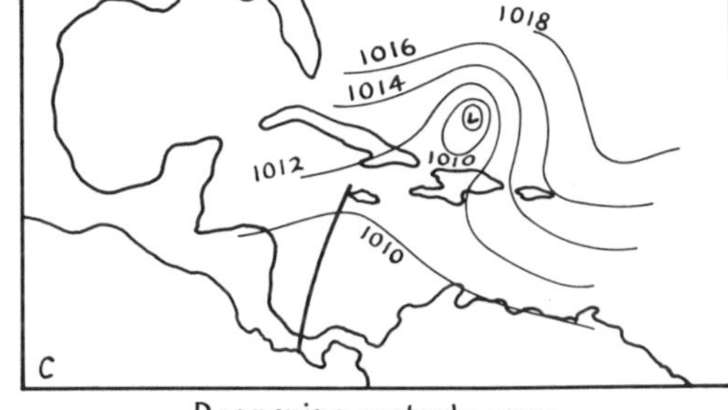

than the latter. The perturbations in the wind field are greater in amplitude than those of the pressure field, especially in the middle troposphere. The variations in pressure and wind, in both time and space, in the vicinity of an easterly wave can be explained if there is a pressure trough that moves from east to west and is open toward the equator (see Figure 4.33). Air moves through the trough from east to west (i.e., the air particles move more rapidly than the trough), and, unlike the air particles moving through a long wave in the westerlies, do not follow a constant-vorticity path. At the point of the trough, Coriolis parameter and relative cyclonic vorticity simultaneously reach their maximum value, so that convergence and divergence fields are an essential feature of an easterly wave. The lower wind field is divergent ahead of the trough and convergent behind it; the axis of the trough slopes eastward with increasing height. This field exists not only for streamline divergence, but also for the divergence of the total horizontal velocity. The resulting wind field includes northeasterlies ahead of the trough in the northern hemisphere, and southeasterly winds behind it.

Although the surface circulation in winter in the Caribbean is easterly, westerlies often exist aloft. Perturbations in the lower easterlies move from west to east against the surface current; they are caused by the movement of upper troughs in the overlying westerlies, which form the periphery of the circumpolar vortex. Such perturbations are termed *polar troughs*, and are continuous both with the upper trough in the tropical westerlies (i.e., the antitrades) and with a surface trough in the middle-latitude westerlies.

Easterly waves occur in early winter in southeast Asia, to the south of the subtropical high-pressure belt. In the New Guinea area, a special type of easterly wave is found (the *Freeman wave*), which is apparent in the windspeed field but not in the field of wind direction. Freeman waves form maxima or minima of windspeed that travel downstream in the lower layers of the tropical easterlies.[161]

Equatorial waves are found in the tropical Pacific. They are very similar to easterly waves, but occur in very low latitudes, so that their crests and troughs may be on opposite sides of the equator. Thus both crest and trough circulations may be cyclonic, which may result in the weather pattern sometimes described as the "double equatorial front."[162]

The equatorial wave fails, as a progressive model, to explain the observed precipitation distribution in East Africa, for which three quite distinct types of equatorial disturbance have been recognized.[163] These include (see Figure 4.34): the *equatorial duct*, an easterly airflow channeled between the subtropical anticyclones of the northern and southern hemispheres, and the *cross-equatorial drift*, in which a cross-equatorward flow from the north converges with a flow from the south, because there is a subtropical anticyclone north of the equator at the same time as a trough or cyclone in low latitudes south of the equator and in the same longitude. *Tropical waves* are recognized in East Africa as high-level disturbances, of approximately the same size as medium-scale troughs or depressions in middle latitudes. They bring precipitation, which, because of their slow rate of movement, may affect one area for several days in succession. Even though the general airflow may be easterly at all levels, these tropical waves move from west to east, against the current in which they are embedded. Their movement is a consequence of the instability or convergence created by an upper trough forming in the tropics while a cold front or cold-front wave is forming in middle latitudes.

Tropical upper troughs are important circulation features in Southeast Asia. They develop above 400 mb in the southwesterlies, in late spring and winter, and travel eastward across southern India and Burma to become stationary over Thailand and the Indo-China region in general. East of the troughline, disturbances develop along the polar front over the China Sea, producing extensive precipitation areas.[164]

Waves similar to easterly waves may be induced in the easterlies by the development of a upper cyclone or trough moving slowly from west to east. Such waves are usually found beneath upper cold-core cyclones in the summer-hemisphere easterlies, or beneath upper troughs in the westerlies overlying surface easterlies in the winter hemisphere.

Linear disturbances are very typical tropical circulation features. These are synoptic systems in which divergence, vorticity, or both are concentrated in a zone whose length is much greater than its width. They usually include areas of both cyclonic vorticity (associated with convergence and hence violent weather) and anticyclonic vorticity (associated with divergence and hence quiet weather), and are often very elongated, comparable in length with extensive active cold fronts in middle latitudes. Examples are the *disturbance lines* of West Africa, which move from west to east and, although producing no perturbation in the wind or pressure fields, are associated with violent squalls and other instability phenomena.

Tropical *shearlines* are linear zones characterized by cyclonic wind shear. They may be due to variations in windspeed across an air current of nearly constant direction, or to variations in wind direction. Shearlines are commonly found between two warm anticyclones in the upper troposphere in summer, when they give rise to extensive

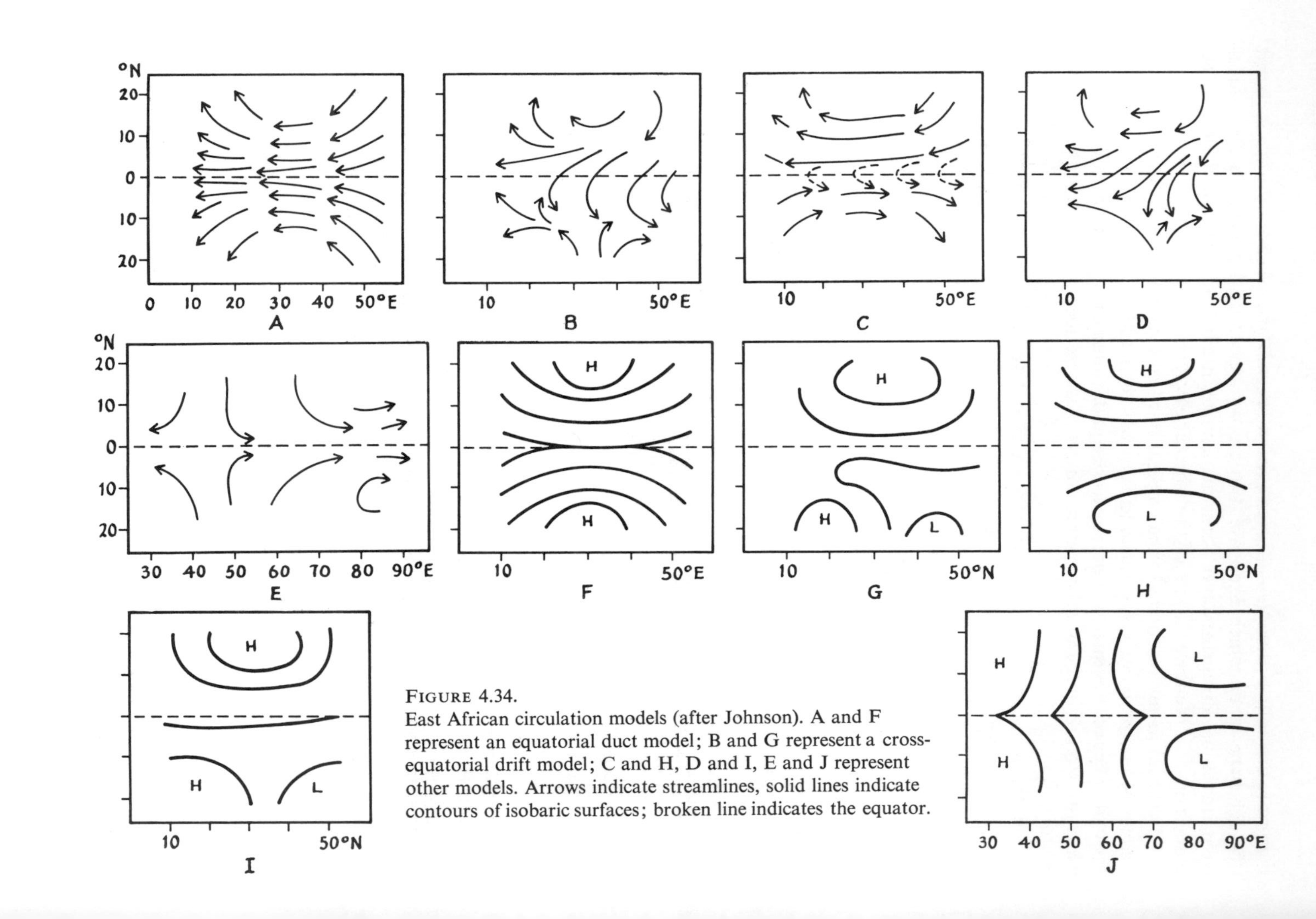

FIGURE 4.34.
East African circulation models (after Johnson). A and F represent an equatorial duct model; B and G represent a cross-equatorial drift model; C and H, D and I, E and J represent other models. Arrows indicate streamlines, solid lines indicate contours of isobaric surfaces; broken line indicates the equator.

medium and high clouds. They also occur in the lower troposphere near the equatorial trough, when the latter is far from the equator; a series of small, low-level cyclonic vortices is usually present on the equatorward side of the shearlines, forming one (statistical) component of the equatorial westerlies.

The expression "equatorial front" was once the name given to long lines or narrow zones of cumulonimbi or cumulus congestus in or near the equatorial trough, which usually produced showers. Later it was recognized that the observed weather was due due to *asymptotes* in the streamline field. These are defined as streamlines from which all neighboring streamlines diverge (positive asymptotes), or toward which all neighboring streamlines converge (negative asymptotes). Negative asymptotes in or near the equatorial trough give rise to the weather phenomena formerly regarded as constituting the equatorial front. Asymptotes, of course, are characteristic features of most streamline fields; they occur frequently in extratropical regions, where they are nearly always accompanied by fronts, whereas the tropical variety are not.

Pressure surges, representing the rises in pressure accompanying the movement of cold polar air into tropical regions, sometimes form important circulation features in the tropics. The pressure rise moves equatorward as a wave at a much greater speed than the cold air. Strong polar outbreaks occasionally reach 20° of the equator (particularly in the South Indian Ocean, and also in winter in the Caribbean and Southeast Asia) still retaining the characteristics of middle-latitude cold fronts. These cold outbreaks, which are recognized locally in South America as *friagems*, *northers*, and *pamperos*, do not move along isobaric paths, and may result in unexpected local temperature falls at sea level to as low as 50°F in a comparatively few minutes.

Occasionally a middle-latitude cold front moves equatorward and penetrates the subtropical high-pressure belt; this is frequent in winter in the Caribbean and in Southeast Asia, when one of the northern hemisphere subtropical anticyclones intensifies and builds up more than its corresponding cell to the east, and is also fairly frequent in the area between the East African coast, Madagascar, and Mauritius. The equatorward extremity of such a cold front, in the lower atmosphere, is then carried along in the tropical easterlies. Its effect on weather is soon masked by local orographic and convection effects, but is presence remains evident as a pressure change (troughlike, but easily obliterated by the strong diurnal pressure wave of the tropics), accompanied by a weak wind-shift line that is usually associated with a line of fairly continuous precipitating clouds, and occasionally with a narrow zone of continuous precipitation. The upper portion of the cold front on the margin of the tropics moves as part of the upper westerlies, and may induce the formation of a trough in the tropical easterlies. Such a trough is particularly common in northern India and Pakistan and in southern China, slightly less common in North Africa and the Caribbean.

Our discussion of the tropical circulation has concentrated so far on zonal phenomena, but there is also an appreciable mean meridional circulation in tropical latitudes, for example, across the equator (see Figure 4.35). In all seasons except the northern hemisphere winter, the momentum flux across the equator amounts to more than 10 per cent of the total westerly momentum produced within the northern hemisphere. The cross-equator momentum flow is strongest above 250 mb. and all the modes of momentum transport (i.e., transport by local transient eddies, standing eddies, and by the toroidal circulation) are in the same direction, which gives a momentum flux directed toward the summer hemisphere. The standing-eddy flux is particularly

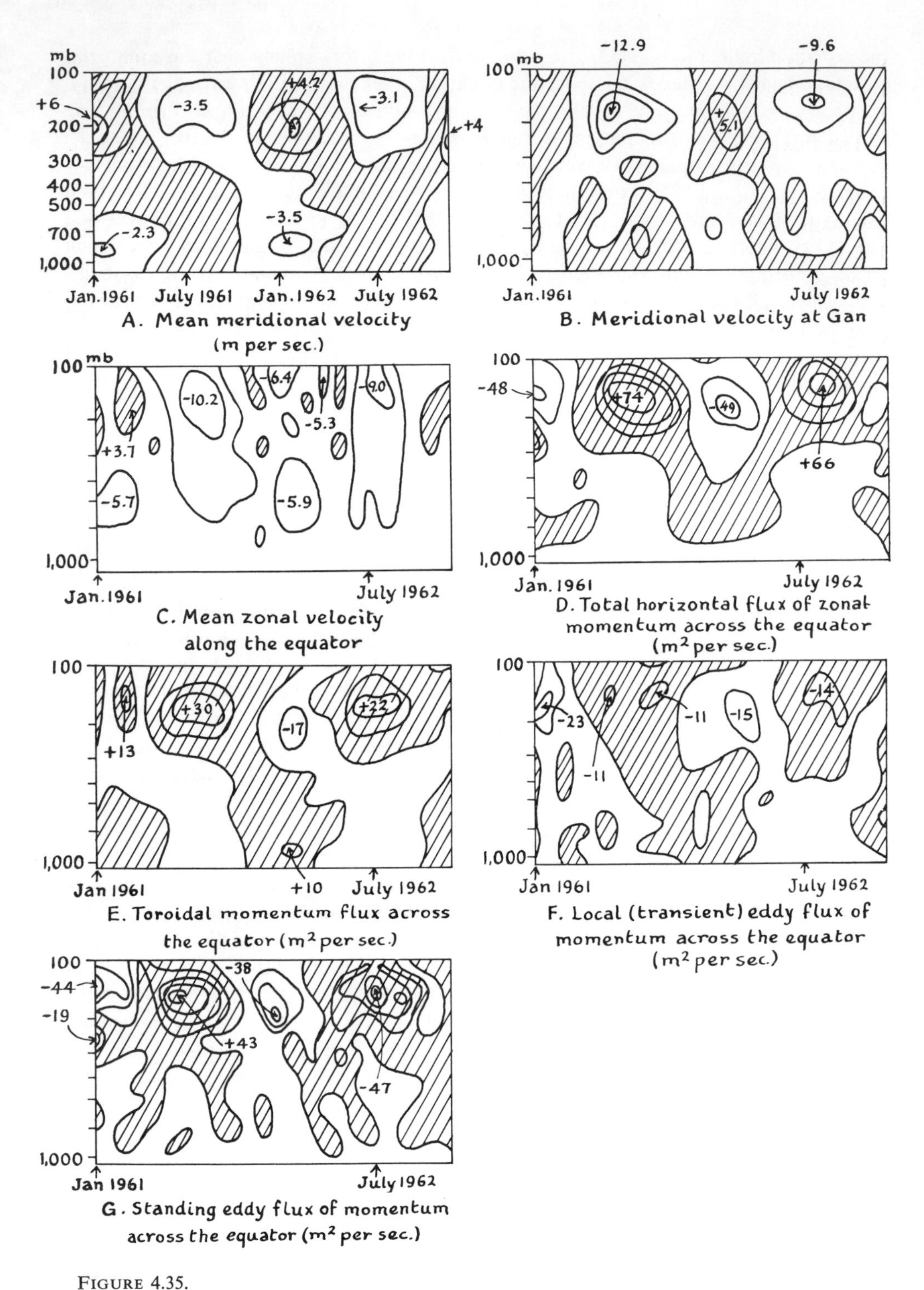

FIGURE 4.35.
Wind regime of the equatorial troposphere (after Tucker). Isopleth interval is 2 m per sec for A, 4 m per sec for B and C, 20 m per sec for D, 10 m per sec for E, F, and G. Positive values are southerly in A and B, westerly in C, northerly in D to G.

strong, because of the upper monsoon flow over the Indian Ocean–East Pacific area. In the equatorial Pacific, momentum is transported mainly by the meridional circulation below 500 mb, mainly by eddies above.[165]

The flux of atmospheric mass across the equator, integrated over all longitudes, amounts to 5.25×10^{11} grams per second; approximately 0.1 per cent of the total mass of the atmosphere (i.e., 5.45×10^{12} tons) is transported seasonally (twice a year) across the equator. The average mass flux of air across one centimeter of the equator is 131 grams per second; the average velocity component required to bring this flux about (i.e., the average meridional component of wind velocity, at the equator, averaged from the surface of the Earth to the top of the atmosphere) is only 0.127 centimeter per second.[166]

The mean meridional circulation is the major transporter of heat across the equator; the cross-equator eddy flux of heat is very small. The meridional heat flow across the equator is opposite in direction to the general planetary heat flow; during the northern winter, the northeast trade wind of the central Pacific extends some degrees of latitude south of the equator, so that its mean northward component carries heat and water vapor southward into the summer hemisphere, and the situation is reversed during the northern summer. The heat flow is maintained in this manner because in the lower atmosphere the I.T.C.Z. acts as a cold source, thus reversing the direction of the heat flux in its vicinity.

The model of mean meridional circulation devised by Flohn (see Figure 4.36) illustrates the actual complexity of the apparently simple thermal cell in tropical latitudes. The model shows that the Hadley cells have internal circulations that are partly driven by surface friction. The antitrades cannot exist as a simple current, because in large areas of the undisturbed tropical easterlies, meridional components shift between 850 and 700 mb (i.e., just above the trade-wind inversion), which is near the level of greatest intensity of the trades. In fact, there seem to be two types of tropical meridional-circulation cell.

Type I, *oceanic*, is the normal cell giving easterly trades over the oceans. It is a modified Hadley cell in which easterlies prevail from the surface up to 6 or 10 km, are separated by spurs of the middle-latitude westerlies (with traveling troughs and ridges) from 6 or 10 to 17 km, and prevail as stratospheric easterlies above 18 km.

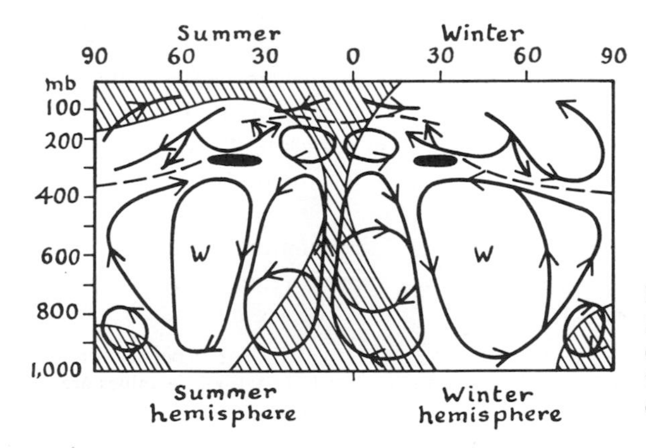

FIGURE 4.36.
Flohn's model of the mean meridional circulation. Broken line indicates the tropopause; *W* indicates west winds, east winds are shaded.

Type II, *continental*, is only found in regions in which extensive monsoonal surface currents are directed from the winter hemisphere to the summer hemisphere, and in which large areas have westerly wind components, which extend from the surface to at least 2 km, or to as much as 6 km over India during the summer monsoon. In the continental cell, the monsoonal (equatorial) westerlies are quasigeostrophic winds driven by the poleward decrease in pressure; they form a shallow layer completely encircled by the tropical easterlies and extending to 25 km or more. Seasonal heating of the subtropical parts of the continents results in wide movements of the I.T.C.Z. in both northern and southern hemispheres, but apparently the distance between the equator and the I.T.C.Z. must remain small enough (estimated as rather less than 10° of latitude) that the pressure gradient between the I.T.C.Z. and the equator can generate the equatorial westerlies.

Explanation of the General Circulation

This chapter has so far presented a huge collection of facts. An obvious question now is whether or not these facts can be ordered and explained. Of course, the description of the facts of wind, temperature, and humidity distribution throughout the atmosphere in terms of a postulated general circulation of the atmosphere is an attempt to order them, but then the postulated general circulation must be explained.

It is obvious that the general circulation is driven by the temperature contrast between polar and equatorial regions, and that the higher temperature of the latter is maintained by the Earth's receiving energy from space. The atmosphere can be regarded as being driven by a thermodynamic heat engine. The efficiency of this engine is about 60 per cent, which is considerably higher than one would expect for a "natural" engine.[167] The atmospheric engine operates as follows.

The total kinetic energy of the atmosphere is about 3×10^{20} joules; it is derived from the absorption of solar radiation, and is dissipated by the atmospheric turbulence and viscosity. Solar radiation is absorbed by the Earth at 1.1×10^{17} watts per sec averaged for the entire globe, and energy is dissipated by turbulence and viscosity at 2.6×10^{15} watts per sec averaged for the entire Earth. Therefore conversion of just over 2 per cent of the absorbed solar radiation into kinetic energy is enough to compensate for the loss of energy by turbulent dissipation. The conversion is accomplished by the atmosphere and the Earth's surface working in conjunction with each other, as a heat engine, whose working substance is air that contains water vapor, and whose maximum efficiency is given by the second law of thermodynamics. The "boiler" of the atmospheric engine is the Earth's surface in the tropics; the "condenser" is the high atmosphere in middle and high latitudes. If the mean temperature of the boiler is 30°C and that of the condenser is −30°C, the Carnot fraction comes to 20 per cent; this is the maximum fraction of the heat transported from boiler to condenser that can be converted into mechanical energy.

Heating within the atmosphere maintains its available potential energy, but in an unstable physical state, so that eddies appear within the smooth flow. This connection between energy and eddies led Lorenz to propose a theory of the general circulation in which the potential energy of the mean atmospheric flow maintains the eddies against the dissipative effects of turbulence and friction, and the eddies maintain the kinetic energy of the mean flow against dissipation.[168] In terms of Fultz' rotating-saucepan analogy of the atmospheric circulation, the present rate of rotation of the Earth and the existing poleward temperature gradient at the Earth's surface are at the critical level that results in an unstable regime. In such a regime, the existing combination of rotation and heating of the Earth-atmosphere system cannot maintain large-scale symmetric convection or even stable symmetric global-scale airflow. This baroclinic instability results in the formation of eddies. If the zonal atmospheric flow is barotropically stable, processes that suppress eddies convert their kinetic energy directly into kinetic energy of the zonal flow, and potential energy (which represents, in other words, the poleward gradient of temperature at the Earth's surface) is not involved. If, on the other hand, the zonal flow is baroclinically unstable, then the eddies are maintained by the transfer of available potential energy (i.e., not kinetic energy) from the zonal circulation to the eddies, which derive their kinetic energy from the available potential energy they thus acquire. In the Lorenz theory, therefore, it is shown thermodynamically that the zonal flow of an atmosphere that is bounded at its lower limit by a surface which exhibits a poleward temperature gradient must be unstable and must generate baroclinic disturbances. Such a theory provides the ultimate explanation of the nature of the general circulation, but, as it stands, is not readily useful for the prediction of air movements. For such predictions one must turn to hydrodynamic theory.

Possibly the best-known of the classic theories of the general circulation, that of Hadley in 1735, attempted to explain the tradewind circulation of the tropics as the direct result of convection between an axially symmetric heat source and two symmetric cold sources to north and south. The basic convection circulation was modified by the influence of the Earth's rotation, but remained symmetric provided the effects of land and sea (with their resulting monsoon systems) were disregarded. The Hadley theory implies that cyclones and anticyclones must be disturbances whose main function in the general circulation is to consume the kinetic energy created by the primary convective cells.[169]

Extending Hadley's concept of directly driven convection cells to the entire atmosphere, Bjerknes estimated theoretically in 1933 how intense a direct meridional circulation would provide the required upward and poleward heat transport for maintenance of Hadley cells of hemispheric extent. He found that the maximum velocity of the meridional circulation would be 2.5 m per sec, which cannot be reconciled with the existence of surface westerly winds in middle latitudes. Thus the contribution of large-scale eddies to the meridional transport of energy and zonal angular momentum must be very important, as Exner had recognized in 1925. Jeffreys showed hydrodynamically in 1926 that there must be a net transfer of angular momentum from polar and subtropical latitudes to middle latitudes. Widger in 1949 and Mintz in 1955 showed that the movements of cyclones and anticyclones account for the flow of angular momentum into middle latitudes to balance that lost by friction between the Earth's surface and the atmosphere.[170]

In effect, Jeffreys proved that a symmetric convection regime, such as that envisaged by Hadley, must break down into an asymmetric regime in which northerly and southerly air currents exist side by side. This can be seen from the following. The Earth's rotation will cause a horizontally moving airstream to be deflected in the direction opposite to the rotation; thus high pressure (due to the accumulation of air) will be built up to the right of the airstream (if the observer is looking downstream) in the northern hemisphere, and low pressure will develop to its left. Pressure will continue to distribute itself in this manner until the resulting pressure gradient across the air current overrides the deflection produced by the Earth's rotation. A symmetric convection cell of the Hadley type, incorporating equatorward winds at the surface and poleward winds aloft, must be accompanied by air pressures that increase westward at the Earth's surface and eastward aloft. However, since pressure must of necessity be continuous in a cyclical sense, an axially symmetric pressure adjustment would not be possible, and so a breakdown of the convection cells would result.

Purely hydrodynamic explanation of the general circulation, with some thermodynamics added, has proved very successful. For example, Eady set out to deduce the observed mean pattern of the general circulation directly from the equations of motion without the intervention of additional hypotheses. He assumed that within the atmosphere there is an upward and poleward transfer of heat, and an angular momentum transfer that is poleward for most of the Earth, but equatorward in high latitudes. He then inferred, from the equations of motion, that the atmospheric circulation must be maintained by turbulent rather than steady currents. He then had to explain how large-scale turbulence develops, and how it transfers heat and angular momentum. The mean flow in a baroclinic atmosphere may be shown to be always unstable, and to generate turbulence continually. The instability may be shown to be continually reestablished by a process connected with the observed heat transfer. Eady found that the transfer of angular momentum is a secondary (rather than primary) feature of the circulation, and is associated with the number of mathematical constraints (boundary conditions, for example) on the complete set of equations. Traveling cyclones at the Earth's surface are observed to have radii of about 10^3 km. Eady was able to show that the amplitudes of individual disturbances—and the amplitude of the zonal currents resulting from the transfer of angular momentum—are limited by surface friction. He was also able to infer that there are *driven* meridional circulations, which explain the main zonal precipitation belts of the world, and are secondary effects of heat transfer.[171]

Petterssen set himself the following problem. Given the existing location of the major heat and cold sources, is it possible to account for the following from dynamic and thermodynamic principles: (a) the maintenance of the known semipermanent pressure systems, and (b) the tendency of traveling cyclones and anticyclones to follow (in a statistical sense) their observed patterns? With the knowledge in 1950 of the zonal distribution of incoming and outgoing radiation, and of the criteria for dynamic stability of zonal motion in the atmosphere, Petterssen employed a generalization of vorticity theory to show that frictional and nonadiabatic thermal components of the vorticity sources and sinks must balance for each global isentropic sheet, although, in general, local balance cannot be attained. The result agreed quite well with the facts of cyclone and anticyclone behavior during 1899–1939, and indicated that many of the perturbations that develop into the migratory cyclones and anticyclones of middle latitudes originate within the equatorial belt and are carried into the westerlies with the airflow around the western extremities of the subtropical anticyclones.[172]

Petterssen expressed the vorticity equation for an isentropic surface in the form:

$$\nabla_A \cdot (Q\mathbf{V}) = F - \frac{\partial Q}{\partial \theta}\frac{d\theta}{dt},$$

in which $\nabla_A \cdot (Q\mathbf{V})$ represents the three-dimensional divergence of the vector $Q\mathbf{V}$ over the area A, and equals the export of vorticity through the boundary of the volume of the atmosphere above area A, the latter being chosen to lie in an isentropic surface. The air motion is assumed to be adiabatic and (internally) frictionless, Q represents the component of absolute vorticity of the motion normal to area A, $\mathbf{V}$ is the three-dimensional velocity vector of the motion of the atmosphere over area A, F is the vorticity of the frictional force opposing the air motion, θ represents potential temperature, and t represents time. The equation specifies the combined dynamic and thermodynamic controls of any stationary circulation. The term on the left-hand side of the equation is a measure of the export of vorticity along the isentropic sheets; i.e., it measures the intensity of the vorticity source. The equation states therefore that the intensity of a vorticity source is equal to the vorticity of the (external) frictional force supplemented by a term that depends on the rate of heating or cooling. The latter term represents the local intensity of the source of heat or cold through which the air is moving. Anticyclonic systems represent frictional sources of vorticity and cyclonic systems represent frictional sinks of vorticity, because if the frictional force is to act against the motion (which it must by definition), then the sign of the vorticity of the frictional force must be opposite to that of the vorticity of the relative motion. The sign of the thermal component of the equation varies with the spatial distribution of vorticity and with the degree of static stability of the atmosphere.

For the atmosphere in its mean state, Petterssen showed that the intensity of a vorticity source must be positive where the air is cooled in a field in which vorticity increases or where it is heated in a field of decreasing vorticity in the direction of increasing entropy, and it must be negative where the air is heated in a field where vorticity increases or where it is cooled where vorticity decreases in the direction of increasing entropy. Application of these inferences leads to a deductive model of the general circulation of the lower troposphere based on vorticity theory. The model has two major features, the first being that there are three main circulation regimes.

The *low-latitude circulation* is mainly a thermal regime, in which air is heated in a field where absolute vorticity decreases in the direction of increasing entropy, and in which the thermal component of the vorticity source is positive. Since the lower tropospheric circulation in low latitudes is mainly cyclonic, the frictional force must be anticyclonic. Therefore thermal and frictional components tend to balance, and as a consequence the circulation systems of low latitudes should be steady systems.

The *high-latitude circulation* is essentially thermal, in which air is cooled in a field where absolute vorticity increases with increasing entropy. Both frictional and thermal sources of vorticity are positive, therefore stationary circulation features can only be maintained by the export of vorticity to middle latitudes.

The *middle-latitude circulation* is dominated by the presence of migratory cyclones and anticyclones, in which separation of vorticity occurs, such that cyclonic eddies are shed towards the poles and anticyclonic eddies are shed towards the equator. Therefore the subpolar low-pressure troughs and the subtropical high-pressure belts tend to be maintained.

The second feature of the Petterssen model is that various modifications are imposed on the above circulations by mountain ranges, inland water bodies, the annual cycle of heating and cooling, and snowfields. The annual variation of heating and cooling of the Earth's surface leads to a zonal transfer of vorticity from the continents to the oceans in winter and from the oceans to the continents in summer. The land surfaces of the continents in middle and high latitudes in winter, and snowfields in particular, act as cold vorticity sources, and, because these cold sources are associated with higher atmospheric pressures than adjacent heat sources, they reduce the efficiency of the atmospheric circulation. Consequently, extensive inversions would be expected to develop over the snowfields and continental interiors in winter, with deep layers of stagnant air.

The Petterssen model is an effective way of explaining the mean state of the general circulation, but it does not enable predictions to be made for short periods. Phillips conducted an experiment at Princeton in 1955 that was a real breakthrough in theoretical climatology, made possible by the development of electronic computers with large, rapid-access internal memories. Commencing with an atmosphere at rest, Phillips integrated in steps the equations describing the motions and energy relationships of a two-level model of the atmosphere. Phillips' model was derived from the equations of motion, the equation of continuity, and the thermodynamic energy equation (see Appendix 4.5). The model was very primitive; in it the flow pattern of the upper half of the atmosphere was represented by the airflow at 250 mb, the flow in the lower half by that at 750 mb. Geostrophic flow was assumed. Friction and nonadiabatic effects were incorporated into the mathematical model (the nonadiabatic effects as a linear function of latitude), and empirical elements were added to represent intensity of heating, degree of vertical stability, and the type of frictional dissipation of energy. The computer was set in motion with the "atmosphere" initially at rest with respect to the Earth's surface. After 130 "days" of operation (30 hours computer time), a zonal flow had developed with a fairly concentrated belt of westerlies in the upper atmosphere. Random disturbances at all scales were then introduced into the computer, and after a time it was found that eddies of a certain range of sizes came to dominate. Finally, waves resembling the long waves in the westerlies made their appearance. Thus the model had successfully explained the existence of zonal surface westerlies in middle latitudes, the Polar Front jetstream, and the growth of a large disturbance, purely as dynamic and thermodynamic properties of atmospheric flow. Also, the index cycle was approximated to, despite the unrealistic "atmosphere" adopted, which was bounded by rigid mathematical walls to the north and south, and was heated in the south and cooled in the north through a net supply of heat that varied linearly with latitude but was constant with time. The extreme simplicity of the equations made the predicted relative strength of subtropical easterlies and polar easterlies incorrect, and the predicted mean latitudinal gradient of temperature within the atmosphere was too large.[173]

Smagorinsky and his co-workers have developed a comprehensive model of the general circulation that does not have the defects of the Phillips model. Phillips used the geostrophic assumption, and as a consequence was unable to account for the essentially nongeostrophic dynamics of the Hadley cells in tropical latitudes, and for the interactions between quasigeostrophic motions and inertial-gravitational motions in extratropical latitudes. In addition, the quasigeostrophic character of the general circulation, an observed fact accepted as such by Phillips, must itself be explained. Smagorinsky avoided these complications by making direct use of the primitive equations of motion (see Appendix 4.6). His initial model was confined to latitudes between the equator and 64.4°; over this zone an integration was performed for 60 days, commencing with an initial state in which a random pattern of temperature disturbances was superimposed on a zonally symmetric flow regime that was baroclinically unstable. The final result was a circulation pattern in the form of a series of waves (wave-numbers five to six) exhibiting an index cycle and an energy cycle (with a period of 11 to 12 days for the first 40 days of the experiment, then one of 17 days and less amplitude for the rest of it), plus a Hadley cell in low latitudes and a Ferrel cell in middle latitudes. The transport and balance of angular momentum were both realistically reproduced, but the mean profile of zonal velocity, although the right shape, was too intense.[174]

The initial Smagorinsky model of 1963 was a two-level model; i.e., it assumed a model atmosphere consisting of two layers, and thereby took a grossly oversimplified view of the vertical structure of the atmosphere. A more realistic model was produced in 1965. This model assumed a nine-layer atmosphere, and was intended to resolve the problem of boundary-layer fluxes, as well as to incorporate the effects of radiative transfer of energy by carbon dioxide, ozone, and water vapor. As in the initial model, the primitive equations of motion were employed, and the lower boundary of the model was assumed to be a kinematically uniform land surface without any heat capacity. The stabilizing effect of moist convection was incorporated by introducing a term that required an adjustment of the lapse-rate within the model atmosphere whenever it exceeded the moist adiabatic value. The model was applied to the prediction of mean annual conditions over a hemisphere, commencing with an isothermal atmosphere at rest. The model reached an equilibrium, in which a stable circulation pattern exhibited a cyclic energy variation with an irregular period of about two weeks. The predicted circulation pattern was fairly realistic below 30 km, including a three-cell meridional circulation in the troposphere and tending toward a two-cell meridional circulation in the stratosphere. The height of the main jetstream in the model agreed with the observed facts, as did the level of maximum poleward momentum transport, but the predicted intensity of the jet was too great. The predicted rate of change of temperature in the stratosphere was in the direction of increasing latitudes, as is the actual rate of change, but it was too small. The predicted energy balance of the model was realistic, apart from the fact that the ratios of eddy to zonal kinetic energy and eddy to zonal available potential energy were too small.[175]

The nine-level model was improved by having a hydrological cycle added to it. The hydrological cycle adopted involved: advection of water vapor by the large-scale atmospheric motion; evaporation from the surface of the Earth; precipitation; and when the relative humidity reached 100 per cent and the lapse rate exceeded the moist adiabatic value, an artificial adjustment intended to simulate the process of moist convection. The lower boundary of the model was assumed to be a completely wet surface without any heat capacity, and the radiative flux of heat was computed from the observed mean distribution of water vapor throughout the atmosphere. Commencing from an initial completely dry, isothermal atmosphere, a state of quasiequilibrium was reached after integrations covering 187 days (see Figure 4.37). The results indicated that incorporating moist convection in the model increased the intensity of the meridional circulation in the tropics, and decreased the rate of poleward transport of total energy in middle latitudes. The predicted mean rate of precipitation over the hemisphere amounted to 1.06 mm per year, which is close to the actual value; in the tropics, the predicted rainfall exceeded the predicted evaporation, and in the subtropics, predicted evaporation exceeded predicted precipitation, in agreement with observation. The predicted distribution of relative humidity increased with decreasing altitude in the troposphere and was very low in the stratosphere (except at the tropical tropopause), as in the actual atmosphere. However, the predicted rate of export of water vapor by the meridional circulation from the dry subtropics to the wet tropics was too great. The model shows that the general effect of condensation is to increase the wave-number of the tropospheric circulation and of the surface pressure field.[176]

Detailed analysis of the tropical atmosphere as predicted by the nine-level model with a hydrological cycle indicates that many of the features of the actual tropical circulation are realistically reproduced. For example, the I.T.C.Z., weak tropical cyclones, and shearlines with strong convergence are generated, and humid towers develop within the central core of regions of strong upward motion and sometimes

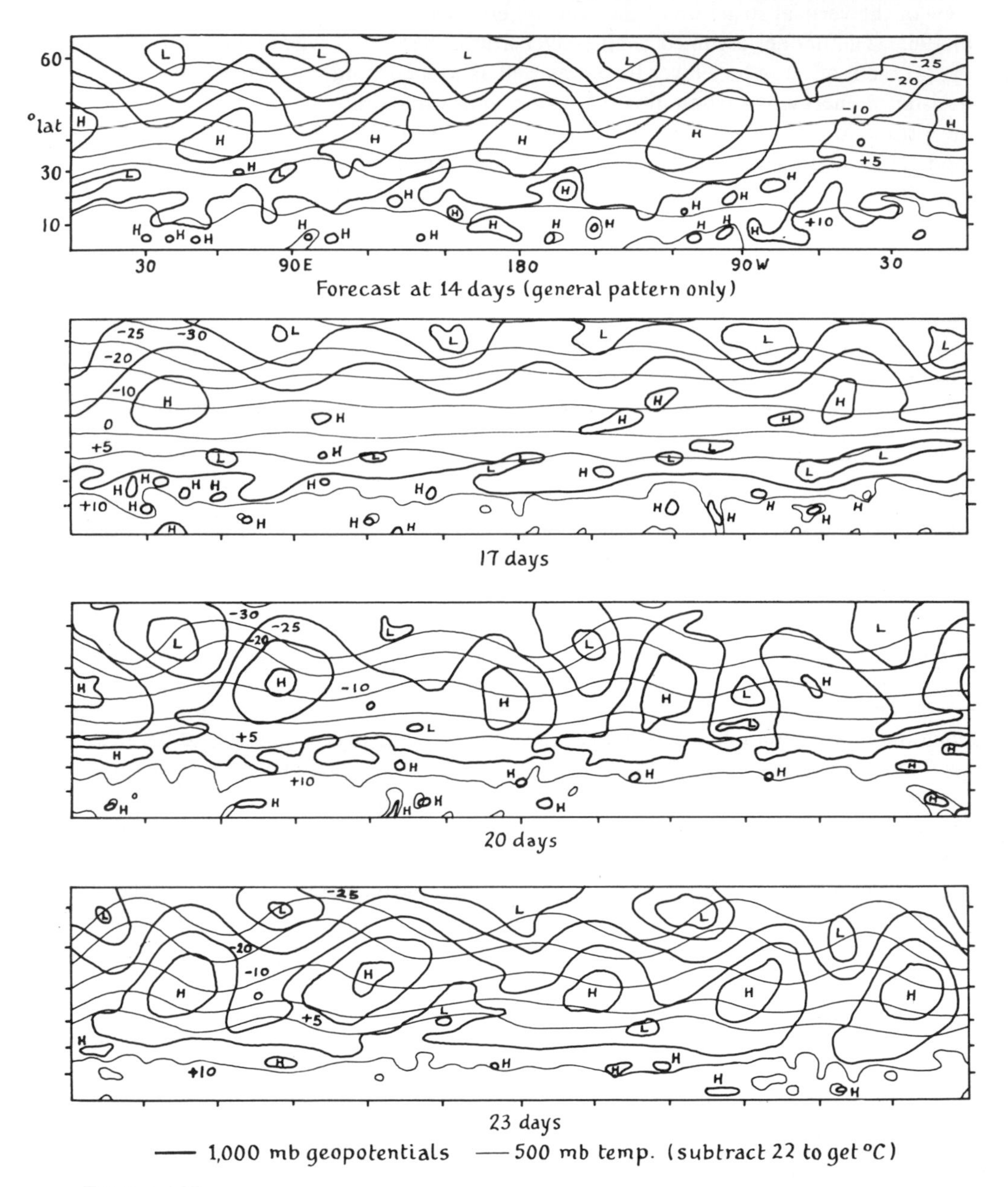

FIGURE 4.37.
Smagorinsky's simulation model of the general circulation.

reach the tropical tropopause. The humid towers result from condensation and the incorporation of the moist convective adjustment in the model (see appendix 4.7); they heat the upper troposphere of the tropics and thereby compensate for cooling due to radiative effects and the meridional circulation. The results of the tests of this model indicate that the release of eddy available potential energy—which is mainly generated by the release of latent heat of condensation—is the major source of eddy kinetic energy for the generation of tropical disturbances.[177]

The foregoing are the best-known attempts to explain the general circulation of the atmosphere in terms of its inherent properties. Others are, for example, that of Sheppard, which is based on the principle of energy transactions (see Appendix 4.8), and that of Adem, which incorporates dynamic and thermodynamic processes and involves equations giving the temperature, excess of radiation, and mean zonal wind for each day of the year. The success of some of these models, in particular the Smagorinsky models, represents an appreciable scientific achievement. It also indicates that the observed facts of the macroscale distribution of weather and climate can be explained in terms of a few simple properties of the atmosphere as a whole, for which geographical influences need not be considered, with very occasional (and very significant) exceptions.[178]

Implications of the Study of the General Circulation

Quite apart from the major scientific problem of how to explain the atmospheric general circulation as it exists today, there are two important questions implicit in general circulation theory: (a) How can we explain the known past fluctuations of the general circulation? (b) How can we use our knowledge of the existing and past general circulation to explain the future pattern; i.e., how can we use it in long-range weather (or climate) forecasting?

Painstaking study of climatic data and weather charts for past periods establishes quite clearly that there have been important fluctuations of the general circulation in recent times; in fact, the pattern of the general circulation is always changing. For example, the periods 1901–30 and 1921–50, which are taken as "normal" for purposes of establishing climatic means, were actually highly abnormal in terms of the general circulation pattern. The week-to-week and month-to-month variations in the general circulation, from a strongly zonal (i.e., high index) to strongly cellular (i.e., low index) character, are essentially similar to the secular variations of world weather patterns that have periods measured in centuries. Since the Pleistocene Ice Age, the basic characteristic of the general circulation has been the expansion or contraction of the circumpolar vortex of the northern hemisphere. Important climatic phases of postglacial times include: (a) the postglacial optimum, a warm period culminating between 5000 and 3000 B.C.; (b) the colder epoch of the early Iron Age, culminating 900–450 B.C.; (c) a secondary climatic optimum in the early Middle Ages (A.D. 1000–1200); and (d) the "Little Ice Age," a very cold period between A.D. 1430 and 1850. Phases (a) to (c) inclusive fit in with the concept of an expanding or contracting circumpolar vortex: for example, during (a) the climatic zones were displaced toward high latitudes, and the equatorial and monsoon rains belts widened (i.e., corresponding to a contracting vortex). During phase (b), the climatic zones were displaced toward low latitudes and the equatorial rain belt narrowed (i.e., an expanding vortex); (c) followed the same pattern as (a), but to a lesser degree. During phase (d), the climatic zones of the northern hemisphere shifted equatorward, and those of the southern hemisphere shifted poleward (although the movements south of the equator were less than those to the north), followed by return movements in both hemispheres during the nineteenth century.[179]

During the Little Ice Age, the atmospheric circulation as a whole was one or two per cent weaker than today (probably more in the northern hemisphere), but during the early Middle Ages, particularly A.D. 1000–1200, it was stronger than today. A general increase in circulation intensity has been apparent from 1800 to the present day, and the northern hemisphere circulation belts have moved poleward; there have been some exceptions, mainly caused by extension of the area occupied by sea ice and cold water generally. The increase in strength of the zonal circulation peaked about 1930 in the northern hemisphere, and between 1900 and 1910 in the southern hemisphere. The atmospheric circulation at 500 mb in the northern hemisphere was probably rather weaker in the year 1800 than it is today, although it probably intensified where cold land and sea surfaces were more extensive than today. Around 1800, the long waves in the westerlies had shorter lengths and larger amplitudes than today.

The period of maximum intensity of the North Atlantic atmospheric circulation, from 1920 to 1940, coincided with the period of most rapid shrinkage of the Arctic ice. The circulation intensification resulted in increased oceanic influences (particularly in winter, when very low surface temperatures caused by inversions over the cold continents became rare), decreased gradients of surface temperture in winter concomitant with increased gradients of upper-air temperature and atmospheric thickness, and increased surface temperature gradients in summer. The winter atmospheric circulation over the North Atlantic began to weaken just before the year 1940, and by the 1950's the extent of polar ice began to increase around Iceland and off the coast of northern Europe, although sea temperatures still remained relatively high.

In general, atmospheric activity as a whole has increased since 1900. There has been a greater rate of exchange of air between equator and poles, turbulence has been more intense, the zonal currents have increased in intensity, and the preferred locations of blocking action have shifted slightly, so that increased cyclonic activity has resulted over the oceans and increased anticyclonic activity over the continents. The interchange of air between northern and southern hemispheres has been modified. These effects have not progressed in a regular manner: although phase-lags have differed with region, the effects have usually been more pronounced (particularly for temperature) in high latitudes than in low, and more marked in the northern hemisphere than in the southern. Despite the circulation changes since 1900, four distinct, although composite, circulation types may be recognized in the northern hemisphere during the period 1899–1954 (see Figure 4.38). Type (I) involves a well-formed polar anticyclone, with a zonal circulation in high latitudes. Zonal transfer of atmospheric properties dominates most of the hemisphere, but two or three intrusions of southerly cyclones from low latitudes to high are typical. Type (II) involves a violation of zonality, i.e., a single intrusion of Arctic air over the northern hemisphere. Type (III) involves two to four simultaneous intrusions of Arctic air, forming a group of meridional circulation subtypes. Type (IV) involves the development of cyclonic activity in high latitudes and in particular over the Arctic Ocean, the cyclones often crossing the North Pole region. A very important point is that the transit from one type to another is very rapid: in fact, the four major types occupied more than 97 per cent of the time.[180]

Comparatively small changes in the pattern of the "normal" (i.e., 30-year mean) circulation chart can have profound consequences for weather and climate over a restricted area. Variations in the mean circulation pattern indicate that for Europe the pressure gradient increased between 1780 and 1930, together with a southerly shift in the positions of the mean European ridge and the westerlies to the north. Such a sequence of events has many similarities with the atmospheric conditions required for the onset of a glacial epoch. The increased vigor and prevalence of the westerlies in the present century has been accompanied by a general rise in winter temperatures over Europe and a slight fall in summer temperatures.[181]

The North Atlantic was colder in 1800 than in 1930, with more extensive sea ice and cyclones following more southerly tracks. At the same time, the South Atlantic

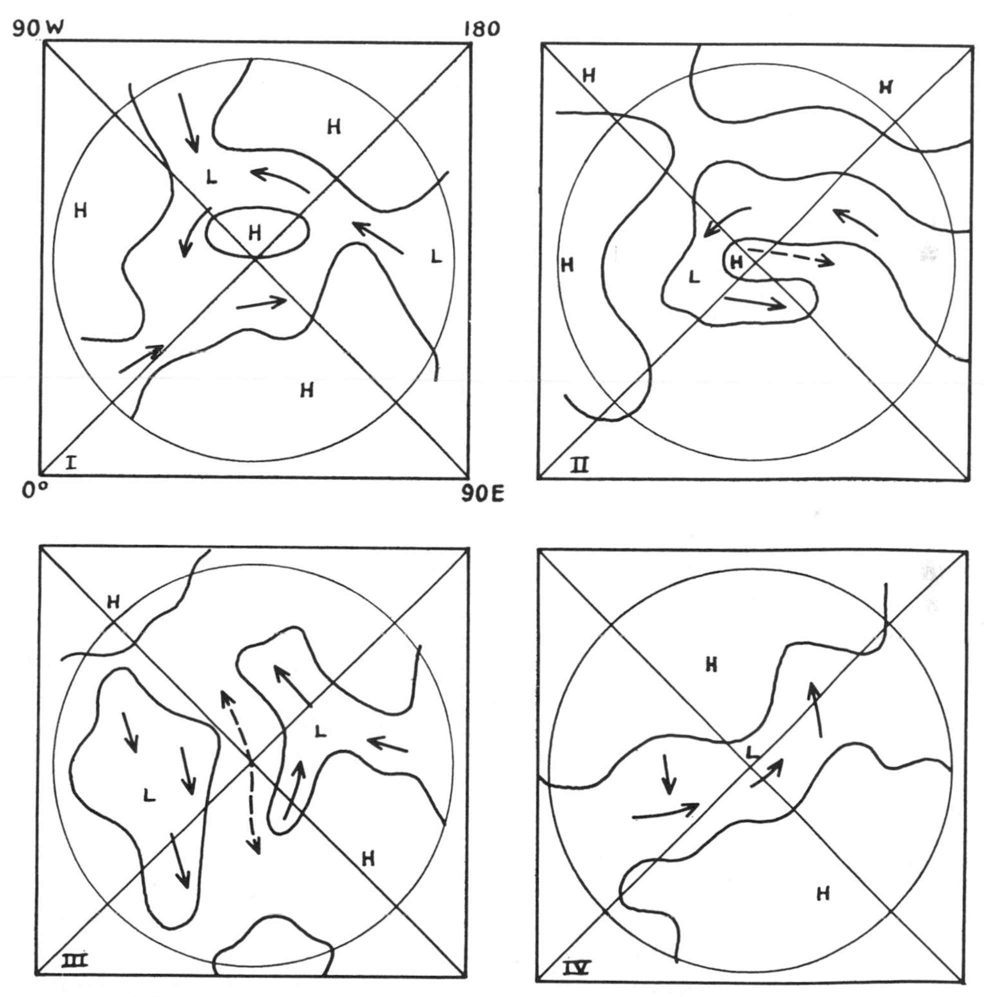

FIGURE 4.38.
Types of circulation in the nothern hemisphere. *L* indicates the area of cyclonic activity, *H* of anticyclonic; solid-line arrows indicate cyclone tracks, broken-line arrows anticyclone tracks. (After Dzerdzeevskii.)

was warmer than today, and the belt of Antarctic sea ice was narrower. In 1800 the Gulf Stream was located farther south than it is now, and its extension tended to turn southward before reaching the coast of Europe. The Polar and Labrador Currents were wider than they are today, and the Equatorial Current from the South Atlantic recurved southward along the Brazilian coast more than today, so that less warm water was supplied to the Caribbean and the North Atlantic. The increase in the zonal strength of the atmospheric circulation over the North Atlantic between 1800 and 1930 was between 5 and 10 per cent. Between 1922 and 1938, temperatures over most of the tropical Atlantic increased about 0.6–0.9°C, the greatest increase being observed in latitude 7°N. In the vicinity of the Cape Verde Islands, the strength of the northerly wind component decreased 0.8 m per sec during this period, and off the coast of central Brazil the amount of cloud cover decreased by 5 to 7 per cent. In the regions subject to northeasterly trades in the tropical Atlantic, wind and temperature anomalies were inversely related; along the Brazilian coast the anomalies of temperature and northerly wind components were directly related, whereas the anomalies of temperature and easterly wind components were inversely related. This illustrates the complications involved in predicting a weather change from a circulation change.[182]

Reconstructions of past atmospheric circulation patterns suggest that a synoptic study of the general circulation is possible. In fact, the broadcasting and publication of CLIMAT data by international agreement has opened up a new phase in climatology. This involves the use of monthly mean-pressure patterns to predict the general climatic conditions likely during the succeeding month (or months) for a fairly extensive area. This is genuine "climate forecasting," and, just as ordinary day-to-day weather forecasting uses synoptic meteorology to translate the weathermap patterns it has predicted by applying the concepts of dynamic meteorology to an observed weather pattern, so this approach involves the application of synoptic climatology to the general circulation pattern that has been predicted by application of the concepts of dynamic climatology to an observed mean-weather pattern. Early studies of this type made use of five-day means, but today 30-day or monthly means are usually employed. As with short-period forecasting, the pressure pattern is first predicted, and then the expected weather is fitted to the pressure pattern. The technique has been particularly well developed in the United States, and comparisons between predicted and actual weather patterns have brought to light many interesting climatic facts and correlations.[183]

One difficulty in extended-range forecasting by mean circulation patterns is that the latter completely overlook the existence of singularities (or *calendaricities*) and periodicities in general. Singularities pose a very difficult problem for the general-circulation theorist. Following Rebman, who paraphrased R. A. Fisher, a singularity is the result of a mechanism for generating a high degree of improbability. Primary singularities are caused by solar-terrestrial interactions, and reflect characteristic modes of the general circulation. They are responsible for generating secondary singularities, which are in effect physicostatistical servomechanisms (see Appendix 4.9). Secondary singularities are usually local; they are determined mainly by the climatological properties of the region in which they occur and its location relative to the semipermanent centers of action; and they may be reinforced or suppressed, depending on the characteristic "note" of the general circulation at the time of their appearance.[184]

Identical causes may result in secondary singularities of opposite nature in different regions. For example, the northward movement of the subtropical anticyclones of the northern hemisphere, and their southward return after January 24, results in a warm spell over the coast of eastern North America but brings the coldest spell of winter to western Europe. The reason is, of course, that the American singularity is brought by warm southwesterly airflow on the northwestern margin of the anticyclone, whereas the European singularity results because the influx of warm maritime air is blocked from the continent by the northeastward extension of the anticyclone.

On statistical evidence alone, one cannot be certain whether or not a given singularity is real. For example, calendaricities may be suppressed if the data are taken from too large an area, because different regions react differently and at different times to the same cause (e.g., the passage of a given threshold of solar radiation). Thus an outbreak of cold air in one locality must be accompanied by an outbreak of warming and moistening to the east. Since the only "fixed" climatic influence we yet know is the position of the sun in the sky, which is the same every year at a given time in a given location, it would appear that singularities are bound up with solar influences.[185]

If a singularity can be linked with events in the fluctuating cycle of the general circulation, then its reality is much more certain. For example, a considerable increase in precipitation is characteristic of the few days before July 1 in much of Arizona. This results from a rather sharp transition from one dominant airmass to another, according to the following sequence. Between June 1 and 15, there is usually a broad, flat, subtropical high-pressure belt south of latitude 25°N, with a semipermanent upper trough that makes a southward bulge in the northern boundary of that portion of the belt which is over Arizona. Between June 15 and 20, the trough usually intensifies and takes on a northeast-southwest orientation to the west of Arizona, with the result that the northwesterly upper flow over the state is replaced by a southwesterly airstream. During the whole of the period June 1–20, the main jetstream is normally south of the state, and heavy precipitation is experienced when the trough is intense. During June 20–24, the trough fills, and the circulation pattern returns to that which prevailed during June 1–15, with northwesterly winds over Arizona. On June 25 or 26, the northwest extremity of the Bermuda high breaks off to form a small migratory anticyclone that moves northwest over Arizona at the same time as the jetstream moves rapidly north to latitude 45°N. The upper trough then reforms over the southeastern United States, with the result that Arizona lies under a region of strong subsidence and usually undergoes a dry spell from June 25 to 30. On July 1 the anticyclone normally moves northeast, and the upper trough reforms again, but in weakened form, off the California coast. Arizona is then covered by a deep flow of air from the Gulf of Mexico; the air originates on the southwestern side of the westernmost extension of the Bermuda high. As a consequence, persistent and widespread rain falls over the state.[186]

Other examples of singularities linked with the general circulation may be quoted. Denver, Colorado, is characterized by a sudden increase in the probability of snowfall during October, associated with a widespread singularity that coincides with the transition from summer-type to winter-type pressure patterns. Although the mean monthly charts suggest that this transition is smooth, it is in fact usually very sudden. The January thaw of New England is a singularity of a similar nature, which coincides with the changeover from westerly to northwesterly circulation over the northeastern United States. The normal winter circulation in this part of the United States involves a trough off the east coast and an anticyclone over the Midwest. The northward movement of the Bermuda high disturbs this pattern, resulting in the formation of a trough over the Midwest that moves eastward during January 17–22, bringing to New England a warm spell that has long been recognized in folklore. The warm spell often brings about a thaw, which is followed by an intense frost as the trough fills and the normal winter circulation pattern is restored.[187]

Explaining climatic change and fluctuation in terms similar to those used to explain the general circulation is quite feasible in theory, but difficult in practice. Because of this, such explanations, regarded as too complicated by dynamicists and physicists, were left to the empirical climatologist. The problem is to explain the behavior of an atmosphere that is a thermodynamic system defined by properties that, although few in number, range from global to microscopic in size and from minutes to millenia in period. Each scale, of both time and size, defines a distinct physical problem, and the totality of these interdependent subsystems is in a state of constant commotion. Thus instabilities and irreversible pseudoperiodicities and oscillations are characteristic of the system. Although this system may be completely described by sets of nonlinear differential equations, the equations cannot be solved explicitly.[188]

The first aim is to determine whether or not year-by-year variations in the general circulation may be explained as a built-in characteristic of the system, i.e., as an intrinsic property of the atmosphere. If they can, then extraneous factors, such as variations in solar radiation, volcanic dust, carbon dioxide, and so on, need not be invoked. It would seem that the year-by-year variations are in fact built-in characteristics of the atmosphere, which, as a system, appears to control its own evolution, but conclusive statements on empirical grounds are not yet possible, mainly because ten years seems to be the shortest sampling period needed to discover a climatic trend that is not merely a "statistical entity" (i.e., a spurious result from random sampling of individual years)

Climate may be regarded as the product of a thermodynamic system that includes both the atmosphere and the Earth's surface. The latter—in particular its oceans and ice-cover—controls the lower boundary condition of the atmosphere and most of its energy input. One feedback mechanism is that the state of the atmosphere to a large extent controls some physical properties of these parts of the Earth's surface, which are of great significance in the system that constitutes climate. The heat capacity of the oceans is 1,000 or more times greater than that of the atmosphere, and the natural fluctuation of the permanent ice-fields has a period in the tens of millennia, rather than in centuries like the oceans. An important feature of the general circulation is that it is not uniquely specified by figures representing seasonal or even monthly values. The effects of snow cover, sea-surface temperatures, and soil variations are all of crucial importance in deciding the magnitude of the feedback that to some extent controls the monthly and seasonal variations in the circulation.[189]

Air-sea exchanges in the North Atlantic Ocean provide a good illustration of the importance of such surface interactions in the study of the general circulation. Physical theory indicates that the effect of windstress upon the surface of the oceans, away from coastal barriers, should be to produce a wind drift of water. The angle between the direction of the drift and the direction of the wind varies with depth beneath the sea surface, but at the surface, in stationary conditions, the surface waters should move at an angle of 45° to the right of the surface wind in the northern hemisphere and at 45° to the left in the southern hemisphere. The wind drift will bring about a transport of water, the *Ekman transport*, which is directed at an angle of 90° to the right of the direction of windstress at the surface in the northern hemisphere. The Ekman transport produces mass divergence of water in regions of cyclonic vorticity of windstress, and mass convergence where the windstress vorticity is anticyclonic (see Appendix 4.10). The effect of mass divergence is to make surface water diverge from a center, and colder deep water must rise to the surface or near to replace it; the result is to widen the area over which deep water is in contact with the atmosphere, to reduce the depth of any preexisting warm surface layer, and to lower its surface temperature.

The initiation and subsequent maintenance of climatic extremes around 1800 was explained by Bjerknes in terms of the above mechanism. A low-index type of atmospheric circulation gives rise to less cyclonic windstress than normal in the North Atlantic, and after some time leads to a weakening of the northward-flowing branches of the warm North Atlantic current. As a result, the Sargasso Sea becomes warmer and the ocean north of latitude 40°N becomes colder. The new state can be maintained as a steady state, because the low-index circulation brings about a slightly increased meridional flux of atmospheric heat in latitudes where the oceanic heat flux was reduced.

The sequence of events when the Bjerknes mechanism operates may be described as follows. First, the atmospheric circulation tends toward the low-index stage. The semipermanent Newfoundland trough shifts westward and connects with the col in the high-pressure belt near Bermuda, and the wavelength of the stationary long waves in the westerlies becomes shorter, causing the next downwind trough to move eastward toward western Europe. The central and eastern portions of the North Atlantic are then between the Newfoundland and western Europe troughs, and therefore come under the influence of westerlies with less cyclonic (possibly even slightly anticyclonic) stress vorticity. Second, the immediate effect of the change to low index is a reduction in the rate of oceanic cooling exerted by winds from arctic America. Consequently, the ocean surface in the westerlies belt becomes warmer. Third, if the low-index circulation persists, the weakened cyclonic windstress causes a quasipermanent weakening of the geostrophic part of the northward-flowing branches of the North Atlantic current over the central and eastern portions of the ocean (see Appendix 4.11). This, in turn, leads to the establishment of new (and lower) surface temperature normals for the sea surface north of latitude 40°N. Fourth, feedback from sea to air becomes important.

Sea-surface temperature constitutes the "thermostat" of the air-sea heat-interchange system. The oceanic response in the northern portion of the North Atlantic to both long- and short-period atmospheric trends is to set this thermostat lower in the Labrador–Irminger Sea area the deeper the Icelandic cyclone becomes. Its response in the southern portion of the ocean is to set the thermostat higher the more intense is the Azores-Bermuda subtropical anticyclone. At low-index stage, the thermostat will therefore be set very low in the north and very high in the south, with the result that the meridional temperature gradient in the surface layers of the water of the mid-Atlantic between 40° and 50°N is strengthened; i.e., the baroclinicity of the ocean surface is increased. Increasing oceanic baroclinicity automatically induces increasing baroclinicity in the lower layer of the atmosphere through convective exchanges between sea and air. Increasing atmospheric baroclinicity within the westerlies increases the rate of vertical wind shear and the maximum for windspeeds near the tropopause, and so more kinetic energy becomes available in the upper troposphere as input into cyclones. Young frontal cyclones must always move to the north of the zone of maximum baroclinicity, i.e., either toward the British Isles or toward Iceland. Those moving toward Iceland are in the majority, deepening and retarding as they move west; they finally stagnate near Iceland, thus maintaining the Icelandic low. As the low deepens, the magnitude of the Ekman divergence increases, so that the temperatures of the ocean surface are reduced. The final result is to set the oceanic thermostat in the Labrador–Irminger Sea area still lower, thereby further increasing the baroclinicity. The ocean-air chain reaction so continues.

The Bjerknes model assumes a stationary Icelandic low with a calm central area, and is a fair description of what has actually happened since 1800. The oceanic baroclinicity has intensified, taking a long time (from 1894–98 to 1920–24) to extend from latitude 70°W to latitude 20°W. The rises or falls in sea-surface temperature between 1894–98 and 1920–24 coincided with the areas of warm or cold isallobaric advection. The changes in sea-surface temperature were in phase with the transport of sea-surface isotherms by the isallobaric wind, both north and south of the Azores anticyclone. The Icelandic low weakened after 1920–24, which resulted in increasing seasurface temperatures in Icelandic waters, and was later associated with a general warming of the lower atmosphere in high latitudes. In general, above-normal sea-surface temperatures have characterized the latitudes of the westerlies during low-index years, and high-index periods have been associated with below-normal sea-surface temperatures. The areas of maximum rates of heating or cooling of the sea surface coincide with the area of greatest increase or decrease in the strength of the westerlies.

There are many general theories of climatic change, but none of them are conclusive. Some recent attempts do point the way toward a quantitative explanation. One approach is to estimate the magnitude of the terms in the atmospheric energy-balance equations. This really means estimating the probable variation in the magnitude of these terms. For example, evaporation is an important term in the water-balance equation. Although the rate of evaporation for a large area varies only slightly from year to year, it is extremely sensitive to human interference. Thus the conversion of savannah and steppe to farmland, and the clearing of tropical forests from large areas of the Earth's surface, must alter the actual rate of evapotranspiration very considerably, while the rate of potential evapotranspiration must remain constant. The carbon-dioxide content of the atmosphere is also very susceptible to human influence. The CO_2 content of the atmosphere has probably increased from 293 parts per million to 325–330 ppm since the end of the nineteenth century, partly because of the burning of fossil fuels (accumulated for 250 million years) in a relatively short space of time in industrial areas, partly because the CO_2 produced by ground bacteria is not immediately taken up by plants, since the plant-covered area of the world has been greatly reduced. The increase in CO_2 has resulted in an increase in back radiation and a corresponding temperature rise of 0.011°C per annum. Man-made large aerosols, from pollution sources and from bush fires, have also increased the back radiation.[190]

The possible connection between solar radiation fluctuations and climatic change has long been a subject of contention, and the disadvantages of nonquantitative arguments are soon apparent from a perusal of Simpson's classic papers. There is definite statistical evidence that *some* elements of the general circulation *sometimes* show a correlation with solar events that, if it continued, could generate a climatic fluctuation. For example, a spotted sun favors longer persistence of wave-number six of the atmospheric circulation at the expense of the stability of wave-number five. Since number six is a winter-type pattern and number five is a summer type, the solar correlation might have an influence on local weather. There is also statistical evidence that high-level Pacific tropical cyclones originate over the Marshall Islands as the tropospheric response to sudden increases in solar radiation. When such cyclones move poleward, they bring about changes in the middle-latitude circulation. The evidence suggests that local sources or sinks of radiative heat mainly influence the ultralong waves in the atmospheric circulation, and hence affect the areas of greatest cyclonic or anticyclonic development. Changes in the horizontal gradient of solar radiation affect the short (i.e., frontal) and the moderately long waves of the circulation, and the sequence of transition from one ephemeral period to another.[191]

Mathematical modeling of the general circulation can point to the influences that must be important in climatic fluctuation or change. For example, Smagorinsky showed that large-scale, longitudinal heat sources produce virtually the same effect on the mean zonal atmosphere flow as do major mountain barriers. Since land-sea temperature contrasts reverse between summer and winter, there may be resonant interactions that depend partly on the zonal circulation ,which is in itself a consequence of the horizontal radiation-balance gradient. He also showed that nonadiabatic heating rather than radiative processes is the main control of low-level perturbations of the zonal flow, so that sea-surface temperature distributions will be an important control of local climate.[192]

Some aspects of local climate appear to be directly related to radiation, the primary driving force of the general circulation; others lag behind the radiative changes, except on a geological time scale. Changes in roughness and other physical properties of the Earth's surface seem to have little general effect on the atmospheric circulation, because the surface windspeed over a large area is primarily determined not by the degree of roughness of the underlying surface but by the wind's need to adjust itself to the flux of angular momentum that is necessary to maintain the circulation. This angular momentum flux is controlled mainly by the poleward rate of change of atmospheric energy. If the poleward gradient of radiation balance at the Earth's surface increases, the poleward temperature gradient will increase, and so will the poleward rate of heat transfer, by an increase in the ratio of eddy available to zonal available potential energy. Thus the period of the index cycle must become shorter. If the meridional gradient of temperature is increased, the length of the long wave of maximum instability is increased, and the character of the general circulation must change. The consequence is that any increase in the amount of radiation received at the Earth's surface would be at least partially balanced by an increase in the surface temperature and by an increase in the rate of upward long-wave radiation. Any increased poleward transport of sensible heat would be accompanied by increased poleward transport of latent heat, which would be released as precipitation in middle latitudes.

The conclusion is that one does not have to deal with variations in the receipt of solar radiation in order to explain short-period fluctuations in climate. For periods on the order of days or, say, up to five years, the observed fluctuations in the general circulation should be explainable by dynamic reasoning. For periods on the order of tens of years, storage and feedback effects involving the oceans must be taken into account.

A major difficulty attending the explanation of the general circulation and its fluctuations in mathematical terms is the inherent turbulence of the atmosphere. The restlessness of the atmosphere sets a theoretical limit to its predictability. Even if we knew all the laws governing the atmosphere's motion, even if we had completely accurate initial grid-point data for the entire atmosphere, and even if the truncation error of the prediction equations could be reduced to negligible proportions, we still could not make error-free long-range forecasts, because the continuous character of the turbulence spectrum of atmospheric motions means that, however fine the grid-point network, some smaller-scale atmospheric motions will remain unobserved; i.e., some turbulent energy will not be accounted for. It will therefore always be possible for part of this energy to masquerade as energy of the large-scale flow, and appear as an "observation error" in the latter. Such a manifestation could be minimized by reducing the size of the grid mesh to that of smallest eddy permitted by viscosity, but then the grid pattern would be too fine to be workable.[193]

The unpredictability of the atmosphere therefore gives rise to the following problem. Although the atmosphere contains periodic and quasiperiodic components (for example, the lunar and solar tides, and the sunspot cycle), in fact its motions are aperiodic and therefore unstable. A given "error," interpreted as a perturbation of the smoothed general circulation, will therefore grow. The problem is to determine how much time must elapse before this error has grown so much that the perturbed motion differs from the unperturbed motion by as much as two randomly chosen flows. This question was investigated by a comparison of mathematical models of the circulation devised by Leith, Mintz and Arakawa, and Smagorinsky. Successive predictions were produced by computer until the predicted state bore no discernable relation to the initial state of the model atmosphere. A sinusoidal "error" perturbation was then introduced in the temperature field, and the prediction process was continued for a further 30 days or more. The results indicate that the temperature error grows exponentially, doubling every five days or so in each hemisphere and at each atmospheric level, and that the pressure and wind errors grow in a like manner. The limit for mathematically precise predictability of the atmosphere thus seems to be about two weeks in winter and perhaps a little longer in summer.[194]

Notes to Chapter 4

1. Good accounts of the general circulation, as understood at one particular time, may be found in: C.-G. Rossby, *HM*, pp. 501–29; B. Bolin, *AGP*, no. 1 (1952); P. A. Sheppard, *SP*, 40 (1952), 89; R. G. Fleagle, *SP*, 48 (1960), 72. See also *CM*, pp. 541–66, for articles by C. P. Starr, J. Namias, and P. F. Clapp.

2. The classic sources of information are H. Heastie and P. M. Stephenson, *MOGM*, vol. 13, nos. 103 and 104 (1960), parts I and II, and G. B. Tucker, *MOGM*, vol. 13, no. 105 (1960). See also G. B. Tucker, *QJRMS*, 83 (1957), 290, and 85 (1959), 209, for details.

3. E. T. Eady, in D. R. Bates, ed., *The Planet Earth* (New York, 1957), pp. 113–51, presents a simple explanatory description of this model.

4. E. Palmén, *QJRMS*, 77 (1951), 337.

5. A. D. Belmont, *General Mills Science Report*, no. 1 (1960).

6. W. W. Kellogg and G. F. Schilling, *JM*, 8 (1951), 222; R. J. Murgatroyd, *QJRMS*, 83 (1957), 417; E. S. Batten, *JM*, 18 (1961), 283. The Kellogg-Schilling model was the first modern model of the stratospheric circulation. For reviews of later models,

see R. J. Murgatroyd *et al.*, *WMO Tech. Note*, no. 70, p. 19.

7. R. A. Ebdon and R. G. Veryard, *N*, 189 (1961), 791, and *MM*, 90 (1961), 125. R. A. Ebdon, *QJRMS*, 87 (1961), 322.

8. C. R. Faulkner, *N*, 187 (1960), 926. J. S. Sawyer, *QJRMS*, 87 (1961), 24. For details of the mathematics, see J. S. Greenhow and E. L. Neufeld, *loc. cit.*, p. 472.

9. A. W. Brewer, *QJRMS*, 75 (1949), 351. G. M. B. Dobson, *Proc. Roy. Soc.*, 236 (series A, 1956), 187.

10. J. Spar, *Sonderausschmuss Radioactivität der Deutscher Bundesrepublik* (Bad Kreuznach, 1959). See also L. Machta, *BWMO*, 9 (1960), 64, for a review of general circulation models inferred from the movements of radioactive fallout through the atmosphere.

11. P. Goldsmith and F. Brown, *N*, 191 (1961), 1033.

12. P. Storebø, *JM*, 17 (1960), 547. On the mean meridional circulation, see: P. A. Gillman, *T*, 17 (1965), 277, on the southern hemisphere; Y. P. Rao, *T*, 19 (1967), 483, on its maintenance; E. O. Holopainen, *T*, 17 (1965), 283, on its role in atmospheric energy balance; J. M. Wallace, *QJRMS*, 93 (1967), 176, on its role in stratospheric biennial oscillation; R. J. Murgatroyd, *QJRMS*, 95 (1969), 194, on its role in heat and momentum balances at solstices; and D. G. Vincent, *QJRMS*, 94 (1968), 333, on the stratosphere in the northern hemisphere. On the theory of the meridional circulation, see A. Wiin-Nielsen and A. D. Vernekar, *MWR*, 95 (1967), 723, and A. D. Vernekar, *MWR*, 95 (1967), 705. For proposed models of the tropical circulation in which one or two "equatorial cells" separate the Hadley cells, see S. L. Hastenrath, *JAS*, 25 (1961), 979. According to G. C. Asnani, *JAS* 25 (1968), 133 and 935, the direct thermal drive of the Hadley cell may be questioned. On the response of the Hadley circulation to anomalies in equatorial oceanic temperature, see J. Bjerknes, *T*, 18 (1966), 820. On diurnal and semidiurnal cycles in winds at 30 to 60 km over White Sands, New Mexico, see B. T. Miers, *JAS*, 22 (1965), 382. For a climatological picture of mean winds to 67 km, see H. S. Appleman, *JGR*, 69 (1964), 1027, and for zonal and meridional winds to 120 km (equator to 75°N), see A. J. Kantor and A. E. Cole, *JGR*, 69 (1964), 5131.

13. For a review of circulation indices, see A. G. Forsdyke, *MM*, 80 (1951), 156.

14. E. J. Rebman, *MWR*, 81 (1953), 386.

15. J. Namias and W. A. Mordy, *JM*, 9 (1952), 180.

16. R. M. White, *JM*, 7 (1950), 349. C. O. Jenista, *BAMS*, 34 (1953), 10.

17. E. Palmén, H. Riehl, and L. A. Vuorela, *JM*, 15 (1958), 271. For estimates of the seasonal variation in the general circulation on the basis of cyclone-anticyclone movements, see N. J. MacDonald, *T*, 20 (1968), 300. On the Global Atmospheric Research Project, involving intensive worldwide measurements of the general circulation, see G. D. Robinson, *QJRMS*, 93 (1967), 409, and P. A. Sheppard, *W*, 23 (1968), 262.

18. For normal charts, see G. A. Tunnell, *Marine Observer*, 30 (1960), 72. The 12-year charts are from J. F. O'Connor, *MWR*, 89 (1961), 211.

19. For a discussion of the general circulation as a synoptic problem, see R. C. Sutcliffe, *QJRMS*, 75 (1949), 417.

20. H. H. Lamb, *QJRMS*, 85 (1959), 1.

21. W. Schwerdtfeger, *MWR*, 88 (1960), 203. See N. E. La Seur, *JM*, 11 (1954), 43, for an account of the asymmetry of the middle-latitude circumpolar current. On the stability of an expanded circumpolar vortex, see J. Namias, *JAS*, 22 (1965), 728.

22. W. H. Klein, *MWR*, 80 (1952), 203. W. H. Klein and J. S. Winston, *MWR*, 86 (1958), 344.

23. H. K. Saylor and J. R. Caporaso, *MWR*, 86 (1958), 447.

24. H. Riehl, *JM*, 6 (1949), 207. P. M. Wolff, *JM*, 12 (1955), 536, gives full details about the quantitative determination of long waves. On the climatology of 500-mb long waves in latitude 50°N, see P. D. de la Mothe, *MM*, 97 (1968), 333. On long waves in the southern hemisphere, see E. C. Anderson, *NOTOS*, 14 (1965), 57. For fluctuations of long-wave patterns in the northern hemisphere in terms of spherical harmonics, see R. Eliasen and B. Machenhauer, *T*, 17 (1965), 220, and 21 (1969), 149. For maps showing the positions of monthly mean 700-mb troughs and ridges in the northern hemisphere, 1949–63, see L. P. Stark, *MWR*, 93 (1965), 705.

25. *BWMO*, 5 (1956), 103 and 6 (1957), 135.

26. *BWMO*, 7 (1958), 122.

27. On the distribution of jetstreams over Europe, the Mediterranean and the North Atlantic, see A. F. Crossley, *J. Institute of Navigation*, 14 (1961), 432. On subtropical jetstreams, see R. H. Clarke, *JAS*, 23 (1966), 516, for Woomera, Australia. On subtropical and polar front jets in Southwest Asia, see J. M. Walker, *MM*, 96 (1967), 161. For details of a tidal jetstream in the stratosphere, see W. L. Webb, *JAS*, 23 (1966), 531.

28. J. K. Angell, *MWR*, 92 (1964), 203, noted that average speeds of 103 knots in autumn, 123 knots in winter, and 101 knots in spring were measured between 300 and 250 mb above Japan. For a detailed discussion of jetstream climatology, see H. Riehl *et al.*, *AMM*, vol. 2, no. 7 (1954). See also J. S. Sawyer, *W*, 12 (1957), 333.

29. E. Palmén, *JM*, 6 (1949), 20. E. Palmén and K. M. Nagler, *JM*, 5 (1948), 58.

30. E. R. Reiter, *JM*, 15 (1958), 27; R. M. Endlich and G. S. McLean, *AFSG*, no. 121 (1960). E. E. Austin and J. K. Bannon, *MM*, 81 (1952), 321. J. K. Bannon and M. P. Jackson, *MM*, 82 (1953), 100.

31. E. Gold, *loc. cit.*, p. 194. For a general discussion of the nullschicht concept, see H. H. Lamb, *MM*, 86 (1957), 142.

32. W. R. Gommel, *JAM*, 2 (1963), 105.

33. S. Petterssen, *JM*, 12 (1955), 58, and *WAF*, I, 337.

34. S. M. Daniels and R. H. Johnson, *QJRMS*, 80 (1954), 212. R. Murray and S. M. Daniels, *QJRMS*, 79 (1953), 236.

35. D. A. Richter and R. A. Dahl, *MWR*, 86 (1958), 368.

36. J. Badner and M. A. Johnson, *MWR*, 85 (1957), 62.

37. T. C. Yeh, J. E. Carson, and J. J. Marciano, *AMM*, vol. 1, no. 3 (1951).

38. C. J. Boyden, *MM*, 92 (1963), 259.

39. G. S. McLean, *BAMS*, 38 (1957), 579.

40. C. J. Boyden, *MM*, 92 (1963), 319.

41. For general information about the relationship between cloudiness and jetstreams, see R. Murray, *MOGM*, no. 97 (1956); G. S. McLean, *op. cit.* (1957); R. M. Endlich and G. S. McLean, *AFSG*, no. 121 (1960); A. I. Eltantawy, *GPA*, 46, no. 2 (1960), 352.

42. For photographs of clouds typically associated with jetstreams, see G. J. Jefferson, *MM*, 93 (1964), 91; V. J. Shaefer and W. E. Hubert, *T*, 5 (1953), 27. On jetstream cirrus, as photographed from TIROS satellites, see W. Viezee, R. M. Endlich, and S. M. Serebreny, *JAM*, 6 (1967), 929.

43. T. A. M. Bradbury, *W*, 19 (1964), 87.

44. L. H. Clem, *BAMS*, 36 (1955), 53. For photographs, see J. H. Conover, *JM*, 17 (1960), 532.

45. For examples, see: H. Riehl *et al.*, *AMM*, vol. 2, no. 7 (1954); G. T. Trewartha, *Erdkunde*, 12 (1958), 205–14.

46. C. W. Newton, *ASM*, p. 288.

47. C.-G. Rossby, *BAMS*, 28 (1947), 53.

48. E. Palmén, *QJRMS*, 77 (1951), 337.

49. For details, see H. Riehl *et al.*, *op. cit.* (1954).

50. For details, see *BAMS*, 28 (1947), 255. J. Namias and P. F. Clapp, *JM*, 6 (1949), 6. C. J. Boyden, *MM*, 92 (1963), 287. C.-G. Rossby, *JM*, 2 (1945), 187.

51. On orographic influences, see B. Bolin, *T*, 2 (1950), 184. For the theory of the effect of the Alpine chain, see H. Reuter and H. Pichler, *T*, 16 (1964), 40. For the theory of the effects of the Himalayas on the general circulation, see A. M. Chaudhury, *T*, 2 (1950), 56, and B. Bolin, *op. cit.* (1950).

52. H. H. Lamb, J. Fleming, H. D. Hoyle and J. Robinson, *MM*, 86 (1957), 97.

53. R. Lee, and W. L. Godson, *JM*, 14 (1957), 126.

54. H. Landers, *BAMS*, 36 (1955), 371. R. R. Dickinson, *BAMS*, 36 (1955), 195.

55. E. Palmén and C. W. Newton, *JM*, 5 (1948), 220. H. Riehl, F. A. Berry and H. Maynard, *JM*, 12 (1955), 26. R. M. Endlich and G. S. McLean, *JM*, 14 (1957), 543. K. C. Brundidge and J. L. Goldman, *JAM*, 1 (1962), 303.

56. G. S. Mclean, Jr., *AFSG*, no. 140 (1962), p. 221. Endlich and Mclean, *AFSG*, no. 121 (1960). E. R. Reiter, *GPA*, 46, no. 2 (1960), 193.

57. J. Briggs and W. T. Roach, *QJRMS*, 89 (1963), 225. G. S. McLean, Jr., *op. cit.* (1962).

58. S. M. Serebreny, E. J. Wiegman and R. G. Hadfield, *JAM*, 1 (1962), 137. For observations of the structure of jetstreams over North America, see: D. R. Johnson and W. C. Shen, *MWR*, 96 (1968), 559, on temperature and radiation profiles; R. M. Endlich and G. S. McLean, *JAM*, 4 (1965), 83, on aircraft observations in the central United States; C. W. Newton and Y. Omoto, *T*, 17 (1965), 449, on energy distribution and associated wave-amplitudes. See also V. J. Oliver, R. K. Anderson, and E. W., Ferguson, *MWR*, 92 (1964), 441, on detection of jets from TIROS photos over eastern North America. On jetstreams over South America, see J. Alvarez, *NOTOS*, 11 (1962), 67. For details of 500-mb jetstream characteristics associated with heavy precipitation over the British Isles, see G. R. R. Benwell, *MM*, 96 (1967), 4. On 300-mb jetstreams and precipitation in Cyprus, see R. M. Morris, *MM*, 95 (1966), 366. On the annual variation of jets across the Greenwich meridian, see W. G. Ritchie and C. J. M. Aanensen, *MM*, 94 (1965), 337.

59. L. C. Chamberlin, M. L. Barad, R. Ely, and H. H. Lettau, *EAFM*, p. 276.

60. R. W. Longley, *BAMS*, 40 (1959), 190.

61. K. L. Pitchford and J. London, *JAM*, 1 (1962), 43.

62. Y. Izumi and M. L. Barad, *JAM*, 2 (1963), 668. J. R. Gerhardt, *JAS*, 19 (1962), 116. Y. Izumi, *JAM*, 3 (1964), 70.

63. W. H. Hoecker, *MWR*, 93 (1965), 133.

64. J. R. Gerhardt, *JAM*, 2 (1963), 49.

65. L. L. Means, *JAM*, 1 (1962), 588. On the climatology of low-level jetstreams in the United States, see W. D. Bonner, *MWR*, 96 (1968), 833. On the physical characteristics of southerly low-level jets over the western plains of the United States, see W. H. Hoecker, *MWR*, 93 (1965), 133. On low-level jets in Kenya that cross the equator, see J. Findlater, *MM*, 95 (1966), 353, and 96 (1967), 216. According to J. Findlater, *QJRMS*, 95 (1969), 362, the low-level jets over Kenya are merely one part of an extensive low-level air current covering the western Indian Ocean during the northern summer. For case studies, of individual low-level jets, see: L. J. Rider, *JAM*, 5 (1966), 283, on White Sands, New Mexico; J. C. Kaimal and Y. Izumi, *JAM*, 4 (1965), 576, and Y. Izumi and H. A. Brown, *JAM*, 5 (1966), 36, on Cedar Hill, Texas; L. J. Rider and M. Armendariz, *JAM*, 5 (1966), 733, on Green River, Utah; and W. D. Bonner, *MWR*, 94 (1966), 167, on thunderstorm correlation with such jets in the south central United States. For a mathematical explanation of diurnal wind variations observed above low-level jets, see

G. L. Darkow and O. E. Thompson, *JAS*, 25 (1968), 39. For a theory of diurnal oscillation of boundary-layer winds above sloping terrain that explains the amplitude of the nocturnal low-level jet over the American Great Plains, see J. R. Holton, *T* 19 (1967), 199.

66. C.-G. Rossby, *J. Marine Res.*, 2 (1939), 38. H. C. Willett, *TAGU*, 29 (1948), 803.

67. J. Namias, *JM*, 7 (1950), 130; *ASM*, p. 240.

68. F. Defant, *SPIAM*, p. 302.

69. J. Namias, *ASM*, p. 240. For a cross-spectral analysis of northern hemisphere index-cycle data, see P. R. Julian, *MWR*, 94 (1966), 283.

70. J. Namias, *JM*, 4 (1947), 125.

71. R. D. Elliott and T. B. Smith, *JM*, 6 (1949), 67.

72. M. K. Miles, *W*, 10 (1955), 237.

73. E. J. Sumner, *QJRMS*, 80 (1954), 402; *MM*, 88 (1959), 300.

74. R. A. Sanders, *MWR*, 81 (1953), 67. C. P. Mook, *BAMS*, 35 (1954), 379.

75. J. Namias, *T*, 16 (1963), 394.

76. Sumner, *op. cit.* (1954).

77. C. M. Woffinden, *MM*, 89 (1960), 236.

78. R. Berggren, B. Bolin, and C.-G. Rossby, *T*, 1 (1949), 14. D. F. Rex, *T*, 2 (1950), 196 and 275.

79. J. Namias, *BAMS*, 33 (1952), 279.

80. J. Namias, *BAMS*, 35 (1954), 112. I. Enger, *BAMS*, 36 (1955), 36.

81. R. Shapiro, *JM*, 15 (1958), 435. W. H. Munk, *JM*, 17 (1960), 92.

82. Y. Tu-cheng, D. Shih-yen, and L. Mei-ts'un, *ASM*, p. 249.

83. I. I. Schell, *HMS*, no. 8 (1947); *JM*, 13 (1956), 471. On blocking patterns during hot weather in Britain, see A. Perry, *W*, 22 (1967), 420. For seasonal rainfall sequences over England and Wales in relation to winter blocking patterns, see P. M. Stephenson, *MM*, 96 (1967), 335. For details of the blocking pattern of February 1965 in the northern hemisphere, see R. Murray, *W*, 21 (1966), 66. For the relation between blocking to the northwest of Britain and central England temperatures during the following winter, see R. F. M. Hay, *MM*, 96 (1967), 167. On dynamic persistence, see I. Schell, *AMGB*, supplement 1 (1966), p. 152. On seasonal persistence and blocking, see J. Namias, *T*, 16 (1964), 394. For the effect of persistence on the accuracy of numerical forecasts at 500 mb, see N. J. MacDonald and R. Shapiro, *JAM*, 3 (1964), 336.

84. For a review of the methods adopted and results obtained, see R. J. Murgatroyd, *QJRMS*, 83 (1957), 417.

85. A. von Berson, *Bericht über die Aerologische Expedition nach Östafrica im Jahre 1908*, *Ergebn. Preuss. Aero. Obs.* (1910). W. Van Bemmelen, *Verh. Magn. Meteorol. Obs. Batavia*, nos. 1 (1911), 4 (1916), 6 (1920). L. S. Clarkson and L. W. Littlejohns, *MM*, 87 (1958), 105.

86. C. J. Brasefield, *JM*, 7 (1950), 67. A. D. Belmont, *General Mills Science Report*, no. 1 (1960). P. B. Storebø, *JM*, 17 (1960), 547.

87. A. R. Goldie, *CPRMS*, p. 156. C. E. Palmer, *W*, 9 (1954), 341.

88. H. S. Muench, *AFSG*, no. 141 (1962).

89. W. L. Webb, *JAS*, 21 (1964), 582. H. T. Mantis, *JAM*, 2 (1963), 427. H. T. Mantis, *JM*, 17 (1960), 465.

90. T. J. Keegan, *JGR*, 67 (1962), 1831.

91. H. Riehl and R. Higgs, *JM*, 17 (1960), 555.

92. R. J. Reed and D. G. Rogers, *JAS*, 19 (1962), 127.

93. A. D. Belmont and D. G. Dartt, *JAS*, 21 (1964), 354.

94. R. A. Ebdon, *W*, 18 (1963), 2.

95. J. E. Laby, J. G. Sparrow, and E. L. Unthank, *JAS*, 21 (1964), 249. J. G. Sparrow and E. L. Unthank, *loc. cit.*, p. 592.

96. J. K. Angell and J. Korshover, *loc. cit.*, p. 479.

97. R. E. Newell, *loc. cit.*, p. 320. For a review, see R. J. Reed, *BAMS*, 46 (1964), 374.

98. R. Scherhag, *JM*, 17 (1960), 575. R. A. Craig and W. S. Hering, *JM*, 16 (1959), 91. F. G. Finger and S. Teweles, *JAM*, 3 (1964), 1. There is an extensive literature dealing with climatological problems of the stratospheric circulation. For the climatology of zonal winds in the equatorial stratosphere, see R. J. Reed, *JGR*, 71 (1966), 4223. For stratospheric winds over Panama, see B. T. Miers, *JGR*, 72 (1967), 5149. On small-scale wind variations in Florida, see R. E. Newell, J. R. Mahoney, and R. W. Lenhard, Jr., *QJRMS*, 92 (1966), 41. On large-scale disturbances in the northern hemisphere, see H. S. Muench, *JAS*, 25 (1968), 1108. On day-to-day variations in stratospheric winds in Australia, see J. E. Laby and E. L. Unthank, *JAM*, 24 (1967), 98. On contour and temperature patterns at 100 mb in the northern hemisphere, see R. A. Ebdon, *W*, 20 (1965), 91. On correlations between 50-mb circulation change and surface weather in Britain, see R. A. Ebdon, *W*, 21 (1966), 259. On winter disturbances in the stratosphere, see B. H. Williams, *MWR*, 96 (1968), 549. On stratospheric disturbances during a sudden warming, see J. E. Morris and B. T. Miers, *JGR*, 69 (1964), 201, and on the tropospheric circulation during midwinter stratospheric warmings, see P. R. Julian, *JGR*, 70 (1965), 757. On the dynamic climatology of the stratospheric circulation, see: J. S. Sawyer, *QJRMS*, 91 (1965), 407, on the lower stratosphere; F. K. Hare, *QJRMS*, 91 (1965), 417, for north of 40°N; and H. S. Muench, *JAS*, 22 (1965), 349, on the northern hemisphere winter. On the climatology of the quasi-biennial oscillation, see: G. B. Tucker and J. M. Hopwood, *JAS*, 25 (1968), 293, for the southern hemisphere; R. J. Reed, *JAS*, 22 (1965), 331, for Ascension Island; G. M. Shah and W. L. Godson, *JAS*, 23 (1966), 786, for harmonic analysis; J. K. Angell and J. Korshover, *JGR*, 70 (1965), 3851, for

height variation; A. D. Belmont and D. G. Dartt, *MWR*, 96 (1968), 767, for longitude variation; J. K. Angell, J. Korshover, and T. H. Carpenter, *MWR*, 94 (1966), 319, for its period; and J. M. Wallace and R. E. Newell, *QJRMS*, 92 (1966), 481, for momentum and heat fluxes; see R. S. Lindzen and J. R. Holton, *JAS*, 25 (1968), 1095, and J. M. Wallace and J. R. Holton, *JAS*, 25 (1968), 280, for theories of the oscillation. See A. D. Belmont and D. G. Dartt, *T*, 18 (1966), 381, for the climatology of the resultant wind arising from the interaction of the quasibiennial and the annual waves. On evidence for a *tropospheric* biennial oscillation, see J. K. Angell and J. Korshover, *MWR*, 96 (1968), 778, and P. W. Wright, *W*, 23 (1968), 50, on the Indian Ocean area; also see A. J. Miller, H. M. Woolf, and S. Teweles, *JAS*, 24 (1967), 298, for evidence that both stratosphere and troposphere in middle and high latitudes show a biennial oscillation in angular momentum transport. For details of a semiannual wind variation in the equatorial stratosphere (easterlies in summer and winter, westerlies at the equinoxes), see R. S. Quiroz and A. J. Miller, *MWR*, 95 (1967), 635. On periodic features of tropical zonal motion at 50 mb, see D. G. Dartt and A. D. Belmont, *JGR*, 69 (1964), 2887.

99. For a general discussion of water balance and the atmospheric general circulation, see R. C. Sutcliffe, *QJRMS*, 82 (1956), 385. On the climatology of atmospheric water balance, see: J. Adem, *T*, 20 (1968), 621, on mean water budget in the northern hemisphere; L. A. Vuorela and I. Tuominen, *GPA*, 57, no. 1 (1964), 167, on northern hemisphere moisture flux in relation to mean zonal and meridional circulations; E. M. Rasmusson, *MWR*, 95 (1967), 403, and *MWR*, 96 (1968), 720, on North America; and G. B. Tucker, *JGR*, 67 (1962), 3129, on the convergence/divergence of horizontal water-vapor flux over the North Atlantic. On drought in the northeastern United States in relation to the general circulation, see J. Namias, *MWR* 94 (1966), 543, and 95 (1967), 497.

100. See, in particular, the various publications containing water-balance data of the Laboratory of Climatology, C. W. Thornthwaite Associates, Centerton, N.J.

101. Quoted in *MM*, 85 (1956), 153. R. C. Sutcliffe, *QJRMS*, 75 (1949), 417.

102. R. C. Sutcliffe, *QJRMS*, 82 (1956), 385.

103. W. J. Gibbs, *QJRMS*, 79 (1953), 121.

104. H. van Loon, *AM*, p. 274.

105. U. Radok and A. M. Grant, *JM*, 14 (1957), 141.

106. M. J. Rubin and H. van Loon, *JM*, 11 (1954), 68. S. T. A. Mirrlees, *MM*, 83 (1954), 33.

107. E. Palmén, *QJRMS*, 81 (1955), 459. C. H. B. Priestley and A. J. Troup, *CTP*, no. 1 (1954).

108. Y. P. Rao, *QJRMS*, 90 (1964), 190, and 86 (1960), 156.

109. J. Namias, *MWR*, 91 (1963), 482; see also *MWR*, 92 (1964), 427.

110. Y. P. Rao, *op. cit.* (1960).

111. L. F. Hubert, *JM*, 6 (1949), 216.

112. G. B. Tucker, *QJRMS*, 91 (1965), 140. For northern hemisphere zonal winds, see G. R. R. Benwell, *MM*, 95 (1966) 33, on harmonic analysis. On half-yearly oscillations in zonal flow in the southern hemisphere tropics, see H. van Loon and R. L. Jenne, *JAS*, 26 (1969), 218. On the southern hemisphere circulation during the IGY, see H. van Loon, *JAM*, 4 (1965), 479, and 6 (1967), 803. On zonal winds in the southern hemisphere stratosphere, see J. S. Theon and J. J. Horvath, *JGR*, 73 (1968), 4475, and for a cross section of southern hemisphere zonal winds, see J. G. Sparrow, *JAM*, 4 (1965), 635. On cyclones, anticyclones, and cyclogenesis in the southern hemisphere, see J. J. Taljaard and H. van Loon, *NOTOS*, 11 (1962), 3, and 12 (1963), 37, and J. J. Taljaard, *NOTOS*, 13 (1964), 31, and 14 (1965), 73; see also, for south of 30°S, excluding the Pacific sector, N. A. Streten, *MWR*, 97 (1969), 193. For zonal winds in the free atmosphere over South America (10°N–80°S), see W. Schwerdtfeger and D. W. Martin, *JAM*, 3 (1964), 726. On the interhemispheric circulation in the lower troposphere over the Indian Ocean, see J. Findlater, *QJRMS*, 95 (1969), 400.

113. For general reviews of this topic, see H. H. Lamb, *MM*, 87 (1958), 364, and "M. O. discussion," *MM*, 86 (1957), 130. R. M. Holcombe, *PAS*, p. 9.

114. Lamb, *op. cit.* (1958).

115. F. K. Hare, *W*, 17 (1962), 256.

116. W. Ferrel, *Nashville J. Med. Surg.*, 11 (1856), 287 and 375.

117. For general accounts, see: H. G. Dorsey, Jr., *CM*, p. 942; A. D. Belmont, *AGP*, 7 (1961), 249; and F. K. Hare, *QJRMS*, 94 (1968), 439.

118. J. Namias, *PAS*, p. 45. H. Flohn and G.Seidel, *PAS*, p. 62.

119. See R. Fjørtoft, *PAS*, p. 87, for details.

120. For maps, see H. G. Heastie, *MM*, 85 (1956), 368; 86 (1957), 198; and 87 (1958), 301.

121. F. K. Hare, *JM*, 17 (1960), 36; see also P. R. Julian, *JM*, 18 (1961), 119.

122. B. W. Boville, C. V. Wilson, and F. K. Hare, *JM*, 18 (1961), 567.

123. E. P. McClain, *JAM*, 1 (1962), 107.

124. B. W. Boville, *JM*, 17 (1960), 329.

125. C. V. Wilson and W. L. Godson, *QJRMS*, 89 (1963), 205.

126. F. K. Hare, *QJRMS*, 86 (1960), 127.

127. W. H. Hobbs, *JM*, 2 (1945), 143. See also F. K. Hare, *CM*, p. 952, for a general review of climatological problems of the arctic and subarctic.

128. H. G. Dorsey, *JM*, 2 (1945), 135.

129. H. H. Lamb, *MM*, 81 (1952), 33 and 282.

130. "M. O. discussion," *MM*, 86 (1957), 130.

131. T. J. Keegan, *JM*, 15 (1958), 513.

132. R. J. Reed and B. A. Kunkel, *JM*, 17 (1960), 489.

133. R. Scherhag, *PAS*, p. 101.

134. For general reviews of the subject, see G. P. Brittan and H. H. Lamb, *W*, 11 (1956), 281 and 339; see also "M. O. discussion," *op. cit.* (1957). On atmospheric pressure and circulation variations in the Antarctic, see F. Loewe, *QJRMS*, 93 (1967), 373. For the climatology of Antarctic stratospheric winds, see R. C. Taylor, *NOTOS*, 10 (1961), 3, and for details of severe winter weather in South Africa due to an outbreak of Antarctic air, see D. Stranz and J. J. Talgaard, *NOTOS*, 14 (1965), 17.

135. H. H. Lamb, *MOGM*, no. 94 (1956).

136. G. C. Simpson, *British Antarctic Expedition, 1910–1913*, vol. XI (1919).

137. For a general account, see A. Court, *CM*, p. 917.

138. W. J. Gibbs, *AM*, p. 84. F. A. Berson and U. Radok, *AGP*, 9 (1962), 193.

139. Simpson, *op. cit.* (1919).

140. J. C. Langford, *AGP*, 9 (1962), 256. H. M. Treloar, *AM*, p. 176. A. K. Hannay, *AM*, p. 153.

141. E. B. Kraus, *AM*, p. 145.

142. N. J. Ropar and T. I. Gray, *MWR*, 89 (1961), 45.

143. W. B. Moreland, *AGP*, 9 (1962), 394.

144. A. Court, *BAMS*, 23 (1942), 220.

145. For discussions of tropical circulation problems, see K. R. Ramanathan, *SPIAM*, p. 317, and A. G. Forsdyke, *TMA*, p. 1. For long-term mean values of the general circulation of the tropics, see J. W. Kidson, D. G. Vincent, and R. E. Newell, *QJRMS*, 95 (1969), 258. For details of climatological problems of tropical oceanic areas, see: S. L. Hasthenrath, *MWR*, 20 (1968), 163, on the Caribbean and Gulf of Mexico; C. A. van den Berg, *W*, 23 (1968), 462, on the I.T.C.Z. and weather in the Caribbean; C. S. Ramage, *JAS*, 23 (1966), 144, on the Arabian Sea, which is covered in summer by the Earth's most extensive heat low; D. E. Pedgley, *W*, 21 (1966), 350 and 394, on the Red Sea convergence zone; A. H. Gordon, *JAS*, 23 (1966), 712, on Aden and Bahrein; C. S. Ramage, *W*, 23 (1968), 28, on the Indian Ocean; and B. Ramsey, *MM*, 95 (1966), 47, on Nairobi, a high-level equatorial station on the fringe of the Indian Ocean monsoon system. According to C. S. Ramage, *MWR*, 96 (1968), 365, Indonesia and the Caroline islands constitute a "maritime continent" that generates a considerable amount of heat for export. According to J. G. Charney, *JAS*, 26 (1969), 182, synoptic-scale motions in the lower tropical troposphere are nearly uncoupled from those of the upper tropical troposphere except when joined through deep cumulus convection. On tropical circulation features, see: H. Riehl, *JAM*, 4 (1965), 149, on easterly waves; W. H. Quinn and W. V. Burt, *JAM*, 6 (1967), 988, on equatorial troughs at Canton Island; and F. A. Godshall, *MWR*, 96 (1968), 172, on the Pacific I.T.C.Z. On the climatology of tropical disturbances, see: W. M. Gray, *MWR*, 96 (1968), 669, on their origin; J. G. Lockwood, *W*, 20 (1965), 279, for equatorial ones; and F. E. Lumb, *MM*, 95 (1966), 150, on rainy periods on the Kenya coast. For characteristic circulation types, see J. Sissons, *W*, 21 (1966), 228, 298, and 319, on East Africa, and J. G. Lockwood, *W*, 21 (1966), 325, on Southeast Asia. For case studies, see: J. Simpson, M. Garstang, E. J. Zipser, and G. A. Dean, *JAM*, 6 (1967), 237, on a nondeepening disturbance in the West Indies; C. O. Erickson, *MWR*, 95 (1967), 121, on a hurricane with extratropical and tropical features; R. W. Fett, *MWR*, 96 (1968), 106, on a typhoon forming and intensifying within the I.T.C.Z.; and R. H. Simpson *et al.*, *MWR*, 96 (1968), 251, on satellite observations of Atlantic tropical disturbances in 1967. See J. M. Wallace and V. E. Kousky, *JAS*, 25 (1968), 900, for empirical (synoptic) evidence of Kelvin waves in the equatorial stratosphere. For an "inverted V" cloud pattern found from satellite photographs to be characteristic of easterly waves in the tropical Atlantic, see N. L. Frank, *MWR*, 97 (1969), 130. For observations of the trade-wind inversion on the slopes of Mauna Loa, Hawaii, see B. G. Mendonca and W. T. Iwaoka, *JAM*, 8 (1969), 213.

146. W. H. Portig, *BAMS*, 44 (1963), 79; see also W. H. Portig, *BAMS*, 45 (1964), 104.

147. For a general review, see C. E. Palmer, *CM*, p. 859, and *QJRMS*, 78 (1952), 126. For details and references concerning the climatological school, see *CM*, pp. 860–63.

148. J. G. Lockwood, *MM*, 92 (1963), 75.

149. For details and references concerning the airmass school, see *CM*, pp. 863–68.

150. The perturbation school originated with G. E. Dunn, *BAMS*, 21 (1940), 215. For other references and details, see *CM*, pp. 868–78.

151. H. Riehl and J. S. Malkus, *QJRMS*, 83 (1957), 21.

152. For an account of the trade-wind inversion, see H. Riehl, *Tropical Meteorology* (McGraw-Hill, 1954), pp. 53–70. Piazzi-Smyth, *Trans. Roy. Soc. London*, 148, series A (1858), 465.

153. H. Flohn, *SPIAM*, p. 431.

154. For details, see H. Flohn, *TMA*, p. 244.

155. C. E. Palmer, *CM*, p. 859. C. L. Jordan, *JM*, 9 (1952), 376. R. D. Fletcher, *JM*, 2 (1945), 167.

156. H. Flohn, *TMA*, p. 253.

157. C. S. Ramage, *JM*, 9 (1952), 403.

158. A. G. Forsdyke, *TMA*, p. 14, and N. E. La Seur, *TMA*, p. 47, give a detailed review of these models.

159. H. Riehl, *JM*, 5 (1948), 247. See also B. W. Thompson, *QJRMS*, 77 (1951), 569.

160. H. Regula, *Ann. Hydrogr.*, 64 (1936), 107. G. E. Dunn, *BAMS*, 21 (1940), 215. H. Riehl, *Waves*

in the Easterlies and the Polar Front in the Tropics (Chicago, 1945). See also E. S. Merritt, *JAM*, 3 (1964), 367) for a reappraisal of easterly waves in the light of TIROS photographs.

161. C. S. Ramage, *JM*, 12 (1955), 252. J. C. Freeman, Jr., *JM*, 5 (1948), 138.

162. C. E. Palmer, *QJRMS*, 78 (1952), 126.

163. D. H. Johnson, *SP*, 51 (1963), 587.

164. C. S. Ramage, *op. cit.* (1955).

165. G. B. Tucker, *QJRMS*, 91 (1965), 140.

166. H. Flohn, *SPIAM*, p. 431.

167. O. R. Wulf and L. Davis, Jr., *JM*, 9 (1952), 79.

168. E. N. Lorenz, *SPIAM*, p. 457.

169. For a simple introduction to the application of hydrodynamic theory to the problem of the general circulation, see C.-G. Rossby, *ASM*, p. 18. G. Hadley, *Phil. Trans. Roy. Soc.*, 29 (1735), 58.

170. V. Bjerknes, *et al.*, *Physikalische Hydrodynamik* (Berlin, 1933). F. Exner, *Dynamische Meteorologie* (Vienna, 1925). H. Jeffreys, *QJRMS*, 52 (1926), 85. W. Widger, *JM*, 6 (1949), 291. Y. Mintz, *BAMS*, 35 (1954), 208.

171. E. T. Eady, *CPRMS*, p. 156.

172. S. Petterssen, *CPRMS*, p. 120, and *WAF*, I, 257–66.

173. N. A. Phillips, *QJRMS*, 82 (1956), 123.

174. J. Smagorinsky, *MWR*, 91 (1963), 99.

175. J. Smagorinsky, S. Manabe, and J. L. Holloway, Jr., *MWR*, 93 (1965), 727.

176. S. Manabe, J. Smagorinsky, and R. F. Strickler, *MWR*, 93 (1965), 769.

177. S. Manabe and J. Smagorinsky, *MWR*, 95 (1967), 155.

178. J. Adem, *T*, 14 (1962), 102. For a general review of theories of the general circulation, see O. G. Sutton, *The Observatory*, 80 (1961), 169, and for a review of E. N. Lorenz's important survey for the WMO of the problem of the general circulation, see R. V. Garciá, *BWMO*, 17 (1968), 116. For a mathematical theory explaining the maintenance of atmospheric zonal flows in terms of planetary waves propagated from some other region of the atmosphere, see R. E. Dickinson, *JAS*, 26 (1969), 73. For explanations by means of mathematical models, see: D. J. Davies and D. R. Davies, *QJRMS*, 95 (1969), 148, for a computationally economical model; Y. Kurihara and J. L. Holloway, Jr., *MWR*, 95 (1967), 509, for a 9-level primitive equation model for global use; E. B. Kraus and E. N. Lorenz, *JAS*, 23 (1966), 3, for one with large-scale seasonal forcing due to continental/oceanic effects; S. Manabe and B. G. Hunt, *MWR*, 96 (1968), 477 and 503, for the stratosphere; and A. Kasahara, *JAS*, 23 (1966), 259, for one incorporating large-scale orographic influences; for mathematical explanations of zonal wind patterns, see R. E. Dickson, *JAS*, 25 (1968), 269, and 26 (1969), 73.

179. For general reviews of the subject, see: G. Manley, *QJRMS*, 79 (1953), 185; H. H. Lamb, *W*, 14 (1959), 299; the symposium *World Climate from 8000 to 0 B.C.* (Roy. Meteorol. Soc., 1966). H. C. Willett, *JM*, 6 (1949), 34.

180. R. G. Veryard, *USCC*, p. 97. B. L. Dzerdzeevskii, *USCC*, p. 285.

181. R. Fay, *MWR*, 86 (1958), 467.

182. G. I. Roden, *JAS*, 19 (1962), 66.

183. For monthly resumés of weather events in the United States, January to June 1967, in relation to the general circulation of the northern hemisphere, see: W. R. Winkler, *MWR*, 95 (1967), 227; J. W. Posey, *loc. cit.*, p. 311; J. F. Andrews, *loc. cit.*, p. 383; R. A. Green, *loc. cit.*, p. 491; L. P. Stark, *loc. cit.*, p. 587; and A. J. Wagner, *loc. cit.*, p. 650. For an early study, see H. C. Willett, *PPOM*, vol. 9, no. 1 (1941). For later studies making use of five-day means, see: W. H. Klein, B. M. Lewis, and J. Enger, *JM*, 16 (1959), 672; A. I. Cooperman and H. S. Rosendal, *MWR*, 91 (1963), 337. R. L. Haney, *MWR*, 89 (1961), 391, analyzed the harmonics of five-day mean 500-mb charts. For monthly mean charts based on CLIMAT data, see H. H. Lamb and A. I. Johnson, *W*, 15 (1960), 83; for charts based on CLIMAT/TEMP data, see R. A. Ebdon, *W*, 20 (1965), 91. For examples of new correlations, see: J. Namias: *JM*, 8 (1951), 251, on the great Pacific winter anticyclone of winter 1949–50; *JM*, 14 (1957), 235, on Scandinavian seasonal anomalies; and *AMM*, vol. 2, no. 6 (1953).

184. E. W. Wahl, *JM*, 10 (1953), 42, E. J. Rebman, *BAMS*, 35 (1954), 498.

185. W. H. Portig, *JAS*, 21 (1964), 462.

186. R. A. Bryson and W. P. Lowry, *BAMS*, 36 (1955), 329.

187. E. W. Wahl, *BAMS*, 35 (1954), 351, and 33 (1952), 380.

188. R. C. Sutcliffe, *USCC*, p. 277.

189. J. Namias, *USCC*, p. 345, gives full details.

190. For classic reviews of climatic-change theories, see: C. E. P. Brooks, *Climate Through the Ages* (Benn, 2nd ed., 1948); H. Shapley, ed. *Climatic Change* (Harvard, 1953). On atmospheric energy-balance equations, see H. Flohn, *TMA*, p. 270, and *USCC*, p. 339.

191. G. C. Simpson, *QJRMS*, 83 (1957), 459. A. H. Glaser, *JM*, 14 (1957), 375. C. E. Palmer, *JM*, 10 (1953), 1. W. L. Godson, *USCC*, p. 323.

192. J. Smagorinsky, *QJRMS*, 79 (1953), 342, and *TAGU*, 41 (1960), 590.

193. For details of a worldwide experiment to demonstrate this unpredictability, see *BAMS*, 47 (1966), 200.

194. *Ibid.* On possible fluctuations in the atmospheric circulation, see J. S. Sawyer, in the Roy. Meteorol. Soc. symposium, *World Climate from 8000 to 0 B.C.* (London, 1966), p. 218, and for known fluctuations, see H. H. Lamb, R. P. W. Lewis, and A. Woodroffe, *loc. cit.*, p. 174, also H. H. Lamb,

ANYAS, 95 (1961), 124. For climatic fluctuations and the general circulation as a problem in dynamic cliamtology, see B. L. Dzerdzeevskii, *T*, 14 (1962), 382, and 18 (1966), 751. For speculations on climatic change, see R. A. Bryson, *WW*, 21 (1968), 56, on causes, and D. M. Ludlum, *WW*, 21 (1968), 62, on early and late snowfall evidence. On singularities, see: J. Namias, *T*, 18 (1966), 731, for a weekly periodicity in precipitation in the eastern United States; A. S. Dennis, *QJRMS*, 93 (1967), 522, for weekly periodicities in radiosonde pressure/temperature data for Rapid City, South Dakota; and G. W. Brier, R. Shapiro, and N. J. MacDonald, *JAM*, 3 (1964), 53, for a periodicity of 18 cycles per year in precipitation data for the United States. According to G. Nicholson, *W*, 20 (1965), 322, Thursday is the wettest day of the week at Teddington, England; but according to J. Maybank and M. M. Qureshi, *JAS*, 23 (1966), 13, rainfall singularities in the Canadian prairies result from day-to-day inhomogeneities in the long-term average pressure pattern. For effects of volcanic dust on the stratosphere, see R. A. Ebdon, *W*, 22 (1967), 245, and for the global spread of dust from the Bali eruption of March 17, 1963, see A. J. Dyer and B. B. Hicks, *QJRMS*, 94 (1968), 545. On the climatology of ocean-atmosphere interaction, see: J. Namias, *MWR*, 97 (1969), 173, on seasonal interactions between the North Pacific Ocean and the atmosphere; J. Bjerknes, *MWR*, 97 (1969), 163, on the response of westerlies over the northeastern Pacific to anomalies in equatorial sea temperature, in relation to the Southern Oscillation; E. B. Kraus and R. E. Morrison, *QJRMS*, 92 (1966), 114, for local ones in the North Atlantic; J. Namias, *JGR*, 70 (1965), 2307, for winds and mean monthly sea-surface temperatures; and N. Marshall, *W*, 23 (1968), 368, on sea-ice synoptics. On long-range forecasting as a circulation problem, see: J. Namias, *BAMS*, 49 (1968), 438, and *MWR*, 92 (1964), 449, on United States seasonal outlooks based on 700-mb circulation anomalies; also J. M. Craddock, *MM*, 93 (1964), 88, and M. H. Freeman, *W*, 22 (1967), 72, on British Meteorological Office procedure and results. On the use of 5-day mean circulation patterns, see J. F. O'Connor, *MWR*, 92 (1964), 303, on 700-mb northern hemisphere patterns. See R. A. S. Ratcliffe, *MM*, 97 (1968), 258, for the use of monthly mean 500-mb charts in forecasting the rainfall of the following month over England and Wales. For a review of the relation between the oceanic circulation and climatic fluctuation, see S. J. Rasool and J. S. Hogan, *BAMS*, 50 (1969), 130.

CHAPTER 5

Scientific Inference in Climatology

In physical and dynamic climatology, "explanation" involves demonstrating how the observed behavior of the atmosphere either is a necessary consequence of some of its physical properties, or follows a sequence of developments that conform to mathematical principles. Either interpretation is, of course, in terms of the "laws" of nature, and constitutes scientific inference. The first interpretation is that of the physical meteorologist, the second that of the dynamic meteorologist, but they do overlap.

The classic literature on the dynamic and physical analysis of meteorological phenomena is that by Napier Shaw and Vilhelm Bjerknes. Shaw's work is best illustrated by his four-volume *Manual of Meteorology*, the standard source for developments in both climatology and meteorology up to about 1939. Under the sponsorship of the Carnegie Institute of Washington, Bjerknes worked for over forty years on the application of physical hydrodynamics to meteorological problems. The Polar Front theory was only one of the many achievements of his researches. At various stages in this work, Bjerknes summarized his results in volumes that have become meteorological classics. In 1948, with the help of Norwegian, Swedish, and American meteorologists, his son, J. Bjerknes, surveyed the whole field of physical and dynamic meteorology in a comprehensive work that is now a standard reference. Unfortunately, the book was not published until 1957, and omits many topics of crucial importance today, but it is nevertheless the most convenient modern, authoritative account of both the synoptic and the theoretical sides of meteorology and climatology by some of their original founders.[1]

In the introduction to the latter volume, it is emphasized that the best way to study nature scientifically is to simplify the complex phenomena we observe therein by the use of mental pictures or "images." Atmospheric phenomena are so complicated, however, that it is rarely possible to arrive at models that can both be treated mathematically and yield a correct picture of the conditions as observed. Therefore two different kinds of image must be used: dynamic models and synoptic models. Dynamic models can be interpreted mathematically, since they are derived from the fundamental equations of motion by successive approximation. Synoptic models, on the other hand, are obtained by simplification of the complex motions of the atmosphere as actually observed and represented on weather maps. Many geography students become acquainted only with these latter models, and so are unaware of at least half the work of Bjerknes and his collaborators. Ultimately, the two types of model will approach each other, producing dynamic-synoptic models. The models derived from the Sutcliffe development theory provide possibly the best examples of this trend.

Empirical explanation in climatology nearly always means arguing in a circle, rarely leads to possible predictions, and is not intellectually satisfying. The alternative to empirical description, which is to follow the procedure outlined in the previous paragraph, would involve a considerable portion of the field of theoretical meteorology, but since my intention is to address readers interested in the practical analysis of weather and climate, rather than the purely mathematical aspects, I will only discuss those topics that have a definite practical value. For a more rigorous treatment of the whole field of theoretical meteorology, see the standard meteorological textbooks.[2] In this chapter and the following one will be found derivations of most of the major theoretical concepts and techniques used in everyday weather analysis.

Four topics in theoretical meteorology provide the "first principles" that are a necessary preliminary to the scientific study of climate. These are the thermodynamics, statics, dynamics, and kinematics of the atmosphere. Thermodynamics is the study of the variations in the temperature, pressure, and density of gases and other bodies that are, on the whole, at rest, caused by variations in the supply and extraction of heat. The thermodynamic conditions for each particle of gas or other substance are assumed to be unaffected by adjoining particles. The classic model is provided by a gas in a cylinder whose volume can be altered by moving a piston.[3] In the study of the statics of the atmosphere, it is not assumed that the air particles are isolated from each other, or that changes of state for given particles take place independently of one another. Statics assumes that pressure, temperature, density, and humidity vary continuously in space, and then investigates the relation between the space distributions of these variables, i.e., their atmospheric fields. The atmosphere is assumed to be in a state of relative equilibrium, following the motion of the Earth. Meteorological dynamics studies atmospheric motions by means of the purely mathematical concepts of classical hydrodynamics, and meteorological kinematics is the study of these motions independently of their thermodynamic and dynamic causes and consequences. Before we discuss these topics in detail, we must undertake some preliminary consideration of units, and how they are related by the theory of dimensions.

The basic units employed in theoretical climatology are usually those in the centimeter-gram-second (c.g.s.) system, although for larger quantities the meter-ton-second (m.t.s.) system is more convenient. The c.g.s. temperature scale, with freezing point at 0°C and boiling point at 100°C, is now termed the Celsius scale instead of the Centigrade scale, although, strictly speaking, the original Celsius scale had 100°C as the freezing point and 0°C as the boiling point (see Appendix 5.1).

The fundamental physical dimensions are length L, mass M, time T, and temperature θ. Any physical quantity may be represented by algebraic combinations of these dimensions. Thus area is length multiplied by length or L^2, volume is L^3, and density is mass divided by volume or M/L^3, usually written ML^{-3}. Specific volume, i.e., the volume occupied by unit mass, is L^3M^{-1}, velocity is LT^{-1}, acceleration is LT^{-2}, and angular velocity is T^{-1}. The dimensions of derived quantities used in theoretical climatology are MLT^{-2} for momentum. T^{-1} for divergence or vorticity, and $L^2T^{-2}\,\theta^{-1}$ for specific heat capacity.

The procedure for studying physical properties and relationships in terms of these fundamental dimensions is set forth in the study known as *dimensional analysis*,[4] which is particularly useful for verifying the form of equations and for determining the correct units to be assigned to derived quantities or parameters. It is not permissible —i.e., physically logical—to add together or otherwise combine two quantities of different dimensions, if one expects the result to represent the true dimension of some physical phenomenon. A climatic index based on temperature and precipitation, which have dimensions θ and L, respectively, cannot delimit a real feature of nature, although it may provide a useful convention.

Elementary Thermodynamics of the Atmosphere

Atmospheric thermodynamics is concerned basically with three concepts: first, the equation of state, which describes the relationship between the physical variables defining the state of the atmosphere at a given time and place; second, the first law of thermodynamics, an empirical principle concerning the conservation of energy; third, the second law of thermodynamics, a concept concerned with the degradation of energy that specifies the direction in which heat may flow during thermodynamic processes.

The study of the thermodynamics of the atmosphere produces graphs or thermodynamic diagrams that are absolutely vital in the study of both local and regional weather developments. Accordingly, the following account deals specifically with the derivation of these diagrams and their uses.[5]

It is necessary first of all to define the hypothetical atmosphere with which we are going to deal. The model atmosphere normally used is one in which the density is uniform, so that the surfaces of constant pressure (the isobaric surfaces) and the surfaces of constant density (the isopycnic surfaces) within it are coincident. Such an atmosphere is *barotropic*. An atmosphere in which the isobaric and isopycnic surfaces are at an angle to one another is a *baroclinic* atmosphere. In actuality the greater part of the atmosphere is usually baroclinic; a barotropic atmosphere would eventually come to rest because it could not generate sufficient kinetic energy, and friction with the ground surface would soon use up whatever energy it originally possessed. However, the atmosphere above large areas of the tropical lands is approximately barotropic; and in the polar-night vortex of the Arctic winter, a warm, barotropic, anticyclonic airflow exists above a cold low near the pole, around which move baroclinic waves.[6]

Other initial assumptions are that (a) atmospheric air is homogeneous and compressible, and (b) we can regard the atmosphere as consisting of a hypothetical gas known as "clean dry air" plus water vapor and various impurities, such as smoke particles. We assume that we can discuss this hypothetical gas separately. Actually, although nitrogen and oxygen make up 99 per cent of the atmosphere, they are quite passive in most meteorological processes. The water-vapor content is much more important, but only accounts for at most 4 per cent of the volume of a sample of moist atmospheric air. Having defined the "atmosphere" we are going to deal with, we shall define our basic concepts and laws. First, pressure, defined as force per unit area, and density, defined as mass divided by volume, are self-evident quantities. Second, three laws, those of Boyle, Charles, and Avogadro, which can of course be simply verified in a school laboratory, are combined into the equation of state for our hypothetical gas.[7] This equation may be written as either

$$p = \rho \frac{RT}{m} \quad \text{or} \quad pV = \frac{RT}{m}, \qquad [5.1]$$

where p and V are the pressure and specific volume of the gas, ρ and T are its density and absolute temperature, R (the universal gas constant) is 8.315×10^7 ergs per °C, and m (the gram-molecular weight of dry air) is 28.9; p is in dynes per square centimeter, V in cubic centimeters per gram, T in degrees absolute, and m in grams.

The equation of state refers to clean dry air, and in order to produce an equation applicable to the real atmosphere, the effect of water vapor must be included. This is done by using Dalton's Law and introducing a fictitious temperature, the *virtual temperature* T', which varies with the amount of water vapor present in the air. In effect, the virtual temperature is the temperature of dry air having the same total pressure and density as the moist air. The equation of state for moist air is then

$$P = \rho \frac{RT'}{m} \quad \text{or} \quad PV = \frac{RT'}{m}, \qquad [5.2]$$

where P is the total pressure, i.e., $p + e$, where e is the pressure exerted by the water vapor in the atmosphere. T and T' do not differ by more than 2 or 3°C, but the difference must be taken into account in all accurate calculations. (See Appendix 5.2.)

The energy changes taking place in the atmosphere must now be considered. According to the kinetic theory of gases, if pressure remains constant, then heating a gas will increase its kinetic energy, and if the gas is compressed but no heat is added, then its kinetic energy will be increased and its temperature must rise. It follows from the first law of thermodynamics that an increase in the internal energy of a gas can be brought about by the addition of heat to it, by performing work on it, or by both. In other words, the change in internal energy equals the heat added plus the work done on the gas, or, for unit mass,

$$dq = C_p \, dT - \frac{RT}{m} \frac{dp}{p}, \qquad [5.3]$$

where dq is the heat added to unit mass, dp is the change in pressure, dT is the change in absolute temperature, and C_p is the specific heat of the gas at constant pressure. (See Appendix 5.3.)

If we specify that the change in the internal energy of the gas be due entirely to work done on the gas (i.e., compression) or work done by the gas (i.e., expansion), then the

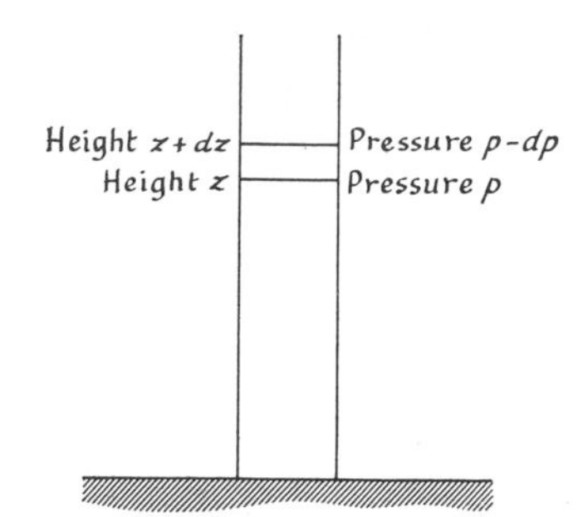

FIGURE 5.1.
Pressure difference in the vertical.

processes involved are adiabatic (see Appendix 5.4). By adopting the adiabatic model of the atmosphere, i.e., by assuming that a small mass or "parcel" of air moving through the atmosphere neither gains heat from nor loses heat to the surrounding air, we can calculate exactly how the temperature of this parcel of air will vary as it moves vertically. To do this, it is necessary to consider the variation of pressure and temperature with height above the Earth's surface.

Consider a vertical air column of unit cross section, and let the pressure difference between a height z and a height $z + dz$ be dq (see Figure 5.1). This pressure difference will equal the weight of the air column of height dz and unit cross section, and if dz is sufficiently small then neither the density, ρ, of the air nor the acceleration due to gravity, g, will vary with height through it. Hence

$$dp = -g\rho dz, \qquad [5.4]$$

where the minus sign signifies that pressure decreases upward. This is the *hydrostatic equation*, one of the most fundamental relationships in meteorology, and describes how pressure varies with height (see Appendix 5.5).

Since we specify that the motion of the air must be adiabatic, then dq in Equation 5.3 will equal zero. Hence if we write

$$\frac{C_p}{C_v} = \gamma,$$

where C_v is the specific heat of the gas at constant volume, then

$$\gamma\frac{dT}{T} - (\gamma - 1)\frac{dp}{p} = 0. \qquad [5.5]$$

For dry air, γ is approximately equal to 1.41, and therefore

$$T = kp^{0.286},$$

where k is constant. This expression, *Poisson's equation*, describes how temperature must change if the pressure of the gas parcel is varied. Then by making use of the hydrostatic equation, and writing T^* to represent the temperature of the parcel of air, as distinct from T, the temperature of the surrounding atmosphere (i.e., the environment), it can be shown that

$$\frac{dT^*}{dz} = \frac{-g}{R}\left(\frac{\gamma - 1}{\gamma}\right)\frac{T^*}{T}. \qquad [5.6]$$

This equation describes how the temperature of a parcel of air moving vertically varies with height. If we take a case in which the temperature of the air parcel differs infinitesimally from the temperature of the surroundings, i.e., $T^* = T$, then we can substitute known values for g, R, and γ, and find that $\frac{dT}{dz} \approx -9.86 \times 10^{-5}$ °C per cm = −1°C per 100 meters, or 5.4°F per 1,000 feet. This is the *dry adiabatic lapse-rate*, which means that, if the adiabatic model can be applied to the real atmosphere, then a mass of rising dry air must lose heat at the rate of 5.4°F every 1,000 feet, and a mass of descending dry air must gain heat at the same rate. Observation tells us that the adiabatic model does apply in nature, on most occasions, except for large cumulus clouds, where air from the environment is entrained into the rising air currents and so heat is brought in (entrained) from outside the parcel.[8]

Adiabatic processes of saturated air are more difficult to deal with than those of dry or moist air, because latent-heat release or absorption is involved. Although water vapor obeys the same laws as other gases, it both solidifies and liquifies within the normal range of temperatures found in the atmosphere. When water changes phase, the total heat content of the system is changed (see Appendix 5.6).

Only a certain amount of water vapor can exist in a given volume of air; the exact amount depends on the air temperature. If water is introduced into a volume of air, it will first of all evaporate and form water vapor, which will exert a pressure, known as the vapor pressure e. After a time the air volume will contain the maximum amount of vapor possible at its current temperature: its vapor is then exerting the saturation vapor pressure, e_s. Any further amounts of water injected into it will not evaporate, but will remain in liquid form, and the vapor pressure will remain the same, at its maximum value, e_s. The saturation vapor pressure is thus the pressure of water vapor at which water and water vapor can exist side by side. It depends only on temperature, increasing as temperature increases. The exact relationship is given by the Clausius-Clapeyron equation (see Appendix 5.7).

Two types of adiabatic expansion of saturated air are defined. In a *moist adiabatic* or *saturated adiabatic expansion*, the condensation products are retained in the air. If the condensation products are regarded as falling out of the atmosphere immediately after formation, the expansion is *pseudoadiabatic*, since the products carry heat away from the system (see Appendix 5.8).

A simplified expression describing the adiabatic process for saturated air, analogous to Equation 5.3 for dry air, is given by

$$-Ldr_s = C_{pd}\,dT - R_d T\frac{dp}{p}, \qquad [5.7]$$

which describes the change undergone by a parcel of $1 + r_s$ tons of saturated air of original state (pressure p, temperature T, saturation mixing ratio t_s) expanded adiabatically to state $p + dp$, $T + dT$, $r_s + dr_s$. The assumption is made that the latent heat involved, $-Ldr_s$, is used exclusively to heat the dry air, and hence heating of the water vapor is ignored.[9] In Equation 5.7, C_{pd} and R_d represent the specific heat of dry air at constant pressure, and the gas constant for dry air, respectively.

The expression defining the adiabatic lapse-rate for saturated air, the *saturated adiabatic lapse-rate*, is more complicated than that for dry air. Substitution of known values shows that near the Earth's surface the saturated adiabatic lapse-rate is approximately 2.7°F per 1,000 feet, but increases with height, coming very close to the dry adiabatic value in the upper layers of the atmosphere.[10]

Adiabatic lapse-rates are the basic concepts in practical atmospheric thermodynamics. They were first explicitly described by William Thomson (Lord Kelvin) in 1862.[11] Their main use is in the assessment of local atmospheric stability.

An air parcel is said to be stable, unstable, or neutral (indifferent) with respect to its environment if, on being given an initial impulse in a vertical direction, it returns to its original position, continues its movement in the direction given by the impulse, or remains in its displaced position, respectively. If parcel and atmosphere are in hydrostatic equilibrium, then the vertical acceleration of the former will be zero. If they are not in hydrostatic equilibrium, then the displaced parcel will be accelerated. If the air parcel is lighter than its environment, then this vertical acceleration will be directed upward; if the parcel is heavier than its surroundings, the vertical acceleration will be downward (see Appendix 5.9). As a general rule, the following relationships apply.

If $\gamma_e > \Gamma_d$ or $\gamma_e > \Gamma_s$, the atmosphere is *unstable.*

If $\gamma_e = \Gamma_d$ or $\gamma_e = \Gamma_s$, the atmosphere is *neutral* or indifferent.

If $\gamma_e < \Gamma_d$ or $\gamma_e < \Gamma_s$, the atmosphere is *stable.*

If $\Gamma_s < \gamma_e < \Gamma_d$, the atmosphere is *conditionally unstable*, i.e., stable for dry air but unstable for saturated air.

Here γ_e is the normal rate of decrease of temperature with height in the atmosphere, i.e., the environmental lapse-rate, which averages 3.6°F per 1,000 feet, although actual lapse-rates may be considerably different; Γ_d is the dry adiabatic lapse-rate; and Γ_s is the saturated adiabatic lapse-rate.

If $\gamma_e < \Gamma_d$, the atmosphere is said to be *absolutely stable*; if $\gamma_e > \Gamma_d$, it is *absolutely unstable*; if $\gamma_e = \Gamma_d$, the atmosphere is *dry neutral*, and if $\gamma_e = \Gamma_s$ it is *saturated neutral*, Thus moisture content is a very important factor in stability considerations.

The moisture factor is particularly important for conditional instability. In this condition, the atmosphere is stable below a particular level (the *lifting condensation level*, the level to which unsaturated air has to be lifted in dry adiabatic expansion before condensation results), but if a parcel is forced to rise above this level, it becomes unstable.

Anomalous adiabatic cooling rates have been observed in the lowest 30 feet or so of some clouds: an air parcel may even be *heated* during ascent in this region.[12] Thus the saturated adiabatic lapse-rate does not always come into operation immediately above cloud base, i.e., immediately the lifting condensation level is reached.

In terms of weather, stable air is characterized by clear skies, absence of precipitation, and generally quiet conditions because of lack of vertical movement. Unstable air is associated with rapid cloud development, heavy shower formation, and violent vertical motions. In mountainous regions in particular, transition from stable to unstable weather can take place in a few minutes where orography enables ascending air to rapidly reach the lifting condensation level.

The stability considerations so far outlined refer only to small bubbles or parcels of air. The *parcel method* of stability analysis assumes that an air parcel retains its identity during ascent or descent, and does not mix with its surroundings. It further assumes that the atmosphere is in hydrostatic equilibrium, and that no compensating vertical movements occur in the environment as the parcel moves through it. The basic theory then shows that a parcel displaced vertically in a stable atmosphere will oscillate about its original position, but one displaced in an unstable atmosphere will increase its displacement distance indefinitely.[13] The parcel method usually predicts excessive buoyancy when compared with observed data.

The *slice method*, by contrast, allows for the existence of compensatory vertical movements in the environmental air, as the parcel or column of air rises. It is useful when studying convection, but is otherwise difficult to apply in practice.[14]

If a layer of air of appreciable horizontal and vertical extent is considered instead of a small parcel, then the problem becomes more complex.[15] This layer will have its own lapse-rate, which will become modified during vertical motion. Assuming that both the amount of air in the layer and its horizontal extent remain unchanged during its displacement, then the *difference* in pressure between the top and bottom of the layer will be unaltered. Changes will occur, however, in pressure and temperature within the layer: hence the specific volume and the vertical thickness between its upper and lower boundaries will be modified. If γ_e represents the lapse-rate within the air layer before displacement, and γ'_e its lapse-rate after undergoing ascent or descent, then

(a) if $\gamma_e > \Gamma_d$, (i.e., the layer is *superadiabatic*), then $\gamma_e > \gamma'_e$;

(b) if $\gamma_e = \Gamma_d$, (i.e., the layer is adiabatic), then $\gamma_e = \gamma'_e$;

(c) if $\gamma_e < \Gamma_d$, (i.e., the layer is *subadiabatic*), then $\gamma_e < \gamma'_e$.

In a superadiabatic layer, ascent decreases and descent increases its lapse-rate; in a subadiabatic layer, ascent increases and descent decreases its lapse-rate. Ascent or descent of an adiabatic layer brings about no change in its lapse-rate. In general, if no saturation occurs during lifting, then ascent will destabilize a stable layer and descent will stabilize it still further.

Superadiabatic lapse-rates are very persistent and common in the lowest few inches or feet of the atmosphere immediately adjacent to the Earth's surface. During daytime, the environmental lapse-rate may be several thousand times the dry adiabatic value in this shallow layer, and in many locations the *average* lapse-rate is superadiabatic during the greater part of the year. This layer is laminar, and theory shows that in such a layer one centimeter deep, the bottom could be 6°C warmer than the top without instability developing.[16]

Quite frequently, widespread layers of air of considerable thickness have a vertical stratification of moisture content such that the bottom part of the layer reaches its lifting condensation level before the upper part does so. The lapse-rate of the layer is then considerably steepened, and instability develops with extreme rapidity. Such a layer is said to be *convectively unstable*. Otherwise stable layers of air (even inversions) may be convectively unstable because of a marked moisture stratification within them.

Two stability parameters are of special use in geographical studies: static stability and the stability index. *Static stability* is a measure of the resistive force offered to vertical air movements by the density stratification of the atmosphere, which is, of course, controlled by gravity. In its simplest form, it may be defined as the difference between the actual (environmental) lapse-rate and the dry adiabatic lapse-rate at some specific place.[17]

The *Showalter stability index* assesses the degree of instability due to convergence: a positive value indicates stability and a negative value instability (see Appendix 5.10). Positive values of 3 or less are usually associated with showers, and values between +1 and −2 with increasing probability of thunderstorms. Values of −3 to −5 normally give severe thunderstorms; tornadoes are found with values of −6 or less. Unfortunately, not all cases of convective instability give a negative index.

A line enclosing all stations with negative Showalter indices demarcates an *instability area.* These instability areas cannot be extrapolated, or followed consistently from one weather map to another. Galway's index, which assesses the instability due to low-level heating, corrects this to some extent.[18]

Other stability indices have been produced by Rackliff, Jefferson, and Boyden (see Appendix 5.11). The *Boyden index*, which is especially useful geographically, is plotted on a 700-mb chart, and the closed *S* (stable) and *U* (unstable) centers so revealed can be traced from day to day in the normal synoptic fashion. The index shows the likelihood of air becoming stable (below the 700 mb-level) when subjected to both surface heating and horizontal convergence. Thus air of high index may not produce an unstable type of weather over the sea, but will give thunderstorms when it moves over a heated land surface.

It is useful for many purposes to define a temperature that is unaffected by variations in pressure for all adiabatic changes. This can be done by using Equation 5.5; the temperature so obtained is the *potential temperature* (see Appendix 5.12). Equation 5.5 can be written

$$\frac{T^{\gamma}}{p^{\gamma-1}} = \frac{\theta^{\gamma}}{p_0^{\gamma-1}} \qquad [5.8]$$

where p_0 is a standard pressure (usually 1,000 mb) and θ is a standard temperature, i.e., the potential temperature; θ can be calculated from

$$\theta = T\left(\frac{p_0}{p}\right)^{0.288} \qquad [5.9]$$

From the potential temperature equation and the gas equation, it can be shown that

$$C_p \log \theta = C_p \log T - R \log p = \phi, \qquad [5.10]$$

where ϕ is the change in entropy; the change in entropy of the parcel of air is proportional to the logarithm of its potential temperature (see Appendix 5.13). This relationship can be used to plot a special type of graph, the *T-ϕ* graph, usually known as the *tephigram.* First introduced by Napier Shaw, the tephigram enables us to solve very easily most of the thermodynamic equations so far discussed, and many more. Since it is in daily use among meteorologists and climatologists, familiarity with its use is a necessity for serious students. The *T-ϕ* graph is constructed as follows.

First, a normal rectangular coordinate graph is plotted, of log θ as ordinate against T as abscissa (see Figure 5.2, A), and the plotted points are labeled with the appropriate values of p from Equation 5.10. Isopleths of equal values of p (i.e. isobars) are then drawn, giving Figure 5.2, B. In this graph, horizontal straight lines are lines of constant log θ, i.e., lines of constant entropy, and vertical straight lines are lines of constant T, i.e., isotherms. Thus at any point on the graph we can read off values of the pressure of

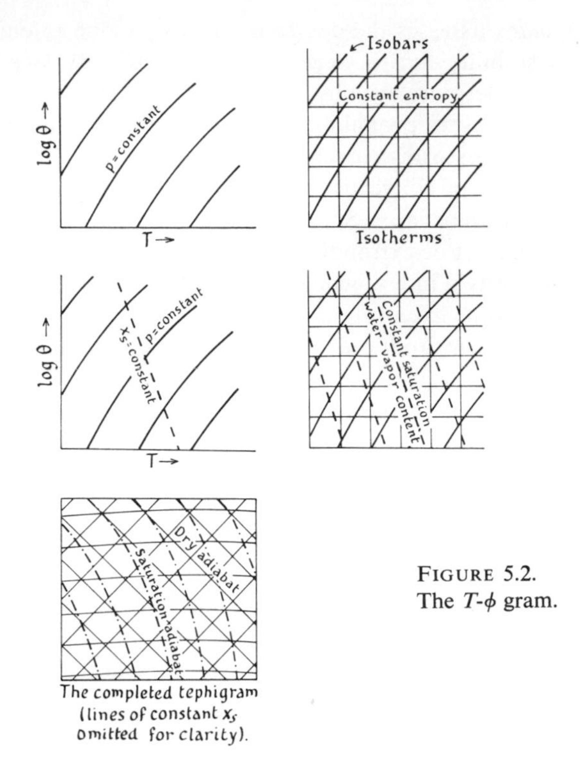

FIGURE 5.2.
The T-ϕ gram.

the air, its temperature, and its potential temperature. To be able to use it practically, we need some way to represent the water content of the air on the graph. This involves introducing some new definitions.

The *specific humidity* of the air is defined as the weight of water vapor contained in a unit weight of *natural* air (i.e., dry air plus water vapor), expressed in grams per kilogram. The *humidity mixing ratio* is defined as the weight of water vapor contained in a unit weight of *dry* air, in g per kg. If we denote the specific humidity by q, and the humidity mixing ratio by x, then there will be x grams of water vapor in one gram of dry air, and q will be equal to $x/(1 + x)$. If e denotes the vapor pressure, it can then be shown that

$$x = \frac{0.622e}{p - e}, \qquad [5.11]$$

or, for saturated air,

$$e_s = \frac{x_s(p - e_s)}{0.622}. \qquad [5.12]$$

Saturation vapor pressure, e_s, is a function of temperature T; therefore the saturation mixing ratio is a function of p and T.[19] Water-vapor content can now be added to Figure 5.2, because e_s may be found for given values of T from empirical tables, and values for e_s and p may then be substituted in Equation 5.12 to obtain x_s. Lines of constant x_s on the graph will represent values of constant water-vapor content for for saturation.

Figure 5.2, D, now represents the basic form of the tephigram. Other features are added for convenience, and the completed tephigram is shown in Figure 5.2, E. Notice that, because pressure levels form the datum lines for atmospheric analysis, for convenience isobars are given as horizontal lines in the routine tephigram, so that Figure 5.2, D has in effect been rotated some 45° clockwise from Figure 5.2, D.

THERMODYNAMIC DIAGRAMS

Before we consider in detail the use of the tephigram, we must discuss some points concerning thermodynamic diagrams in general. Very briefly, thermodynamic diagrams may be defined as graphs on which are plotted isopleths representing the relation between various thermodynamic quantities.[20] They have the important property that areas on them enclosed by continuous isopleths represent energy. In classical thermodynamics, this recalls the Clapeyron diagram (see Figure 5.3, A), in which the specific volume V of a gas is plotted as a function of pressure p. On the pV diagram, changes in the temperature of an ideal gas, brought about by keeping the volume constant and varying the pressure, are represented by curved lines. If a cyclic series of changes is plotted (see Figure 5.3, B), the work done on an air parcel moving through the atmosphere may be found. If the cycle is enclosed in a clockwise direction, the enclosed area (shaded) is said to be positive, and the air parcel does work, i.e., expends energy. If the cycle is closed in an anticlockwise direction, the enclosed area is negative, and the environment does work on the air parcel, i.e., the latter acquires energy. The area enclosed represents the net amount of heat transferred during the cycle, the scale of energy per unit area depending only upon the scales adopted to represent p and V.

From a curve of p against V for different levels in the atmosphere, it is possible to determine the *geopotential*, which is defined as the potential energy of a unit mass in the Earth's gravitational field (see Appendix 5.14). The unit of geopotential is the dynamic meter, which is equal to 10^5 ergs in the classic system of energy units. Geopotential is taken as zero at sea level, and then the geopotential between z_1 and z_2 is given (see Figure 5.4) by the expression

$$-\int_{z_1}^{z_2} V dp.$$

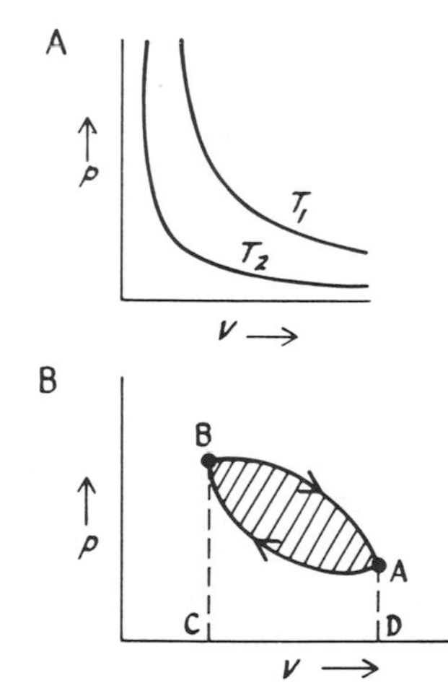

FIGURE 5.3.
A shows the Clapeyron diagram, which indicates the state of a gas for two temperatures, T_1 and T_2. The temperature of the gas may be changed either by moving along a horizontal line (i.e., at constant pressure), or by moving along a vertical line (i.e., at constant volume). B shows the thermodynamic cycle. If the gas changes from state A to state B along the path indicated work will be done on the gas. If gas changes from state B to state A along path indicated, the gas will do work against the outer pressure forces. In either case, the work involved is given by the area $ABCD$, where AB is along the appropriate path.

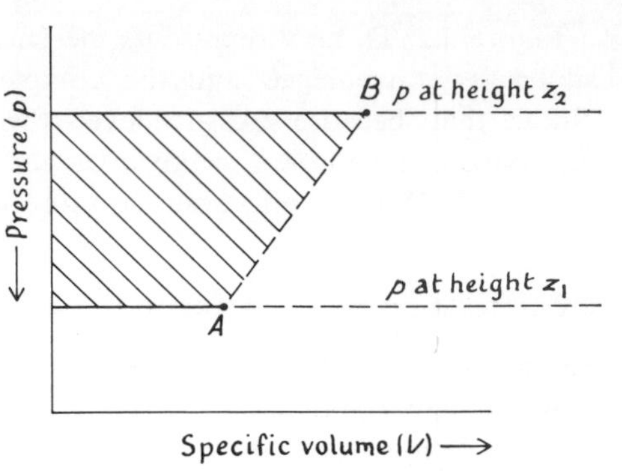

FIGURE 5.4.
The geopotential of the atmospheric layer between heights z_1 and z_2 is given by the shaded area; *A* and *B* are specific points at the bottom and the top of layer, respectively, with point *B* vertically above point *A*.

In other words, the geopotential of a unit mass of the atmosphere at a height of z meters above mean sea level is $0.981z$ dynamic meters. The concept of geopotential is of great importance, since all upper-air data are now quoted in terms of geopotential.

The pV diagram is little used in meteorology and climatology, because it is not possible to measure specific volume directly. Instead, different thermodynamic diagrams are produced by making equal-area transformations of the pV diagram. In other words, other coordinates than p and V are chosen, so that an area of A square centimetres, which is enclosed by some particular isopleth on the pV diagram, will remain exactly A cm^2 when it is replotted on the new diagram. The shape of the area may change, but its magnitude must not. A large number of transformations are possible, but only four have gained widespread acceptance, resulting in the tephigram, the aerogram, the emagram, and the Stüve (or pseudoadiabatic) diagram. The first thermodynamic diagram, devised by Hertz in 1864, and revised by Neuhoff in 1900, is little used today.

In the tephigram (Figure 5.2, D) originally introduced in 1928, the abscissa is $C_p \log \theta$ and the ordinate is $-T$.[21] Vertical lines, equidistantly spaced, are isotherms, from $-130°$ to $+98°$F. Horizontal lines, also equidistant, are lines of constant entropy or constant potential temperature, which also form *dry adiabatics* (*dry adiabats*) or path curves for rising unsaturated air. Isobars are represented as slightly curved lines sloping upward from left to right, from 1,050 mb on the extreme right of the diagram to 170 mb on the extreme left. Saturation water-vapor content of the air is represented by nearly vertical pecked lines, labeled from 36 grams per kilogram on the extreme right to 0.01 gm per kg on the extreme left. Curved full lines sloping upward from right to left are the saturation adiabatics (*saturation adiabats*), representing path curves for rising saturated air. Finally, figures printed midway between isobars at 100-mb intervals represent the vertical thickness of air between those pressure surfaces for various temperatures.

The tephigram is much used in the United Kingdom and in Canada, but in the United States the *pseudoadiabatic* or *Stüve diagram* is favored (see Figure 5.5). An adiabatic diagram is produced quite simply by plotting temperature in °C as abscissa and pressure in mb as ordinate on rectangular coordinates. The temperature scale is usually linear, but the pressure scale is logarithmic, decreasing upward. If, instead of $\log p$, $p^{0.286}$ was used for the ordinate, the potential temperature lines, i.e., the dry adiabatics, would be straight lines. $\log p$ is not an unreasonable approximation to $p^{0.286}$, and so the diagram in which T is plotted against $\log p$ is referred to as the pseudoadiabatic diagram. In the Stüve version of this, the abscissa is given by RT,

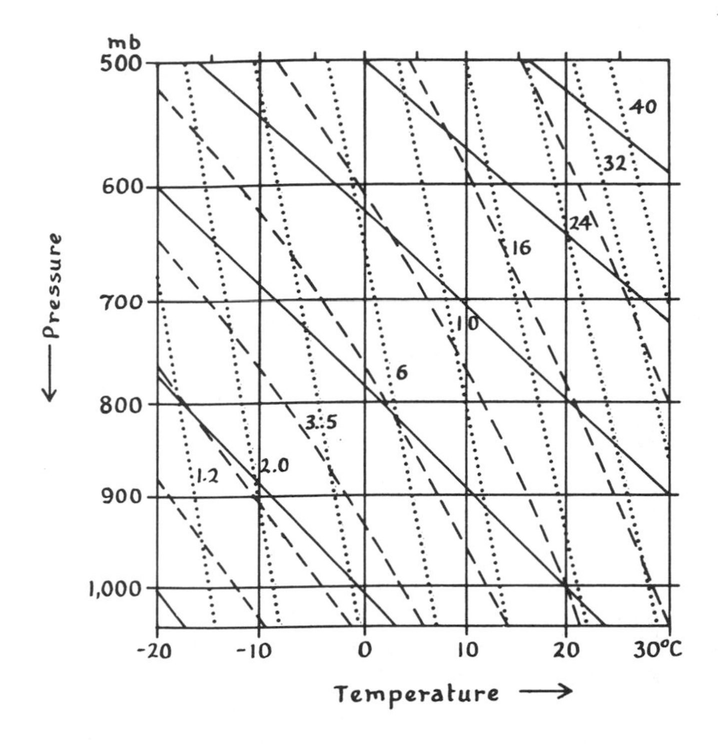

FIGURE 5.5.
Pseudoadiabatic (Stüve) diagram. Basic lines only are shown. The dotted lines show water-vapor content, in grams per kg.

and the ordinate by $-p^{R/C_p}$. Strictly speaking, this diagram does not truly preserve areas from the pV graph, although it is referred to as a thermodynamic diagram.[22]

In the pseudoadiabatic diagram, horizontal lines are isobars spaced at intervals of R/C_p, from 1,050 mb at the bottom and usually up to 400 mb at the top. Vertical lines, equidistantly spaced, are isotherms, usually increasing in value from $-45°$C on the extreme left to $+40°$ on the extreme right. Energies are represented on the chart by areas, the exact relation being

$$energy = area \times K, \quad \text{where} \quad K = 0.937\ \theta/T.$$

By chosing a suitable value for K, the diagram can be made to yield energy values for any level. If K is made unity at 800 mb, then areas on the chart will indicate energy values at 800 mb; they will underestimate energy values if p is less than 800, and overestimate if p is greater than 800 mb.

Lines of constant potential temperature (the dry adiabats) are straight lines, sloping upward from right to left at different angles, the slope of a line being given by $100^{R/C_p}/\theta$. If the diagram is extended to zero temperature and zero pressure, the dry adiabats diverge radially from the upper left-hand corner of the graph. The water-vapor content lines, representing the saturation mixing ratio, slope upward from right to left, more steeply than the dry adiabats, and are usually labeled from 40 gm per kg (extreme right) to 0.1 gm per kg (extreme left). They are derived from the ratio $0.622\ e_s/(p - e_s)$, where p is the total pressure (i.e., the pressure plotted on the diagram) and e_s is the saturation vapor pressure. The saturation adiabatics are curved lines, sloping steeply upward from bottom right to top left. They are very steeply curved at higher temperatures, but approach the dry adiabats at low temperatures.

Both the tephigram and the pseudoadiabatic diagram are based on rectangular coordinates. In the *skewed diagram*, of which there are various types, the coordinates are made nonrectangular, while the equal-area property is preserved.[23] For example, in the USAF type, log p remains the ordinate as before, but the temperature lines are rotated clockwise almost 45°. This increases the angle of intersection between the isotherms and the dry adiabats, and facilitates plotting.

In Scandinavia, France, and Germany, the *aerogram* (see Figure 5.6) is preferred. Introduced by Refsdal in 1935, this diagram has log RT as abscissa, and $-RT \log p$ as ordinate.[24] Isotherms are shown as straight, equidistant, vertical lines. Isobars are curved, and only slightly deviate from horizontal, sloping very slightly up from left to right. The dry adiabats are curves sloping from top left to bottom right, while the saturation adiabats are curves sloping in the same direction, but more steeply than the dry adiabats. The water-vapor content lines slope very steeply from top left to bottom right. In this diagram, the angles between the isotherms, the dry adiabats, and the saturation adiabats are less than with the tephigram, so that it is not so suitable as the latter for making stability assessments. It is, however, superior to the tephigram for the display of aerological data.

Basically, the *emagram* is an adiabatic chart that indicates the difference between actual temperature T and virtual temperature T' at saturation for different pressures.[25] To find the actual difference, the saturation value must be multiplied by the relative humidity. The basis of this diagram is that, since $T' - T$ at saturation is a function of temperature and humidity mixing ratio alone, and the mixing ratio at saturation is a function of temperature and pressure, then $T' = T$ at saturation must also be a function of temperature and pressure. In some versions of the tephigram, this information is already incorporated, so that an additional chart is not necessary.

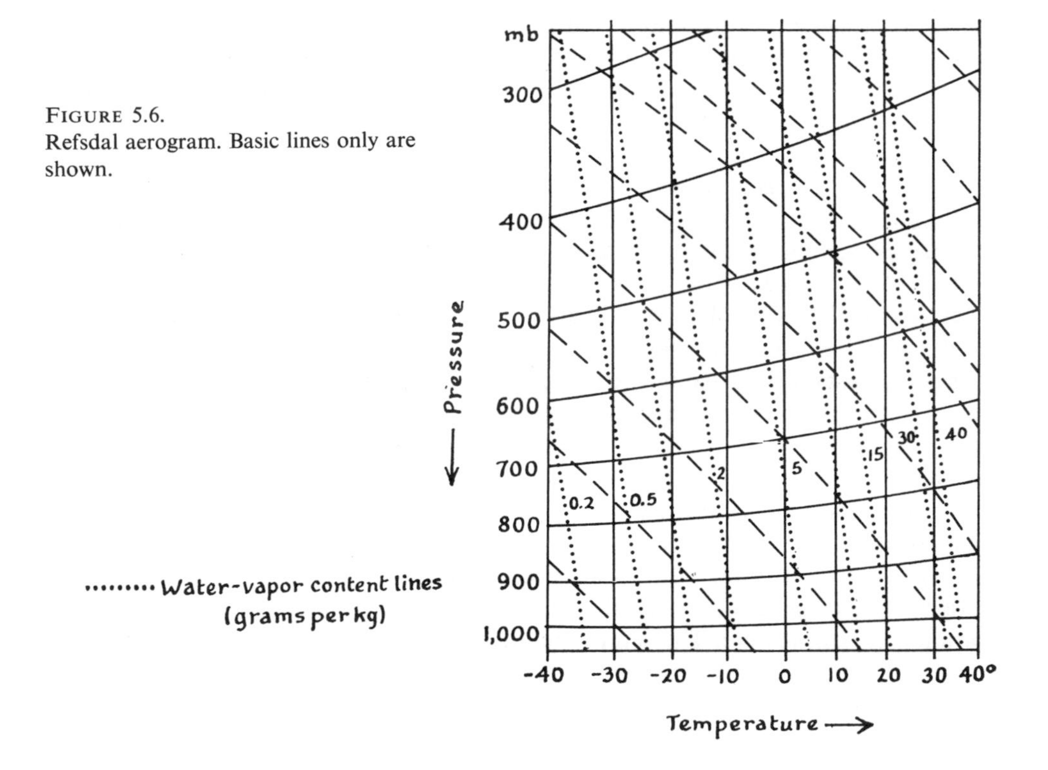

FIGURE 5.6.
Refsdal aerogram. Basic lines only are shown.

USE OF THERMODYNAMIC DIAGRAMS

All thermodynamic diagrams enable the following information to be derived: identity of air-masses; presence of fronts; occurrence of inversions; degree of stability or instability of the atmosphere; likelihood of convection-cloud formation; possibility of thunderstorm development; and estimation of maximum temperature likely during the day. The first step usually involves first plotting the pressure, temperature, and humidity data derived from the nearest radiosonde ascent. This is plotted as a series of points, which are joined by straight lines to form a sounding for the region of the atmosphere traversed by the radiosonde. The sounding usually falls between the isotherms, representing conductive equilibrium, and the *isentropes*, representing adiabatic equilibrium.[26] In other words, a number of different physical processes must be operating in the atmosphere simultaneously. The following discussion will be illustrated specifically from the tephigram, but it applies with minor modifications to the other diagrams too.

First, thermodynamic diagrams enable the properties of moist air to be assessed quantitatively. Given the pressure and temperature, the water-vapor content line through the intersection of the appropriate isobar and isotherm gives the water-vapor amount required for saturation. Once we know the pressure and actual water-vapor content of the air, the value of the isotherm through the appropriate isobar and water-vapor content line will give us the *dew point*, which is defined as the temperature at which the amount of water-vapor present would be just sufficient to saturate the air. Dew points are usually found for each pressure level, and joined up by straight dotted lines. Thus the complete sounding consists of two "curves": a full one for dry-bulb temperatures, and a dotted one for dew points.

Given pressure, temperature, and relative humidity, both water-vapor content and dew point may be found as follows. Find the value of the water-vapor content line through the intersection of the appropriate isobar and isotherm (Y in Figure 5.7, B); this value is the saturation value. Multiply this value by the relative humidity to derive the actual water-vapor content (X in Figure 5.7, B). Then find the isotherm that cuts through the intersection of the actual pressure and actual water-vapor content lines; this isotherm will give the dew point.

Knowing the dry-bulb temperature of the air, the wet-bulb temperature, and the pressure, one can find the dew point, water-vapor content, and relative humidity easily on the tephigram, by making use of *Normand's theorem*, which states that the dry adiabatic through the dry-bulb temperature (T in Figure 5.7, C), the saturation adiabatic through the wet-bulb temperature (T_W), and the water-vapor content line through the dew point (T_d), always meet in a single point P.[27] Hence, knowing any two of the quantities T, T_W, and T_d, one can find the third. This theorem is very useful in routine work, for it obviates the need for hygrometric tables.

It is necessary on all thermodynamic diagrams to carefully distinguish between the sounding plotted from radiosonde data, which is referred to as the *environment curve* or the actual lapse-rate of the atmosphere at a certain time and place, and the curve that indicates the changes in state a small parcel of air undergoes when moving vertically through the atmosphere. Figure 5.8, A, makes this clear. In it AB represents the environment curve. If the air at point A, which represents ground level, is unsaturated (i.e., if its temperature T is less than the dew point temperature, T_d), then if a small parcel of air is forced to rise from A, its condition will be given by moving along the dry adiabatic AX. When its temperature has fallen to that of the dew point, i.e., point Y,

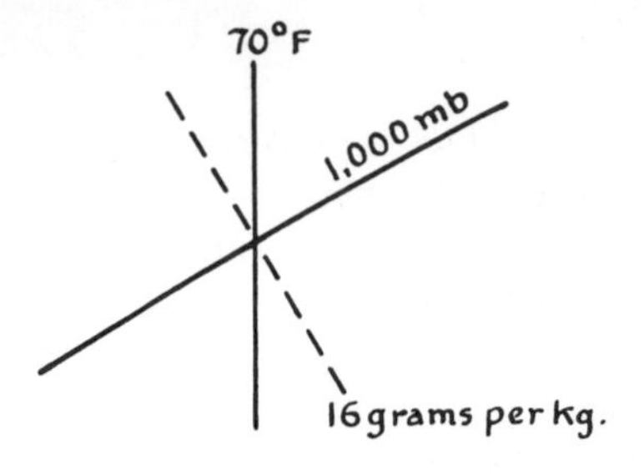

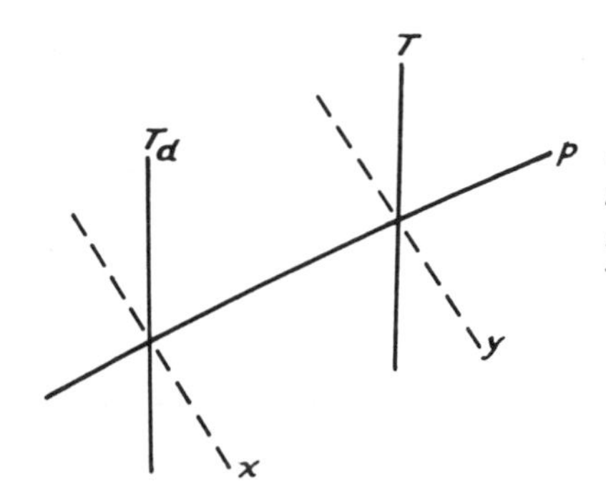

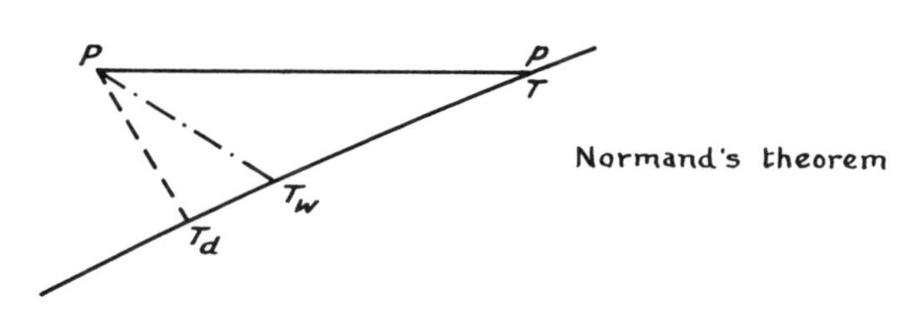

FIGURE 5.7.
Use of the tephigram.

it then cools at the saturated adiabatic lapse-rate. The latter is less than the dry adiabatic lapse-rate because of the release of latent heat, and so the condition of the air parcel at different levels in the atmosphere will be given by moving along the saturation adiabatic *YZ*.

Three possible relationships exist between the path curve and the environment curve. If the path curve indicates that at every pressure level the rising air is cooler than its environment, i.e., the environment curve lies wholly to the right of the path curve on the tephigram (Figure 5.8, B), then the atmosphere is stable, because at any given level the rising parcel will be denser than its surroundings, and so, if the cause of its ascent is removed, it will not continue to rise, but will sink back toward the ground. If the path curve shows that the rising parcel is warmer than its environment at all levels, i.e., the environment curve lies wholly to the left of the path curve (Figure 5.8, C), then the atmosphere is unstable, because at any given level the rising air will be warmer and lighter than its surroundings. (If path and environment curves approximately coincide,

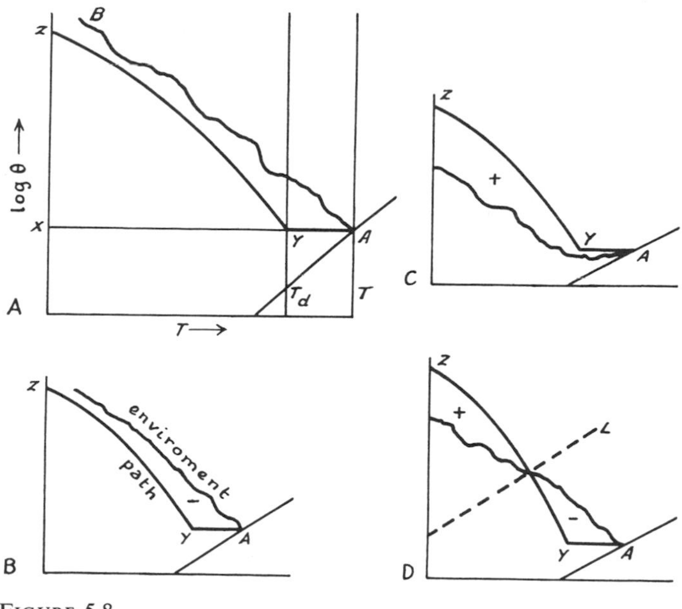

FIGURE 5.8.
Environment curves
and lapse-rates.

which is a very rare occurrence, the atmosphere is said to be in neutral equilibrium.)

A quite common situation (Figure 5.8, D) is for the atmosphere to be stable near the ground, but unstable above a certain level *L*. This represents conditional instability; the atmosphere is stable so long as air is not forced to rise above the level *L*. All these stability considerations, incidentally, refer to small parcels of air, not to layers of air. It is not permissible to apply them to föhn (chinook) conditions, as is done in many elementary textbooks; for the föhn, convective instability is the important consideration. Convective instability is a potential condition, whereas conditional instability refers to the instantaneous conditions; an air layer may be conditionally stable and yet convectively unstable at the same time.

In the stable situation, the area between the environment and path curves is said to be a *negative area*; in the unstable one, it is said to be a *positive area*. Large positive areas are necessary for intense convection. If the air is conditionally unstable, convection may occur if there is only a small negative area above the ground, because a rising air parcel may acquire sufficient momentum to penetrate the negative area. If a large negative area occurs at ground level, however, strong mechanical lifting up to the top of the negative layer is necessary before convection can occur.

Early morning soundings are frequently used to predict the possibility of convection clouds or thunderstorms during the ensuing day.[28] For this, the maximum temperature likely during the day must first be estimated. The procedure for this on the tephigram (Figure 5.9, A) is as follows. If the atmosphere is stable, draw a horizontal line PT_m on the early morning sounding such that the area PT_mT_e does not exceed the *Gold Square value* for the month (see Appendix 5.15). Gold squares, enclosed by 10 degrees of temperature and 10 degrees of potential temperature, represent the maximum solar energy that can be received in a particular latitude per day; T_m will then represent the maximum temperature likely to occur. If the atmosphere is unstable (Figure 5.9, B)

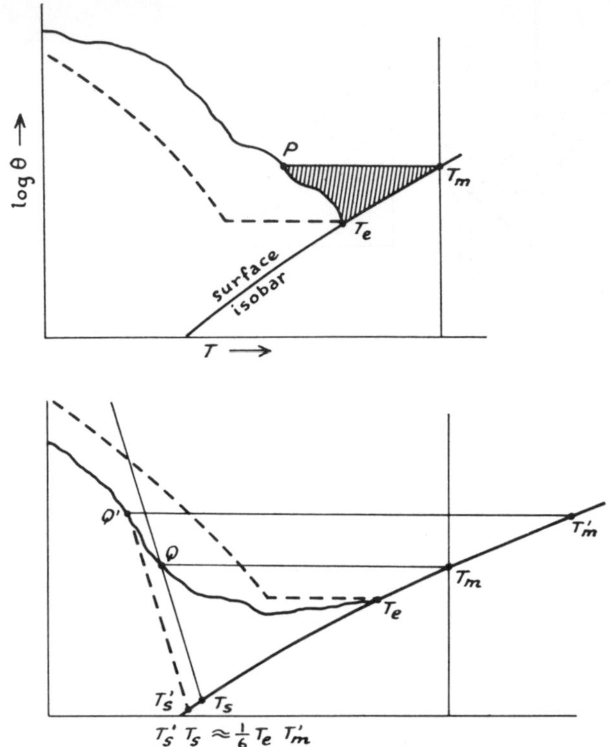

FIGURE 5.9.
Predicting thunderstorms or convection clouds from the tephigram.

a horizontal line $T_m Q$ is drawn through the point Q where the water-vapor content line through the early morning dew point T_s cuts the sounding; T_m will then be the maximum temperature, provided the dew point does not also increase during the day. If it is likely that the dew point will rise, $T'_s Q'$ and $T'_m Q'$ are drawn, such that $T'_s T_s$ equals one sixth of $T_e T'_m$, this value being found empirically to give the required allowance for a change in the dew point.

After estimating the maximum temperature likely during the day, convection is predicted as follows. The point C, where the water-vapor content line through the dew point T_s cuts the sounding, gives condensation level; T_m is found from one of the methods already outlined. If T'_m occurs as in Figure 5,10, A, dry convection will occur, with no clouds; if T_m is as in Figure 5.10, B, convection will be just sufficient for rising air to reach condensation level, and fair-weather cumulus will develop, with a flat base at level C and flat tops if the air is stable. If the upper part of the sounding indicates instability, as in Figure 5.10, C, a large positive area will exist; convection will be unrestricted and cumulonimbus should form. If there is a large positive area, but the atmosphere becomes stable above level Z, then towering cumulus will develop with bases given by C and tops by Z. If Z is above freezing level, showers are likely. Cloud amount may be forecast approximately by Poulter's empirical formula,

$$C = \frac{R - 6}{6},$$

where C is the cloud amount in tenths, and R is the relative humidity at condensation

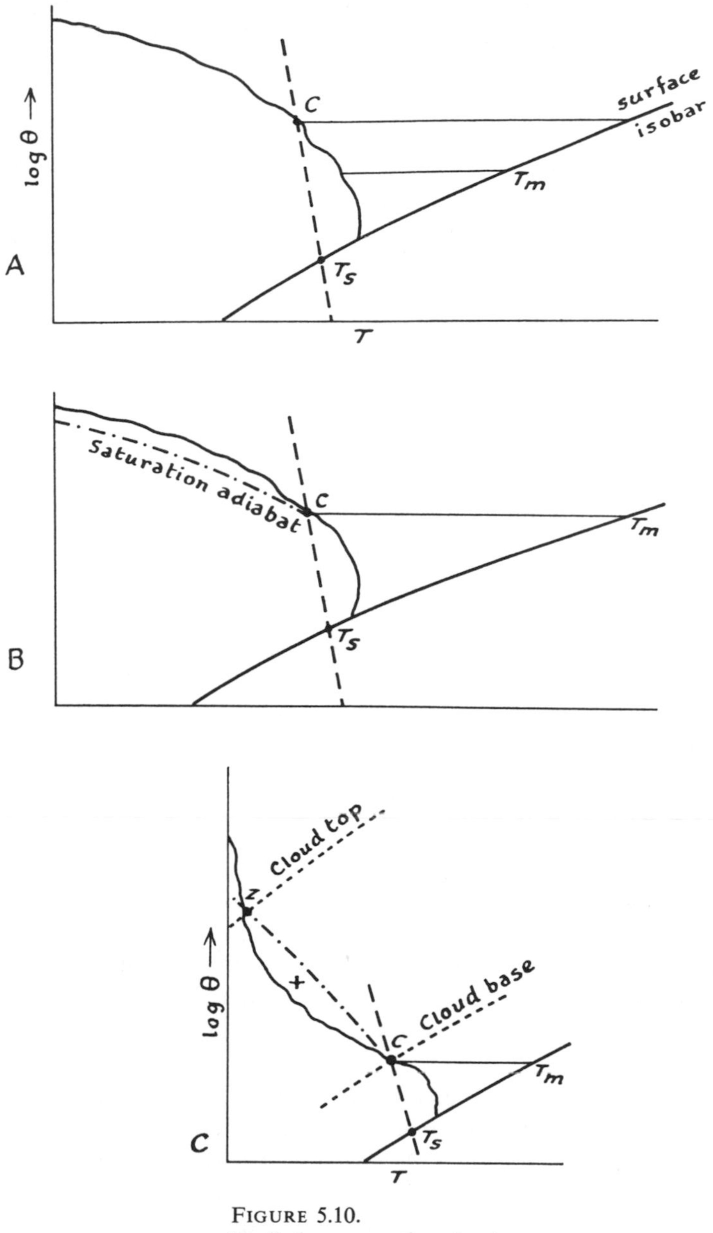

FIGURE 5.10.
Predicting convection clouds from the tephigram.

level.[29] If a very large positive area occurs, extending vertically at least 10,000 feet above condensation level, thunderstorms are probable.

Thermodynamic diagrams are very useful for identifying types of air mass, and they often indicate that several air masses occur, one above the other, over a point on the Earth's surface. For the identification of thermodynamic states, a number of different properties may be used. Two of them, the potential temperature and the dew-point temperature, have been already defined. The potential temperature is found from the thermodynamic diagram by following the dry adiabatic, from the intersection of the appropriate isobar and isotherm, to the 1,000-mb pressure line, and reading off the temperature there. The dew-point temperature is given by the isotherm through the point where the line representing the actual water-vapor content intersects the isobar representing the observed pressure. Two others are the *wet-bulb potential temperature* and the *equivalent potential temperature*.[30] The former is defined as the temperature a standard parcel of air would attain if brought pseudo-adiabatically to a standard pressure of 1,000 mb. It is found by following the appropriate saturation adiabatic back to 1,000 mb. The equivalent potential temperature is defined as the maximum potential temperature a parcel of air can attain by adiabatic motion. It is found by following the saturation adiabatic back as far as the dew-point temperature, then following the dry-adiabatic to the standard pressure of 1,000 mb.

All of these different properties have various advantages and disadvantages as tracers for following the horizontal and vertical movements of air parcels through the atmosphere. The most useful properties are the *conservative* ones, i.e., those that do not vary appreciably during a few hours or a few days. The wet-bulb potential temperature and the equivalent potential temperature are the only properties remaining constant both before and after saturation; they are conservative for all adiabatic processes so long as no moisture is added to the air or subtracted from it. However, they remain unchanged with altitude over large areas, so that an air parcel displaced from its original location will not, from these properties alone, be observed to have distinctive thermodynamic qualities that differentiate it from its surroundings. The humidity mixing ratio has a very nonuniform distribution, and enables an air parcel to be easily followed. The specific humidity is also useful in this respect. Before condensation (i.e., for all dry adiabatic ascents) the relative humidity increases, dry-bulb temperature, virtual temperature, vapor pressure, dew-point temperature, absolute humidity, equivalent temperature, and wet-bulb temperature all decrease, and the specific humidity, mixing ratio, equivalent potential temperature, and wet-bulb potential temperature remain constant. After condensation (i.e., for all saturation-adiabatic ascents), the potential temperature increases, the wet-bulb potential temperature, equivalent-potential temperature, and relative humidity remain constant, and all the other properties decrease. A convenient diagram for indicating the conservative properties of air masses is the Rossby diagram, introduced in 1932, which plots mixing ratio and potential temperature as abscissa and ordinate, respectively, on rectangular coordinates.[31]

Thermodynamic diagrams also enable the *resilience* of the atmosphere to be determined, i.e., the energy available within it to restore a parcel of air to its original position if the parcel is displaced vertically.[32] Atmospheric resilience varies with the seasons—for example, it reaches a maximum in winter at Larkhill and a minimum in summer, and a maximum in summer at Ocean Weather Stations I and J with a minimum in winter—and provides a measure of the potential energy of the atmosphere as the latter fluctuates in space or time. Regions of minimum resilience indicate locations where synoptic development is most likely; more cyclones originate and deepen over the North Atlantic in winter than in summer, the former being the season of least resilience.

Elementary Dynamics of the Atmosphere

The basis of the study of atmospheric dynamics is the Newtonian system. For practical purposes in climatology, Einsteinian relativity theory need not be considered. The foundations of the Newtonian system are the three laws of motion: (1) bodies at rest remain at rest, and bodies in motion remain in motion with the same velocity, unless acted on by unbalanced forces; (2) the rate of change of momentum of a body with time equals the vector sum of all the forces acting on the body in the same direction; (3) to every action or force there is an equal and opposite reaction. The nature of these laws is interesting. They are *initial assumptions* in a theoretical system, not inductive laws based on experiment. In fact, they cannot be verified by experiment without arguing in a circle, because our definitions of *force* and of *mass* are derived from the first two laws, and it is impossible to verify either law without assuming that either force or mass is already known. Although Newton's laws of motion represent a deductive theory, they are not the dynamic equivalent of Euclid's axioms of geometry, which "neither require, nor are capable of proof," because, although incapable of proof, they are not self-evident. Their value is that they represent the simplest possible description of the relations between moving bodies in the physical universe, and enable predictions to be made that are in accordance with experience; i.e., the laws "work" in the everyday world.[33]

Newton's second law, expressed in the form $P = ma$, where P is the force producing an acceleration a on a mass m, is the basis of all quantitative studies in dynamics. However, it cannot be applied directly to the atmosphere, because, as it stands, the equation is only valid in an *inertial coordinate system*, whereas the coordinate system required for the atmosphere is noninertial. The essential difference is that inertial coordinate systems do not accelerate, or accelerate negligibly, whereas noninertial systems accelerate appreciably. The geographical system of coordinates, which are latitude, longitude, and height above sea level, is an accelerating one, because it rotates about the Earth's axis (see Appendix 5.16). The effects of its acceleration are negligible for small-scale motions, such as those observed in a laboratory or measured in microclimatology, but they become appreciable for medium- and large-scale motions, such as extensive land-sea breezes or the global wind belts. The distinction between inertial and noninertial coordinate systems is important, because Newton's second law does not apply in a noninertial system unless a new force is introduced into its expression.[34]

Various coordinate systems are used in dynamical climatology, and for some purposes it is convenient to have a system of mechanics that is valid in all coordinate systems. The latter system is provided by the tensor calculus, by means of which the laws of motion may be expressed in a general form in terms of arbitrary coordinates that can be applied to any desired coordinate system. However, it is usual in theoretical climatology to use the standard meteorological system of rectangular coordinates, in which the horizontal axes run east-west and north-south, and the third axis is vertical; the latter may be graduated in either height or pressure units, the horizontal axes are graduated in length units.[35]

Before it may be used in theoretical climatology, Newton's second law must be transformed first from an inertial to a rotating coordinate system, and second to a form that is applicable to a spherical Earth. The transformations may be illustrated as follows. Figure 5.11 represents horizontal coordinates at right angles, in which OY, OX, are at rest, and OY', OX', are those same axes in a coordinate system that is rotating about the origin O of the fixed system at a constant angular velocity Ω with respect to the fixed system.

Introducing Newton's system of calculus notation, in which one dot placed above a quantity means one differentiation with respect to time, two dots means two differentiations, and so on (see Appendix 5.17), then Newton's second law for a particle of

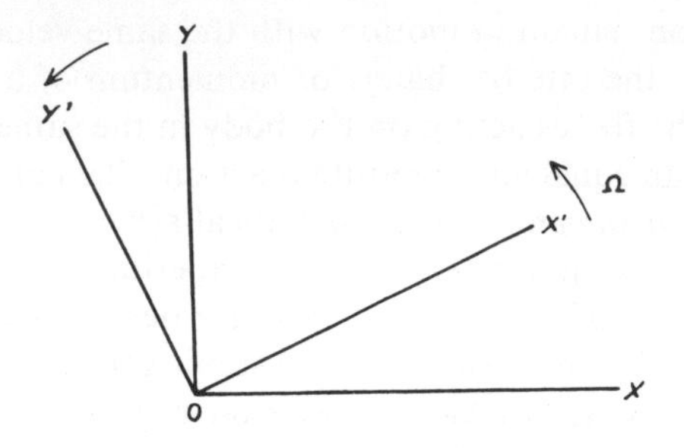

FIGURE 5.11.
Fixed and rotating coordinate systems.

constant mass located in the fixed coordinate system is expressed by three equations,

$$\ddot{x} = \frac{F_E}{m}, \ \ddot{y} = \frac{F_N}{m}, \ \ddot{z} = \frac{F_Z}{m},$$

where x, y, z, represent the distance moved by the particle in a given time along the OX, OY, and OZ axes, respectively; F_E, F_N, F_Z, represent the net force acting on the particle (referred to the approximate axes) during that time, and m is the mass of the particle. Then $\ddot{x}$ represents the acceleration of the particle in the OX direction, $\ddot{y}$ its acceleration in the OY direction, and $\ddot{z}$ its acceleration in the OZ (upward) direction. The equations, when transformed to the Y', X', system become

$$\ddot{x}' - \frac{F_E}{m} = x'\Omega^2 + 2\dot{y}'\Omega,$$

$$\ddot{y}' - \frac{F_N}{m} = y'\Omega^2 - 2\dot{x}'\Omega,$$

$$\ddot{z}' - \frac{F_Z}{m} = 0.$$

The first terms on the right-hand sides of the first two equations represent the *centrifugal accelerations*, i.e., the accelerations produced by the outward-acting centrifugal force that is necessary to maintain any body in uniform motion in a circular orbit; the magnitude of the combined centrifugal components in the horizontal plane is $r\Omega^2$, where r is the distance of the particle from the axis of rotation, which is located at the origin O (see Figure 5.12). The second terms on the right-hand sides represent the *Coriolis accelerations*, which arise from the combined effects of the rotation of the coordinate system and the motion of the particle relative to the coordinate system. It is important to note that the Coriolis accelerations involve only those components of the motion that are in the plane perpendicular to the axis of rotation of the coordinates, i.e., the horizontal plane; they are directed to the right of these components, with a magnitude of $2\Omega V$, where V is the total velocity of the particle in the horizontal plane. The Coriolis accelerations represent the effect of a force, the *Coriolis force*, which is a deflecting force (unlike the force F), because the Coriolis accelerations always act perpendicularly to the velocity of the particle, and consequently cannot change its speed of movement but can only alter its direction of motion.[36]

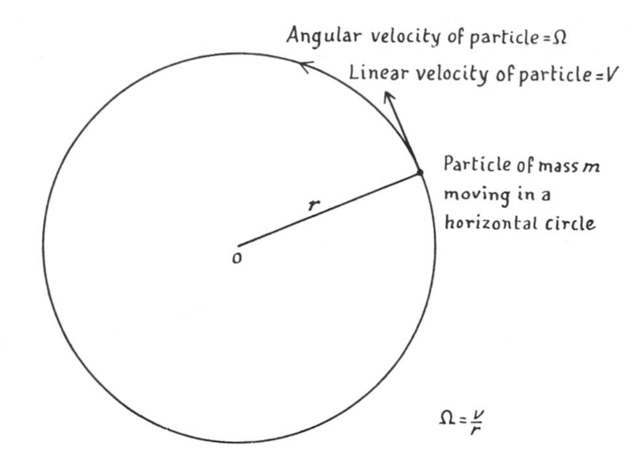

FIGURE 5.12.
Centrifugal acceleration.

The importance of the Coriolis force in dynamic climatology cannot be overemphasized. It causes considerable difficulty and mistaken concepts in elementary teaching, because some writers feel they must explain things in "concrete" terms. It cannot be overstressed that the Coriolis force is a *mathematical* concept, and it is not possible to explain such a concept by means of chalk lines drawn on rotating gramophone turntables and so forth. The apparent *existence* of the Coriolis force may easily be demonstrated, because it is merely a consequence of the rotation of the Earth. However, the purely qualitative demonstration of it in most elementary textbooks on meteorology and climatology is not scientific because it does not enable verifiable predictions to be made.

As an example of such a qualitative demonstration, let us consider the following. If a gun is fired at a distant target and the bullet travels southward in a straight line, then, because the surface of the Earth is rotating from west to east beneath the bullet during its passage through the air, the bullet will not hit the target but will travel to the right of it (left for a northward-moving bullet), provided the bullet's passage takes enough time for the Earth's rotation to be appreciable. If so, then the path of the bullet across the Earth, *relative to a stationary observer on the Earth's surface*, will not be a straight line but a curve. The deviation of this path from a straight line is ascribed, by the stationary observer in his noninertial coordinate system, to the effect of a fictitious force. But to an observer stationary in space, in an inertial coordinate system, it would be evident that the bullet had in fact traveled in a straight line, so that he would not require a fictitious force to explain its behavior.

The existence of the Earth's diurnal rotation may be demonstrated by a Foucault pendulum,[37] and simple reasoning will indicate that to an observer on the Earth this rotation must give rise to a deflecting force. However, such reasoning does not get us very far, because it does not enable us to make any quantitative predictions that may be verified. Consequently, we have to return to the Coriolis force in its original mathematical conception. There is, in fact, a considerable gap between qualitative discussions of bullets moving over the face of the Earth and the mathematical derivation of the Coriolis force as a necessary consequence of a rotating coordinate system. This gap is the first of the hurdles the serious student of climatology must jump, and it is impossible to make the jump without going through the necessary mathematics.

The aim is to derive an expression that enables us to quantitatively determine the magnitude of the Coriolis acceleration at any point on the Earth's surface. The crux of the problem is that there is an inherent asymmetry in Newton's second law, such that the properties of the coordinate system we use to make measurements of any atmospheric motions may, if the scale of motion is sufficiently great, influence the acceleration term in the equations stating the law but not the force term.[38] Complications arise because the expressions for Newton's second law of motion obtained for a rotating coordinate system must now be transformed to a spherical Earth, and because of the existence of gravitational attraction.

Figure 5.13 shows a rectangular coordinate system with its origin O at the Earth's surface in latitude ϕ, and with its horizontal (X, Y) axes in a plane tangent to the surface at O and in the east-west and north-south directions, respectively. If Ω is the angular velocity of the Earth, then the equations representing Newton's second law transformed to a spherical Earth are

$$\ddot{x}' = 2\Omega(v \sin\phi - w \cos\phi) + \frac{F_E}{m}, \qquad [5.13]$$

$$\ddot{y}' = -2\Omega u \sin\phi - r\Omega^2 \cos\phi \sin\phi + \frac{F_N}{m}, \qquad [5.14]$$

$$\ddot{z}' = 2\Omega u \cos\phi + r\Omega^2 \cos^2\phi + \frac{F_Z}{m}, \qquad [5.15]$$

in which u, v, w, represent the speed of the particle (i.e., $\dot{x}'$, $\dot{y}'$, $\dot{w}'$) in the east-west, north-south, and vertical directions, respectively.[39]

Gravitation is the universal attractive force between bodies in the Newtonian system of dynamics.[40] If a particle is at rest on a rotating Earth, and is subject to no external

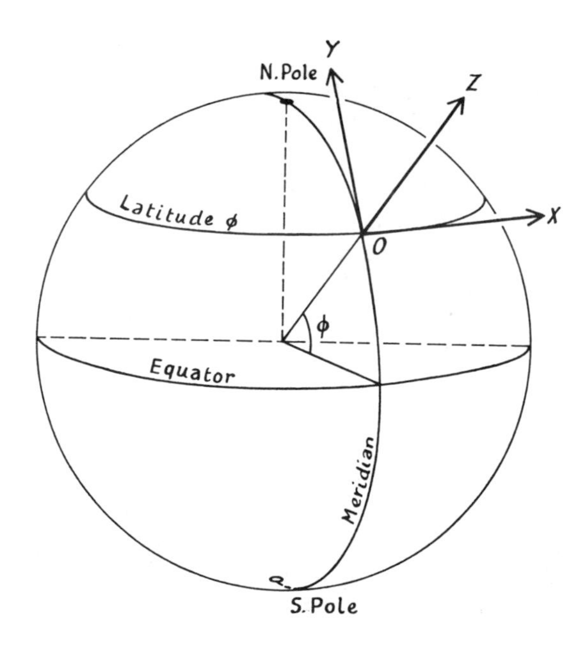

FIGURE 5.13.
A rectangular coordinate system on a rotating Earth.

force other than that of gravitation, then Equations 5.13 to 5.15 become simplified to

$$\ddot{x}' = 0,$$

$$\ddot{y}' = -r\Omega^2 \cos\phi \sin\phi$$

$$\ddot{z}' = r\Omega^2 \cos^2\phi - G,$$

where G is the universal gravitational constant. The accelerations have disappeared for east-west motions, but not for north-south or vertical motions. Consequently the body will not continue in its initial state of rest, but will move towards the equator. If the consideration is extended to include all particles initially at rest on the surface of a rotating Earth, the result of the remaining accelerations will be to cause these particles to move equatorward, so that the Earth's surface itself is in north-south motion and the shape of the Earth becomes deformed from a perfect sphere into an *oblate spheroid.* The oblate spheroid has the familiar shape of a sphere flattened at the poles, and its surface represents an equilibrium position: the surface is perpendicular to the vector sum of the centrifugal force of the Earth's rotation and the gravitational forces. An observer on the rotating Earth can only measure the combined centrifugal and gravitational forces (which he terms *gravity*), and he cannot distinguish between them. In other words, if P is a particle at the surface of the Earth (see Figure 5.14), and if G_T represents the true attraction on the particle due to gravitation, then when the Earth is rotating a centrifugal force C on the particle comes into operation. The *apparent* gravitational attraction g will then be the resultant of C and G_T. The Earth adjusts its shape so that no net force exists along its surface (otherwise the latter would be constantly changing), i.e., so that g is perpendicular to this surface. In this case, the equatorward component of centrifugal force C is exactly balanced by the poleward

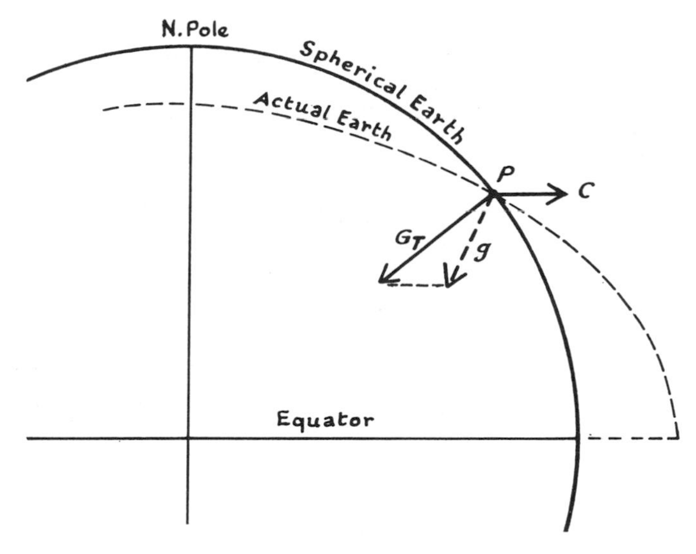

FIGURE 5.14.
Gravity versus gravitational force. G_T, true direction of gravitational attraction; g, apparent direction of gravitational attraction due to the rotation of the Earth (i.e., gravity); C, centrifugal force on particle P.

component of the true gravitational attraction G_T. The level parts of the Earth's surface correspond to *equipotential surfaces*, i.e. surfaces along which a particle is not subject to any component of gravity. Equipotential surfaces are thus equilibrium surfaces, and the direction of the acceleration due to gravity, g, is perpendicular to them; particles on this surface have constant potential energy. In practice, the angle between the directions of G_T and g is very small, with a maximum value of 0.1° in latitude 45°, from which it decreases both northward and southward until it becomes zero at the equator and the poles. The oblateness of the Earth may therefore be taken into account by taking the Z axis in the rectangular coordinate system to be along the direction of the g vector instead of along the G_T vector. The nonsphericity of the Earth's surface may then be neglected with negligible error; the difference between the vectors is not observationally significant in most studies of atmospheric motion. With allowance for the oblateness of the Earth, the equations forming the counterpart of the simple formula $P = ma$ for the expression of Newton's second law become

$$\ddot{x}' = 2\Omega(v \sin \phi - w \cos \phi) + \frac{F_E}{m}, \tag{5.16}$$

$$\ddot{y}' = -2\Omega u \sin \phi + \frac{F_N}{m}, \tag{5.17}$$

$$\ddot{z}' = 2\Omega u \cos \phi - g + \frac{F_Z}{m}. \tag{5.18}$$

F_Z represents all external forces in the new Z direction other than gravity. It will be noted that the centrifugal accelerations have disappeared, and the first terms on the right-hand side of the equations represent the Coriolis accelerations. Equations 5.13 through 5.18 form the *equations of motion* for a system of Cartesian coordinates tangent to the Earth's surface in latitude ϕ. We have used the "dot" system of notation for simplicity; it is more usual to adopt Leibniz's form of notation, in which differentials are employed, [41] and in which the equations of motion are written

$$\frac{du}{dt} = 2\Omega(v \sin \phi - w \cos \phi) + X, \tag{5.19}$$

$$\frac{dv}{dt} = -2\Omega u \sin \phi + Y, \tag{5.20}$$

$$\frac{dw}{dt} = 2\Omega u \cos \phi - g + Z, \tag{5.21}$$

in which X, Y, Z, represent the forces per unit mass in the OX, OY, and OZ directions, respectively; $\frac{du}{dt}$ represents the acceleration of the particle in the OX direction, $\frac{dv}{dt}$ its acceleration in the OY direction, and $\frac{dw}{dt}$ its acceleration in the OZ direction. If the air motion is purely horizontal, Equation 5.21 will have no effect and Equations 5.19 and 5.20 reduce to

$$\frac{du}{dt} - 2\Omega v \sin \phi = X, \tag{5.22}$$

$$\frac{dv}{dt} + 2\Omega u \sin\phi = Y. \qquad [5.23]$$

The second terms of these equations then provide the required definitions of the Coriolis force. The magnitude of the Coriolis force in the east-west direction is $2\Omega v \sin\phi$, and its magnitude in the north-south direction is $2\Omega u \sin\phi$. These expressions may be derived quite easily, and they show that the magnitude of the Coriolis force is zero at the equator (where $\sin\phi = 0$) and a maximum at the poles (where $\sin\phi = 1$). Such a derivation, however, only applies to zonal or meridional motion; for motion in *any* direction a derivation using vector algebra is required.[42]

The equations of motion as given in Equations 5.22 and 5.23 still cannot be applied immediately to the atmosphere, because: (a) the terms $\frac{du}{dt}$ and $\frac{dv}{dt}$ represent the acceleration of an element of mass, i.e., differentiation following the motion of the air, and atmospheric quantities are normally described in terms of their rate of variation with time at a given point; and (b) the force terms X and Y require further elucidation. It will be convenient to consider (b) first.

The force terms X, Y, and Z (returning to air motion that is not necessarily horizontal) are defined as the external forces acting on an element of mass of the air. They include the effects of pressure, friction, and turbulence. It is quite easy to take into account the effect of pressure distribution: consider the case of a parallelepiped whose edges are parallel to the axes of the rectangular coordinate system (see Figure 5.15). It may then be shown that the resultant force in the OX directions due to the variation in pressure is equal to $-V\frac{dp}{dx}$, where V is the volume of air within the parallelepiped, and $\frac{dp}{dx}$ is the rate of change of pressure with distance in the OX direction. The minus sign indicates that the resultant force acts in the direction of lower pressure. This resultant force is obviously equivalent to one of magnitude $-\frac{dp}{dx}$ per unit volume in the OX direction, or a force of $-\frac{1}{\rho}\frac{dp}{dx}$ per unit mass, where ρ is the air density. A resultant force of magnitude $-\frac{1}{\rho}\frac{dp}{dy}$ per unit mass will also exist in the OY direction, due to the distribution of pressure, and one of magnitude $-\frac{1}{\rho}\frac{dp}{dz}$ in the OZ direction. The resultant of the forces acting in the horizontal plane on a unit volume, i.e., the resultant of $-\frac{dp}{dx}$ and $-\frac{dp}{dy}$, is termed the *pressure-gradient force*, which acts in a direction perpendicular to the isobars. This pressure-gradient force is usually expressed in the form $-\frac{1}{\rho}\frac{dp}{dn}$ per unit mass, where n is the perpendicular distance apart of the isobars.[43]

The question of notation now becomes important. Pressure is a function of several variables, in this case x, y, and z, and we must distinguish between the *total* rate of change of pressure with distance, and the *partial* rates of change of pressure with

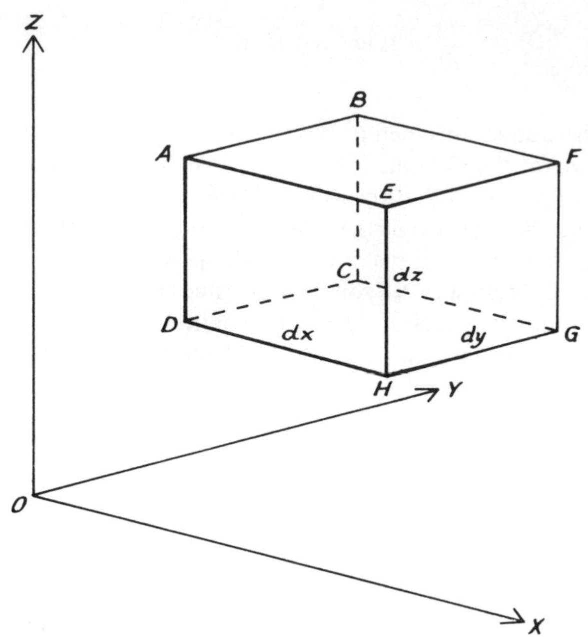

FIGURE 5.15.
Pressure-gradient force. If the force on face *ABCD* due to the surrounding air is *pdydz*, where *p* denotes atmospheric pressure, then the force on face *EFGH* must be $-(p + dp/dx \times dx) \times dy \times dz$, where $dp/dx \times dx$ represents the variation in pressure over the small distance *dx*. *ABCDEFGH* is an imaginary parallelepiped in the atmosphere, in which $dx \times dy \times dz = V$.

distance in the *OX*, *OY*, and *OZ* directions.[44] Using partial derivative notation, the correct expression for the pressure-gradient force thus becomes $-\frac{1}{\rho}\frac{\partial p}{\partial n}$ per unit mass, and the components of this force in the *OX*, *OY*, and *OZ* directions become $-\frac{1}{\rho}\frac{\partial p}{\partial x}$, $-\frac{1}{\rho}\frac{\partial p}{\partial y}$, and $-\frac{1}{\rho}\frac{\partial p}{\partial z}$ respectively. The equations of motion, taking into account the pressure-gradient force, may now be written

$$\frac{du}{dt} = 2\Omega(v \sin\phi - w \cos\phi) - \frac{1}{\rho}\frac{\partial p}{\partial x} + X', \qquad [5.24]$$

$$\frac{dv}{dt} = -2\Omega u \sin\phi - \frac{1}{\rho}\frac{\partial p}{\partial y} + Y', \qquad [5.25]$$

$$\frac{dw}{dt} = 2\Omega u \cos\phi - g - \frac{1}{\rho}\frac{\partial p}{\partial z} + Z', \qquad [5.26]$$

in which X', Y', Z', represent the components of all forces not due to pressure and gravity. The forces involved include friction between the ground surface and the air moving over it or between adjacent layers of air moving with different velocities, turbulence effects, and centripetal forces when the air particles are moving in a curved path with respect to the surface of the Earth. Confining our considerations to horizontal motion and neglecting turbulence effects, we may then write the equations of motion for *frictionless flow in a straight path* as

$$-\frac{1}{\rho}\frac{\partial p}{\partial x} = -2\Omega v \sin\phi,$$

$$-\frac{1}{\rho}\frac{\partial p}{\partial y} = 2\Omega u \sin\phi,$$

assuming that the motion of the air is steady, so that there is no acceleration; i.e., $\frac{du}{dt} = \frac{dy}{dt} = 0$. The acceleration terms are, indeed, usually negligible in comparison with the pressure-gradient and Coriolis terms, except when atmospheric pressure is changing very rapidly during the formation of a cyclone or anticyclone, so that the assumption of steady motion is a valid one. The equations in effect state that the pressure-gradient force is equal and opposite to the Coriolis force. Since Ω is known (the angular velocity of the Earth's rotation is 7.29×10^{-5} radians per second), the value of the pressure gradient at any point can be measured, ϕ is known, and p can be easily found from the temperature of the air, the equations may be solved for u and v. The resulting values are the velocities of the *geostrophic wind* in the east-west and north-south directions. If we measure the pressure gradient in the direction perpendicular to the isobars at the point in question, we then obtain the total geostrophic wind, instead of merely its components, as

$$-\frac{1}{\rho}\frac{\partial p}{\partial n} = 2\Omega c \sin\phi \qquad [5.27]$$

where c is the actual velocity of the geostrophic wind. Equation 5.27 is the *geostrophic wind equation*, which may also be expressed in the form

$$c = \frac{-P_G}{\rho\lambda},$$

where P_G is the pressure gradient and λ (equal to $2\Omega \sin\phi$) is termed the *Coriolis parameter*. The geostrophic wind equation is the basic of a scale that enables us to determine the windspeed corresponding to any given isobar separation.[45] Together with the tephigram, this wind scale is the most useful tool in theoretical or synoptic climatology. Since one can observe pressure with much greater precision than the steady component of wind velocity, it is customary to plot pressure distributions generally and then to adjust the isobars locally to fit in with the observed winds by means of the geostrophic wind scale. A given geostrophic wind scale is only applicable to maps of a specific scale, in a limited range of latitudes, and for air with a specific range of temperatures. For other scales, latitudes, or temperatures, new scales must be computed, or a special type of all-purpose scale must be used. Geostrophic-distance circles plotted on a map provide a convenient way of relating pressure and wind distributions.[46]

The direction of the geostrophic wind is the same as would be obtained by the application of Buys-Ballot's empirical law. Thus, the geostrophic wind blows parallel to the isobars with low pressure to its left and high pressure to its right in the northern hemisphere, and with high pressure to the left and low pressure to the right in the southern hemisphere. Figure 5.16 makes this clear: the geostrophic wind represents the resulting airflow when the pressure-gradient force P_G is equal and opposite to the Coriolis force C; since P_G acts at right angles to the isobars, and C is a deflecting force acting to the right of the air motion in the northern hemisphere, a flow in the direction illustrated is the only one that satisfies these conditions. The actual airflow is geostrophic in areas of straight isobars above the frictional layer, i.e., above 2,000 feet or so. In the middle layers of the atmosphere, the geostrophic wind has a standard vector error of 5.5 knots for a mean wind speed of 40 knots. Attempts have been made to improve the theory of the geostrophic wind, but in general it is a very satisfactory concept in both climatology and meteorology, except possibly for the determination of air-particle trajectories. Various methods of obtaining geostrophic winds very rapidly have been introduced, and the effect of such complications as a transverse windspeed gradient within the geostrophic wind in relation to its stability have been considered.[47]

The geostrophic wind refers to straight isobars; the corresponding theoretical wind for frictionless flow in a system of curved isobars is the *gradient wind.* This wind has the same direction as the geostrophic wind, but its speed differs. Figure 5.17 shows that for curved isobars, an additional (centrifugal) force must be introduced if a balanced flow is to be obtained. The magnitude of this force will be $\frac{V^2}{r}$ per unit mass, where V is the velocity of the air at point P, which is distant r from the center of curvature of the path of the air particle. In the general case, the equations for balanced flow will be

$$\frac{P_G}{\rho} = \lambda V + \frac{V^2}{r} \text{ for a cyclone,} \qquad [5.28]$$

$$\frac{P_G}{\rho} = \lambda V - \frac{V^2}{r} \text{ for an anticyclone,} \qquad [5.29]$$

in which V has been substituted for c in the geostrophic wind equation. These equations indicate that to obtain the speed of the gradient wind, one should obtain the geostrophic wind c at the point in question by means of a geostrophic wind scale, and subtract the value $\frac{c^2}{\lambda r}$ from the geostrophic wind so obtained for cyclonic isobar curvature, and add $\frac{c^2}{\lambda r}$ for anticyclonic curvature. This means that, for systems of equal size, the gradient wind in an anticyclone will be greater than the gradient wind in a cyclone, which may appear to be contrary to experience. However, anticyclones are normally very much larger than cyclones, and the value r in the term $\frac{V^2}{r}$ will consequently be much greater. The gradient wind may be found graphically from the geostrophic wind quite conveniently.[48]

The term "gradient wind" was first introduced by Gold to denote the theoretical wind whose direction and speed were such that the forces due to the Earth's rotation and the curvature of the path of the air particles exactly balanced the force due to the horizontal gradient of pressure. The effect of the curvature of path is of a lower order of magnitude than the effect of the pressure gradient in many synoptic situations, and is in any case of greater uncertainty. This led Shaw to introduce two new terms

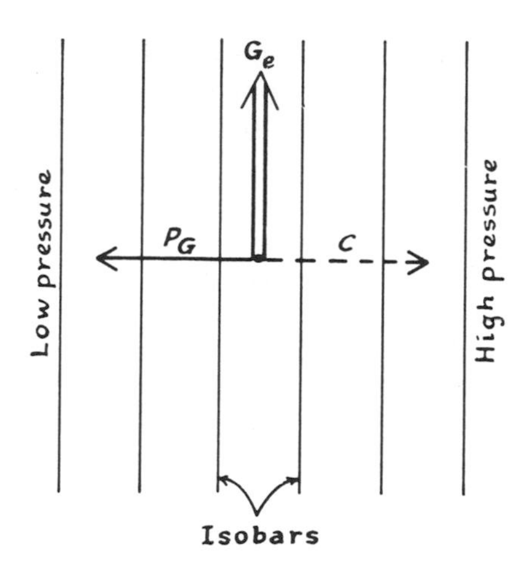

FIGURE 5.16.
Geostrophic balance; G_e represents the geostrophic wind.

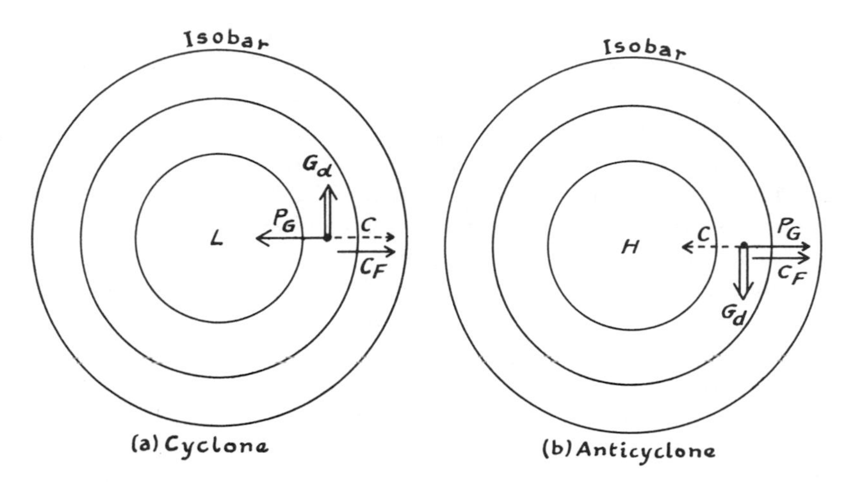

FIGURE 5.17.
Gradient balance; C_F represents the additional centrifugal force for motion in a curved path, and G_d represents the gradient wind.

for the components of the gradient wind, i.e., *geostrophic* for the effect of the Earth's rotation, and *cyclostrophic* for the effect of the path curvature. The *cyclostrophic correction* is therefore the adjustment that must be made to the geostrophic wind to correct for isobar curvature. The latter may be difficult to measure accurately because of map distortion when the system covers a large range of latitude, and the errors in isobar curvature arising from this effect must sometimes be allowed for, for example, when a large anticyclone is in relatively low latitudes.[49]

Equations 5.28 and 5.29 are quadratics, and to solve either of them for V to obtain the gradient wind will result in two values identical except for signs. The positive value is taken to be the gradient wind, but little attention has been given to the negative root until recently. The negative component has been termed the *anomalous wind*,[50] and its effect can be important. For an anticyclonic airflow, the anomalous wind represents a clockwise rotation in space that is opposite to the Earth's rotation; this may provide a dynamic mechanism for the triggering of hurricane formation and the deepening of troughs downstream from intense ridges.

The main problem in the use of the gradient-wind concept is the accurate measurement of isobar curvature, but even with poor measurements windspeeds, divergences, and vorticities computed by means of the gradient-wind assumption should be considerably more accurate than those computed under the geostrophic assumption. Sometimes the gradient-wind concept is invalid, for example, for hurricanes. Deepening or steady-state hurricanes must have pressure-gradient forces that are considerably greater than the corresponding centrifugal and Coriolis forces, particularly the latter. There is thus a lack of radial balance between these three forces in hurricanes, of about 25 to 30 per cent of the pressure-gradient force, which is probably countered by internal friction.[51]

The question of friction is important, particularly friction between the airflow and the Earth's surface. The solid part of the Earth's surface is usually rough or at least irregular, and will reduce the speed of movement of the air by frictional retardation. Therefore the terms on the right-hand sides of Equations 5.27 to 5.29 will be reduced, because these involve windspeed, but the pressure-gradient force will be unaffected. The result of friction is thus to reduce the magnitude of the Coriolis force without altering the pressure-gradient force; consequently, the direction of the balanced airflow will be toward the lower pressure rather than parallel to the isobars. Figure 5.18 shows the relation between actual and geostrophic or gradient winds. The angle θ increases as the irregularity of the surface increases: for a relatively smooth surface, such as the sea, it averages 10° or so; for a land surface, it averages 20–30° for a site of standard exposure. These figures are very general averages, and considerable deviations from them may occur locally. At Gorleston, the more cloudy is the sky at midday, the more nearly does the offshore wind approach the geostrophic wind. On the average, the surface geostrophic wind at Gorleston is at an angle of 30–40° to the isobars with offshore winds, but with onshore winds in winter the angle is only 24°. The speed of onshore winds at Gorleston is four-fifths of the geostrophic speed in winter, and two-thirds in summer; this ratio varies diurnally in summer, when the offshore winds are relatively weak at night and relatively strong during the day, but not in winter.[52] Obviously, local influences such as land- and sea-breeze effects, which are not embraced by the concept of strophic balance if they are very small, and which are not depicted on the synoptic chart, can cause the actual wind to deviate locally from the surface geostrophic value.

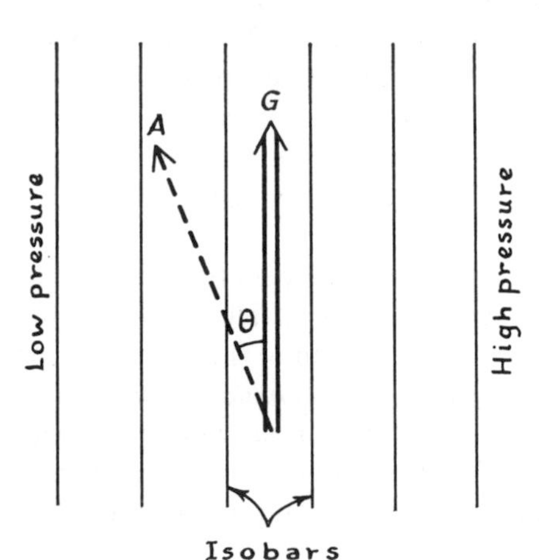

FIGURE 5.18.
Surface wind versus geostrophic/gradient winds; *G* represents the geostrophic or gradient wind, and *A* represents the actual wind at the Earth's surface.

The *ageostrophic wind* is the difference between the actual wind and the gradient or geostrophic winds. Consequently, the smaller the magnitude of the ageostrophic wind, the more closely the geostrophic/gradient wind approximates the true wind. The geostrophic wind generally gives a poor approximation to the actual winds for geostrophic speeds exceeding 50 miles per hour, in which case the gradient wind gives a better result, provided the trajectory curvature can be accurately measured; but in latitudes below about 30°, neither the geostrophic nor the gradient wind provide reliable results. The "ageostrophic wind" may also include a subjective component if the geostrophic/gradient winds are computed from map analyses.[53]

The *isallobaric wind* bears the same relation to a map of pressure tendencies as the geostrophic wind does to a map of isobars. According to Brunt and Douglas, the actual wind velocity consists of two parts: the geostrophic wind, proportional to the pressure gradient, and the isallobaric wind, proportional to the gradient of pressure tendency. Thus the isallobaric wind is part of the ageostrophic wind. The validity of the isallobaric wind concept has been questioned; nevertheless, the measured isallobaric wind provides a useful first approximation to the ageostrophic wind.[54]

The basis of the concept of strophic balance is that there must be a connection between wind and pressure, which can be expressed in terms of the laws of dynamics. It is worthwhile considering to what extent the observations indicate that the correlation between wind and pressure is a precise one. Personal factors must enter from the start if synoptic charts are to be used, as they must, instead of single-station data, for the heights of pressure surfaces must usually be interpolated for the point at which the test is made, and a typical (average) subjective error in locating the height of a pressure surface in this manner is 15 meters. The windspeed must then be computed from the interpolated pressure heights, by means of the geostrophic wind scale; the personal factor leads to errors of up to 25 per cent here, the error decreasing as windspeeds increase. Once the wind values have been computed, it is generally found that the *variabilities* of wind and pressure are closely related. Thus the square of the standard deviation of height or pressure and the square of the standard vector deviation of wind are in an approximately linear relationship over large areas and through deep layers of the atmosphere.[55]

Another interesting question is how a state of strophic balance is achieved. Since the pressure distribution is continually changing, so is the pressure-gradient force. How long does it take the wind field to adjust itself to the pressure field? The fact that the pressure-gradient force is almost never static may seem to imply that balance is never attained, yet synoptic experience proves that it frequently is. It may be shown that the sensitivity of the wind field to pressure variations depends on the lateral shear of the geostrophic wind and the tangential and orthogonal curvature of the streamlines of the geostrophic wind.[56] The latter factor is the dominant one when the geostrophic windspeed is high. Crossisobaric flow, i.e., airflow that has not come into strophic balance, once it is established over an extensive area, can continue over that area for some time, can cross the equator, and can even penetrate to higher latitudes. Thus the pressure-gradient force by itself is not sufficient to determine the horizontal motion of the air. Subsiding cold air in the tropics is a good example of this; in the West Indies, for example, shallow outbursts of cold air produce strong winds with a marked cross-isobaric component that penetrates rapidly toward the equator independently of the pressure gradient.

In tropical regions in general, the geostrophic/gradient wind theory is of no practical consequence in the interpretation of air movements. Consequently, some basis other than isobars is necessary for synoptic analysis; it is provided by *streamline analysis*. Streamlines are lines drawn tangent to the instantaneous wind vectors; a streamline map gives an instantaneous picture of the field of motion. In contrast, a map of *trajectories* shows the actual paths followed by individual air parcels or particles during some period of time. Streamlines may be obtained directly from the wind observations, or may be computed from the pressure distribution; when computed they sometimes give better indications of actual winds than the geostrophic winds do.[57] Although streamlines indicate wind direction, not wind velocity, by means of the *stream function* streamline patterns may be used to estimate wind velocities. The stream function bears much the same relationship to streamlines as the geostrophic wind does to isobars (see Appendix 5.18). In order to understand the stream function, one must deal with the hydrodynamics of the atmosphere.

Elementary Kinematics and Hydrodynamics of the Atmosphere

Dynamics may be defined as the branch of mechanics that studies the motion of bodies in terms of forces and mass, and kinematics as the branch of mechanics that studies the motion of bodies without reference to force or mass. It is possible to deal with certain atmospheric motions by purely kinematic methods, for example, the prediction of changes in the surface pressure field, but usually some reference to force, especially, and to mass is necessary, so that an elementary consideration of hydrodynamics is essential for the climatologist who wishes to understand the phenomena he is dealing with.[58]

The mechanics of fluids is more complicated than the mechanics of solids because a fluid will yield to any shear stress on it. The atmosphere in motion is subject to many shear stresses, and under these conditions the elements of fluid motion within it are being continuously deformed. The two general methods for describing fluid motion are the *Eulerian* and the *Lagrangian* methods: in the former, the velocity field is described in terms of time and space coordinates; in the latter, the trajectories of the fluid elements are described. For the atmosphere the Lagrangian method can be regarded as describing the variations recorded by instruments moving with the wind; the Eulerian method describes the variations recorded by an instrument in a fixed location. The Eulerian system is the one most commonly used in meteorology, and the Lagrangian equations can be considered integrals of the Eulerian equations.

The equations of motion, Equations 5.19 to 5.26, contain the expressions $\frac{du}{dt}$, $\frac{dv}{dt}$, and $\frac{dw}{dt}$, which represent the accelerations experienced by an element of mass; i.e., they represent differentiation following the motion. These equations may be written in hydrodynamic form, in the Eulerian system and neglecting the pressure-gradient force, as

$$\frac{\partial u}{\partial t} + u\frac{\partial u}{\partial x} + v\frac{\partial u}{\partial y} + w\frac{\partial u}{\partial z} = 2\Omega\,(v \sin\phi + w\cos\phi) + X, \qquad [5.30]$$

$$\frac{\partial v}{\partial t} + u\frac{\partial v}{\partial x} + v\frac{\partial v}{\partial y} + w\frac{\partial v}{\partial z} = -2\Omega u \sin\phi + Y, \qquad [5.31]$$

$$\frac{\partial w}{\partial t} + u\frac{\partial w}{\partial x} + v\frac{\partial w}{\partial y} + w\frac{\partial w}{\partial z} = 2\Omega u \cos\phi - g + Z. \quad [5.32]$$

The left-hand sides of these equations represent the expanded forms of $\frac{du}{dt}$, $\frac{dv}{dt}$, and $\frac{dw}{dt}$, respectively (see Appendix 5.19). The expression $\frac{du}{dt}$ represents the instantaneous rate of change of velocity with time that an observer following the motion would note, whilst $\frac{\partial u}{\partial t}$ represents the instantaneous rate of change of velocity with time that a stationary observer, past whom the air particles are moving, would note.

A new concept has now to be introduced: the *equation of continuity*. This is a development of the law of conservation of matter, and is simply the mathematical statement of the fact that the mass of air flowing into a given volume in a given time must equal the mass of air flowing out, unless a change in density has occurred. The mass of air flowing into face A of the cube in Figure 5.19 must equal the mass of air flowing out of the opposite face, provided the airflow is horizontal and parallel to the shaded face of the cube, if the density of air in the cube remains unchanged. The equation of continuity is

$$\frac{\partial \rho}{\partial t} + \frac{\partial}{\partial x}(\rho u) + \frac{\partial}{\partial y}(\rho v) + \frac{\partial}{\partial z}(\rho w) = 0,$$

which may be expanded to

$$\frac{\partial \rho}{\partial t} + u\frac{\partial \rho}{\partial x} + v\frac{\partial \rho}{\partial y} + w\frac{\partial \rho}{\partial z} + \rho\frac{\partial u}{\partial x} + \rho\frac{\partial v}{\partial y} + \rho\frac{\partial w}{\partial z} = 0, \quad [5.33]$$

the first four terms of which represent $\frac{d\rho}{dt}$, the rate of change of density with time following the motion, expanded as in Equations 5.30 to 5.32. Equation 5.33 may

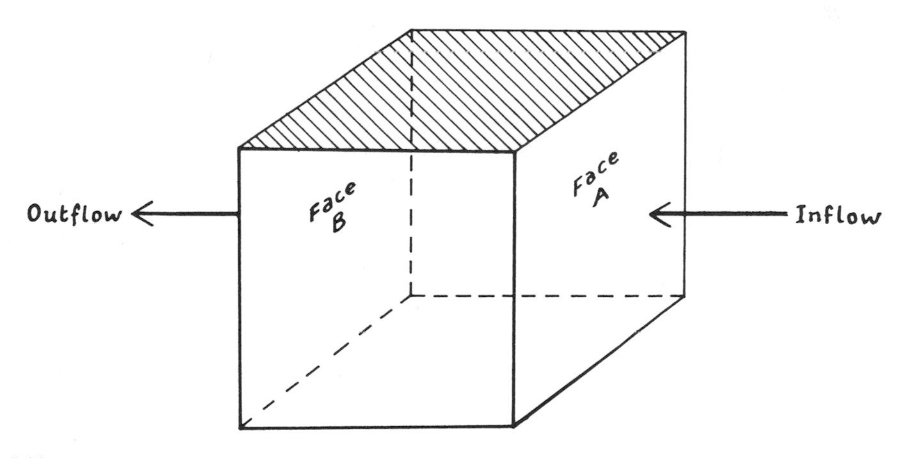

FIGURE 5.19.
Equation of continuity.

therefore be written in the form

$$\frac{d\rho}{dt} + \rho\left(\frac{\partial u}{\partial x} + \frac{\partial v}{\partial y} + \frac{\partial w}{\partial z}\right) = 0,$$

or, more usually,

$$\frac{1}{\rho}\frac{d\rho}{dt} + \left(\frac{\partial u}{\partial x} + \frac{\partial v}{\partial y} + \frac{\partial w}{\partial z}\right) = 0. \qquad [5.34]$$

Equation 5.34 is the normal way of writing the equation of continuity. The terms in parentheses have an important meaning: they define the *divergence* of the air.[59] If their sum is positive, then by definition divergence is occurring; i.e., a depletion of mass is occurring within a given volume of air. If their sum is negative, then *convergence* is taking place; i.e., an accumulation of mass is occurring within the given volume. If their sum comes to zero, then any lateral divergence (or convergence) is being compensated by vertical shrinking (or expansion), and the flow is said to be incompressible. Atmospheric flow is normally divergent; and nondivergent (i.e., incompressible) flow is comparatively rare, except at a certain level in the upper half of the atmosphere. If the flow were always incompressible, so that vertical convergence was always balanced by horizontal divergence and vice versa, no surface pressure changes would be possible. There can be no accumulation or depletion of mass if the airflow is truly geostrophic; divergence is a common feature of the atmosphere, and so the actual air flow is usually not geostrophic, but *quasigeostrophic*.

Divergence and convergence as just discussed are hydrodynamic concepts and should not be confused with topographic convergence and divergence, or with confluence and diffluence, which refer to the channeling effect of valleys on the airflow and the coming together or spreading apart of streamlines, respectively.[60] Confluence (diffluence) does not necessarily also imply hydrodynamic convergence (divergence). Instantaneous values of hydrodynamic divergence may be estimated from a synoptic chart by replacing the differentials in Equation 5.34 by finite differences, and mean values of divergence may be approximated by simply resolving the mean vector winds into their north-south and east-west components at four points separated by 5° of latitude (see Appendix 5.20). In the lower and middle troposphere, mean divergence values are on the order of 1 to 2×10^{-5} per sec, the larger value applying to the smallest scale features.[61]

Divergence and convergence fields may be readily mapped, and cross sections showing their distribution may also be plotted, by means of objective techniques. Divergence is a very significant control of weather and climate at all scales. At the local scale, for example, there is a connection between the magnitude of divergence and the occurrence or absence of precipitation in the tropics, precipitation being heavy with a divergence of -10^{-4} per sec, light with a divergence of -10^{-5} or -10^{-6}, and almost nil with a divergence of -10^{-8}. With divergence values on the order of $+10^{-6}$ to $+10^{-5}$ per sec, moderate subsidence occurs. On a larger scale, the diurnal variation of precipitation over the plains of the central United States is related to the diurnal variation of horizontal divergence. Nighttime rains are very frequent in this area because of low-level convergence and ascending air over the plain, accompanied by upper-level divergence.[62]

On a still larger scale, consider a sinusoidal pattern of isobars covering an area the size of an extensive continent (Figure 5.20). If the gradient wind blows through this system of troughs and wedges, cyclonic curvature will occur at the trough line *B* and anticyclonic curvature at the wedge line *A*. From the gradient-wind equation, the wind velocity at A must be greater than the wind velocity at B; the wind velocity must therefore be decreasing downstream between A and B, resulting in horizontal convergence. Mass will thus be taken from A and added to B, decreasing the pressure at A and increasing it at B. Horizontal divergence is occurring between B and C, so that pressures at C must be falling. Therefore the entire wave pattern must be displaced from left to right. In the middle-latitude westerlies, the long waves in the upper air should therefore move from west to east more slowly than the gradient wind through them.[63] Since it may easily be demonstrated that convergence is associated with ascending air, cloud formation, and precipitation, and divergence with subsiding air and clearing skies, we should expect the weather in middle latitudes to consist of periods of bad weather separated by periods of good weather, successively following each other across the continent from west to east (see Appendix 5.21). This state of affairs is obviously very frequent. Connections between divergence and numerous other weather phenomena have been demonstrated.[64]

In an important theorem published in 1845, Stokes showed that the motion of a small element of fluid, during a short period of time, may be regarded as consisting of four parts: (a) a translation, i.e., a movement of the entire fluid system as a whole; (b) a solid rotation of the element without deforming it; (c) a volume expansion or contraction of the element; and (d) a shear.[65] Part (b) is of especial interest at this stage of our discussion. A basic quantity in hydrodynamics is *circulation*, involving movement around a closed path within the fluid. Circulation is defined as the line integral of the velocity of the fluid around this closed path (see Appendix 5.22). The simplest form of circulation is uniform motion around a circular path r with tangential velocity v; the circulation is then given by

$$C = 2\pi rv.$$

The concept of circulation is the basis of an extremely important notion, that of *vorticity*. According to Stokes' theorem, vorticity may be defined as *circulation per unit area*. For a circular disc of fluid rotating as a solid with angular velocity ω, the circulation around its outer boundary will be equal to $2\pi R$ multiplied by ωR, where R is the radius of the outer boundary; that is circulation equals $\pi R^2(2\omega)$ or area times 2ω. Dividing through by area gives us circulation per unit area, and we

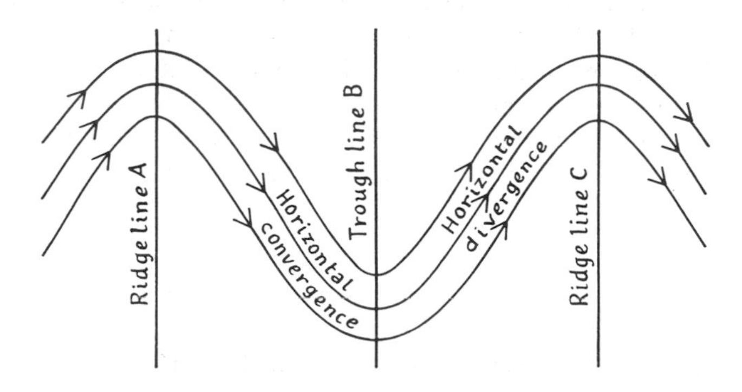

FIGURE 5.20.
Airflow through a sinusoidal pattern of isobars. Arrows indicate the direction of the geostrophic/gradient wind.

find that for a disc rotating as a solid, the vorticity equals twice the angular velocity of rotation. It may be shown that in the general case, i.e., whatever the form of the area and whether or not the fluid rotates as a solid,

$$\zeta = \frac{\partial v}{\partial x} - \frac{\partial u}{\partial y}, \qquad [5.35]$$

$$\xi = \frac{\partial w}{\partial y} - \frac{\partial v}{\partial z}, \qquad [5.36]$$

$$\eta = \frac{\partial u}{\partial z} - \frac{\partial w}{\partial x}. \qquad [5.37]$$

These equations give the components of vorticity referred the three orthogonal axes of the Cartesian system: ζ is the vorticity component about the vertical axis, i.e., vorticity in the xy plane; ξ is the vorticity component in the yz plane; and η is the component of vorticity in the xz plane.[66] The ζ component is the most important in most investigations.

Vorticity is thus a measure of rotation or spin. Any fluid motion that has vorticity is termed *rotational motion*, and fluid motion without vorticity is *irrotational motion*. Vorticity is a function of two parameters: curvature and shear. It clearly must depend on the curvature of the path around which the element travels, because it is measured in terms of rotation or angular velocity. Isobar curvature consequently is an important determinant of the magnitude of vorticity at a given instant. *Shear* is defined as the rate of change of velocity in a direction at right angles to the direction of motion; i.e. a wind shear occurs if the windspeed varies in the horizontal but the wind direction remains constant. An air current may have vorticity even if it is a straight current flowing for a thousand miles or more without curving, provided it has a shear.[67] The convention is to take cyclonic shear and curvature that goes counterclockwise as positive (in the northern hemisphere), and anticyclonic shear and curvature that goes clockwise as negative. Equations 5.35 to 5.37 therefore give positive vorticity values for counterclockwise motion in the northern hemisphere, which is the direction of rotation of the Earth in that hemisphere.

Since the Earth is rotating, it must have a vorticity: this vorticity is always positive and cyclonic. The Earth's atmosphere rotates with the Earth, and its mean vorticity, regarding the atmosphere as a whole in solid rotation, is 2Ω, where Ω is the angular velocity of the Earth. However, local variations in atmospheric vorticity occur, and these are of prime significance for both climate and weather.

A distinction must be made between *relative vorticity*, which is the vorticity of the airflow relative to the Earth's surface, and *absolute vorticity*. The latter is equal to the sum of the relative vorticity and the vorticity of the Earth itself:

$$\zeta_A = \zeta + f,$$

where f is the vorticity of the Earth's surface and ζ is the relative vorticity. It will be noted that f is the same as the Coriolis parameter, because it is apparent from the equations of motion that a point on the surface of the Earth in latitude ϕ is rotating with angular velocity $\Omega \sin \phi$ about a vertical axis through that point, and it has already been shown that the vorticity of a rotating disc is 2ω.[68] If we take the hypothetical case of a column of air that is the size of a cyclone or anticyclone and that is rotating as a solid with an angular velocity ω with respect to the Earth's surface (counted as positive for cyclonic rotation and negative for anticyclonic rotation), then the absolute velocity of this air column is $\omega + \Omega \sin \phi$. Assuming that the air is homogeneous and incompressible, the equation of continuity implies that volume will be conserved; i.e., lateral convergence will be compensated for by vertical stretching of the air column and lateral divergence by vertical shrinking. Assuming that the principle of conservation of angular momentum (see Appendix 5.23) applies to the air column, then it may be shown that

$$\frac{2\Omega \sin \phi + 2\omega}{H} = k$$

where H is the height of the air column and k is a constant. This may be stated in the form.

$$\frac{\lambda + \zeta}{\Delta p} = k, \qquad [5.38]$$

where λ is the Coriolis parameter, ζ is the relative vorticity of the air column, and Δp is the pressure difference between the top and bottom of the column. Equation 5.38 is a simplified version of the *theorem of conservation of absolute vorticity*.[69] It indicates that for a given latitude, i.e., for a given value of λ, any increase in ζ must be accompanied by an increase in Δp, and vice versa. The implications are as follows. (see Figure 5.21).[70]

A. At a given latitude, if cyclonic vorticity is increasing (or anticyclonic vorticity decreasing), then lateral convergence must be occurring, leading to cyclogenesis for a straight air flow.

B. At a given latitude, if cyclonic vorticity is decreasing (or anticyclonic vorticity increasing), then lateral divergence must be occurring, leading to anticyclogenesis for a straight air flow.

C. If there is no pronounced lateral divergence, then, if there is an equatorward airflow, it must develop increasing cyclonic or decreasing anticyclonic vorticity.

D. If there is no pronounced lateral convergence, then if there is a poleward-moving air flow, it must develop decreasing cyclonic or increasing anticyclonic vorticity

Thus the final effect of a change in latitude will be to create a low-pressure trough aloft, just to the west of the surface position of a cold front, and an upper high-pressure wedge above the slope of the warm front, because anticyclonic vorticity must develop in the warm air above the warm front and cyclonic vorticity above a cold front. In terms of actual weather, the following rules emerge.

1. Clouds and precipitation will develop under cyclonically curved isobars aloft, independently of the presence or absence of surface fronts.
2. Clouds and instability showers in a cold mass of air can only occur in that portion of the air flow moving in a cyclonically curved path.
3. If an air current has a component from the south, clouds and precipitation are likely everywhere except where the current turns sharply anticyclonically. This rule particularly applies in mountainous areas, over the oceans, and in the warm sectors of frontal depressions.

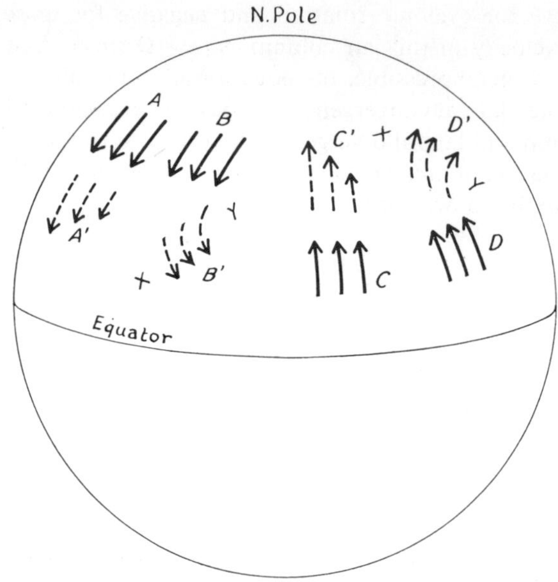

FIGURE 5.21.
Consequences of conservation of absolute vorticity.
A and *A′* are narrow air currents moving south, forced to follow a straight path; *B* and *B′* are similar currents, but are free to seek their own path. *A* develops cyclonic shear (*A′*); *B* develops cyclonic curvature (*B′*).
C and *C′* are narrow air currents moving north, forced to follow a straight path; *D* and *D′* are similar currents, but are free to seek their own path.
C develops anticyclonic shear (*C′*); *D* develops anticyclonic curvature (*D′*).

4. Clear skies occur where equatorward-moving air is following a straight or anticyclonically curved path, or where poleward-moving air turns sharply anticyclonically.
5. Elongated V-shaped upper troughs should be associated with clouds and precipitation in the equatorward-moving current in advance of the trough-lines, with clearing at and behind the latter.

The importance of the concept of vorticity cannot be stressed too much; it is, more than any other idea perhaps, the keynote of dynamic climatology.[71] Its importance is both theoretical and practical. On the theoretical side, Gordon has shown that there is a connection between vorticity and divergence.[72] The total instantaneous rate of change of absolute vorticity with time, $\frac{d(\zeta + \lambda)}{dt}$, which equals the instantaneous rate of change of relative vorticity with time, $\frac{d\zeta}{dt}$, plus the rate of change of the Coriolis parameter, $\frac{d\lambda}{dt}$, is given by

$$\frac{d(\zeta + \lambda)}{dt} = -(\zeta + \lambda)\left(\frac{\partial u}{\partial x} + \frac{\partial v}{\partial y}\right). \qquad [5.39]$$

The importance of Equation 5.39 is in connection with the estimation of divergence. The method of estimating divergence given in Appendix 5.20, that is, by computing it from the synoptic chart, is not entirely successful because it is based on actual winds, which may vary considerably between stations. The divergence cannot be computed from geostrophic winds, which would of course provide a sound method of interpolation between stations, because geostrophic flow is nondivergent; i.e. there can be no accumulation or depletion of mass if the flow is geostrophic. However, geostrophic values may be used to obtain ζ in Equation 5.39, giving the *geostrophic vorticity*, and the divergence may be estimated from this equation. On the practical side, numerous connections between vorticity and various meteorological phenomena have been demonstrated, because the type of current (i.e., its vorticity) is as important as heating and cooling from below for the modification of air masses. Lapse-rates in cyclonic and anticyclonic vorticity areas often differ as much as for warm and cold air masses; anticyclonic vorticity is associated with subsidence, and deep convection (i.e., with the tops of convective clouds above 600 mb) with cyclonic vorticity aloft. Furthermore, vorticity is associated with rapid cyclogenesis; spectacular cyclonic development occurred downstream just after the build-up of strong cyclonic vorticity off the Pacific coast of the United States on April 22, 1958. Vorticity transfer can create strong zonal currents within about three weeks, and the deepening of wave depressions is related to the advection of geostrophic vorticity. Vorticity considerations are very enlightening in case studies of upper-wind distributions and in investigations of, for example, advection, vertical motion, the generation of horizontal wind shear, the growth of troughs and wedges aloft, the generation of upper-air fronts, and the development of large-scale circulation pattern anomalies. In terms of the weather elements, connections with vorticity distribution have been shown, for example, for showers, precipitation generally, and zonal wind profiles.[73]

Geostrophic vorticity may be estimated from isobaric charts by means of finite differences: upper-air contour charts are usually accurate enough to give the vorticity correctly to within 30 per cent. The method involves superimposing a rectangular grid on to the chart and reading off height values for the pressure contours at intersections of grid lines; the grid size has a marked effect on the magnitude of the computed vorticity (see Appendix 5.24). Observed winds give a better approximation to the true vorticity than do geostrophic vorticities for levels above 300 mb, provided the winds have been determined by radar or by equally precise methods. The geostrophic vorticity may differ appreciably from the true vorticity in storms of moderate intensity, and the computed advection of the two vorticities may also differ significantly.[74]

The magnitude of vorticity varies from 75×10^{-5} per sec for the vertical component of relative vorticity in small-scale wind systems to 2.4×10^{-5} per sec for large-scale systems. At all scales, mean vorticity values are about 50 per cent larger than the mean divergence. The mean relative vorticity in the lowest 25,000 feet of the atmosphere over the eastern United States is 2.4×10^{-5} per sec, where the mean absolute vorticity is 11.6×10^{-5} per sec. Charts that show the changing synoptic patterns of vorticity distribution may be constructed; relative vorticity charts enable climatological studies of the cyclonic or anticyclonic "energies" of the atmosphere over different parts of the Earth to be made, and absolute vorticity charts are of considerable use in weather forecasting on a global scale.[75]

Synoptic experience indicates that horizontal convergence within the lower troposphere in developing cyclones is usually balanced by horizontal divergence in the upper sphere, and that horizontal divergence near the Earth's surface in developing anti-cyclones is usually balanced by horizontal convergence aloft. Consequently, an intermediate level must exist at which the divergence is zero. The level of nondivergence is, on the average, near 600 mb, but not necessarily everywhere at the same time. The existence of a level of nondivergence is of considerable significance: from Equation 5.39 it follows that, at the level of nondivergence, the vertical component of absolute vorticity, i.e., $\zeta + \lambda$, must be constant following the motion of the air. Air moving poleward at the level of nondivergence must therefore curve anticyclonically back toward the equator, because λ increases as it travels poleward and therefore ζ (i.e., its positive, or cyclonic, vorticity) must decrease. Similarly, air moving equatorward at this level must curve cyclonically back towards the pole because λ decreases as it moves toward the equator and ζ must consequently increase. It follows that the pattern of airflow in the upper air of middle latitudes must be in the form of sinusoidal waves (see Figure 5.22), which as we have already seen from the application of divergence considerations to the gradient wind, must move from west to east. This is the basis of *Rossby's theory of long-waves in the westerlies*. Writing Equation 5.39 in the form

$$\frac{d(\zeta + \lambda)}{dt} = 0,$$

Gordon shows that

$$c = U - \frac{\beta L^2}{4\pi^2}, \qquad [5.40]$$

where c is the speed of movement of the long-wave system as a whole from west to east, U is the velocity of the westerlies (which are assumed to be invariant with latitude), β is the instantaneous rate of change of the Earth's vorticity at a given point in the

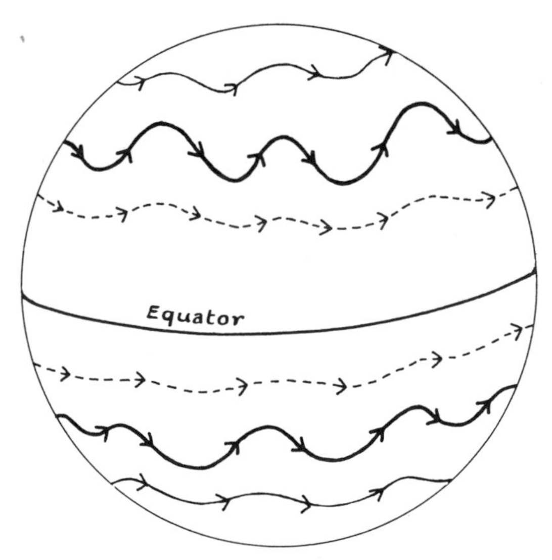

FIGURE 5.22.
Rossby waves. The heavy lines indicate the most pronounced, the broken lines the least pronounced, Rossby wave in each hemisphere.

airflow (i.e., the rate of change of the Coriolis parameter with latitude), and L is the wavelength of the sinusoidal disturbances. Equation 5.40 is the *Rossby wave equation*, which describes the velocity of the long-wave system in terms of the wavelength and the mean zonal windspeed of the middle-latitude westerlies. It shows that if the decrease in vorticity between a trough and the next ridge in a wave just balances the poleward rate of increase in the Coriolis parameter, then the wave system as a whole must become stationary. The wavelength of this stationary wave varies from 40° to 120° in longitude, depending on latitude and windspeed. Reinforcement and interference between different wave systems must be allowed for.[76]

The Rossby wave equation is actually a solution of the equation of constant absolute vorticity, which is derived from a set of six simultaneous differential equations. These equations are: (a) the three equations of motion in three coordinates; (b) the equation of continuity; (c) the physical equation; (d) an equation relating the amount of energy received to that used and emitted by the air particle. Equation (c) is derived from thermodynamics, and describes the consequences of the variation in atmospheric density on air motion. The equations are in the form of nonlinear differential equations, whose solution is mathematically very difficult. In fact, no standard mathematical methods were available to Rossby to solve the equations, and he adopted the *perturbation technique* for dealing with them. This technique involves linearizing the equations by assuming that the atmospheric variables consist of basic values unchanging with time plus small perturbations whose second-order terms may be neglected. The perturbation technique is the standard method for dealing with the meteorological equations of motion. The *method of characteristics* is another way of dealing with a related problem, that of hyperbolic differential equations for the atmosphere. Rossby applied the perturbation technique by assuming that the perturbations of the motion are infinitely small, but the vorticity equation can be solved without making this assumption.[77]

There have been numerous extensions of the vorticity equation since Rossby, for example, to westerlies which do not vary in speed with latitude, and to wave patterns that change with time.[78] In the literature the vorticity equation and the equations of motion are written completely differently from the forms given in this chapter; the usual mode of writing equations in atmospheric dynamics is in terms of vector algebra. It will be noted that the important dynamic equations given in the present chapter, i.e., Equations 5.24–5.26, 5.30–5.32, 5.34–5.37, and 5.39, either deal with atmospheric motion components in different directions separately, or ignore all motions other than horizontal. This is unsatisfactory, because atmospheric motion is three-dimensional. It is also unnecessary, because three-dimensional motion can be dealt with by *vector analysis*, which not only is a very concise form of description, but also provides considerable insight into the physical motions discussed. Here the reader who is not a mathematician may begin to experience some considerable difficulty, because vector terminology looks very unfamiliar at first, but if he is serious, he should persevere. Without a knowledge of vector notation, he will find much of the literature in dynamic climatology incomprehensible, and in any case vector algebra is the basic method of description in most branches of physical science and engineering. The vector form of the dynamic equations is as follows.[79]

The Coriolis force in vector terminology is $-2\mathbf{\Omega} \times \mathbf{V}$ where the symbol $\times$ represents the *vector cross product* (see Appendix 5.25). The expression is read "minus two omega cross vee." The pressure gradient in vector terminology is $-\nabla p$. The symbol ∇ is called *del.* It is not a number or a quantity, but an *operator* in differential calculus. In Cartesian coordinates,

$$\nabla = \frac{\partial}{\partial x} + \frac{\partial}{\partial y} + \frac{\partial}{\partial z}.$$

The pressure gradient in three dimensions is therefore given by

$$\nabla p = \frac{\partial p}{\partial x} + \frac{\partial p}{\partial y} + \frac{\partial p}{\partial z}$$

which is sometimes written **grad** p. The vector idea of a gradient is an intrinsic property of a scalar field, therefore, and is *invariant*; i.e., it is entirely independent of the coordinate system, and is unaffected by a rotation of the coordinates. The concept of divergence in vector form, i.e., div **V**, is given by the *vector dot product* (see Appendix 5.26) of del and the velocity **V** (read "del dot V"):

$$\text{div } \mathbf{V} \equiv \nabla \cdot \mathbf{V}. \qquad [5.41]$$

Therefore the idea of divergence must also be invariant.[80] The concept of vorticity in vector form, known as the *curl* of the vector, is given by

$$\text{curl } \mathbf{V} \equiv \nabla \times \mathbf{V}. \qquad [5.42]$$

Curl or vorticity is also independent of rotation of the coordinate system. In Equations 5.41 and 5.42, the Cartesian forms of divergence and vorticity are given by

$$\text{div } \mathbf{V} = \frac{\partial \mathbf{V}_x}{\partial x} + \frac{\partial \mathbf{V}_y}{\partial y} + \frac{\partial \mathbf{V}_z}{\partial z},$$

$$\text{curl } \mathbf{V} = \left(\frac{\partial \mathbf{V}_z}{\partial y} - \frac{\partial \mathbf{V}_y}{\partial z}\right) + \left(\frac{\partial \mathbf{V}_x}{\partial z} - \frac{\partial \mathbf{V}_z}{\partial x}\right) + \left(\frac{\partial \mathbf{V}_y}{\partial x} - \frac{\partial \mathbf{V}_x}{\partial y}\right),$$

where $\mathbf{V}_x$, $\mathbf{V}_y$, and $\mathbf{V}_z$ represent the resolved parts of the velocity vector **V** in the OX, OY, and OZ direction, respectively, of the orthogonal coordinate system.

The operator div grad, i.e., the operator $\nabla \cdot \nabla$ or "del dot del," usually written ∇^2 (which is read "del two" and not "del squared" because it is not a quantity), has an important meaning: ∇^2 is known as Laplace's operator, and to operate by ∇^2 on any quantity is to find its *Laplacian.* The Laplacian of the quantity ϕ is given by.

$$\nabla^2 \phi = \frac{\partial^2 \phi}{\partial x^2} + \frac{\partial^2 \phi}{\partial y^2} + \frac{\partial^2 \phi}{\partial z^2}.$$

For nondivergent flow, vorticity can be represented as the Laplacian of the stream function. This enables vorticities to be found from streamlines more accurately than they may be determined from isobars. The equation $\nabla^2 \phi = 0$ is known as *Laplace's equation*, which has been described as the most important equation in the whole of physical science; it has profound consequences for the construction and interpretation of valid isopleth maps of any quantity.[81]

The importance of the ∇^2 operator in dynamic climatology arises from the fact that, for geostrophic flow on a pressure surface, and ignoring small variations in the Coriolis parameter λ, it may be shown that the relative vorticity about a vertical axis is given by

$$\zeta_z = \frac{g}{\lambda} \nabla^2 z,$$

where z is the height of the constant-pressure surface and $\nabla^2 z$ is the Laplacian of height z; i.e.,

$$\nabla^2 z = \frac{\partial^2 z}{\partial x^2} + \frac{\partial^2 z}{\partial y^2}.$$

This two-dimensional form of the Laplacian of contour height can be estimated from a contour chart by finite-difference approximations (see Appendix 5.27).

The *Jacobian* operator J is defined by

$$J(P_1, P_2) = \frac{\partial P_1}{\partial x} \cdot \frac{\partial P_2}{\partial y} - \frac{\partial P_1}{\partial y} \cdot \frac{\partial P_2}{\partial x},$$

where $J(P_1, P_2)$ is the Jacobian of properties P_1 and P_2, and represents the advection of property P_2 along isopleths of constant P_1.[82]

The Laplacian and Jacobian operators appear in the equations of motion and the vorticity equation as follows:

$$\frac{\partial \mathbf{V}}{\partial t} = -\alpha \nabla p - 2\mathbf{\Omega} \times \mathbf{V} - g, \qquad [5.43]$$

$$\frac{1}{\alpha} \frac{d\alpha}{dt} = \nabla \cdot \mathbf{V}, \qquad [5.44]$$

$$p\alpha = RT, \qquad [5.45]$$

$$dQ = C_p \, dT - \alpha dp, \qquad [5.46]$$

$$\nabla^2 p \frac{\partial z}{\partial t} = -J(z, F). \qquad [5.47]$$

Equation 5.47 is the vorticity equation, in which $\nabla^2 p$ is the Laplacian operator referred to a coordinate system that has pressure p instead of height z as the vertical coordinate, and F represents the quantity $\left(\frac{g}{\lambda}\right)\nabla^2 pz + \lambda$. Equation 5.47 is the form of the vorticity equation adapted for weather forecasting by means of the barotropic model. Equation 5.43 is the *vector equation of motion*, omitting friction; Equation 5.44 is the *vector equation of continuity*; Equation 5.45 is the equation of state; and Equation 5.46 represents the first law of thermodynamics. In each of these equations, α represents specific volume, i.e., the inverse of density. The economy effected by making use of vector notation is clear: Equation 5.43 is equivalent to the three scalar equations 5.30 through 5.32.[83]

Equations 5.43 through 5.47 provide the fundamental basis for the study of dynamic climatology. It may therefore be advisable at this point to stress what these equations in fact are: *they represent the atmospheric equivalent of Newton's second law of motion.* In other words, Equation 5.43 is the meteorological equivalent of $P = ma$, and the other equations represent mathematical identities that must be taken into account when using this equation because of the physical nature of the atmosphere.

Our treatment of atmospheric motion is not yet complete, because the effect of temperature variation has not so far been discussed. The equations of motion we have described apply only in a barotropic atmosphere. In a baroclinic atmosphere (see Figure 5.23), the isopleths of density (or specific volume) and of pressure do not coincide with each other as they do in a barotropic atmosphere, but intersect each other. Whether or not the atmosphere in some region is barotropic can be determined from the synoptic chart: if, on a pressure surface, isotherms can be drawn, the atmosphere is baroclinic; if there is no horizontal gradient of temperature, the atmosphere is barotropic. In a baroclinic atmosphere, the intersection of the isobaric and the isosteric (i.e., constant-density) surfaces may be regarded as forming a number of tubes or *solenoids* (see Figure 5.23, B). It is shown in *Bjerknes' circulation theorem* that the number of solenoids measured at fixed unit intervals is a function of the intensity of the atmospheric circulation.[84] Thus the greater the degree of baroclinicity of the atmosphere, the more intense are the resulting circulations. We now have to consider the influence of baroclinicity on atmospheric dynamics.

The geostrophic and gradient wind equations, Equations 5.27 through 5.29, may be used directly on pressure-contour charts, but not on constant-level charts, i.e., only on maps that depict variations in the height of a specified pressure surface. The reason is that the geostrophic wind equation includes density, and a geostrophic wind scale for a constant-level chart must be constructed for a specific density, whereas for pressure-contour charts it may be shown that the geostrophic wind at a given latitude depends only on the slope of the isobaric surface, and not on the density.[85] In a barotropic atmosphere, the wind direction must of necessity be constant with height (because in a barotropic atmosphere there can be no thermal wind). In the actual atmosphere, the wind direction and the windspeed usually vary with height: up to the middle or high troposphere, windspeeds normally increase with height. In a vector diagram (see Figure 5.24) depicting the geostrophic wind at two different levels in the atmosphere, for a baroclinic atmosphere a *thermal wind*, $\mathbf{V}_T$, must exist, which is defined as the vector difference between these two geostrophic wind vectors; i.e.,

$$\mathbf{V}_T = \mathbf{V}_{500} - \mathbf{V}_{1,000},$$

where $\mathbf{V}_{500}$, $\mathbf{V}_{1,000}$ are the geostrophic wind vectors at 500 and 1,000 mb respectively. It may be shown that the east-west and north-south components of the thermal wind are given by

$$\frac{\partial v}{\partial z} = \frac{g}{\lambda T}\frac{\partial T}{\partial x} + \frac{v}{T}\frac{\partial T}{\partial z} \qquad [5.48]$$

and

$$\frac{\partial u}{\partial z} = -\frac{g}{\lambda T}\frac{\partial T}{\partial y} + \frac{u}{T}\frac{\partial T}{\partial z}, \qquad [5.49]$$

where u and v represent the horizontal components of the geostrophic wind, $\frac{\partial T}{\partial Z}$ represents the instaneous rate of change of temperature with height, λ is the Coriolis

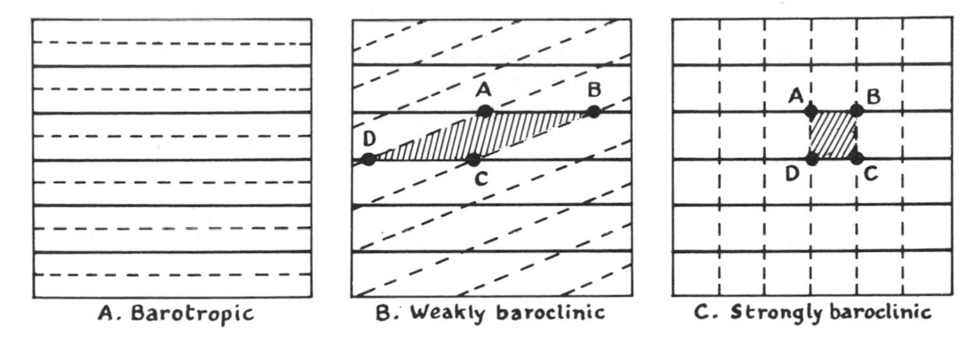

FIGURE 5.23.
Barotropic and baroclinic atmospheres. Isobars shown by solid lines, isosteres or isotherms by broken lines. *ABCD* indicates one solenoid. Squares represent horizontal or vertical areas.

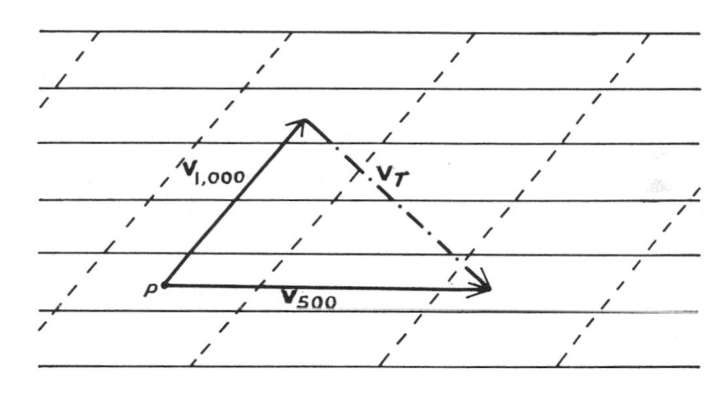

FIGURE 5.24.
Thermal wind. Broken lines show contours of 1,000-mb pressure surface, solid lines contours of 500-mb pressure surface. $V_{1,000}$ is the wind vector for surface wind at point P; V_{500} is the wind vector for 500-mb wind at point P; and V_T is the thermal wind vector for the 1,000-to 500-mb layer.

parameter, g is the acceleration of gravity, T is the actual temperature at the level at which $\delta Z \to 0$ (i.e., T is the temperature of a layer of air of negligible thickness), and $\frac{\partial T}{\partial x}, \frac{\partial T}{\partial y}$, are the horizontal gradients of temperature on a level surface in the east-west and north-south directions, respectively.[86] These equations show that if v is to increase with height, temperatures must increase horizontally toward the east (in the northern hemisphere), and if u is to increase with height, temperatures must increase horizontally toward the south. The vertical shear of the geostrophic wind is a vector $\left(\text{i.e., the thermal wind, whose components are } \frac{\partial u}{\partial z} \text{ and } \frac{\partial v}{\partial z}\right)$ whose direction is parallel to the isotherms on a level surface, such that low temperatures are to the left in the northern hemisphere and to the right in the southern hemisphere. The observed increase in strength of the middle-latitude westerlies with height must therefore be the result of the normal decrease of temperature with height toward the poles that is a characteristic feature of the troposphere. The equations also indicate that if the geostrophic wind veers (goes clockwise) with height, warm-air advection must be in progress, whereas if it backs (goes anticlockwise) with height, cold-air advection is taking place.

If Equations 5.48 and 5.49 are rewritten in terms of pressure coordinates instead of height coordinates (see Appendix 5.28), the terms involving the vertical temperature gradient $\frac{\partial T}{\partial z}$ drop out, and the thermal wind equation is considerably simplified but just as accurate. This is an important reason for using constant-pressure charts rather than constant-level charts in dynamic climatology.

The thermal wind equation bears the same relation to isotherm distribution as the geostrophic wind equation does to isobar distribution, and thermal wind analogies to geostrophic gradient winds exist for both straight and curved isotherms. These show that:

1. Geostrophic air flow around a low-pressure center will increase in strength with height if the low has a cold core, and will decrease if it has a warm core.
2. Geostrophic flow around a high-pressure center will decrease in strength with height if the high has a cold core, and will increase if it has a warm core.
3. If the closed centers of the isobaric and isothermal distributions do not coincide, the cyclonic centers must be displaced toward the colder air with increasing height, and the anticyclonic centers must be displaced toward the warmer air with increasing height.

There must clearly be a connection between the wind and temperature fields in the atmosphere, i.e., between its dynamics and its thermodynamics. This connection is described quantitatively in the *Sutcliffe development theory*.[87] A developing depression may be defined as one that is deepening, i.e., one in which pressure is falling with time; a developing anticyclone may be defined as one that is intensifying, i.e., one in which pressure is increasing with time. In a developing cyclone, isallobaric and frictional effects combined are directed inward, in the lower part of the atmosphere, so that convergence must be occurring. In a developing anticyclone, the isallobaric and frictional effects are directed outward, so that divergence must be occurring in the lower part of the anticyclone. For a cyclone to deepen, a process must exist that removes air faster than the rate of inflow due to convergence in the lower layers, and that exceeds the inflow by an amount equal to the drop in surface pressure tendency. For an anticyclone to intensify, a process must exist that accumulates air faster than the rate of outflow due to divergence in the lower layers, and that exceeds the outflow by an amount equal to the rise in surface pressure tendencies. These processes are upper-level divergence and upper-level convergence, respectively (see Figure 5.25). The combination of low-level convergence with high-level divergence in a developing cyclone obviously indicates ascending vertical motion, and the combination of low-level divergence with high-level convergence in a developing anticyclone implies descending vertical motion. The intensities of development in either case depend on the arithmetic difference between the convergence and divergence at the two levels, i.e., on the *relative divergence*. Sutcliffe was able to show that

$$-\lambda \operatorname{div} \mathbf{V}_T = \mathbf{V}_T \cdot \nabla(\lambda + \zeta_T + 2\zeta_0), \qquad [5.50]$$

in which div $\mathbf{V}_T$ represents the divergence of the thermal wind $\mathbf{V}_T$, ζ_T represents the thermal vorticity (i.e., the vorticity computed from isotherms instead of from isobars), and ζ_0 represents the vorticity of the 1,000-mb wind field. The term λ is the Coriolis force, i.e., the *latitude term* in the development equation; the term ζ_T is the *development term*, and the term $2\zeta_0$ is the *steering term*. Equation 5.50 is usually written in the form:

$$-\text{div}\ \mathbf{V}_T = \frac{1}{\lambda}\mathbf{V}_T \cdot \frac{\partial}{\partial s}(\lambda + \zeta_T + 2\zeta_0), \qquad [5.51]$$

where $\frac{\partial}{\partial s}$ means instantaneous rate of change measured along the isotherms in the direction of the thermal wind. The development term $\frac{\partial \zeta_T}{\partial s}$ consists of two components, one due to the curvature of the isotherms and the other due to the shear of the isotherms. The type of development likely to occur depends on the sign of these components, since the relative divergence is in part determined by them.[88] Equation 5.51 is the basis of thickness-chart analysis, which will be considered in detail in the next chapter.

The Sutcliffe development theory has been enlarged upon by Petterssen, who showed that cyclonic development at sea level occurs where and when an area of positive vorticity advection in the upper troposphere becomes superimposed upon a baroclinic zone along which thermal advection is discontinuous (i.e., a frontal zone in the lower layers of the atmosphere).[89]

One topic still has not been given its due importance in our discussion of atmospheric dynamics: the problem of baroclinic instability, or of hydrodynamic stability in general. The concept of *hydrodynamic stability* should be clearly distinguished from that of thermodynamic stability or of static stability. Under certain conditions, a statically stable atmosphere may be unstable for quasihorizontal perturbations of the zonal flow. In other words, under these circumstances the long-wave pattern of the upper westerlies does not remain a fluctuating one of approximately sinusoidal waves, but breaks down into separate closed cyclonic and anticyclonic centers. The results in terms of surface weather and climate are obviously very profound. As another example, hydrodynamic instability may be induced by anomalous winds in the upper atmosphere, which contribute to hurricane development.[90]

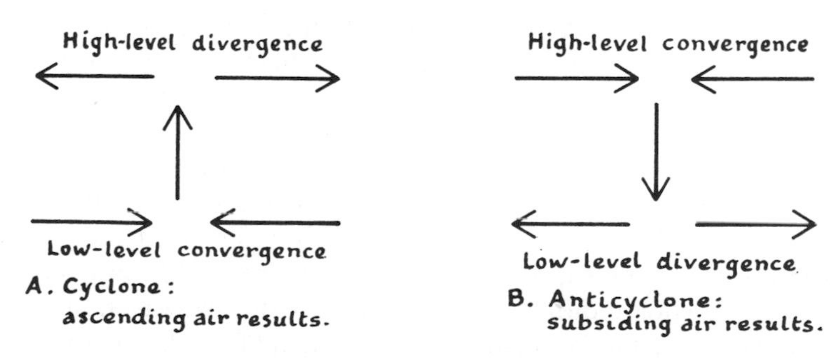

FIGURE 5.25.
Motion in the Sutcliffe development models.

In general, hydrodynamic instability refers to the stability of wind currents. Thus a hydrodynamically stable airflow will persist as a distinct current, whereas a hydrodynamically unstable airflow will degenerate into small, separate wind systems. To consider this formally, consider an air parcel embedded in a steady geostrophic wind current. The particle is initially in a state of hydrodynamic equilibrium, in which it maintains the same position relative to the geostrophic flow. It is then given an impulse in an arbitrary direction. If the parcel then tends to return to its initial path, the current is said to be *hydrodynamically stable*; if the parcel acquires an acceleration away from its initial position, the current is said to be *hydrodynamically unstable.* If the displacement is vertical, then the concept of hydrodynamic stability reduces to that of static stability. If the displacement is quasihorizontal, the concept of hydrodynamic instability reduces to that of *inertial stability.* Thus hydrodynamic stability is essentially a combination of two concepts: static stability and inertial stability. The criteria for the latter are as follows:

If $\lambda - \left(\frac{\partial V_g}{\partial y}\right) > 0,$ the current is inertially stable.

If $\lambda - \left(\frac{\partial V_g}{\partial y}\right) = 0,$ the current is inertially indifferent.

If $\lambda - \left(\frac{\partial V_g}{\partial y}\right) < 0,$ the current is inertially unstable.

This assumes a steady-state zonal current with geostrophic wind speed V_g, the current being constant in the east-west direction and invariant with time; λ is the Coriolis parameter. It is assumed that the current remains undisturbed as an air parcel moves isentropically through it.[91]

The term *baroclinic instability* refers to hydrodynamic instability in a baroclinic atmosphere, and is especially important for climate in relation to the stability of long waves in the circumpolar vortices. It may be shown that, neglecting the effect of friction, waves shorter than a critical wavelength, L_c, must all be unstable, and that there must exist a "most unstable" wave of length L_k. Both L_c and L_k depend on the magnitude of vertical wind shear and static stability in the zonal flow, and the critical wavelength for any given wind-shear value varies considerably with the shape of the velocity profile. The theory indicates that the very long waves in a baroclinic atmosphere must be limited to the upper levels of the latter, and the very short waves must be essentially low-level phenomena. In general, unstable baroclinic waves should only occur poleward of a critical latitude, which is near 30°. The amplification of baroclinic waves is also predicted by the theory: a certain latitude exists—exactly which latitude depends on the static stability—below which the waves should not amplify. There is also a connection between wave-number and baroclinic stability; the waves in an $n = 3$ pattern (where n is the number of waves in a latitudinal belt completely circling the Earth) appear to be stable if they occur at latitudes below 60°. Frictional influence is important too. Friction reduces the amplitude of baroclinic waves, but some of the waves on the fringes of the latitudinal band may amplify with friction. An increase in the drag coefficient affects amplitude in the same way as a decrease in the thermal wind.[92]

Mechanics of Some Atmospheric Circulations

As illustrations of the dynamic and kinematic concepts discussed, the mechanisms of atmospheric circulations at different scales will be discussed briefly, and some thermodynamic concepts will be introduced as well. Our purpose here is to show how inadequate purely empirical description is for climatic phenomena, if we wish to understand quantitatively any observations made of these phenomena.

The middle-latitude westerlies are the best example of a climatic phenomenon whose mechanics are fairly well understood, and whose behavior can be predicted from first principles with a fair degree of success. A study of the mechanism of these wind belts is really a study of the development of our ideas of atmospheric dynamics and kinematics. The main problem has always been that of cyclones and anticyclones in relation to the mean westerlies: do the latter merely represent the statistical average of the passage of large numbers of cyclones and anticyclones generally from west to east, or do the disturbances represent perturbations of a real zonal current? The success of the theoretical studies indicate quite definitely that the latter is the correct interpretation. The development of the theoretical studies is interesting.[93]

In 1916, V. Bjerknes advanced the theory, based on hydrodynamics, that cyclones originate as dynamically unstable, wavelike disturbances of a global westerly current; in 1919, J. Bjerknes was able to provide empirical confirmation of this with the discovery of a strongly baroclinic zone in middle latitudes, which was named the Polar Front. In 1921–22, J. Bjerknes and H. Solberg provided empirical evidence to show that cyclones develop from unstable waves originating along the Polar Front, and in 1928–1931, Solberg was able to demonstrate mathematically the theory of frontal waves. In all these studies, the emphasis was on the surface aspects of the baroclinic zone, i.e., on the wind shear at the frontal surface.[94]

In 1937, J. Bjerknes demonstrated that the general baroclinicity characteristic of the atmosphere in middle latitudes implies a vertical wind shear throughout the westerlies, which in turn causes horizontal displacement of an upper-air wave relative to a surface cyclone, resulting in deepening of the latter. The reason for the intensification of a given pressure system was therefore to be sought in a wide region of the entire troposphere rather than in the evolution of its system of surface warm and cold fronts.[95]

In 1939, Rossby published the theory of long waves for the special case of a constant zonal air current in a homogeneous, incompressible atmosphere. This theory was extended in 1940 by Hauwitz, who incorporated the effect of the Earth's curvature and the effect of finite lateral extent of the waves, and in 1944 by Holmboe, who obtained a long-wave equation for a more general type of barotropic atmosphere. A barotropic atmosphere with no wind shear cannot generate hydrodynamic instability, because there is no source of potential energy within it that can provide the energy for unstable wave motions. In 1947, Charney generalized the Rossby wave equation to cover a baroclinic atmosphere, and proved that dynamic instability within the westerlies increases as wind shear, lapse-rate, and latitude increase, and decreases as wavelength increases. He also showed that the Rossby barotropic wave equation applies if the constant value assumed for the zonal wind is equal to the 600-mb windspeed.[96]

In 1944, Bjerknes and Holmboe proved that with a sufficiently strong westerly current, horizontal convergences and divergences occur in such a manner, if the zonal flow is wavelike, to cause the wave pattern to move from west to east, the distribution of divergence/convergence reversing if the speed of the westerlies drops below a certain critical value. If the speed of the westerlies is above this critical speed, and if their velocity increases with height, then incipient waves must become thermally asymmetric and will intensify. A similar situation must hold for tropical easterlies in summer and autumn, when they increase in velocity with height. Thus the middle-latitude westerlies in all seasons, and the trade winds in summer and autumn, are regions of hydrodynamic instability, in which depressions and tropical cyclones, respectively, are generated.[97]

Later developments in the theory of the westerlies have included splitting the observed long waves into their harmonic components and studying the dynamics of each component separately, and investigating the influence of the Earth's orography on the zonal flow. Thus Queney, Charney, and Eliassen demonstrated theoretically that orographic obstacles of the size of the Rocky Mountains are capable of generating a wave pattern in their lee that resembles the observed long waves in magnitude: in particular, the "stationary profile" computed from the Rossby wave equation for 500 mb agrees with the observed mean wind profile. Differences in the physical properties of land and sea surfaces were recognized as important influences on the wave pattern too; James showed that the shape of the vertical profile of geostrophic westerly winds over North America is invariant with time or space, but over the sea the profile is much more variable from month to month and has a lower ratio of high-level to low-level winds than over the continent.[98]

On the generation of the mean zonal flow by the movement of disturbances, Kuo proved that the effect of the Coriolis parameter is to produce a mean poleward transport of westerly momentum, and a tilt of the axes of the troughs and ridges constituting the long-wave pattern toward the southwest-northeast direction in the northern hemisphere, throughout the westerly belt except near its northern boundary. Maximum transport of momentum occurs to the south of the middle of the zonal belt occupied by the disturbances, so that mean westerly and easterly currents are created in the central and outer regions of the belt, respectively. Therefore the Earth's rotation may have either an amplifying or a damping influence on a disturbance, depending on the asymmetry of the profile of mean flow; for example, if the latter is symmetric, a slow damping of the disturbance must result.[99]

All these studies of the zonal westerlies have in effect followed the lines of the classical Lagrangian theory of atmospheric waves initiated by V. Bjerknes. Other approaches are no doubt possible, and in fact the atmosphere contains features that are not included in the classical theory. For example, the perturbations of the westerlies in the density, pressure, and velocity fields show nodal surfaces. At Valentia, the zonal wind component of the westerlies has nodal surfaces at 4 and 12 km, but the meridional wind component reaches its maximum at 9 km.[100]

The mechanics of large-scale atmospheric disturbances obviously provide the key to a quantitative study of macroclimate, but the disturbances cannot be separated from the basic zonal flow without introducing some degree of artificiality. There are several ways of approaching the problem of cyclone dynamics. For example, attention may be focused on the mechanics of pressure change in a deepening or filling cyclone, or the generation and dissipation of cyclones and cyclone families may be traced by studying to what extent they are controlled by the dynamics of upper long waves in the westerlies. An alternative approach is to start with an isobaric chart at a given instant in time, and then to predict by means of the equations of motion, etc., how that chart will look at some later time; i.e., one starts with a given field of motion, and then determines the final field of motion mathematically after making appropriate boundary assumptions. If the predicted field of motion contains disturbances in approximately their true positions, then our equations must describe the mechanics of cyclones and anticyclones fairly well. The mechanics of anticyclones are difficult to study, because, even more than with cyclones, these disturbances cannot be treated as dynamically separate entities. Tropical cyclones also pose complex dynamic problems, not only because of their inherent characteristics, but also because they normally occur between middle latitudes, where the Coriolis parameter is an important term in the equations of motion, and the equatorial atmosphere, in which the Coriolis term is no longer predominant in balancing the pressure term in the equations.[101]

In all studies of the mechanics of the disturbances of the macroclimatic field, the central interest of the dynamic climatologist is the influence of the Earth's variegated surface on the behavior of cyclones and anticyclones. By the application of universal principles to a rotating body whose physical attributes approximate as closely as possible the Earth and its atmosphere, the dynamic meteorologist can predict the pattern of movement, temperature, and so on that the atmosphere ought to exhibit if it obeys all the known physical laws. But he cannot approximate *all* the physical attributes of the atmosphere and the Earth in his dynamic models, and there may be atmospheric properties about which he knows nothing. Consequently, the behavior of disturbances may not be as predicted. When this irregularity of behavior proves to be due to geographical influences and not to inadequate dynamic concepts, the climatologist is particularly interested, for his concern is with actuality, and he must know when and where he can rely on physics and mathematics to supplement his observational data, and when and where he can only rely on empirical information. The surprising thing is that physics and mathematics often provide the right answer, even when oversimplified terrain models have been assumed, so that any need for pure empiricism recedes more and more into the background.[102]

A good example of a climatologically important study in cyclone dynamics is that of cyclogenesis in the lee of mountain ranges. In the Rocky Mountains, a cyclone developed during November 16–21, 1948, at first as described by Petterssen's development theory, but later becoming abnormally destructive. This cyclone did not go through the classic "wave-depression" stage, and did not acquire a frontal system until *after* the cyclonic circulation was well under way. The intense deepening occurred when the region of strongest upward motion at 500 mb, directly beneath the Polar Front jet stream, became superimposed over the zone of descending surface air in the lee of the Rockies. Thus the center of low-level development was located where a combination of ascending air at 500 mb and descending low-level air produced the greatest vertical stretching of the atmosphere, i.e., where the jet axis ahead of the lee trough crossed the zone of descending air due to orography. Terrain-induced developments such as this may be very complex, because the terrain contours induce ageostrophic wind components parallel to their trend, which may result, for example, in net horizontal convergence even though the main component of the atmospheric flow is divergent.[103]

The mechanics of anticyclones pose an interesting geographical problem. Simple considerations of the theory of baroclinic instability suggest that in middle latitudes we should find a belt of "synoptic turbulence," in which cyclones and anticyclones occur in equal numbers and develop alternately. In fact, cyclones predominate, because of four main influences. First, it may be shown from vorticity considerations that intensification of the surface circulation should be more rapid in the center of cyclones than in the center of anticyclones (see Appendix 5.29). Second, cyclonic development is favored when appreciable quantities of latent heat are being released into the atmosphere. The intensification of cyclones (or anticyclones) is retarded by the cooling by ascent (or warming by subsidence) that is typical of these systems, and any process that reduces this effect will increase the liberation of latent heat and hence will favor cyclonic development. Third, "scale effects" are important. The relative importance of vertical stability in synoptic development varies inversely with the square of the dimensions of the system, and becomes negligible when the radius of the latter exceeds about 1,000 km in middle latitudes. Therefore when cyclones and anticyclones are separated by distances of about 2,000 km, there should be little difference evident in their rates of growth, but when they are closer together, the cyclones should grow rather than the anticyclones. Development theory indicates that cyclones should grow more rapidly than anticyclones in wave systems whose wavelength does not exceed about 5,000 km. Waves whose length is greater are not unstable, and at a wavelength of 5,000 km, anticyclones should develop at approximately the same rate as cyclones, which agrees with synoptic experience.[104]

The fourth influence is purely geographical. Except when the waves in the upper westerlies are unusually long, the southern hemisphere is always favorable for cyclonic development in middle latitudes, because it is almost entirely water-covered in these regions. There is no tendency for a quasistationary long-wave pattern to develop in these latitudes. In the northern hemisphere middle latitudes, in contrast, there is a definite tendency, mainly because of the Rocky Mountains, for a quasistationary pattern to occur, the waves being usually only slightly longer than the limit of baroclinic instability (i.e., 5,000 km). Therefore, anticyclonic tendencies exist in this latitudinal band, particularly in winter, when North America and Eurasia are largely snow-covered. At such a time, thermal influences override dynamic ones in determining the long-wave pattern; because there is very little water vapor in the air over the snow surfaces, very little latent heat is available for cyclone development. Under such conditions, therefore, dynamic anticyclogenesis is favored, and not infrequently the resulting anticyclones become persistent and built right across the westerlies to block their flow.

Tropical cyclones are very intense cyclones, almost perfectly circular, that develop over the tropical oceans. Dynamically, they provide a number of problems. For example, it is a well-known characteristic of tropical cyclones that they decay soon after moving over land. Although frictional dissipation of energy increases considerably when a hurricane or typhoon moves over land, it can be shown that frictional effects alone are too small to dissipate the dangerous winds of the cyclone. When a tropical cyclone moves over land, surface roughness and wind stress must increase, causing the surface wind to blow across the isobars at an increased angle and thereby increasing the convergence of air within the surface friction layer. Despite this effect, it is the energy budget of the cyclone that controls its dissipative tendency.[105]

Asymmetry of the wind field is another characteristic feature of tropical cyclones. This has long been recognized, and was explained by Ferrel in the following terms. Consider a rotary wind system of 40-knot winds, the system as a whole moving at 10 knots. If the system is moving from east to west in the northern hemisphere, then the speed of translation of the storm and the speed of its winds will combine in the northern quadrant, producing a net speed of 50 knots, but will be opposed in the southern quadrant, producing a net speed of 30 knots. A "dangerous semicircle" of

50-knot winds must then exist, with a semicircle of winds of only 30 knots on the opposite side of the storm. In fact, the difference in windspeeds on opposite sides of the cyclone is often *more* than twice the speed of translation of the storm as a whole, so that Ferrel's explanation is only partial. It can be demonstrated, by means of non-fluence lines, that a windspeed asymmetry must be an inherent feature of a tropical cyclone, as would be evident to an observer moving with the storm.[106]

The wind field in a tropical cyclone, although apparently very simple on a conventional weather chart, is actually quite a problem in kinematics to represent correctly. The wind system of a typhoon or a hurricane—normally described by means of streamlines, because the geostrophic wind approximation is not a good one in tropical regions—is characterized by two *singular points*. One of these points, taken as positive, is the cyclonic-indraft point, usually taken to be the cyclone center; the other, taken as negative, represents the hyperbolic point (see Figure 5.26, A). Location of the hyperbolic point in a hurricane is difficult, because it generally occurs in an extensive area of very light winds, but this point must be located if a complete picture of the wind system of a cyclone is to be built up. The existence of singular points is crucial for the theory of equatorial waves according to Palmer: when a singular point forms, the wave begins to be transformed into a vortex, and so a tropical cyclone may be originated (see Figure 5.26, B). Only a small proportion of these cyclonic vortices attain the intensities of hurricanes or typhoons, but an essential accompaniment of an incipient hurricane is that the two singular points within the cyclonic vortex become separated.[107]

Tropical cyclones are much rarer than extratropical cyclones, yet they are the most powerful low-level wind systems found anywhere in the world. These two facts in themselves pose dynamic problems, which have been studied in various ways. For example, the vorticity approach has provided some useful correlations, particularly concerning the transformation of an equatorial wave into a vortex. It may be shown that a region of negative relative vorticity must exist around a tropical cyclone, and

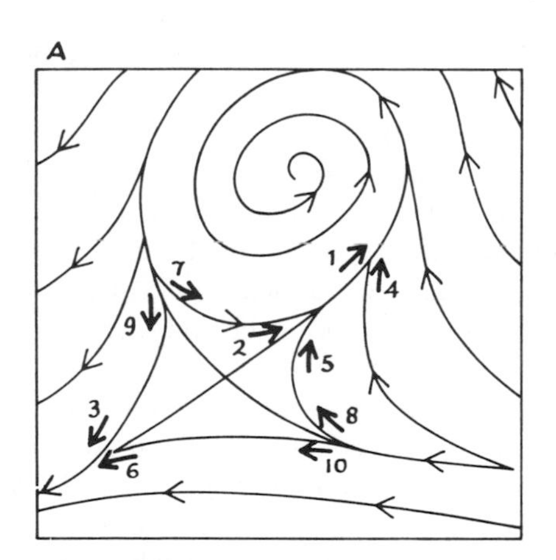

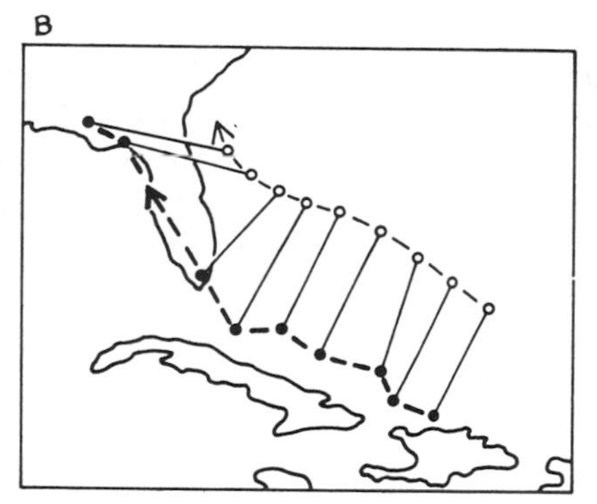

FIGURE 5.26.
Hyperbolic and singular points. A shows the typical cyclonic indraft to a hurricane, embedded in an easterly current. Note that wind flows at stations 1, 2, 3, are converging with those at stations 4, 5, 6, toward an asymptote of convergence. Stations 7, 8, 9, 10, indicate the asymptote of divergence. B shows Hurricane Able, August 30, 1952. The empty circles show the hurricane's surface position at three-hour intervals; the filled circles show the surface positions of the hyperbolic points at those times. (After Sherman and Carino.)

that hurricane development is associated with a steady change from anticyclonic to cyclonic absolute vorticity. The initial impulse for cyclone development comes during the early part of the anticyclonic absolute vorticity stage, but the main increase in energy necessary for hurricane-intensity winds occurs during the cyclonic vorticity stage.[108]

Tornadoes are very intense cyclones on a very local scale. The main dynamic problems are, first, to explain the source of potential energy and how it is converted into kinetic energy, and second, to explain why the low-pressure center within the tornado does not disappear because of the strong surface inflow. An important correlation with vorticity and stability may be noted: tornadoes normally develop within an area of conditionally unstable air under the influence of increasing cyclonic vorticity tendency. Conditions such as these exist for a short time before the tornado occurs, and the latter develops only after their combined intensity reaches a critical value.[109]

Orographic influences on atmospheric circulations are of special interest to the climatologist, particularly if it can be shown that the circulations would not exist at all if the orography were different. We shall here be concerned with macroinfluences, not with very local microclimatological phenomena; with the latter the friction of the Earth's surface has an important influence on the "skin" of air in contact with it, but the study of this effect is one in atmospheric turbulence rather than in general atmospheric mechanics. For now we are concerned with the *form drag* of orography on airflow, rather than the surface friction influence. The essential point is that the dynamic equations show, when applied to synoptic-scale phenomena, that the macroorography of the Earth's surface sometimes exerts more influence on atmospheric circulations than atmospheric dynamics.[110] In other words, the atmospheric circulations in a given area at a given time represent a balance between the dynamically induced climatology for that area and its orographically induced climatology. The exact form of that balance depends also on the energy-balance conditions at the Earth-atmosphere interface. For example, tropospheric airflow above the friction layer always contains wave motions. The effect of an orographic obstacle is to cause certain of these waves to become stationary, and sometimes to amplify them considerably, creating the orographic and lee waves considered in an earlier chapter. It is important to note that mountains are not necessary for the formation of orographic and lee waves; any kind of permanent stationary interference with the flow pattern may induce them, for example, the change in surface friction between land and sea on either side of a coastline. Our earlier consideration of orographic waves was largely two-dimensional; there are certain orographically induced atmospheric circulations that must be considered in three dimensions.

We have now to consider various wind systems that are not more than 100 km in extent. These winds are distinguished dynamically by the fact that they blow approximately in the direction of the pressure gradient. They are known as *antitriptic winds*, in which the friction term in the strophic balance equation is very large, and the Coriolis term is very small. They contrast with *Eulerian winds*, in which both the Coriolis and the friction terms are negligible, so that the acceleration of the air is measured in terms of the pressure gradient, and with geostrophic winds, in which the Coriolis term is very large in comparison with the friction and acceleration terms. Antitriptic winds occur when the pressure gradient balances the frictional force; they are consequently controlled by the irregularities of the Earth's surface and by surface heating effects, and are essentially local winds. Other types of winds, theoretically possible but rarely important in practice, include the *antibaric wind*, in which the pressure-gradient force is in the same direction as the Coriolis force, and the *inertial wind*, in which the horizontal pressure gradient is zero. Normal horizontal frictionless flow in the atmosphere is baric flow, defined as flow in which the pressure-gradient flow opposes the Coriolis force. The *baric wind* is therefore equivalent to the geostrophic wind.[111]

Examples of antitriptic winds include land and sea breezes, mountain and valley winds, gravity winds, and (in part) föhn and bora winds. All these circulation systems are, of course, familiar features in descriptive climatology texts, but it may be shown that all the elementary qualitative descriptions of them are inaccurate in terms of dynamics.

Land and sea breezes are usually described as the consequence of the different heating rates of land and water surfaces. For quantitative studies of actual land and sea breezes, it is necessary to know the strength of the breeze corresponding to a given land-sea temperature difference, and this involves hydromechanics. Four controls are important in the generation of land and sea breezes: (a) the pressure-gradient force resulting from the difference in temperature between land and sea surfaces; (b) vertical heat exchange by turbulence; (c) the internal friction of the air motion due to turbulence; and (d) the Earth's rotation.

The thermal theory of the origin of land and sea breezes devised by V. Bjerknes is very different from the elementary thermal explanation.[112] According to Bjerknes, the land-sea breeze circulation originates as a periodic current around solenoids in which the wind is 90° out of phase with the change in air density. During the day, when the land surface is relatively warm and the sea surface relatively cool, if the air over the coast region is initially at rest, the isobars (P_1 and P_2 in Figure 5.27) will be horizontal, but the isosteres (ρ_1 and ρ_2) will be inclined because of the differential heating at the surface. The relationship between pressure, density, and air movement may then be expressed quantitatively by means of Bjerknes' circulation theorem, from which it appears that circulation around the contour *ABCD* must be in the direction shown and must increase with time, thus giving rise to an onshore breeze at the surface and an offshore breeze aloft (see Appendix 5.30). During the night, radiative cooling of the land surface lowers its temperature below that of the sea surface, and the horizontal temperature gradient is reversed. The effect of this is to reverse the solenoidal circulations, producing an offshore surface current and an onshore current above. The circulations are thus in a direction that tends to rotate the isosteric surfaces towards the isobaric surfaces, with lighter air accumulating over heavier air.

According to the Bjerknes theory, the sea breeze should commence at the time of greatest heating of the land surface, and should reach a maximum at the time of smallest land-sea temperature difference in the evening, but this does not agree with observational experience. Haurwitz was able to make the theory fit the facts more satisfactorily by taking frictional influences into account, and showing that the land- and sea-breeze circulation should keep on increasing until the difference in temperature between land and sea disappears, and also showing that the diurnal variation of the circulation is a Coriolis effect.[113]

An alternative theory, an adaptation of Rayleigh's convection-cell theory, takes into account vertical as well as horizontal heat transfer, and also the effect of turbulent friction. This theory proceeds from the equations of motion, continuity, and heat transfer to give a solution for the velocity of the land-sea breeze circulation that is in good accord with experience.[114] Other theories exist, and numerous observational studies stress the fact that land and sea breezes can be quite complicated, particularly

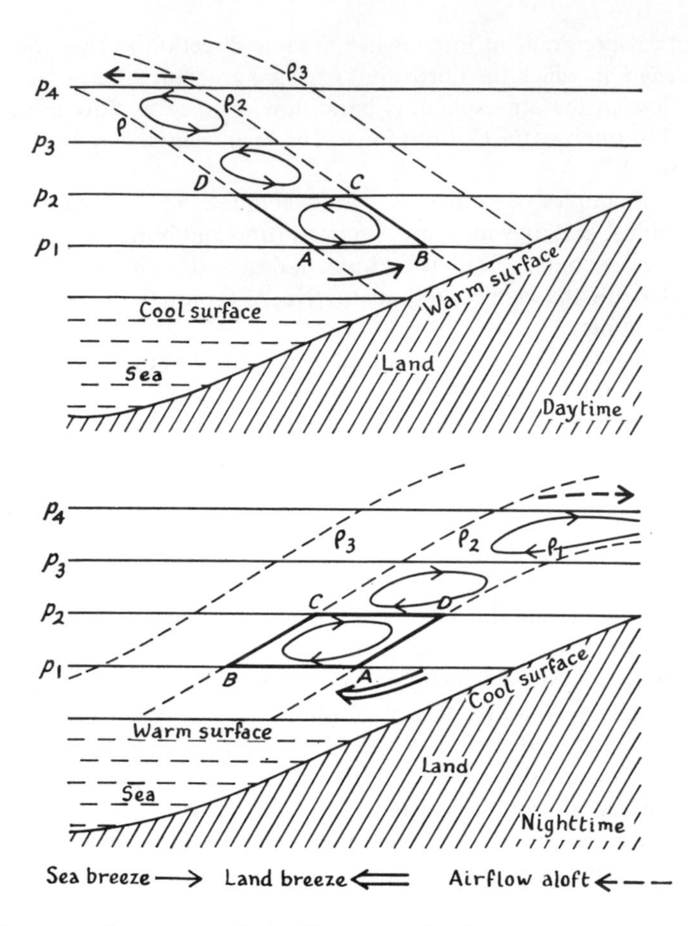

FIGURE 5.27.
Land and sea breezes.
Sea breeze → Land breeze ⇐ Airflow aloft ← – – –
p_1, p_2, p_3, p_4, isobars.
ρ_1, ρ_2, ρ_3, isosteres.

when they interact with the general synoptic winds. For example, the sea breeze takes on frontal characteristics when it is opposed by the gradient wind. Sometimes the mathematical equations describing a given situation become too complicated for an exact solution to be possible, in which case approximation techniques and indices may be used. Local orography is not usually taken account of by the land-sea breeze equations, but it can be a major influence.[115]

Mountain and valley winds (see Figure 5.28) are more complex phenomena, because they represent a combination of two distinct effects. First, there is the mountain/valley wind proper, involving an up-valley surface current (the *valley wind*) between 9 or 10 A.M. and sunset, and a down-valley surface current (the *mountain wind*) from after sunset until soon after sunrise. Second, there is the cross-valley wind, the *thermal slope wind*, which is caused by differential heating of the valley side slopes by the sun. The latter is easier to deal with dynamically, and fundamental theoretical treatments of convection motion on a slope by Prandtl and of cross-valley winds by Gleeson are available. The theory of mountain and valley winds devised by Defant combines the two effects, and enables estimates to be made of the length of time required for the mass of air initially over the valley bottom to be replaced in a closed circulation by the heated air rising from the slopes, moving transversely, and ultimately descending to the valley bottom. For a valley 1,500 m wide and 1,750 m deep, the theory shows that a slab of air 1 meter thick will be completely replaced in 4.5 hours. This replacement

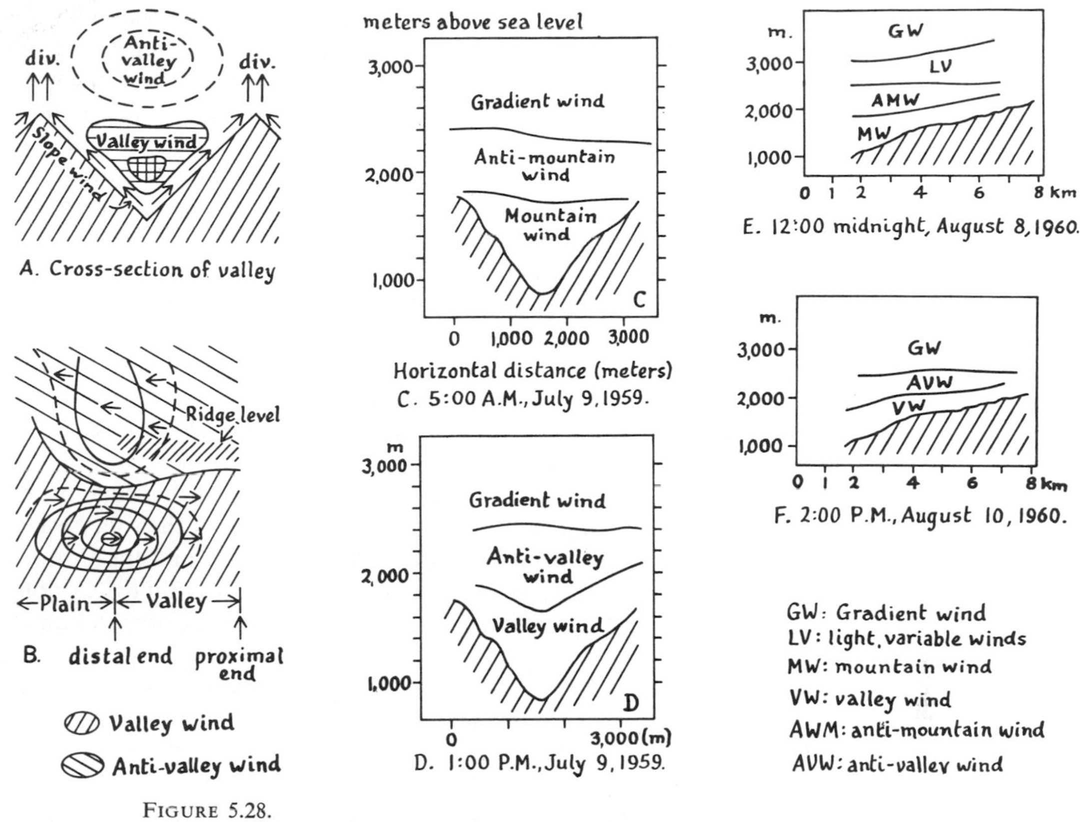

FIGURE 5.28.
Mountain and valley winds in the Mt. Rainer area, Cascade Mountains, Washington State. A, model of valley wind in ideal *V*-shaped valley. B, atmosphere cross section along long profile of valley. C–F, Carbon River valley wind structure. (After K. J. K. Buettner and N. Thyer.)

of air over the center of the valley influences the surface pressure gradient considerably and so is an important control of the antitriptic wind.[116]

The Defant theory is based on classic observations by Wagner in the Alps.[117] These observations indicated that, at a certain height above the valley, roughly equal to that of the surrounding ridge summits, the otherwise different diurnal variations of pressure over the head and the mouth of the valley become equalized. This height is termed the "effective ridge altitude" and sets a limit to the depth of the mountain- and valley-wind circulation. The observations also showed that the diurnal variations in air temperature within the valley, up to the height of the ridge summits, are more than twice those within a layer of similar depth over the adjacent plain. A pressure gradient from plain to valley must therefore exist during the day, and reversed at night. The mountain- and valley-wind circulation is most developed in wide, deep valleys. It is particularly characteristic of cloudless anticyclonic weather in summer, but also appears, usually as a modification of the general wind arising from the synoptic situation rather than as a distinct circulation, in winter and under cloudy skies. The circulation is essentially thermodynamic, and neither the longitudinal profile of the valley floor nor the shape of the valley cross section appear to influence it to any marked degree.

During daytime, air rises along the upper slopes of the valley sides because of the difference in temperature between the relatively warm air over these slopes and the relatively cool air over the center of the valley at the same altitude. This upslope wind commences 15 minutes to an hour after sunrise, continues to increase in intensity until it peaks at the time of maximum insolation, and then reverses in direction 15 minutes to one hour after sunset. It is particularly prominent on south-facing slopes, almost nonexistent on north-facing slopes, and its intensity depends on local exposure and degree of slope. The thickness of the daytime upslope wind in the Alps is between 100 and 200 meters, the nighttime downslope wind being usually somewhat thinner. The time of commencement of the upslope flow varies locally, depending on the time at which insolation first falls on slopes of different exposure, and the circulation is in general very sensitive to shading of the slopes by clouds during even very short periods. The speed of the thermal-slope circulation increases with distance above the slope, and the presence of the current may be indicated by an isolated cumulus above a summit or by lines over cumuli above ridges.

Valley winds in general can be described dynamically in terms of the slope of the valley floor, the pressure force resulting from the constraining effect of the valley walls, the diurnal temperature variation of the air and its eddy viscosity, and the Coriolis force. The theoretical treatment starts from the equations of motion, and the Coriolis force is—somewhat surprisingly, at first—necessary to give a good representation of actual valley winds.[118]

Local orography can result in unusual forms of the mountain wind. Thus the *Maloja wind* is a mountain wind that blows down-valley both day and night. Originally described from the area between Bergell and the Engadine in Switzerland, it arises because the wind from one valley spills over a pass into another valley.[119]

Nocturnal air drainage is an apparently simple phenomenon, involving the gravity flow of shallow masses of cold air down from radiation-cooled upper slopes, yet its mechanics are complex. This type of air drainage may occur on an extensive scale along cold slopes, for example, in Antarctica, or on a very localized scale along valleys. In both cases, the cold air flows are usually very shallow, and the *katabatic wind* they give rise to is soon destroyed by friction and by turbulent mixing with warmer air.[120]

The *föhn wind* is an example of an atmospheric circulation whose mechanics can be explained only by a combination of dynamics and thermodynamics. The föhn is a local wind blowing down the leeward side of a mountain and causing abnormally rapid temperature rises and dryness in the adjacent lowlands. It is normally a very gusty wind, bringing extremely good visibility, and is usually characterized by lenticular cumulus clouds in the general airflow, and by a mass of flat cumuliform cloud (the *föhnwall*) above the mountain summits and the upper parts of the windward slopes. The föhnwall is usually a spectacular sight, and remains stationary despite the strong wind, continuously forming and dissolving to windward and leeward of the summit, respectively.

The classical explanation of the föhn devised by Hann is that precipitation on the windward side of an orographic barrier will release latent heat, so that the air descending the leeward slope will gain heat at the dry adiabatic rate, whereas air ascending the windward slope will lose heat only half as fast. It is clearly an inadequate explanation, because it does not provide a mechanism that would cause the air to descend in the lee. Neither does it fit all of the facts: in the chinook in Montana, for example, which is essentially a föhn wind that covers much more geographical range than the föhn of the Alps (which is more or less restricted to valleys), latent-heat release by precipitation on the windward slopes is unnecessary for the initiation of the föhn, which actually gains heat rather than loses it during its passage across the mountains, because of evaporation occurring in its descent down the leeward slopes.[121]

Hann did later conclude that precipitation on the windward slopes was unnecessary, and showed how the temperature rise could be explained by replacement of air in the lee, which has the usual wintertime lapse-rate, by air descending from above, which is warmed at the dry adiabatic lapse-rate. The general condition for Hann's modified thermodynamic explanation to be applicable is that the airflow must be maintained at right angles to the orographic barrier by a pressure gradient parallel to the ridge.[122] Several synoptic situations may produce such a condition.

The problem of the descent of the warm föhn current in the lee was explained by von Ficker by means of three phases in the development of the föhn. The *preliminary phase* involves general subsidence over the area, with the air increasing in stability and the development of a temperature inversion. During the *anticyclonic phase* the underlying cold air moves out of the valleys under the influence of the pressure gradient, and the subsidence inversion descends but does not reach the ground. Finally, in the *stationary föhn phase*, the föhnwall forms on the windward side of the mountain and the horizontal pressure gradient is increased in the direction of the mountains. The normal Hann thermodynamic effect then comes into operation, and the resulting föhn causes changes in the pressure distribution and in air-mass properties, which in turn influence the thermodynamic balance so that the stationary phase ultimately comes to an end.[123]

The essence of the Hann–von Ficker theory of the föhn is that the descent of the warm air is the immediate consequence of the removal of cold surface air from the valleys in the lee. The orography is such that air can only be replaced from above, in which case dynamic heating must occur. Other theories have emphasized effects that must also occur on occasion. Küttner showed that warm air penetrates to the valley floors when lee waves are set up by the föhn-inducing orographic barrier; Frey demonstrated how föhns could be due to a solenoidal circulation set up in the lee when a considerable isobaric temperature gradient exists in the free atmosphere from the region above the barrier to that above the foothills; and Willett regarded the föhn as an upper warm current that has been forced to descend to the surface by frictional drag.[124]

Although the typical area of föhn occurrence is the Alps, they have been observed in many other regions. Comparatively modest mountain masses can produce föhn effects, for example the Cairngorms and the Grampians in Scotland.[125] In some areas, geographical influences may severely modify the föhn effect. Thus *bora* winds are föhns in which the lee descent is insufficient to heat the air appreciably. When a very cold air mass moves out from a continental interior in winter and descends steep slopes, it can produce extremely violent winds with intense cold, as on the Dalmation coast of the Adriatic. The bora may be explained dynamically by the theory of Defant.[126] *Mistral* winds are boras that have been intensified during a passage through a constricted orographical gap; for example, the mistral of Provence and adjacent regions in southern France is an intrusion of deep polar air into the western Mediterranean basin that has been forced through the gap between the Pyrenees and the western part of the Alps. *Northers* are similar winds on the Pacific Northwest coast of North America.

A study of actual föhns, land and sea breezes, and mountain and valley winds will prove to anyone just how complicated the mechanics of atmospheric circulations are. Dynamic climatology works by simplifying the observed phenomena into systems that have a physical meaning. For local wind phenomena, this simplification requires introducing a whole range of new concepts from the mathematical field of turbulent diffusion, because the air particles constituting a local wind are subject not only to the dynamic and thermodynamic laws already discussed, but also to dispersion in terms of certain probability laws that are dealt with in Chapter 3. Therefore, the quantitative study of local winds is difficult, and if the investigator is not inclined to study turbulent wind fluctuations, he may be able to find only empirical explanations for his observations.[127]

Very large-scale wind circulations, for example, the Asiatic monsoon, provide completely different dynamic problems. Empirical study of monsoons in relation to the general atmospheric circulation suggests that they are consequences of the seasonal shifts of the planetary wind systems, amplified by continental heating and cooling, particularly over southern Asia and Africa. Empirical study further suggests that there are connections between the onset of the monsoon and blocking anticyclones over northwestern Europe in late spring, the formation of a midtropospheric warm anticyclone over Tibet in early summer, and the Southern Oscillation.[128] The question therefore arises whether the monsoon may not be a dynamic rather than a geographical problem; i.e., would the monsoon occur where it does even if Asia, Africa, and Australia were not in their present positions? The question cannot be completely answered yet, but it is clear that the dynamic part of the monsoon is of much greater importance than was formerly thought.

The classic description of the Asiatic monsoon is based on interpreting the geographical distribution of mean pressure in terms of Buys-Ballot's law.[129] For example, the Siberian anticyclone in winter produces a clockwise circulation of surface air that gives rise to the *northeast monsoon* over China and the Indian subcontinent. In summer, the surface low-pressure center over northwestern India and Pakistan causes a continuous increase in surface pressure southward to the center of the subtropical anticyclone in the South Indian Ocean. This pressure gradient results in a broad air current covering the whole of the Indian Ocean; the current commences as a southeasterly wind south of the equator, becomes deflected into a southwesterly wind after crossing the latter, and picks up considerable quantities of water as it moves toward India as the *southwest monsoon.* The summertime surface low-pressure center is also regarded as providing a southerly airflow over the China seas (the *southerly monsoon*) and a northerly airflow over the eastern part of the Mediterranean (the *Etesian wind*). Although synoptic studies show that the actual airflows in the northeast and southwest monsoons are not simple currents, but contain disturbances, it is useful to regard them as very simple flows. Continuity requires that there be compensatory return currents. A dynamic theory devised by Jeffreys explained the simplified system very well, and indicated that the return circulation should be at a little above 2.1 km. Jeffreys' theory of the monsoon commenced with an equilibrium condition, in which the surfaces of constant pressure and constant density are horizontal, and then determined what effects surface heating and cooling would have on air motion. The theory made use of the equations of motion, expressed in terms of polar coordinates.[130]

A more modern approach by Kraus and Lorenz develops a theory of the summer monsoon circulation in which the equations are written in terms of Laplacians and Jacobians, so that a more truly three-dimensional picture is obtained.[131] The Kraus-Lorenz theory includes the effects of advection, nonadiabatic influences, radiation effects, and static stability, as well as the usual dynamic and surface-heating effects. The maps computed from the theory show that, assuming a latitudinal surface-heating differential between land and water that is zero at the equator and poles and 25°C in latitude 30°, and a meridional temperature gradient giving a difference of 75°C between equator and pole, then, starting with the isobaric and isosteric surfaces at rest, the final steady state will be a pattern with low-pressure centers over the continents and high-pressure centers over the sea. However, the exact location and intensity of these pressure centers depends very much on the value of the static stability parameter as imposed by external influences. These external influences consist of radiational processes in the upper atmosphere, in particular, those due to ozone variations, supplemented by the effect of atmospheric humidity variations.

In principle, the Asiatic monsoon might be considered, as a first approximation, a gigantic land- and sea-breeze phenomenon, in which the southwest monsoon corresponds to the sea breeze and the northeast monsoon to the land breeze. However, superimposed upon the relatively simple mechanism of the land-sea breeze would be dynamic effects that are internal to the atmosphere and do not appear important with land and sea breezes proper, and radiational effects from processes in the upper atmosphere. At this point the importance of empirical studies, not necessarily linked with a dynamic model, becomes evident: Wagner showed in 1931, from upper-air observations, that no "antimonsoon" flow occurred aloft. Thus Jeffreys' theory, although it undoubtedly explains some features of the real monsoon, is not the complete explanation. The main result of Wagner's work was his proof that the Indian monsoon is linked with the atmospheric circulation to the north of Tibet.[132]

But the monsoon phenomenon is not necessarily related to the distribution of land and sea, and upper-atmosphere dynamics may well be its controlling feature. For example, the formation of the polar-night vortex in the winter stratosphere and its replacement in summer by an anticyclone at the same level is a monsoon phenomenon on an even larger scale than the Asiatic one; geographical influences could only play a small part in these stratospheric phenomena. In any case, synoptic studies of the Asiatic monsoon in depth show the variety of airflows involved, and seem to imply that terrain influences may be very minor in the maintenance of the monsoon circulation, although the Himalayas seem to be essential for the present form of the Indian monsoon. The rest of the monsoon explanation lies in upper-air dynamics.[133] The sequence is as follows.

First, because of the existence of the Himalayan-Tibetan highlands massif, the main Polar Front jet over eastern Asia develops a double structure, one branch locating itself north of the Himalayas, the other south. In conformity with the confluence theory of jet formation, the southern branch increases considerably in strength during November because of the confluence of warm air from the south and cold air from the north. The confluence zone extends as far upstream as northern India and Pakistan, giving the characteristic subsidence and winter dryness of these regions. The split in the jet is a consequence of winter cooling of the Tibetan plateau: the east-west baroclinic zone over the highlands is strengthened by the cooling effect of the plateau on its southern border, but is weakened over the plateau itself. The opposite state of affairs prevails during summer: a quasistationary extension of the subtropical high-pressure belt, thermally generated, lies over the whole Himalayan-Tibetan region, producing easterly winds over northern India. The crucial factor for dynamic climatology is the coincident locations of this anticyclone and the Himalayan-Tibetan massif. As demonstrated by Bolin, the influence of a large circular mountain barrier on a uniform westerly airflow without lateral shear is to lead to the development of a critical velocity. If the airflow velocity exceeds the critical value, the current is not able to cross the mountain but is forced to curve back anticyclonically. Vorticity considerations decide that the split in the current must occur to the west of the mountain, as is actually observed.[134]

The nonmathematical student who is dismayed at the apparent complexity of the first principles of dynamic and physical climatology as portrayed in the present chapter should not be discouraged. The mathematics is not difficult once the notation is understood, and one cannot expect to follow the movements of our restless atmosphere without some complication. Climate is essentially a complex atmospheric phenomenon that is still more complicated by its interaction with the irregular surface of the Earth, which involves boundary-layer and turbulence effects that are not nearly as important in the free atmosphere. To approach the study of climate by electing to ignore the complexities of atmospheric dynamics and to concentrate upon the apparent "simplicities" of surface weather is quite wrong. The correct approach, the simplest one, is via atmospheric dynamics: the real complications come later.

A final point to be emphasized is that our discussion of atmospheric dynamics has been essentially Newtonian. Yet, as stated by Wiener, the number of particles in the atmosphere is immense, and if all the readings made at all the meteorological stations on Earth were taken simultaneously, they would not provide a billionth part of the data necessary to characterize the actual state of the atmosphere from a Newtonian point of view. The terms "cloud", "temperature," "wind," and so on, refer to a distribution of possible situations of which only one is realized; they do not describe a single physical situation. However good is the system of mechanics developed for atmospheric study, it can only predict the probability distribution of the constants of the system at some time in the future. Dynamic meteorology, and therefore dynamic climatology, is still a semiexact science; indeed, it may well be that there is a definite limit to the predictability of weather and climate.

It must also be admitted that approaches completely different from the Newtonian one may be developed for atmospheric dynamics in the future. For example, a thoroughgoing application of modern thermodynamics has much to commend itself, and recent developments in mathematical and statistical theory could no doubt be applied to climatology.[135]

Notes to Chapter 5

1. Of the four volumes of Napier Shaw's *Manual of Meteorology*, vol. III, *The Physical Processes of Weather*, and vol. IV, *Meteorological Calculus: Pressure and Wind*, are particularly relevant to this chapter. V. Bjerknes' major works are *Dynamic Meteorology and Hydrography: Vol. I, Statics* (Washington, 1910); *Vol. II, Kinematics* (Washington, 1911); and *Physikalische Hydrodynamik* (Berlin, 1933). See also *DMWF.*

2. The most elementary treatment is *EDM*; a more advanced treatment, that of *ITM*, avoids the use of vector algebra. The standard work, which assumes

a knowledge of vector algebra, is G. J. Haltiner and F. L. Martin, *Dynamical and Physical Meteorology* (McGraw-Hill, 1957). *PDM* is still a classic.

3. See *HM*, p. 316.

4. For a standard account, see P. W. Bridgeman, *Dimensional Analysis*. On dimensional analysis applied to the wind distribution in the planetary boundary layer, see A. B. Bernstein, *MWR*, 93 (1965), 579.

5. The account is based primarily on *DMWF*, pp. 27–33, and *HM*, pp. 361–71, but also makes use of H. R. Byers, *General Meteorology* (New York, 1944), and B. Haurwitz, *Dynamic Meteorology* (New York, 1941).

6. F. K. Hare, *JM*, 17 (1960), 36. On atmospheric density, see: R. S. Quiroz, *JAM*, 7 (1968), 969, for synoptic maps for the 30 to 40-km level; P. F. Nee, *JAM*, 3 (1964), 175 for hourly variation, surface to 25 km, at Bedford, Mass.; J. W. Smith, *JAM*, 3 (1964), 290, for the climatology, surface to 31 km, from pole to pole near longitude 80°W. See P. F. Nee *et al.*, *JAM*, 4 (1965), 300, for monthly mean vertical profiles for the United States and Greenland of density variation at constant altitude and at constant pressure. See V. S. Whitehead, *MWR*, 93 (1965), 559, for a method of inferring 500-mb heights and temperatures from surface pressure and temperature and the variation of density with height.

7. See Byers, *op. cit.* (1944), p. 131, for the derivation of this equation.

8. For a discussion of entrainment, see J. M. Austin, *CM*, p. 694. On diabatic (i.e., nonadiabatic) properties of the atmosphere from TIROS III radiation measurements, see P. A. Davis, *MWR*, 93 (1965), 535. On the hydrostatic equation for an autobarotropic atmosphere, see M. Yanowitch, *GPA*, 64, no. 2 (1966), 169.

9. This follows *EDM*, p. 35.

10. See B. Haurwitz, *op. cit.* (1941), p. 54, for the derivation; also see J. E. McDonald, *BAMS*, 44 (1963), 203, for details of early developments in the theory of the saturated adiabatic process. *PDM*, p. 66, gives a graph of the situation.

11. W. Thomson, *Mem. Lit. Phil. Soc. Manchester*, 2 (1865), 125. The paper was read in 1862, but not published until 1865.

12. J. E. McDonald, *JAS*, 19 (1962), 309. On the estimation of air temperatures at the lifting condensation level, see R. L. Inman, *JAM*, 8 (1969), 155. For a rapid method of estimating temperature and pressure at the lifting condensation level, see S. L. Barnes, *JAM*, 7 (1968), 511.

13. *ITM*, pp. 96–100, gives the basis of the theory. On the parcel method, see J. S. A. Green, F. H. Ludlam, and J. F. R. McIlveen (*QJRMS*, 92, 1966, 210) who show, from the parcel theory, how the speed of a jetstream may be estimated from areas on a thermodynamic diagram.

14. See J. Bjerknes, *QJRMS*, 65 (1938), 325, for the basis of the slice method. On the use of the slice method in explaining the areal extent of convective precipitation as observed by radar, see J. N. Myers, *JAM*, 5 (1966), 832.

15. See *ITM*, pp. 100–103, for criteria involved in the assessment of layer stability.

16. W. A. Baum, *JM*, 8 (1951), 196.

17. See W. L. Gates, *JM*, 18 (1961), 526, for details of static stability measures in the atmosphere.

18. J. G. Galway, *BAMS*, 37 (1956), 528. On the stability of layers of air in a baroclinic airflow, see F. P. Bretherton, *QJRMS*, 92 (1966), 325.

19. For Equations 5.11 and 5.12, see Byers, *op. cit.* (1944), p. 152. For a formula for the conversion of vapor pressure over water to equivalent vapor pressure over ice, see H. R. Glahn and R. A. Allen, *MWR*, 92 (1964), 509. On the computation of saturation vapor pressure, see F. W. Murray, *JAM*, 6 (1967), 203. For an approximate method for the computation of saturation vapor pressure over sea water, see W. E. Langlois, *JAM*, 6 (1967), 451.

20. *DMWF*, p. 27. See also *HM*, pp. 361–409, for a discussion of thermodynamic diagrams, and J. C. Bellamy, *JM*, 2 (1945), 1, for details of the pastagram, which combines many of their properties.

21. Shaw, *op. cit.* (1928), III, 269.

22. G. Stüve, *Beitr. Phys. Freien Atmos.*, 13 (1927), 218.

23. N. Herlofson, *Met. Ann.* 2 (1947), 311.

24. A. Refsdal, *Meteorol. Z.*, 52 (1935), 1.

25. *HM*, p. 367.

26. *Isentropic analysis* was much in vogue as a method of synoptic study of atmospheric moisture during the 1940's. See J. Namias, in *WAF* (1940), pp. 351–77, for an account of this technique. There are signs of renewed interest in isentropic analysis; see F. H. Ludlam, *QJRMS*, 91 (1965), 529. On isentropic charts, see J. S. A. Green, F. H. Ludlam, and J. F. R. McIlveen, *QJRMS*, 92 (1966), 210, for a study of the rise of tradewind air over a front into a jetstream (Caribbean-North Atlantic); see also A. K. Betts and J. F. R. McIlveen, *QJRMS*, 95 (1969), 639.

27. C. W. B. Normand, *QJRMS*, 64 (1938), 47 and 71.

28. See S. W. C. Pack, *Weather Forecasting* (New York, 1948), for details. See also D. W. Johnston, *MM*, 87 (1958), 265.

29. R. M. Poulter, *QJRMS*, 66 (1940), 84.

30. See *ITM*, pp. 56, 62, for mathematical derivations.

31. C.-G. Rossby, *Thermodynamics Applied to Air-Mass Analysis* (Cambridge, Mass., 1932). For tephigram soundings of a "mean hurricane," see R. C. Sheets, *JAM*, 8 (1969), 134. For vertical profiles and charts of total energy content of the atmosphere, which supplement tephigram analysis, see J. G. Lockwood, *MM*, 97 (1968), 145. For maps of average monthly dewpoints and their standard deviations

in the contiguous United States, see A. V. Dodd, *MWR*, 93 (1965), 113. For details of routine automated analysis of aerological soundings, see J. D. Stackpole, *JAM*, 6 (1967), 464.

32. For details, see A. H. Gordon, *MM*, 81 (1952), 333.

33. See *ITM*, pp. 161–73, whose treatment I follow, for a discussion of the Newtonian basis of atmospheric dynamics. On the nature of the laws, see the article "Mechanics," by Sir R. V. Southwell, in *Encyclopaedia Britannica* (1951).

34. See *ITM*, p. 161, for the useful analogy of a moving elevator in clarifying the distinction.

35. G. C. McVittie, *JM*, 8 (1951), 161, gives a review of systems. See *PDM*, p. 160, for the use of polar coordinates. For a discussion of the use of tensor calculus in atmospheric mechanics, see P. Defrise, *AGP*, 10 (1964), 261.

For examples of specialized coordinate systems, see: M. S. Rao, *GPA*, 65, no. 3 (1966) 196, for equations for global monsoons and toroidal circulations in a coordinate system in which the ground forms a coordinate surface; and W. D. Bonner, S. Esbensen, and R. Greenberg, *JAM*, 7 (1968), 339, for a study of low-level jetstreams.

36. See *EDM*, p. 68, for mathematical details, and see C. G. Coriolis, *Traité de la mécanique de corps solides* (Paris, 1844). As to who should be given credit for the discovery of the Coriolis force, see: C. L. Jordan, *BAMS*, 47 (1966), 401; B. Haurwitz, *loc. cit.*, p. 659; H. E. Landsberg, *loc. cit.*, p. 887; H. L. Burstyn, *loc. cit.*, p. 890; F. R. Leslie, *BAMS*, 48 (1967), 103; and D. G. Baker, *loc. cit.*, p. 458.

37. See almost any elementary textbook of dynamics for such a demonstration.

38. For an amplification of this viewpoint, see J. E. McDonald, *BAMS*, 34 (1953), 192.

39. It is important to realize that Ω varies, because of fluctuations in the atmospheric and oceanic circulations. See W. H. Munk and R. L. Miller, *T*, 2 (1950), 93.

40. See *ITM*, pp. 166–69, for the basic mathematics.

41. D. T. Perkins and A. F. Gustafson, *JM*, 8 (1951), 418.

42. For the derivation, see J. M. Leaver, *BAMS*, 23 (1942), 298. For an elementary derivation of the Coriolis acceleration using vectors, see J. G. Breiland, *BAMS*, 26 (1945), 69.

43. See *PDM*, p. 169, for the full derivation.

44. For an elementary discussion of the physical meaning of partial differentiation, see P. Abbott, *Teach Yourself Calculus* (London, 1946), pp. 319–24.

45. For illustrations of such a scale, see *Admiralty Weather Manual* (London, 1941), p. 177; also, for complete scales, *HM*, pp. 50–62.

46. For a direct-reading scale, see L. S. Matthews, *MM*, 85 (1956), 263. J. C. Bellamy and R. E. Nordstrom, *BAMS*, 34 (1953), 132, give details of the circles.

47. S. Petterssen, *JM*, 7 (1950), 76. E. J. Aubert and J. S. Winston, *JM*, 8 (1951), 126. A. F. Gustafson, *BAMS*, 34 (1953), 196. A. F. Crossley, *MM*, 85 (1956), 311. G. Arnason, G. J. Haltiner, and M. J. Frawley, *MWR*, 90 (1962), 175. C. S. Durst and N. E. Davis, *MM*, 86 (1957), 138. S. L. Hess, *BAMS*, 39 (1958), 228. For a theory of instability in a geostrophic wind with a transverse windspeed gradient, see S. D. R. Wilson, *QJRMS*, 91 (1965), 132. For probability predictions of geostrophic winds, based on a barotropic model, (North American examples), see T. A. Gleeson, *JAM*, 6 (1967), 355. On the direct computation of geostrophic winds from observed winds in the United States by means of the *balance equation*, see R. M. Endlich, *JAM*, 7 (1968), 994. The balance equation,

$$\frac{1}{\rho}\nabla^2 p - f\zeta + \beta u - 2J(u, v) = 0,$$

where the notation is as in the text, is a differential relationship between horizontal wind and pressure that is different from the geostrophic and gradient wind equations, and is obtained from the divergence equation by neglecting certain terms. It describes that state of air motion in which the airflow is always adjusted to the pressure field, under the control of gravity and Coriolis forces. For a technique for solving the balance equation, see K. Miyakoda and R. W. Moyer, *T*, 20 (1968), 115. For a study of a deepening wave cyclone over North America, in which vertical motions are obtained via the balance equation, see T. N. Krishnamurti, *MWR*, 96 (1968), 197 and 208. For a review *in* geostrophic motion, see N. A. Phillips, *Revs. in Geophysics*, 1 (1963), 123. For a mathematical model describing the adjustment of an airflow from a nongeostrophic initial state to a final geostrophic equilibrium, see W. M. Washington, *T*, 16 (1964), 530.

48. For details, see R. Silvester, *MM*, 84 (1955), 348.

49. E. Gold, *Proc. Roy. Soc. London*, 80, series A, (1908), 436, and *BAMS*, 44 (1963), 249. N. Shaw, *Meteorological Glossary* (London, 1916). M. H. Freeman, *MM*, 91 (1962), 255.

50. M. A. Alaka, *MWR*, 89 (1961), 482.

51. R. M. Endlich, *MWR*, 89 (1961), 187. W. M. Gray, *QJRMS*, 88 (1962), 430. On whether gradient flow is balanced or unbalanced, see: C. P. Arnold, Jr., *BAMS*, 48 (1967), 715; also O. E. Thompson, *BAMS*, 49 (1968), 24; L. F. Graves, *loc. cit.*, p. 25; W. J. Saucier, *loc. cit.*, p. 26.

52. C. S. Durst, *MM*, 78 (1949), 157.

53. A. Kochanski, *JM*, 15 (1958), 84. J. K. Bannon, *QJRMS*, 75 (1949), 131. W. L. Godson, *BAMS*, 30 (1949), 342. M. Neiburger *et al.*, *JM*, 5 (1948), 87.

54. D. Brunt and C. K. M. Douglas, *Mem. Roy. Meteorol. Soc.*, vol. 3, no. 2 (1928). B. Haurwitz,

JM, 3 (1946), 95. M. K. Miles, *QJRMS*, 76 (1950), 99.

55. C. E. Buell, *BAMS*, 38 (1957), 47, and *JM*, 15 (1958), 309. J. W. Hutchings, *JM*, 14 (1957), 566. R. Berggren, *W*, 14 (1959), 181. Neiburger *et al.*, *op. cit.* (1948).

56. S. Petterssen, *T*, 2 (1950), 18.

57. See: W. J. Saucier, *Principles of Meteorological Analysis* (Chicago, 1955), p. 305, for details of the determination of streamlines from wind observations; R. C. Bundgaard, *JM*, 13 (1956), 569, on computing them from pressure distribution. On the theory of the relation between wind and pressure in low latitudes, see S. L. Rosenthal and R. W. Reeves, *MWR*, 95 (1967), 11. For a simple formula of confluent/diffluent streamline patterns, analogous to the gradient-wind formula, see P. Groen, *QJRMS*, 90 (1964), 194. For details of the computation of stream functions, see: R. M. Endlich and R. L. Mancuso, *T*, 16 (1964), 32, for a direct method; H. F. Hawkins and S. L. Rosenthal, *MWR*, 93 (1965), 245, on computing from the windfield; and R. L. Mancuso, *JAM*, 6 (1967), 994, on computing from the vorticity of the horizontal windfield.

58. See Petterssen's rules for prediction of the pressure field in *WAF* (1940), 378–440. See *HM*, pp. 412–98 for a comprehensive review of the field of atmospheric hydrodynamics. For examples of the use of Lagrangian concepts, see: A. H. Gordon, *W*, 22 (1967), 455, on tropical meteorology; S.-K. Kao and L. L. Wendell, *JGR*, 70 (1965), 769, on large-scale eddy frictional forces; J. K. Angell, D. H. Pack, and C. R. Dickson, *JAS*, 25 (1968), 707, on helices or longitudinal roll-vortices in the planetary boundary layer at Idaho Falls. See A. H. Gordon *AMGB*, 17, series A (1968), 125, for a Lagrangian solution to the following problem. Consider a baroclinic atmosphere initially at rest, with a uniform pressure field at sea level. Given data for the atmosphere's baroclinicity during some one month, how will its mass readjust itself (as a function of time and latitude) if the initial pressure distribution is retained, and baroclinic changes are ignored? For the mathematics of changes in the lapse-rate of temperature moving with the wind, see D. O. Staley, *QJRMS*, 92 (1966), 147.

59. For a precise derivation of the general equation of horizontal mass divergence in the atmosphere, see E. H. Hsu, *JM*, 8 (1951), 395; see also L. Sherman, *JM*, 9 (1952), 359. On the computation of divergence, see A. Eddy, *QJRMS*, 90 (1964), 424, for an objective analysis of horizontal wind divergence. For a method of adjusting wind data to yield zero divergence of total atmospheric mass, using Florida examples, see A. Gruber and J. J. O'Brien, *JAM*, 7 (1968), 333. For a method of altering observed winds to give the wind pattern in the United States for nondivergence, see R. M. Endlich, *JAM*, 6 (1967), 837. For case studies of divergence, see G. J. Haltiner, R. F. Alden, and G. C. Rosenberger, *MWR*, 93 (1965), 297, on migratory nondeveloping cyclones in North America, and M. A. Lateef, *MWR*, 95 (1967), 778, who also discusses vorticity in the Caribbean.

60. For examples of topographic divergence in the New Guinea area, see A. H. Glenn, *BAMS*, 30 (1949), 50. A simple formula for confluent or diffluent currents, with application to winds, is given by P. Groen, *QJRMS*, 90 (1964), 194.

61. H. Landers, *JM*, 13 (1956), 121.

62. For mean charts of the normal fields of convergence and divergence at the 10,000-foot level in the northern hemisphere, see J. Namias and P. F. Clapp, *JM*, 3 (1946), 14. For objective maps and cross sections of the horizontal divergence field, see A. Eddy, *QJRMS*, 90 (1964), 424. On the tropics, see K. Bryant, *MM*, 88 (1959), 270, and on the central plains, see W. Bleeker and M. J. Andre, *QJRMS*, 77 (1951), 260. See J. S. Sawyer, *QJRMS*, 90 (1964), 395, for evidence that airflow over northern North America at the 30-mb level is nondivergent for wavenumber two.

63. This was first demonstrated by V. Bjerknes, *Astrophys. Norv.*, 2 (1937), 262.

64. R. G. Fleagle, *JM*, 3 (1946), 9. G. P. Cressman, *loc. cit.*, 85. D. C. House, *BAMS*, 39 (1958), 137, and 42 (1961), 803.

65. G. Stokes, *Camb. Phil. Trans.*, vol. 7 (1845).

66. See *ITM*, pp. 198–201, 208–12, for the mathematical derivation.

67. H. Riehl, *Tropical Meteorology* (New York, 1954), pp. 199–200. For objective criteria for distinguishing between irrotational and rotational motions in the atmosphere, see F. Di Benedetto, *GPA*, 67, no. 2 (1967), 193.

68. See Haltiner and Martin *op. cit.* (1957), p. 164, for details.

69. See Byers, *op. cit.* (1944), p. 363, for a derivation. See also C.-G. Rossby, *QJRMS*, supp. to vol. 66 (1940).

70. The discussion here follows *HM*, pp. 813–48.

71. For a simple account, see R. S. Scorer, *W*, 12 (1957), 72. See also "M. O. discussion" on vorticity, *MM*, 80 (1951), 169.

72. *EDM*, p. 118.

73. S. Petterssen, *BAMS*, 30 (1949), 191. R. F. Shaw and E. J. Joseph, *MWR*, 86 (1958), 141. H. L. Kuo, *JM*, 8 (1951), 307. R. Dixon, *QJRMS*, 90 (1964), 175. C. W. Newton and J. E. Carson, *T*, 5 (1953), 321. H. Landers, *JM*, 13 (1956), 511. L. Sherman, *JM*, 10 (1953), 399. B. Haurwitz, *JM*, 3 (1946), 24. F. W. Ernst, R. L. Fox, and H. E. Hutchinson, *BAMS*, 34 (1953), 441. R. J. Reed and F. Sanders, *JM*, 10 (1953), 338. R. D. Elliott, *BAMS*, 37 (1956), 270. J. P. Jenrette, *BAMS*, 41 (1960), 317. H. Riehl, K. S. Norquest, and A. L. Sugg, *JM*, 9 (1952), 291. A. M. Grant, *JM*, 9 (1952), 439.

74. E. Hovmöller, *T*, 4 (1952), 126. S. Petterssen,

T, 5 (1953), 231. R. C. Bundgaard, *BAMS*, 37 (1956), 465.

75. H. Landers, *op. cit.* (1956), and *JM*, 13 (1956), 511. On charts, see: B. G. Wales-Smith, *MM*, 88 (1959), 4; E. M. Carlstead, *JM*, 10 (1953), 356; G. P. Cressman, *JM*, 10 (1953), 17. For a vorticity budget of the wintertime lower stratosphere over North America, see R. A. Craig, *JAS*, 24 (1967), 558. For details of how to construct wind fields that correspond to specific vorticity distributions, see R. M. Endlich, *JAM*, 6 (1967), 837.

76. *EDM*, pp. 130–32. On reinforcement and interferences, see A. V. Carlin, *BAMS*, 34 (1953), 311. See also "M. O. discussion," *MM*, 82 (1953), p. 148, for a review of wavelength concepts.

77. R. A. Craig, *JM*, 2 (1945), 175. For a discussion of the perturbation technique and the method of characteristics, see *CM*, pp. 401–33.

78. H.-I. Kuo, *JM*, 6 (1949), 105. S. A. M. Hassanein, *T*, 1 (1949), 58. For a review of recent work on Rossby waves, see G. W. Platzman, *QJRMS*, 94 (1968), 225. For a theory of the effect of variations in the Coriolis parameter (i.e., Rossby's β term) on planetary circulations, see A Huss, *JAS*, 21 (1964), 507. On the theory of vertical propagation of stationary planetary Rossby waves through regions of weak westerly winds forming "wave guides," see R. E. Dickinson, *JAS*, 25 (1968), 984. For details of atmospheric wave motions other than Rossby waves, see: R. J. Deland and K. W. Johnson, *MWR*, 96 (1968), 12, on westward-moving planetary waves, wave-numbers 1 to 3, which appear to be present at all seasons between latitudes 22 and 87°N in the 1,000 to 10-mb layer; R. J. Deland, *MWR*, 93 (1965), 307, on spherical-harmonic waves at 500 mb in the northern hemisphere; N. K. Balachandran and W. L. Donn, *MWR*, 92 (1964), 423, on short- and long-period internal gravity waves producing a rapidly moving pressure rise, at the Palisades, New York; F. Press and D. Harkrider, *JGR*, 67 (1962), 3889, on acoustic-gravity waves, using evidence from thermonuclear explosions; and D. Stranz, *GPA*, 64, no. 2 (1966), 243, on Helmholtz waves near the tropopause, producing wave clouds in southwestern Germany.

79. For a more detailed discussion, see e.g., D. E. Rutherford, *Vector Methods* (Edinburgh and London, 1951), and *HM*, pp. 219–31.

80. See *ITM*, pp. 198–201, for a discussion of this concept.

81. If a large number of Laplacian operations must be performed, a computer method is useful; See S. L. Hess, *BAMS*, 38 (1957), 67. See *EDM*, pp. 155–56, for the mathematics of vorticity as the Laplacian of the stream function. See Rutherford, *op. cit.* (1951), p. 85, on Laplace's equation.

82. The Jacobian is a type of determinant. See P. G. Hoel, *Introduction to Mathematical Statistics* (New York, 1962), pp. 381–83, for a discussion in terms of probability theory.

83. See Haltiner and Martin, *op. cit.* (1957), p. 350, for the derivation of the vorticity equation, expressed in this form, from the conservation of vorticity theorem, and pp. 366 and 390 for the derivation of the other equations. For expressions of the other equations of atmospheric dynamics in a coordinate system that has p instead of z as the vertical coordinate, see *ITM*, pp. 219–31. For discussions of the use of the vorticity equation in weather forecasting, see: J. Van Mieghem, *T*, 6 (1954), 170, and *SPIAM*, p. 415; G. W. Platzman, *JAS*, 19 (1962), 313, where the equation is developed in spherical surface harmonics for nondivergent flow; and B. Saltzman and M. S. Rao, *JAS*, 20 (1963), 438, who apply it to the mean state of the atmosphere. For the basic equations enabling the long-term statistical properties of the atmosphere (i.e., its mean climate) to be deduced analytically, see E. N. Lorenz, *T*, 16 (1964), 1.

84. V. Bjerknes, *Videnskapsselsk Skrifter*, no. 5 (1898).

85. See *EDM*, p. 138.

86. See *ITM*, pp. 189–96, for details. On energy exchange between the barotropic and baroclinic components of the atmospheric circulation during the abnormal winter of 1962–63 in the northern hemisphere, see A. Wiin-Nielsen and M. Drake, *MWR*, 94 (1966), 1. For the climatology of barotropic/baroclinic energy exchange in the total atmosphere, see A. Wiin-Nielsen and M. Drake, *MWR*, 93 (1965), 67. On wind shear, see: M. Armendariz and L. J. Rider, *JAM*, 5 (1966), 810, for thin atmospheric layers, at White Sands, New Mexico; A. F. Crossley, *MOSP*, no. 17 (1962), for extreme values; and O. Essenwanger, *AMGB*, 15, series A (1966), 50, for persistence of specific shear values; see also G. H. R. Reisig, *JAM*, 6 (1967), 961. For an example of wind shear in the planetary boundary layer at Little America, Antarctica, produced both thermally by the temperature gradient at the ice edge and frictionally, by the normal Ekman spiral, see B. Lettau, *MWR*, 95 (1967), 627.

87. R. C. Sutcliffe, *QJRMS*, 73 (1947), 370, and 74 (1948), 178.

88. See the table in *EDM*, p. 180, for details.

89. S. Petterssen, *JM*, 12 (1955), 36. See also J. Spar, *JM*, 13 (1956), 123.

90. W. L. Godson, *JM*, 7 (1950), 268. M. A. Alaka, *MWR*, 90 (1962), 49.

91. For further details, see *CM*, pp. 434–52, and Haltiner and Martin, *op. cit.* (1957), p. 381. The principle of hydrodynamic instability was devised by H. von. Helmholtz, *Meteorol. Z.*, 5 (1888), 329; see J. Bjerknes, *Proc. Toronto Meteorol. Conf. 1953* (London, 1954), p. 138, for a historical review of ideas.

92. H.-L. Kuo, *JM*, 10 (1953), 235. J. Spar, *JM*, 14 (1957), 136. R. G. Fleagle, *T*, 7 (1955), 168. G. Arnason, *T*, 15 (1963), 205. G. J. Haltiner and D. E. Caverly, *QJRMS*, 91 (1965), 209. On baroclinic

instability, see R. G. Fleagle, *QJRMS*, 83 (1957), 1, and, for practical application of Fleagle's work, A. E. Parker, *MM*, 97 (1968), 340; see also F. P. Bretherton, *QJRMS*, 92 (1966), 335. For a theory of baroclinic instability of the zonal wind, see J. W. Miles, *Revs. in Geophysics*, 2 (1964), 155, and *JAS*, 21 (1964), 500 and 603. According to A. Wiin-Nielsen, *MWR*, 95 (1967), 733, baroclinic instability of the zonal wind is a function of its vertical profiles. According to P. H. Stone, *JAS*, 26 (1969), 376, the existence of a single jetstream in middle latitudes cannot be explained solely by momentum transport in baroclinic waves; the latter can only intensify an existing jet. See R. T. Williams, *JAS*, 22 (1965), 388, for a mathematical model which shows that small-scale distortions of planetary wave patterns (i.e., a sharpening of troughs, a flattening of ridges, and the development of frontal zones) result when the amplitude of an unstable baroclinic wave exceeds a critical value.

93. See J. G. Charney, *JM*, 4 (1947), 135, for a historical review.

94. V. Bjerknes *et al.*, *Physikalische Hydrodynamik* (Berlin, 1933), p. 785. J. Bjerknes, *Geofys. Publ.*, vol. 1, no. 2 (1919). J. Bjerknes and H. Solberg, *Geofys. Publ.* vol. 2, no. 3 (1921), and vol. 3, no. 1 (1922). H. Solberg, *Geofys. Publ.*, vol. 5, no. 9 (1928), and *Proc. Third Int. Congr. Appl. Mech.* (Stockholm, 1931), 121.

95. J. Bjerknes, *Meteorol. Z.*, 54 (1937), 462.

96. C.-G. Rossby, *J. Marine Res.*, 2 (1939), 38. B. Haurwitz, *J. Marine Res.*, 3 (1940), 35. See J. Holmboe, G. E. Forsythe, and W. Gustin, *Dynamic Meteorology* (New York, 1945), Chapter 12. J. C. Charney, *JM*, 4 (1947), 135. On the mathematics of Charney's theory of long-wave instability, see J. W. Miles, *JAS*, 21 (1964), 451. On baroclinic waves, see: G. J. Haltiner, *T*, 19 (1967), 183, on the effects of sensible heat exchange; G. J. Haltiner and D. E. Caverly, *QJRMS*, 91 (1965), 209, on the effects of surface friction; and A. Wiin-Nielsen, A. Vernekar, and C. H. Yang, *GPA*, 68, no. 3 (1967), 131, on the effects of friction and heating. On the stability of Polar Front waves (mathematical model), see I. Orlanski, *JAS*, 25 (1968), 178.

97. J. Bjerknes and J. Holmboe, *JM* 1 (1944), 1.

98. E. Eliasen, *T*, 10 (1958), 206. H. J. Stewart, *JM*, 5 (1948), 236. B. Bolin, *T*, 2 (1950), 184. R. W. James, *T*, 3 (1951), 258. For the dynamics of orographic influences on large-scale atmospheric circulations, see H. Reuter and H. Pichler, *T*, 16 (1964), 40, on the Alps and vorticity effects, and M. Sankar-Rao, *GPA*, 60, no. 1 (1965), 141, on the influence of continental elevation on the stationary harmonics of atmospheric motion.

99. H.-L. Kuo, *T*, 5 (1953), 475. On the dynamics of the stratospheric mean zonal flow at the equator, see G. B. Tucker, *QJRMS*, 90 (1964), 405.

100. M. Doporto, *T*, 6 (1954), 32.

101. *CM*, pp. 454–63. J. Bjerknes, *CM*, p. 577. E. Palmen, *CM*, p. 599. E. T. Eady, *CM*, p. 464. H. Wexler, *CM*, p. 621. H. Riel, *CM*, p. 909. A. Grimes, *CM*, p. 883. On the dynamics of cyclones, see: D. W. Stuart, *JAM*, 3 (1964), 669, on the baroclinic features of a cutoff low in the western United States; C. E. Wallington, *W*, 24 (1969), 42, on depressions as moving vortices; R. Dixon, *QJRMS*, 90 (1964), 174, on deepening of wave depressions in relation to vorticity advection; C. B. Pyke, *BAMS*, 46 (1965), 4, on the role of air-sea interaction in cyclone development; and M. D. Danard, *JAM*, 3 (1964), 27, on the effect of latent heat release in the United States. For a numerical-dynamic model simulating the life-cycle of a tropical cyclone, see K. Ooyama, *JAS*, 26 (1969), 3.

102. See *BAMS*, 47 (1966), 200, for the feasibility of a global observation and analysis experiment tied to dynamics.

103. See *DMWF*, p. 600, for an account of leeward vortices. On the cyclone, see C. W. Newton, *JM*, 13 (1956), 528. For case studies of lee cyclones in the Colorado Rocky Mountains, see E. B. Fawcett and H. K. Saylor, *MWR*, 93 (1965), 359. For examples of the complexity of mountain influences on atmospheric circulations and weather, see D. E. Pedgley, *W*, 22 (1967), 266.

104. For an elaboration of these factors, see R. C. Sutcliffe, *Proc. Toronto Meteorol. Conf. 1953* (London, 1954), p. 139.

105. E. J. Sumner, *QJRMS*, 76 (1950), 384. L. F. Hubert, *BAMS*, 36 (1955), 440. For studies of the energy budgets of tropical cyclones, see, e.g., B. I. Miller, *MWR*, 92 (1964), 389.

106. W. Ferrel, *A Popular Treatise on the Winds* (New York, 1890). L. Sherman, *JM*, 13 (1956, 500).

107. See L. Sherman, *BAMS*, 34 (1953), 256, for details. For examples of hyperbolic points, see L. Sherman and I. Carino, *BAMS*, 35 (1954), 220. L. Sherman and N. E. La Seur, *BAMS*, 36 (1955), 152. C. E. Palmer, *QJRMS*, 78 (1952), 126.

108. S. Syono, *JM*, 8 (1951), 103. R. W. James, *JM*, 8 (1951), 17. On the dynamics of tropical cyclones, see: R. J. Renard, *MWR*, 96 (1968), 453, on their motion, making use of a numerically derived steering current; M. Yanai, *Revs. in Geophysics*, 2 (1964), 367, for a review; and K. Ooyama, *JAS*, 26 (1969), 3, for a numerical model simulating the life cycle of a tropical cyclone. On the dynamics of hurricanes, see H. F. Hawkins and D. T. Rubsam, *MWR*, 96 (1968), 617, on the vorticity and mass budgets of Hurricane Hilda, 1964.

109. E. M. Brooks, *CM*, p. 676. D. S. Foster, *MWR*, 92 (1964), 339.

110. E. T. Eady and J. S. Sawyer, *QJRMS*, 77 (1951), 531. For the dynamics of airflow over heated coastal mountains (Santa Ana Mts., southern Calif.),

involving three distinct components (ridge-top convection, transition flow, and wave flow), see M. A. Fosberg, *JAM*, 8 (1969), 436. For examples of the influence of mesoorography on synoptic-scale phenomena, see: R. S. Scorer, *QJRMS*, 78 (1952), 53, on mountain-gap winds at Gibraltar; A. Ward, *MM*, 82 (1953), 322 on the local wind at North Front, Gibraltar; and J. Meiklejohn, *QJRMS*, 81 (1955), 468, on the very strong local westerly wind found at the foot of the Cheviot escarpment, Northumberland, Eng.

111. For a review of local winds, see F. Defant, *CM*, p. 655. For the dynamics of antibaric and inertial winds, see Haltiner and Martin, *op. cit.* (1957), p. 178. The term *geotriptic* has been suggested by W. B. Johnson, Jr., *BAMS*, 47 (1966), 982, for all winds arising from the effects of the Earth's rotation and surface friction. A *geotriptic wind* is an unaccelerated airflow in the planetary boundary layer, which represents a balance between pressure-gradient, Coriolis, and frictional forces: it thus has both geostrophic and antitriptic characteristics.

112. V. Bjerknes *et al.*, *Physikalische Hydrodynamik* (Berlin, 1933).

113. B. Haurwitz, *JM*, 4 (1947), 1, and 7 (1950), 164.

114. The original theory devised by Lord Rayleigh, *Phil. Mag.*, 32 (1916), 529, was developed in a meteorological context by F. Defant, *AMGB*, 2 (1950), 404. For details of the mathematics, see *CM*, p. 661.

115. F. H. Schmidt, *JM*, 4 (1947), 9. J. K. Angell and D. H. Pack, *MWR*, 93 (1965), 475. M. A. Estoque, *JAS*, 19 (1962), 244. J. A. Frizzola and E. L. Fisher, *JAM*, 2 (1963), 722. W. G. Biggs and M. E. Graves, *JAM*, 1 (1962), 474. C. W. Frenzel, *JAM*, 1 (1962), 405. On land-and-sea breezes in relation to diurnal wind oscillations in Hawaii, see R. L. Lavoie, *T*, 19 (1967), 354. On sea breezes, see: J. K. Angell and D. H. Pack, *MWR*, 93 (1965), 475, for Atlantic City, New Jersey; A. Elliott, *W*, 19 (1964), 147, for Porton Down, Wiltshire, Eng.; D. S. Gill, *MM*, 97 (1968), 19, for northeastern Scotland; and M. J. Schroeder *et al.*, *BAMS*, 48 (1967), 802, for the Pacific coast of North America. Dynamic theory indicates that a disturbance (the *sea-breeze forerunner*) should be generated at a coastline, propagating both inland and out to sea, prior to the development of the sea breeze: see J. E. Geisler and F. P. Bretherton, *JAS*, 26 (1969), 82. For a numerical model of the lake breeze on the eastern shore of Lake Michigan, see W. J. Moroz, *JAS*, 24 (1967), 337. For the effect of the Great Lakes on winds, see T. L. Richards, H. Dragert, and D. R. McIntyre, *MWR*, 94 (1966), 448. On the interaction between lake breeze and urban heat island (Toronto), see R. E. Munn, M. S. Hirt, and B. F. Findlay, *JAS*, 8 (1969), 411. On sea breezes on the Texas coast, see A. Eddy, *W*, 21 (1966), 162.

116. *M*, p. 268. T. A. Gleeson, *JM*, 8 (1951), 398. F. Defant, *AMGB*, 1 (1949), 421, and 2 (1950), 404. For a mathematical model of mountain and valley winds, see N. H. Thyer, *AMGB*, 15, series A (1966), 318. For a mathematical model of mountain upslope winds in the Santa Catalina Mountains of Arizona, see H. D. Orville, *JAS*, 21 (1964), 622. For examples of valley winds, see: W. Koutnik, *WW*, 21 (1968), 186, for the San Fernando Valley, Calif.; L. B. MacHattie, *JAM*, 7 (1968), 348, for the Kamanaskis Valley, Alberta; and M. A. Fosberg, *JAM*, 6 (1967), 889, for Wolfskill Canyon in southern Calif. For details of the valley wind in the Mt. Rainier area of Washington, see K. J. K. Buettner and N. Thyer, *AMGB*, 14, series A (1966), 125.

117. A. Wagner, *Meteorol. Z.*, 49 (1932), 209.

118. T. A. Gleeson, *JM*, 10 (1953), 262.

119. A. Defant, *CM*, p. 664.

120. See R. G. Fleagle, *JM*, 7 (1950), 227, for the mathematical theory. On Antarctic katabatic winds, see *AM*, pp. 3, 9, 52, and 317. G. S. P. Heywood, *QJRMS*, 59 (1933), 47. For further information on katabatic winds, see C. E. Cornford, *QJRMS*, 64 (1938), 553, and R. Geiger, *The Climate near the Ground* (Cambridge, Mass., 1950), pp. 437–39. For examples of katabatic winds, see: M. E. López and W. E. Howell, *JAS*, 24 (1967), 29, on the equatorial Andes; and N. A. Streten, *JAM*, 7 (1968), 46, on coastal east Antarctica.

121. J. von Hann, *Z. Öst. Ges. Meteorol.*, 1 (1866), 257. For meteorological details of the Montana chinooks, see A. W. Cook and A. G. Topil, *BAMS*, 33 (1952), 42, and E. P. McClain, *loc. cit.*, p. 87.

122. J. von Hann, *Handbuch der Klimatologie* (Stuttgart, 1908), p. 357.

123. H. von Ficker, *Denkschr. Akad. Wiss. Wien*, 78 (1906), 33; 85 (1910), 113; 80 (1907), 131; and *Föhn und Föhnwirkungen der Gegenwartige Stand der Frage* (Leipzig, 1943).

124. J. Kuttner, *Beitr. Phys. Freien Atmos.*, 25 (1938), 79 and 251. K. Frey, *Experientia*, 4 (1948), 36. H. C. Willett, *Descriptive Meteorology* (New York, 1944), p. 274.

125. W. D. S. McCaffery, *MM*, 81 (1952), 151. E. N. Lawrence, *MM*, 82 (1953), 74.

126. A. Defant, *Denkschr. Akad. Wiss. Wien*, 80. (1907), 107.

127. A. Defant, *CM*, pp. 671–72 gives further references.

128. Sir C. Normand, *QJRMS*, 79 (1953), 463. See: the symposium *Monsoons of the World* (New Delhi, 1960); P. R. Pischaroty, *PSTM*, p. 373, on monsoon pulses; R. Ananthakrishnan *et al.*, *PSTM*, pp. 62, 89, and 144; G. C. Asnani and A. U. Rao, *PSTM*, p. 207, on the atmospheric circulation over India; *PSTM*, p. 278, for evidence of mesovariations in the monsoon rains.

129. W. G. Kendrew, *Climates of the Continents* (Oxford, 3rd ed., 1937), and *PDM*, p. 13.

130. See E. R. Biehl, *Climatology of the Mediterranean area* (Chicago, 1944), for basic facts concerning

Etesians. H. Jeffreys, *QJRMS*, 48 (1922), 29, and 52 (1926), 85; see *PDM*, p. 409, for details of the mathematics.

131. E. B. Kraus and E. N. Lorenz, *USCC*, p. 361. According to K. Saha, *T*, 20 (1968), 601, a climatological picture of instantaneous vertical velocities in the Asiatic summer monsoon indicates that only a three-dimensional thermal convection model can describe adequately the immensity of the monsoon circulation of Asia and the northern Indian Ocean.

132. A. Wagner, *Gerlands Beitr. Geophys.*, 30 (1931), 196.

133. F. K. Hare and S. Orvig, *McGill Univ. Artic Meteorol. Res. Group Publ.*, no. 12 (1958). E. B. Reiter and H. Hengerber, *GA*, 42 (1960), 17. See also P. F. McAllen, *MM*, 91 (1962), 157, for the synoptic climatology of these different airflows over Singapore.

134. J. Namias and P. F. Clapp, *JM*, 6 (1949), 330. A. M. Chaudhury, *T*, 2 (1950), 56. H. Flohn, *J. Meteorol. Soc. Japan*, 35 (1957), 180, and *Erdkunde*, 12 (1958), 294. B. Bolin, *T*, 2 (1950), 184. For dynamic theories of monsoons in general, see M. Sankar-Rao, *MWR*, 93 (1965), 417, on the dynamic influence of stationary heat sources and sinks; see also M. Sankar-Rao and B. Saltzman, *T*, 21 (1969), 308, on the steady-state theory of global monsoons, and a northern-hemisphere verification of it. On the dynamics of the Indian summer monsoon, see R. N. K. Murty, *MWR*, 96 (1968), 23, on the maintenance of westerly flow in the lower troposphere and easterly flow in the upper troposphere; on the uniqueness of the Arabian Sea southwest monsoon compared to summer monsoons in other parts of the world, see B. N. Desai, *JAS*, 24 (1967), 216. On the relation between the onset of the Indian southwest monsoon and trough–ridge patterns at 500 mb over extratropical Eurasia, see P. D. de la Mothe and P. B. Wright, *MM*, 98 (1969), 145; on the relation between rapid changes in atmospheric circulation at 200 mb in May or June each year, and the development of the Indian southwest monsoon, see P. B. Wright, *MM*, 96 (1967), 302. For a review of ideas concerning the Indian monsoon, see J. G. Lockwood, *W*, 20 (1965), 2. For a review of the problem of marine air invasions of the Pacific coast of the United States, involving sea breezes, monsoons, and airflow over coastal mountains, see M. J. Schroeder *et al.*, *BAMS*, 48 (1967), 802.

135. N. Wiener, *Cybernetics* (Cambridge, Mass., 3d ed., 1961), pp. 30ff. O. G. Sutton, *W*, 6 (1951), 291. For reviews of the predictability of the atmosphere, see E. N. Lorenz, *BAMS*, 50 (1969), 345, and J. Smagorinsky, *BAMS*, 50 (1969), 286, the latter also dealing with deterministic extended-range forecasting. On the predictability of an airflow possessing many scales of motion, see E. N. Lorenz, *T*, 21 (1969), 289, and T. A. Gleeson, *JAM*, 6 (1967), 213. On the updating of predictions by optimum use of current synoptic data, see I. G. Tadjbakhsh, *JAM*, 8 (1969), 389 (theory).

CHAPTER 6

The Synoptic Method

Synoptic meteorology, the study concerned with obtaining a general, synchronic, three-dimensional view of atmospheric conditions,[1] is essentially a form of geographical exploration, because the thickness of the atmosphere is practically infinitesimal in comparison with the size of the Earth: in round figures, the vertical thickness of air between mean sea level and 10 mb (30 km), which includes more than 99 per cent of the atmosphere, is less than one four-hundredth of the average diameter of the Earth (12,735 km). In a vertical cross section of the atmosphere from mean sea level to 10 mb, pole to pole, the horizontal axis (about 20,000 km) would have to be almost 700 times as long as the vertical axis to be drawn true to scale. In other words, for geographical and practical purposes the atmosphere can be regarded quite realistically as part of the Earth's surface. Experience of flying or climbing in high mountains will soon convince the sceptic that this statement is true. It is therefore misleading to attempt a complete geographical picture of any region of the Earth without describing conditions in the atmosphere above it as far up as accurate data will permit.

The synoptic meteorologist has the task of geographically exploring the atmosphere, and representing his results by cartographical methods. The synoptic climatologist is concerned with charting the relatively permanent features in the atmosphere's ever-changing topography, as revealed by the synoptic meteorologist's maps and sections.

In 1855, the electric telegraph made possible the construction of the first synoptic maps. Since then, there have been developed numerous types of map, graph, synoptic model, and other aids. Although weather-chart analysis and weather forecasting were at first carried out by the same person or persons, there is now a tendency to separate the two functions. Aviation requirements during the Second World War demanded a decentralization of forecasting services. The preparation of weather charts became the job mainly of large-area weather centrals, which also produce forecasts of the overall pressure field. Local forecasting became the task of regional offices, which fit the local weather into the prognostic pressure pattern supplied from the central source. There is thus a distinction to be drawn between *weather analysis* and *motion analysis*. Bjerknes and his collaborators were mostly concerned with motion analysis, whereas weather analysis was the particular field of Bergeron from 1928 on.[2]

The motion field of the atmosphere is the only quantity shown as an explicit continuous function on weather maps. In middle and high latitudes, motion is represented by isobars, in tropical latitudes it is described by streamlines. Winds and temperatures are shown implicitly by the geostrophic and thermal wind relationships. Other quantities are shown either by dynamic models, which correlate various meteorological elements, or by conventional display models of "spot observations" to be mentally integrated by the analyst. The main dynamic models used are the hydrostatic equation, which correlates pressure, temperature, and height, and the geostrophic (or gradient) wind equation, which correlates the pressure, temperature, and wind fields.

The international display models for synoptic work were mainly designed to exhibit weather reports significant for air-mass and frontal analysis, and therefore are less suitable for general use today than when the Polar Front theory was the only synoptic model used. The conventional barbed-arrow presentation of wind is, for example, unsuitable for showing the continuous variation of wind with time. A more useful method involves graphs showing the east-west and north-south components of the wind as functions of height or time, in place of the wind arrows.[3]

There are two sides to the synoptic problem: how to represent the field distribution of physical properties, i.e., how to chart the geographical distribution of pressure, temperature, and humidity in three dimensions; and how to deal with the fine structure of the atmosphere, the phenomena too small to be observed by the synoptic network, that is, scattered showers, local cloudiness variations, and so on. The first problem is solved by various types of synoptic chart: the synoptic weather map; the pressure-contour map; the cross section, either in space or in time; the thickness chart; and the hodograph, or vector wind diagram. The second problem can sometimes be solved by applying the laws of micrometeorology to specific situations shown in the synoptic charts. If no physical theory is available of direct relevance to the problem, statistical methods can be used on the basis of past occurrences, although this amounts, of course, to scientific cheating.

Several assumptions are made in constructing a synoptic chart. First, pressure is assumed to have the hydrostatic value everywhere, a good assumption, which enables the heights of upper isobaric surfaces to be computed from observed surface pressures and observed upper-air temperatures. Second, wind and pressure are assumed to be represented perfectly by the geostrophic relationship, despite the fact that divergence, convergence, and vertical motion (and hence clouds, precipitation, and "weather") can only occur if the actual wind differs from the geostrophic wind. Third, vertical motion and friction are neglected in explaining the weather shown on the chart, because they are too complex to be included in the standard observations. Fourth, the development or decay of pressure systems is inferred from horizontal divergence as determined by the rate of change of vorticity. The measure of vorticity employed is the geostrophic vorticity, obtained from the isobars on the chart. In other words, divergence is obtained by applying the geostrophic equation to a series of pressure charts, yet if the geostrophic relation applied perfectly there could be no divergence.

Synoptic Maps

Weather-analysis centers usually prepare hemispheric maps once or twice per day, and more detailed maps covering smaller areas much more frequently. Thus in the United Kingdom, maps for the northern hemisphere are prepared every 12 hours on a scale of 1:30,000,000, and maps covering the British Isles every hour on a scale of 1:3,000,000.[4] For areas the size of the United States, maps on scales of 1:5,000,000 to 1:12,500,000 are usually preferred.

For hemispheric and regional maps the choice of a suitable map projection is critical. The properties requiring accurate reproduction are area, direction, relief, scale and shape. Obviously it is impossible to preserve all these perfectly in one map.[5] Figure 6.1 demonstrates that erroneous conclusions may be drawn from a map if the properties of particular projections are not understood. The Mercator map in Figure 6.1 shows a large anticyclone (over Greenland) in the extreme north, with a low of approximately the same size to the south. The sinusoidal equal-area map, however, shows that in area the Greenland high is insignificant, and the Icelandic low is only one-fifth the size of the Azores anticyclone.[6]

In 1951 the Lambert conformal conic projection, with standard parallels at 30° and 60°, was recommended as most suitable for middle-latitude weather analysis. In 1959 the WMO recommended three standard projections for meteorological use. These are *Mercator's projection* for maps of areas near the equator, with scale true in latitude $22\frac{1}{2}$°; *Lambert's conformal conic projection* for middle latitudes, with standard parallels at 30° and 60° or 10° and 40°; and the *Polar stereographic projection* for circumpolar maps, with scale true in latitude 60°. All three are orthomorphic projections. Although their scales vary with latitude, the scale at any one point on the map is the same in all directions, so that shape is preserved within small areas, with no local distortion of direction. In addition, the *Gnomonic projection* is normally used for thunderstorm location. Usually the oblique case is required, so the tangential point can be located anywhere on the Earth's surface. Climatological maps often need different projections from those just mentioned, depending on the element displayed. Thus winds require conformal maps, precipitation areas and air-mass source regions equal-area maps.[7]

The basic synoptic maps are the surface weather map, and the pressure-contour maps for 1,000, 850, 700, 500, 300, 200, 100, and 10 mb. The surface weather map depicts mean sea-level isobars; the other maps show the height above mean sea level of the various pressure surfaces. The 1,000-mb contour map is identical with the surface map, apart from the fact that the isobars are relabeled as heights of the 1,000-mb surface, making use of the relationship that, near sea level, 4 mb of pressure corresponds to 100 feet of vertical height in the atmosphere. Thus the 1,000-mb isobar on the surface weather map is relabeled to form the zero contour on the 1,000-mb map, the 996-mb isobar becomes the −100-foot contour, the 1,004 mb isobar the +100-foot contour, and so on.

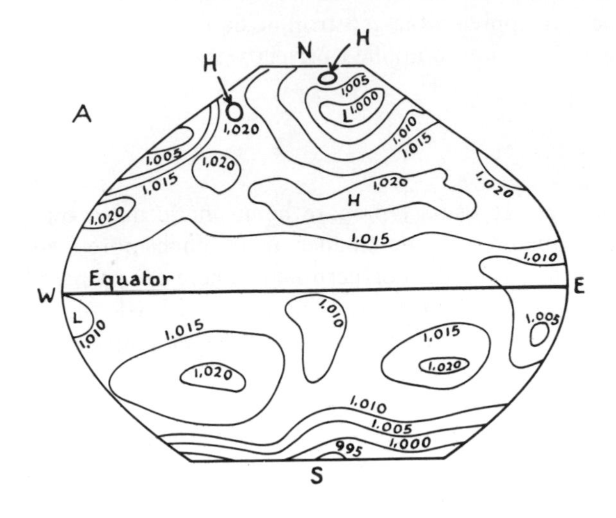

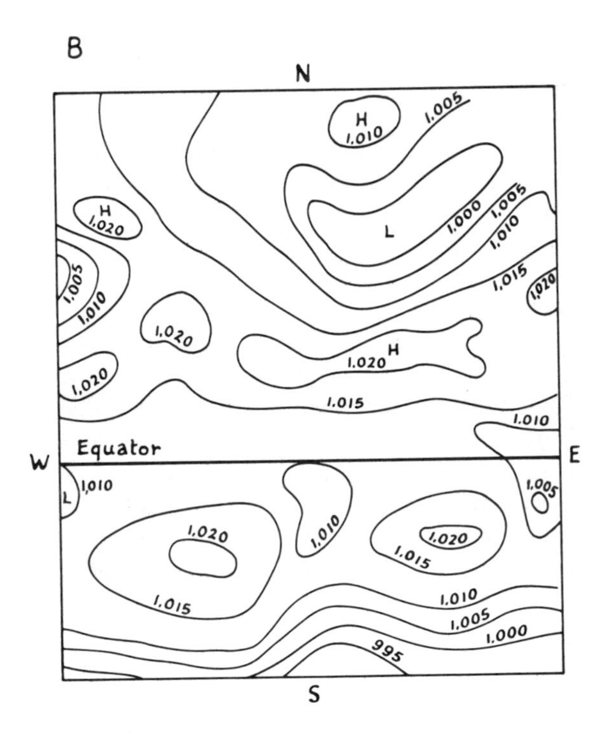

FIGURE 6.1.
Comparison of map projections (after J. B. Rigg). A shows the January mean sea-level pressure for 80°N to 60°S, 150°W to 30°E, plotted on the sinusoidal equal-area projection. B shows the same information plotted on Mercator's projection.

The maps for the higher levels of the atmosphere present simpler patterns than those for levels near the ground. This does not necessarily mean that the actual wind patterns at the higher levels are very simple. Neither does it imply, that, because the isobar or contour pattern is one of large, smooth curves in the form of (usually open) waves, the curvature of the isobars can be neglected in computing windspeeds from the contours. For the 200-mb level at Crawley, for example, the gradient wind was found to give the closest approximation to the actual wind for winds from SE through NNE to west, and for anticyclonic isobar curvature. For winds from SE through south to west, however, the curvature could be neglected and the geostrophic wind scale used directly.[8]

The smooth pattern of airflow shown by upper-air maps does seem to be real, not just the result of paucity of observations. A comparison of observed winds at 100 mb over Britain, with those inferred from the slope of the 100-mb surface, shows that the standard error of measurement of the 100-mb heights is only about 145 feet.[9]

The highest level for which charts can at present be fairly confidently constructed is 30 km (10 mb). IGY data enabled the completion of a series of these charts for 1958. They indicate (see Figure 6.2) some remarkable circulation changes throughout the year over North America and adjacent seas. Easterlies prevailed in summer 1957 until September, and then westerlies increased from October to mid-November. The Aleutian high exhibited pulsation from then until early January 1958, after which explosive warming occurred until early February. The circumpolar flow then re-established itself and persisted until the middle of April, after which the summer easterlies returned.[10]

Weather-map construction and analysis by hand is a tedious procedure, and often somewhat subjective. Attempts have therefore been made to speed up the work, and at the same time provide a more objective approach. *Numerical prediction techniques* using electronic computers, in particular, have necessitated this development. In the Cressman technique, a method of objective map analysis developed at the Joint Numerical Weather Prediction Unit, data from observations at irregularly spaced points are transformed into data at points in a grid pattern. The latter can easily be fed into a computer, with the result that objective maps for the entire northern hemisphere can be obtained in about 15 minutes. No hand analyses for 500 mb have been made at the JNWPU since April 1958, as a consequence of this development.[11]

An objective scheme of analysis must interpolate data in regions of few observations, remove errors, smooth the data, and maintain internal consistency. The Cressman method essentially involves using reported data to make successive corrections to an initial guess. In other words, a "first guess" chart must be provided for input to the computer, which then adjusts the data on the basis of current conditions. If a numerical prognostic chart is used as a first guess, i.e., a forecast chart produced by computer on the basis of a numerical model, the procedure should be completely objective. The adjustments made to the "first guess" map involve weighting each observation according to its distance from the nearest grid point. For northern hemisphere maps, a grid of almost 2,000 points is used, at the intersections of grid lines 200 nautical miles apart (see Appendix 6.1).

In Britain, experiments in objective analysis by the Meteorological Office show that the resulting analyses are smoother than the subjective maps. Comparison of maps derived by the two techniques proves that the objective method systematically

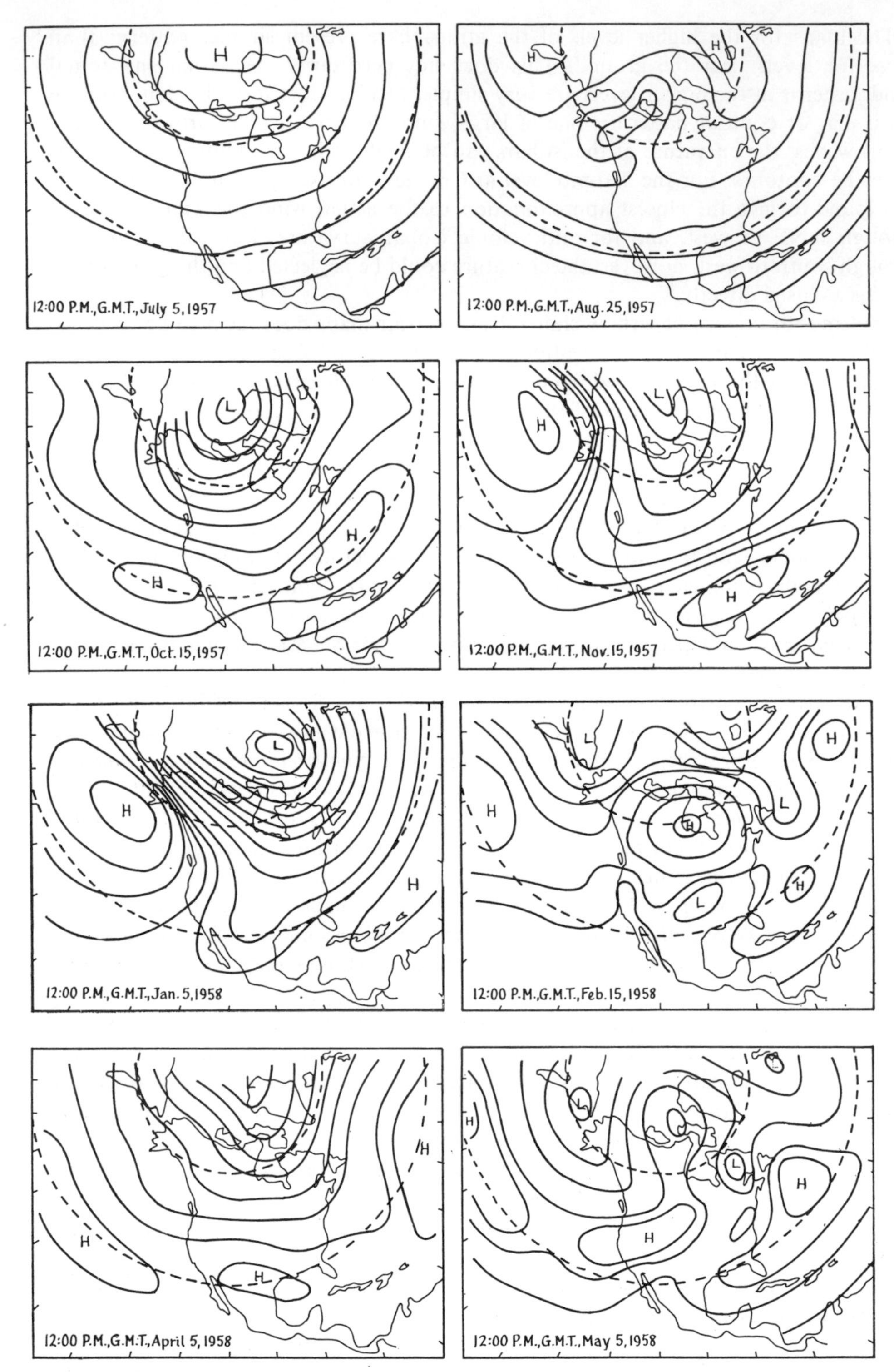

FIGURE 6.2.
Examples of weather patterns at 10 mb (after Teweles, Rothenberg, and Finger).

underestimates the depth of intense depressions over the Atlantic, largely because of the use of the geostrophic approximation in the numerical technique.[12]

Experiments have also been made involving the direct transmission of weather maps, in order to eliminate any need to plot the observations on a chart prior to analysis. Instead, the data is fed into a computer, and the final map is displayed on the screen of a cathode-ray tube.[13] This method is undoubtedly of great value in enabling a general picture of the instantaneous atmospheric circulation over a large area to be rapidly obtained, but it does not, of course, provide the detailed type of map required for local weather study.

Atmospheric Trajectories

Synoptic maps show conditions at certain selected instants of time: for published serial maps, these instants are separated by 12 hours for hemispheric charts, and by either three or six hours for maps covering smaller areas. The question arises: how is it possible to determine conditions at other times?

Normal synoptic charts present lines of constant pressure. These lines are not streamlines; i.e., they are not tangent curves to the wind throughout an instantaneous flow pattern. Streamline charts are in common use in tropical regions, but they are not regularly drawn for middle-latitude or polar regions. Instead, isopleths of constant wind direction (*isogons*), drawn from scalar analysis of wind-direction observations, and isopleths of constant windspeed (*isotachs*), are frequently superimposed on the same chart. It is not always easy to mentally integrate the two fields, however, and so a third technique is commonly used, that of trajectories.

Trajectories are curves exhibiting the successive locations of a given air parcel or air particle at different times; i.e., they indicate its track over the Earth's surface. This cannot be done for a complete chart without excessive computation, so it is normal practice to concentrate on significant points in space or time to commence the analysis. Trajectory analysis supplies information otherwise impossible to read directly from the charts.

Although isotach and isogon analyses show wind as a continuous function of space everywhere except at sharp fronts, shear lines, and at the ground surface, we may need to trace the movements of air particles across these very zones. Streamlines may fork or join, diverge or converge, and approach one another asymptotically; hence at first sight they seem to be very suitable for tracing air particles. However, *singular streamlines* occur, i.e., lines of streamline divergence or convergence, which are not necessarily lines of maximum horizontal velocity divergence or convergence, respectively. In addition, streamlines show *singular points*, marking the meeting or diverging point of two or more streamlines; if rotation is superimposed, the streamlines form spirals. Isogons and isotachs also exhibit singular points, as well as appearing and disappearing at discontinuities. It is thus clear that, merely employing streamlines, isotachs and isogons alone—even discounting isobars, in which the geostrophic assumption is involved—will mean that it will frequently be impossible to trace satisfactorily the movement of an air parcel from one chart to another.

Assuming geostrophic flow, the ordinary isobaric chart represents trajectories approximately correctly only when the pressure systems are stationary. If the pressure pattern shown on the chart can be regarded as stable, a simple measurement of the

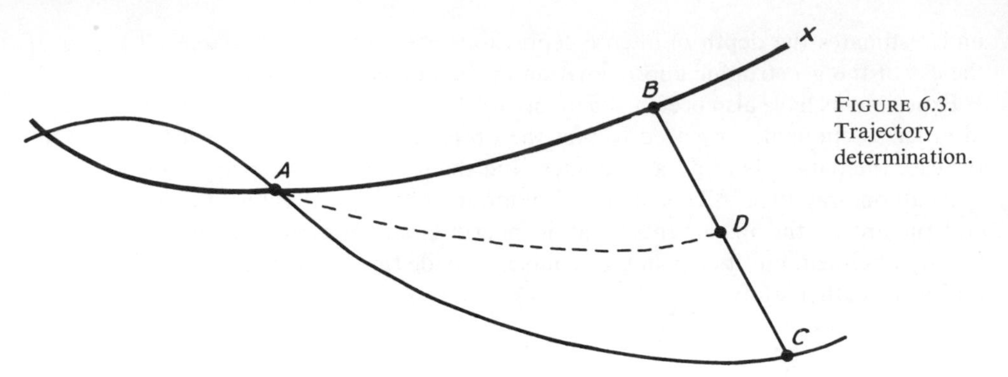

FIGURE 6.3. Trajectory determination.

isobar curvature enables the trajectory to be determined (see Appendix 6.2). Unfortunately, the pressure pattern usually cannot be considered stable.

A simple trajectory may be plotted by superimposing one chart upon another. In Figure 6.3, if AX represents the streamline through point A at time T, let B represent the point which the parcel of air at A will reach in time t if it continues to move with its present speed. Then if AC represents the streamline through A on the subsequent chart, the displacement of A through the same time interval with the speed shown for A on the second chart takes it to point C. The midpoint D of BC may then be assumed to represent the approximate position of the parcel on the second chart. AD then represents the trajectory of the air parcel between time T and time $(T + t)$.[14]

Trajectories of air particles have been computed from actual winds, from geostrophic and gradient winds, from CAV trajectories, from isentropic charts, and by dynamic methods. *Isobaric trajectories* (i.e., using geostrophic or gradient winds) cannot take vertical motions or velocity shears into account. If a large amount of vertical motion occurs, isobaric trajectories can deviate 1,000 km or more in 12 hours, in a horizontal plane, from the true trajectory (see Figure 6.4). *Isentropic trajectories*, determined using the stream function, deviate less from the actual path for air of average stability. At low stabilities, however, the isentropic trajectory is impossible to obtain because the air parcel loses its identity by mixing.[15]

In general, the errors of horizontal trajectories computed from isobaric maps may

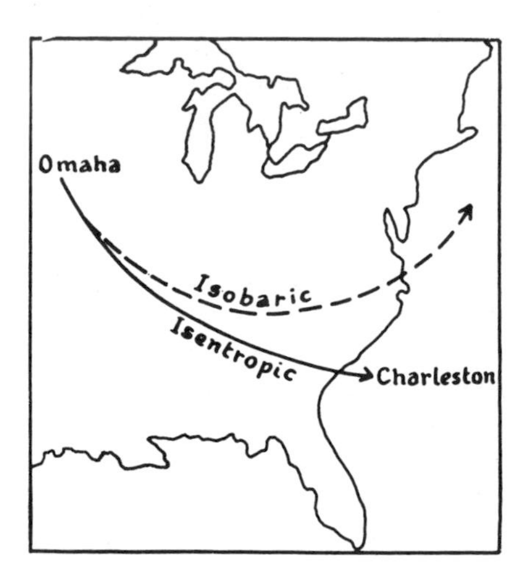

FIGURE 6.4.
Comparison of 12-hour isobaric and isentropic trajectories originating at 435 mb at midnight G.M.T. on January 2, 1958 (after Danielson).

be up to 20 per cent of the trajectory length in a 12-hour period, because of the geostrophic approximation alone. Deficiencies in the density of observing stations result in errors of up to 10 per cent for regions with relatively large numbers of stations, and up to 30 per cent for the oceans, for periods of 12 hours.[16]

Trajectories may be accurately computed from dynamic theory. These computations show that actual trajectories are very different from what one would expect from simple inspection of the charts. In a pressure field in which the isobaric or isentropic surfaces have a constant slope, the particle trajectories take the form of waves, cycloids, or loops, depending on whether the magnitude of the geostrophic velocity is greater than, equal to, or smaller than that of the initial velocity of the particle. For wave-shaped isobars, the trajectories show a wave pattern whose amplitude increases linearly with time under certain conditions, even though the isobars remain stationary. Although actual trajectories are of course three-dimensional, horizontal trajectories, as determined by dynamic theory, can be found quite easily by a graphic method.[17]

Atmospheric Cross Sections

Synoptic charts may be classified into three types, all of which make use of the Cartesian coordinate system, with a vertical z plane, through a line joining the observing point to the center of the Earth, and an xy plane tangent to the Earth's surface at the place of observation. If the z coordinate is constant, and x,y are variable, a synoptic map results. If z is variable, and x,y, are constant, an upper-air sounding chart results. If the x (or y) coordinate is constant, and y (or x) and z are variable, an atmospheric cross section results.

If t represents time, then a t,z, cross section displays the sequence of events at a fixed station, and a t,x, or t,y, cross section shows the progression of weather along a line of closely spaced observing points, the data for each point being plotted as functions of time and distance from an arbitrary origin.

Atmospheric cross sections must normally be plotted with highly exaggerated vertical scales. Sometimes a linear height scale is used, in which case the section will describe conditions accurately only if the atmospheric conditions approximate those in the appropriate Standard Atmosphere. More usually, a logarithmic pressure scale is used for the vertical axis (see Appendix 6.3).

Figures 6.5 to 6.9 present some examples of atmospheric cross sections. Figure 6.5 gives time-sections showing conditions up to 80 km over Churchill, Manitoba, based on IGY data. These indicate the occurrence of several distinct layers in the atmosphere, each layer having its own characteristic types of horizontal circulation. This stratification is mainly due to the vertical variation in static stability brought about by the variable ability of atmospheric constituents to absorb radiation. The stratospheric circulation systems are large and develop slowly, relative to systems nearer the ground. Figure 6.5,B, shows that during summer 1957, irregular easterly flow occurred in the middle stratosphere, with highly variable westerly airstreams above 75 km. Intense, rapidly moving cellular circulations were present near the mesopause. By winter 1957–1958, the westerly circulations extended downward to occupy the entire mesosphere in January. Very strong westerlies are shown developing in the midstratosphere in winter, with a temperature minimum at 30 km.[18]

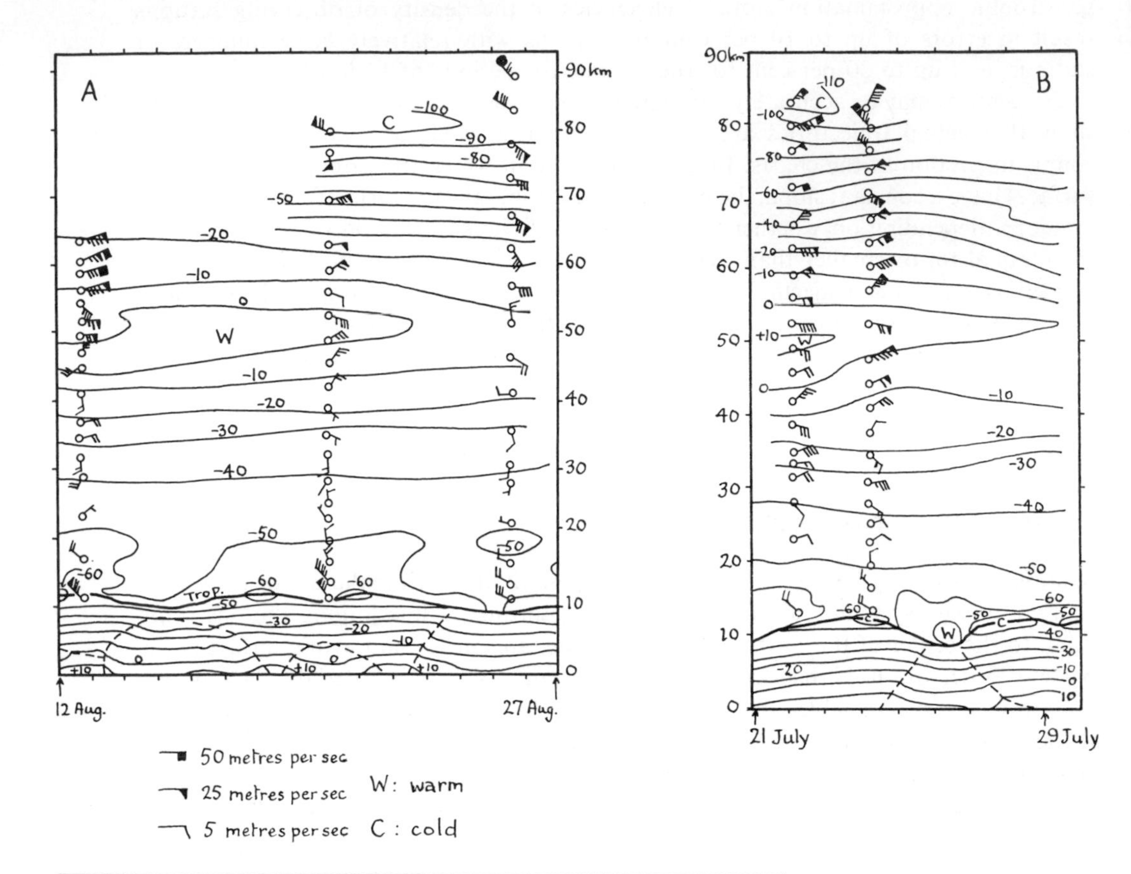

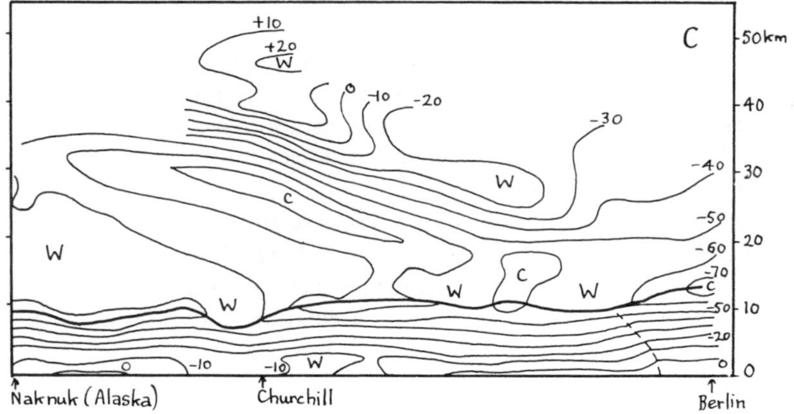

FIGURE 6.5.
A,B, Time cross sections through Churchill, Manitoba, August 12–27 and July 21–27, 1957.
C, Space cross section near 55°N, January 29, 1958. (After S. Teweles.)

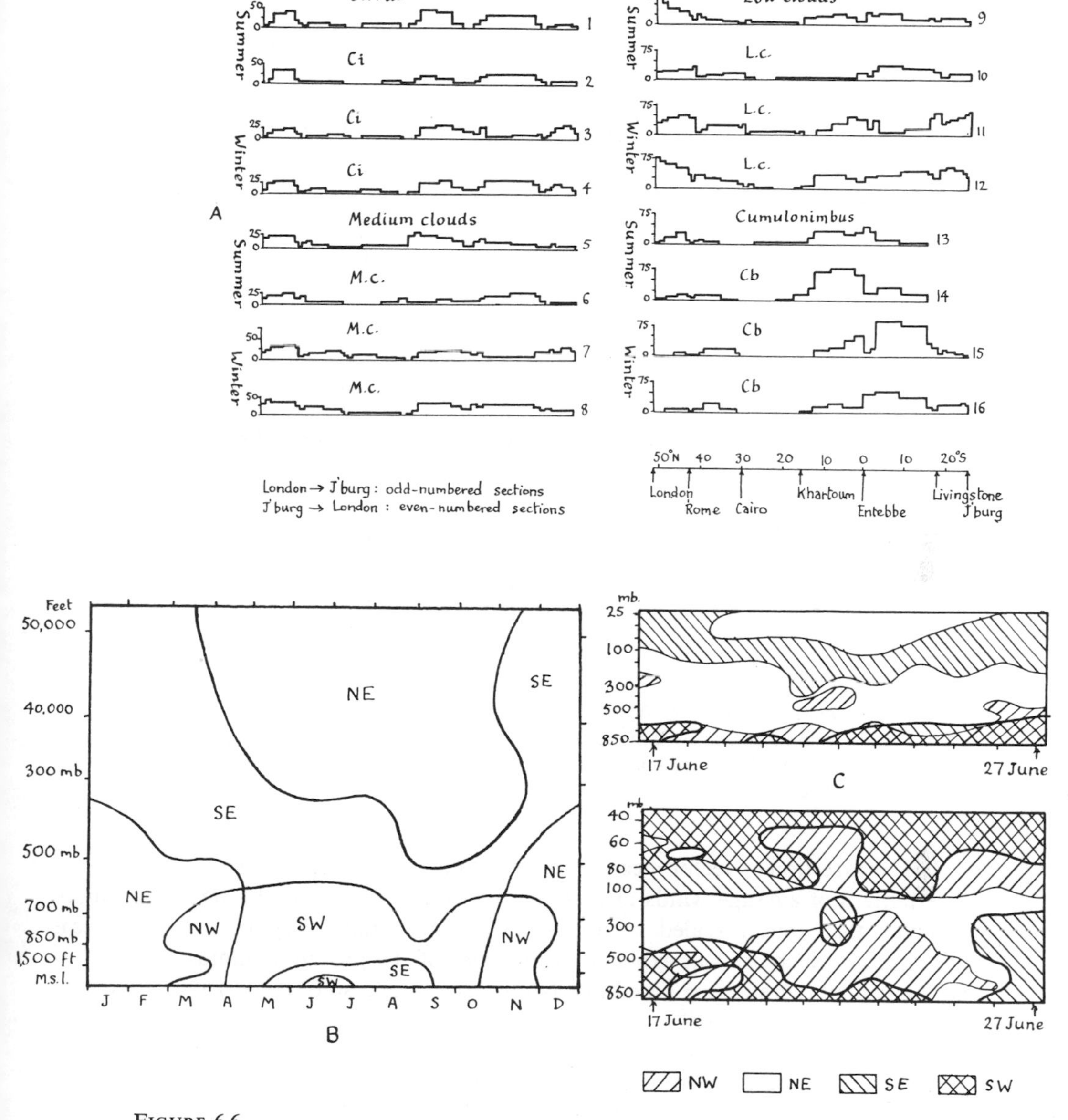

FIGURE 6.6.
Examples of atmospheric cross sections (after Kerley). A, percentage occurrence of clouds along London-Johannesburg route. B, Time cross section of wind direction, June 17–27, 1958, at Aden (upper) and Nairobi (lower). C, Vertical cross section over Singapore of vector mean winds in quadrants (after McAllen).

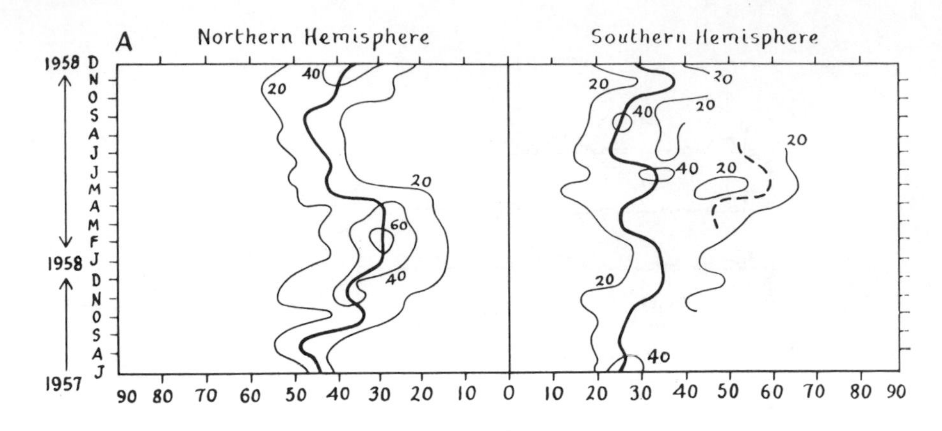

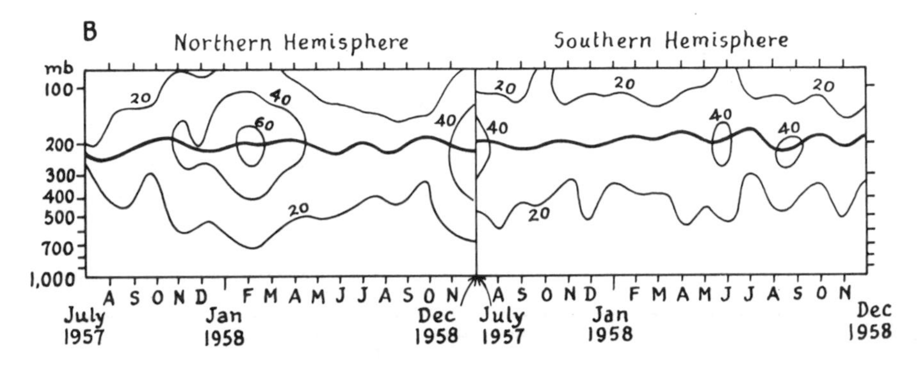

FIGURE 6.7.
Zonal wind field along 80°W for I.G.Y. (after W. M. McMurray).
A, Time cross section by latitude. B, Time cross section by height. Heavy lines show monthly mean west-wind maximum near the tropopause, lighter lines are isotachs in m per sec.

The cross section of vector mean winds over Singapore (Figure 6.6) indicates that the true pattern of average winds at a station or over an area may be very much more complex than that revealed by the surface winds alone. Regional climatological description, which aims at explanation in terms of the prevailing atmospheric circulation, cannot therefore be based purely on surface data, since these may not indicate representative conditions.[19]

The zonal wind field along 80°W for the IGY period in Figure 6.7 shows that the monthly mean maximum westerly wind components ranged over 20° of latitude and 1.5 km of altitude in the northern hemisphere, but over 17° and 1.7 km in the southern hemisphere. The height of the mean position of the westerly maximum oscillated about the 11.5-km (220-mb) level in the northern, and about the 12-km (200-mb) level in the southern hemisphere. Clearly, therefore, there must be a basic difference between the dynamics of the atmospheric circulation in the two hemispheres, quite apart from the difference due to the differing extents of land and sea.[20]

Aircraft observations are very conveniently summarized in the form of atmospheric cross sections. Figure 6.8 presents a summary based on more than 500 scheduled flights by the Scandinavian Airways System over Greenland. The ice cap was crossed

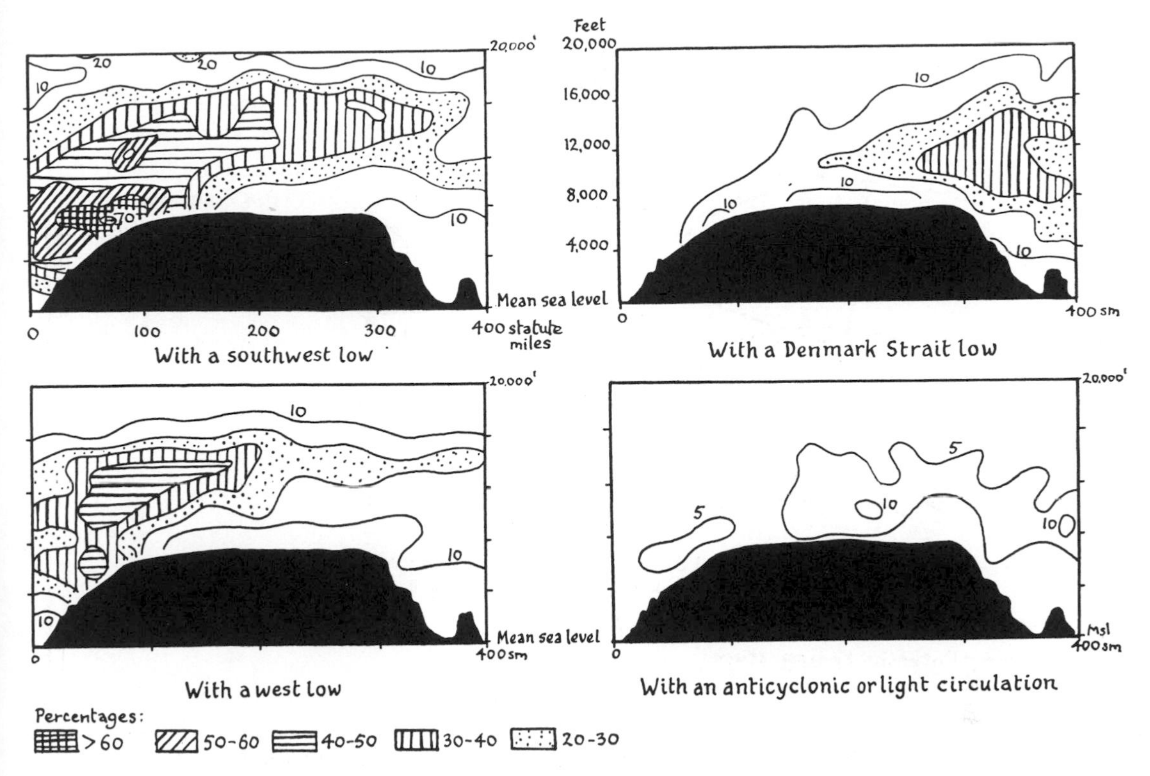

FIGURE 6.8.
Annual cloudiness over the Greenland ice cap (after C. A. Carpenter).

at a height of 17,000 feet above mean sea level, and cloud estimates were made at grid points 40 miles apart, every 2,000 feet. These cross sections indicate that Hobbs' "glacial anticyclone" effect only occurs when the ice cap is covered by an ordinary anticyclone, or when the air circulation is very light. When the synoptic situation is dominated by a low to the southwest or west of Greenland, or a low is centered over the Denmark Strait, there is no evidence at all of the presence of an anticyclone over the ice cap. When a high incidence of cloudiness occurs on one side of the ice cap, a low incidence occurs on the opposite side; strong upslope motion on one side results in downslope motion (hence less cloud) on the other.[21]

Figure 6.6 shows some data produced by high-altitude flights between Britain and South or East Africa. The time-section for Nairobi and Aden demonstrates how considerable was the range of variation in speed and height of the easterly wind belt in the middle and upper troposphere during a period of 10 days. The graphs depicting the percentage occurrence of clouds over the London-Johannesburg route for 1952–1953 indicate, as would be expected, considerable cloud amounts in low latitudes, small amounts between latitudes 20 to 30°N, and larger amounts again in temperate latitudes. Cumulonimbus amounts show a marked maxima at 5 to 10°S in winter, and at 5 to 10°N in summer. All cloud amounts in temperate latitudes show little seasonal variation, but the tropical "no cloud" areas move northward in summer and southward in winter.[22]

Finally, cross sections can obviously be employed to produce tentative synoptic

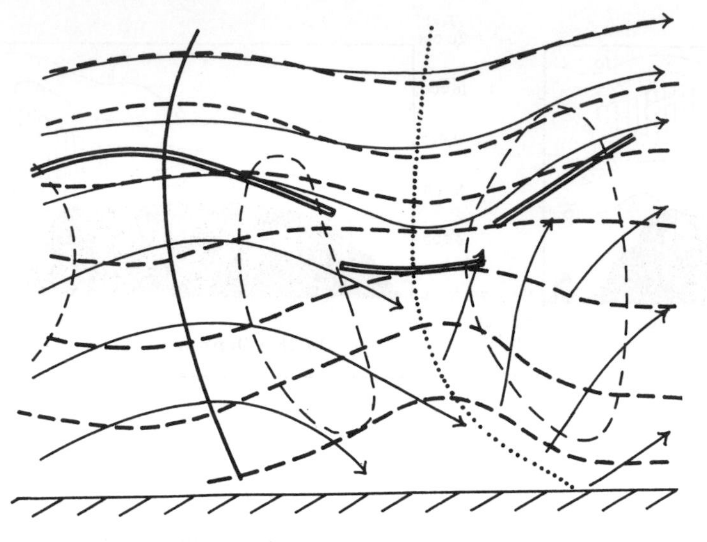

FIGURE 6.9.
A vertical cross section model (after Fleagle).

models that are useful in indicating empirically the probable connections between various synoptic quantities. The difficulty then is to account for these connections by dynamic reasoning. Figure 6.9 presents idealized conditions in a wedge and trough line, based on observations from Colorado and the eastern United States.[23]

The charts so far discussed that are designed to present the results of geographical exploration of the atmosphere are: (a) maps representing the actual or mean pattern of instantaneous conditions in any chosen plane parallel to, or coincident with, mean sea level; and (b) cross sections for vertical planes through the atmosphere in space or time. These charts are normally integrated mentally by the analyzer to produce a complete picture, although there have been attempts to make charts in three dimensions.[24] Another class of charts do not depict features that are readily visualized, and these will be considered in more detail.

Thickness Charts

Although the advent of upper-air observations completely changed ideas about the development and decay of weather phenomena, the technique of upper-air analysis really involved only applying the long-established methods of synoptic cartography to various levels of the atmosphere instead of merely to mean sea-level conditions. With the introduction of the thickness chart, however, a new technique of major importance was made available. Thickness charts are of first-rate importance for the geographical study of both weather and climate, and are particularly interesting in that they are based on very sound physical theory.

The term "thickness" was first introduced by V. Bjerknes in 1910, and in Part I of his *Dynamic Meteorology and Hydrography* the fundamental principles were given, together with synoptic charts at 100-mb intervals up to 300 mb, with both absolute and relative topographies of these pressure surfaces. Thickness charts were first put into routine use by the Germans before the Second World War. They were adopted in Britain in 1941, and recommended by the IMO for worldwide use in 1947.[25] Real utilization of the potentialities of the thickness chart had to await the introduction of the Sutcliffe development theory. The physical significance of this theory was dealt with in Chapter 5. In the following pages, only its practical, synoptic significance will be described.

The main achievement of the Sutcliffe theory was the production of fundamental formulae linking the temperature and motion fields of the atmosphere. The basic formula gives the relative isobaric divergence between two pressure levels in terms of the wind shear between the levels, the Coriolis parameter, and the vorticities at the two levels (see Appendix 6.4). It indicates where subsidence or ascent, hence anti-cyclonic or cyclonic development, respectively, can be expected. Anticyclonic "development" means the generation of anticyclonic vorticity, but this does not necessarily mean that an anticyclone will form, because sometimes not enough vorticity will be generated. If the process of subsidence is continued long enough, an anticyclone must eventually result.

Two approximations were made in deriving the basic formulae, and these may modify the results on occasion. First, the geostrophic approximation is used to determine the magnitude of windspeed and vorticity from the pressure distribution, and then these quantities are used to estimate other quantities that would not exist in purely geostrophic motion. The validity of this approximation depends on the relative magnitude of the terms involved in the calculation. Second, certain factors that may modify the thermal pattern, adiabatic heating and cooling, for example, are for simplicity ignored.

Exceptions to the Sutcliffe theory do occur. Vigorous cyclogenesis is often present off the east coast of North America under straight isobars, and diffluences occur in which there is no fall in surface pressure. Occasionally, surface pressure is found to fall on both sides of a diffluence, and rise all over a confluence. Such situations are relatively infrequent.

The *thickness chart* depicts cartographically the vertical height difference (in geopotential units) between the standard isobaric surfaces. The height difference between the 1,000-mb and 500-mb surfaces is the "total thickness," and that between for example, 1,000 and 700 mb, or 700 and 500 mb, is the "partial thickness" (see Table 6.1).

TABLE 6.1.
Equivalence Relationships Between Total and Partial Thickness over Europe, the Mediterranean, and the North Atlantic.

Total thickness (1,000 to 500 mb) in geopotential feet*	Partial thickness (1,000 to 700 mb) in geopotential feet*
17,400	9,200
17,800	9,400
18,200	9,600

*Data is from J. C. Gordon, *MM*, 90 (1961), 352.

With the hydrostatic assumption, the thickness value is a function of only the virtual temperature distribution in a vertical air column between the appropriate pressure surfaces. Since the thickness is proportional to the average specific volume of the air column, the isopleths of equal thickness are in effect isotherms depicting the distribution of mean air temperature in the layer between the two pressure surfaces. Thus the isopleths of 1,000 to 500-mb thickness are also isotherms for the lower half of the atmosphere.

Variations in the thickness pattern are due solely to temperature variations, which depend on physical processes (such as advection), adiabatic changes, and nonadiabatic changes (such as the phase changes of water in the atmosphere, radiation, conduction, and turbulence). The thickness lines are therefore more useful than isotherms for various levels, because they indicate more than temperature alone. For example, they show at a glance where the atmosphere as a whole is warm or cold, where isolated tongues or pools of cold or warm air exist, and where the thermal gradients and curvatures are large or small. They aid air-mass and frontal analysis, indicating whether a given front is of major or minor significance, depending on whether the thickness (and hence temperature) gradient across it is steep or slack. In a sequence of charts, changes in the thickness pattern over a particular region indicate changes in the distribution of temperature that emphasize nonadvective processes, the latter being very difficult to assess directly from the ordinary pressure chart.

Figure 6.10 gives the mean positions of the thickness isopleths in the northern hemisphere. Any departure of an isopleth from its mean location obviously means that physical processes are operating to give a temperature anomaly, and the thickness pattern provides a means of isolating these processes to some extent.

The local rate of change of thickness (i.e., $\frac{\partial h}{\partial t}$, the rate of change at a given point on the Earth's surface) is given by

$$\frac{\partial h}{\partial t} = A + D + N,$$

where A represents the effect of advection, D the effect of dynamic changes, and N the nonadiabatic effects.[26] The term represented by A shows that the thickness isopleths are advected with the 1,000-mb geostrophic wind; i.e., the thickness pattern moves as if embedded in the surface pressure pattern.

The term D represents the changes in temperature brought about by dynamic heating (i.e., adiabatic compression) and dynamic cooling (adiabatic expansion), mainly resulting from subsidence or ascent of the air. If subsidence occurs, thickness increases; if ascent occurs, thickness decreases, assuming a lapse-rate less than the adiabatic one. If the lapse-rate exceeds the saturated-adiabatic lapse-rate, however, ascent produces warming, leading to an increase in thickness. Warming by subsidence is often considerable, averaging about 1°C in six hours up to 500 mb, and this may distort the thickness pattern if long continued. An increase in thickness due to the ascent of saturated air does not occur often, because it is rare for an air mass to become saturated and unstable throughout a very deep layer, without becoming broken into deep convective cells.

The term N includes radiation, latent heat liberated by precipitation, and heat transfer by convection and turbulence, the latter being the normal processes by which heat gained at the surface of the Earth by an excess of incoming over outgoing radiation is spread through the atmosphere. Short-period radiation changes are reflected in the thickness pattern, because the rate of cooling of the atmosphere by radiation amounts to 2 or 3°C per day, which is equivalent to 200 feet difference in thickness, or a surface pressure change of 7 mb. On a global scale, the net interchange of atmospheric heat by advection between low and high latitudes balances their different rates of gaining heat from and losing heat to space by radiation. At least in extratropical latitudes, synoptic disturbances are essential features of the general circulation, and many nonadiabatic processes operate in these systems. For example, although cold air masses moving to warmer regions are generally supposed to be warmed mainly by subsidence, and warm air masses moving into colder regions to be cooled by ascent or lifted (i.e., occluded) in frontal situations, cyclonic polar air-mass circulations penetrate into low latitudes with little or no subsidence, and warm air masses sometimes move well into polar regions without convergence. Therefore N must be of the same order of magnitude as A and D.

It is difficult to assess the exact contribution of A, D, and N in any given situation, but each may be significant under certain conditions, and may contribute to the thickness tendency by an amount equal to the mean tendencies observed.

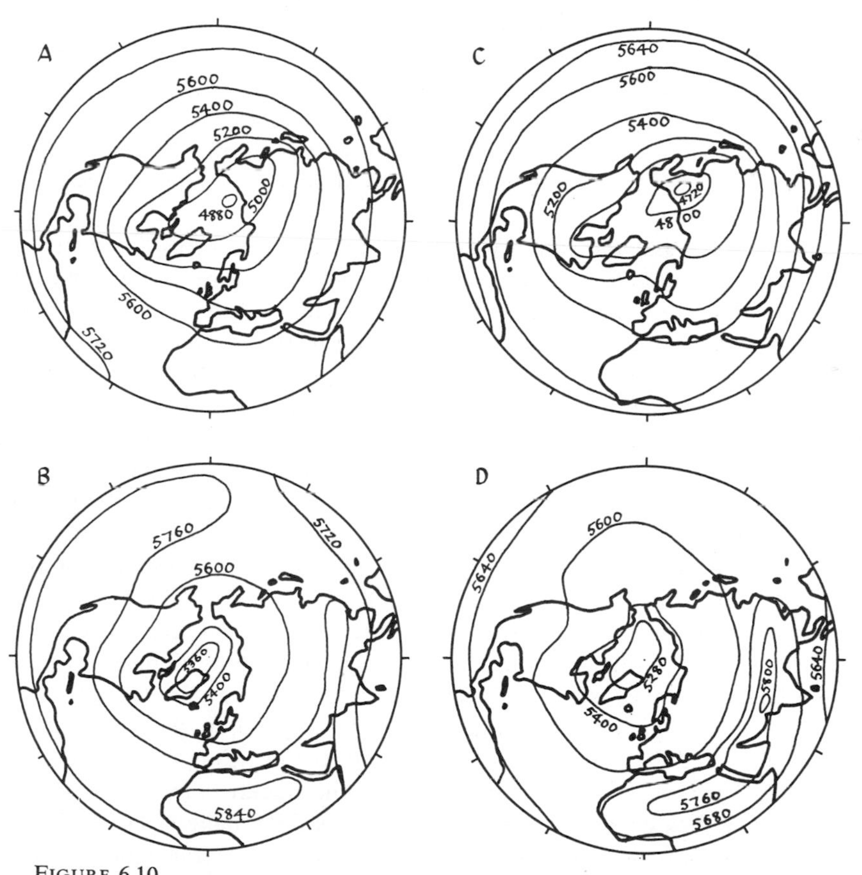

FIGURE 6.10.
A and B show mean absolute 500-mb topographies; C and D show mean relative (1,000- to 500-mb) topographies. A and C are for January, B and D for July. (After R. C. Sutcliffe.)

Algebraically subtracting actual thicknesses from the mean values produces the thickness anomaly pattern, which indicates both the generalized temperature departure from normal over a large area, and the circulation anomaly. Five-day mean thickness anomalies, for example, provide large, slow-moving, and long-lived patterns that persist much longer than the pressure anomalies for the same period. New thickness anomalies sometimes indicate temperature advection in the expected direction according to the previous pressure chart, but the temperature anomaly persists after the pressure pattern completely changes. At other times, the temperature (thickness) anomaly develops first, and later the advective flow to maintain it makes its appearance. When large-scale thickness patterns remain quasistationary for long periods, very abnormal weather conditions frequently result.[27]

EXAMPLES OF THICKNESS PATTERNS AND THEIR SIGNIFICANCE

On a global scale, thickness patterns provide a means of elucidating some fundamental problems. A comparison of mean thicknesses and mean upper-pressure contours for the northern hemisphere (see Figure 6.10) demonstrates the value of thickness patterns as indicators of geographical influences. According to Bolin and others, the observed mean pattern of upper troughs and ridges in the pressure field could be set up by purely dynamic processes acting in a barotropic atmosphere. The secondary trough over Europe, which cannot be explained by the thermal influence of the Earth's surface, is regarded as a consequence of the wave pattern set up by the Rocky Mountains. Obviously, the summer and winter flow patterns should be similar if this theory is correct.

Figure 6.10 shows that the January and July patterns are broadly similar. Bolin regarded the slight differences as due to the baroclinic nature of the atmosphere. Sutcliffe considered these small differences as the significant features; even near the Rockies, the summer and winter patterns are more nearly antiphase than in phase. If this is so, the mean 1,000 to 500-mb thickness pattern (i.e., the mean temperature of the lower half of the atmosphere) in the northern hemisphere mainly reflects the distribution of land and sea. In turn, this pattern decides the mean upper contours of pressure. Thus the western mountain barriers of the Americas, and the southern barriers in Eurasia, prevent free advection of lower tropospheric air, thereby producing sharp temperature and moisture differences, which result in the formation of lee depressions. The latter then affect the mean upper flow, resulting in a sinusoidal pattern of contours instead of purely east-west isobars.[28]

On a smaller scale, thickness patterns enable the development and decay of synoptic disturbances to be both explained and anticipated in geographical terms. This is accomplished with the assistance of a series of models (see Figure 6.11).[29]

The areas of favoured anticyclonic and cyclonic development in the model situations are denoted by *A* and *C*, respectively. According to the principle of *thermal steering*, cyclones and anticyclones must move parallel to the thickness isopleths, and hence the patterns also indicate the direction of movement of disturbances. Thermal steering is essentially a process of development, as distinct from the mere translation of pressure systems. It works well for depressions and anticyclones associated with straight, well-defined thermal patterns, but less well for more complex patterns.

Basically, the Sutcliffe development theory postulates that upper-level divergence

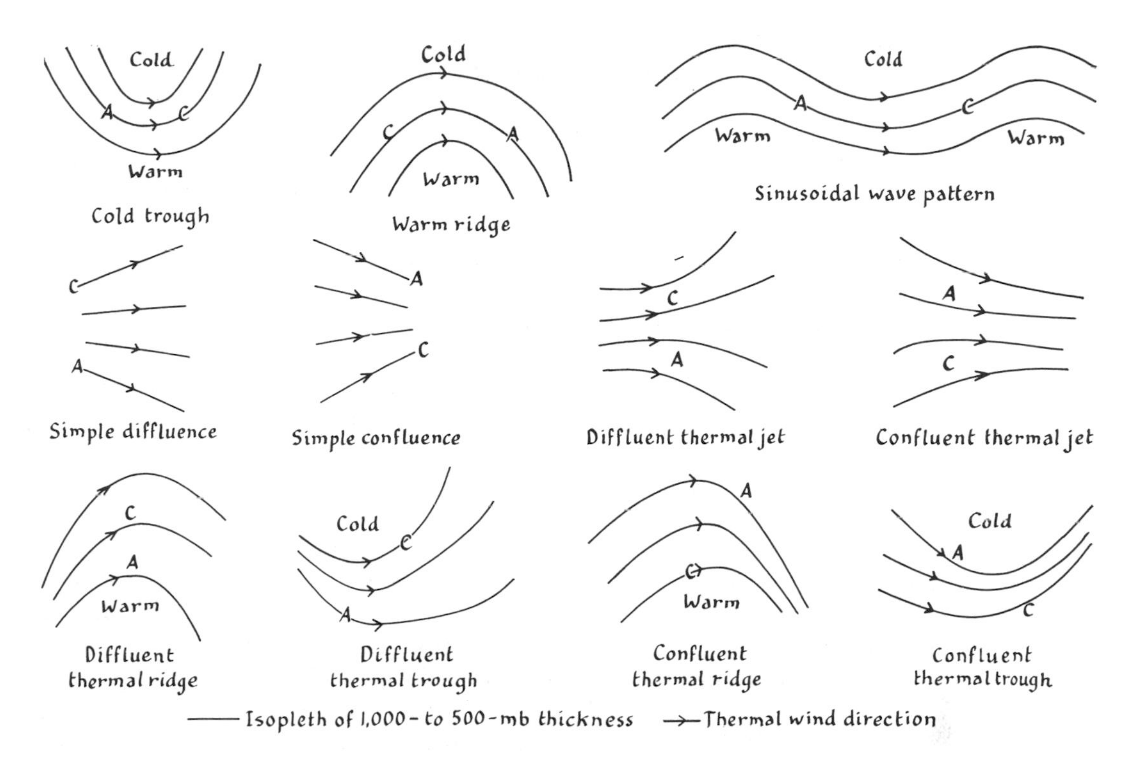

FIGURE 6.11.
Basic models of 1,000- to 500-mb thickness patterns. *A* is an area of anticyclogenesis; *C* is an area of cyclogenesis. (After Sutcliffe and Fordsdyke.)

must be balanced by low-level convergence, and vice versa. The process of attaining this balance produces the vertical motion necessary for precipitation and the relatively small pressure changes observable at the ground. The thickness pattern provides a means of changing the sign of the divergence with height. It operates either by making windspeeds differ at different heights, or by changing the curvature of the isobars with height. The former is, in effect, thermal steering. It means, for example, that a cyclone should move in parallel to the thickness isopleths over its center, at a speed proportional to the closeness of the isopleths. The latter implies, from the vorticity theorem, that if advection of thermal cyclonic vorticity is taking place, divergence must be occurring aloft and convergence near the ground, leading to cyclogenesis.

Empirically, small lows with open warm sectors are found to move parallel to the thickness isopleths at four-fifths of the speed of the thermal wind above them. (The thermal wind blows parallel to the thickness isopleths; standing with one's back to this wind, low temperatures, i.e., low thickness values, occur to the left.) Waves on cold fronts move at four-fifths of the speed of the warm-sector geostrophic wind; they usually form when advection of positive thermal vorticity occurs up to 500 or 600 miles east of the thermal trough line, provided the thermal wind is at least 25 knots along at least 1,200 miles of the cold front. About one cold front in every four crossing the United Kingdom has a wave develop on it, and whether a wave forms or not decides whether a particular day will be brilliantly fine or completely rainy.[30]

The model thickness patterns shown in Figure 6.11 have been amply confirmed by day-to-day analysis. In addition, empirical studies have been made of the relationship between thickness patterns and surface synoptic features.

Figure 6.12 illustrates the relation of thickness isopleths to surface fronts, and the consequent rate of movement and life history of the fronts. The relative position of surface and thickness lows indicates the relative age of a cyclone. Figure 6.12 gives examples of two typical methods of formation of secondary depression.[31] The thickness pattern clarifies why the secondaries form thus.

For warm occlusions, the necessary conditions for secondary formation are a slow moving primary depression, with a strong thermal gradient ahead of the primary, and a thermal wind of 40 to 80 knots. For cold occlusions, for a secondary to form, a strong thermal gradient must be present several hundred miles ahead of the primary center, with a weak gradient over the latter, and thermal diffluence must be present. Thus secondary depressions on warm occlusions form just to the right of a thermal confluence, and secondaries on cold occlusions just to the left of a diffluence. The explanation is that both of these locations are areas of cyclonic development according to the Sutcliffe-Forsdyke models in Figure 6.11.

Waves form on warm fronts with the same type of thermal pattern as for a warm-occlusion secondary; slow-moving secondary depressions also form on cold fronts with thermal patterns similar to those for cold-occlusion secondaries. Once formed, both warm-occlusion secondaries and warm-front waves move rapidly away from the primary at speeds of from 30 to 50 knots, in directions closely following the thickness lines over the primary center, but slightly inclined toward the cold side of the strong thermal belt that existed before the secondary appeared, the angle of inclination varying from 10° for warm-front waves to 30° for warm-occlusion secondaries. These secondaries and waves form *break-away depressions*, which are often responsible for bringing rain into a high-pressure area.

Secondary depressions forming on cold occlusions behave less regularly, being more dependent on details in the thickness pattern, and although deepening rapidly, they

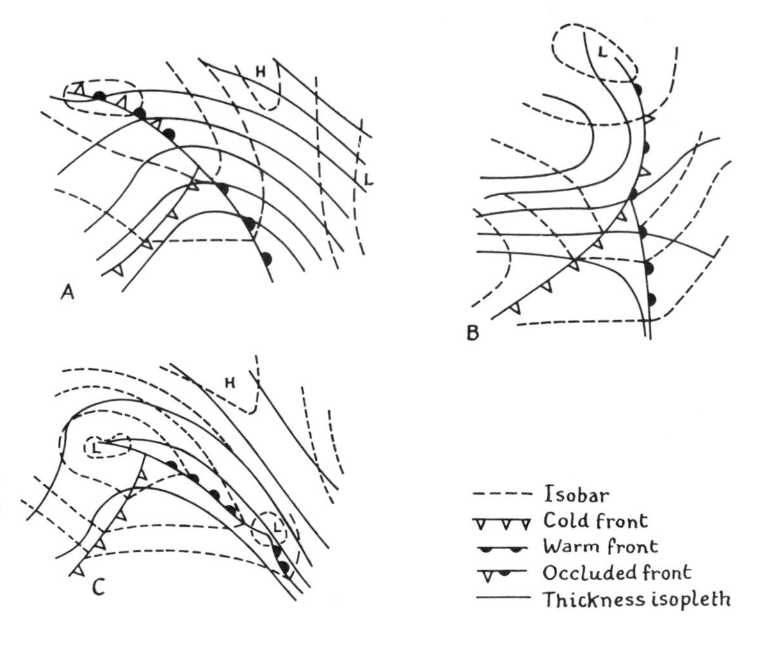

FIGURE 6.12.
Formation of secondary depressions (after Sawyer). A, Secondary forming at point of a warm occlusion. B, Secondary forming at point of a cold occlusion. C, Formation of a wave on a warm front

seldom become major synoptic features. They usually are associated with symmetric diffluence in the thermal pattern, moving slowly (10 to 20 knots) in the direction of the strongest thermal winds (see Appendix 6.5).

Thermal steering works less well with anticyclones than with cyclones, but their associated thickness patterns enable anticyclones to be separated into four types quite distinct from the classical separation into warm and cold anticyclones (see Figure 6.13).[32]

In *open-wave anticyclones*, the high-pressure center moves rapidly toward the thermal trough with an average speed of 25 knots, usually showing some tendency to move toward the region where the thermal wind is strongest. The central pressure may rise or fall slowly as the system progresses through the thickness pattern, but a rise in pressure is more usual.

In *distorted-wave anticyclones*, the center moves slowly toward the thermal trough, averaging 10 to 15 knots, in a direction that is inclined to the right of the perpendicular to the trough line, but is displaced across the thickness isopleths into the cold air. Central pressures in anticyclones of this type change very little, or show a very slow increase.

Warm anticyclones have very weak thermal gradients near their centers, and show

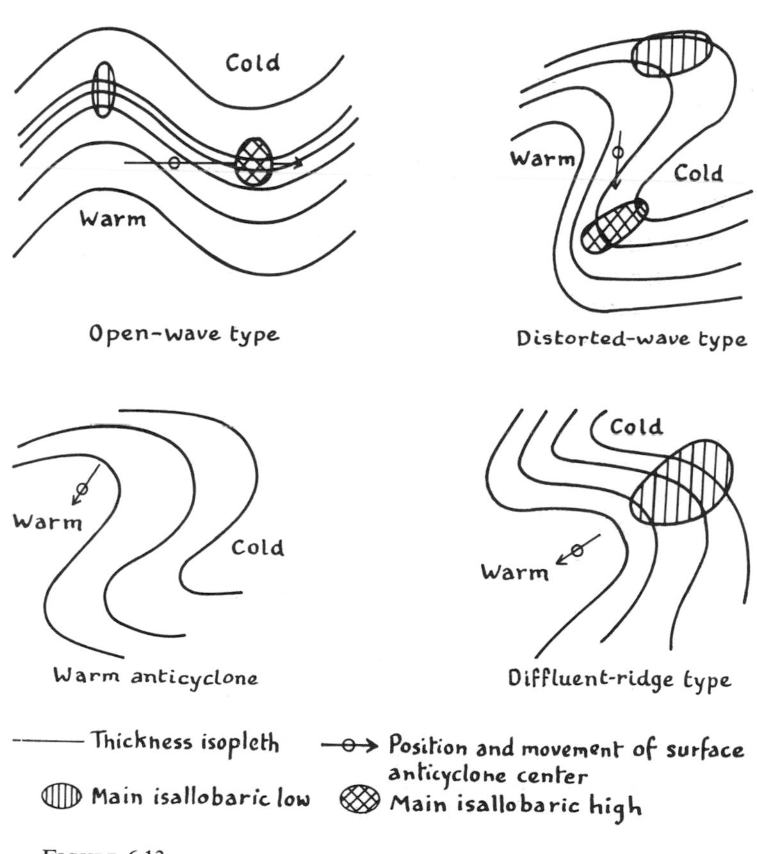

FIGURE 6.13.
Anticyclones and thickness patterns (after Sawyer).

very little change in central pressure as they move through the thickness pattern. They move slowly (less than 10 knots) in a direction between the axis of the warm tongue and a line from their center to the apex of the cold trough, i.e., toward the higher temperatures.

Diffluent-ridge anticyclones move more rapidly than warm anticyclones, at 15 to 20 knots, and their central pressures rapidly decrease as they move through the thermal pattern.

Figure 6.13 does not apply to cold polar anticyclones, continental anticyclones, or subtropical anticyclones in their normal positions, with all of which, other thickness patterns are found. The four types distinguished, however, indicate the value of thermal patterns in explaining the different behavior of anticyclones that appear, from the surface weather map, to be exactly similar. The sequence commencing with an open-wave anticyclone, and passing through the distorted wave type to the warm anticyclone, illustrates the transformation of a rapidly-moving cold anticyclone into a slow-moving warm one, a process difficult to follow without the aid of the thickness chart.

A closed thickness line, cutting off an area relatively small by wave-pattern standards, defines features known as *cold pools* and *warm pools*, depending on whether the cutoff area contains air colder or warmer than its surroundings. Cold pools, in particular, have proved to have very significant weather connections, and they explain curious weather phenomena termed cold drops, which have been recognized since the 1930's.[33]

Cold drops, which are really special cases of cold pools, were defined by Buschner as pools of cold air in the middle or upper troposphere that are associated with little or no pressure irregularity at ground level. They occur over the European continent with easterly winds, and rapidly die out over the sea, although they may result in violent cyclonic activity before they finally disappear. Cold drops are steered in the direction of the surface gradient wind, moving in winter at three-fifths of the speed of this wind. The maximum cooling in them is found at 650 mb (15°C), and above 400 mb cold drops are actually warmer than their surroundings. The weather ahead of a cold drop is cloudless, with much cloud and (in winter) snowfall behind the drop. Occasionally, precipitation may occur ahead of a cold drop moving from the east. Temperature and pressure show a symmetric vertical cross section through the drop, but the moisture field shows a tongue of moister (rising) air behind it.

Cold pools may be defined as deep masses of cold air entirely surrounded by relatively warm air, which appear as one or more closed thickness lines on the chart.[34] They may be associated with any surface pressure pattern. The weather connected with them is usually cloudy, with or without precipitation, although both fine and overcast conditions may occur at ground level under cold pools in different areas. In other words, although the surface weather found beneath a cold pool will always be different from that outside the cold pool, these features are not always associated with any one kind of weather, although generally, anticyclonic conditions occur ahead of a cold pool, and cyclonic conditions behind it.

In the Mediterranean region, cold pools always seem to be associated with rain, and over the Atlantic with storms and hail. Slow-moving surface depressions in the Azores area, bringing long periods of bad weather involving frequent showers, thunderstorms, gales, or drizzle with very low stratus, have their severe-weather conditions much intensified when associated with cold pools. These spells of very bad weather usually occur four to six times a year, each lasting one to two weeks, but are very rare in the summer. Their life is often prolonged by renewal of the cold air in the pool by the arrival of polar lows, or old occluded lows from Florida, but they normally end when the low fills or moves northeastward.

Cold pools are most frequent where latitudinal temperature gradients are weak, and under such conditions they are very important causes of unexpected weather phenomena. The British Isles usually have low temperature gradients for their latitude, and hence experience a high incidence of cold pools. In comparison with the North Atlantic area, the continent of Europe has a relatively high concentration of cold pools in all seasons, and other areas of high frequency are found between northeastern Greenland and northern Scandinavia in spring and summer.

Well-marked cold pools always form upper-air depressions, but the low pressure center and the closed thickness line do not always coincide, and not all upper depressions are necessarily cold pools. The formation of a cold pool is usually due to an outbreak of polar air. A thermal trough forms, then the supply of cold air becomes cut off and the northern end of the trough is warmed by advection and possibly subsidence, so that its southern end becomes a cold pool. Deep cold air is much more limited in area than is shallow cold air, since it tends to subside and spread out laterally. Thus a weak latitudinal temperature gradient is a very favorable area for cold pool formation.

Once formed, cold pools are usually fairly slow moving, covering less than 500 miles per day, and they very rarely move rapidly for longer than one day. Over Europe and the North Atlantic, they generally travel eastward or southward, but oscillate considerably. In winter, they sometimes move westward from Siberia. They tend to be more frequent over land than over sea in winter and autumn, and vice versa in spring and summer. Their usual length of life is three to four days, but if a balance is achieved between warming by convection (for example, by the cold air passing over a warm sea surface) and cooling by radiation to space at high levels, they may persist for nine or ten days.

The presence of a cold pool often explains an otherwise inexplicable motion in a surface pressure system. For example, a cold pool moving south or southeast across Europe or the North Atlantic may control the movement of a sea-level depression further east, causing it to travel in a direction opposite to that expected from its frontal and thermal wind structure. New cyclones tend to form at the edge of a cold pool, and surface heating may cause the cyclonic development to move closer to the center of the cold pool, so that the latter may in time come to coincide with the center of the surface low.

Cold and warm pools exert a steering effect on small, fast-moving surface cyclones. The upper air features control the motion of the sea-level features in all situations except that of a very marked sea-level pressure gradient. If the latter occurs in the area of a cold pool, there is often a tendency for the latter to move in the direction of the sea-level geostrophic wind. Warm fronts cannot penetrate cold pools, even if the surface pressure distribution indicates that they should: they become delayed, finally dissipating or becoming overtaken by the following cold front.

Instability weather phenomena, showers for example, occur beneath cold pools when no particular instability features appear on the surface chart. Thundery outbreaks take place in the afternoon and evening beneath small, fast-moving cold pools that develop in the lower layers of the atmosphere. If a cold pool of this type extends from the ground surface to well above 700 mb, thunderstorms often develop by surface heating in the central area, although they are rarely associated with upper cold pools. If surface heating is absent, however, lower cold pools give rise to fine weather. Southerly weather types in Britain in summer are frequently associated with a cold pool to the west, and if the pool moves, the weather ahead of it is usually much worse than that behind the pool.

Thermal troughs and *thermal ridges* are very common features of the thickness chart, the former consisting of relatively cold air and the latter of relatively warm air. At least four distinct types of thermal trough have been recognized over the British Isles.[35]

The *mature thermal trough* (see Figure 6.14) is a feature normally fully developed when it enters the region, and is fairly long-lived. The *northerly type trough* is formed by the advection of cold air within the British Isles region, whereas the *cyclonic tongue* or *thermal involution* is associated with katafronts, and is more frequent in winter than in summer, because of the existence of deeper cyclones in the former season. The short-lived *break-away trough* is less common than the other types: it forms by splitting off from larger thermal troughs, and then moving away at a relatively high speed.

Some features of cold-front structure and precipitation can be interpreted in terms of the type of the accompanying thermal trough. Cold fronts without precipitation in the cold air (i.e. *katafronts*) occur with all break-away troughs. Both katafronts and *anafronts* (i.e., cold fronts with precipitation in the cold air) are found in mature and northerly type thermal troughs with equal likelihood.

In blocking situations, thermal troughs of all types tend to become distorted. For example, during April 1954, there were eight mature troughs over the Atlantic, yet only one of these reached the British Isles: the others were too distorted by the blocking of the normal westerly current to be recognizable as troughs. In summer, thermal troughs tend to be propagated downward, so that a trough develops in the surface pressure pattern, coincident with the thickness trough. In winter, this tendency is not so noticeable.

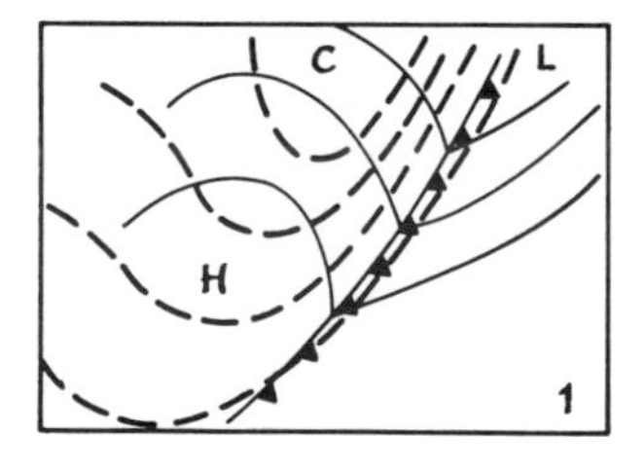

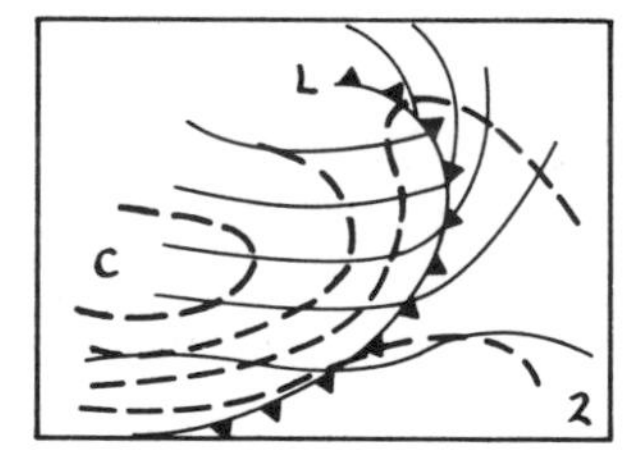

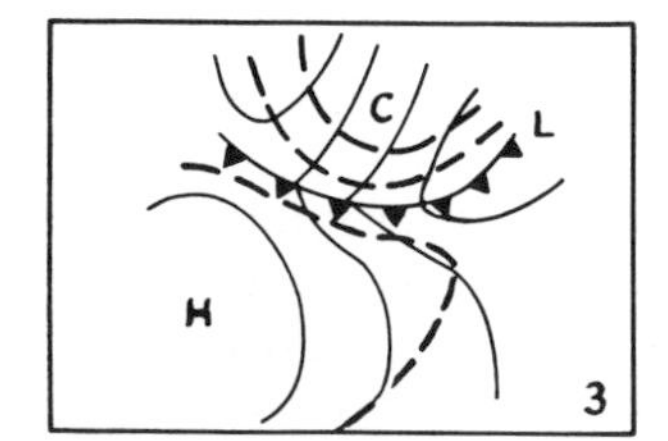

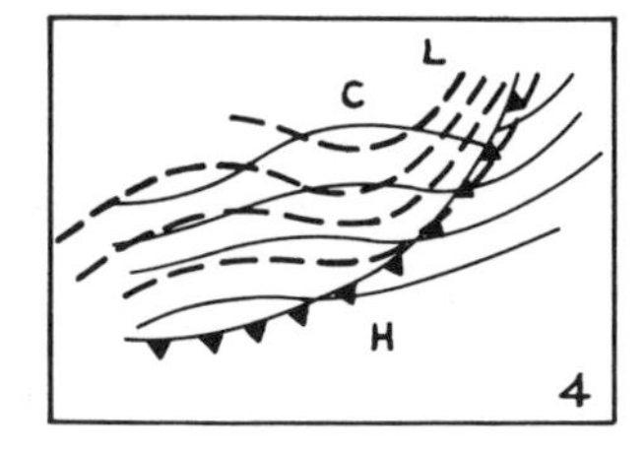

FIGURE 6.14.
Main types of thermal trough over the Atlantic Ocean and the British Isles. 1, Mature thermal trough; 2, thermal involution or cyclonic tongue; 3, Northerly trough; 4, break-away trough. C denotes cold air. Broken lines are thickness isopleths; full lines are isobars. (After M. K. Miles.)

Small-scale thermal troughs occasionally develop, superimposed on a main trough. A thermal long-wave pattern became stationary over the North Atlantic between February 25 and March 9, 1953, and small-scale cold troughs with wavelengths of 20° of longitude moved through it. These miniature troughs moved rapidly (30° of longitude per day) in the direction of the mean thickness lines, traveling up to the crest of a large warm ridge. Dynamic warming taking place during this movement ultimately led to their disappearance.[36]

Extension of thermal troughs in a meridional direction is nearly always accompanied by a substantial fall in surface pressure to the southeast of the trough, and very definite weather changes result. For example, meridional extension in the sector between 20°W and 20°E gives rise to an extensive area of rain or snow to the southeast of the thermal trough, the precipitation area moving southeastward as the trough extends further south (see Appendix 6.6).

The possibilities for empirical geographical study of the relationship between surface weather phenomena and thickness patterns are almost infinite, and even taking only a few years' data one can usually obtain consistent results, although the results obtained from examination of a very long series of charts can sometimes be misleading. One correlation may work well for some one area in certain years, but a different correlation may apply in other years. At times, apparently normal synoptic features may prove to have most unusual thermal structures. For example, a system that formed near Greenland on October 4, 1952, started off as a normal depression, but later developed a well-marked cold trough in phase with its center. The depression did not fill up or become stationary, as would be expected in such circumstances, but moved erratically for a time and then continued on a southeastward track without any pressure change at its center.[37]

Cyclogenesis is normally accompanied by distortion of the thickness pattern, a cold trough developing behind, and a warm ridge ahead of, the surface depression. The cold trough often overtakes the surface low when the occlusion process is well under way, and the thermal gradients usually weaken by that time, a cold pool gradually being formed in association with the surface low. In other words, even with the aid of the thickness pattern, it is not possible to design a cyclone model that will fit all depressions.

In the depression forming near Greenland on October 4, 1952, the movement of the low-pressure center was in conformity with the 500-mb to 300-mb thickness pattern. Very frequently, in fact, weather systems whose motion and behavior cannot be explained by the total thickness pattern prove to be explicable in terms of thickness features within the upper half of the atmosphere.

The pattern of airflow at 100 mb, for example, changes comparatively little with time, and exhibits a slow advection as its predominant feature. The pattern at 200 mb shows much larger changes with time, which changes are advected at approximately the same speed as those on the surface chart. The 100-mb to 200-mb thickness pattern appears to be very significant in the control of weather and climate at ground level; its features are closely associated with surface fronts, in an inverse sense, and it shows a close relation to the profile of the tropopause. This is important, because it indicates that the 100-mb to 200-mb pattern—which results from essentially dynamic processes, with negligible (if any) influence by surface geography—may hold the key to explaining weather systems that do not move in conformity with either surface or total thickness patterns.[38]

OTHER USES OF THE THICKNESS CHART

Attention has so far been focused on the geographical patterns assumed by thickness isopleths, and how these may be employed to explain weather and climate. In addition to this type of work, thickness charts have many other uses, including: the isolation of synoptic features not caught by the normal charts; the assessment of stability; the delimitation of the freezing level in the atmosphere; distinguishing between the likelihood of liquid or frozen forms of precipitation; and the study of surface temperature variations.

Thickness charts are useful, for example, in the location of upper cold fronts in the Mediterranean area. These fronts are not infrequent, and are the cause of periods of unsettled weather unexplainable from the surface chart. Abrupt changes in the thickness gradient indicate the presence of the fronts, and therefore it may be unwise to smooth out an apparently irregular gradient in the thickness isopleths.[39]

Inspection of 1,000-mb to 500-mb and 1,000-mb to 700-mb thickness values on successive charts indicates whether the atmosphere in the area concerned is stable or unstable. This procedure is based on whether or not instability showers or thunderstorms occur within three hours of the time of the sounding.[40] In general, if the total thickness increases relative to the 1,000-mb to 700-mb thickness, the air is becoming more stable; if the total thickness decreases relative to the 1,000-mb to 700-mb thickness, it is becoming more unstable (see Table 6.2).

The height of the freezing level above the ground may also be estimated from thickness charts. For polar air over the British Isles, the freezing level can be quite easily determined from the 1,000-mb to 700-mb chart, provided there is no front within 200 miles, and no anticyclonic center within 400 miles.[41] Table 6.3 gives the height of the freezing level above the 1,000-mb surface for various 1,000-mb to 700-mb thicknesses.

A definite relationship would obviously be expected between thickness values and temperatures at the Earth's surface, in view of the equation connecting the vertical depth of air between two pressure surfaces and the mean temperature of this air column, which, as indicated elsewhere in this book, can be established from first principles (i.e., the total thickness in feet equals 66.5 times the mean absolute virtual temperature of the 1,000-mb to 500-mb layer). The question is: how close is the actual relationship?

Mean air temperatures would be expected to bear a closer relation to thickness

TABLE 6.2.
Relationship Between Partial Thickness and Stability.
If a rectangular coordinate graph of 1,000- to 700-mb thickness (abscissa) against 700- to 500-mb thickness (ordinate) is plotted, the critical zone within which partial thicknesses provide a useful guide to stability is defined as follows:

Upper limit of zone					
1,000- to 700-mb thickness in feet	9,000	9,200	9,600	9,800	10,000
700- to 500-mb thickness in feet	8,060	8,230	8,570	8,750	9,000
Lower limit of zone					
1,000- to 700-mb thickness in feet	9,000	9,200	9,600	9,800	10,000
700- to 500-mb thickness in feet	7,980	8,150	8,490	8,660	8,830

(After Murray)

TABLE 6.3.
Height of the Freezing Level Over the British Isles Above the 100-mb Surface for Various Values of 1,000- to 700-mb Thickness.

Thickness (in feet)	9,000	9,050	9,100	9,150	9,200	9,250	9,300	9,350	9,400	9,450
Height of freezing level (in feet)	100	700	1,300	1,900	2,500	3,200	3,800	4,400	5,000	5,600

Data from R. Murray, *MM*, 78 (1969), 349.

values than actual air temperatures, due to the wide variations in local surface temperatures because of surface heat transfer, the latter being determined in turn by windspeed and direction, cloud cover, and time of day. The theoretical relationship is a linear one. At London Airport, however, the actual relation between daily mean temperature and daily mean thickness is a power curve. Two distinct thickness "seasons" occur at Crawley and Larkhill, in relation to London surface temperatures: during the summer season (April to August), the monthly mean temperature is higher than would be expected from the monthly mean thickness; the remaining months form the "winter" season, during which monthly mean temperatures are low for the monthly mean thickness (see Appendix 6.7).

A critical thickness isopleth would be expected to occur, separating liquid from frozen precipitation. From the tephigram, it is possible to deduce the temperatures corresponding to various total thicknesses, for different degrees of humidity (see Table 6.4). These figures in Table 6.4 indicate a theoretical value for the critical thickness, which varies with the humidity content of the air. Empirically, the critical thickness for the changeover from liquid drizzle and rain to snow and ice is 17,300 feet on the average (see Appendix 6.8). This critical thickness has a higher value over established snowfields, and a lower value over sea surfaces: it averages 17,400 to 17,600 feet for the former, and for the latter a figure of 17,150 feet is representative over a surface water temperature of 10°C. For cold pools, the change-over takes place at 16,900 to 17,150 feet for shower precipitation, and at 17,200 feet for other types of precipitation. Figure 6.35 gives the probability of snow and sleet for given surface temperatures and total thicknesses.[42]

TABLE 6.4.
Relationship Between Total Thickness and Temperatures at 1,000 mb.

Total thickness in feet	Temperature in air saturated throughout 1,000 to 500-mb layer (°F)	Temperature in air unsaturated below 850-mb (°F)
18,600	68	79
18,000	55	64
17,400	43	49
16,800	29	35

(After Lamb)

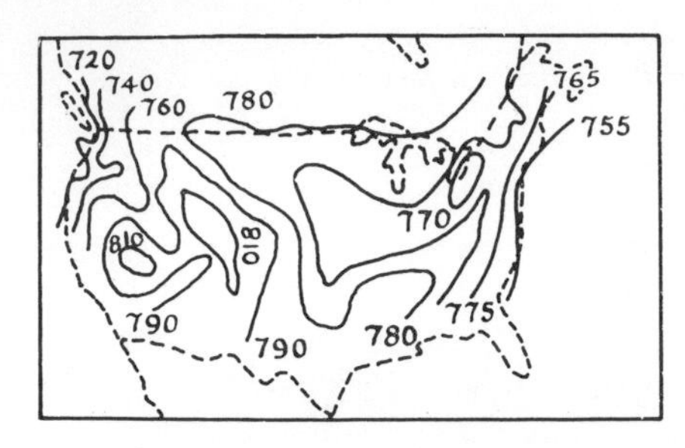

FIGURE 6.15.
Equal probability thickness in hundreds of feet; tens of thousands digit omitted (after Wagner).

In general, for low-level inland locations in winter, the chances of rain or snow are about equal for total thicknesses of 5,280 m, which thickness value corresponds to a surface temperature of 38°F, not, as might be expected, with the 32°F isotherm. The 0°C mean isotherm in North America and Asia coincides approximately with the 17,507-feet thickness isopleth.

With an established snow cover, snow flurries can be expected from low clouds (e.g., stratocumulus beneath a subsidence inversion) with a thickness value of up to 5,400 m, which thickness value will, however, give rain from medium-level frontal cloud.[43]

The critical thickness for a changeover from snow to rain over maritime areas, islands, and seas is 5,250 m, in all seasons except summer. Since this value is, of course, only a general figure intended as a first approximation, and since the equivalent figure for land areas already mentioned is only 30 m different, by taking a thickness value of about 5,260 or 5,270 m, one obtains a very useful climatological parameter for charting purposes. Owing to the smaller size of the unit, it is more difficult to give a generally applicable single figure if total thicknesses are expressed in feet instead of meters.

The change in critical thickness resulting from the formation of a snow cover is of great importance in the determination of the severity of European winters. For a severe, prolonged winter to occur in the British Isles, a continuous snow cover must first form across Europe north of the Alps and the Carpathians, with abnormal cooling of the North Sea. An extensive continental snow cover results in the development of a broad cold trough over the snow mantle, and a great warm ridge just outside the western limit of the snow. Both trough and ridge extend well above the middle layers of the atmosphere. Anticyclogenesis must then occur over the northwestern portion of the snow-covered area, giving persistent anticyclones over the British Isles, and cyclogenesis will develop to the west of the warm ridge. Depressions are then steered well to the northwest of Britain, and are unable to penetrate into the continent and ultimately cause the melting of the snow cover. For the snow to melt, a very vigorous pulse of major warm-air advection is required.

A sprinkling of snow, such as that forming over Britain and the northern fringes of western Europe during northerly outbreaks of Arctic maritime air, for example, is insufficient to initiate a long, cold winter, because there is not enough snow to cool the atmosphere appreciably so that the eastward advance of the ridges and troughs in the upper westerlies is not restricted. If a snow mantle forms in the extreme west of Europe only, any snow lying in Great Britain usually disappears abruptly by midwinter, and is followed by mild westerly spells.

In the United States, it has been found feasible to map the "equal probability thickness," i.e., the critical thickness value for which frozen and unfrozen precipitation forms are equally likely (see Figure 6.15). This critical value not only has a coherent geographical distribution, but also increases with altitude, from 17,700 feet at sea level to 18,050 feet at 6,000 feet. The considerable variations in equal-probability thicknesses from place to place contrast very remarkably with the situation in Europe.[44]

A very good estimate of the geographical pattern of surface minimum temperatures in the United States can be obtained by using the departure of total thicknesses from their average values.[45] From theory, a 5.4°F change in the mean virtual temperature of the 1,000-mb to 500-mb layer corresponds to a change of 200 feet in the thickness of this layer. Under favorable conditions, the conclusion may reasonably be made that every 200-foot change in thickness should bring about a temperature change of 5.4°F. Many local factors combine to reduce the frequency of favorable occasions, so that it cannot be expected that the relationship will always apply precisely, or even approximately. However, the American work proves that areas where thickness departs from normal are closely related to the areas where minimum temperatures depart from normal.

The correlation works particularly well when factors such as cloudiness, precipitation, or radiation do not vary to any great extent locally, for example, in fresh polar outbreaks, when the departure from normal thickness is at a maximum between a surface high-pressure area and a surface low-presure area. Typical conditions favoring a good correlation involve moderate air movement of 10 to 20 knots with an absence of warm (frontal) clouds. Poor correlation occurs when radiation effects are great, under clear skies and light winds.

Other useful correlations between thickness patterns and weather phenomena have been discovered. For example, in the midwestern and southern United States, the 18,600-foot isopleth of 1,000-mb to 500-mb thickness is roughly parallel to the normal tracks of tornadoes. Multiple tornadoes are usually found in the vicinity of the mean 18,600-foot isopleth.[46]

Synoptic Phenomena

Certain meteorological entities or weather systems are so large that they appear as phenomena in their own right on the synoptic chart, for example, cyclones, anticyclones, and tropical storms. Other phenomena of the synoptic chart—fronts, for example—do not always exist as physical entities in the actual atmosphere, and their representation on the chart is more an assemblage of conventional signs. Still other phenomena, such as tornadoes, are too small to be followed by the usual synoptic-station network, and therefore the orthodox synoptic chart presents a somewhat unreal picture when they are present within the area it depicts. The synoptic picture of weather as implicit in the chart is therefore by no means complete. Nevertheless, studying a series of successive charts does enable one to build up a valuable mental model of certain associations of meteorological conditions. "Synoptic experience," i.e., the knowledge gained from carefully studying synoptic charts and comparing the picture they present with the actual weather, is very much to be desired as training for the climatologist. Without this experience, it is easy to be misled by the chart; in particular, it is easy to acquire a totally incorrect concept of synoptic phenomena.

DISTURBANCES

The most obvious type of synoptic phenomenon is the *disturbance*. For thousands of square miles, the isobars may run smoothly, with little deviation from a straight path. But every so often, this smooth pattern will be interrupted by a disturbance. The undisturbed areas of the atmosphere usually occur where it is barotropic, and the disturbances where it is baroclinic, but disturbances develop in barotropic air in tropical regions, and baroclinicity may develop anywhere with great rapidity. Composite maps of the frequency of cyclonic and anticyclonic disturbances show that the regions of greatest cyclone frequency are located to the left of the axis of the climatological Polar Front jetstream at 700 mb, i.e., in the areas of maximum cyclonic relative vorticity. The regions of greatest anticyclone frequency are situated just to the south of the jet axis, in the areas of strong anticyclonic wind shear and vorticity. The average life of all types of pressure center is on the order of five days, but anticyclones tend to persist, on the average, one day longer than cyclones. At least three cyclones form to every two anticyclones in the northern hemisphere. The frequency of disturbances at times of high zonal index is significantly different from that at times of low index, especially over the oceans. Regardless of the zonal-index effect, geographical influences such as mountain ranges, coastline configurations, and large lakes or inland seas appear to exert the same effect on atmospheric disturbances whatever the stage reached in the index cycle.[47]

The main facts of synoptic development, i.e., the observed patterns of formation, movement, and disappearance of cyclones and anticyclones, may be explained fairly satisfactorily by a combination of vorticity considerations with the Sutcliffe development theory and its extension by Petterssen.[48] However, unlike the problem investigated by the dynamic climatologist, which involves mean motions in which geographical influences may cancel themselves out during a period of time, the synoptic problem is such that effects of the surface form of the Earth must receive careful attention.[49]

The two main large-scale influences on synoptic development are the location of vorticity sources and sinks and the effect of major orographic barriers. It may be shown that a large source of vorticity is created in the lee of a mountain barrier when a pronounced upper trough crosses the range; the form of the resulting lee depression—indeed, whether or not one will be generated—depends on the vertical profile of the ground surface, the thermal structure of the atmosphere, and the airstream's speed and the angle of incidence it makes the crestline of the barrier. For the location of vorticity sources and sinks relative to nonorographic surface features, it may be shown that hot or cold areas have an effect that depends on whether the vorticity increases or decreases in the direction of increasing potential temperature. If the magnitude of absolute vorticity increases with height, a cold source makes a positive contribution to synoptic development; and if the absolute vorticity decreases with height, then a hot source will make a positive contribution.[50]

In theory, polar regions should generally be sources of relative cyclonic vorticity, and the subpolar low-pressure belts or cells should be sinks. The theoretical model works quite well for Antarctica. The surfaces of constant potential temperature may be considered to form domes over an approximately circular continent, giving a maximum concentration of isotherms at the coasts. Writing the vorticity equation as it applies to a surface of constant potential temperature shows that absolute vorticity must be generated above the cold pole and then exported along the potential temperature surfaces towards the warm coastal waters surrounding the continent (see Appendix 6.9). Because the vorticity is being exported in the direction of decreasing Coriolis parameter, an even greater export of relative cyclonic vorticity must occur down to the lowest layers of the atmosphere in the zone where the sea-surface isotherms are packed most closely together. Furthermore, if the winds are to be maintained in geostrophic balance in this coastal zone, a semipermanent low-pressure belt must be maintained here to balance the high values of the Coriolis parameter. If the surface winds in the low-pressure belt are to be opposed by friction, then the vorticity of the frictional component of the air motion there must be negative (i.e., anticyclonic), so that vorticity is transferred from the atmosphere to the Earth's surface in this belt, which therefore becomes a sink of relative cyclonic vorticity. This contrasts with the central portion of the continent: the average sea-level pressure distribution is (in the model) anticyclonic, and therefore the vorticity of the frictional component must be cyclonic if it is to oppose the resulting motion.

The picture in the Arctic is more complex than the simple model allows, because it is an almost land-locked sea. For example, not only does the North Pole provide the required cold source—as postulated by the model—in winter, but additional cold sources are produced by the Eurasian and North American coasts, so that the surfaces of constant potential temperature slope downward from the continental interiors toward the coast, as well as sloping downward from the cap over the North Pole. The result is that relative cyclonic vorticity is exported zonally as well as meridionally in Arctic regions; the zonal export is mainly toward the northern parts of the North Atlantic and the North Pacific, which act as frictional vorticity sinks in winter. In general, the northern oceans—and also nonfrozen inland seas, such as the Black and Caspian Seas, the Mediterranean, the Baltic, and the Great Lakes, which are partially or completely surrounded by cold land masses—favor cyclonic synoptic development in winter, particularly along boundary zones between open ocean and frozen areas. During the summer, the northern continents are relatively warm, and the vorticity export is from sea to land or from the air layers above cold ocean currents to those above warm surface currents.

Synoptic development in middle latitudes is difficult to explain by reference to the average location of cold or hot sources, but must be studied in relation to the motion of individual disturbances. For example, even though the northern portion of the North Atlantic acts as a heat source for Arctic air moving south and as a cold source for tropical air moving north, the rule provided from the thermal development theory, i.e., that a disturbance moves in the general direction of the gradient of the Laplacian of thermal advection, proves to be a valid one.* In general, in these latitudes anticyclones move in the direction of warm to cold advection, and cyclones move in the direction of cold to warm advection. Liberation of latent heat may give rise to important local modifications in the movement of a disturbance as predicted from this rule.

The equatorward portions of the subtropical high-pressure belts provide vorticity sources for synoptic development in the intertropical zone. Vorticity produced here is exported in three directions: (a) westward around the subtropical anticyclones and thence to the westerlies; (b) via the trades to the doldrums, which form a frictional

---

* Advection is usually described by the two-dimensional advection equations

$$\frac{\partial A}{\partial t} = -u\frac{\partial A}{\partial x} - w\frac{\partial A}{\partial z}, \quad \text{or} \quad \frac{\partial A}{\partial t} = -J(\psi, A),$$

where A is the quantity being advected, x and z are position coordinates, u and w are velocities in the x, z, directions, t is time, ψ is the stream function, and J is the Jacobian operator. On the accuracy of finite difference methods for solving the advection equation, see C. R. Molenkamp, *JAM*, 7 (1968), 160.

vorticity sink; and (c) to the continental monsoons over Asia and to the thermal low-pressure areas over southern Africa, South America, and Australia, which form important vorticity sinks in summer.

Maps of the frequency of disturbances show clearly the importance of various types of geographical influence (see Figure 6.16). The frequency of cyclones at 500 mb in the northwest quadrasphere shows a maximum at all seasons in the Labrador–Quebec–Hudson Bay region, over the Gulf of Alaska, and in cellular areas extending down the Atlantic and Pacific coasts of North America. At 500-mb cyclones are very rare over the interior of the United States in summer and winter, but occur frequently during spring and fall. Cyclone frequencies in general over the United States show great variability from year to year, and the variability showed an increase between 1905 and 1954 that was probably real, not merely due to fashions in synoptic analysis.[51]

Certain generalizations can be made about favored orographic areas for synoptic development. For example, maxima of cyclonegenesis are found in the lee of major mountain barriers in all seasons: three maxima occur in the United States, associated with the Alberta and Colorado mountain ranges and the Sierra Nevada. Cyclogenesis is also at a maximum in winter over large inland water bodies that remain unfrozen. Contrasts between land and sea are important in causing cyclone modifications and even cyclogenesis. For example, cyclones moving over the European continent in winter from the sea are cooled from below, and their intensity decreases, but those moving from land to the North Sea or the Baltic do not lose, and sometimes increase, their intensity. A cyclone centered over southern Scandinavia in summer usually deepens during the day and fills at night, more or less. This effect is partly due to land-sea breeze processes, but more important is the "corner effect," which comes into operation where the coastline is cyclonically curved in the same manner as the isobars.[52]

Numerous examples of geographical influences on cyclonic development may be found in the literature: most of the studies are concerned with the analysis and explanation of the movement of an individual cyclone. It is in such studies that the synoptic method proves its worth, for it enables conclusions of general application to be arrived at from only a few days of observation. Without the physical understanding implicit in the synoptic chart, many years of daily situations would have to be accumulated before similar conclusions could be obtained by means of simple inspection and intuition. In every case study, there are two problems of special interest to the climatologist: (a) to what extent the development and movement of the given cyclone is related to mean patterns of atmospheric motion; and (b) whether or not the synoptic pattern of the storm represents a dynamic or a geographical influence predominating. For example, the most damaging storm of the present century along the Atlantic coast of North America, which occurred in March 1962, proved to be related to a northwest-southeast trending ridge over the North Atlantic, as revealed by 5-day mean contours. When this ridge is weak (as in March 1962), the easterly winds in the lower atmosphere have an unusually long fetch over the ocean, extending from Britain to the American coast. On the second problem, in the absence of orographic influences, the area of maximum precipitation in an extratropical cyclone is located on the forward, left-hand side of the axis of the area of maximum water-vapor content. The limit of the rain area in a cyclonic storm, excluding thunderstorms, on its warm side nearly coincides with the isopleth of zero mean vorticity. Even with anticyclonic circulation at sea level, widespread cloudiness and precipitation can occur if a cyclone is present in the upper troposphere. For example, a cold upper low over the southeast United

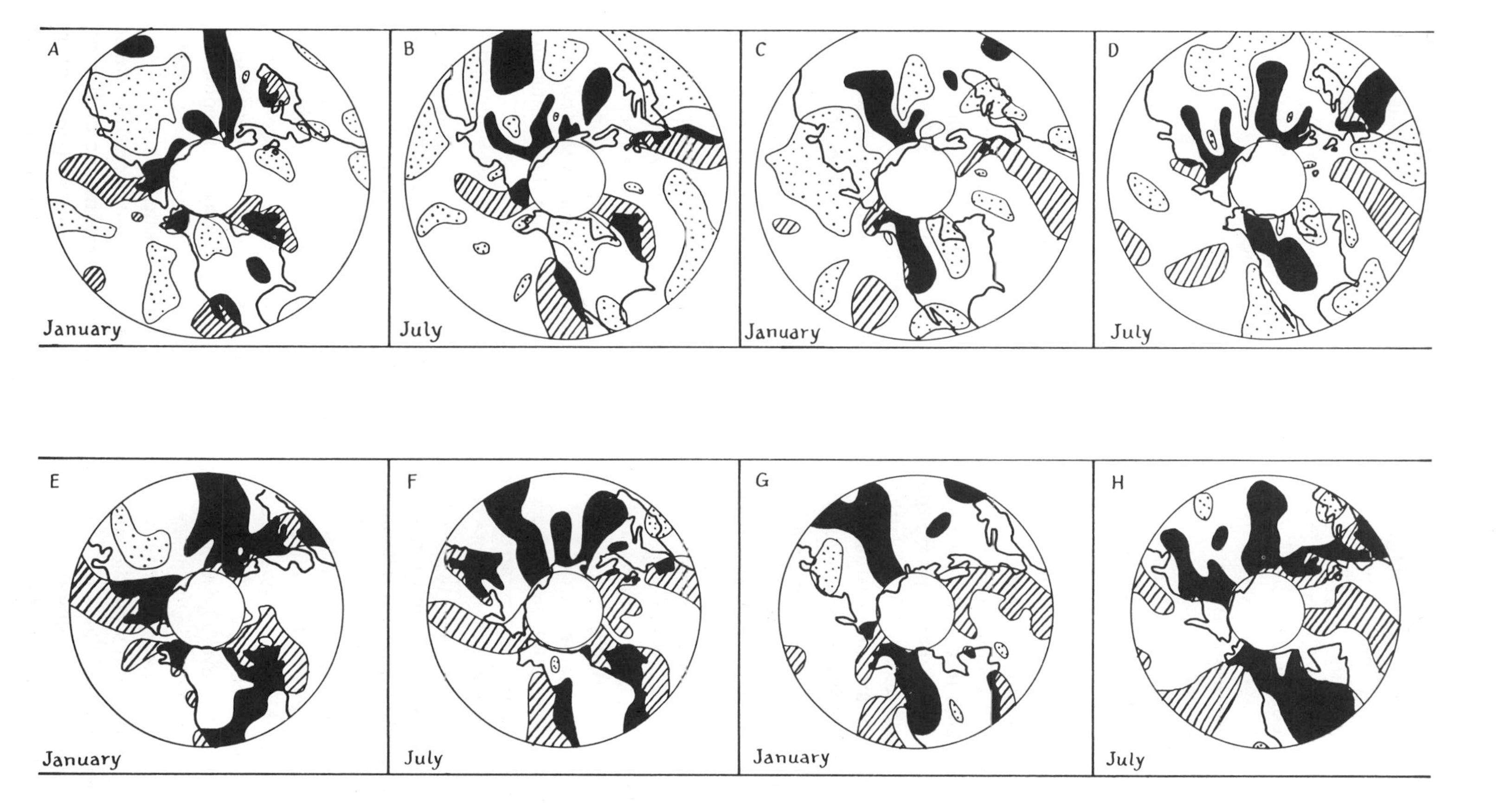

FIGURE 6.16.
Occurrence of mean troughs and ridges at 700- mb (after Klein). Mean troughs: 30-day means at A and B; 5-day means at E and F. Mean ridges: 30-day means at C and D; 5-day means at G and H.

States during the period July 1–6, 1958, resulted in showers and thunderstorms at the ground despite the prevailing weakly anticyclonic surface circulation. The precipitation falling over the intermountain states of the western United States in association with upper cold lows is relatively greater at valley stations than at mountain stations, in comparison with the precipitation resulting from surface fronts. The precipitation associated with a cold cyclone in the upper troposphere over these states is little influenced by orographic uplift, and usually is a result of very pronounced upward motion in the upper atmosphere arising from appreciable advection of vorticity in strongly baroclinic synoptic situations.[53]

Synoptic case studies show that cyclogenesis may develop locally as a result of processes not usually considered in dynamic theories. For example, cyclogenesis may arise mechanically as the effect of a surge of cold air on a quasistationary cold front. The development of secondary cyclones over the east coast of North America is usually preceded 12 hours before by a characteristic anomaly in the broad-scale distribution of temperature at 200-mb. Upper troughs moving eastward in the westerlies have been observed to undergo cyclogenesis without the existence of surface fronts or a surface low-pressure center; the resulting cyclone forms first at intermediate levels in the atmosphere and then extends both upward and downward, finally reaching sea level as a closed low-pressure center.[54] Thus, although most surface cyclones move parallel to the upper flow, giving rise to weather that will be roughly in conformity with the mean zonal pattern of the general circulation, at times the disturbances of the upper flow have by far the greater influence on surface weather.[55]

The classic Norwegian model of the cyclone has been considerably modified as a result of TIROS and later satellite photographs, resulting in a new series of models of cloud distributions associated with a developing extratropical cyclone.[56] The major feature to note is the spiral pattern of cloud typical of such a storm. Radar precipitation echoes indicate that different precipitation areas may be distinguished within this spiral pattern. With a large extratropical maritime cyclone, only a very small percentage of the clouds associated with it proved to be precipitating at a given time. Minor cloud vortices, which were centers of heavy and very local precipitation, were found to persist in the zones of confluence between the different air masses on the periphery of the spiral cloud pattern, and rotated with the latter.

Extratropical cyclones as represented on the synoptic chart have obvious differences from tropical cyclones; in particular, the symmetry and lack of fronts in the latter make the middle-latitude depression appear to be a very much more complex phenomenon than the hurricane or typhoon. In fact, the tropical disturbance presents the greater problem. The synoptic pattern of tropical weather, in the vertical sense, is the reverse of that for middle latitudes. In middle latitudes, the majority of disturbances are surface phenomena that are replaced by relatively smooth wave flow in the upper troposphere. In tropical latitudes, the airflow of the lower troposphere is predominantly wavelike—except when interrupted by a hurricane—but the trades and the easterly waves are replaced above 30,000 feet by rotating cyclonic or anticyclonic eddies of about 500 to 1,000 miles in diameter, which move slowly westward. The synoptic problem posed by the tropical disturbance is thus an interesting one.[57]

The energy to maintain a middle-latitude cyclone is derived mainly from the release of potential energy as cold air sinks and warm air rises, but that required to maintain a tropical cyclone comes mainly from the release of latent heat during condensation. The trade-wind cumuli pump heat and moisture upward from the sea surface into the high troposphere, and the cyclone is maintained by tapping the latent energy stored in the water vapor. Despite the intensity of convection associated with the trade cumuli, the water content of a tropical cyclone is usually well below that theoretically possible for an adiabatic process. The fractional area of a hurricane within which convection is active varies with the size of the cyclone, and is usually at a maximum when the storm has achieved its greatest intensity. The maximum intensity (i.e., the minimum pressure within its eye) that can be achieved by a tropical cyclone is related to the temperature of the sea surface over which it moves, and the processes taking place within this central eye region are the most important ones deciding the synoptic structure of the whole cyclone.[58]

Synoptic charts indicate that tropical cyclones have very intense pressure gradients in comparison with extratropical cyclones of comparable size. This is a result of the warm-core structure of the tropical cyclone; in fact, it was recognized by Shaw in 1922 that tropical storms are warm-core cyclones that decrease in intensity with height and are replaced by anticyclonic circulations by the time the 45,000-foot level is reached.[59] The synoptic situations favorable to hurricane or typhoon development have been analyzed in detail, but no single situation is an infallible guide to tropical cyclone prediction.

The annual frequency of tropical cyclones in the different oceanic regions of the world varies considerably from year to year, but there is no apparent evidence of cycles, trends, or significant space-correlations. The mean frequency of cyclone occurrence also varies a great deal from region to region—for example, typhoons are three times as likely to develop in the western part of the North Pacific than anywhere else—and in general the frequency and preferred sites of tropical cyclogenesis fluctuate in accordance with variations in the form and position of the planetary wave pattern. The frequency of the tropical cyclones is a function of horizontal wind shear in middle latitudes, and of the strength of the westerlies in lower and middle latitudes. There is thus a relationship between long-period circulation anomalies and the development and motion of tropical storms. In the Atlantic, for example, hurricanes tend to move northward of their average path when a marked trough persists near the Appalachians, and may reach New England; however, when the trough occurs well off the eastern seaboard, the hurricanes recurve northward or northwestward well out to sea and do not reach New England.[60]

Although some Atlantic hurricanes develop out of disturbances that originate over the continent of Africa, hurricanes in general tend to form over relatively warm ocean surfaces and then to move along tracks following the areas of warmest surface water, finally weakening when they move over colder waters. Detailed analysis in the Caribbean indicates that for occurrences averaged over 10-day periods and 5° latitude-longitude squares, hurricanes develop downstream from the geographical centres of regions of maximum sea-to-air heat transfer, and then move along the axes of maximum positive anomaly of this quantity. If averages for periods of five years or more are used, there is some correlation between the frequency of hurricanes and their tracks and the location of monthly and annual sea-surface temperature isopleths.[61]

Tropical cyclones in general tend to move in parabolic tracks, involving an initial leg from east to west, then a clockwise (northern hemisphere) or an anticlockwise (southern hemisphere) track to north or south respectively, and a final leg directed towards the northeast or the southeast, respectively. The mode of recurvature of an individual tropical cyclone is related to the height of the base of the polar westerlies that overlie the trades within which they are embedded. If the base lowers considerably to the west of a tropical cyclone, and remains low, then the cyclone usually recurves poleward; if the base does not lower, then recurvature often does not occur and the cyclone continues to move due westward. On a very local scale, vectors showing the magnitude and direction of the barometric pressure-gradient may be useful in indicating the direction of movement of a hurricane, particularly when it reaches middle latitudes.[62]

A synoptic fact of great importance was noted very early in the modern period of hurricane and typhoon investigation. Despite the apparent convergence of the northeast and southeast trades, as exhibited on global maps, it soon became apparent that tropical cyclone development commences as a consequence of instability *within* one of the trade currents, and not as a result of the interaction between the trades of the northern and southern hemispheres. A tropical storm originates as a local disturbance within a homogeneous trade wind, and if the storm happens to be located under an upper ridge, it usually grows into a hurricane or a typhoon. Poleward flow at 200 mb (i.e., as in the eastern portion of troughs in the upper westerlies) is more conducive to the development of tropical cyclones than is equatorward flow, and upper flow with anticyclonic vorticity is more favorable for the growth of a hurricane than is cyclonic vorticity.[63]

Despite the regular symmetry of a tropical cyclone on the synoptic chart, individual hurricanes and typhoons show marked asymmetry in the pattern of their latent-heat release and hence in their precipitation pattern (see Figure 6.17). Although some features common to large numbers of cyclones have been described, the resulting rules are by no means infallible and the precipitation pattern associated with each tropical cyclone must be studied as an individual problem. The same conclusion applies to cloud patterns. Although satellite photographs show that changes in the intensity of a tropical storm are accompanied by corresponding changes in the appearance of its associated cloud pattern as viewed from above, no simple generalizations have been formulated. An interesting result of the TIROS series of observations was the discovery that typhoons in the eastern part of the North Pacific are three times as frequent as the conventional synoptic observations lead one to suppose. An important influence appears to be a ridge separating easterlies and westerlies at 200 mb. When this ridge is strongly developed, many tropical cyclones are prevented from moving into the central Pacific.[64]

The subtropical anticyclones appear on almost every synoptic chart that covers sufficient latitudinal extent, but the atmosphere does not appear to possess cyclonic counterparts to these quasipermanent features. However, upper-level cut-off cyclones are usually termed *subtropical cyclones* when they are present in relatively low latitudes. Such a cyclone is a very remarkable synoptic feature, which may reach sea level and persist for days or even weeks. A complete subtropical cyclone system comprises a region of ascending air in the upper troposphere overlying a region of subsiding air in the middle troposphere, which in turn overlies a trade-wind type of circulation in the lower troposphere. The system does not decay—surface friction has a negligible effect on it—and usually persists until it is finally absorbed by a large-amplitude trough extending from the upper westerlies of middle latitudes. During its relatively long life, a subtropical cyclone exports a considerable amount of energy, and can produce spectacular amounts of rainfall. A good example is provided by the *kona storm*, which produces very heavy rainfall in Hawaii two or three times each winter half-year. The kona storm is a subtropical cyclone that originates as a normal cold-core

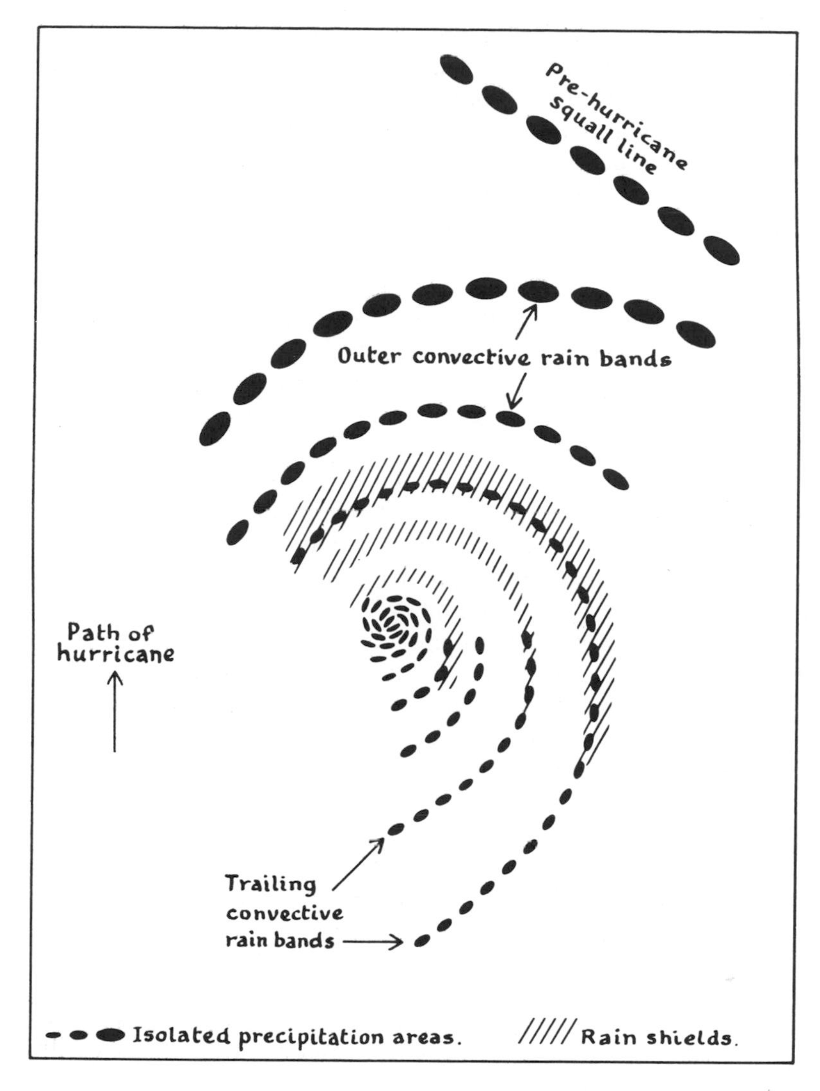

FIGURE 6.17.
Composite radar picture of a hurricane. (After V. D. Rockney, *Munitalp Foundation Proceedings*, Nairobi, 1960.)

cyclone in middle latitudes, intensifies, and then acquires warm-core cyclone properties as its wind and precipitation become similar to those of a tropical cyclone.[65]

As one would expect, maps and cross sections of cyclonicity and anticyclonicity show that cyclonic and anticyclonic activity over a large area are interdependent.[66] Because of the concentration of dynamic and synoptic studies on cyclonic development, i.e., on "bad weather," studies of anticyclones have tended to lag behind. Nevertheless, anticyclones pose very important problems both meteorologically and geographically. Anticyclones in general appear and disappear on the synoptic chart in rough conformity with the Sutcliffe development theory, but many extraneous influences may exist in individual cases. For example, anticyclonic development is much more sensitive to nonadiabatic processes (i.e., those connected with the differing thermodynamic properties of different parts of the Earth's surface) than is cyclonic development; and in any case the quantities involved in dynamic considerations—such as vorticity relative to the Coriolis parameter—are very much smaller in the anticyclonic case than in the cyclonic, and therefore their magnitudes are much more critical in synoptic considerations. Despite these qualifications, a simple rule is that most anticyclones at sea level develop downwind from upper wedges, particularly polar continental anticyclones developing in the lee of the Rocky Mountains and the Scandinavian mountains.[67]

In middle latitudes, a linear relation exists between the mean central pressure of anticyclones (and cyclones) and the sine of their latitude of occurrence. The data imply that the mean "potential intensity" of anticyclones is independent of the latitude in which they occur, a result that presents some problems, since, in terms of vorticity, it is possible to envisage a *maximum anticyclone*, i.e., one in which the absolute angular momentum is zero. TIROS observations have revealed an upper troposphere anticyclone that in magnitude approached this limiting case.[68]

Yearly averages of the positions of the 1,000-mb and 500-mb contours show that the northern hemisphere is dominated by three anticyclones.[69] These anticyclones—which will obviously not necessarily appear on the synoptic chart but which may be regarded as "circulation poles" controlling or at least representing the day-to-day movement of individual anticyclones—are located over the Azores, in central Siberia, and in the eastern Pacific. Their axes are tilted southwestward with increasing height, so that at 500 mb the anticyclones coincide with the warmest regions of the world, i.e., the Caribbean and Central America, the area between Central Africa and India, and the West Pacific east of the Marshall Islands. The locations of these mean anticyclones are significant in the movement of individual "cold poles" (i.e., zones within which the temperature at 500 mb is less than −30°C in summer or less than −40°C in other months) that develop in a lifting process within Arctic cyclones and have a great deal of influence on Arctic weather. The movement of the cold poles (see Figure 6.18) controls the height of the 500-mb surface over the North Polar region, which can delay considerably the rise in surface temperature in this region.

The manner in which anticyclones persist from day to day on the synoptic chart has several implications. First, "mean anticyclones" must have synoptic importance as entities in the physical sense, unlike "mean cyclones," which are primarily statistical features. As an example, the great Pacific anticyclone of winter 1949–50 may be cited. This warm anticyclone persisted from early December 1949 to late March 1950, during which time it moved along an arc from the southeast Pacific to the Bering Sea and Canadian Yukon, exerting a controlling influence on the movement of cyclones in the North Pacific and the large-scale climatic anomalies over the United States. A second implication of the persistence of anticyclones as synoptic features is that certain areas must become very prominent as centers of anticyclogenesis. The Alaska–Northwest Canada area, for example, is subject to very rapid anticyclogenesis, of about 7 mb in 24 hours, which is related to sunspot variations.[70]

Excluding the circumpolar vortices, the anticyclone which develops in summer at 100 mb over southern Asia is the most intense and persistent synoptic feature of the northern hemisphere at this level of the atmosphere. The influence of this anticyclone extends from the Atlantic coast of Africa eastward into the Pacific, and it is connected in some fashion with the southwest monsoon of Asia; it also is responsible to some extent for the longitudinal location of wave-number one of the planetary wave pattern.[71]

Scandinavia is a favored location for persistent winter anticyclones, but for an individual anticyclone to persist for more than three days, it should be located some 600 nautical miles to the east of a warm ridge of large amplitude, and this ridge should continue to develop for at least 24 hours after the appearance of the anticyclone. Winter anticyclones developing over northwestern Canada and then moving eastward or southeastward show an interesting change in form. Their appearance on the synoptic chart changes as they approach the east coast of the continent: they become oval in shape, with their longest axis parallel to the coast, as a consequence of a retardation they suffer as heating over the sea begins to transform their vorticity from anticyclonic to cyclonic. Their intensity gradually decreases as they move out over the Atlantic.[72]

Detailed study of conditions within an individual anticyclone indicates that considerable variety in weather conditions may be associated with anticyclones that are very similar on the synoptic chart. For example, the amount of stratocumulus within easterly airstreams over the British Isles may vary from nil to 8 oktas at a given station even for identical seasons, time of day, geostrophic trajectory, and wind conditions. The variations are due to the existence of distinct regions of cold or warm air circulating in the lower levels of anticyclones. These regions represent successive pulses of cold or warm air on the flanks of an anticyclone, and may be regarded as the remnants of different air masses after the fronts separating them have dissipated or been dropped from the synoptic analysis. The cold or warm masses have depth-to-length ratios of about one to 700 and they move with the speed of the air at their level. If the anticyclone moves over an extensive water surface, stratocumulus forms within the cold-air regions, with convection resulting up to the inversion level.[73]

AIR MASSES AND FRONTS

Air-mass and frontal analysis is the basis of routine synoptic forecasting in middle and high latitudes, and will probably continue to be so for many years. Consequently, the climatologist concerned with climate as the synthesis of daily weather must become acquainted with the techniques of air-mass and frontal analysis if he is not to be content with taking his "primary documents," i.e., the published daily weather maps of the

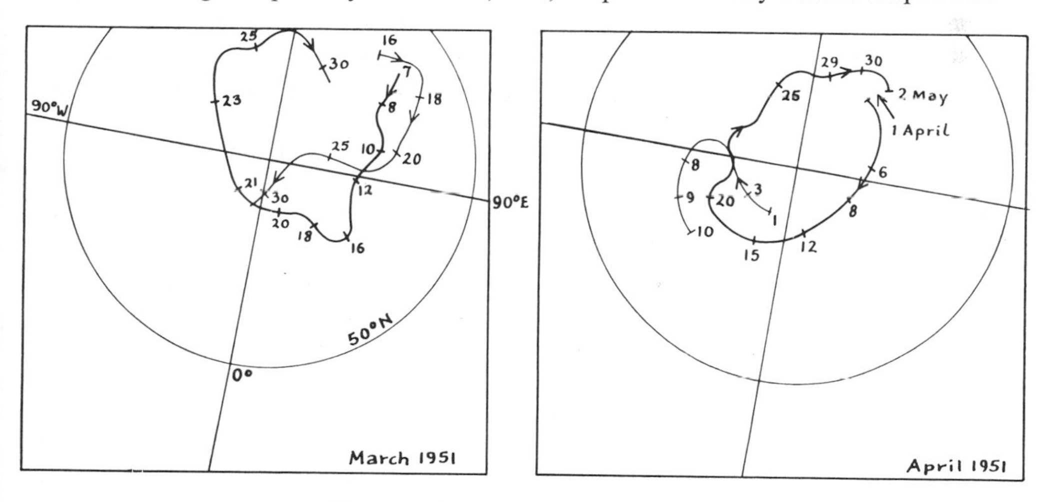

FIGURE 6.18.
Tracks of cold poles (after Scherhag).

TABLE 6.5
Classification of Air-Masses (following T. Bergeron).

Basic types	Geographical subtypes	Thermodynamic subtypes
Equatorial (E)	mE = Maritime equatorial[1]	mEk = mE colder than surface over which it is moving
	M = monsoon air, over southeast Asia in summer[6]	mEw = mE warmer than surface over which it is moving
Tropical (T)	mT = Maritime tropical[4]	mTk = mT colder than surface over which it is moving
		mTw = mT warmer than surface over which it is moving
	cT = Continental tropical[3]	cTk = cT colder than surface over which it is moving
		cTw = cT warmer than surface over which it is moving
Polar (P)	mP = Maritime polar[5]	mPk = mP colder than surface over which it is moving
		mPw = mP warmer than surface over which it is moving
	cP = Continental polar[2]	cPk = cP colder than surface over which it is moving
		cPw = cP warmer than surface over which it is moving
Arctic (A)	mA = Maritime arctic (or antarctic)[5]	mAk = mA colder than surface over which it is moving
	cA = Continental arctic (or antarctic)[2]	cAk = cA colder than surface over which it is moving
		cAw = cA warmer than surface over which it is moving.

Source regions:
1. Oceanic areas between subtropical anticyclones of northern and southern hemispheres.
2. Snow-covered continents under anticyclonic wind circulations.
3. Hot deserts of continents.
4. Oceanic areas on equatorial side of subtropical anticyclones.
5. Oceans in middle and high latitudes.
6. Equatorial areas of Indian Ocean.

state meteorological services, at their face value. Many textbooks give instruction in these techniques, and therefore they will not be outlined here.[74] Instead, attention will be focused on the deficiencies of air-mass and frontal analysis, and on some of the physical and geographical problems involved. The only general point to be made at this stage is that the climatologist is advised *not* to take published daily weather maps at their face value, if he is interested in precise information. The exceptions to this are provided by such series as the I.G.Y. synoptic charts, which are published after months of careful analysis.

Air masses may be defined as large masses of air that are more or less horizontally homogeneous in such physical properties as heat and moisture content. A disadvantage of the concept is immediately apparent: the definition excludes vertical conditions, which are of prime importance in local weather. The concept of an air mass is therefore essentially one in which the atmosphere is studied in terms of its horizontal pattern, and vertical variations are neglected. Nevertheless, useful generalizations concerning vertical conditions may safely be made from a knowledge of the air mass prevailing at a particular place and time.

If an air mass remains over one portion of the Earth's surface for some time, its lower layers will tend to acquire, from turbulent diffusion and other vertical-exchange processes, some of the physical properties of the surface on which it is resting, and will develop a characteristic lapse-rate. For example, an air mass resting on an ice surface will tend to become cold and dry, but one resting on a warm water surface will tend to become warm and moist; each will have a characteristic type of stability. Thus *source regions* may be recognized that give an air mass its physical character. The most obvious source regions are land areas and sea areas, each divided according to whether it falls into arctic, polar, tropical, or equatorial latitudes. On this basis, a classification of air masses may be built up, each air mass being described in terms of its source region (see Table 6.5). Obviously, an air mass in its source region will be characterized by specific type of weather, depending on the season and the time of day, and when the air mass begins to move its associated weather will become modified as the nature of the underlying surface changes.

Several different air masses may be present at a given time at a station, superimposed one above the other, but conventional air-mass analysis restricts itself to the air mass at the surface of the Earth. Orographic features give rise to complications: for example, the Rocky Mountains and the Appalachians form a vast funnel that channels cold air masses southward, so that Arctic air masses have been recognized in Central America.[75] However, in dynamic weather studies the emphasis should be on the transitory nature of air masses, i.e., on the transformation of one type of air mass into another.

The transformation of polar continental air to polar maritime as it moves from land to over the sea may be explained by the theory of transfer processes operating at the sea surface. The theory enables the surface air temperature within the air mass to be found at any point over the sea from the initial surface air temperature, the initial lapse rate within the air mass, the sea-surface temperature, and the distance the air has traveled over the sea. The depth to which polar continental air in westerly airstreams over North America is modified when it crosses Hudson Bay depends on the physical state of the water surface. If the water is frozen, then the air mass suffers negligible modification, but if the water is not frozen, the air mass is modified to a depth of 5,000 to 10,000 feet in early winter. The rate of warming of cold Arctic air masses moving in a southerly or southeasterly direction over the eastern North Atlantic is related to the difference between the temperature of the surface layer of the sea and the surface air temperature of the air mass. Warm maritime air masses moving over the cold European continent in winter are cooled at the rate of 194 cal per cm^2 every 12 hours during the night—by long-wave radiation and evaporation—and at the rate of 120 cal per cm^2 every 12 hours during the day. These cooling rates produce a decrease in the total thickness of 41 m every 12 hours during the night and 25 m every 12 hours during the day, the nighttime decrease corresponding to a drop in the mean temperature of the 1,000-mb to 500-mb layer of 2°C.[76]

Too few air masses may be identified on the conventional synoptic chart to explain observed weather variations. A good example of this is the *marine layer* of the coastal regions of southern California. The marine layer is the thin lower layer of the extensive air mass moving around the northeastern portion of the subtropical anticyclone of the North Pacific. Unlike the main body of the air mass, which is warm and dry because of the prevailing subsidence, the marine layer is cool and humid, owing to moisture and heat exchange with the sea surface. The marine layer persists for most of the year and the type of weather experienced by the coastal districts depends largely on its depth. It owes its existence to two processes of heat transfer that act in opposite senses: gradient diffusion, i.e., diffusion of physical properties in a turbulent manner, down the gradient of concentration of the property concerned, and bulk convection, in which buoyancy effects completely swamp the normally dominant gradient diffusion.

The point of synoptic interest is the extent to which the marine layer is capable of penetrating inland. The layer is usually short-lived when it moves inland, because convective mixing transforms it from its normal near-neutral temperature stratification to a stable stratification. This is an interesting effect in itself, because convective mixing usually results in a neutral stratification. Provided the upper boundary of the marine layer is impervious to convection, the layer can persist as an entity. Inland, however, the air within the marine layer warms so much that convection elements developing in contact with the ground may rise completely through the layer and escape into the warm air above. In such a situation, the marine layer breaks down and pure air-mass weather exists at ground level.[77]

In terms of air-mass analysis, *fronts* are simply the zones between different types of air mass; unlike air masses, which are normally associated with quiet weather (or at least the same type of weather) over extensive areas, fronts in the synoptic sense are usually associated with violent forms of weather. Although maps showing the locations of the main frontal zones of the Earth have been available for many years, it was only in 1960 that maps depicting actual fronts for the entire northern hemisphere were published for the first time. The standard textbooks presented maps showing the average positions of the main frontal zones as long ago as the 1930's, but these maps are misleading in that they indicated the locations in which the basic fronts *ought* to occur, rather than where they actually occur, because the frontal locations on the maps were based on the distribution of mean temperature and mean pressure, plus the assumption that fronts should occur in regions where the deformation of the mean atmospheric flow pattern, and the horizontal gradient of mean temperature, are of appreciable magnitude and suitably oriented to each other. This assumption is not always justified.[78]

The maps in Figure 6.19 give the actual preferred locations of fronts at the Earth's surface, based on counts of daily frontal occurrences within standardized areas. They were constructed from data over a five-year period, but it is believed that they would be little altered by averaging over a much longer period. The positions indicated by these new maps differ significantly from those inferred from mean pressures and mean temperatures.

In winter, three axes of high frontal frequency occur, forming the principal frontal zones. These define the Pacific and Atlantic Polar Fronts, and the Eurasian Polar Front. A fourth belt of high frequency runs parallel to the Western Cordillera of North America, and there is also some evidence for a weak Arctic Front over the North Atlantic. Four zones of high frontal frequency occur in summer, delineating the Pacific and Atlantic Polar Fronts, the Eurasian Polar Front, and the Siberian-Canadian Arctic Front. In all cases, the axes of high frontal frequency shown in Figure 6.19 occupy nearly the same positions if maps are drawn for warm-front and cold-front frequencies separately.

The winter map agrees quite well with those in some of the standard texts, but does not confirm the double frontal zone in the Pacific shown in other works.[79] The summer map differs considerably from the earlier maps. The latter show a Polar Front in the vicinity of Alaska and eastern Siberia, which is 20° to 30° of latitude too far north, and do not recognize the existence of the Siberian-Canadian Arctic Front at all. The earlier maps incorrectly show a Polar Front crossing southern Canada and dipping southward along the west coast of the United States as a trade front. They do not show a Polar Front in the east Pacific, in the same latitude as in the winter, and they either omit the

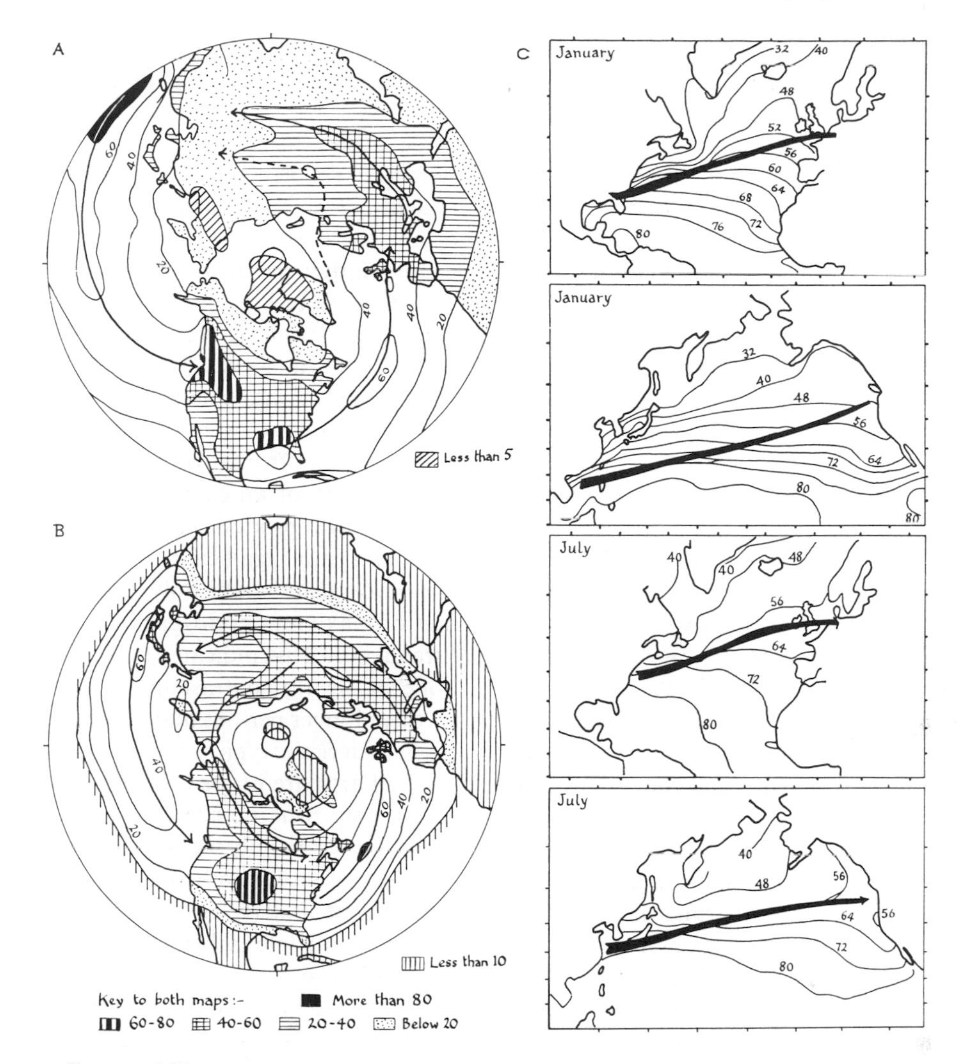

FIGURE 6.19.
Occurrence of fronts in the northern hemisphere (after R. J. Reed).
A, Frontal frequencies, December–February.
B, Frontal frequencies, June to August.
C, Mean sea-surface temperatures compared with mean location of Polar Front zones.

Eurasian Polar Front completely or carry it too far northward, to merge with the western portion of the Siberian-Canadian Arctic Front.

The locations and intensities of the frontal zones prove to be closely related to the distribution of sea-surface temperatures. The regions of very high frontal frequency are found on the western sides of the oceans in both seasons, approximately coinciding with the zones of maximum concentration of the isotherms of sea-surface temperature. It has long been recognized that fronts do not represent true temperature discontinuities in the atmosphere, but instead are narrow transition zones. In order to secure uniformity in plotting fronts on synoptic maps, therefore, the convention is to draw them at the warm boundaries of the frontal zones. It is noteworthy that the axis of the highest frequency of fronts lies along the warm edges of the zones of maximum thermal contrast in the underlying surface, corresponding to this convention. The Siberian-Canadian Arctic Front is due to the strong thermal contrast developing along the borders of the Arctic seas.

Except in the Atlantic in winter, the frontal frequencies decrease appreciably eastward, as the sea-surface temperature gradient weakens. The asymmetry of the frontal distributions is partly due to the influence of ocean currents, in particular the Gulf Stream–Labrador Current system in the Atlantic and the Kuroshio-Oyashio system in the Pacific. More detailed analysis shows, for example, that the greatest frequency of frontal passages over the North Atlantic is over the line marking the initial contact of cooler air with that associated with the Gulf Stream.[80] High frontal frequencies in winter also result from orographic wave effects in the lee of the Rocky Mountains. The Atlantic Arctic Front in winter is produced both by the thermal contrast developing near the edge of the northern pack ice, and by the channeling of cyclones through the oceanic corridor between Greenland and Scandinavia. This frontal zone dips southeastward over Russia, associated with the development of an extensive upper-air trough, which is probably a resonance effect set up by waves produced by the Rockies.

The frontal zones shown in Figure 6.19 are dynamically significant, because they coincide in general with areas of high frequency of cyclogenesis. The only exception is in the western Pacific in winter, where the region of maximum cyclogenesis is 10° of latitude north of the region of greatest frontal frequency. This indicates that shallow cold fronts, which enter the western Pacific zone of strong contrast in sea-surface temperatures, are maintained on the synoptic charts for longer periods than in other regions for some reason, despite the absence of wind characteristics typical of frontal zones.

The maps indicate that the surface features of the Earth are of great importance in deciding the location of the main frontal zones. This empirical induction is in conflict with that derived from both theory and experiment. Dynamic theory shows that fronts necessarily arise in the atmosphere as a consequence of its inherent properties, whether or not any thermal or other contrasts exist in the underlying surface. Experiments with rotating cylinders also indicate the same principle. Obviously, we have here a very interesting geographical problem for investigation.

The possible occurrence of fronts in tropical regions has received some attention. When fronts enter the tropics, the temperature contrast across them usually disappears because of the uniformly warm sea surface and the prevailing atmospheric subsidence. However, they may persist for days as *shear lines*, moving slowly equatorward.[81]

Fronts in the Arctic are similar in general to those in middle latitudes. In the Antarctic, however, lack of synoptic stations has meant that a synoptic model has had to be developed if weather changes are to be interpreted in terms of frontal concepts. The normal practice is to regard Antarctic fronts as "line" phenomena that are usually associated with cyclones or with cols between two high-pressure centers and two low-pressure centers, and that often have a very limited life and geographical extent. The fronts are drawn on the charts as either zonal or meridional phenomena, depending on the viewpoint of the analyst.[82] The meridional "fronts" are frequently not fronts in the hydrodynamic sense; they are merely the conventional way of interpreting certain assemblages of weather conditions that are observed in the deformation field between migratory anticyclones in the subtropical high-pressure belt of the southern hemisphere. Although these anticyclones are normally separated by north-south extending troughs, many of the troughs do not contain fronts.

Notice that there is some difference of opinion as to what exactly constitutes a

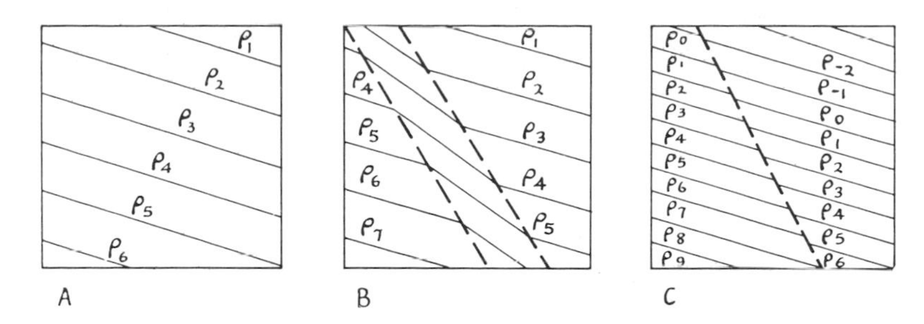

FIGURE 6.20.
Fronts as density discontinuities. A shows a continuous distribution of density. B shows a continuous density distribution containing a narrow transition zone. C shows a possible appearance of B on a synoptic chart; because of map-scale effects, C. shows a zero-order discontinuity in place of the actual transition zone.

"front". Essentially, the original conception of a front involved some form of *discontinuity*. Several orders of discontinuity are possible. This may be illustrated by considering the distribution of the variable X in the horizontal plane (see Figure 6.20). If X is discontinuous along a line, then the discontinuity is said to be of *zero order*. If X is continuous over the whole area of the plane, but its first derivatives are discontinuous along a line, then the line is said to represent a *first-order discontinuity*. If both X and its first space derivatives are continuous over the whole area, but its second space derivatives are discontinuous along a line, then the latter is said to be a *second-order discontinuity* (see Appendix 6.10). Real zero-order discontinuities are very rare in the atmosphere, because diffusive, radiative, and turbulent processes tend to produce zones of transition rather than to maintain sharp boundaries. However, at the usual scales adopted for synoptic charts, such transition zones may appear to be very narrow or even to decrease to the thickness of a line on the chart. Consequently, what in actuality is a narrow transition zone may appear on the chart to be a zero-order discontinuity.

The atmospheric feature—as distinct from the synoptic entity—known as a "front" was first introduced in 1918 by V. Bjerknes as a three-dimensional concept, but it could only be related to surface weather conditions owing to the absence of upper-air observations. From 1950 onward, the availability of numerous radiosonde reports have enabled the *actual* three-dimensional structure of fronts to be compared with the Bjerknes model. Two important conclusions result.[83]

First, the frontal "surface" shown in the classic frontal models is actually a zone, rarely less than 40 miles (and averaging 130 miles) in width. This frontal zone is most marked, and most intense, at 500 mb. Here it is also at its narrowest width (30 to 50 miles), and usually has sharp boundaries. The frontal zone is most marked underneath the jetstream, and frontogenetic processes in the upper troposphere only operate where the jet is intensifying or becoming more cyclonically curved along its length. The Polar Front is clearly, therefore, a primary feature of the atmosphere, due to *dynamic* processes, and is *not* formed by the undercutting of warm tropical air by cold polar air, except in a very local and restricted sense.

In the middle levels of the atmosphere the primary Polar Front is very pronounced. The average horizontal temperature contrast across it averages 15°F, spread over a distance of 600 miles, but 9°F of this contrast is actually concentrated into the frontal

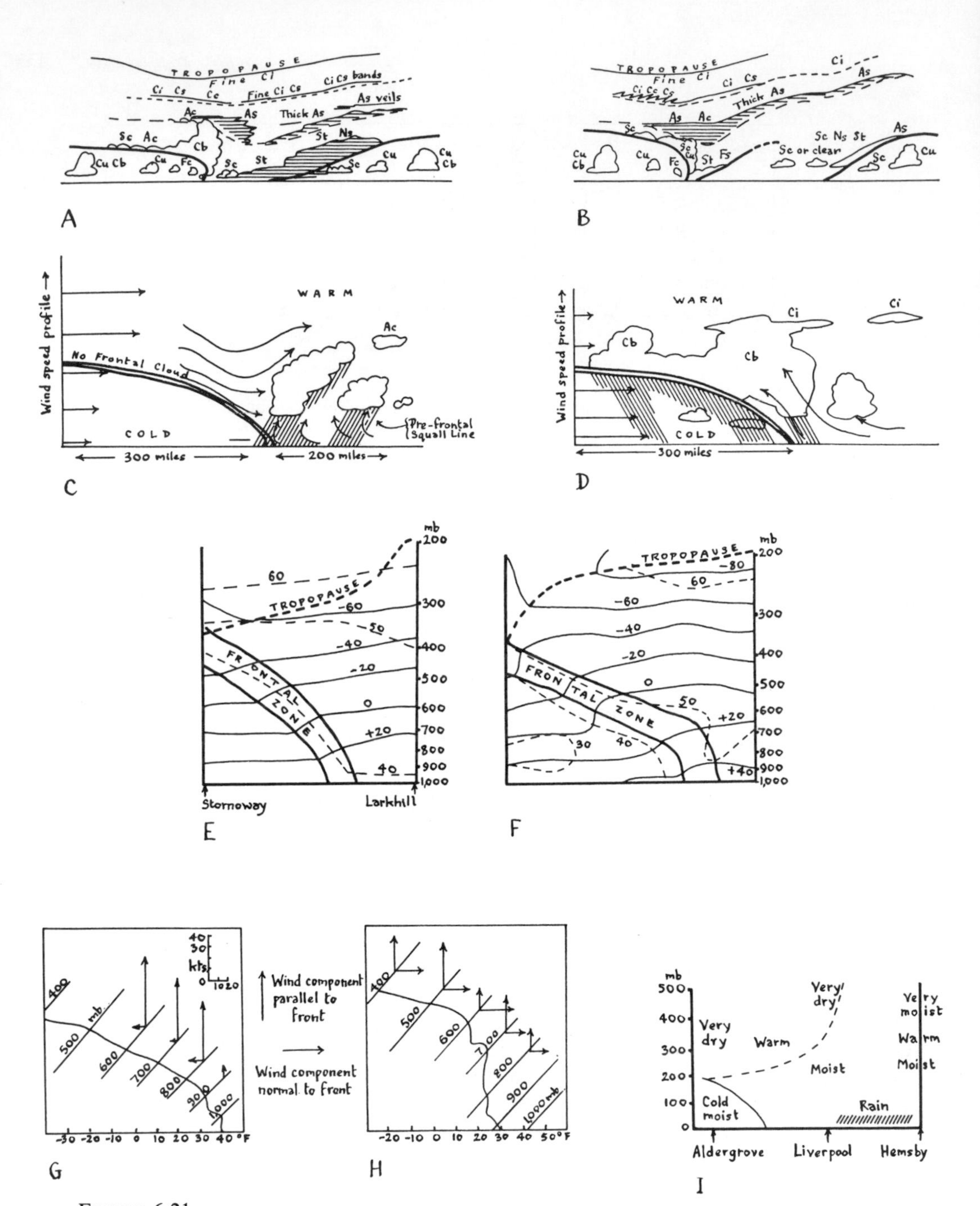

FIGURE 6.21.
Some typical frontal structures. A, B, Lamb's frontal models. C, D, The Oliver's models. E-H, Sansom's models. In E and F, solid line shows dry-bulb temperature, broken line, wet-bulb potential temperature. G and H show dry-bulb temperature soundings and wind components. A, Common type of cloud structure in a warm sector. B, Cloud structure in a warm sector of more complex origin. C, Cross section through an inactive cold front (precipitation areas shaded). D, Cross section through an active cold front (precipitation areas shaded). E, Section of an anafront, January 3, 1945. F, Section of a katafront, March 1, 1945. G, Anafront, January 3, 1945. H, Katafront, March 1, 1945. G and H are tephigrams for Liverpool at noon G.M.T., speed of front was 23 knots in each case, I, An unusual cold front, July 6, 1953. (After H. H. Lamb, V. J. and M. B. Oliver, and H. W. Sansom.)

zone. The air in and near the frontal zone is very dry, less than 5 per cent relative humidity in the middle of the zone, down to 10,000 feet above the ground; often, a tongue of dry air occurs in the vicinity of the frontal zone, extending downward, and tilted in the direction of the frontal slope. This dryness of the air is due to a considerable amount of subsidence during the formation of the front, for example, in the left entrance to the jetstream.[84] Generally, the humidity structure of a front is very complex, while the temperature structure is fairly clear-cut.

These humidity and temperature distributions must obviously have profound consequences for cloud development. The boundary of the mass of frontal cloud is found to slope more steeply than the front itself, and to intersect the warm boundary of the frontal zone at around 600 mb. With warm and cold fronts, the high and medium clouds more often than not are found to form separate systems, distinct from the lower clouds, with lanes of clear blue sky between the systems. These clear lanes are indicated by stable layers in the upper-air soundings.[85] The normal situation is for quite distinct upper and lower cloud decks to occur, but these merge into a single system if one (or more) of the following conditions occurs: the uplifted air mass has a high moisture content; it is extremely unstable; or heavy rainfall occurs, raising the humidity of the layers it falls through.

Figure 6.21,A,B, illustrates Lamb's models of cloud distribution in a warm-sector depression. The "common model" shows banded cirrostratus parallel to the advancing cold front, and some 250 miles ahead of it. The cloud layers are usually entirely ahead of the cold front. The residual discontinuity surfaces are the remnants of former frontal surfaces; i.e., they represent all that remains of former fronts, which have been imperfectly frontolyzed. Such residual surfaces are best preserved in stable air masses, for example, in warm air and in uplifted warm sectors, and they preserve part of their former cloud systems. This is recognized as the "front within front" structure, and the difficulties of frontal analysis on the synoptic chart are much increased. However, the resulting model will bear a much closer approximation to reality than did the old models. The "more complex model" shows that the *warmest* air (also the most humid air, with the lowest clouds) occurs just before the *cold* front. This happens because the cold front was formerly an occlusion, or was preceded by relics of the warm front of an occluding depression. This type of warm sector is common in Australia.

Thus one fact brought to light by upper-air data is that a given front may have cloud and precipitation features associated with it that come from a previous stage in its existence as a front. This agrees with the experience gained by viewing fronts from the air. Their cloud structure, on many occasions, will be seen to be of truly fantastic complexity, and one then realizes how greatly the "fronts" plotted on the synoptic chart from surface-weather analysis omit many of the really important facts about the weather. Lamb's models are empirical models, presenting the observed distributions of clouds and interpreting them as due to multiple frontal structures. In contrast to this are the "active" and "inactive" frontal models, which are theoretical models, distinguished on the basis of what one would expect to happen if the wind varied with height. Figure 6.21,C,D give the models of V. B. and J. Oliver for these features, with the associated cloud distributions. These models have been found to work well in everyday weather analysis.[86]

The second conclusion derived from a study of upper-air data is that the classic two-front model of an extratropical cyclone, i.e., the warm-sector depression with warm and cold fronts, changing to an occluded front when the warm sector is lifted off the ground, has been found wanting. In its place has appeared the *three-front model*, developed by Canadian meteorologists, largely because it was found that an occlusion in the classic sense rarely occurred in nature. In other words, instead of carrying on the evolution of a weather sequence in terms of the Norwegian models, they decided to be on the continual lookout for the formation of new fronts from the weather observations. The result was that a double-frontal structure, or front-within-front situation in the horizontal geographical sense, was found to be the rule, not the exception. The reader can easily prove this for himself by following the evolution of a long series of weather situations on a sequence of synoptic charts. If, instead of trying to make the classic Polar Front depression evolution fit the actual weather by "forgetting" about those cases that do not fit, he studies each situation on its own merits if it fits the criteria for a front, then he will find that the frontal picture more often than not is a multiple one.

The basic difficulty in applying the concept of the front to atmospheric events has been the standpoint of the analyst. If he was an essentially practical meteorologist, then "front" to him meant clouds and precipitation. If a theoretician, a front was a first-order discontinuity in density in the atmosphere, and hence a discontinuity in temperature and winds. Bjerknes, Solberg, and Bergeron combined the two, and showed that, on many occasions, this discontinuity was accompanied by a distinct sequence of clouds and precipitation. Later workers have examined the cases where a discontinuity surface was *not* accompanied by clouds and precipitation, and have introduced the concepts of "inactive" fronts and "dry" fronts (described later). One should always be wary, therefore, of the bias of the analyst when one is examining a weather situation. The theoretician's front is now a "three-dimensional hyperbaroclinic zone with a first-order discontinuity in the temperature and wind fields."[87] There is no reference to weather in this definition, but the concept can be used in investigations and predictions by means of mathematical and physical models. The weatherman's front is a line of extensive cloud, usually with precipitation, which extends for some hundreds of miles longitudinally and is up to 100 or 200 miles wide. There is no reference in this definition to the original criterion of a density discontinuity, as specified by Bjerknes and his collaborators. The definition has value in that it *always* refers to an actual feature of the weather; the difficulty with it is that this feature may or may not be a front in the original sense. Both theoretical and practical meteorologists agree, however, that a front must be reasonably continuous on the chart both in space and time, and it must also be a quasisubstantial surface that moves with the general airflow.

Fronts recognized by different people, therefore, tend to differ very much both in location and extent, and also in their times of appearance or disappearance. This is one reason why purists are not too happy with the synoptic method. The three-front model seems to be the concept most useful, to both theoretical and practical meteorologists, of those at present available.

Developed in Canada from 1950 onward by Godson and his co-workers, the three-front model (see Figure 6.22) introduces the following modifications to the Bjerknes' concept.[88] First, the idea of a front as an interface between two air masses is replaced by the idea of it forming a hyperbaroclinic zone of transition. The convention is then introduced that the frontal surface is taken as that between the warmer air mass and the hyperbaroclinic zone, which therefore, with its maximum wind and temperature change with height, lies on the cold side of the front.

Second, the isobars in the hyperbaroclinic zone must be cyclonically curved, not kinked as in the classic model, with the maximum cyclonic curvature actually in the

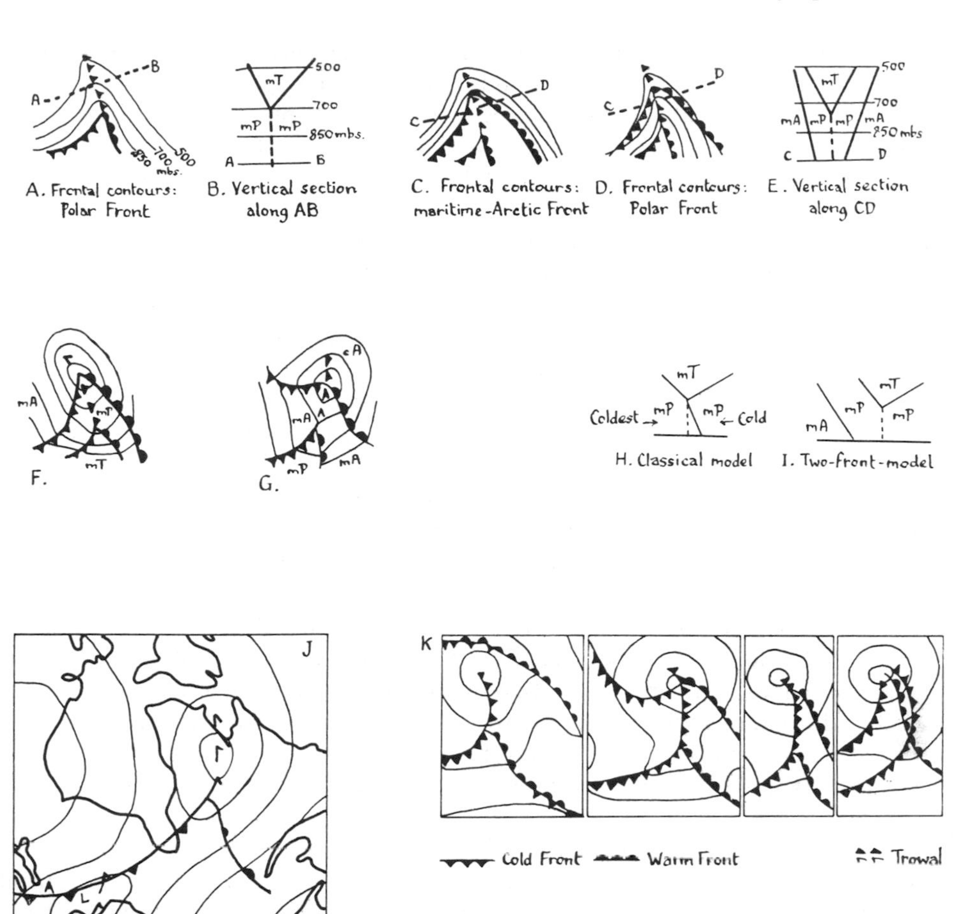

FIGURE 6.22.
The three-air-mass cyclone model (after C. M. Penner and J. L. Galloway). A, B, Classical polar-front model of a cyclone wave. C-E, Two-front model of a cyclone wave. F, G, The three-air-mass cyclone, two common cases. H, I, Interpretations of the cold occlusion. J, Surface chart for 12:30 G.M.T., May 9, 1956, interpreted by means of the 3-front model. K, Possible interactions between polar and Arctic fronts. See Figure 6.23, C, for frontal contours.

zone. The true position of the fronts at the ground surface, in this concept, must be with the warm front to the rear of the trough formed by the cyclonically-curved isobars, and the cold front in advance of the trough.

Third, a new synoptic feature is introduced, the *trowal*: a trough of warm air aloft. The reason for the introduction of this new model is that, in Canada at least, the classic occlusion model very rarely occurs in practice. In other words, there appears to be no significant temperature difference between the cold air in advance of a system after occlusion, and the cold air to its rear. The trowal is a distinct feature, a type of upper front, that moves with the upper flow and is accompanied by (a) a positive pressure-tendency discontinuity at the ground; (b) a change from prefrontal precipitation to air-mass showers; and (c) a rapid clearance of middle and high clouds. No significant temperature change occurs at ground level during the passage of a trowal.

The position of the trowal in the common three-front (i.e., four air mass) cyclone is illustrated in Figure 6.22,F. The classic occlusion, very rare indeed in Canada, is a two-air-mass phenomena. To account for the weather found in *actual* occlusions in Canada by surface observations, three air masses are necessary. And finally, to take into account the upper-air observations, four air masses must be distinguished. These air masses are distinguished in general on the basis of their wet-bulb potential temperatures; the exact frontal position is found from the wet-bulb lapse-rate on the tephigram. The air masses recognized over North America by the Canadians are tropical maritime (mT), polar maritime (mP), arctic maritime (mA), and arctic continental (cA), separated by the Maritime Front (M-front) between mP and mA, the Polar Front (P-front) between mT and mP, and the Arctic Front (A-front) between mA and cA. Typical values for wet-bulb potential temperature and tropopause level are given in Table 6.6.

A new type of synoptic chart, the *frontal-surface contour chart*, has been found to be very useful in following the evolution of fronts. Figure 6.23,A,B, shows a situation for one day in North America; the situation provides a model that fits 95 per cent of all occluded frontal waves crossing the west coast. The maps show an occluded wave on the west coast associated with an upper cold front but *not* with a surface front or an occluded front. The frontal system thus must represent *both* Arctic and Polar Fronts; i.e., the Polar Front stretching across the continent in this example must be a combination of *two* different frontal systems.[89]

Figure 6.21,K, shows models of three-air-mass systems, in which an upper cold front (of the Polar Front system) is approaching the Arctic Front. The upper cold front sets off a wave on the Arctic Front. Three distinct possibilities can (and do) occur then: the frontal systems may amalgamate, or reach an equilibrium position with a close, in-phase, double-frontal structure; the cold fronts may amalgamate, the warm fronts remaining separate; or the warm fronts may coalesce, the cold fronts remaining separate. Only in this last event is the occlusion concept valid, and in this case the Polar Front wave usually outstrips the Arctic Front wave in its speed of movement.

TYPES OF FRONT

Apart from the classic cold and warm fronts, we now have dry fronts, coastal fronts, sea-breeze fronts, and other varieties of geographical fronts. In addition, we have the

TABLE 6.6
Criteria for Air Masses

	Winter				Summer			
Air mass	mT	mP	mA	cA	mT	mP	mA	cA
Wet-bulb potential temperature, in °C	13–18	8–12	2–7	<2	>15	10–14	4–10	4
Tropopause in mb	200	275	350	400	125	250	350	—

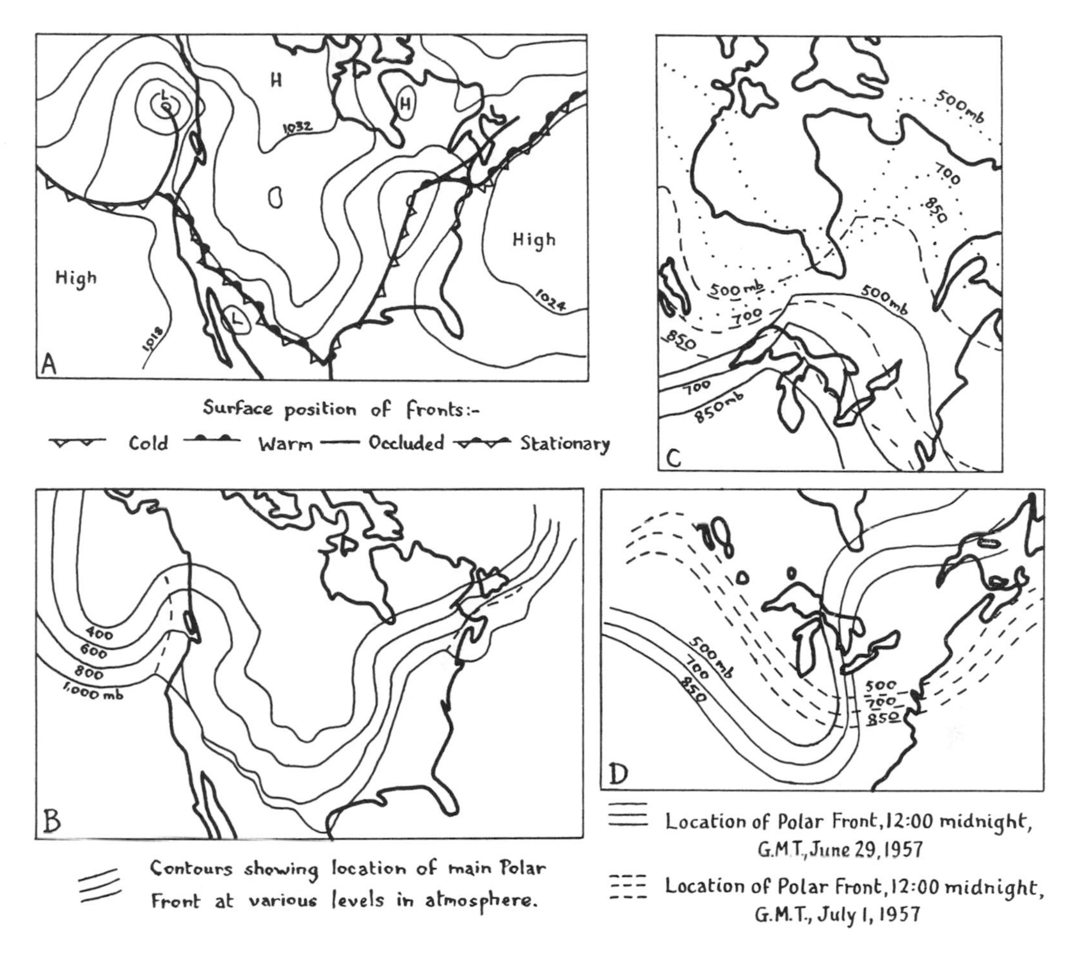

FIGURE 6.23.
Examples of frontal contour charts (after W. L. Godson and J. L. Galloway). A, B, Charts for 3:00 P.M. G.M.T., April 4, 1950; surface chart at A, frontal at B. C, Frontal contour chart for 3:00 P.M. G.M.T., May 9, 1956. D, Frontal contour charts for two separate occasions superimposed.

important upper-air fronts, which produce appreciable weather changes at the surface of the Earth, even though no density discontinuity occurs there.

The upper front that has been best studied so far is the large-scale Polar Front of the middle and upper troposphere. Figure 6.24 illustrates the effect of this front on temperatures at 500 mb on one occasion in the northern hemisphere. The temperature gradient is concentrated into a band of isotherms from $-28°$ to $-30°C$, and marks the southern limit of cold polar air.[90] Clearly such a front will have profound consequences for the weather over extensive areas in middle latitudes, mainly because of its effect on traveling depressions and anticyclones.

The extremely strong temperature gradient found in the upper Polar Front in the middle layers of the atmosphere can be shown to arise dynamically from the differential subsidence of air on the cold side of the jetstream, which leads to the tilting of the isentropic surfaces (see Figure 6.25,C).[91] Thus the large-scale tropospheric Polar Front is a dynamic feature of the atmosphere, the direct consequence of the existence of the jetstream. It owes nothing to the geographical differentiation of the Earth's surface, a fact that sets it apart from all the medium-scale and small-scale types of front.

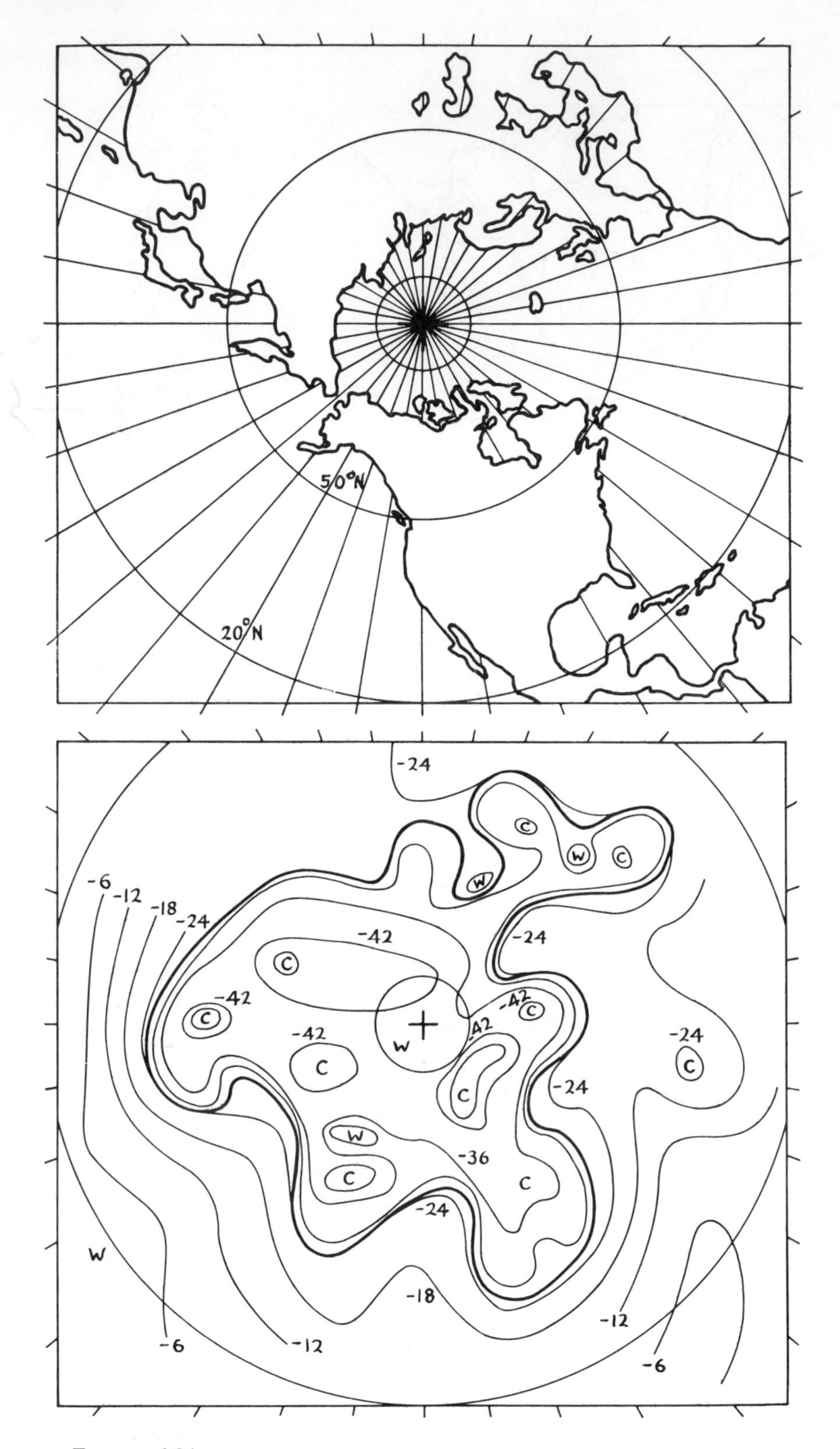

FIGURE 6.24.
Polar Front jetstream of northern hemisphere at 500 mb on February 6, 1952, at 3:00 A.M. G.M.T. (After D. L. Bradbury and E. Palmén.) Lighter lines indicate isotherms (°C), heavier lines the southern limit of polar air (including frontal zone). *W* indicates a warm pool, *C* a cold pool.

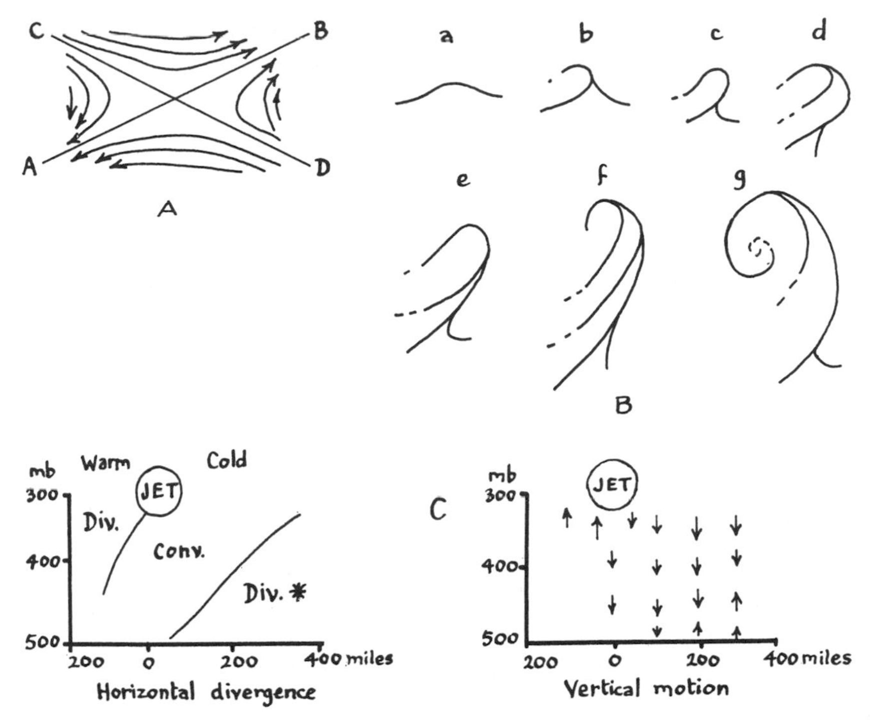

FIGURE 6.25.
Theories of modes of formation of fronts (after H. H. Lamb and J. S. Sawyer). A, Petterssen's deformation field, showing streamlines with inflow (CD) and outflow (AB) axes. B, Refsdal's theory, showing stages in development of a frontal pattern in a major depression. C, Sawyer's theory of the formation of midtroposphere fronts; Distribution of horizontal divergence and vertical motion on the cold side of an intensifying jetstream.

Dynamic theory shows that the jetstream, with strong wind shear on its cold side, flowing from regions of relatively low speed to regions of relatively high speed, must develop a zone of general subsidence on its cold side that is accompanied by an intensification of the horizontal and vertical gradients of potential temperature. The latter, of course, form the 500-mb frontal zone. The same result occurs if the jet flows into a slow-moving upper trough, but in this case the frontal zone is built up at a smaller horizontal distance from the jet axis. The theory also applies in reverse, so that frontolysis occurs as the air particles move into weaker sections of the jet or into the upper ridges.

The very sharp and well-defined nature of the Polar Front zone in the middle atmosphere is probably due to the intensification and dynamic tilting of preexisting boundaries of lapse-rate; very occasionally, the tropopause may do likewise, and subsidence may then be strong enough to bring stratospheric air into the frontal zone.

In general, then, the Polar Front zone of the middle and upper troposphere is intimately linked with the jetstream. The frontogenetic processes in the upper troposphere only operate when the jet is intensifying or becoming more cyclonically curved along its length (see Appendix 6.11). The Polar Front probably never completely disappears, but it may exist in a less intense form, and its characteristic structure may be carried forward into weaker sections of the jet. A permanent Polar Front zone *does* completely encircle the Earth, as envisaged in the original Bjerknes' Polar Front theory, *but* this zone is actually in the upper atmosphere, not at the ground as the original theory postulated. There is no justification for attempting to make all the ground observations fit in with the concept of a surface Polar Front that can be traced completely around the world, and along which all the depressions of middle latitudes are believed to originate.

The notion of upper fronts was not accepted by meteorologists with enthusiasm. Because of their years of practical experience with charts, in which the effects of the differing geographical conditions at the Earth's surface were used to conjure fronts out of apparently homogeneous air masses, the idea that fronts on a very extensive scale could apparently still exist in the atmosphere, whatever the nature of the Earth's surface, was not easy to accept. As late at 1954, British meteorologists were still skeptical of the existence of upper fronts, although Berggren claimed that stratospheric fronts occur on the average once a month in the vicinity of the British Isles.[92]

Berggren produced a model in 1953 that distinguished fronts in the upper troposphere and lower stratosphere. In this model, both tropical and polar air are regarded as high-reaching air masses separated by a Polar Front believed to reach well into the stratosphere. Where the Polar Front is a warm front in the troposphere, it is a cold front in the stratosphere, and vice versa. The temperature difference between tropical and polar air in the model increases with height up to 400 mb; the polar air is isothermal above the tropopause; both air masses have the same temperature at 250 mb, while above this level the polar air is warmer than the tropical air. In a vertical cross section, the tropical air invades the polar air as a wedge, with its axis at 9 km. The Polar Front is vertical at the level of zero temperature difference between the tropical and polar air, and at this level the horizontal wind shear across the front reaches its maximum values of 40 to 50 m per sec per 100 km. Hence the Berggren model considers the tropopause to be a phenomenon localized within one air mass, and not the boundary between different air masses, as was assumed to be logical in many theories about the tropopause.

Upper fronts cannot always be directly linked with surface weather changes, but that does not mean that the meteorologist or geographer interested only in weather at the surface of the Earth can ignore them. In the first place, such direct links may well be found in the future. Second, and more importantly, both weather and upper fronts are phenomena existing in the atmosphere, and to understand either we must understand the atmosphere. If we don't completely understand upper fronts, our understanding of the atmosphere is incomplete. Knowledge of what fronts really are and what their function is in the atmosphere can only come from study of upper fronts, for only in the upper atmosphere are we relatively free from the immensely complicating influence of the varied geography of the Earth's surface. At the Earth's surface, fronts change so rapidly that it is almost impossible to explain them completely. We may be able to do so in a general qualitative sense, but such "explanation" is of little or no use in the practical world. For example, a front passing through Dunstable as a well-marked cold front became a warm front over the North Sea, as it approached southern Scandinavia, because the air to the west of the warm front rapidly subsided, becoming "warm air," in the frontal model, then picked up moisture from the sea.[93] But why did this frontal transformation occur when and where it did? To answer this we have to investigate the mechanics of the subsidence, and soon we have to delve into large-scale atmospheric processes, which brings us to the dynamic features of weather, and so back to upper fronts.

The discovery of upper fronts and their dynamic nature was a very fortunate event for synoptic meteorology, for if it had not been discovered that "some" fronts were dynamic, real features of the atmosphere, then in the not too distant future, fronts would gradually have disappeared from the synoptic chart as the synoptic analyst became more and more versed in meteorological theory. This would mean that historical geographers of a meteorological bent would have weather maps covering three-quarters of a century—and no more—with which to describe the weather experienced over the world, and its importance in our lives, forming a sort of "Domesday Book" of twentieth-century weather. No doubt such a project would make interesting reading, and to see the events of human history accounted for in terms of fronts would be a welcome change, even if not everybody would go so far as to follow Petersen, who ascribed many later events in American history to the passage of a cold front on December 31, 1840, which brought about an attack of, literally, "cold feet" in Abraham Lincoln.[94] Before going ahead with investigations in this manner, it is just as well to convince oneself that fronts actually occur in nature, and are not just conventional lines on a chart.

The characteristic feature of the classic *cold-front* model, the wedge of advancing cold air with an upsliding of warm air above the frontal surface, does not necessarily occur in the actual atmosphere. For example, typical cold fronts encountered in south-eastern England show no evidence of an upslide of warm air above a wedge of cold air; sometimes even the wedge of cold air is absent. In many types of cold front, the main prefrontal precipitation area is often completely unrelated to the warm air moving up over the cold wedge. The drop in temperature behind the cold front, which is characteristic of the classic model, is not always observed in practice. It is not unusual for the passage of a weak surface cold front to be followed several hours later by the drop in temperature representing the contrast between the air masses on either side of the front. Such an occurrence is typical of situations where the horizontal component of wind at right angles to the front increases with height; this permits the development of a very marked invasion of cold air near the center of the depression, in which case the classic model of a wedge-shaped mass of cold air is completely unrealistic.[95]

The sequence of cloud development in a cold front is illustrated by a front that moved northeastward over Arkansas in a winter cyclone. The first cloud to form was cirrus, which thickened to form altostratus. The latter then lowered to merge with altocumulus, which appeared as the cirrus thickened. The cloud ceiling then rapidly lowered to form nimbostratus. As the cyclone developed, the highest cirrus moved on ahead of the surface position of the cold front, and the altocumulus layer became very extensive. A lower layer of cloud also formed at the frontal surface near the ground, and this later extended over the occlusion area and grew to form a medium-level cloud. Clearly, such a picture of cloud development does not fit in with the classic model of a cold front.[96]

Two distinct types of cold front have been recognized over the British Isles, the *katafront* and the *anafront* (Figure 6.21,E,F).[97] The katafront is characterized by descending motion in the warm air (mean vertical velocity, 2.5 cm per sec, forming a subsidence inversion). There is a slight backing of wind with height, and the wind component at right angles to the front increases considerably with height. The upper wind direction is inclined at a large angle to the front (40° average), and the mean slope of the frontal surface above the friction layer is 1 in 300. There is usually fairly rapid occlusion of the pressure trough at the front, as is reflected in the weather usually associated with katafronts: a rapid clearance of cloud, followed by fine weather. At the passage of a katafront, there is only a gradual and very slight temperature change, but often a considerable, very sharp decrease in relative humidity. A gradual wind veer occurs, but little change in windspeed. The precipitation is usually slight, even nil, but latent or convective instability is often present at the nose of a katafront, where strong upcurrents occur.

The anafront is characterized by ascending motion in the warm air (mean vertical velocity, 10 cm per sec), and, unlike the katafront, it gives no sharp discontinuity in an upper-air sounding plotted on a tephigram. The wind rapidly backs with height, and the component at right angles to the front decreases with height, while the upper wind direction is usually inclined at a small angle (15°) to the line of the front. The changes in wind and temperature are much more abrupt than with the katafront, and much more precipitation falls. There is a large and sudden temperature fall, while the relative humidity remains at a high level, with little change. A sharp wind veer usually occurs, followed by a marked decrease in windspeed; heavy rain falls during the passage of the front, and steady rain behind it. The clouds slowly clear behind the anafront, with the cloud types resembling a warm front in reverse. The slope of the frontal surface is much greater than with the katafront, averaging 1 in 70. Delays in the clearance of clouds behind cold fronts in Britain are usually associated with relaxing troughs rather than with extending ones.[98] Generally, the anafront is normally the initial state of a cold front in a warm-sector frontal depression, and develops into a katafront as the depression becomes more occluded. Despite this sequence of formation, katafronts are more common over the British Isles than anafronts, especially during the summer.

On July 6, 1953, a remarkable katafront crossed England, in which warm, very dry air moved in at medium levels ahead of the surface front. The cold front was very well-defined at ground level (see Figure 6.21,I), but despite this the warm dry air aloft moved forward independently of the shallow cold wedge beneath, probably under the control of an intensifying 300-mb trough some 30° of longitude to the west.[99]

A very unusual cold front was discovered by radar observations over Texas on December 10, 1957. A distinct, although slight, pressure jump was found to occur nearly coincident with the radar-echo line. The front had no cloud or precipitation whatsoever associated with it; many "dry" fronts of this type have been found in northern Texas and the Great Plains. With no cloud or precipitation particles to reflect the radar pulses, it was very puzzling exactly how the front could be picked up on a radar screen at all. The explanation apparently is that the radar echoes were produced by microwave scattering by the gradient of the index of refraction.[100] In other words, the mixing of air-masses of different temperature and moisture characteristics, in the frontal zone, resulted in bubbles or filaments of air of slightly different refractive index becoming embedded in the general environment with a "normal" refractive index. Such a variation in refractive index necessary to give a radar echo means that the movement of the air in the vicinity of the cold front must have been very complex indeed. Figure 6.26,K, shows the way in which large amounts of air of contrasting refractive properties could be rapidly and effectively mixed. At a is shown the typical velocity profile to be expected in a tongue of advancing cold air, under the influence of surface friction and viscous drag alone. At b, in the transition zone, air must be moving away from the advancing nose at the top and bottom of the cold air tongue, and toward the nose in the intermediate layer. The resulting pattern of

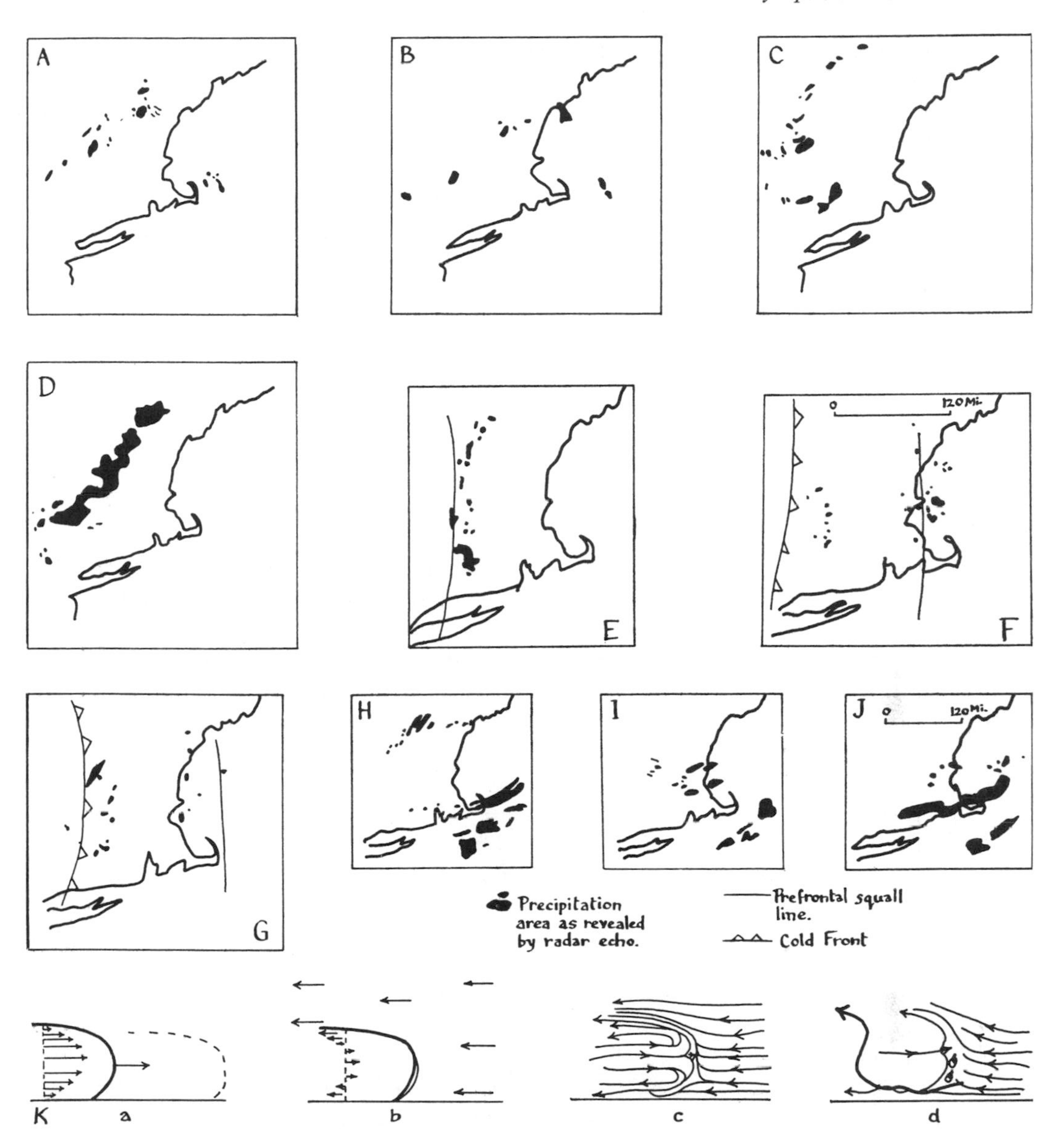

FIGURE 6.26.
Cold front precipitation patterns (after J. M. Austin, R. H. Blackmer, M. G. H. Ligda, and S. G. Bigler). New England precipitation areas from radar observations. A–D, June 17, 1952. E–G, June 2, 1954. H–J, August 24, 1949. A, B, Prefrontal precipitation. C, D, Cold frontal precipitation. E-G, Cold frontal and prefrontal precipitation close together. H-J, Precipitation areas associated with a slow-moving cold front. K, Development of a dry front.

streamlines is shown in c, which indicates that two circulation cells should form in the advancing tongue, hence some of the warm air must be overrun by the latter, and mixing will occur, as at d.

Cold fronts are generally associated with showers, falling from cumuliform cloud. Only since the advent of radar has it been possible to examine the geographical distribution of showers associated with specific cold fronts. Radar pictures of summertime cold fronts in New England show that the precipitation occurs not only in the form of showers, but also in bands and in extensive masses (see Figure 6.26). Most of the frontal precipitation was in the form of bands; most of the postfrontal precipitation consisted of showers; and the prefrontal precipitation was in the form of bands or showers, but with a few precipitation masses also. No correlation was found between the general synoptic pattern and precipitation type with these American cold fronts. The frontal precipitation pattern was continually changing: bands of rainfall changed into isolated showers and vice versa within an hour. Even in frontal situations, local influences, such as low-level fields of divergence or convergence produced by local heating and cooling, were more important than synoptic considerations in deciding the type of cloud and precipitation likely to be associated with a particular cold front.[101] The geostrophic wind appears to be totally inadequate as an index to specify the small-scale topographic and land-sea influence important in deciding such developments. This obviously means that each individual cold front must be treated on its own merits —a conclusion that somewhat weakens the validity of a cold-front "model" applicable to all occasions.

Cold fronts can and do arise in the middle of apparently homogeneous air masses, far from low-pressure areas and the "normal" location for a cold front. These "free-lance" cold fronts can be extremely persistent. For example, the southward journey of the whaling ship "Balaena" in 1946 kept pace with the cloud system associated with such a cold front. The front, with its associated clouds, persisted from Finisterra as far south as 8.5°N, which journey took nine days. Although the front became quiescent for a time in the high-pressure belt near the Canary Islands, it later continued its equatorward progress. Over the oceans, at least, fully developed frontal weather in the intertropical convergence zone may be caused by the arrival of cold fronts from higher latitudes in a still-recognizable form. Cold fronts are also known to advance northward right across the equator near the east coast of Africa, and also move northward through tropical regions in the vicinity of Ceylon. One cold front penetrated from the southern hemisphere middle-latitude westerlies, across the equator, and as far as 5°N, bringing heavy rain to southern Venezuela.[102]

The existence of these isolated cold fronts within the tropics, embedded in the trade winds, means that some cause other than the undercutting of warm air by cold air, as envisaged in the Polar Front theory, must be invoked to explain them. Lamb's suggestion is that each time a subtropical high-pressure cell is rejuvenated by the arrival of a former polar anticyclone from middle latitudes, a cold front will be pushed toward the equator as the trade-wind airstream acquires a new impulse of momentum. This suggestion stems from Flohn's idea that the trade winds are driven toward the intertropical convergence zone by the subtropical anticyclones, so that the intertropical convergence zone is the indirect result of whatever features behind the general circumstances result in the development of the subtropical anticyclones. The observation of the movement of a cold front from middle latitudes, through the subtropical anticyclone, and right into the intertropical zone, means that some of our simple ideas about the general circulation must be reappraised, since the old concept that the belt of equatorial bad weather and rains (i.e., the intertropical convergence zone) is associated with the zone of maximum solar heating clearly cannot apply if Flohn and Lamb are correct.

Less variation has been noted in *warm fronts* than in cold fronts. A debatable point concerns the speed of movement of warm fronts. Cold fronts generally move with the speed of the geostrophic wind in the cold air immediately behind the front, and one can see intuitively why this would work. Warm fronts, however, are found in practice to move more slowly than associated cold fronts in equal geostrophic winds. Generally the warm front is found to move with two-thirds the speed of the geostrophic wind component at right angles to, and measured at, the front. There are slight variations from this figure, depending on geographical location, so that it is perhaps more correct to say that warm fronts move with 50 to 80 per cent of the speed of the geostrophic wind. For the British Isles and North America, the two-thirds rule works well in practice, but for the North Atlantic five-sixths is more accurate (see Appendix 6.12). The discrepancy is important because it can result in a difference in forecast position of more than 100 miles in 24 hours.

Why do warm fronts not move with the speed of the geostrophic wind through them? Putting the question differently, why, if cold fronts move as though embedded in the geostrophic wind field, do not warm fronts behave similarly? If the frontal surface is a substantial surface of discontinuity, with the cold air below it moving horizontally, then it would be expected that the speed of the warm front would equal the component of the actual windspeed that is at right angles to the front above the friction layer (say, at 900 mb), on the cold side of the front. Dynamic reasoning, however, shows that this is not strictly true (see Appendix 6.13). From empirical considerations, if u_f is the speed of movement of the warm front, u_g is the geostrophic wind component at right angles to the front as measured at the front, and u_g' is the geostrophic wind component at right angles to the front as measured for warm fronts moving across the British Isles, 75 miles ahead of the front, then,

$$u_f = 0.7\, u_g' + 3.0, \text{ or } u_f = 0.6\, u_g + 2.4,$$

and, for warm fronts moving across the eastern North Atlantic,

$$u_f = 0.91\, u_g' + 2.6, \text{ or } u_f = 0.75\, u_g + 3.7.$$

These formulae give an accurate estimation of the speed of movement, based on dynamic theory, but the two-thirds rule often gives an approximation that is close enough for practical purposes.

Although the relatively slower movement of warm fronts can be explained dynamically by using the 900-mb wind, there is still the movement of the front at sea level to be explained. Obviously the latter must be connected with friction and turbulence near the ground, since the 900-mb level is above this atmospheric ground layer. A combination of friction and turbulence thus appears to be the most likely cause of the observed speed of movement of warm fronts, and of their deviation from the two-thirds rule, particularly over the sea. A strong thermal wind exists parallel to and just ahead of a warm front, so that turbulence originating at ground level in this region could influence the wind up to a greater height than normal. Owing to the existence of the nonfrontal thermal wind (i.e., the thermal wind always present in the atmosphere whether or not a front exists), a strong forward wind shear also exists a few thousand feet above the surface front, and this, combined with the rapid surface heating of the shallow cold air just ahead of the front, may easily affect the rate of movement of the warm front, especially over the sea.

Double warm fronts occasionally occur, but they are not as frequent as double cold fronts. A tricky problem in synoptic study is concerned with the likelihood of the development of a small wave on a warm front. Such a wave is a comparatively insignificant feature on the synoptic chart, but it is associated with a narrow, rapidly moving belt of very low cloud and heavy precipitation that can penetrate into a high-pressure area. Development of a wave usually slows down the rate of movement of the warm front for a time. On the average, only four waves form on warm fronts in the vicinity of the British Isles each year. They normally develop when the front is near the right-entrance region of a jetstream at 300 mb and to the east of a 300-mb ridge.[103]

Many frontal types other than warm fronts, cold fronts, and occluded fronts have been recognized. Some of these soon outlived their usefulness, and have been dropped from routine synoptic analysis, but others have persisted, for example: fronts tied to specific geographical locations, such as the Gulf Stream (or Gulf Coast) front and the east coast "backdoor" front in the United States; fronts, such as the sea-breeze front, that are connected generally with major geographical differences; and "unusual" fronts, such as dry fronts.

The *Gulf Stream front* (see Figure 6.27,A–E) has been suggested as the cause of the very intense and persistent wintertime stratus along the lower Atlantic coast of the United States.[104] It has an interesting human connection in that it is caused by the very contrast that leads many residents of the northeastern states to fly southward to escape the winter weather, with the result that the peaks of the worst weather and the heaviest air traffic in Florida coincide.

Persistent winter stratus has long been a rather enigmatic feature of the east coast of the United States. Its mode of formation is as follows. The cold high-pressure center that is usual in the eastern states at this time of year causes a flow of cold air down the Atlantic coastal plain. A pronounced wedge develops east of the Appalachian Mountains parallel to the coast, but this high-pressure ridge only occurs north and west of the Gulf Stream, because of the heating of the air overlying the latter and the consequent lowering of pressure. Thus the coastal isobars trend southwest and northeast to parallel the coastline, and adjacent parts of polar airflow encounter vastly different conditions. The air over the central axis of the Gulf Stream is much heated, then carried aloft and spread westward by easterly winds that occur often at 1,000 feet or just below. Thus the air aloft is 5 to 15°C warmer than the surface air in Florida, and a warm frontal surface develops.

The entire lower layer of cold air has temperatures well below those of the precipitation drops falling from the warm front, and evaporation of this falling rain could be expected to build up an extensive stratus layer right down to the ground. However, not much precipitation falls from the Gulf Stream front, and so the evaporation process is not of major importance in the formation of the stratus. The main factor involved is the trajectory of the air that leaves the coast in the neighborhood of Charleston: this is from the north-northeast at the surface. The cold air picks up much heat and moisture during its short passage over the relatively warm water surface; it becomes unstable, and convection is set up. The warm frontal surface that forms over the Gulf Stream offshore of the part of the Florida coast that is concave eastward, provides a cap for the convection, so that the convection cloud rapidly spreads out at low levels, forming an extensive sheet of stratus that blankets the coast from Charleston to St. Augustine, Florida. Ceilings as low as 200 to 800 feet are common with this stratus, which soon spreads as far as Tampa, Florida, and Atlanta, Georgia, and may persist for a week or more.

The *east coast "backdoor" front* is one that from time to time moves in from the Atlantic across the coastlands of the eastern United States, even though the general westerly atmospheric circulation suggests that such a frontal movement is very unlikely. Backdoor fronts are usually associated with the stagnation and intensification of an anticyclone to the northeast. The high-pressure area often reorients itself from a nearly circular cell to one with its major axis from east to west as the front forms. The first heat wave of late spring in the mid-Atlantic states is frequently broken by the passage of a backdoor cold front, moving from an easterly or northerly direction instead of from the usual westerly direction. A good example was during May 16 to 20, 1951 (see Figure 6.27,F–J).[105] Such backdoor cold fronts usually occur over the eastern slopes of the Appalachians, or the eastern slopes of the Rockies. The cold air, moving from the east, banks up against the mountains, and is forced southward or southwestward at a wide angle across the isobars, so that it penetrates much further south than it would if the orographic barriers were absent.

That the onset of a sea breeze has frontal characteristics has led to the recognition of the *sea-breeze front* which forms a very narrow transitional zone between the incoming marine air of the sea breeze, and the air inland. This front appears to show a friction head similar to that exhibited by a density surge in the flow of a stratified liquid. It moves inland in a series of pulses, so that two or more parallel belts may exist simultaneously, each with frontal characteristics. A repeating mechanism probably exists in airstreams flowing parallel to the coast, so that "early" and "late" sea-breeze fronts develop. After the wind change that occurs at the passage inland of the first front, the low-level winds near the coast gradually revert to a direction parallel to the coastline, and the synoptic situation is then favorable for a second front.[106]

Figure 6.28 shows some examples of sea-breeze fronts. These fronts are favored by deep convection, the main factor in their development being not so much the contrast between land and sea-surface temperatures, as the difference between the mean temperatures of the lowest kilometer or so of the sea-breeze air and the air warmed by inland convection. The sea breeze is, in effect, a developing front, which may be quite a significant mesoscale feature inland in moderate to fresh convectively unstable airstreams that parallel the coast. In light winds, the front may thus be more important inland than at coastal stations. Across the front, important wind and humidity changes occur, but little temperature variation. The thin, steeply inclined wafer of ascending sea air in the front is marked by wisps of cloud that form a tattered cloud "curtain." From the south coast of England, the sea-breeze front moves well inland, and sometimes reaches the Thames estuary as a "sea breeze" from a landward direction.

A *dry front* is one that has no precipitation associated with it, and little or no cloud. One over the eastern slope of the Rockies on June 11, 1956, was observed from aircraft. The frontal surface was tilted eastward at an inclination of 1 in 30—much steeper than the frontal surface of a warm front—and separated very dry air to the west from moist air to the east. Convective activity was very pronounced to the east of the leading edge of the dry air. The wind direction in the moist air was parallel to the front, while that in the dry air to the west was perpendicular to the front, which moved at approximately the speed of the wind in the dry air. An increase in dew-point temperature of 6°C occurred on entering the moist air, and the maximum dew-point temperature occurred at the same times and places as the minimum dry-bulb temperature. The structure of the dry front indicated that the moist air that was replaced by the pushing dry-air mass

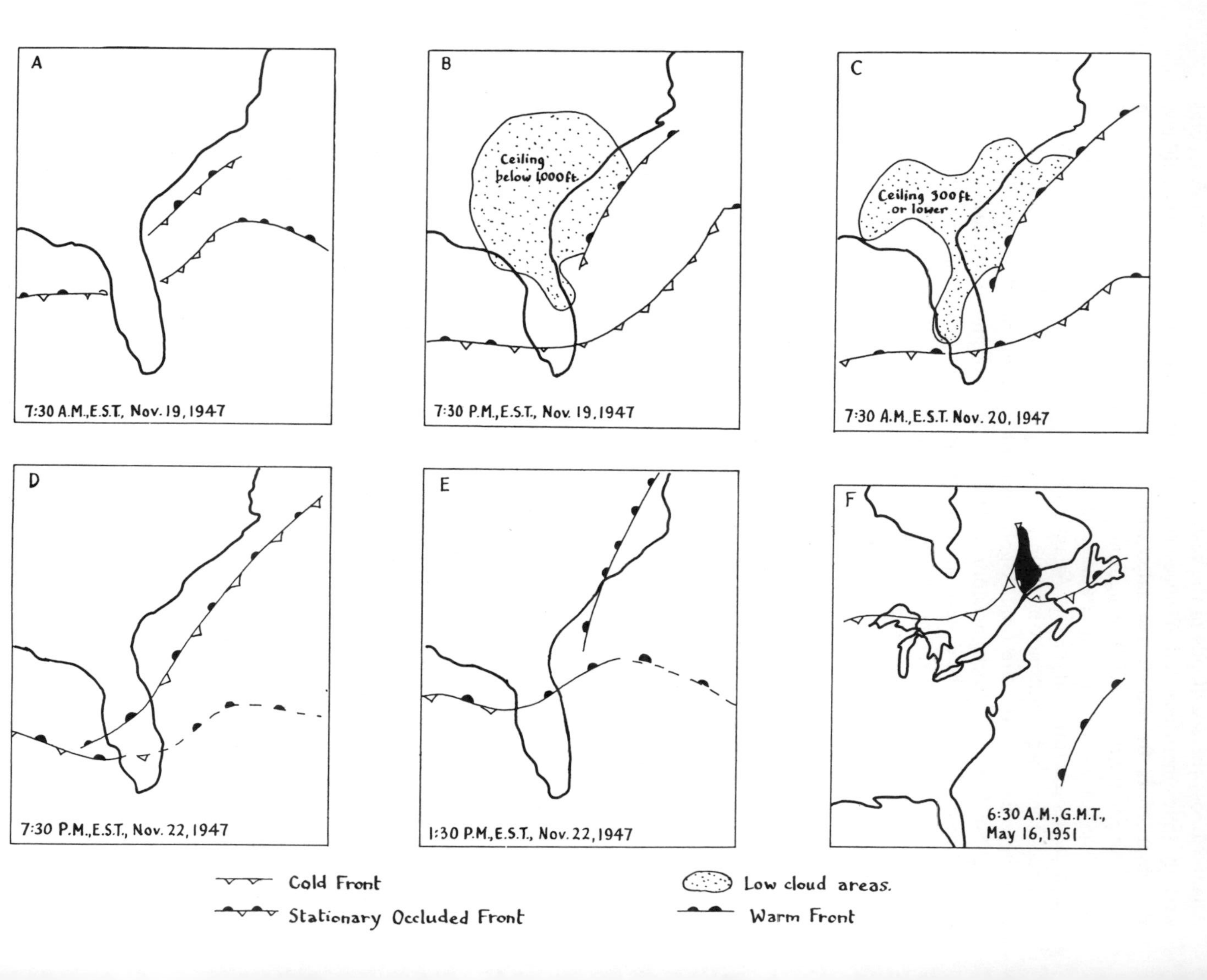
A
7:30 A.M., E.S.T., Nov. 19, 1947
B
Ceiling below 1000 ft.
7:30 P.M., E.S.T., Nov. 19, 1947
C
Ceiling 300 ft. or lower
7:30 A.M., E.S.T. Nov. 20, 1947
D
7:30 P.M., E.S.T., Nov. 22, 1947
E
1:30 P.M., E.S.T., Nov. 22, 1947
F
6:30 A.M., G.M.T., May 16, 1951
Cold Front
Stationary Occluded Front
Low cloud areas.
Warm Front

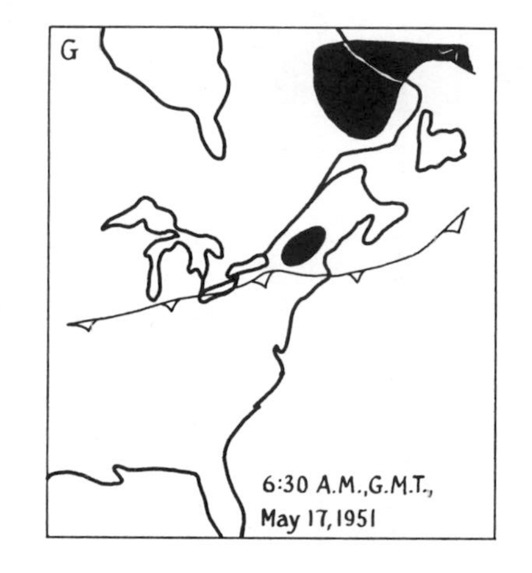

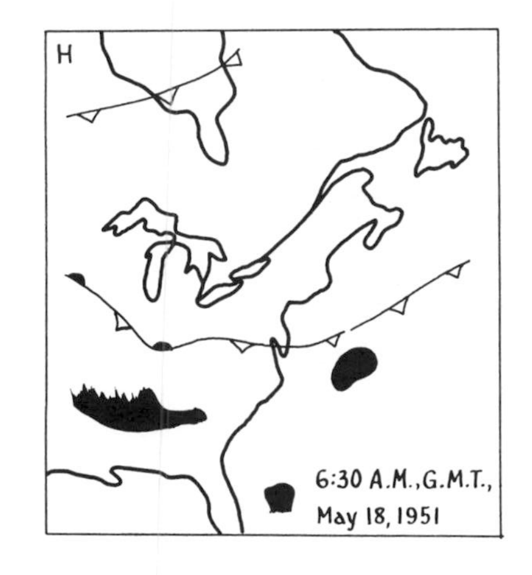

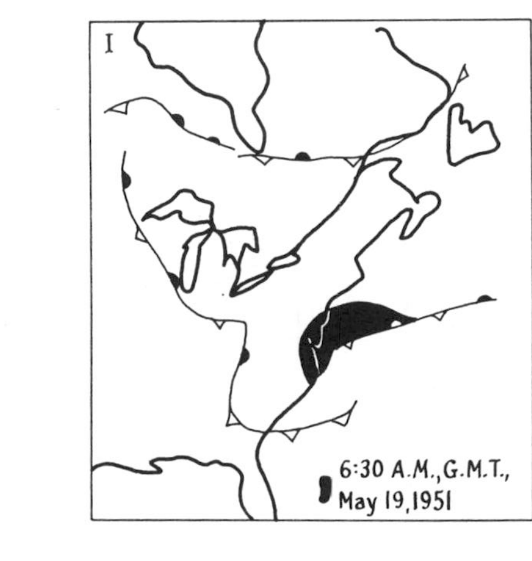

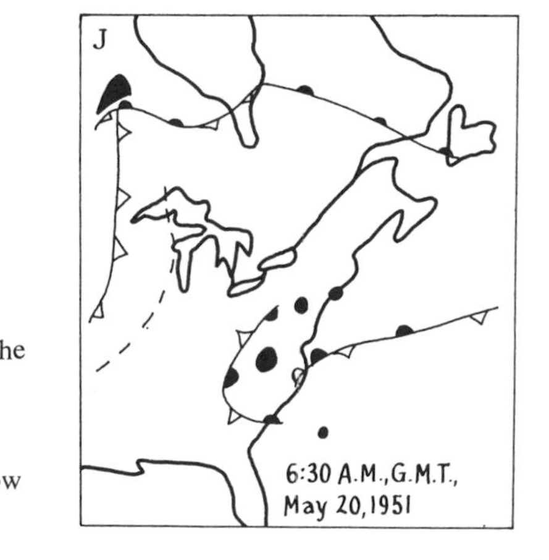

FIGURE 6.27
Some regional peculiarities in front formation in the United States
(after R. B. Carson and J. A. Carr).
A–E, The gulf coast front.
F–J, The East coast backdoor front. All maps show surface situations.

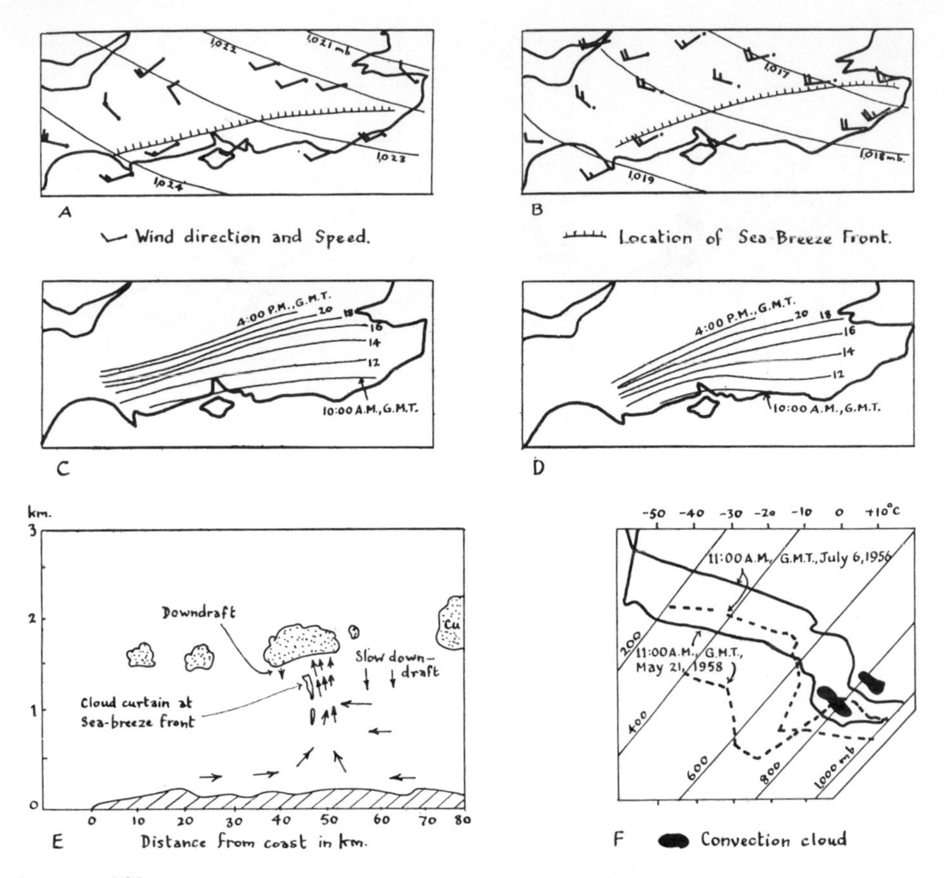

FIGURE 6.28.
The sea-breeze front in southern England. (After C. E. Wallington). A, Surface chart for noon G.M.T., July 6, 1956. B, Surface chart for 3:00 P.M. G.M.T., May 21, 1958. C, Isochrones for sea-breeze front, July 6, 1956. D, Isochrones for sea-breeze front, May 21, 1958. E, Air motion associated with sea-breeze front at 5:00 P.M. G.M.T., July 6, 1956. Arrows show air ascending at more than 2 m per second in a belt 100 to 250 m. wide along the sea-breeze front. F, Tephigrams for Crawley for the two sea breezes.

must move upward, and under certain conditions might move up enough to break the conditional instability inside the moist air east of the dry front.*

Since fronts are generally associated with bad weather, or at least with the less-settled forms of weather, a knowledge of the typical precipitation patterns associated with different types of front should be invaluable in the study of the geographical distribution of rainfall. Early studies of frontal weather concentrated on the weather resulting from cooling and condensation processes during the upslope motion of warm air over a wedge of cold air. These upslope processes as represented in the classical warm-front model are deceptively simple, and detailed investigation of upslope weather in actual situations show that it is very complex.

It is obvious that the location of a front in a certain area on a specific occasion

* R. Fujita, *BAMS*, 39(1958), 574. For an example of a dry front, see D. R. Grant, *MM*, 96(1967) 33, on Farnborough, Hants.

will often be critical in deciding the short-period variations in precipitation in that area. But frontal occurrences may also be very important in explaining the long-period (i.e., the climatic) variation in the distribution of rainfall. For example, 60 per cent of the variance of the average rainfall of England can be explained by only two variables, pressure and frontal contrast, the latter being represented by the change in mean total thickness at 250 miles ahead of the front. In other words, despite the undoubted close resemblance between the map of average rainfall and the orographic features of England, nongeographical factors are actually *more* important than orography in determining the geographical distribution of rainfall.[107]

Although fronts are so often accompanied by precipitation, not all heavy and organized precipitation belts are associated with fronts; some precipitation that has all the geographical characteristics of the frontal variety occurs far from any front. A good example was on May 12–13, 1957, when heavy convective rain, of the type usually associated with cold fronts, fell over Texas, although no fronts were in existence and the surface isobars were anticyclonically curved. In this case, wind shear between 500 mb and the ground surface was caused by vorticity changes in the lowest layer of the atmosphere. These vorticity changes led to low-level convergence, the trajectories of the air parcels converging over north-central Texas in the lowest air layer, although the isobars showed anticyclonic curvature.[108]

The geographical pattern of the precipitation areas associated with fronts is often not at all like the broad zones of continuous rainfall one is led to expect by the classic frontal models. Although observations from regular synoptic stations may suggest that these broad belts actually occur, the closer networks provided by rainfall stations show that, instead of broad belts, there are several tongues of rainfall separated by dry zones. These tongues are not due to local topographic effects, and maintain continuity over relatively long periods of time. They are often mesoscale features, and are usually associated with waves forming on warm fronts in a confluent thermal ridge. In such a situation, cyclonic development occurs close to the warm front, some distance from the depression centre. The main criterion for the development of such a wave on a warm front is the stability, both static and hydrodynamic. If the air is hydrodynamically stable, then any vertical motion produced by the broad-scale fields of temperature and pressure will be rapidly damped out.[109]

Hydrodynamic stability depends on both temperature and wind, while static stability depends only on temperature. If the airstream is such that the windspeed increases with height, then, even if the air is statically stable, an air parcel displaced upward will not immediately return to its original level, because the displaced air will have a lower windspeed than that required for geostrophic flow at the new level. Hence the air parcel will be diverted toward lower pressure. The windspeed increases with height because temperature decreases toward the region of lower pressure, hence a transverse displacement of the parcel takes it into a progressively colder environment. Therefore the effective hydrodynamic stability is less than the static stability in such a situation. If the upward motion, after damping by the hydrodynamic stability, is still sufficient to produce saturation, then the saturation will immediately reduce stability, so that the upward motion receives a sudden impetus. In other words, in certain airstreams, the effective of *increasing* stability may actually result in *increased* precipitation. A region of cyclonic development may contain several separate cells of rising air; the largest of these may distort the frontal pattern into small waves, and the most intense cells may generate tongues of frontal precipitation. In general, the tongue-like form of the precipitation areas associated with warm fronts is due to the fact that the slowly rising air of the frontal zone is often augmented in places of instability, which may either be local hydrodynamic instability or be due to the existence of unstable frontal waves of length 100 to 300 miles.

Although, with rapidly moving and deepening cyclones, the movement of the isohyetal patterns may be closely correlated with the movements of the associated fronts,[110] the precipitation areas associated with different types of front do not usually move with the same speed as the front, except for warm fronts. The precipitation area associated with a cold front moves with the speed of the mean 600-mb wind perpendicular to the front. Precipitation areas associated with open frontal waves move with half the speed of the 500-mb airflow, and in the same direction; while extensive rain areas not associated with fronts usually move with the 700-mb flow.

It has been usual for many years to divide precipitation into two general types, frontal and nonfrontal, both of which may be intensified locally by orographic uplift. Frontal precipitation is usually regarded as ordered and capable of systematic analysis, but nonfrontal precipitation, particularly shower activity, is generally supposed to be too sporadic for accurate explanation and prediction to be possible. However, there is a definite geographical pattern to showers. Winter showers over land are not formed in situ by local convection, because insolation is too weak, but they may be advected in from the sea. If the wind from the sea is sufficiently strong, the showers may be maintained all through the winter night. At Belfast, for example, a westerly gradient wind of 50 knots or a northwesterly gradient wind of 25 to 30 knots will keep the showers up all night, and a mere 20-knot wind from 330° will maintain snow showers all night.

Showers are not common over the sea in spring and summer, because the water surface is relatively cold. Showers may be active inland, but absent on windward coasts in what is usually a showery airstream. Showers in coastal areas can also be removed by the sea breeze in the warmer parts of the year. Northwesterly airstreams in the western parts of Great Britain are usually showery; at Aldergrove, they bring showers early in the day in summer, but the showers disappear for the rest of the day, probably because the condensation level is lower in the morning than in the afternoon because of the evaporation of dew. In the afternoon, the drier ambient air is entrained in the convection currents, and the resulting upcurrents are too dry for condensation to result.

Various criteria can be used to decide whether showers will fall, apart from indications of local convection made, for example, with a tephigram. In maritime districts in Britain, the lower the value of the total thickness, the greater the chance of showers. For snow showers over hill regions, the height of the wet-bulb freezing level in the cold air is a useful indication of the lower limit of snow. The drier and colder the air, the lower down will snow fall.

Local topographical influences are very important in the production of showers. Mildenhall and Lerwick are both exposed to northerly airstreams, yet winter showers are ten times more frequent at Lerwick than at Mildenhall, presumably because of differences in their local geography. Rainfall variations of quite large magnitude within short distances are more likely to be due to geographically induced differences in shower incidence than to frontal effects. In January 1958, for example, Kew recorded 49 mm of precipitation, while Golders Green, only eight miles distant, received 69 mm. At Cardington, considerable spatial and temporal variations in the *rate* of rainfall have been noted in an area only two miles square.

Apart from frontal and shower activity, a third contribution to precipitation comes from lee waves. In 1957 in the Belfast area, for example, places to the east of the hills received 40 to 44 inches of rain, those to the west recorded 34 to 38 inches, the prevailing airstream being westerly. The areas in the lee of the hills received more precipitation than those to windward, apparently because of the existence of standing waves. The lee wave formed by the hills was several miles downwind from the hill crest, so that uplift of the airstream and the resulting rainfall was leeward of the hills. Nutts Corner experienced showers with northwesterly airstreams, as did places upwind, yet Aldergrove to the northwest had no precipitation. This may be due to large lee waves produced by the Sperrin Hills, and these waves also probably explain the relatively low annual average rainfall at Aldergrove (34 inches) compared with adjacent areas. In general, it may be shown theoretically that prefrontal precipitation bands may be wave phenomena, associated with gravitational wave oscillations set up originally at a frontal surface. An important feature is that these waves are not sinusoidal in form: each "wave" is, in fact, a complex group of waves itself.[111]

The pattern of precipitation, as presented in elementary textbooks in terms of frontal rainfall belts and "air-mass showers" in the areas between the fronts, appears to be very simple and easy to appreciate. This impression is illusory. The geographical distribution of precipitation is a very complex problem. As an example, a cool, damp period in the eastern United States from July 18 to 25, 1956, may be cited. In the Gulf coast states, fronts are normally very ephemeral in July, but during this period there was a well-defined polar front, bringing heavy precipitation from the Gulf coast right to the Atlantic states. The regions of maximum precipitation amount and greatest precipitation intensity coincided in a narrow band that tapered from its maximum width of 60 miles near Washington, D.C., to its northward and southward limits at Philadelphia, Pa., and Florence, S.C., respectively.[112] Although cursory examination might suggest that this exceptionally heavy rainfall was merely due to an unusual, well-defined front being in the region, in fact the precipitation was due to a combination of at least ten factors.

First of all, there was obviously the active front positioned over the Gulf states, along which stable (i.e., nondeveloping) waves traveled, each leading to local increases in precipitation. Second, a block of high pressure intensified in the east directly ahead of the developing frontal wave, both at the ground and aloft, and provided a definite limit to the movement of these stable waves. The remaining factors are all events taking place in the upper-air, and the surface chart was inadequate to isolate any of them.

A cold low centered near Chicago was responsible for the movement of comparatively very cold air southward at 500 mb; these recurring impulses of cold air assisted in the development of the frontal waves. The developing frontal wave was also associated with an upper trough, which rapidly intensified and sharpened as it approached the high-pressure block building up over New England and along the eastern seaboard. The intensification of the trough created an area of very pronounced cyclonic isobar curvature and convergence. An area of marked cyclonic vorticity also existed at 500 mb, and moved in phase with the surface frontal wave in its northeastward progress from the Gulf coast to Washington, D.C. The concentration of cyclonic vorticity allowed the intensification of the trough to take place, and this in turn assisted in the development of the frontal wave and its increasing instability.

Factors that were of more importance in the local intensification of the precipitation in this case than they are in the general development of the frontal wave, included the following: advection of near-saturated air over the entire region between 500 and 700 mb, the exact cause being uncertain; the movement of a zone of very marked vertical motion across the northern part of the area; and the development of an area of anticyclonic vorticity in the total thickness pattern just west of the region at the beginning of the period of the rains. This area reached its maximum value at 3:00 A.M. G.M.T. on July 20, and during the next 24 hours the thickness pattern showed extremely strong anticyclonic vorticity being advected across the southern part of New England. This anticyclonic thickness area coincided with the region of heaviest precipitation.

This example illustrates the complexities of frontal rainfall. The rainfall pattern indicated in the Polar Front model is only a very general one, and the *actual* precipitation associated with a specific front can only be explained by a detailed study of the charts for that front. It may then appear that such influences as convergence or divergence, orographic or other waves, or areas of cyclonic or anticyclonic vorticity production, are present in critical areas, so that the actual precipitation pattern may be much modified from the model pattern.

Developments in the upper air can be of crucial importance. For example, the 300-mb chart is useful as an indicator of situations in which inactive fronts suddenly develop new precipitation areas. The mechanism requires the movement of an upper trough over an inactive frontal system in the lower troposphere. The resulting combination of high-level divergence and low-level convergence ahead of the trough gives rise to ascending air, with consequent formation of precipitation. The very rapid development of precipitation over England on April 7, 1961, may be attributed to such a cause.[113]

Geographical effects may also be important. The occurrence of thunder along a warm front is an interesting case in point. Thunderstorms in a warm-front situation are extremely rare in Britain but are common in North America and, in particular, in the Persian Gulf area. The storms develop in polar-maritime air (instead of in the more usual tropical-maritime air), which is sufficiently unstable locally to give rise to cumulonimbus if forced to rise over a colder air mass.[114]

DYNAMICS OF FRONTS

The first front was charted in 1918, when Bjerknes introduced the concept into synoptic meteorology, as a hydrodynamic model. In 1928, Bergeron introduced the indirect aerology technique of analysis, by means of which frontal structures (i.e., the appropriate model to apply in a particular situation) could be inferred from the distribution of clouds and precipitation as observed from the ground surface. This indirect analysis remained standard until after the Second World War, with the result that the formation and dissolution of fronts was a problem little considered by the synoptic analyst. Fronts were formed by man, when the analyst thought that the observed cloud and precipitation patterns warranted them, and then disappeared in some unspecified manner by moving into areas with which he was little concerned. But the advent of upper-air analysis on a large scale meant that frontogenesis and frontolysis became important problems to solve. Palmén was able to show by 1951 that frontal zones were well defined at 500 mb, and by 1958 Sawyer was able to demonstrate from dynamics that the Polar Front in the midtroposphere was linked with the existence of the jet-stream.[115] The very strong temperature gradient across the frontal zone at 500 mb arises dynamically, from differential subsidence of the air on the cold side of the jet-stream, which leads to tilting of the isentropic surfaces.

Upper-air analysis has shifted the emphasis from the rapidly varying minor fronts on the daily chart, which often are nothing more than conventional signs synthesizing several weather phenomena, to the major frontal systems of the Earth. These major frontal systems are not easily formed, and once formed they do not easily die away.[116] They usually travel for great distances across the Earth's surface between birth and decay, and are subject to successive frontogenetic or frontolytic influences that alternately strengthen or weaken their activity. They undergo continual distortions and changes of slope, as cyclonic disturbances form and move along them and as various air masses are continually fed into them from either side. In other words, the activity of the major fronts of the Earth changes with time and place, but they do not necessarily lose their identity.

Examples of this change in activity of a major front are as follows. Occlusions moving eastward across northern Europe arrive in China as cold fronts. Other occlusions moving south over the western parts of the U.S.S.R. reach the Eastern Mediterranean as cold fronts, in all seasons except summer. The warm sectors of North Pacific storms become occluded against the Rocky Mountains, and in winter advance eastward of the Rockies as warm occlusions, followed by chinooks. These occlusions become rejuvenated by the deformations in them caused by disturbances in the wind flow induced by the great topographical contrasts. Finally, occlusions moving north through the Bering Strait from the Pacific, or rounding the Polar Basin from the North Atlantic, move southward or southeastward across Canada east of the Rockies as cold fronts. It is noteworthy that fronts usually emerge from transformations, such as those just described, as cold fronts. The major fronts between the chief anticyclonic areas normally advance eastward in successive stages as cold fronts and cold occlusions, because the air masses in their rear are drawn from higher latitudes than the air masses ahead of them.

Although the formation of fronts may be demonstrated physically by means of a heated rotating water-tank model, the process of frontal formation and dissipation is normally described in essentially mathematical terms. The first mathematical proof that adjacent currents of air, at different temperatures and moving with different velocities, may remain in steady motion with a discontinuity surface separating them, was devised by Helmholtz. He was able to show that the angle of incidence of the discontinuity surface to the horizontal depends on the differences between air temperatures and velocities on the opposite sides of the surface. A classic treatment of the slope of a frontal surface was provided by Margules. In general, the slope of a discontinuity surface in steady motion is given by

$$\tan \alpha = \frac{2\omega \sin \phi T(u_1 - u_2)}{g(T_1 - T_2)}$$

where u_1, u_2 are the velocities in the OX direction (see Figure 6.29) in the air masses on opposite sides of the surface; T_1, T_2, are the corresponding temperatures, and T is the mean of T_1 and T_2; and α is the angle of inclination of the discontinuity surface. It is assumed that the motion of the air in each air mass follows the law

$$2\omega u \sin \phi = -\frac{1}{\rho}\frac{\partial p}{\partial y}$$

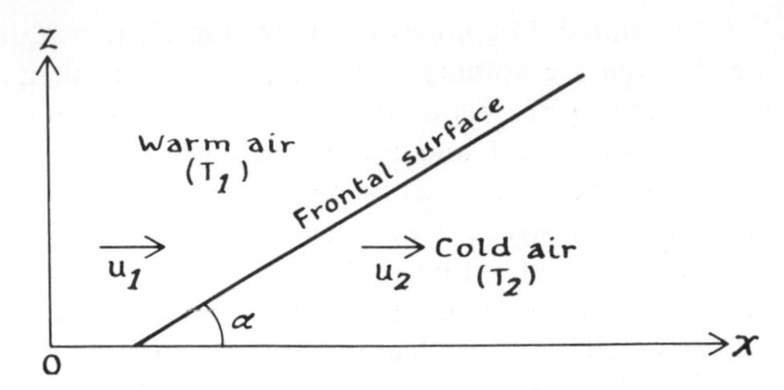

FIGURE 6.29.
Slope of a frontal surface. The *OY* axis is taken to be along the front; *OX*, *OZ*, plane is taken to be perpendicular to the frontal surface.

where the letters have their usual meaning in the equations of motion. The equation for tan α indicates that the slope of the discontinuity surface increases in direct proportion to the difference in wind velocity across it, and in inverse proportion to the difference in temperature across it.[117]

Frontogenesis may be defined as a positive tendency toward the formation of an atmospheric discontinuity or the intensification of an existing discontinuity zone. *Frontolysis* is a negative tendency toward either of these. The theory of frontogenesis which has been standard for many years is that of Petterssen, who envisages fronts as arising in a hydrodynamic *deformation field.* Strictly speaking, the theory requires the existence of an *axis of dilatation*, which acts as a collector of isotherms (see Appendix 6.14).

The circulation in an isobaric col approximates a deformation field. The streamlines in Figure 6.25,A, show that *AB* is an outflow axis and *CD* an inflow axis, so that the isotherms parallel to *AB* in an initially weak thermal gradient will be brought progressively closer together along the axis *AB*, which will thus become a line of frontogenesis. By similar reasoning, *CD* will become a line of frontolysis. The angle between *AB* and *CD* determines whether the effect of the col will be predominantly frontogenetic or predominantly frontolytic. Unfortunately, frontogenetic cols seldom remain long enough in one location for a front to be formed when none existed before, so that the Petterssen theory seems to provide a mechanism for the intensifying of already existing weak fronts and old occlusions, rather than for the production of completely new fronts.

The frontogenetic col is quite effective in certain areas, if the inflow axis lies over belts of the Earth's surface where a steep temperature gradient induces the formation of a rapid transition zone in the overlying air. All these geographical belts show rapid temperature transitions, for example, from a cold land mass to a warm ocean surface in a comparatively short distance, and among them may be noted the following areas: (a) the North Atlantic off the east coasts of North America and Greenland, near the boundaries of the polar ice, the Labrador Current, and the Gulf Stream; (b) the North Pacific off the east coasts of Asia and the Aleutians; (c) the fringe of the permanent ice in the Polar Basin along the northern coasts of Asia and North America in late summer and autumn, replaced in late autumn and winter by a belt trending southeastward through Alberta along the flank of the Northern Rockies and by a northwest-southeast belt near the White Sea and the Northern Urals; (d) the Baltic Sea and the St. Lawrence valley; (e) the northwest African coast, through the Mediterranean and across Turkey to Mongolia and Tibet; (f) a belt around the Antarctic continent and, in winter, the edge of the pack ice; and finally (g) a series of subsidiary frontogenetic zones near the coasts of the warmer southern continents. The importance of these various geographical belts for frontogenesis varies with the seasons. Belts (a) and (b) are strongest in winter; belt (d) is well-developed in spring and autumn; belt (e) in winter and early spring; and the zones included under (g) are important in all seasons.

Large-scale orographic effects may be important frontogenetic influences. For example, cold fronts appearing in troughs in the lee of mountain barriers the size of the Rocky Mountains differ from normal cold fronts in that their horizontal temperature field is that of a sinusoidal thermal ridge. The cold front in the lee trough is a *pseudofront*, i.e., one whose associated surface trough does not exhibit the abrupt low-level temperature discontinuity characteristic of normal cold fronts, although quite a pronounced wind shift may be present. In the pseudofront, the thickness isopleths are almost symmetrically disposed about the surface trough, with maximum air temperatures occurring at the front. The origin of this uncharacteristic type of temperature pattern associated with a cold front is orographic. It may be proved dynamically that the thermal ridge develops when the surface airflow produces large-scale descent of air in the lee of the Rockies, usually when a 500-mb ridge approaches from the west. The thermal ridge intensifies, remaining stationary relative to the mountains, then moves eastward after the passage of the 500-mb ridge, with a Pacific cold front embedded in it.[118]

Conditions are very favorable for frontogenesis if an already existing air-mass boundary, or an old frontal system, moves into one of these specified geographical belts, and there comes under the influence of a frontogenetic col. If a frontogenetic col appears in a homogeneous air mass over one of the geographical belts, then embryonic fronts may form. Thin sheets of altostratus or altocumulus develop, with occasional stratus sheets and slight rain or drizzle, well before there is any sign of discontinuities in surface temperature and humidity. For a full-fledged front to develop, the col must persist in the same position for many hours, even several days. Usually, however, the col moves away to a fresh position before development can go further than the production of an embryonic front.

The importance of Lamb's empirical frontal models, which recognize the existence of front-within-front structures, has been emphasized already. These models show the normal situation in a frontogenetic col as including distinct upper and lower cloud decks with clear sky in between. If the uplifted air mass is extremely unstable, these merge into a single layer, the main frontal cloud becomes altocumulus castellanus, and turbulence results, which destroys minor upper discontinuities and produces the "chaotic sky" recognized by the category C_M9 in the synoptic code. Such chaotic sky is common with fronts approaching the British Isles from Spain.

Frontolytic cols are rare, and fronts are usually dissolved by assimilation of the frontal zone into the adjacent air masses. This air-mass assimilation may be brought about in three different ways. Two air masses stagnating side by side over a uniform terrain become subject to the same influences from the underlying surface and from direct radiation, so that they tend to merge. Air masses moving along adjacent parallel tracks, with little difference in velocity, also tend to merge into each other. Finally, the arrival of a front in a geographical area where the orientation of the ground temperature gradient is unfavorable for its continued existence—i.e., where the "cold" air mass fed into the cold side of the front becomes as warm as the "warm" air mass on the warm side of the front, and vice versa—will ultimately lead to the disappearance of the front. Assimilation of very shallow, cold air masses also takes place in regions of great surface heating. In North Africa, for example, or over the Iberian peninsula in summer, a dry-adiabatic lapse-rate is set up from the ground surface upward, and any shallow masses of cold air arriving over the surface are soon assimilated. Relics of fronts remain as isolated cloud decks, but if subsidence occurs these frontal cloud sheets are soon destroyed, and any decks of turbulence cloud—undulating stratocumulus or altocumulus layers, for example—are largely cleared away.

Apart from the Petterssen deformation-field theory of front formation and dissolution, other theories have been advanced from time to time to explain specific aspects of frontal development. Many of these theories involved the *bent-back occlusion*, a feature that appears much more frequently on the synoptic chart since the advent of thickness-pattern analysis. In Refsdal's theory of frontogenesis, advanced in 1930, the development of a bent-back occlusion is followed by reocclusion of the pseudo-warm sector (see Figure 6.25,B), the process often repeating itself several times in one depression. This process is capable of generating new discontinuity lines in a horizontal field of flow, which usually form secondary cold fronts. The association of one of these secondary cold fronts with a frontogenetic col will produce a major front. Another source for the formation of new fronts is provided by *breakaway depressions*, which form at the point of occlusion in an occluded low, and then move on ahead of the parent depression. The result is the formation of a single, long frontal line.

Sawyer's dynamic theory linking frontal formation in the middle and upper troposphere with the jetstream is a major advance, since the Petterssen theory only dealt with the surface situation, and fronts are essentially three-dimensional phenomena. Sawyer showed that jetstreams, which flow from regions of relatively low windspeed to regions of relatively higher speed, with strong wind shear on the cold side of their axis, must develop a region of general subsidence on their cold sides, which is accompanied by an intensification of both horizontal and vertical potential-temperature gradients, such as are observed in frontal zones at 500 mb. [119] The same reasoning applies if a jet flows into a slow-moving upper trough, but in this case the frontal zone is built up at a smaller horizontal distance from the jet axis. Figure 6.25,C, presents sections that both apply to an intensifying jet stream. The vertical stability increases in the divergence area marked *, and the vertical potential-temperature gradient intensifies. At the same time, tilting about a horizontal axis increases the horizontal temperature gradients, and is accompanied by an increase in cyclonic vorticity about the vertical axis.

The Sawyer theory does not explain why the boundaries of frontal zones in the mid-troposphere are so sharp and well-defined, but suggests that it is due to the intensification and dynamic tilting of already existing boundaries of inversions or stable layers. In other words, it presupposes the existence in the atmosphere of initial discontinuities of lapse-rate, and then applies dynamic concepts to show how a frontal zone must arise as a necessary consequence of the existence of a jetstream that is intensifying. Very occasionally, the tropopause may provide the initial discontinuity, in which case subsidence may be strong enough to bring stratospheric air into the frontal zone. The dry air in the frontal zone must owe its dryness to subsidence; it is very unlikely that the frontal zone as high as 500 mb owes its origin to different geographical origins of the air on either side of it.

The front-forming processes in the upper and middle troposphere only operate where the jet is intensifying or becoming more cyclonically curved along its length. But the characteristic structure may be carried forward into weaker sections of the jet, and probably does not completely degenerate. The theory also acts in reverse, so that frontolysis occurs if the air passes into weaker sections of the jet, or into the locality of an upper ridge.

The basis of the Sawyer theory is the derivation of a formula that links wind velocity with the rate of change of vorticity along a particular streamline (see Appendix 6.11). It thus explains how the rotational energy of an airstream changes from point to point through the axis of the jetstream and in its vicinity. It ignores the local change in vorticity, in comparison with the advective change, and is of course only applicable to *major* frontal zones. No suitable dynamic theory is yet available for small-scale or local fronts, and the tendency is usually to treat each occurrence on its own merits.

Perusal of a long series of daily weather charts will show that the great variation in fronts therein displayed has reflected the current fashions in synoptic analysis rather than the true vagaries of the weather. The American synoptic charts, for example, have undergone a succession of dramatic changes in appearance as one idea after another has dominated the minds of the analysts.[120] The first major change occurred during the years 1938 to 1943, which saw the introduction, and then the wholehearted acceptance, of the conventions of air-mass analysis. The ultimate in this development was the "spoked wheel" stage of frontal analysis, in which numerous fronts radiated out in all directions from every cyclone. The next stage involved the elimination of spurious fronts, and the adoption of a policy of "logical moderation" in the recognition and delimitation of fronts on a synoptic chart.

The second major change in American chart analysis involved the adoption of the idea of the *instability line* or the squall line to explain the organized bands of bad weather frequently found in the warm sector ahead of the cold front. The idea was accepted reluctantly, since it radically departed from the classic Norwegian models, and so for a time organized bands of thunderstorms or showers were all accounted for by means of surface fronts or upper cold fronts. Then ideas reversed, and all lines of thunderstorms were regarded as instability lines, even those that were fronts in the classic sense. Instability lines are very much more versatile synoptic features than are fronts, since they are not tied to the boundaries between different air masses, and can move backward or forward in a homogeneous air mass, accelerate or decelerate, disappear and reform, and even result in the formation of tornadoes or other severe weather. By the mid-1950s, overenthusiasm for instability lines had settled in, so that a stage equivalent to the "spoked wheel" stage in frontal analysis had arrived.

Two trends may now be observed. First, there is a trend toward a complete revision of frontal concepts. Especially for summer cold fronts, instability lines are drawn at the forward edge of a transition zone, with a cold front at the rear of the zone, so that the cold front is regarded as coincident with the rear edge of the precipitation area, and is placed there because a dew-point discontinuity must exist between the rainy and rainless areas. In the classic Norwegian model, the cold front lies by definition at the *forward* edge of the transition zone. Thus the American revision of the frontal depression model tends to conform with the three-front model introduced by the Canadians from different types of reasoning.

The second trend is toward objective frontal analysis. For example, the U.S. Navy

Fleet Numerical Weather Facility has developed a numerical system for the objective location of the boundaries of frontal zones and of the region of maximum baroclinicity within that zone. The parameter employed is the second derivative of almost any air-mass property that is conservative, such as potential temperature.[121]

A recent discovery of great importance is a proof by Kirk that fronts may be generated dynamically and thermodynamically without any reference to air masses. The essential result is that fronts prove to be dynamic consequences of cyclonic development, and the air-mass concept is not essential to the theory of frontogenesis or frontolysis. In fact, the air-mass theory only appears to be valid if it is interpreted in terms of vorticity transfer. The theory shows that frontogenesis is fundamentally linked with concentrations of vorticity, and the appropriate atmospheric discontinuities across a frontal zone may be located by means of the Laplacian operator (see Appendix 6.15).

Kirk's theory, in which fronts are generated thermodynamically within an initially homogeneous body of air, is a great advance. Before it, the existence of two non-homogeneous air masses had to be assumed; i.e. the air-mass on either side of an incipient front was assumed to possess horizontal gradients in both temperature and wind. Once these assumptions have been made, it is possible to show theoretically that a discontinuity in vertical velocity must develop at the incipient frontal surface; i.e., upsliding or downsliding of air must occur at this surface. Observation confirms the deduction from the thermal wind equation that vertical windshear is discontinuous at the boundaries of a frontal zone.[122]

A difficulty in the designing of new frontal models of universal application is that local weather effects can completely change the structure of a front from its ideal form. Rainfall can modify the temperature difference across a front, causing it to accelerate toward the warm air, and acquire the characteristics of an instability line. This modification of a front by rainfall intensifies with increasing dryness of the air mass into which the rain falls. Evaporation of the falling rain, and consequent release of latent heat, is the connecting factor in the process. For example, the storm of May 9–13, 1953, in the American Midwest produced widespread heavy rain and numerous tornadoes, while the rain-cooled air masses both intensified and modified existing fronts and also resulted in frontogenesis (see Figure 6.30).

Local or regional airflow idiosyncracies also result in the formation of miniature fronts. For example, a discontinuity surface is sometimes produced by topographic winds over the upper Snake River Plain in Idaho, particularly in summer. The discontinuity surface usually occurs in the center of the 60-mile-wide plain, where mountain and valley winds are quite pronounced, even against an opposing pressure gradient (see Figure 6.31). Southwesterly (valley) winds predominate during the day, and northeasterly (mountain) winds at night. Convergence of the katabatic wind down the slope of the continental divide with the valley wind produces a miniature front, even with opposing geostrophic winds of 20 miles per hour. Temperature contrasts of up to 20°F have been measured across the front, contrasts of 10°F being frequent. The front is of the "fair-weather" variety, no pressure variation being found with it, and no cloud, but its existence may be proved by means of smoke plumes. It is usually nocturnal, occurring embedded in the surface temperature inversion, and its intensity depends on the depth of the mountain wind, which is usually less than 500 feet thick, but very occasionally reaches 3,000 feet in thickness.[123]

INSTABILITY LINES AND SQUALL LINES

Instability line is the term usually applied to the line of intense thunderstorms that frequently precedes, and parallels, the surface position of a cold front. It is associated with low-level horizontal convergence, which in turn is usually associated with a layer of strong confluent winds in the lower troposphere. In other words, contrary to what one would expect, instability lines are not necessarily associated with dynamic instability in the atmosphere.[124]

No generally applicable dynamic theories of instability-line formation have yet been formulated, and the rules for extrapolating the movements of instability lines are empiric, based mainly on the associations between surface and upper airflows. Examples of these rules, for the southeastern United States in April and May only, are as follows. An instability line usually forms if a cold tongue at 700 mb and a warm tongue at 850 mb develop well to the rear of a trough line, the instability line usually taking at least 24 hours to form after this coincident development occurs. The temperature difference from the axis of the warm tongue to that of the cold tongue, measured along a circle of latitude, must be at least 7°C for the instability line to develop; if it drops below 5°C, the instability line dissipates. Once formed, the instability line moves ahead of the cold front, and remains parallel to it. Instability lines also develop along the axis of 850-mb warm tongues after strong cold-air advection at 700 mb has reached the fore part of the associated trough, and is advancing at a more rapid rate than the warm tongue.[125]

A *squall line* is defined as a narrow zone of weather activity usually parallel to the surface position of a cold front, and some distance in front of it. The "weather activity" includes heavy rain, thunderstorms, brief wind shifts and gusty winds, abrupt pressure rises, and temperature falls. Squall lines occasionally develop north of warm fronts, north of the crests of the frontal waves, and in areas far from fronts. Surface wind shifts are considered to be good criteria for tracking them on the synoptic chart: the formation of a squall line is indicated if two or more stations in the same area report thunderstorms with pronounced wind shifts and pressure rises.[126]

Southerly surface winds usually occur ahead of a squall line, and shift almost instantaneously to westerly or northerly at the onset of the squall line. The peak wind gust occurs at the time of the squall-line passage, reaching 75 miles per hour or more for a period of one or two minutes. The pressure rise behind a squall line is occasionally enough to appear in the pressures recorded on the synoptic chart, and sometimes may be so great that a pressure wave results, in which case pressure falls abruptly several hundred miles to the rear of the squall line, so that dense cirrus forms in the crest of the pressure wave, because of the strong vertical currents that result at 20 to 30 thousand feet.

Thunderstorms are often critical factors in the formation of a squall line. The formation, linear structure, propagation, and dissolution of squall lines are primarily due to horizontal variations in warm air advection in the layer below 10,000 feet above sea-level.[127]

In the north-central United States, squall lines form in the vicinity of fronts, in areas of maximum moisture content of the layers of the atmosphere up to 700 mb, but they often move far from their associated fronts later in their history.[128] They also form: on the western and central parts of moist and warm tongues; on the eastern

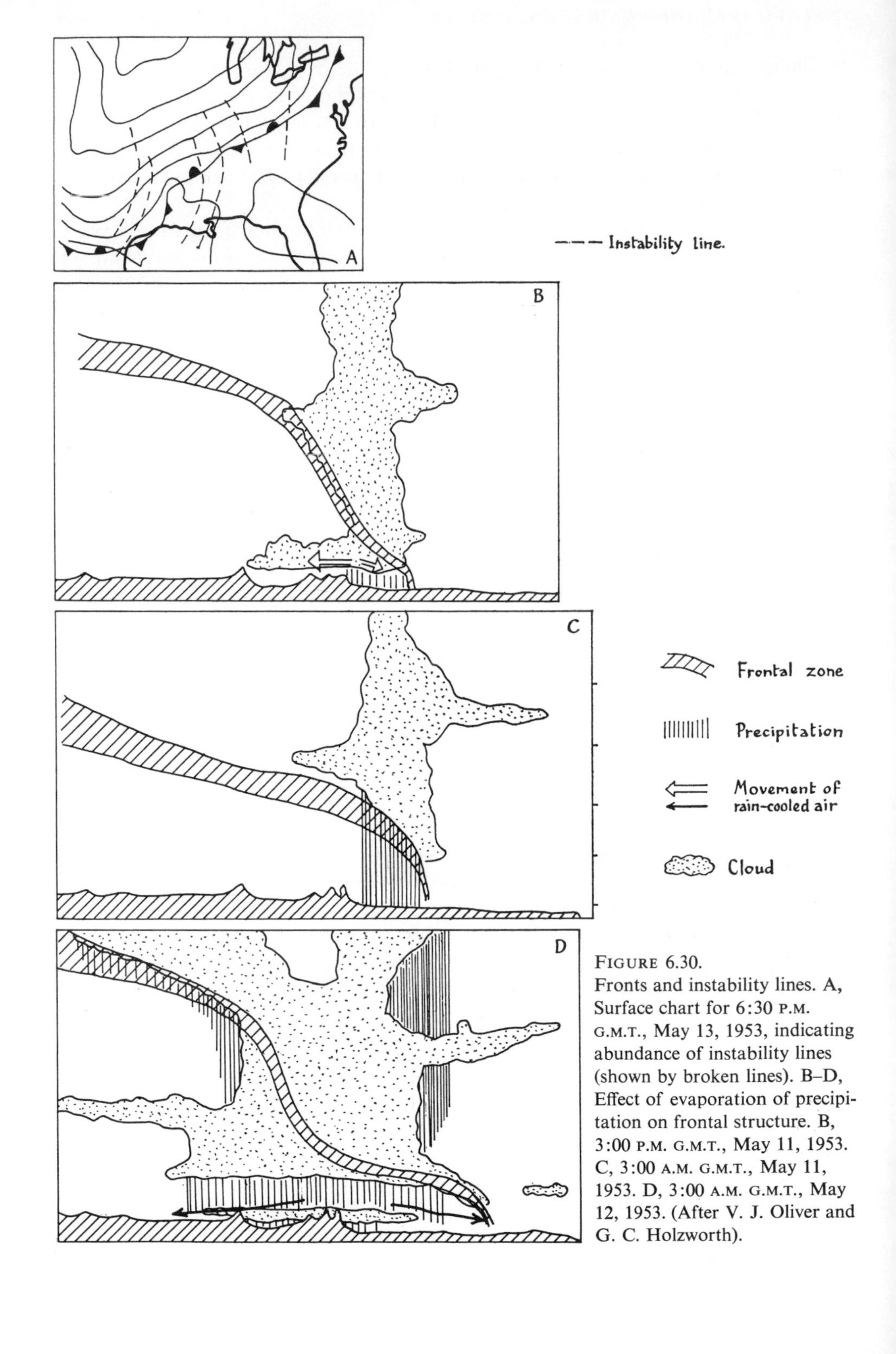

FIGURE 6.30.
Fronts and instability lines. A, Surface chart for 6:30 P.M. G.M.T., May 13, 1953, indicating abundance of instability lines (shown by broken lines). B–D, Effect of evaporation of precipitation on frontal structure. B, 3:00 P.M. G.M.T., May 11, 1953. C, 3:00 A.M. G.M.T., May 11, 1953. D, 3:00 A.M. G.M.T., May 12, 1953. (After V. J. Oliver and G. C. Holzworth).

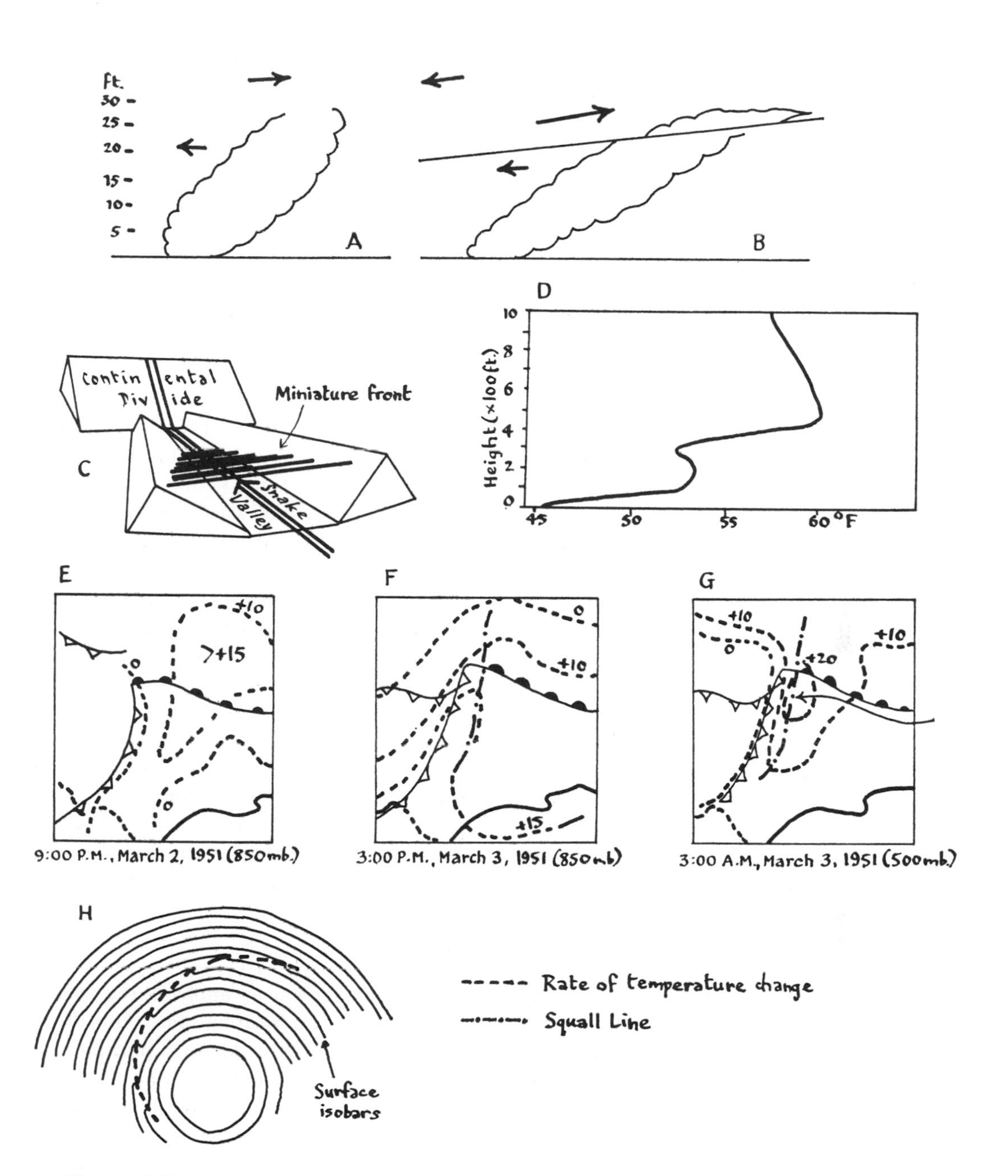

FIGURE 6.31.
Local fronts and squall lines. A, B show Snake Valley front: smoke-plume cross sections distant 100 feet (A) and 500 feet (B) from source of smoke. C, model of Snake valley. D, temperature sounding, April 23, 1952. E–G, squall-line development, showing instantaneous rates of temperature change due to advection, six hours prior to squall-line development. H, hurricane with single outer precipitation band. (After E. M. Wilkins, J. Macdonald, and M. G. H. Li da).

or southeastern sides of areas of maximum warm-air advection; in the area of cyclonic shear associated with the 850-mb jet; in warm sectors, and in advance of cold fronts generally, when a sharp 500-mb ridge is followed upstream by a sharp cyclonic bend on the eastern side of the succeeding trough; in similar surface situations just east of areas of marked concentrations of cyclonic vorticity, in the vicinity of an inflection point of the upper contours; in spring, just east of the northward-curving tip of the jet-maximum; along warm fronts or stationary fronts associated with broad, flat 500-mb ridges, in which case they occur under the right front quadrant of the jet maximum, with one end of the squall line near the tip of the 80-knot isotach, curving away from this point towards the 50-knot area. With a double jet core, particularly in spring, a squall line frequently extends from the right rear quadrant of the northern jet maximum to the left front quadrant of the southern jet maximum. Southward-moving bursts of squall-line activity also develop below the northeastern sector of 500-mb high-pressure areas, where strong northwesterly wind components occur, usually to the right of a jet oriented from northwest to southeast. Their movement is usually closely connected with the mean temperature of certain of the lower layers of the atmosphere. The squall lines are steered by the mean 850-mb to 700-mb thickness pattern, and continue to move as long as cross patterns occur downstream in the direction of movement of the disturbances. They then dissipate if they move into areas beneath regions of lighter winds (i.e., less than 50 knots) between the 500-mb and 200-mb layers, or if they move into areas where isobars have more anticyclonic curvature.

In the southeastern United States, squall lines are associated with hurricanes and tornadoes. With hurricanes, the usually concentric pattern of the isobars is sometimes disturbed by slight irregularities that coincide with precipitation bands due to squall lines (see Figure 6.31,H). The outer precipitation bands coincide with dips in the barograph trace, and are similar in some ways to instability lines found in extratropical cyclones. They are convective in nature, giving rise to thunderstorms and tornadoes in which the wind gusts are sometimes equal to twice the speed of the gusts in the eye of the hurricane, and they are nonfrontal. Squall lines also occur in thunderstorm cells, for which evidence has been found in Illinois and Massachusetts that the right rear quadrant of these cells is a preferred region for the generation of tornadoes.[129]

EXTENSIVE NONFRONTAL PRECIPITATION

The basic practical feature of synoptic analysis as followed by applying the models of the Polar Front theory is, of course, the division of a weather chart into broad areas of relatively uniform conditions (air-mass areas), where extensive precipitation and thick cloud is the exception, not the rule, and which are separated by narrow zones of intense "weather" activity (fronts). We tend to think of all bands or zones of precipitation as coinciding with fronts, instability lines, or squall lines. However, there can be extensive precipitation areas with an organized form that are not connected with these synoptic features, such as large convective rainstorms, of 20 miles or more in diameter, whose movement is completely unrelated to fronts, and which can occur in the middle of homogeneous air masses. In the United States, such rainstorms move with an appreciable component of movement toward the right of the 700-mb wind, i.e., the wind in the cloud layer. Thus the storms do not move directly downwind, but in a direction which deviates 25° on the average from the downwind direction, because of the veer of wind with height (see Appendix 6.16).

In general, in typical squall-line situations in which the wind veers strongly with height, individual convective storms move in directions up to 60° to the right and up to 30° to the left of the direction of the mean wind in the cloud layer. This is explained by the water-budget requirements of the storm.[130] For a given precipitation intensity, the amount of water precipitated from the storm is proportional to its area, i.e., to the square of its diameter (for roughly circular storms). The amount of water vapor intercepted is proportional to both the diameter of the storm and its velocity relative to the winds in the moist layer of the lower troposphere. A large storm must intercept more water vapor, in proportion to its diameter, than a small storm, and therefore must move with a higher velocity relative to the moist layer. This requirement is satisfied when the large storm moves in a direction to the right of that of the mean wind in the moist layer, provided the general wind veers with height. In addition to this form of movement, the individual storms in a squall-line situation tend to migrate from the southwestern end of the squall line towards its northeastern end.

There is some degree of correlation between the occurrence and movement of severe local storms and the circulation at 500 mb.[131] Although such storms may not be depicted on the synoptic chart because of their relatively small size, the movement of the precipitation areas they produce may be understood by reference to very large-scale synoptic parameters. An interesting contrast to this situation is found with very large cyclonic storms. The distribution of precipitation associated with such storms is explained, in the classic model, in terms of a gentle lifting of air masses over frontal surfaces throughout a very extensive area. Detailed observations, for example, of storms over the Pacific coast of North America, show that the actual rainfall in such an area is far from uniform, and much of it is nonfrontal. Organized bands of convective precipitation within the storm can be 20 to 40 miles wide and 30 to 60 miles apart, and persist for at least three hours and over distances of 100 miles or more. Such bands are not necessarily related to fronts, but are located along the shear vector between the upper portion of the layer of convective cloud and the air layer above this. They usually move in the direction of the low-level shear vector.[132]

Extensive areas of nonfrontal precipitation may sometimes be tracked on the synoptic chart: for example, the rain areas under stratus decks over the English Midlands can be followed by means of isochrones. More usually, however, the actual precipitation amounts show variations that are very difficult to represent synoptically. On Guam, for example, the actual precipitation consists of the rainfall due to synoptic disturbances plus a rainfall component (which has considerable very local spatial variation) due to shower activity. The latter represents the "background noise" in the time-curve of rainfall fluctuation produced by the presence of the island, and the latter's influence extends tens of miles out to sea, particularly to leeward.[133]

Diurnal variations in precipitation may occur over large areas and introduce a complication in interpreting the synoptic picture. Data from weatherships suggest that rainfall in maritime areas is most frequent at night. The effect varies with season and with latitude, and is at least as large for nonconvective as for convective rain. It appears to be related to the absorption of solar radiation by water vapor in deep maritime clouds: during the day, the absorbed solar radiation is used, to a large extent, to evaporate the cloud droplets, so that less liquid water is available to form precipitation. Over land areas, diurnal precipitation variations may depend partly on surface heating effects. For example, the rainfall in summer over eastern China, Korea, and Japan shows a morning maximum at all stations because of monsoon effects, but on this is superimposed, at some stations, an afternoon maximum due to surface heating. The morning maximum occurs at or just before dawn inland and over the middle-latitude coasts, but just after dawn over tropical and subtropical coasts. It is produced by the rapid development of cumulus clouds during the morning, which are the consequence of extensive overturning in the lower layer of the atmosphere. The overturning is generated during the low-level advection of warm air between midnight and sunrise, a process favored by the limited depth of the monsoon current and its lack of turbulence.[134]

Over tropical land areas, diurnal variations in precipitation may result in very serious synoptic problems. The rainfall of much of East Africa, for example, is largely convective, and isochrones depicting its time of onset are usually not practicable. In such cases, it is useful to map daily rainfall in terms of distribution rather than amounts; i.e., rain areas are delimited that are more or less homogeneous in the numbers of stations reporting rain, and that are bounded by wetter or drier areas. Rain-area maps of this type reveal the existence of large-scale rain systems that develop and persist for days, but are not mobile features on the synoptic chart. The rain areas also change type, the whole process taking place against a background pattern of seasonal change. In Rhodesia, where the distribution of rainfall in both area and time cannot be related, in detail, to synoptic features, it has been found convenient to describe the development and decay of precipitation patterns in terms of "progressions," defined as areas of precipitation that move (on the radar scope) continuously by a process of new echo formation.[135]

Unusually high rainfall amounts or rain intensities are usually confined to small areas within an extensive precipitation cover, and explaining them in synoptic terms is often problematic. Some record rainfalls may be explained synoptically without recourse to geographical influences. For example, a cloudburst over Tulsa, Oklahoma, on July 27, 1963, was produced by a combination of low-level convection with nearly calm upper winds. Precipitation falling in an extensive area of clear skies is another difficult synoptic problem. When a strong wind exists, it is reasonable to expect that fine rain may fall some miles to the leeward of a precipitating cloud, but cases are on record of light rain falling from a cloudless sky thirty minutes after the dispersal of the cloud.[136]

Particularly during the winter, it is of obvious importance to be able to determine, from the synoptic chart, just which portions of an extensive area of precipitation may be characterized by snow instead of rain or drizzle. The total thickness chart gives a first indication, but conditions within the lowest few hundred feet of the atmosphere may differ from those implied by the thickness value. For the United Kingdom the best synoptic indicators of snow are (a) the height of the freezing level, and (b) the 1000-mb to 850-mb thickness combined with the sea-level pressure. A combination of total thickness, height of freezing level, and surface temperature also permits a determination of the probability of snowfall.[137]

In applying such predictors to determine whether precipitation will be liquid or solid in a certain area, one should not overlook orographic effects. For example, one case study showed that the forced uplift of an airstream over the Cairngorms (4,084 feet) caused the freezing level—which at the time was at 6,400-7,000 feet over the northern parts of Great Britain—to descend 2,500 feet over the Cairngorm area. Other geographical effects are of importance in other areas. For example, the passage of an airstream across the southern portion of Lake Michigan was sufficient to bring snow to its southwestern shore in the early hours of January 19, 1963. The airstream consisted of very cold Arctic air, to which energy was added in the form of latent heat of vaporization during its passage across the lake. This surplus moisture was then precipitated in the form of snow. If a single indicator is desired, then the wet-bulb temperature provides a useful criterion for the extent of the downward penetration of snow. During continuous precipitation of moderate intensity, it may be shown that, provided the lapse-rate is approximately saturated-adiabatic, precipitation should be in the form of snow down to the level at which the wet-bulb temperature is 1.5°C. Snow is unlikely to descend below the 3.5°C level in areas of moderate precipitation, or below the 4.5°C level in areas of heavy precipitation.[138]

Synoptics of Some Weather Phenomena

Some weather phenomena that rarely appear on the synoptic chart are of very great practical significance, and their study has involved the development of a new scale of synoptic analysis. Tornadoes, thunderstorms, and hailstorms are good examples of such phenomena.

For many years, *tornado* study was mainly confined to purely climatographic considerations. For obvious reasons, accurate observations of the standard meteorological elements within a tornado were lacking. Movie films of tornadoes and measurements of debris used as "tags" have now provided some accurate figures. For example, the Dallas tornado of April 2, 1957, was found to have attained tangential and upward windspeeds of 170 and 150 miles per hour, respectively, and unexpected variations in angular momentum were found at different heights above ground level (see Appendix 6.17). Application of the cyclostrophic equation to the observed tangential windspeeds enables the total fall in surface pressure at the tornado center to be computed: this proved to be 60 mb. Variations in the geometry of the tornado "funnel" were related to changes in the moisture content of the air in the vicinity of the tunnel, and to changes in pressure in a relatively large area around the tornado. From these observations, it is clear that there ought to be *some* indications on the synoptic chart of incipient tornado formation.[139]

Maps of tornado occurrences reveal that one is justified in searching for synoptic indicators, because the reports show there is a certain pattern to tornado locations. Maps for individual American states have been compiled, showing the movements and frequency of tornadoes. In Arkansas, for example, between 1879 and 1955, the average width of the path of a tornado was 350 yards, and, although seven tornadoes moved for 100 miles or more, most paths were usually short. Three exceptionally wide paths, of about 2 or 2.5 miles, were recorded. The paths generally lay from southwest to northeast, despite the irregular nature of the Arkansas topography, and there is some evidence that the paths of the tornadoes a century ago were longer than those of today. Some counties in Arkansas proved to be especially likely to have tornadoes; the number of tornadoes reported during the 76-year period ranged from only one in Marion County to 29 in Pulaski County.[140]

Tornadoes are more frequent over the Canadian prairies than is generally believed. During the period 1916–1955, tornadoes were most frequent in an area near the junction of the Manitoba—Saskatchewan—North Dakota borders. From a maximum in

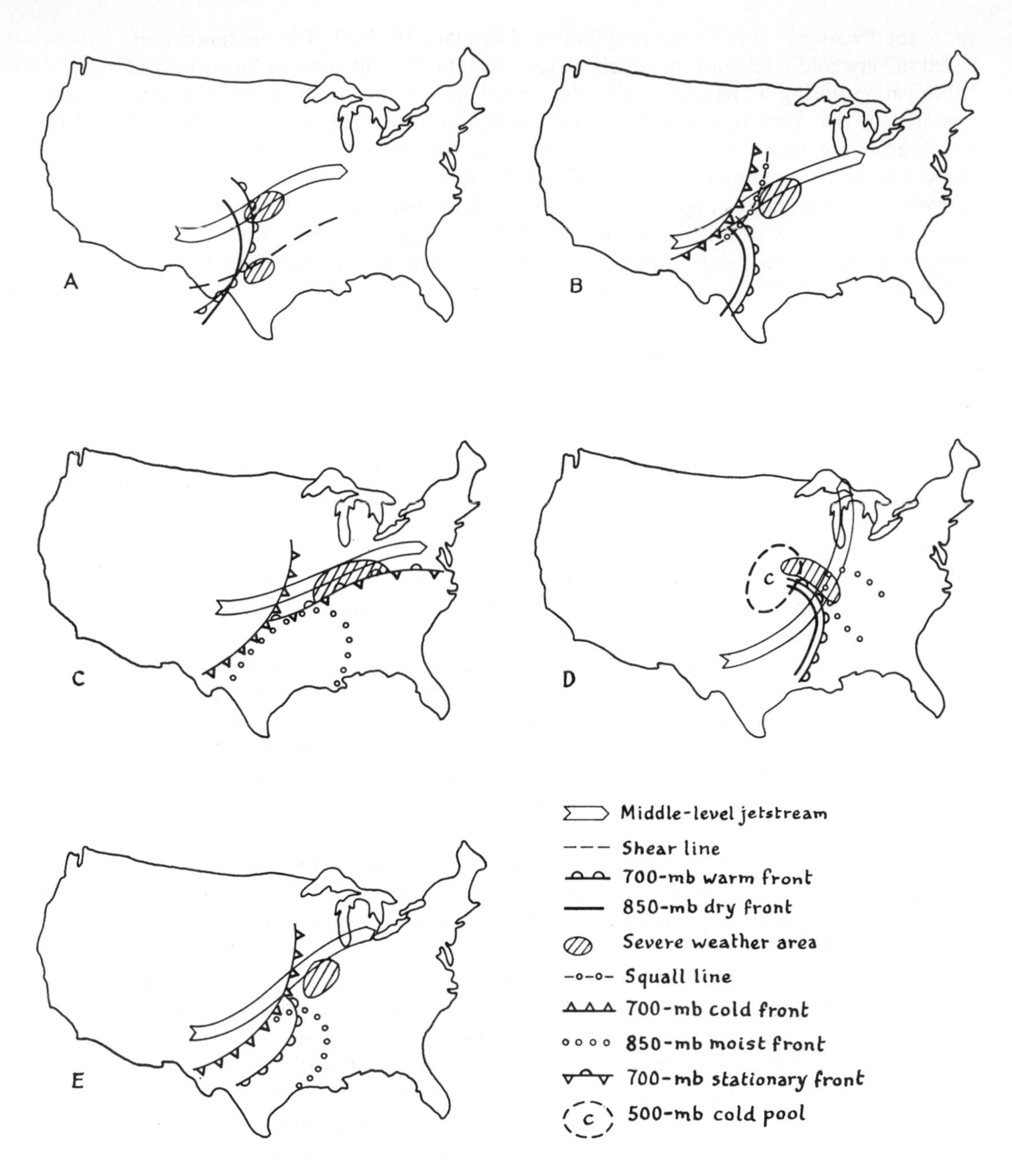

FIGURE 6.32.
Tornado-producing synoptic situations at local noon (after R. C. Miller).

southeast Saskatchewan and southwest Manitoba, their frequency decreased rapidly towards the northeast and gradually decreased to the north and west. The tornadoes occurred over dense forest as well as over grasslands. They occurred as far north as latitude 56°N in Alberta and, although rare in the west, in the foothills of the Rocky Mountains, became frequent again in a small area near Calgary.

The Canadian tornado season extends from mid-April to mid-September, with maximum tornado frequency near July 1. The tornadoes have been reported at all hours of the day, but are most likely to develop in late afternoon and evening. The synoptic situations associated with tornado outbreaks appear to differ in Canada and the United States. In Canada the tornadoes often pick up moisture (and hence acquire additional energy) through evapotranspiration from the prairies and the area to the north of the latter. The most common synoptic situation is for the tornadoes to develop along an instability line that is parallel to, and moving in front of, a cold front. Sometimes a tornado develops in the area where an instability line intersects a diffuse warm front.[141]

American tornado-producing synoptic patterns have received considerable attention. Composite charts have been produced that relate tornado occurrence to certain characteristic combinations of sea-level pressure (or height of an isobaric surface), temperature, and dew point as plotted on a single map. Tornado occurrence has been related to areas of intense warm advection, as indicated by surface isobars and isotherms, and, for Massachusetts tornadoes, a typical location is in the right rear quadrant of a squall-line thunderstorm cell.[142] Typical synoptic conditions favoring tornadoes in the United States involve (a) upper winds exhibiting a maximum between 10,000 and 20,000 feet, the maximum speed exceeding 35 knots in a relatively narrow band; (b) the appearance of a distinct dry tongue in the middle levels of the atmosphere just prior to tornado generation; (c) the movement of an upper dry tongue over a lower moist wedge, the latter with a horizontal moisture distribution such that a distinct maximum is present along a relatively narrow band on the windward side of the area of tornado inception. The synoptic situations conducive to the development of these critical tornado-producing conditions are illustrated in Figure 6.32. In addition, three types of air mass are especially favorable for tornado generation. Soundings taken upstream from an area of incipient tornado formation within warm air are usually characterized by an inversion. However, the inversion normally lifts and has completely disappeared by the time the tornado makes its appearance, and there is a rapid lifting of low-level moisture to a great height simultaneously with the rise and disappearance of the inversion.[143]

Although the literature on tornado synoptics consists mainly of American examples, there are interesting cases of tornadoes elsewhere. For example, very occasional tornadoes have been reported in Britain, and an important study has been made of a tornado observed in Malta on October 14, 1960, indicating the importance of thermodynamic factors in the generation of tornadoes. The typical synoptic concomitants of American tornadoes—i.e., a stable layer or a dry inversion above a moist surface layer, dry air above the inversion, and hail in the vicinity of the tornado—were also present in the Malta example, but with the latter the synoptic-scale wind field was not an important factor. Tornadoes obviously require low-level convergence to maintain them; for the Malta tornado it was provided by the thunderstorm downdraft at the margin of a nearby thunderstorm cell, not by a general convergence of the low-level wind field. That is, since the dynamic conditions associated on the synoptic chart with tornado development may also occur when tornadoes are not generated, the generation of a tornado depends on an additional thermodynamic factor. For Malta, this factor is provided by the Sea of Sidra, which acts as a source of warm moist air. For the United States, where tornadoes are, of course, on a much larger scale and much more frequent than in Malta, the arrival of warm moist air from the Gulf of Mexico is the required thermodynamic "trigger."[144]

Only very recently has the physics of the tornado been given detailed attention. One difficulty has been that, unlike cyclones, tornadoes are so constructed that their physical structure must be inferred rather than directly measured; i.e., observational data concerning tornadoes must usually be deductive. An exception to this is provided by radar observations. Tornadoes characteristically produce a hooked echo (see Figure 6.33) on the radar screen, which represents cloud or precipitation targets spiraling inward, counterclockwise, into the vortex. If the observer faces in the direction toward which the tornado is moving, the hook is normally located in the right rear quadrant of the tornado, and the path of the end of the hook usually coincides with the track of tornado damage along the ground.[145]

The synoptic technique known as *mesoanalysis* is proving very useful in the study of tornadoes. Although the Norwegian meteorologists during the First World War carried out mesoscale investigations, using observations from a close network of

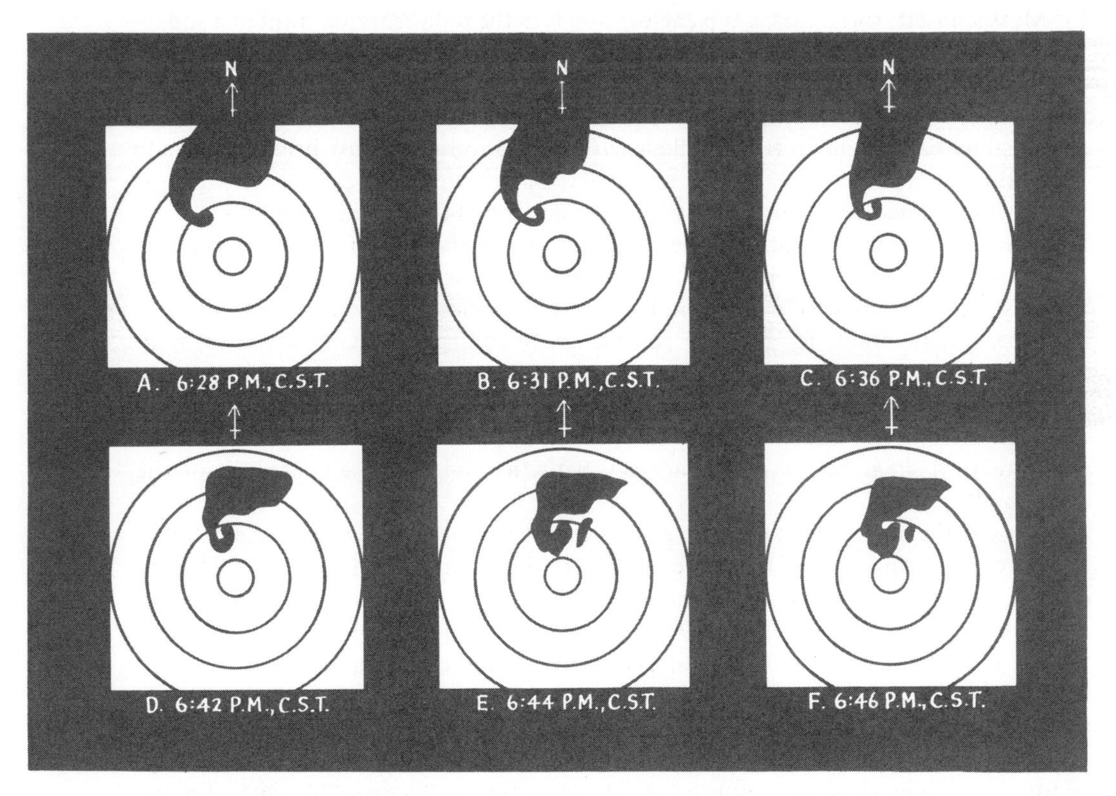

FIGURE 6.33.
Tornado hook echo. Appearance of PPI scope of WSR-3 radar (S-band) at Topeka Weather-Bureau Office, May 19, 1960. PPI shows horizontal distribution of radar echoes. Range between marker circles is 5 nautical miles.

stations, their work was not truly mesometeorological, for the data were interpreted in terms of a synoptic-scale model, the Polar Front theory. Mesometeorology as a formal study was first defined in 1953. The first detailed investigation, conducted in the midwestern United States, involved the setting up of high-speed thermohygrographs and microbarographs, the latter revolving once every 12 hours. The stations were located 20 to 30 miles apart, and since 1952 they have provided routine weather reports that prove the existence of distinct *mesosystems* within airflows that are quite uniform in the normal synoptic sense. The occurrence of mesosystems is not sporadic or random, and the most significant variations in local weather (and, in particular, the most severe weather disturbances) are due to the passage of mesosystems.[146]

Essentially, synoptic meteorology is a study of atmospheric motions described in cartographic terms. The motions are conveniently described by assuming as a convention the existence of certain synoptic systems, the nature and appearance of which depends on the density of reporting stations. If the stations are far apart, say, about 200 to 300 miles, as upper-air reporting stations are, then the synoptic systems that may be distinguished—i.e., the *simplest* systems that may be drawn on the charts so as to fit in with the actual observations—are mainly of the planetary-wave variety. If the stations are closer together, say, 100 to 200 miles, as surface stations are in many parts of the world, then more detailed systems may be distinguished, cyclones, anticyclones, and fronts, for example. All these systems have to do with the macroclimate. If the distance between adjacent stations is less than five miles, then there is no point in recognizing any synoptic "systems" at all, because the observations will all refer to actual weather features, such as tornadoes, thunderstorms, cloud patterns, or showers. All these features may, in principle at least, be described in terms of the laws of micrometeorology. Between the scales of the macroclimate (a synoptic study) and the topoclimate (a microclimatic study), there exists a range of synoptic entities that explain many hitherto unexpected features about local weather. These synoptic entities are the mesosystems, which may be extended to include pressure jumps, pressure surges, and squall lines, as well as typical mesodisturbances, such as thunderstorm highs and pressure couplets.

The techniques of mesoanalysis are quite distinct from those of ordinary synoptic analysis. Of special interest are two techniques: mean-value correction, which is necessary because the raw observations result in very erratic isobars and isotherms if plotted directly; and the conversion of time-sections into space-sections, which has application in many branches of climatology. If desired, the mesoanalysis may be carried out purely objectively, by applying a computer procedure to the data provided by the microbarograms. Once the mesocharts have been completed, several models may be applied in the description of the systems that may appear on them. For the motion of these systems to be detected, the time-interval between successive charts should be not more than two hours, and hourly charts are to be preferred.[147]

Mesosystems may be regarded as horizontal disturbances of the macroflow; typically, these disturbances have diameters ranging from 0.1 miles to just over 6 miles, the range 0.1 to 2.25 miles being the most common. Mesolows include mesodepressions, mesocyclones, and tornado cyclones. Tornadoes are usually associated with mesolows at the intersection of either (a) two instability lines, or (b) a squall line and the northeastern boundary of rain-cooled air. The *mesodepression* usually forms as a result of very pronounced local differential heating, for example, above the heated surface of a tropical island during the daytime, or above the heated slope of a mountain; it does not have closed wind circulation about its center. The *mesocyclone* is characterized by a closed cyclonic wind circulation, although, under certain conditions, a mesocyclone may have an anticyclonic wind circulation about its center. The ratio of convergence-to-vorticity appears to be much larger for mesocyclones than for synoptic-scale cyclones. As with a tropical cyclone, the mesocyclone often exhibits precipitation bands, but these usually cross the isobars at angles of between 45 and 60° in comparison with the 10–30° crossing angle for hurricanes. It is not correct to regard a mesocyclone as a miniature hurricane, because the closed air circulation associated with it can rarely be studied on the synoptic chart, but must instead be inferred from the pattern of radar echoes or from cloud patterns photographed from a satellite. In fact, the first indication of the development of a mesocyclone is usually the appearance of a pendant echo (see Figure 6.33,A) on the radar screen. The echo fairly rapidly begins to curl around the cyclone center, and, for the *tornado cyclone*, becomes transformed into the shape of a hook. Several tornadoes may be associated with a tornado cyclone, which usually averages 30 to 60 miles in diameter; typically, the tornadoes form one to three miles to the right of the path of the center of the cyclone, and then move across the track of the cyclone before finally dissipating. In a tornado cyclone in Illinois on April 9, 1953, the tornadoes were associated with what resembled a very small hurricane, 30 miles in diameter, which produced spiral echo-bands on the radar screen.[148]

Radar is an extremely useful tool for the study of mesosystems, since it enables one to study systems that are not a type of mesolow. Triple-theodolite pilot-balloon ascents may also be used to demonstrate the existence of small-scale features within the mesoscale field. Mesoanalysis has proved to be of considerable value in the synoptic study of sea breezes. For example, each valley and coastal plain along the southern California coast has its own distinct sea-breeze regime. Adjacent pairs of regimes have mutual boundaries that coincide with convergence zones associated with rising air. The synoptics of these very localized regimes must be studied by mesometeorological techniques rather than by normal synoptic methods.[149]

The counterpart to the mesolow is the *mesohigh*. Good examples of mesoanticyclones were observed on July 20, 1956, when five mesohighs developed within an area of 500 by 600 nautical miles in Arkansas, Louisiana, Oklahoma, and Texas (see Appendix 6.18). These mesohighs grew to have diameters of 300 to 400 miles, and were associated with appreciable divergence at ground level.[150] Small mesohighs usually give rise to squall-type weather phenomena, associated, for example, with instability lines, pressure jumps, pressure pulsations, squall lines, and wind-shift lines, and then dissipate as mesodepressions. In particular, mesohighs are the normal synoptic features associated with *thunderstorms*.

Classic observations of thunderstorms in Florida indicated that the storms usually consist of several convective systems, each system constituting a "cell" that is more-or-less independent of adjacent cells, and that goes through a distinct life-cycle. When entrainment occurs in the later stages of the life-cycle, the lapse-rate of the air within the updraft in the thundercloud approaches that of the environment. At this time, falling rain may, by dragging air with it, initiate a downdraft of cold air that reaches the ground and then spreads out radially as the cold core of the rain area associated with the thunderstorm. The thunderstorm downdraft results in the formation of a cold dome beneath the thundercloud, from 1,000 to 5,000 feet in depth, which advances behind a mesoscale cold front that forms a discontinuity zone associated with a pressure surge, a vector wind shift, and rainfall, as well as with a temperature discontinuity. The wind shift usually immediately follows the pressure surge, and the break in temperature occurs one to three minutes later with daytime thunderstorms. The cold dome often spreads out above a thin surface inversion layer with nocturnal thunderstorms, in which case the temperature discontinuity will not be evident in thermograms obtained from instruments exposed at standard screen heights above the ground. A mesohigh usually coincides with the cold dome.[151]

Thunderstorms have other mesoscale associations too; for example, nocturnal thunderstorms along the coasts of the eastern Mediterranean are especially associated with land breezes. The coastline is concave towards the sea, and, particularly in winter, the combination of land breeze and diurnal winds in the friction layer of the atmosphere constitute a field of convergence. In summer, the sea breezes constitute a divergence field, and so are inimical to the thunderstorms characteristic of the Florida peninsula, the interior of which has the highest frequency of thunderstorms in the United States, because the afternoon sea breezes penetrate the peninsula from the ground surface up to 3,000 or 4,000 feet.[152]

Because thunderstorms are of obvious importance in weather forecasting, many attempts have been made to relate them to ordinary large-scale synoptic parameters. For example, thunderstorms in Alaska are associated with low-level synoptic-scale convergence into the low-pressure center that forms over Alaska in summer. Great Britain experiences a relatively high frequency of thunderstorms for its latitudinal range, most of them being between early May and September. The storms usually develop under 500-mb cold pools, especially the ones in winter, and dissipate very rapidly at sunset. A location under or to the left of the Polar Front jetstream favors the thunderstorms increasing in height and vigor, in which case they may persist throughout the night. Provided the air at and below 700 mb has a relative humidity of at least 70 per cent, the occurrence of British thunderstorms is related to the difference in temperature between the 500-mb and 850-mb levels (see Appendix 6.19). In North America, the Showalter index is the normal synoptic indicator for predicting thunderstorms. Provided the Showalter index is known, then frontal thunderstorms may be predicted objectively from a knowledge of the height of the freezing level, the surface dew-point temperature, the surface temperature and wind direction, and the barometric tendency. For nonfrontal thunderstorms, a combination of divergence in the upper troposphere and convergence in the middle troposphere are good synoptic predictors for New England thunderstorms, and the temperature difference between the 700-mb and 500-mb levels proves to be negatively correlated with rainfall for New Mexico thunderstorms.[153]

Tornadoes and thunderstorms are often associated. Radar evidence allows a fourth stage to be distinguished in the life-cycle of a thunderstorm that produces tornadoes. In general, tornadoes occur in the right rear quadrants of thunderstorm cells, but if the life-cycle of a thunderstorm is examined in detail, it appears that the prerequisite for tornado generation is the existence of a quasisteady stage in the mature portion of the life of the storm. This stage involves the development of a very large single convective cell, without subcells, that persists for several hours before dissipating, and that travels to the right of the general wind direction. Tornadoes occur within this cell.[154]

Hail is a fairly frequent accompaniment of thunderstorms, and the synoptician needs to be able to determine whether or not a given thunderstorm is likely to turn into a hailstorm. There is much to be said for considering thunderstorms, hailstorms, and similar types of disturbance together as "severe local storms." Various models of severe local storms have been proposed, and of these the *Browning-Ludlam model* is probably the most well-known. A classic case involving the application of this model was that of a severe hailstorm in southeast England in July 1959. This storm exhibited a most unusual vertical structure (see Figure 6.34,A) when investigated by radar. The right front quadrant of the storm produced no radar echo below 12,000 feet, but an overhanging echo portion was observed. Behind the latter, the echo descended to the ground in an abrupt, nearly vertical wall, the echo-free (hence cloud-free) zone ahead of the wall penetrating for two miles in the direction toward which the storm was moving, and extending upwards to 15,000 feet above the ground. The echo top reached 45,000 feet, forming a single, large dome almost directly above the echo-free vault. A similar echo pattern was observed in a tornadic storm over Oklahoma in 1962. Tornadoes were present beneath the vault, which extended to 24,000 feet and persisted for one hour. Above 24,000 feet, the vault structure was still apparent, in the form of a cylindrical region of low echo-intensity extending to 40,000 feet. In 1963, thunderstorms were observed over Oklahoma in which tornado hooks were present in the vault; the latter proved to be a zone of very strong updrafts.[155]

The Browning-Ludlam model explains the echo structure described in the preceding paragraph in terms of a specific airflow pattern (see Figure 6.34, B). The air within the updraft enters from the region ahead of the storm, to its right. This air passes through the overhang region, where it helps prevent the fall of the moderately intense precipitation, and finally leaves the storm at a high level, moving in the direction of storm movement. The vault is located near the most intense portion of the updraft, which becomes tilted toward the left rear side of the storm as it rises, thus opposing the wind shear. The air within the downdraft enters the storm at middle levels, and becomes cooled (by the evaporation of precipitation falling from the overlying updraft) as it descends. Some of the smaller of the descending hailstones are caught in the updraft within the overhang zone, and pass through a second cycle, thereby growing into large hailstones. In the July 1959 hailstorm in southeastern England, hail fell almost continuously over a swath 130 miles in length at ground level, the maximum hail diameter (greater than one inch) falling during a period of 40 minutes.

Radar provides a very convenient synoptic picture of the initiation and development of a severe local storm, giving information of a type quite distinct from that supplied by conventional synoptic and mesosynoptic techniques. For example, radar focuses the attention of the synoptician on the specific area and time in which thunderstorm development is likely to occur. Each of the following criteria indicate that the development of a severe thunderstorm is possible: (1) the radar echoes are in converging lines, or in line with a wave pattern; (2) the echoes have sharply indented or scallop-shaped edges; (3) echo tops penetrate the tropopause; (4) the echoes move with a speed exceeding 30 knots; (5) equivalent radar reflectivity (see Appendix 1.7 in *Techniques*) exceeds 3×10^5 mm^6 per m^3 anywhere in the storm if measured with a 10-cm radar, or exceeds 10^4 mm^6 per m^3 at 30,000 feet if measured with a 3-cm radar; (6) the radar reflectivity increases above the 0°C level when measured with a 3-cm radar. Furthermore, certain radar criteria are very reliable indicators that the development of a severe thunderstorm is probable. These include: (1) echo tops that rise above tropopause level by at least 10,000 feet; (2) equivalent radar reflectivity exceeding 10^6 mm^6 per m^3 in any part of the storm on 10-cm radar, or exceeding 10^5mm^6 per m^3 at 30,000 feet on 3-cm radar; (3) the presence of a persistent reflectivity maximum at 20,000 feet or more on 3-cm

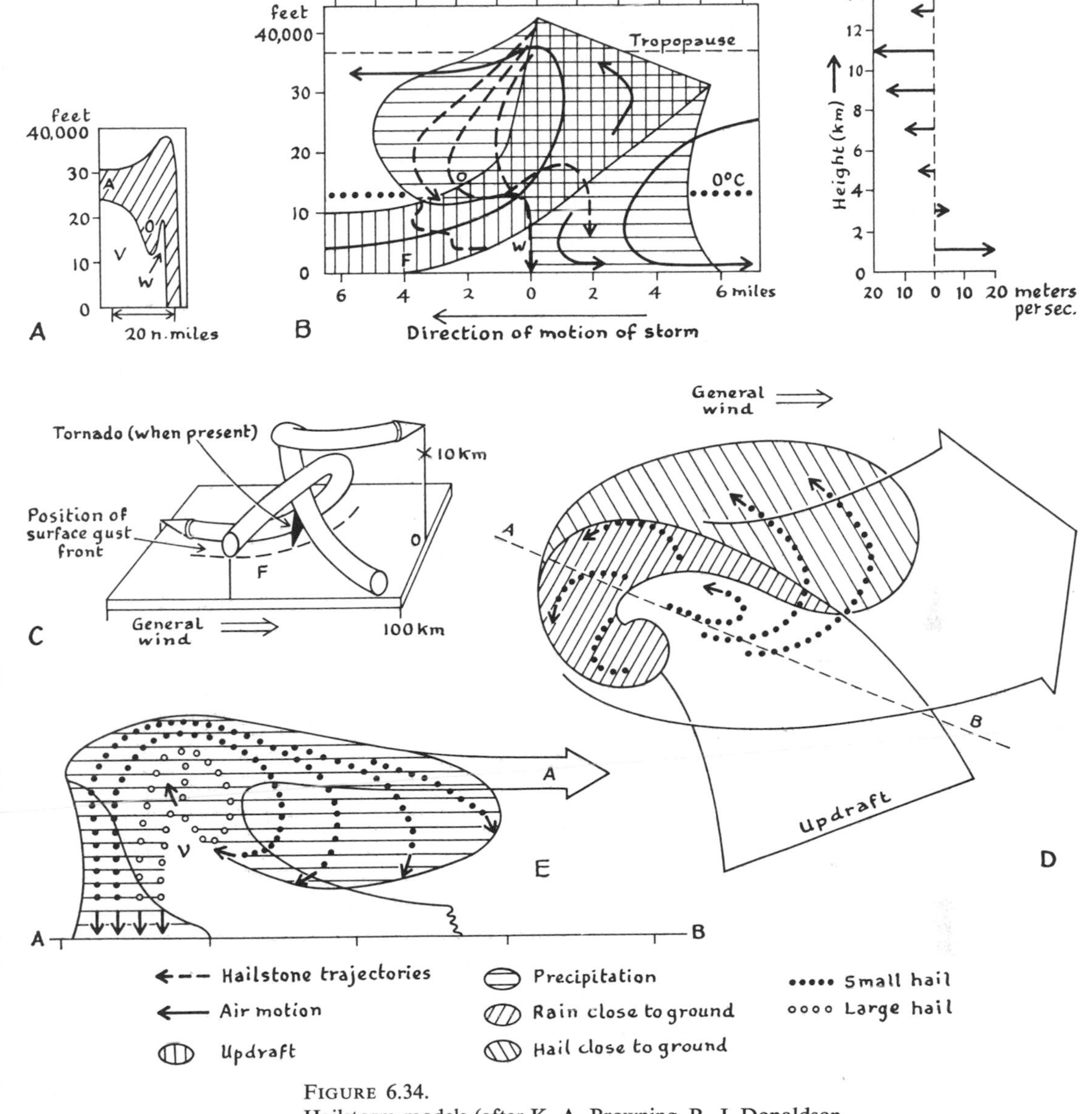

FIGURE 6.34.
Hailstorm models (after K. A. Browning, R. J. Donaldson, Jr., and F. H. Ludlam). A, Structure of Wokingham, England, and Geary, Oklahoma, storms as observed on RHI radarscopes. B, Vertical section through Wokingham storm, with wind-velocity profile in unmodified environmental air. C, three-Dimensional model of airflow in Browning-Ludlam model of hailstorm. D, Plan view of Browning-Ludlam model. E, Cross section of Browning-Ludlam model.
A, Anvil *V*, Vault
W, Wall *O*, Overhang
F, Gust front

radar that is at least 5 db greater than the reflectivity at the level where the 0°C isotherm is; (4) the presence of a tornado hook at low altitudes on the PPI screen; (5) the presence of a persistent vault-and-wall structure beneath the highest echo on the RHI screen. The persistence of a vault-and-wall pattern is important, because it indicates that a more or less steady state has been reached, in which not only is large hail likely to be produced by the recycling of small hail, but tornadoes may be generated. The maintenance of a vault shows that a persistent updraft of high intensity is present (plus the associated downdraft), and also indicates that some form of rotation must exist; i.e., the echo-free vault is partly due to the centrifugal ejection of precipitation and cloud particles from the center of rotation, which most often will be the center of a rotating tornado cyclone.

The tornado hook is a very interesting feature of radar synoptics. First photographed in 1953 in Illinois, it initially appears as a projection from the right rear side of the echo pattern of the storm on a PPI screen. The projection lengthens and curls cyclonically, forming a figure six or a hook, then closes in on itself and fills in, to form a large bulge extending outward from the parent echo. Fine-scale irregularities are usually present at the edges of the hook, and these tend to rotate cyclonically, and spiral inward. The local weather associated with tornado hooks can be of several types. A hook may be present in a severe storm without tornadoes, for example, a hailstorm or a thunderstorm squall. Although hooks are usually associated with large hail (0.5 inch or more in diameter) falling on the outer fringe of a storm, they may occur in storms that do not have hail or in which strong winds are lacking. Usually the severest type of weather is associated with storms whose echo forms a spiral band wound around a cyclonic vortex 5 to 25 miles in diameter, but there are exceptions; for example, a cyclonic-vortex echo was produced by a nonviolent thunderstorm observed in Rhodesia. The instrumental characteristics of the radar appear to be very critical factors in hook occurrence. Hooks produce the most pronounced echoes with antenna beams that are narrow in both azimuth and elevation, and although the closer the hook is to the radar, the stronger the echo, hooks up to 80 miles from the radar may produce echoes. Some tornadoes do not produce echoes on the radar screen, sometimes because of a combination of range and instrumental factors—for example, hooks are rarely detected by radars with wide vertical beams, and even narrow-beam radars rarely detect hooks at more than 50 miles—but often because of the structure of the storm. If a large portion of the radar beam intersects the overhang zone, then the precipitation-free area below it will be masked by the echo from the precipitation in the overhang, and so the presence of any hook in the vault zone will not be detectable.[156]

Radar can determine what type of precipitation is occurring in a given local storm. For example, drizzle and precipitation from stratiform clouds in general display the characteristic "bright band" that indicates the level at which snow is melting to form water drops, but convective precipitation only exhibits a bright band when the convective cells are decaying.[157] The echoes produced by precipitation from cumuliform clouds usually form a cellular pattern, the convective cells being of roughly similar horizontal and vertical extent and forming linear groups. The intensity of echoes received from convective precipitation is usually much stronger than from stratiform precipitation. In cumuliform cloud, the height of the echo top indicates the intensity of convection, and hence the type of precipitation. For example, if the echo top is less than 15,000 feet above ground level, then the cumulonimbus has not developed into a thundercloud, but if it is at 40,000 feet or more, hail is very probable.[158]

However, there are interesting geographical variations. In the U.S.S.R., the $-22°C$ isotherm divides showers from thunderstorms; if the echo tops project above the level of the $-22°C$ isotherm, a thunderstorm is likely, but if the echo tops do not reach this level, showers alone will occur. The echoes in this case are those produced by rainfall with an intensity of 1 mm per hour. For hailstorms in which the hail does not reach the ground, the size of the hail depends on the extent to which the echo top penetrates into the stratosphere. If the echo top reaches 40,000 feet, the probability of hail varies from almost nil in Texas, to 40 per cent in New England and almost 100 per cent in Alberta. For hailstorms which do deposit hail on the ground, there is considerable difference between New England and Alberta in the critical echo-top level. The probability of hail at the surface in New England increases from zero for echo tops less than 20,000 feet high to 48 per cent for echo tops greater than 50,000 feet. The most typical echo-top height is 43,000 feet above ground level for hailstorms, and 38,000 feet for hail-free thunderstorms. The critical echo-top heights in Alberta are 10,000 feet lower than the corresponding heights in New England. Hailstones 4 cm or more in diameter in Alberta are associated with echo-tops over 30,000 feet in height.

Tornado-producing thunderstorms are associated with very high echo-tops. Thunderstorms whose echo-tops rise 10,000 feet or more above the tropopause are very likely to produce a major tornado, and also to produce hail of more than one inch in diameter. Major tornadoes are usually generated in thunderstorms whose echo-tops are 47 to 57 thousand feet in height.

When applying any echo-top criterion, one must remember certain difficulties inherent in the radar technique. Especially serious errors in echo-top height determination may arise because of the power returned in the side lobes and the outer portions of the main lobe of the radar antenna. Because of this effect, echo-top height may be overestimated by as much as 10,000 feet for thunderstorms with very intense cores, if the range of the echo happens to be such that the product of range (in nautical miles) multiplied by the half-power beam width of the antenna (in degrees) equals 100. For greater ranges or beam widths, even larger errors may occur, if the magnitude of the effect is not previously calculated.[159]

After due allowance has been made for instrumental factors, the radar reflectivity (see Appendix 6.20) of a severe local storm provides perhaps the best method of classifying it in terms of the weather it is likely to produce, because there is a critical threshold of reflectivity that separates hail and no-hail occurrences. The actual numerical value of this threshold depends on the wavelength of the radar in use, and also varies very considerably with geographical location. If 3-cm radar is being employed, radar reflectivities exceeding $2 \times 10^6 mm^6$ per m^3 at ground level are associated with hail in Alberta, but reflectivities exceeding $3 \times 10^4 mm^6$ per m^3 at 30,000 feet represent the critical threshold in New England. For 10-cm radars in the latter area, reflectivities exceeding $3 \times 10^5 mm^6$ per m^3 near ground level are associated with hail, and reflectivities exceeding $10^6 mm^6$ per m^3 at 30,000 feet with hail greater than 2 cm in diameter.

Some relatively minor disturbances that cannot usually be detected on the synoptic chart are worthy of mention, in particular, waterspouts, dust devils, and whirlwinds. *Waterspouts* are similar to tornadoes in that both are air vortices that descend rapidly from the base of a large thundercloud, reaching the ground within a minute or so. Both disturbances are made visible by condensation within the vortex, the condensation process progressing downward. Both travel in more or less the same direction as the thundercloud to which they are attached and at the same speed. However, there are very distinct physical differences between waterspouts and tornadoes. *Dust devils* are fairly common in hot arid and semiarid parts of the world. They are similar to tornadoes in their size, high rotation rate, and intense concentration of kinetic energy, but differ in that they form in cloudless air and consequently can derive none of their energy from the release of latent heat. The microbarophone, a device sensitive to pressure variations on the order of 0.1 mb, may be mounted in pairs on a mobile tower, and indicates a distinct pressure drop in a dust devil, which usually has a warm low-pressure core. *Whirlwinds* may be generated by human action. For example, fire whirlwinds develop over hot fires on lee slopes close to a ridge summit line. Provided a steep lapse-rate exists near the ground, the pressure drop produced by the airflow over the ridge crest, which resembles an abruptly terminating aerofoil section, is usually sufficient to generate a vortex motion. However, the fire whirlwind does not usually appear until the rate of heat release from the fire is very great.[160]

Synoptic Climatology

It will be apparent that a series of synoptic charts for an area provides a source of climatic information quite distinct from the physical and dynamic model approach, or the statistical approach, which are considered in *Techniques*. Assuming that the synoptic charts are accurate, the main problem is to find a method that will reveal a pattern in the synoptic vicissitudes the charts depict. The general classes of method available are subjective weather typing, and objective specification. Before discussing them, we should reiterate the basis and purpose of synoptic work. Synoptic meteorology aims to relate all observable weather phenomena at one instant in time, in an area that may vary in size from a few square miles to the entire surface of the Earth, to the instantaneous distribution of pressure. The ultimate purpose is the complete prediction of the type and distribution of weather phenomena at any desired time in the future on the basis of (a) the existing pressure pattern, and (b) synoptic relations known to exist in conformity with the laws of physical and dynamic meteorology. Because of geographical factors, many synoptic relations are known that cannot, as yet, be completely explained in terms of mathematics and physics. *Synoptic climatology* is the formal study of such relations. Given a synoptic chart, we are able—on the basis of our knowledge of synoptic meteorology—to predict what the weather ought to be in the area depicted, provided geographical influences are unimportant. Knowledge of synoptic climatology, on the other hand, enables us to state the actual weather associated with the given pressure pattern in the area concerned, on normal occasions in the past.

The most elementary form of synoptic climatology involves the determination, by simple inspection of the charts, of the most common pressure patterns over an area. In itself, this can be quite an instructive exercise, and is a good way of "getting to know the weather" of new areas. Details of weather types obtained in this manner have been published for many parts of the world; Table 6.7 lists those applying to the British Isles.[161] Various pictorial or graphic methods may be used to bring into prominence how the incidence of each of these types differs from year to year, and so provide a simple way of following the fluctuations of the general circulation.

If the data may be made available on punched cards, a more complete form of subjective weather typing may be carried out. A classic example of this is the synoptic

TABLE 6.7.
Weather Types of the British Isles (after H. H. Lamb).

Type	Synoptic pattern	Monthly frequencies (1938–1961)												Annual Frequencies
		J	F	M	A	M	J	J	A	S	O	N	D	
Westerly	High pressure to the south of Britain, low pressure to the north	32.5[a]	29.8	22.2	27.8	14.3	33.2	32.4	29.8	37.5	31.6	32.2	38.6	27.0[b] 38.1[c] 30.1[d]
Northwesterly	Azores anticyclone displaced northeast toward the British Isles.	4.2	5.5	4.4	5.8	3.9	5.4	7.8	5.2	5.4	3.0	5.6	6.9	6.7[b] 4.8[c] 5.3[d]
Northerly	High pressure to the northwest of Britain, low pressure over the Baltic and North Sea.	9.9	10.0	7.4	10.4	11.4	10.4	8.7	9.1	4.9	7.1	8.1	5.9	6.3[b] 4.8[c] 8.6[d]
Easterly	Anticyclones from Scandinavia toward Iceland, low pressure to the south of Britain.	9.8	13.1	15.1	9.4	16.9	6.3	4.3	6.3	7.2	7.7	8.1	7.4	7.9[b] 8.2[c] 9.3[d]
Southerly	High pressure over central and northern Europe; Atlantic depressions blocked west of British Isles, traveling north.	12.9	12.5	18.0	10.3	11.7	6.8	5.5	8.7	11.0	17.3	15.0	12.4	14.3[b] 9.6[c] 11.9[d]
Anticyclonic	Anticyclones or cols centered over, near, or extending over the British Isles.	11.6	12.5	17.7	20.4	19.5	16.3	15.9	14.8	15.3	16.1	11.4	12.9	16.6[b] 15.0[c] 15.3[d]
Cyclonic	Depressions stagnating over, or frequently passing across, the British Isles.	12.4	11.9	8.6	10.4	13.8	14.4	20.6	19.6	13.5	11.6	13.9	10.2	12.1[b] 15.1[c] 13.4[d]

[a] All figures are percentages. Frequencies do not add up to 100 per cent because of a small residue of unclassifiable days.
[b] Average for 1873–1897. [c] Average for 1898–1937. [d] Average for 1938–1961.

climatology of Hokkaido produced by the Weather Division of the U.S.A.F. during the Second World War.[162] The aim was to provide weather information for bombing operations over Japan that would be more accurate than that supplied by normal synoptic forecasting techniques. The work involved two stages. First, all available daily synoptic (sea-level) charts were classified according to the direction of the gradient wind, the speed of the gradient wind (in terms of three or more velocity groups), isobar curvature, and air-mass type. For this purpose, Hokkaido was divided into areas, each of which could be characterized by a single gradient-wind direction—100,000 to 200,000 square miles is the maximum possible size for such areas—and the synoptic variables were determined separately for each area. Second, all available weather data for the date and time of each synoptic chart were punched on cards. The data could then be summarized mechanically, both singly and combined in terms of the synoptic variables, in the form of frequency distributions. Hence, the probability of a weather condition, or combination of weather conditions, at a given place or over a given area, could be obtained rapidly, provided the synoptic situation was known (see Appendix 6.21). In addition, it became very easy to construct maps showing the probability of each weather condition, given a particular weather type (see Figure 6.36). From a knowledge of the probabilities, it was also possible to obtain (by means of area coupling) the length of time during which a given weather type might persist, the weather type most likely to succeed the prevailing one, and the weather type likely to dominate the region much larger than the area used initially to determine the gradient-wind direction.

Since the Hokkaido example, there have been numerous studies made of such subjective synoptic climatology. The narrower the range of weather events to be studied and the more precise the specifications in terms of which these events are to be described, the more useful the method becomes. For example, the occurrence of dry and

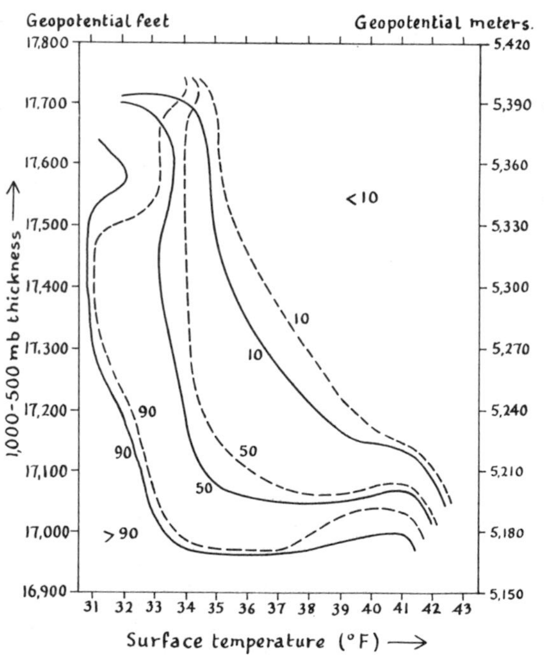

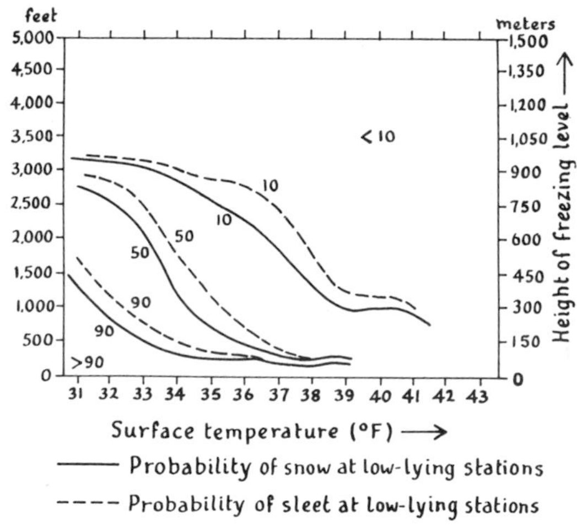

FIGURE 6.35.
Percentage probability of snow or sleet in relation to synoptic parameters (after R. A. Murray). Data are for British stations below 300 feet.

wet trade winds (corresponding to daily precipitation amounts of 0.3 inches or less, and 3 inches or more, respectively) over Hawaii may be determined from the prevailing weather type, defined in terms of the surface pressure pattern, the 700-mb and 500-mb contour patterns, and the 1,000-mb to 500-mb and 1,000-mb to 700-mb thickness patterns.[163] Figure 6.35 gives another example.

The disadvantage of the subjective method of weather typing is that it is impossible to devise a classification of pressure patterns that is completely unambiguous, or that will cover all possible situations. Consequently, objective classification systems are very desirable. The most well-known method in use at present is the *objective specification technique* of pressure-pattern description.[164] The basic idea involved is the representation of the configuration of an isobaric surface (or the horizontal distribution of isobars at constant height) by means of a mathematical equation, instead of by reference to a set of discrete synoptic situations. This procedure can generate a continuous spectrum of patterns, and avoid the difficulties of the subjective approach, especially the difficulty of determining the true airflow direction in some pressure patterns, and the difficulty in matching other patterns against any one discrete weather type.

In effect, the objective specification technique involves splitting the isobaric surface to be described into simple surfaces, each of which may be represented by simple equations, which when added together algebraically approximate the actual surface. The degree of agreement between theoretical and actual surfaces is computed by means of least squares or correlation coefficients. In practice, the equations used are usually Fisher-Tschebyscheff orthonormal polynomials (see Appendix 6.22), but the principle may be illustrated by the expression

$$P_i = \bar{P} + \alpha l + \beta L,$$

in which P_i represents atmospheric pressure (or isobaric-contour height) at point i on a synoptic chart; $\bar{P}$ is the mean pressure over the area covered by the chart; l and L represent the latitude and longitude, respectively, of point i, and α, β, are constants. A rectangular grid is superimposed over the chart, and P_i read off at each grid-line intersection. $\bar{P}$ is determined as the mean of all pressure values at the intersections. Constants α, β, may be determined when values of P_i, l, and L have been read off from adjacent points. The resulting expression forms a description of the pressure surface in the area between the two points. The actual procedure differs from this simplified illustration, but essentially results in equations whose coefficients α, β, etc. form an objective description of the pressure pattern.[165] The greater the number of grid points, the more accurate these coefficients become. The number of terms in the final equations is the same as the number of grid points, so that an electronic calculating machine must be used to obtain the coefficients.

Wadsworth used a grid of 13 by 13 lines to specify the circulation over the entire Arctic at 500 mb level and at sea level for 1955, and Malone and Friedman specified the patterns of sea-level pressure and 700-mb contours, respectively, over North America.[166] The sea-level pattern was found to give the best description of surface temperatures at a specific hour, and the 700-mb pattern gave the best description of 24-hour precipitation amounts. Friedman found that, whereas subjective weather typing of the 700-mb contour pattern accounted for only 22 per cent of the variance of surface temperatures and for only 25 per cent of the precipitation variance, the objective specification technique accounted for 46 per cent of the observed temperature variance, and for 50 per cent of the observed precipitation variance. Thus the objective

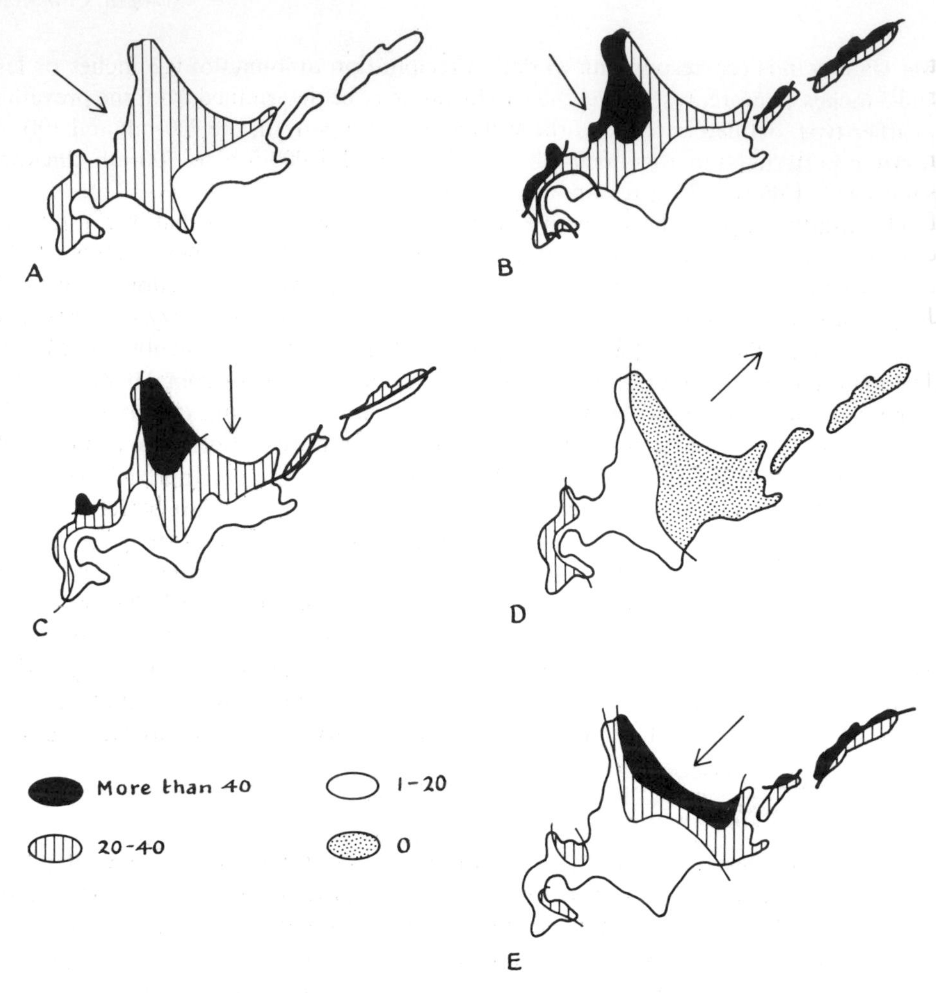

FIGURE 6.36.
Synoptic climatology of overcast skies, Hokkaido (after W. C. Jacobs). A, Mean situation. B, Northwesterly weather type. C, Northerly weather type. D, Southwesterly weather type. E, Northeasterly weather type. Figures give the percentage frequency of days having overcast skies and precipitation at 6:00 A.M. in winter.

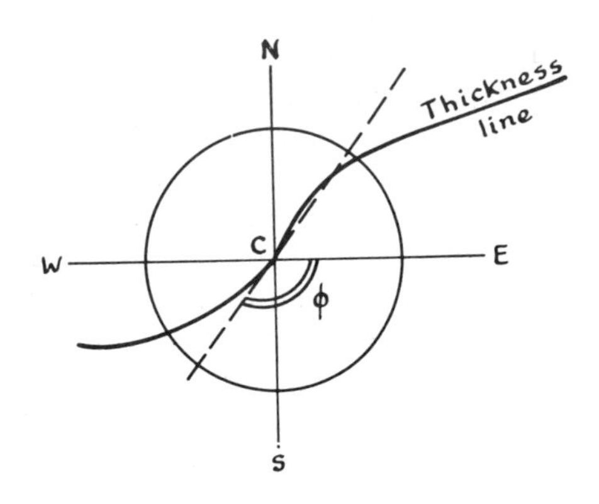

FIGURE 6.37.
C is the center of the storm. Angle ϕ gives the orientation of the thickness line through C.

technique extracts twice as much information from the same data as the subjective method.

Numerous examples of the application of the standard objective specification technique are available; all of them require access to an electronic computer. In addition, studies have been made that do not follow the standard technique but are nonetheless fully objective, and often permit specification of the synoptic information in more detail than the standard approach. For example, in a computer-derived synoptic climatology of precipitation from winter storms over the eastern United States, Jorgensen recognizes that synoptic features are not static (as is assumed in the standard objective technique), but move in some general direction and at the same time rotate. The grid system he adopted moves with the synoptic feature, and permits the latter to rotate about the origin, so that both its directional and its asymmetric characteristics may be included in the description. The initial data involve 6-hourly precipitation amounts at the various stations, related to the sea-level location of the center of the storm that produced them, the latitude and longitude of the stations and of the position of the sea-level low-pressure centers, and the angle (measured clockwise from east) of the orientation of the thickness line through the center of the storm (see Figure 6.37). The grid system is formed by great circles one degree of latitude apart along both abscissa and ordinate, the grid area forming a rectangle 18° of latitude on a side. The result is a network of 324 cells, each one degree of latitude square, the area of a cell becoming slightly less toward the corners of the grid because of the slight convergence of the great circles within the grid area. The 6-hourly precipitation amounts reported by all stations are allocated by a computer program to the appropriate grid cell, and the amounts are then considered to be located at the center of the cell. If more than one report is allocated to a cell, an average amount of precipitation is computed for that cell. The correct allocation of cells is a problem in spherical trigonometry (see Appendix 6.23), but, when completed, enables objective short-period precipitation forecasts to be made rapidly, because the computer printout gives the average precipitation reported in each grid cell during the six hours after the time when the center of the storm was at the given latitude and orientation. On the basis of the printout, probabilities may be computed, and probability maps constructed.

The classification of map patterns by statistical correlation techniques is a useful objective basis for a synoptic climatology. For example, daily sea-level pressure values at 22 stations in the northeast United States were used to describe the map patterns, and each map pattern so defined was correlated with every other map pattern. A number of distinctly different patterns were found to be more frequent than would be expected from chance, and these patterns were then regarded as weather "types." Linear correlation was used, and a correlation coefficient of 0.70 was employed as the lowest acceptable value. Of the ten weather types so determined (see Figure 6.38), types D, E, F, and G were found to be always associated with cloudy weather, and types I and J usually with sunny weather. Type E was usually associated with snow, type D with heavy snow, and types F and G always with precipitation; types I and J never had more than 0.50 inch (usually less than 0.01 inch) of precipitation.[167]

The fitting of a mathematically defined surface to weather observations is a useful method of objective analysis, and can be used to predict data for places where no data exists. Quadratic surfaces are often used.[168]

Whatever the form of synoptic climatology adopted, whether it be subjective or objective, the final result will be a description of the conditions experienced in a certain

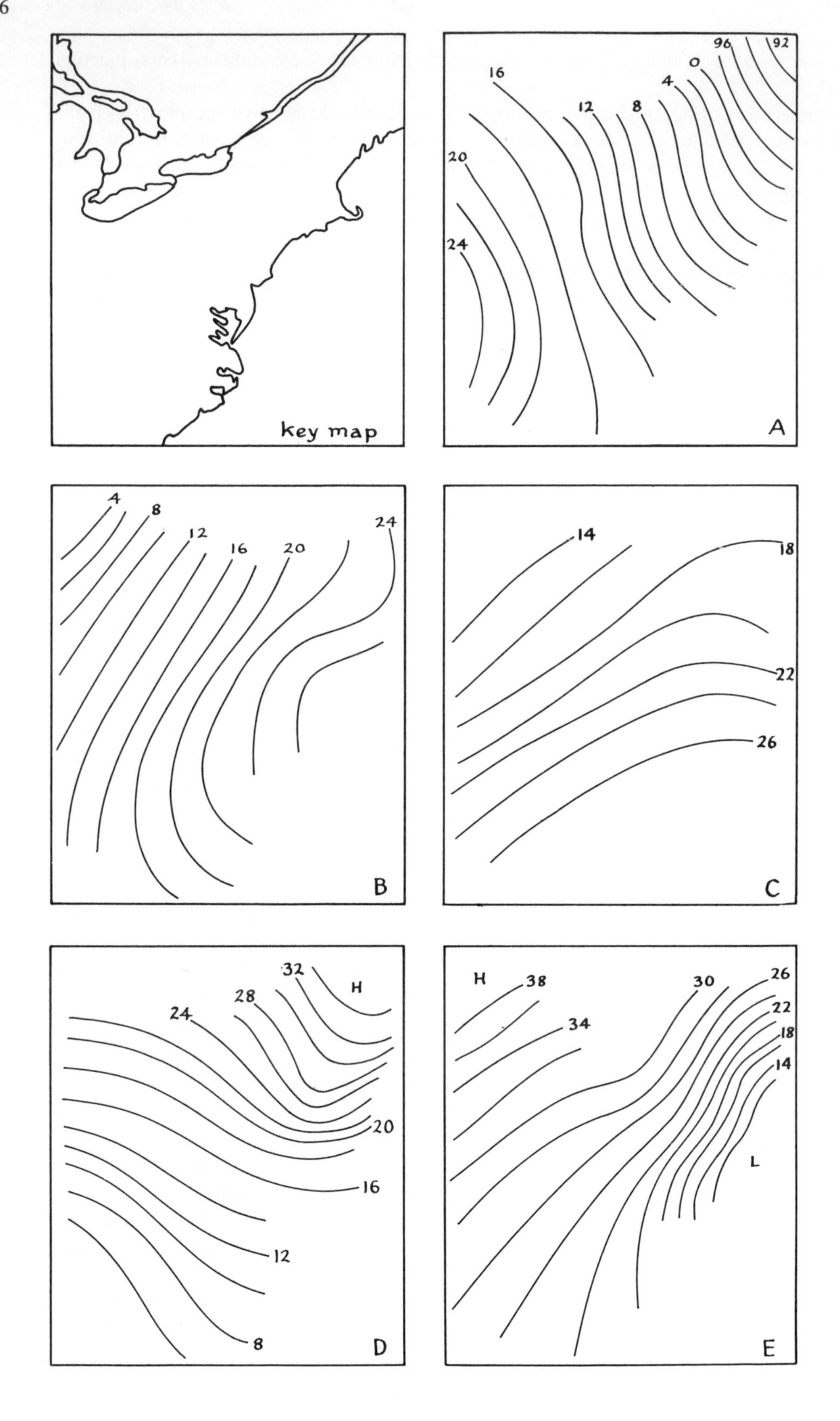

key map
16
12
8
4
0
96
92
20
24
A
4
8
12
16
20
24
B
14
18
22
26
C
32
H
28
24
20
16
12
8
D
H
38
34
30
26
22
18
14
L
E

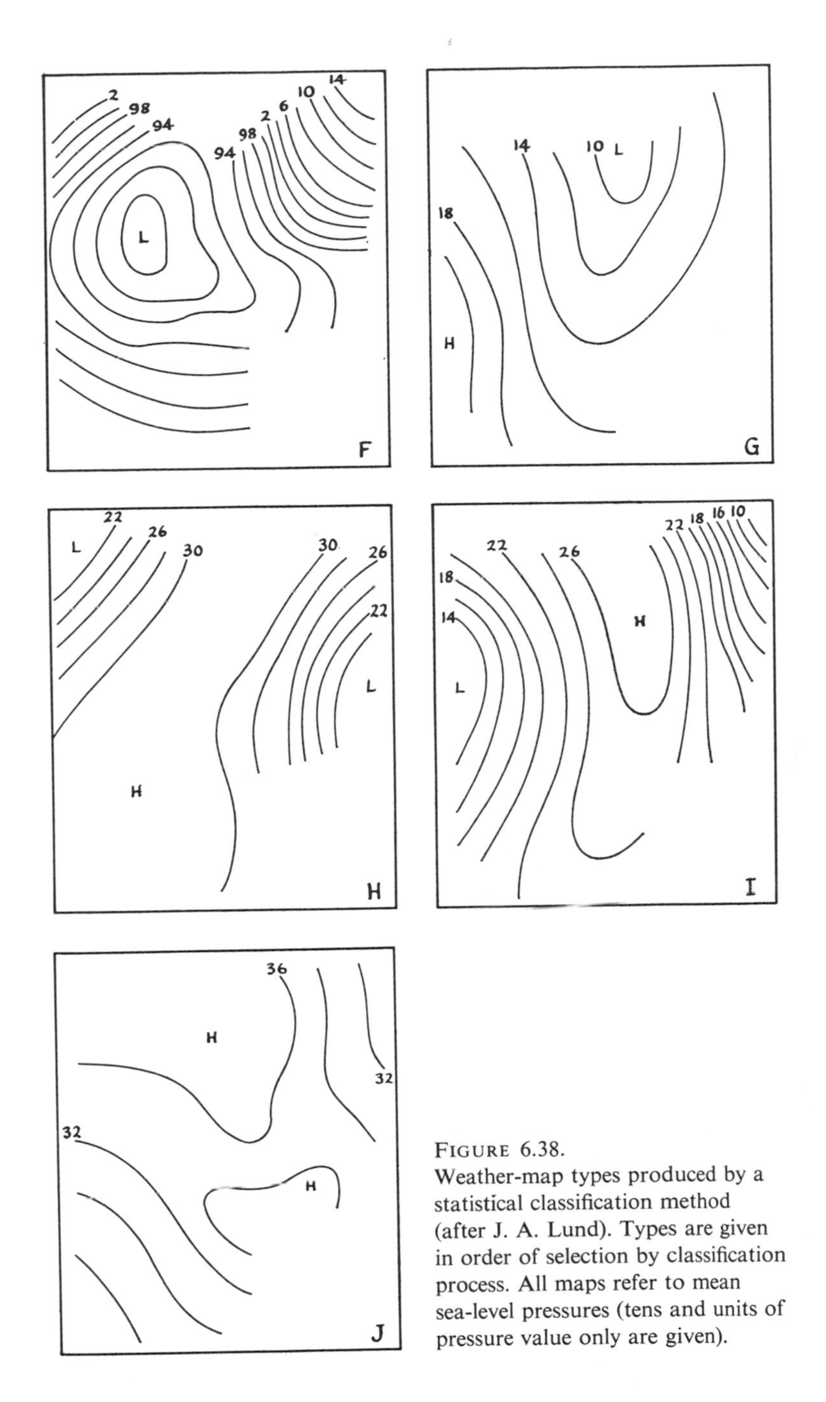

FIGURE 6.38.
Weather-map types produced by a statistical classification method (after J. A. Lund). Types are given in order of selection by classification process. All maps refer to mean sea-level pressures (tens and units of pressure value only are given).

area, for so many years, in terms of pressure maps. To be of practical use, these pressure patterns must be translated into actual weather. Such a translation may be in verbal descriptive terms; it may be in statistical terms, giving, for example, the probability of different ranges of surface temperature associated with a given pressure pattern; it may obtain the desired weather elements from the pressure pattern by an objective procedure; or it may express the relationship between pressure pattern and weather by means of maps.[169]

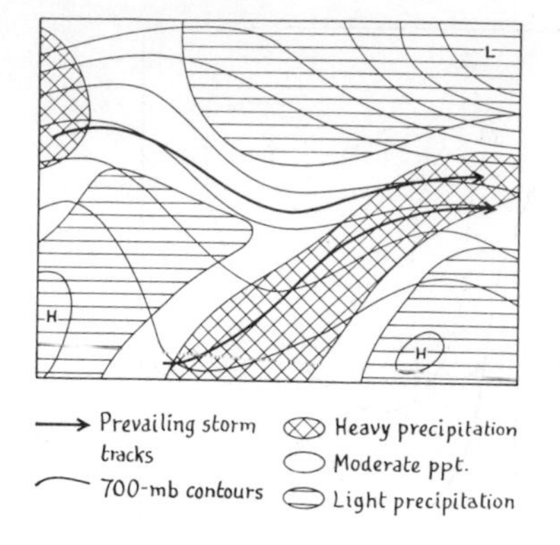

FIGURE 6.39.
Model of winter precipitation distribution in Tennessee Valley area in relation to 700-mb circulation (after W. H. Klein).

Some of the methods devised for translating pressure patterns into weather are especially valuable in that they emphasize very graphically how the weather at a single station is determined by synoptic developments throughout a very wide area. The relation between the 700-mb circulation and winter precipitation in the United States is a good example. Figure 6.39 is a map that indicates the five-day precipitation amounts associated with different locations on a 700-mb contour pattern. The map

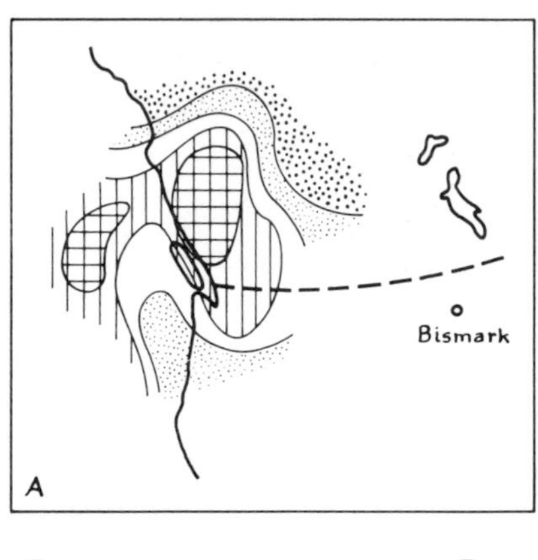

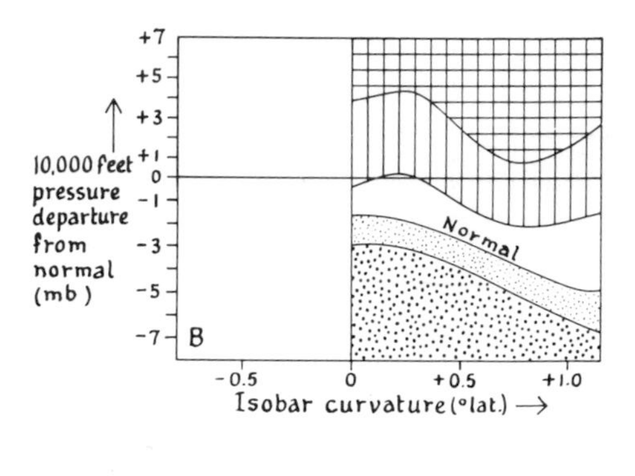

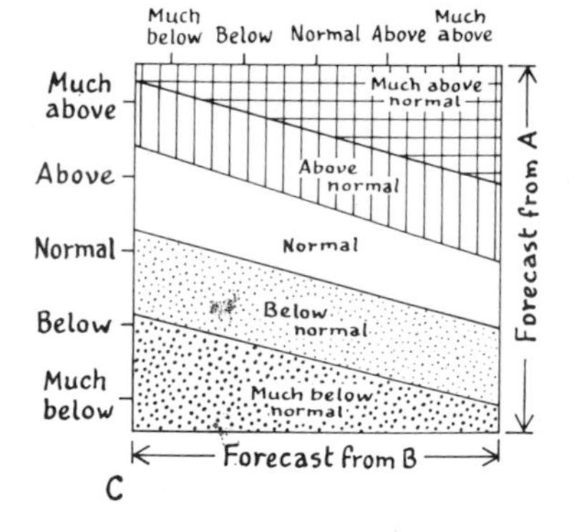

FIGURE 6.40.
Translation of pressure patterns and geostrophic winds into surface temperatures at Bismark, North Dakota (after Martin and Leight). A, First prediction diagram: monthly mean temperature expected at Bismark in winter according to location of source of 48-hour trajectories. B, Second prediction diagram: monthly mean temperature expected at Bismark in winter according to isobar curvature and pressure departure from normal. C, Final prediction diagram: monthly mean temperature expected at Bismark in winter according to trajectories, isobar curvature, and pressure anomaly.

was originally derived from precipitation data for the Tennessee Valley, but was found to apply also to central and eastern North America in general, and to daily and monthly precipitation amounts as well as five-day totals. It appears to give the correct rainfall distribution approximately two-thirds of the time in winter. In summer, the precipitation pattern must be shifted westward relative to the 700-mb pattern to obtain the correct result, the amount of shift being greatest just to the east of the main 700-mb trough line, and least just to the east of the ridge line. A later study, which correlates winter precipitation with departures of 700-mb height from normal, indicates that over most of the United States below-normal 700-mb heights are associated with relatively heavy winter precipitation, and above-normal heights with dry weather. However, in the northeastern states there is a slight tendency for heavy rain to be associated with above-normal heights.[170]

Monthly mean 700-mb patterns may be converted into patterns of surface-temperature anomaly by an objective technique. The required synoptic parameters are the isobar curvature, the departure of pressure from normal, and the air-parcel trajectory. These parameters are determined for the station in question (see Appendix 6.24) and then a comparison between the value of the parameters on different occasions with the observed surface temperature enables a set of correlation graphs to be produced (see Figure 6.40) that are in themselves very valuable contributions to the synoptic climatology of the area concerned.

Numerous examples of the description of actual weather and climate in terms of instantaneous or mean pressure patterns may be cited.[171] These are very useful, indeed, provided two complicating factors are kept in mind. The first of these concerns the existence of singularities. It is sometimes (but not always) possible to distinguish synoptic patterns that seem particularly prone to have singularities.[172] Singularities are, in effect, date-bound synoptic patterns, and as such they depend on the state of the general atmospheric circulation: they are pronounced in middle latitudes at times of low zonal index, but tend to be obliterated during high-index periods. Consequently, the singularity concept provides the link between everyday weather (i.e., the synoptic situation) and the planetary climate (i.e., the general circulation). Pronounced singularities in Europe, which need to be taken into account in all studies of synoptic climate in the continent, include the "ice saints" singularity in mid-May, the monsoon rains at the end of June, the "old wives anticyclone" in summer, and the Christmas warm spell in December. Pronounced singularities in the United States include the January thaw (between January 20 and 23), the early winter invasion on October 15, which occurs almost one month after the equinox and is accompanied by the passage of a major cold front down the whole of the eastern seaboard, with snowfall in most areas east of the Rocky Mountains, and the rainfall minimum ("Indian summer") of late September. The geographical differentiation of the Earth's surface may have something to do with the variations in the incidence of singularities that are apparent in most parts of the world.[173]

The second complicating factor concerns local weather. The literature of meteorology and climatology contains many empirical descriptions of weather in some very restricted area, which is not described in synoptic terms. Sometimes the description is purely empirical because the observer knows nothing about modern meteorology, but more often the synoptic situation proves not to hold the full explanation of the observed weather. In other words, the general synoptic situation explains only part of the weather, and the remaining facts must merely be stated baldly, if the investigator is not prepared to commence a micrometeorological study to explain them. Local weather is thus a distinct field of study, and various examples may be quoted in which the actual weather owes as much to the physical characteristics of the geographical locality as to the physics of the atmosphere.[174]

On occasion it will be found that certain local weather patterns may be explained logically from the synoptic pattern when the nature of the Earth's surface in the locality concerned is taken into account. For example, comparatively minor orographic factors may have an important influence on clouds and precipitation. When the airflow direction is parallel to the coast, with land to the right, a precipitation maximum should be experienced along the coastal strip, even if the latter is a flat plain. The reason for this is simply the difference in surface roughness between the sea surface and a flat land surface: the angular deviation of the wind from the isobars increases as the roughness of the underlying surface increases. A zone of frictional convergence must develop where the wind blows along the coast with the land to the right, and a zone of frictional divergence where the wind blows along the coast with the land to the left. If the air is near saturation, clouds and precipitation will be produced in the rising air of the convergence zone.[175]

The possibility of the existence of local weather completely different from that expected from the synoptic chart means that there is abundant scope for field observation in synoptic climatology. Almost any detailed description of the weather in a specific locality at a given time, provided the observer knows his meteorology, will, after comparison with the weather according to the synoptic chart, provide new knowledge. Whether this knowledge will remain purely local and empirical, or will find application in a much larger field, depends entirely on the observer's understanding of the principles of physical and dynamic meteorology, that is, on how successful he is in relating them in physical (and preferably mathematical) terms to the distributions displayed on the synoptic chart.

The task of the synoptic climatologist in the field is to determine the "normal" weather experienced in a locality under each and every synoptic situation to which the region is subject. To do this successfully, he must distinguish between the "synoptic weather," i.e., the weather that ought to occur according to the chart, and the "geographical weather," which is the local atmospheric state induced by the specific combination of site conditions in the given locality. His starting point must therefore be the state of the atmospheric circulation at the time of his observation, which he must determine from *two* synoptic charts, i.e., the *actual* chart for the preceding synoptic hour, and the *forecast* chart for the succeeding synoptic hour. Although useful

TABLE 6.8
Combinations of Weather Charts.

Actual chart	Forecast chart	Examples
Subjective	Subjective	Old weather forecast procedure; countries with small meteorological depts.
Subjective	Objective	Classic 30-day forecasts.
Objective	Subjective	Outstation practice.
Objective	Objective	Weather centrals.

actual and forecast charts may be constructed from public radio broadcasts, for accurate work the investigator needs all available data and therefore may have to obtain access to a teletype or facsimile, or to a short-wave receiver, preferably fitted to an automatic program timer.[176]

For accurate synoptic fieldwork, the observations should be fitted (by an appropriate statistical forecasting scheme) to an *objectively* drawn pressure map. This pressure map may represent either actual or forecast situations. Four combinations of charts are possible, as indicated in Table 6.8. Depending on the area in which he is working, the field synoptic climatologist may encounter any of these possibilities, and it is important that he should understand the basis on which the forecast chart has been compiled.

When both actual and forecast charts are subjective, the actual chart will have been constructed by fitting fronts and isobars to the station observations by visual estimation, followed by subjective mutual adjustment of frontal and isobaric locations so that continuity with previous actual charts is preserved. The forecast chart will then have been produced by extrapolating the movement of pressure centers and fronts in accordance with empirical rules and kinematic computations (see Appendix 6.25). In themselves, the empirical rules formulated on the basis of past experience in a certain area constitute valuable contributions to the climatography of the area.

Most forecast charts produced by state meteorological services are now, at least to some extent, objective. There is an extensive literature dealing with the preparation of forecast surface (*prebaratic*) and forecast upper-air (*prontour*) charts. The degree of objectivity varies considerably from country to country, depending on the financial and computer resources of the meteorological organization. For example, the forecast charts produced in Britain by the Meteorological Office are now very largely objective, based on the Sawyer-Bushby model (on which see Appendix 3.18 in *Techniques*), but until fairly recently, a semiobjective procedure was followed for the production of prebaratics or prontours, and this procedure is probably the best one for the synoptic climatologist to use in his own forecast studies. The stages are as follows.

1. Extrapolate recent trends in the development or decay of pressure systems, using charts for the last 24 hours to give a first approximation, then subjectively adjust it on the basis of the latest available pressure tendencies. Isallobaric charts (which depict isopleths of equal pressure tendency) are maintained for 3:00 A.M. and 12:00 NOON G.M.T., adjusted for diurnal pressure variation.
2. Locate favorable areas for cyclonic or anticyclonic development as indicated by the latest actual thickness chart. These two steps lead to the production of the prebaratic chart, which depicts the forecast distribution of surface pressure.
3. Allow for nonadvective dynamic and thermodynamic effects, which may affect the future thickness pattern.[177]
4. Produce a *prethickness chart* (i.e., a forecast thickness chart) based on steps 2 and 3, making use of the rule that thickness isopleths move with the speed of the gradient wind through the isobaric layer whose mean temperature they describe; i.e., the thickness lines should move as if they are embedded in the surface wind field.
5. Finally, superimpose the prethickness chart on the prebaratic chart, and add the two together algebraically; i.e., carry out "gridding." The resulting chart is the prontour chart, depicting the forecast distribution of upper-air pressure.

Once a forecast chart has been produced, there still exists the problem of fitting the

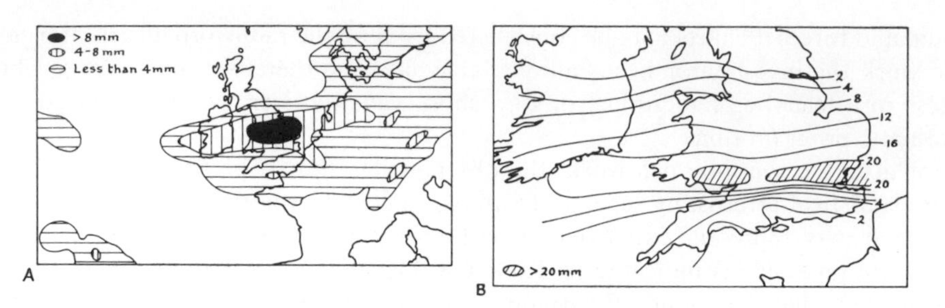

FIGURE 6.41.
Numerical prediction of precipitation by the Bushby-Timpson ten-level primitive equation model (after F. H. Bushby and M. S. Timpson). A, Forecast total rainfall, midnight to midnight G.M.T., December 1, 1961. B, Actual total rainfall during same 24 hours.

weather into it, i.e., translating the pressure pattern into actual weather terms. Obviously, this is where synoptic climatology contributes to weather forecasting. Although the development of better and better mathematical models will undoubtedly enable the objective prediction of certain weather elements to be realized with fair accuracy—for example, the objective short-period prediction of the broad-scale distribution of precipitation in the British Isles is now possible[178]—when local precision is required, the techniques of the synoptic climatologist are absolutely vital (see Figure 6.41).

Synoptic study, because of its contact with actuality, is often unpalatable for the theoretical climatologist. It does, in fact, involve a number of problems quite distinct from those already described. The topics we have discussed in this chapter have been concerned mainly with the logical fundamentals of synoptic work. In addition, there are numerous practical problems.

First, there is the question of the provision of synoptic data. Although the network of synoptic stations is dense enough for global-scale weather-forecasting services in middle latitudes, it is not dense enough in polar and tropical regions, and especially not in the latter. Although more than 60 years of published daily synoptic charts are available for latitudes north of 20°N, only 18 months of records of global synoptics for the tropics existed in 1965.[179] For regional-scale or local-scale forecasting, the reporting networks are not dense enough in many parts of the middle latitudes, and indeed the necessity for recognition of the field of synoptic climatology springs from the forecaster's need for a prediction technique that can be applied in areas that have too few stations.

Extension of synoptic networks in polar and tropical regions is limited not only by financial conditions, but by scientific and operational considerations too. For example, the maintenance of weather stations in polar regions presents very unusual problems for observers accustomed to middle-latitude practice. Drifting snow makes instrument exposure very difficult, and the low temperatures cause the moving parts of the instruments to freeze. The observer has considerable difficulty in making his observations: his breath condenses on the lenses of his theodolite, thus rendering pilot-balloon tracking an even more delicate operation than usual, and his presence near a thermometer when temperatures are low can easily upset the reading. Precipitation is usually in the form of snow, and is therefore measured rather imprecisely. Visual observations, for example, of cloud amount and type, are estimations rather than measurements during the polar night. A common procedure is to assume that clouds are present if the stars cannot be seen, and that such clouds must be cirrus or cirrostratus, because of the low temperatures.[180]

Owing to the peculiar geographical conditions in polar regions, observations made at stations with middle-latitude synoptics in mind may completely overlook very important weather controls. For example, the low-level inversion that blankets most parts of Arctic regions in winter, and covers the Arctic Sea in summer, is a characteristic feature of the polar troposphere that is not represented at all on the conventional synoptic charts used for both middle- and polar-latitude work.[181]

To increase the number of synoptic stations in polar regions, automatic stations mounted in inaccessible land sites or on buoys may be used. An alternative way of increasing the supply of synoptic information from such regions is to develop synoptic models instead of increasing the number of stations. For example, because of a lack of synoptic stations in Antarctica, the Australian synopticians introduced a new series of synoptic models for the area south of latitude 30°S. An essential step was the recognition of fronts as series of significant "lines" of weather, generally associated with cyclones or cols, and often of very limited extent and life. The fronts are drawn as either zonal or meridional features. The meridional fronts, appearing initially in the deformation field between migrating anticyclones in the subtropical high-pressure belt, are rarely fronts in the hydrodynamic sense.[182]

To increase the value of synoptic information provided by existing stations, by means of specially designed synoptic models, is more difficult in the tropics. The equations of dynamic meteorology provide a sure pattern that the models must follow in polar latitudes, but the equations for tropical air motion require special care. Equations of horizontal motion derived for the tropics suggest that absolute vorticity should determine the pressure-wind relationship in tropical latitudes. The equations lead to the concepts of antiflow and a kinematic equator, which are unnecessary in middle-latitude synoptic work, but which must be introduced in tropical work if the analysis is to be carried up to and across the geographical equator.[183] The synoptic difficulties of the equatorial zone may also be circumvented to some extent by means of the *furrow* model. The furrow is, in effect, a synoptic form defined by streamlines that corresponds to a wave-form in a pressure-contour pattern. The equatorial zone, which, synoptically speaking, has a width of about 500 miles, is a zone in which the streamlines of the airflow depart considerably from the pressure contours, partly because the down-gradient component of the wind (i.e., the velocity potential) is a very important element in the balance of forces, and partly because the irrotational component of the wind is usually very pronounced away from the areas of tropical disturbances. The furrow takes the form of a curved streamline pattern astride the equator that seems to resemble a pressure-trough-and-ridge pattern; it corresponds, in isobaric terms, to a cusp in a meridional airstream, also across the equator, that encloses one or more pure indrafts with their associated neutral points. The furrow produces an ageostrophic cross-equatorial curvature of the streamlines, which permits the east-west trending isobars on the poleward margins of the equatorial zone of one hemisphere to transform smoothly to the east-west isobars in the other hemisphere. Furrows are usually associated with pronounced meridional pressure gradients at the equator. Over the East African coast, the furrow enables the easterly trades associated with an anticyclonic circulation to cross the equator and form a cyclonic circulation in the opposite hemisphere. The direction of flow across the equator, i.e., northward or southward along the coast, depends on the season.[184]

Quite apart from what may be termed meteorological difficulties in the provision of synoptic information, there are serious difficulties with the flow of synoptic data.[185] Synoptic stations operated on a routine basis are expensive. In 1963, the cost of a single radiosonde ascent in the United States was 50 dollars, roughly 10 per cent of the monthly income of an average family. There are about 10,000 radiosonde ascents per month in the United States, indicating that the annual expenditure per year for all soundings in that country is 10 cents per family. This figure is much less than the annual American expenditure on private telephone requests for standard local weather forecasts, and it may therefore be taken as a measure of the minimum price that a national community is willing to pay for the provision of general weather information.[186]

Telecommunication problems are acute in the synoptic field. An enormous amount of data must be transmitted, and transmission schedules must therefore be very tight. Delays can be very serious under such conditions. Financial considerations set a definite limit to the number of telecommunication circuits that may be set up, and increases in the amount of data transmitted obviously require expensive modifications. One method is to increase the modulation rate of the circuits: existing meteorological circuits do not operate at modulation rates in excess of 75 bauds, but channels operating at 100 and 200 bauds and over are quite feasible technically.* An alternative is to use standard telephone circuits employing the party-line system. Most meteorological teleprinters transmit 60 to 100 words per minute, whereas up to 3,000 words per minute can be transmitted via telephone. A selective-station scheme may be used; i.e., each station may select the data it wishes to receive by means of a data-selection unit operating on the party-line system. This avoids the difficulties of the existing system, in which data required by only one station are relayed to all other stations on the circuit.[187]

Although facsimile transmissions save time for the synoptician by providing him with an already completed chart, the actual rate of transmission is slow. For example, the U.S. facsimile broadcasts in 1957 took 96 minutes to produce a 1:20,000,000 chart of the entire northern hemisphere analysis. However, it is possible to increase considerably this rate of transmission. Facsimile broadcasts have the advantage over radio-telegraphy or radioteleprinter broadcasts in that they are less subject to fading and interference than the latter, and also in that they enable the transmission of coded five-figure groups three times as rapidly as radioteleprinter transmission, and seven times as rapidly as hand-operated Morse.[188]

Finally, some new lines of development in the provision of synoptic information may be mentioned. The application of the methods of systems engineering can undoubtedly increase the efficiency of the weather information services. Electronic systems can speed up considerably the flow of information; for example, numerical prediction charts can be produced for the entire northern hemisphere in twenty minutes, and transmitted to out-station computers for display in five or six minutes. To introduce a completely different approach to the information problem, statistical theory enables extrapolation formulae (for the movement of isobars, isotherms, etc.) to be generalized in a manner that takes into account the uncertainty about initial states

* One *baud* is a transmission rate of one binary digit per second.

caused by the lack of information between observing stations. In fact, the application of mathematics and logic to the problem of making the most of existing synoptic networks has until recently received insufficient attention. Investigation of the mathematical properties of isolated "holes," i.e., regions for which no synoptic data are available, has much to offer in the provision of synoptic analyses that are logically consistent. It may be shown from the properties of the prediction system in use that the general level of accuracy in a "hole" approaches a definite upper limit that is determined by the dimensions and shape of the "hole," the average rate of flow of atmospheric properties across its boundaries, and the inherent error in the prediction system. Different sizes or hierarchies of "holes" exist, and it is possible to predict the attainable standard of accuracy if the largest "holes" (such as the North Atlantic or the Arctic) are subdivided into a number of smaller "holes."[189]

Notes to Chapter 6

1. R. C. Sutcliffe, *QJRMS*, 78 (1952), 291.
2. T. Bergeron, *ASM*, p. 440, contains full references to the classic literature.
3. For details, see J. C. Bellamy, *CM*, p. 711. For details of the way synoptic information is handled at various centers, see: P. Johns, *BAMS*, 49 (1968), 259, on the Canadian Weather Service; P. G. Kesel, *loc. cit.*, p. 108, on the U.S. Naval Operations center, Monterey, Calif.; W. H. Klein, *BAMS*, 48 (1967), 890, on precipitation, U.S. Weather Bureau. For a suggested solution to organizational problems of synoptic forecasting, see R. Lee, *BAMS*, 47 (1966), 438. On the pictorial representation of wind data over the oceans, see L. Moskowitz, *BAMS*, 47 (1966), 114.
4. For details of British procedure, see Sutcliffe, *op. cit.* (1952), and C. K. M. Douglas, *QJRMS*, 78 (1952), 1.
5. See e.g., J. A. Steers, *An Introduction to the Study of Map Projections* (London, 7th ed., 1949), and more advanced, A. R. Hinks, *Map Projections* (Cambridge, Eng., 2d ed., 1942), for full discussion of the possible solutions.
6. J. B. Rigg, *W*, 12 (1957), 154.
7. G. E. Dunn, *CM*, p. 748. For details of the projections, see P. B. Sarson, *MM*, 91 (1961), 68. For details of the effect of the Polar stereographic projection on the calculation of horizontal curves, see K. Krishna, *MWR*, 96 (1968), 658.
8. R. F. Zobel, *MM*, 87 (1958), 44. On the representation of moisture on synoptic charts, see A. Papež, *MM*, 95 (1966), 210. See F. Mesinger, *GPA*, 64, no. 2 (1966), 178, for details of the vertical interpolation of data from constant-pressure charts to constant-density surfaces.
9. D. H. Johnson, *MM*, 83 (1954), 37.
10. S. Teweles, L. Rothernberg, and F. G. Finger, *MWR*, 88 (1960), 137.
11. See Chapter 3 in *Techniques* for a review of numerical models. On the Cressman technique, see G. P. Cressman, *MWR*, 87 (1959), 367, and S. Teweles and M. Snidero, *MWR*, 90 (1962), 147. The foundations of the method were laid by P. Bergthorssen and B. R. Döös, *T*, 7 (1955), 329. For a review of the results of the first six years of operational objective chart analysis in the United States, see E. B. Fawcett, *JAM*, 1 (1962), 318. See also R. M. Endlich and J. R. Clark, *JAM*, 2 (1963), 66, for details of objective computation of some meteorological quantities, and D. P. Petersen and D. Middleton, *T*, 15 (1963), 387, for a discussion of the smoothing process in objective analysis. For a discussion of the faults of the early JNWPU numerical models, see H. W. Elsaesser, *JAM*, 7 (1968), 153. For an application of the Cressman technique to the analysis of monthly and average rainfall patterns in Victoria, Austr., see R. Maine and D. J. Gauntlett, *loc. cit.*, p. 18.

12. G. A. Corby, *QJRMS*, 87 (1961), 34. On methods of objective analysis of synoptic information, see: H. B. Kruger, *QJRMS*, 95 (1969), 21, for a mathematical approach to the general problem; S. L. Barnes, *JAM*, 3 (1964), 396, on maximization of detail; F. G. Finger, H. M. Woolf, and C. E. Anderson, *MWR*, 93 (1965), 619, on stratospheric constant-pressure charts; D. P. Peterson, *T*, 20 (1968), 673, on optimum mixing of synoptic and prognostic information; M. B. Danard, M. M. Holl, and J. R. Clark, *MWR*, 96 (1968), 141, on correlation assembly; and W. J. Koss, *MWR*, 94 (1966), 237, on pressure-height data for the tropics. For examples of objective analysis, see: H. A. Bedient and J. Vederman, *MWR*, 92 (1964), 565, on upper winds in the tropical Pacific; R. M. Endlich and R. L. Mancuso, *MWR*, 96 (1968), 342, on thunderstorms and tornado environments; D. B. Spiegler, *JAM*, 7 (1968), 736, on the maximum-wind layer in the northern hemisphere; and A. Eddy, *JAM*, 6 (1967), 597 on 500-mb heights and temperatures in the northern hemisphere. See also E. O. Holopainen, *T*, 20 (1968), 128, on errors resulting from the use of a forecast as a "first guess" in the objective system; and L. S. Gandin and K. M. Lugina, *BWMO*, 18 (1969), 86, for a comparison of the accuracy of objective analyses produced by various meteorological services.

13. J. S. Sawyer, *MM*, 89 (1960), 187. On the application of cathode-ray tube contour charts to general circulation experiments, see W. M. Washington *et al.*, *BAMS*, 49 (1968), 822. On automated analysis, see F. H. Bushby, *QJRMS*, 91 (1965), 527.

14. For details, see W. J. Saucier, *Principles of Meteorological Analysis* (Chicago, 1953), p. 313.

15. E. F. Danielsen, *JM*, 18 (1961), 479.

16. D. Djuric, *JM*, 18 (1961), 597.

17. S.-K. Kao and M. Neiburger, *JGR*, 64 (1959), 1283. For the graphic method, see H. V. Goodyear, *MWR*, 87 (1959), 188. On the determination of three-dimensional trajectories by means of a numerical prediction model, see D. C. Barnum and J. W. Diercks, *JAM*, 8 (1969), 3. According to S. G. Cornford, *MM*, 97 (1968), 353, aircraft soundings along a 150-km trajectory over the North Sea indicate that the trajectory concept is invalid at low levels (up to 900 mb), because of differential advection and turbulent exchange.

18. S. Teweles, *MWR*, 89 (1961), 125.

19. P. F. McAllen, *MM*, 91 (1962), 157.

20. W. M. McMurray, *MWR*, 89 (1961), 549.

21. C. A. Carpenter, *BAMS*, 41 (1960), 68.

22. M. J. Kerley, *MM*, 90 (1961), and 89 (1960), 297.

23. R. G. Fleagle, *JM*, 4 (1947), 165. For cross sections of wind shear (surface to 25 km), see D. A. Stewart, *JAM*, 6 (1967), 724. On objective analysis of cross sections, see R. T. Duquet, E. F. Danielsen, and N. R. Phares, *JAM*, 5 (1966), 233.

24. A. V. Carlin, *MWR*, 82 (1954), 97, gives examples. For examples of three-dimensional (anaglyph) layer charts, see J. M. Adams and C. Malone, *W*, 24 (1969), 273.

25. R. C. Sutcliffe, *MM*, 77 (1948), 147. For a discussion of the use of thickness patterns over the Southern Ocean, see J. J. Taljaard and H. van Loon, *AM*, p. 96.

26. The discussion here follows R. C. Sutcliffe and A. G. Forsdyke, *QJRMS*, 76 (1950), 189.

27. For examples, see J. M. Craddock and C. A. S. Lowndes, *MOPN*, no. 126 (1958). On total thickness anomalies in the southern hemisphere, see J. J. Taljaard, *NOTOS*, 16 (1967), 3. For synoptic charts of *dewpoint thickness*, see T. H. Kirk, *MM*, 94 (1965), 214.

28. R. C. Sutcliffe, *QJRMS*, 77 (1951), 435, and A. G. Forsdyke, *MM*, 78 (1949), 127.

29. For details, see Sutcliffe and Forsdyke, *op. cit.* (1950).

30. M. K. Miles, *W*, 16 (1961), 349.

31. For details, see J. S. Sawyer, *MM*, 79 (1950), 1.

32. See J. S. Sawyer, *MM*, 78 (1949), 189, for details.

33. See *MM*, 82 (1953), 81.

34. For details, see C. K. M. Douglas, *MM*, 76 (1947), 225, and E. J. Sumner, *MM*, 82 (1953), 291.

35. M. K. Miles, *MM*, 87 (1958), 1.

36. M. K. Miles, *MM*, 83 (1954), 338.

37. C. A. S. Lowndes, *MM*, 83 (1954), 40.

38. G. A. Howkins, *MM*, 91 (1962), 10.

39. F. E. Lumb, *MM*, 79 (1950), 191.

40. For details, see R. Murray, *MM*, 78 (1949), 132.

41. For details, see *ibid.*, p. 349.

42. R. Murray, *MM*, 88 (1959), 324.

43. "M. O. discussion," *MM*, 87 (1958), 179.

44. A. J. Wagner, *BAMS*, 38 (1957), 584.

45. For maps and details, see C. L. Kibler, C. M. Lennahan, and R. H. Martin, *MWR*, 83 (1955), 23.

46. C. P. Mook, *MWR*, 82 (1954), 160.

47. See W. H. Klein, *JM*, 15 (1958), 98, and D. L. Bradbury, *BAMS*, 39 (1958), 149, for examples. For normal frequency maps for the northern hemisphere, see *WAF*, I, 267.

48. S. Petterssen, G. E. Dunn, and L. L. Means, *JM*, 12 (1955), 58.

49. For a discussion of this factor, see *WAF*, I, 257–90, 320–70, and II, 1–33, 196–224.

50. *WAF*, I, 265 and 259. See also H.-L. Kuo, *JAS*, 26 (1969), 390, for a mathematical model which demonstrates that, in an air current with a non-uniform vorticity distribution, cyclonic or anticyclonic vortices should move into regions of higher or lower absolute vorticity, respectively. This explains why lesser cyclones tend to move into the large, semipermanent Aleutian and Icelandic lows, and also why typhoons and hurricanes tend to move toward higher latitudes.

51. R. T. Duquet and J. Spar, *JM*, 14 (1957), 251. For monthly mean maps, see C. L. Hosler and L. A. Gamage, *MWR*, 84 (1956), 388.

52. *WAF*, I, 325–29. J. E. Miller, *JM*, 3 (1946), 31 and 126, and J. E. Miller and H. T. Mantis, *JM*, 4 (1947), 29 and 104, give examples from the Atlantic coast region of the United States and the Pacific coast region of Asia, respectively.

53. For maps of the March 1962 storm, see A. I. Cooperman and H. E. Rosendal, *MWR*, 91 (1963), 337. D. L. Bradbury, *JM*, 14 (1957), 559. E. A. Di-Loreto and M. Hamada, *MWR*, 86 (1958), 277. P. Williams, Jr., and E. L. Peck, *JAM*, 1 (1962), 343.

54. A. J. Abdullah, *JM*, 6 (1949), 86. E. M. Ballenzweig, *BAMS*, 36 (1955), 318. For an example, see M. F. Grace and V. G. Bohl, *MWR*, 85 (1957), 350.

55. J. M. Austin, *JM*, 4 (1947), 16. Also see R. W. Longley, *loc. cit.*, p. 202. For synoptic studies of cyclones, see: C. Pope, Jr., *MWR*, 96 (1968), 867, on winter cyclogenesis with tropical characteristics over the Gulf Stream; K. G. Mowla, *T*, 20 (1968), 151, on cyclogenesis in the Bay of Bengal and Arabian Sea; I. Dawson, *W*, 20 (1965), 275, on a 500-mb low bringing record rainfall to Malta; T. N. Carlson, *MWR*, 95 (1967), 763, on a steady-state upper cold low in the Caribbean; D. Stranz, *GPA*, 66, no. 1 (1967), 103, on a stationary cold vortex bringing cold spells in South Africa; C. M. Stephenson, *W*, 23 (1968), 156, on polar lows bringing heavy snowfall to Britain; E. V. Jetton, *JAM*, 5 (1966), 857, on a cutoff low in the southwestern United States; K. A. Browning and T. W. Harrold, *QJRMS*, 95 (1969), 288, on a wave depression crossing Britain; E. R. Walker, *W*, 21 (1966), 410, on a warm-sector depression bringing rapid temperature change in Alberta; C. W. Kreitzberg, *JAM*, 7 (1968), 53, on mesoscale winds in an occlusion in Connecticut); L. A. D. I. Ekanayake, *W*, 23 (1968), 195, on an intense cyclone in Ceylon. For studies of severe cyclonic storms, see J. R. Mather, H. Adams, III, and G. A. Yoshioka, *JAM*, 3 (1964), 693, and *JAM*, 6 (1967), 20, on their frequency on the east coast of the United States. For a synthesis of TIROS photos of extratropical vortices, see W. K. Widger, Jr., *MWR*, 92 (1964), 263.

56. For details, see R. J. Boucher and R. J. Newcombe, *JAM*, 1 (1962), 127, also J. W. Deardorff, *JAM*, 2 (1963), 173.

57. For a general account, see J. S. Malkus, *W*, 13 (1958), 75. See J. Namias, *MWR*, 97 (1969), 346, on the 1968 season, which was remarkable for an unusually small number of hurricanes and an unusually large number of easterly waves in the Atlantic; also see K. M. Shamshad, *JAM*, 6 (1967), 199, on problems of monsoon rain due to disturbances in western Pakistan, and R. H. Simpson *et al.*, *MWR*, 97 (1969), 240, on the climatology of Atlantic tropical disturbances.

58. B. Ackerman, *JAS*, 20 (1963), 288. B. I. Miller, *JM*, 15 (1958), 184. N. E. LaSeur and H. F. Hawkins, *MWR*, 91 (1963), 694.

59. For details of this structure, see E. S. Jordan, *JM*, 9 (1952), 340, and B. I. Miller, *MWR*, 92 (1964), 389.

60. C. L. Jordan and Te-Chun Ho, *MWR*, 90 (1962), 157. C. S. Ramage, *JM*, 16 (1959), 227. E. M. Ballenzweig, *JM*, 16 (1959), 121.

61. C. O. Erickson, *MWR*, 91 (1963), 61. E. L. Fisher, *JM*, 15 (1958), 328. C. F. Tisdale and P. F. Clapp, *JAM*, 2 (1963), 358. H. Riehl, *BAMS*, 37 (1956), 413. Also see J. E. O'Hare, *BAMS*, 38 (1957), 226 and 433.

62. H. Riehl and R. J. Shafer, *JM*, 1 (1944), 42. C. H. Smiley, *BAMS*, 37 (1956), 403.

63. H. Riehl, *JM*, 5 (1948), 247. J. A. Colón and W. R. Nightingale, *MWR*, 91 (1963), 329. For synoptic studies of Atlantic hurricanes, see: T. N. Carlson, *MWR*, 97 (1969), 256, on African disturbances that developed into hurricanes; P. Koteswaram, *MWR*, 95 (1967), 541, on structure at time of landfall on a United States coast; B. I. Miller, *MWR*, 92 (1964), 389, on filling in Florida; H. F. Hawkins and D. T. Rubsam, *MWR*, 96 (1968), 701, on degradation while approaching Louisiana coast, and 617, on budgets of mass, heat, moisture, vorticity, and momentum; I. Perlroth, *T*, 19 (1967), 258, on behavior as related to water masses over which the cyclone passed; D. F. Leipper, *JAS*, 24 (1967), 182, on ocean surface conditions in the Gulf of Mexico; S. Penn, *JAM*, 4 (1965), 212, on temperature and ozone structure on the South Carolina coast; J. R. Stear, *MWR*, 93 (1965), 380, on a radiosonde sounding to over 100,000 feet in a hurricane eye over Bermuda; S. Penn, *JAM*, 5 (1966), 407, on tropopause structure; R. D. Fletcher and K. R. Johannessen, *JAM*, 4 (1965), 457, on computation of maximum surface winds; W. M. Gray, *JAS*, 23 (1966), 278, on winds and scales of motion; R. J. Renard, *MWR*, 96 (1968), 453, on predicted motion on the basis of a numerically derived steering current; and J. J. Fernández-Partagás, *MWR*, 94 (1966), 475, on a small hurricane affecting the Cayman Islands, Cuba, and the Bahama Islands, which was not recorded by the normal synoptic network. See M. A. Alaka and D. T. Rubsam, *MWR*, 93 (1965), 673, for details of Hurricane Ella, which developed a most unusual warm-core anticyclonic circulation at lower levels that later spread to the upper troposphere. See A. J. Abdullah, *JAS*, 23 (1966), 367, for a dynamic explanation of spiral precipitation bands in hurricanes; see also R. Wexler, *JAS*, 23 (1967), 441. According to N. J. M. MacDonald, *T*, 20 (1968), 138, hurricane spiral rain bands have characteristics similar to those of troughs in Rossby waves. For satellite photographs of hurricanes, see R. W. Fett, *MWR*, 92 (1964), 43; see also L. F. Hubert and A. Timchalk, *JAM*, 3 (1964), 203, on the accuracy of hurricane

location. For radar features of a hurricane at New Orleans, see W. A. Schultz, Jr., and E. L. Hill, *WW*, 18 (1965), 152. For a study of the genesis of Hurricane Hilda, as revealed by satellite photos and conventional and aircraft data, see H. F. Hawkins and D. T. Rubsam, *MWR*, 96 (1968), 428. For the climatology of hurricane occurrence, see: A. L. Sugg, *MWR*, 94 (1966), 183, on the 1965 hurricane season, and *MWR*, 95 (1967), 131 on the 1966 season; A. L. Sugg and J. M. Pelisser, *MWR*, 96 (1968), 242, on the 1967 season; and A. L. Sugg and P. J. Herbert, *MWR*, 97 (1969), 225, on the 1968 season.

For synoptic studies of Pacific typhoons, see: W. J. Denny, *MWR*, 97 (1969), 207, on the climatology of the 1968 season in the east Pacific; R. W. Fett, *MWR*, 96 (1968), 637, on Typhoon Billie, which had no wall cloud around its eye, but was otherwise normal; S. Y. W. Tse, *QJRMS*, 92 (1966), 239, on prediction of movement from the 700-mb chart; C. O. Erickson and S. Fritz, *MWR*, 93 (1965), 145, on surface ship observations and satellite photos; E. Gherzi, *GPA*, 71, no. 3 (1968), 198, on radar observations of Typhoon Susan in the China Sea, showing five to six spiral rain bands. For a typhoon model, with inner and outer rain bands, see T. Fujita *et al.*, *JAM*, 6 (1967), 3.

64. S. C. Gilbert and N. E. LaSeur, *JM*, 14 (1957), 18. H. R. Byers, *General Meteorology* (New York, 2d ed., 1944), p. 440. R. W. Fett, *MWR*, 91 (1963), 367. J. C. Sadler, *JAM*, 3 (1964), 347. For satellite photographs of tropical cyclones, see R. W. Fett, *MWR*, 94 (1966), 9, on formative stages, and 605, on movement of circulation center during life cycle.

65. C. S. Ramage, *JGR*, 67 (1962), 1401 and 4512. R. H. Simpson, *JM*, 9 (1952), 24. On the synoptic effects of tropical storms, see G. A. Winterling, *WW*, 20 (1967), 72, on cool cycles at Jacksonville, Florida, and K. Raghavan, *W*, 22 (1967), 250. On rainfall associated with double tropical storms in India, see K. Raghavan, *W*, 19 (1964), 106.

66. For details and maps, see S. Karelsky, *AM*, p. 293.

67. R. C. Sutcliffe, *MM*, 82 (1953), 163. *WAF*, II, 338. On stratospheric circulation changes over the British Isles during anticyclogenesis, see M. K. Miles, *MM*, 93 (1964), 365.

68. R. W. James, *JM*, 9 (1952), 243, N. L. Frank, *MWR*, 91 (1963), 355.

69. For maps and details, see R. Scherhag, *PAS*, p. 101.

70. J. Namias, *JM*, 8 (1951), 251. F. T. Bodurtha, Jr., *JM*, 9 (1952), 118.

71. R. B. Mason and C. E. Anderson, *MWR*, 91 (1963), 3.

72. M. K. Miles, *MOSP*, no. 8 (1961). *WAF*, II, 327.

73. J. Findlater, *QJRMS*, 87 (1961), 513. For synoptic studies of anticyclones and their importance for British weather, see: J. Findlater, *MM*, 96 (1967), 69, who also discusses mesoanalysis of cells in anticyclones; P. C. Clarke, *W*, 22 (1967), 15, and R. Murray, *loc. cit.*, p. 16, on Scandinavian anticyclones; and L. C. W. Bonacina, *W*, 20 (1965), 308.

74. For many years, the standard works were J. Namias, *An Introduction to the Study of Air-Mass Analysis* (Boston, 1947), and *WAF*.

75. W. H. Portig, *BAMS*, 40 (1959), 301.

76. C. J. Burke, *JM*, 2 (1945), 94, and 3 (1946), 100. F. E. Burbridge, *QJRMS*, 77 (1951), 365. J. M. Craddock, *loc. cit.*, p. 355. I. J. W. Pothecary, *MM*, 89 (1960), 1.

77. J. G. Edinger, *JAM*, 2 (1963), 706. On air masses, see: H. Flohn, M. Hantel, and E. Ruprecht, *JAS*, 25 (1968), 527, on the dynamics of the Arabian Sea summer monsoon; C. W. Newton, *MWR*, 93 (1965), 101, on the deep southward penetration of a polar air mass; E. C. Jarvis, *JAM*, 3 (1964), 744, on the modification of surface temperatures when cold air masses move over warmer water surfaces; and C. J. Boyden, *MM*, 93 (1964), 138 and 180, on subsidence in the British Isles and surrounding areas. For station summary charts, displaying air-mass changes, see R. L. Ives, *W*, 20 (1965), 123, on the Salt Lake desert. On wave clouds in the marine layer off the California coast, see J. G. Edinger, *JAM*, 5 (1966), 804.

78. See R. J. Reed, *BAMS*, 41 (1960), 591, for comparisons of the maps. For studies of frontal zones, see L. Oredsson, *T*, 16 (1964), 41, on vertical motions and precipitation in Europe; R. D. Elliott and E. L. Hovind, *JAM*, 4 (1965), 198 (heat, water, and vorticity balance in southern California). According to M. Oi and M. Sekioka, *MWR*, 93 (1965), 163, a cold air dome can produce perturbations of an overrunning warm current as large as those produced by the Rocky Mts. in an area east of the Rockies.

79. For comparison, see *DMWF*, pp. 292–93, *WAF*, I, 108 and 211, and H. C. Willett, *Descriptive Meteorology* (New York, 1944).

80. G. C. Whiting, *MWR*, 87 (1959), 409.

81. "M. O. discussion," *MM*, 88 (1959), 113.

82. W. J. Gibbs, *AM*, p. 84.

83. I. Pothecary, *W*, 11 (1956), 147, and J. S. Sawyer, *QJRMS*, 84 (1958), 375.

84. For details, see D. L. Bradbury and E. Palmén, *BAMS*, 34 (1953), 56.

85. H. H. Lamb, *MM*, 80 (1951), 65.

86. *HM*, p. 816.

87. "M. O. discussion," *MM*, 85 (1956), 83, gives this definition by B. W. Boville. See also C. E. Palmer, *JM*, 14 (1957), 403, for a discussion of kinematic aspects of frontal zones.

88. C. M. Penner, *QJRMS*, 81 (1955), 89, and J. L. Galloway, *W*, 13 (1958), 3 and 395. See also J. Clodman, *BAMS*, 35 (1954), 464.

89. W. L. Godson, *QJRMS*, 77 (1951), 633. For

frontal analysis in the southern hemisphere, see J. J. Taljaard, W. Schmidt, and H. van Loon, *NOTOS*, 10 (1961), 25. On objective frontal contour analysis, see W. S. Creswick, *JAM*, 6 (1967), 774.

90. Bradbury and Palmén, *op. cit.* (1953).

91. Sawyer, *op. cit.* (1958). On extraterrestrial factors in the initiation of North American polar fronts, see C. A. Mills, *MWR*, 94 (1966), 313.

92. "M. O. discussion," *MM*, 82 (1954), 83.

93. *Ibid.*

94. W. F. Petersen, *Lincoln-Douglas: The Weather as Destiny* (Illinois, 1943), correlated many critical events in Lincoln's life with the passages of cold fronts.

95. M. K. Miles, *QJRMS*, 88 (1962), 286. For a typical case history, see W. Schwerdtfeger and N. D. Strommen, *MWR*, 92 (1964), 523. For examples of cold fronts, see: K. C. Brundidge, *MWR*, 93 (1965), 587, on wind and temperature structure of the lowest 1,500 feet of nocturnal cold fronts at Cedar Hill, Texas; J. W. Zillman and D. W. Martin, *JAM*, 7 (1968), 708, on a spectacular one at Macquarie Island. On cold front clouds, see *W*, 20 (1965), 165, for an example of dense cirrus to the rear of a cold front, and P. Jackson, *W*, 19 (1964), 383, for an example of roll cloud. On cold frontal changes on the coast of New South Wales coast in summer (the *southerly buster*), see J. Gentilli, *W*, 24 (1969), 173.

96. J. H. Conover and S. H. Wollaston, *JM*, 6 (1949), 249.

97. H. W. Sansom, *QJRMS*, 77 (1951), 96.

98. R. M. Morris, *MM*, 91 (1962), 304.

99. M. K. Miles, *MM*, 83 (1954), 289.

100. M. G. H. Ligda and S. G. Bigler, *JM*, 15 (1958), 494.

101. J. M. Austin and R. H. Blackmer, Jr., *BAMS*, 37 (1956), 447.

102. H. H. Lamb, *MM*, 87 (1958), 76. V. A. Myers, *MWR*, 92 (1964), 513.

103. W. J. Bruce, *MM*, 84 (1955), 251. D. C. E. Jones, *MM*, 91 (1962), 297. For an example of a warm front that exhibited rapid changes in the height of its melting layer (two melting layers existing overhead simultaneously at times), see T. W. Harrold, K. A. Browning, and J. M. Nicholls, *MM*, 97 (1968), 327.

104. R. B. Carson, *MWR*, 78 (1950), 91.

105. J. A. Carr, *MWR*, 79 (1951), 100. On backdoor cold fronts in New England, see W. Hovey, K. Sirinek, and F. Storer, *WW*, 20 (1967), 264. Backdoor warm fronts are very rare.

106. C. E. Wallington, *W*, 9 (1959), 263, and *CD*, p. 119. For examples of sea-breeze fronts, see: J. E. Simpson, *W*, 19 (1964), 208, and C. E. Wallington, *W*, 20 (1965), 140, on Hampshire, Eng.; J. E. Simpson, *W*, 22 (1967), 306, on southeast England; M. A. Fosberg and M. J. Schroeder, *JAM*, 5 (1966), 573, on central California.

107. L. L. Kolb and M. M. Goodmanson, *JM*, 1 (1944), 98. "M. O. discussion," *MM*, 84 (1955), 213.

108. R. O. Cole and D. A. Lowry, *MWR*, 85 (1957), 183.

109. "M. O. discussion," *MM*, 87 (1958), 181.

110. J. Spar, *MWR*, 84 (1956), 291.

111. A. J. Abdullah, *JM*, 10 (1953), 228.

112. H. R. McQueen and R. H. Martin, *MWR*, 84 (1956), 277.

113. T. A. M. Bradbury, *MM*, 91 (1962), 71.

114. J. L. Galloway, *MM*, 88 (1959), 54. For examples of frontal precipitation, see: N. Rutter and J. A. Taylor, *W*, 23 (1968), 94, on western Wales; R. A. Canovan, *W*, 22 (1967), 256, on very localized heavy rain due to a shallow wave developing on an old, inactive cold front at London, Eng.; G. Reynolds, *W*, 22 (1967), 224, and D. E. Pedgley, *loc. cit.*, p. 478, on heavy rain due to a combination of frontal and orographic lifting in western Scotland.

115. J. S. Sawyer, *QJRMS*, 84 (1958), 375.

116. For an essay on frontogenesis and frontolysis, see H. H. Lamb, *MM*, 80 (1951), 35, 65, and 97.

117. For the water-tank model, see A. J. Faller, *JM*, 13 (1956), 1. H. von Helmholtz, *Über Atmosphärische Bewegungen* (Berlin: vol. I, 1888; vol. II, 1889). M. Margules, *Meteorol. Z.*, Hann vol. (1906), p. 293. *PDM*, pp. 203–6.

118. T. N. Carlson, *MWR*, 89 (1961), 163. On frontogenesis, see G. V. Rao, *JAM*, 5 (1966), 377, on the influence of latent heat and baroclinicity, and P. H. Stone, *JAS*, 23 (1966), 455, who describes a mathematical model of frontogenesis by horizontal wind deformation that confirms the original Bergeron theory that frontogenesis is caused by deformation of the horizontal wind field acting on a preexisting horizontal temperature gradient.

119. Sawyer, *op. cit.* (1958).

120. V. J. Oliver and G. C. Holzworth, *MWR*, 81 (1953), 141.

121. P. E. Carlson, J. L. Galloway, and P. C. Haering, *MM*, 94 (1965), 218.

122. J. S. Sawyer, *QJRMS*, 78 (1952), 170. R. J. Reed, *BAMS*, 38 (1957), 357.

123. E. M. Wilkins, *BAMS*, 36 (1955), 397. For a theory of frontal motion in the atmosphere, see A. Kasahara, E. Isaacson, and J. J. Stoker, *T*, 17 (1965), 261. For an example of crossing fronts, presenting a difficult dynamic problem, see G. C. Rider and J. E. Simpson, *MM*, 97 (1968), 24, on radar observations in eastern England. On objective analysis of fronts, see L. C. Clarke and R. J. Renard, *JAM*, 5 (1966), 764, for an evaluation of the U.S. Navy Fleet Numerical Weather Facility system. Also see R. J. Renard and L. C. Clarke, *MWR*, 93 (1965), 547, on numerical objective frontal analysis in the northern hemisphere, and T. H. Kirk, *MM*, 94 (1965), 351, on a parameter for the objective location of frontal zones. On miniature fronts, see D. J. Ride, *W*, 20 (1965), 212, for an

example of a discontinuity in surface wind arising from differential heating in England. See H. H. Lamb, *MM*, 83 (1954), 264, for evidence from a wind discontinuity at Luqa Airport, Malta, that, particularly in the Mediterranean region and North Africa, the wind discontinuity in a front is often the last feature of the front to disappear after frontolysis. See also H. S. Turner, *MM*, 84 (1955), 92, on the effect of topography in this example.

124. J. G. Breiland, *JM*, 15 (1958), 297.

125. M. E. Crawford, *BAMS*, 31 (1950), 351.

126. I. W. Brunk, *BAMS*, 34 (1953), 1.

127. J. D. Macdonald, *BAMS*, 33 (1952), 237.

128. J. M. Porter *et al.*, *BAMS*, 36 (1955), 390.

129. M. G. H. Ligda, *loc. cit.*, p. 340. S. Penn, C. Pierce, and J. K. McGuire, *loc. cit.*, p. 109. For examples of squall lines, see E. R. Lichtenstein and M. L. Schwarzkopf, *W*, 21 (1966), 181, on Argentina. According to W. A. Lyons, *JAS*, 25 (1968), 146, a "square cloud" pattern photographed by a TIROS satellite over Oklahoma was the cirrus canopy of a developing squall line; square-cloud patterns can be due to merging of anvils of adjacent cumulonimbi.

130. C. W. Newton and J. C. Fankhauser, *JAM*, 6 (1964), 651.

131. For maps depicting the correlation patterns, see E. J. Fawbush, R. C. Miller, and L. G. Starrett, *BAMS*, 38 (1957), 115.

132. R. D. Elliott and E. L. Hovind, *JAM*, 3 (1964), 143. On severe convective storms, see C. W. Newton, *AGP*, 12 (1967), 257, for a review. On severe local storms, see: K. A. Browning, *W*, 23 (1968), 429, on their organization; G. L. Darkow, *JAM*, 7 (1968), 199, on their total energy environment in the midwestern United States; K. A. Browning, *JAS*, 22 (1965), 669, on updrafts at Oklahoma City; W. T. Roach, *QJRMS*, 93 (1967), 318, on the nature of storm summit areas in Oklahoma; and T. N. Carlson and F. H. Ludlam, *T*, 20 (1968), 203, on favorable synoptic situations for storms in the midwestern United States and southern England. According to H. Riehl and R. L. Elsberry, *GPA*, 57, no. 1 (1964), 213, there is a semiconstant "noise level" of six inches of precipitation per year over the Colorado watershed, due to small rains; most precipitation in the area comes from storms, which each yield an average of 0.3 to 1.2 inch. For a relation between storm mean precipitation and storm duration in Illinois, see F. A. Huff, *JAM*, 8 (1969), 401.

133. T. M. McElmurry, *BAMS*, 42 (1961), 817. C. L. Jordan, *BAMS*, 36 (1955), 446.

134. E. B. Kraus, *JAS*, 20 (1963), 551. C. S. Ramage, *JM*, 9 (1952), 83.

135. For details and maps, see D. H. Johnson, *QJRMS*, 88 (1962), 1. C. M. Soane and V. G. Miles, *QJRMS*, 81 (1955), 440.

136. U.S. Weather Bureau Staff, *MWR*, 92 (1964), 345. S. E. Ashmore, *MM*, 84 (1955), 156. On synoptic aspects of nonfrontal rainfall, see E. V. Jetton and C. E. Woods, *MWR*, 95 (1967), 221, on heavy rains in New Mexico and Texas, which were preceded by a sharp break in the westward zonal flow near the stratopause, and a concurrent lowering and cooling of the tropopause. On the relation between wind strength and rainfall in the British Isles, see R. F. Zobel, *W*, (1965), 13; see also C. A. S. Lowndes, *MM*, 97 (1968), 226, on westerlies and daily rainfall of two inches or more in northeastern Wales. On the diurnal variation of tropical precipitation, see D. E. Pedgley, *MM*, 98 (1969), 129, on monsoon showers in the Sudan. See also F. A. Huff and W. L. Shipp, *JAM*, 7 (1968), 886, on mesoscale spatial variability of precipitation in the midwestern United States. On the variation of the heavy rainfall of July 10, 1968, over England and Wales, see P. R. S. Salter, *MM*, 98 (1969), 92. On the diurnal variation of tropical rainfall in relation to evaporation in the Sudan, see J. Oliver, *W*, 20 (1965), 58.

137. C. J. Boyden, *MM*, 93 (1964), 353. For a prediction graph, see R. Murray, *MM*, 88 (1959), 324.

138. S. M. Ross, *MM*, 83 (1954), 275. G. C. Williams, *MWR*, 91 (1963), 465. For synoptic study of widespread freezing drizzle over southeast England, see T. H. Kirk, *MM*, 96 (1967), 112; see also A. Perry, *MM*, 96 (1967), 221. For synoptic studies of heavy snowfalls, see R. J. Younkin, *MWR*, 96 (1968), 851, on the western United States. According to R. A. Muller, *WW*, 19 (1966), 248, the complexity of the map of mean seasonal snowfall around the Great Lakes is due to the fact that each of the four types of snowfall affecting the area (warm frontal, cold frontal, orographic, and post-cold-frontal lake squalls or lake bursts) affects different parts of the area. For a synoptic climatology of heavy snowfall over the central and eastern United States, see P. A. Goree and R. J. Younkin, *MWR*, 94 (1966), 663. For synoptic studies of North American blizzards, see: D. Storr, *W*, 20 (1965), 370, on the one of December 15, 1964, in Alberta; R. B. Sykes, Jr., *WW*, 19 (1966), 240, on one in January 1966, in central New York State; H. G. Stommel, *WW*, 19 (1966), 188, on one in early March 1966 in the northern Great Plains; and D. A. Haines, *WW*, 19 (1966), 194, on one in late March 1966 in the middle West.

139. W. H. Hoecker, Jr., *MWR*, 89 (1961), 533.

140. See: M. O. Asp., *MWR*, 84 (1956), 143, and 83 (1955), 117, for Arkansas; H. R. Spohn and P. J. Waite, *MWR*, 90 (1962), 398, for Iowa; and M. O. Asp., *MWR*, 78 (1950), 23, for Oklahoma.

141. A. B. Lowe and G. A. McKay, *JAM*, 1 (1962), 157.

142. R. G. Beebe, *MWR*, 84 (1956), 127. R. M. Whiting, *BAMS*, 38 (1957), 353. S. Penn, C. Pierce, and J. K. McGuire, *BAMS*, 36 (1955), 109.

143. R. C. Miller, *BAMS*, 40 (1959), 465. For typical tephigram soundings see E. J. Fawbush and

R. C. Miller, *BAMS*, 35 (1954), 154. R. G. Beebe, *BAMS*, 39 (1958), 195. On the climatology of tornadoes, see: R. H. Skaggs, *MWR*, 97 (1969), 103, on diurnal distribution in the United States; A. H. Auer, Jr., *MWR*, 95 (1967), 32, on northeastern Colorado; and R. H. Skaggs, *loc. cit.*, p. 107, for a climatological model of association between tornadoes and 500-mb indicators of jetstreams in Kansas, Oklahoma, and Texas. For examples of American tornadoes, see: N. E. Prosser, *MWR*, 92 (1964), 593, on a tornado path in Nebraska; H. W. Hiser, *JAM*, 7 (1968), 892, on Miami; and *WW*, 18 (1965), 122, for the Palm Sunday tornadoes of April 11, 1965, when there were 37 tornadoes in six midwestern states. For an unusual tornado, without a funnel cloud, associated with a large cumulonimbus within a mesolow, at Manhattan, Kansas, see L. D. Bark, *WW*, 20 (1967), 62. On vorticity and air-mass correlations for American tornadoes, see D. S. Foster, *MWR*, 92 (1964), 339.

144. See F. A. Barnes and C. A. M. King, *W*, 7 (1952), 214, and C. K. M. Douglas, *loc. cit.*, p. 311, on the Tibshelf tornado, and T. H. Kirk and D. T. J. Dean, *MOGM*, no. 107 (1963), on the Malta tornado. For examples of British tornadoes, see: T. W. V. Jones, *MM*, 95 (1966), 91, on Wisley, Surrey; E. G. Gilbert and J. M. Walker, *W*, 21 (1966), 211, also on Wisley, Surrey; and W. T. Roach, *W*, 23 (1968), 418, on the Barnacle, Coventry, tornado, which was unusually vigorous for a British one. On tornado occurrences in Britain, 1963–1966, see R. E. Lacy, *W*, 23 (1968), 116. On tornadoes associated with hurricanes, see: J. S. Smith, *MWR*, 93 (1965), 453, on their climatology in the United States; M. I. Rudd, *MWR*, 92 (1964), 251, on Galveston, Texas; A. D. Pearson and A. F. Sadowski, *MWR*, 93 (1965), 461, on case studies of hurricane-induced tornadoes in the United States; E. L. Hill, W. Malkin, and W. A. Schultz, Jr., *JAM*, 5 (1966), 745, on the southeastern United States; and A. Sadowski, *WW*, 19 (1966), 71, for examples. On the life cycle of tornadic storms, see K. A. Browning, *JAS*, 22 (1965), 664. For the wind pattern in a tornado as revealed by tree damage in dense woodland in Elk County, Pennsylvania, see L. J. Budney, *WW*, 18 (1965), 74.

145. F. O. Rossmann, *CD*, p. 167. A. H. Glaser, *CD*, p. 157. R. A. Garrett and V. D. Rockney, *MWR*, 90 (1962), 231. For examples of radar analysis of tornado situations, see: M. Tepper, *MWR*, 78 (1950), 170; G. E. Stout and H. W. Hiser, *BAMS*, 36 (1955), 519; W. F. Staats and C. M. Turrentine, *BAMS*, 37 (1956), 495; H. W. Hiser, *BAMS*, 39 (1958), 353; T. Fujita and H. A. Brown, *loc. cit.*, p. 538; and R. H. Nolen, *BAMS*, 40 (1959), 277. For a physical explanation of the mechanism of formation of the tornado hooked echo on the southwestern edge of eastward-moving cumulonimbus cells in the United States, see T. Fujita, *MWR*, 93 (1965), 67; see H. Arakawa, *JAM*, 6 (1967), 439, for a hook-shaped echo in a mesocyclone associated with a catastrophic rainstorm in Japan. For a theory that the loud roar accompanying tornadoes could be due to circulating acoustic waves, see F. J. Anderson and G. D. Freier, *JGR*, 70 (1965), 2781. For a postulated mechanism to explain the burning and dehydration produced by tornadoes, see P. A. Silberg, *JAS*, 23 (1966), 202. On luminous phenomena associated with tornadoes, see C. M. Botley, *W*, 21 (1966), 318; see also B. Vonnegut and J. R. Weyer, *WW*, 19 (1966), 66. See H. L. Jones, *WW*, 18 (1965), 78, for details of the *tornado pulse generator*, observed for the first time on May 25, 1955, in Oklahoma: a circular area of flashing, pale-blue light at a height of 6 km on the side of the tornado thundercloud. On the role of electrical phenomena associated with tornadoes, see E. M. Wilkins, *JGR*, 69 (1964), 2435 and 5425.

146. For general accounts, see "M.O. discussion," *MM*, 86 (1957), 176, and M. Tepper, *BAMS*, 40 (1959), 56.

147. For a detailed review, see T. Fujita, *AMM*, 5, no. 27 (1963), 77–123. G. J. Dellert, Jr., *MWR*, 90 (1962), 133, gives the computer method. D. C. House, *MWR*, 92 (1964), 589.

148. A. P. Richter, *JAM*, 3 (1964), 339. B. W. Magor, *BAMS*, 40 (1959), 499. L. E. Fortner, Jr., and C. L. Jordan, *MWR*, 88 (1960), 343. T. Fujita, *JM*. 15 (1958), 288. For a small wave disturbance, intermediate in size between mesosystems and synoptic systems, bringing squalls to the lower Great Lakes area, see H. L. Ferguson, *JAM*, 6 (1957), 523. For a mesosynoptic-scale snow belt in eastern England, see D. E. Pedgley, *W*, 23 (1968), 469. For case histories with isohyetal maps, see T. Fujita and H. A. Brown, *BAMS*, 39 (1958), 538.

149. H. Landers, *BAMS*, 38 (1957), 130. For a study of the San Fernando convergence zone in southern California, see J. G. Edinger and R. A. Helvey, *BAMS*, 42 (1961), 626.

150. T. Fujita, *JM*, 16 (1959), 38. On mesoanalysis, see: D. C. House, *MWR*, 92 (1964), 589, on detection of mesosystems; H. Arakawa, *W*, 22 (1967), 229, on a mesocyclone viewed by radar in Japan; R. R. McNair and J. A. Barthram, *MM*, 95 (1966), 304, on a squall line in southern England; D. E. Pedgley, *W*, 20 (1965), 351, and F. G. Thomas, *loc. cit.*, p. 302, on a convectional rainstorm in Dover, Eng.; A. P. Richter, *JAM*, 3 (1964), 339, on horizontal closed circulations, 1 to 2 miles in diameter, in Nevada, Idaho, and Los Angeles; R. J. Taylor *et al.*, *CTP*, no. 18, on forest fires, in southwestern Australia; A. I. Weinstein, E. R. Reiter, and J. R. Scoggins, *JAM*, 5 (1966), 49, on the mesoscale structure of winds in the 11 to 20-km layer at Point Mugu, Calif., and at Cape Kennedy, Florida; and D. Atlas *et al.*, *QJRMS*, 95 (1969), 544, on mesoscale perturbations in the melting layer, induced by precipitation

at Pershore, Eng. See R. L. Peace, Jr., and R. B. Sykes, Jr., *MWR*, 94 (1966), 495, for a mesoanalysis of a snowstorm over Lake Ontario that reveals the existence of "lake-effect bands" due to heating of the air by the lake, and a narrow confluent-convergent wind-shift zone, 0.1 to 1.5 nautical miles wide, beneath the snow bands. On phenomena midway between synoptic and mesoscale, see R. E. Lynott and O. P. Cramer, *MWR*, 94 (1966), 105, on a record windstorm in Oregon and Washington; and H. L. Ferguson, *JAM*, 6 (1967), 523, on a squall line and wave disturbance in the lower Great Lakes.

151. H. R. Byers and R. R. Braham, Jr., *JM*, 5 (1948), 71. DMWF, pp. 596–97. T. Fujita, *AMM*, vol. 5, no. 27 (1963).

152. J. Neumann, *JM*, 8 (1951), 60. H. R. Byers and H. R. Rodebush, *JM*, 5 (1948), 275, and 6 (1949), 289 and 365.

153. W. G. Sullivan, Jr., *MWR*, 91 (1963), 89. A. K. Showalter, *BAMS*, 34 (1953), 250; see also *BAMS*, 37 (1956), 443. For J. G. Galway's "lifted index," see *loc. cit.*, p. 443. C. V. Ardis, Jr., *BAMS*, 42 (1961), 166. S. Penn, *BAMS*, 36 (1955), 278. C. E. Buell, *BAMS*, 35 (1954), 476.

154. K. A. Browning, *JAS*, 22 (1965), 664. Penn, Pierce, and McGuire, *op. cit.* (1955). On the climatology of thunderstorms, see: S. A. Chagnon, Jr., *MWR*, 95 (1967), 209, on thunderstorm and hail day patterns in Illinois; P. J. Feteris, *JAM*, 4 (1965), 178, on thunderstorm situation statistics for the Netherlands; G. Rönicke, *JAM*, 4 (1965), 186, on a thunderstorm center in the Andes of northern Argentina; A. F. Crossley and N. Lofthouse, *W*, 19 (1964), 172, on their distribution over Britain; B. W. Atkinson, *W*, 21 (1966), 203, on thunder outbreaks in southeastern England, and *W*, 22 (1967), 335, on atmospheric structures conducive to thunderstorms in southeastern England; N. E. Davis, *W*, 24 (1969), 166, on the diurnal variation of thunder at London Airport; and M. J. Long, *JAM*, 5 (1966), 467, on thunderstorm penetrations of the tropopause in the United States.

For examples of thunderstorms, see: R. Wexler, *MWR*, 93 (1965), 523, for TIROS photographs and radar observations of a thunderstorm complex over Florida; P. R. S. Salter, *MM*, 97 (1968), 372, on the thunderstorms of July 10, 1968, which gave more than 100 mm of rain in Gloucestershire, Eng.; D. McFarlane and C. G. Smith, *MM*, 97 (1908), 235, on the thunderstorms of July 13, 1967, which brought a record rainfall to Oxford, Eng.; and *MWR*, 92 (1964), 345, for details of an almost stationary thunderstorm that caused a record rainfall at Tulsa, Oklahoma.

On special aspects of thunderstorms, see: W. J. Moroz and E. W. Hewson, *JAM*, 5 (1966), 148, on a mesoscale interaction of a low-level thunderstorm outflow with a lake breeze on Lake Michigan; L. J. Battan, *JAM*, 3 (1964), 415, on vertical velocities and precipitation size in an Arizona thunderstorm; P. L. Kamburova and F. H. Ludlam, *QJRMS*, 92 (1966), 510, on rainfall evaporation in thunderstorm downdrafts; J. C. Fankhauser and C. W. Newton, *WW*, 18 (1965), 68, on movement of thunderstorms as related to winds and moisture supply; and T. Fujita and H. Grandoso, *JAS*, 25 (1968), 416, on the splitting of thunderstorms into anticyclonic and cyclonic storms in Oklahoma and Kansas. On predicting thunderstorms, see W. E. Saunders, *MM*, 95 (1966), 204, and 96 (1967), 85, for a test of thunderstorm forecasting methods, and A. W. Hanssen, *JAM*, 4 (1965), 172, on an objective method for the Netherlands.

155. K. A. Browning and F. H. Ludlam, *QJRMS*, 88 (1962), 117. K. A. Browning and R. J. Donaldson, *JAS*, 20 (1963), 533. R. J. Donaldson, *BAMS*, 46 (1965), 174. On hailstorms, see K. A. Browning, *JAS*, 21 (1964), 634, for a kinematic airflow and precipitation model (for severe local storms that travel to the right of the wind in the middle troposphere) which extends the Browning-Ludlam hailstorm model. For examples of hailstorms, see: M. E. Hardman, *W*, 23 (1968), 404, for Wiltshire, Eng.; A. E. Carte, *QJRMS*, 92 (1966), 290, for the Transvaal; A. H. Auer, Jr., and J. D. Marwitz, *JAM*, 7 (1968), 196, on air and moisture flux in the high plains of the United States; C. J. Neumann, *JAM*, 4 (1966), 161, on mesoanalysis in Florida; and R. M. Brown, *WW*, 20 (1967), 254, for a hailstorm with a frontal system, which produced odd-shaped hailstones, in Long Island, N.Y. For the climatology of hailstorms in Wisconsin, see M. W. Burley, R. Pfleger, and Jen-yu Wang, *MWR*, 92 (1964), 121.

156. R. J. Donaldson, *BAMS*, 46 (1965), 174. C. M. Soane, *BAMS*, 38 (1957), 269.

157. See P. M. Austin and A. C. Bemis, *JM*, 7 (1950), 145, for a quantitative study of the radar bright band in precipitation echoes. On the distribution of summer showers over Florida, as deduced from radar measurements, see N. L. Frank, P. L. Moore and G. E. Fisher, *JAM*, 6 (1967), 309.

158. Donaldson, *op. cit.* (1965).

159. R. J. Donaldson, *JAM*, 3 (1964), 611.

160. For a discussion of the differences between waterspouts and tornadoes, see F. O. Rossmann, *W*, 13 (1958), 259. Waterspouts have been observed to occur in tornadoes; see F. B. Dinwiddie, *MWR*, 87 (1959), 239. L. J. Battan, *JM*, 15 (1958), 235. P. C. Sinclair, *JAM*, 4 (1965), 116, and *MWR*, 92 (1964), 363. H. E. Graham, *BAMS*, 36 (1955), 99. B. R. Morton, *SP*, 52 (1964), 249, discusses fire whirlwinds. For examples of *waterspouts*, see: W. L. Woodley *et al.*, *MWR*, 95 (1967), 799, for aircraft observations in Florida; J. H. Golden, *W*, 23 (1968), 103, for one in Florida; and D. M. Ludlam, *WW*, 20 (1967), 174, for three in Vineyard Sound, Mass., most unusual phenomena in middle latitudes. For examples of *dust devils*, see: J. Hallett, *W*, 24 (1969), 133, for a

dust devil induced by the rotor of a mountain wave in Nevada; P. C. Sinclair, *JAM*, 8 (1969), 32, for a census of occurrences in Tucson, Ariz.; W. D. Crozier, *JGR*, 69 (1964), 5427, on an electric field in New Mexico; R. L. Lamberth *BAMS*, 47 (1966), 522, for measurement of dust-devil parameters in New Mexico; and P. C. Sinclair, *BAMS*, 46 (1966), 388, on their rotation. On dust whirls in northwest Libya, see J. B. McGinnigle, *W*, 21 (1966), 272. For details of *whirlwinds* produced by the eruption of the Surtsey submarine volcanic off the south coast of Iceland, see S. Thorarinsson and B. Vonnegut, *BAMS*, 45 (1964), 440.

161. See the *Aviation Meteorology Reports* of the M.O. for various areas and routes. See *MAB*, vol. 2 (1951), for abstracts on the synoptic climatology of the Middle East, the Near East and northeastern Africa; *MAB*, vol. 3 (1952), for northwestern Africa, Central Africa, and Argentina; *MAB*, vol. 4 (1953), for Australia and the Pacific; and *MAB*, vol. 5 (1954), for southeast Asia and the Arctic. See *MGAB* under "geographical index" and "synoptic climatology" for later references. For the British Isles, see H. H. Lamb, *QJRMS*, 76 (1950), 393, and *W*, 8 (1953), 131 and 176; also R. B. M. Levick, *W*, 10 (1955), 412. See also H. H. Lamb, *W*, 20 (1965), 9, for the frequency of weather types in the British Isles. For regional correlations of British weather types, on the basis of linkage analysis, see A. Perry, *W*, 23 (1968), 325. For details of *synoptic indices*, measuring the occurrence of weather types, see R. Murray and R. P. W. Lewis, *MM*, 95 (1966), 193; and for regional anomalies of temperature, rainfall, and sunshine in Britain, based on synoptic indices, see A. Perry, *W*, 24 (1969), 225.

162. W. C. Jacobs, *AMM*, vol. 1, no. 1 (1947), and *BAMS*, 27 (1946), 306.

163. E.g., see W. E. Howell, *JM*, 10 (1953), 270, on Cuba, and F. M. Ali, *JM*, 10 (1953), 1959, on wet periods in Egypt, who present contrasting approaches to subjective weather typing. See also R. E. Snead, *AMGB*, 16, series B (1968), 316, on weather types in southwest Pakistan. For a more refined example, see W. I. Christensen, Jr., and R. A. Bryson, *MWR*, 94 (1966), 697, on the weather types of Madison, Wisc., and Minneapolis–St. Paul, defined by factor (component) analysis applied to hourly surface observations of *all* synoptic quantities. On Hawaii, see R. K. Siler, *MWR*, 90 (1962), 103, and 92 (1964), 61.

164. See: T. F. Malone, *WAF*, II, 238–56; J. E. Kutzbach and E. W. Wahl, *JAM*, 4 (1965), 542, on the use of orthogonal polar coordinates; and J. M. Craddock and C. R. Flood, *QJRMS*, 95 (1969), 576, on the representation of northern hemisphere 500-mb contour patterns by means of eigenvectors.

165. For details, see Malone, *op. cit.*, and F. K. Hare, *PAS*, pp. 137–50.

166. G. P. Wadsworth, *CM*, p. 849. Malone, *op. cit.* D. G. Friedman, *JM*, 12 (1955), 428. See also D. L. Jorgensen, W. H. Klein, and A. F. Korte, *JAM*, 6 (1967), 782, for an example of the "moving grid" method applied to the climatology of winter precipitation from 700-mb lows over the intermontane areas of the western United States.

167. I. A. Lund, *JAM*, 2 (1963), 56.

168. For details, see S. Penn, B. Kunkel, and W. D. Mount, *loc. cit.*, p. 345.

169. For examples, see: D. E. Martin and W. G. Leight, *MWR*, 77 (1949), 275; W. H. Klein, *BAMS*, 29 (1948), 439, and *WW*, 18 (1965), 252. For examples of verbal (and cartographic) descriptions, see *M.O. Aviation Meteorology Report*, vol. 1, no. 1 (1948), on the Azores, and vol. 1, no. 2 (1949), on South America.

170. W. H. Klein, *op. cit.* (1948), and *MWR*, 91 (1963), 527.

171. J. Namias, *MWR*, 83 (1955), 199, described the mean monthly 700-mb patterns that produce extensive drought in the United States. M. T. Yin, *JM*, 6 (1949), 393, described the onset of the summer monsoon over India and Burma in terms of the 8-km circulation pattern. M. Rahmatullah, *JM*, 9 (1952), 176, was able to show synoptically that even during one month, August 1949, the monsoon current was far from steady over Indo-Pakistan, and in fact five different circulation types could be identified. For examples of British synoptic correlations, see: R. Murray and R. P. W. Lewis, *MM*, 95 (1966), 193, on the use of simple indices; R. A. S. Ratclifie, *MM*, 95 (1966), 98, on fine spells in southwest Scotland; C. A. S. Lowndes, *MM*, 95 (1966), 80 and 248 on showers in northeasterly airstreams over northwest England. For additional examples of British weather, see: R. A. S. Ratcliffe, *MM*, 97 (1968), 258, on monthly rainfall, from monthly mean 500-mb charts, in England and Wales; R. Murray, *MM*, 96 (1967), 65, for September synoptic weather, from cyclonic character of the preceding June, and *MM*, 97 (1968), 321, on autumn temperature and rainfall in England and Wales, from the circulation pattern of the preceding summer; and C. A. S. Lowndes, *MM*, 96 (1967), 212, on southerly weather type and summertime rain in southeastern England. See L. Krown, *JAM*, 5 (1966), 590, on seasonal rainfall in Israel from 500-mb patterns, and J. Namias, *AMGB*, supplement 1 (1966), p. 96, on long-lag correlations between circulation patterns and surface weather, i.e., long-range forecasting based on synoptic climatology.

172. See A. Ehrlich, *BAMS*, 35 (1954), 215, for the synoptic patterns over the northeast Atlantic that are associated with singularities. For details of the relations between circulation patterns and weather spells, which complicate the problems associated with singularities, see: C. A. S. Lowndes, *MM*, 92 (1963), 165, on cold spells at London, and *MM*, 93 (1964), 231 and 273, on winter dry spells in London and southeastern England; P. C. Clarke, *W*, 19 (1964), 366,

on the start of a prolonged cold spell in December 1962 in England and Wales; R. A. S. Ratcliffe, *MM*, 94 (1965), 129, and R. A. S. Ratcliffe and P. Collison, *MM*, 96 (1967), 228, on fine spells in southeastern England, from 500-mb circulation patterns; C. A. S. Lowndes, *MM*, 94 (1965), 241, on summer dry spells in southeastern England, from 500-mb patterns; R. A. S. Ratcliffe and A. E. Parker, *MM*, 97 (1968), 1, on wet spells in southeastern England, from 500-mb patterns. See also W. L. Hofmeyr and V. Gouws, *NOTOS*, 13 (1964), 37, for wet/dry spells in the northwestern Transvaal in relation to circulation patterns.

173. *AMM*, vol. 3, no. 3 (1957). E. W. Wahl, *BAMS*, 33 (1952), 380. L. C. W. Bonacina, *W*, 6 (1951), 79.

174. C. R. Burgess, *W*, 6 (1951), 144 and 163. For examples, see J. Powell and D. E. Pedgley, *W*, 24 (1969), 247, on Termit, Nigeria, and J. Pope, *W*, 24 (1969), 231, on Plymouth, Eng. *Weather indices* provide a useful device for the comparison of weather in different localities. For examples, see: R. M. Poulter, *W*, 17 (1962), 253, and 21 (1966), 109, on a summer-weather index for London; N. E. Davis, *MM*, 96 (1967), 178, for a classification of summers in northwestern Europe based on Poulter's index; G. H. Hughes, *W*, 22 (1967), 199, for a summer-weather index for Manchester; and P. G. Rackcliff, *W*, 20 (1965), 38, for a summer- and winter-weather index for Armagh. On "summer day" indices for Britain, see N. E. Davis, *W*, 23 (1968), 305, and A. Perry, *W*, 23 (1968), 212; for a very simple index, see T. Baker, *W*, 24 (1969), 277. See D. H. Miller, *UCPG*, vol. II (1955), and *AGP*, 11 (1965), 175, and C. P. Patton, *UCPG*, vol. 10, no. 3 (1956), for examples of the effects of local topography on weather and climate. Mountain weather is a classic instance: see D. E. Pedgley, *W*, 22 (1967), 266, for a general account; H. H. Coutts, *W*, 24 (1969), 66, on the rainfall of the Kilimanjaro area of Africa; and D. E. Pedgley, *W*, 21 (1966), 187, on the rainfall of the Mt. Kenya area. Another good example is provided by the weather and climate of islands, on which, see: H. H. Lamb, *MM*, 86 (1957), 73, on St. Helena; A. Elliot, *W*, 21 (1966), 282, on the New Hebrides in the southwestern Pacific; S. G. Irvine, *W*, 23 (1968), 392, on the Shetland Isles; B. R. J. Blench, *W*, 22 (1967), 134, on Jersey; and A. J. Whiten and D. J. Simmons, *W*, 21 (1966), 76, on the Isle of Wight.

175. For details, see *WAF*, II, 205–9. For an example of orographic and frictional convergence effects on the coastal area east of Stockholm, see A. Nyberg and H. Modén, *T*, 18 (1966), 745. For an example of local topographic effects in the Dale peninsula, Pembrokeshire, Wales, see T. M. Thomas, *W*, 11 (1956), 183.

176. For Metmaps, see C. E. Wallington, *W*, 19 (1964), 166, 255, 278, and 320, and *W*, 22 (1967), 2, 50, and 120. R. L. Ives, *BAMS*, 41 (1960), 433, gives details of an automatic timer to switch the receiver on in time for the broadcast. For details of radio-teleprinters suitable for amateur use, see, e.g., the *Newsletter of the British Amateur Radioteleprinter Group*.

177. For a discussion, see R. C. Sutcliffe and A. G. Forsdyke, *QJRMS*, 76 (1950), 189.

178. F. H. Bushby and M. S. Timpson, *QJRMS*, 93 (1967), 1. For details of British weather forecasting procedure, see J. S. Sawyer, *W*, 22 (1967), 350, for a historical review, 1850 to 1967, and D. M. Houghton, *QJRMS*, 91 (1965), 524, on current practice at the Central Forecasting Office, Bracknell. On future aids to centralized forecasting, see J. S. Sawyer, *W*, 22 (1967), 400. For details of the use of three-hour pressure tendencies in tropical forecasting, East Africa, see H. W. Sansom, *QJRMS*, 90 (1964), 298. Corrected for the diurnal variation of pressure, the tendencies may indicate vertical motion.

179. J. C. Sadler, *BAMS*, 46 (1965), 118. On the feasibility of global-scale analysis, see *BAMS*, 47 (1966), 200. For details of an operational system for tropical analysis, see H. A. Bedient, W. G. Collins, and G. Dent, *MWR*, 95 (1967), 942. For a review of the future of synoptic analysis in general, see J. S. Sawyer, *QJRMS*, 90 (1964), 227.

180. For details, see M. J. Toyli, *PTMC*, p. 74, and R. W. Rae, *PTMC*, p. 88. Owing to the effects of altitude and exposure, near-arctic difficulties may also be experienced at middle-latitude stations; see A. Diver, *W*, 24 (1969), 75, for semi-arctic conditions in winter at Great Dun Fell, in the Pennines, Cumberland, Eng.

181. R. J. Reed, *PAS*, p. 124.

182. W. S. Lanterman, *AM*, p. 79. W. J. Gibb, *AM*, p. 84. On the reliability of synoptic analysis in the South Pacific area, see J. J. Taljaard, *NOTOS*, 13 (1964), 15. For southern hemisphere weather maps constructed during the IGY, see J. J. Taljaard, *BAMS*, 45 (1964), 88.

183. For the mathematics, see E. Kruger, *TMA*, p. 168.

184. J. Cochemé, *TMA*, p. 181. On the importance of another synoptic model in equatorial regions, the *meridional trough*, extending from equator to pole, and its vital role in the exchange of heat, moisture, and angular momentum between tropics and higher latitudes, especially in winter, see H. Riehl, *W*, 24 (1969), 288. The fact that easterly waves in the tropical Atlantic are associated with a characteristic "inverted V" cloud pattern, on satellite photographs, is a useful aid to analysis in that area.

185. *BWMO*, 12 (1963), 144, and 13 (1964), 146.

186. P. D. Thompson, *BWMO*, 12 (1963), 146.

187. P. Leclerq, *BWMO*, 12 (1963), 201. P. I. Hershberg, *JAM*, 2 (1963), 531.

188. V. E. Lally, D. H. McInnis, and R. F. Myers,

BAMS, 38 (1957), 62. C. V. Ockenden, *BWMO*, 3 (1954), 30.

189. R. Perley, *BAMS*, 45 (1964), 740. On the statistical theory, see T. A. Gleeson, *JAM*, 2 (1963), 202. P. D. Thompson, *BWMO*, 12 (1963), 146, discusses "holes." One difficulty in the application of modern techniques is that there is, as yet, no generally agreed-on quantitative definition as to what constitutes a weather observation. See M. P. Woodall, *BAMS*, 47 (1966), 111, for one definition. On the design of electronic equipment for meteorological communications, see E. R. Reins and P. M. Wolff, *BAMS*, 46 (1965), 16. See also T. A. Gleeson, *JAM*, 3 (1964), 529, on network sampling theory, and E. S. Epstein, *JAM*, 8 (1969), 190, on the role of uncertainties in describing the initial situation.

Postscript

Foundations of Climatology has been concerned with climate as a physical problem. Its main task has been to demonstrate that climatology is not merely a collection of facts about weather and climate, classified in a manner sometimes scientifically valid, sometimes arbitrary. Instead, it is a physical science with sound foundations. These foundations have formed the subject-matter of the book. The average reader will, no doubt, have received some grounding in the purely descriptive side of climatology, i.e., world climatic regions and climatic types, at school. I hope that such a reader will have concluded, upon reaching this Postscript, that a real understanding of climate requires a knowledge of physics, which reveals that many supposed problems of world climatic regions and climatic types are trivial, if not illusory.

Once the reader has grasped the essentially physical nature of climatology, he will encounter another problem, that of mathematics, for here the climatologist differs from the meteorologist. The latter approaches the study of weather and climate as a physicist, well versed in the use of his primary tool, mathematics, and consequently tends to investigate only those problems that are directly amenable to analysis by mathematical techniques. The climatologist, on the other hand, approaches the study of weather and climate in terms of the phenomena. His inspiration comes from personal experience of the variety of weathers and climates the Earth has to offer, and from a conviction of the importance of meteorological influences on human life and activities. To the climatologist, only those branches of meteorological physics that are relevant to the study of actual weather situations are of interest, and only those parts of mathematics that are necessary to the understanding of these branches of physics are of use to him—hence the highly selective treatment of meteorological physics and meteorological mathematics in these pages.

The geographer who wishes to carry out climatological work that will be a lasting contribution to human knowledge must become proficient in the mathematical analysis of the phenomena that interest him, for without mathematics, prediction is impossible, and without prediction, science cannot advance. The geographical literature contains numerous examples of climatological investigations that came to nought because of a lack of mathematical insight. Likewise, the physicist and the mathematician should not neglect the writings of geographers, naturalists, and other empiricists, for these may reveal climatological problems that the analytical approach may completely overlook. The geographer is very much concerned with problems of balance between different, often conflicting, interests, and may be in a position to outline the most profitable area to which the high-powered techniques of mathematics and physics may be applied. Meteorological literature is full of contributions to pure climatology that, although physically refined and mathematically elegant, are all but useless to mankind, because they deal with situations that rarely, if ever, occur on the real Earth. The actual world of weather and climate is vastly more complicated than the abstract world defined by the computer-generated maps of the numerical weather-prediction expert, even though his contribution does provide a sound foundation for the study of actual weather and climate, and hence for climatology.

APPENDIX I

1.1. The amount of air above or below mean sea level is measured in terms of atmospheric pressure, which is usually expressed in millibars. One mb is equivalent to a weight that exerts a force of 10^3 dynes per square cm multiplied by the local value of gravity (g).

There does not seem to be a suitable general term in existence to describe the disposition of the relatively permanent features of the atmosphere with respect to space. *Aerology* is the scientific study of the free atmosphere, and so *aerography* (not to be confused with areography, the "science of areas" in biogeography) may possibly be used to mean the description of the relatively permanent features of the free atmosphere. *Aeronomy* is the physics of the upper atmosphere.

1.2. Mont Blanc was not climbed until 1787, by de Saussure, and although expeditions were organized to the Alps, the Andes, and the Himalayas, they were not much use scientifically.

1.3. This was radio set SCR–258. A smaller set, SCR–658, operating on a frequency of 400 megacycles, was later developed; after the war it was superseded by set AN/GMD–1, operating on 1,680 mc and incorporating automatic following. Owing to their low price on the surplus market, these early sets are still used in meteorological research.

1.4. The sea-level extremes recorded vary from 887 mb in the Pacific on August 18, 1927, to 1,075 mb in Siberia on December 16, 1877.

1.5. The hydrostatic equation is

$$\partial p = -g\rho\partial z,$$

where an infinitesimal change in height, ∂z, produces an infinitesimal change in pressure, ∂p, ρ being the density of the air. The hypsometric equation is

$$h - h_0 = KT(\log p_0 - \log p),$$

where p, p_0, denote the pressure at heights h, h_0, respectively; T is the mean (absolute) temperature of the air column between h and h_0, and K equals 221 (if h is in feet) or 67.4 (if h is in meters).

1.6. Details of these errors are as follows:

The *index error* represents the cumulative effect of errors in the construction and adjustment of the instrument, and is determined by comparing individual instruments with the readings of a standard instrument whose errors are known exactly. An error that cannot be compensated for in this fashion forms the residual index error, which arises from the fact that a barometer must be graduated on the assumption that a given reading on its scale represents the actual difference in level between the upper and lower mercury surfaces. However, it is not usually possible to adjust the barometer so that the zero of its scale is exactly at the level of the lower mercury surface, so that a small error of up to 0.2 mb will always exist.

The *personal error* is the unconscious error made by every observer in sighting the top of the mercury meniscus. It varies by about 0.07 mb with the position of the vernier on the barometer scale.

The *capillarity error* is due to the fact that the summit of the convex mercury meniscus is lowered slightly from its "true" level by surface tension. The amount of lowering increases as the diameter of the vertical glass tube holding the mercury decreases. The effect is negligible for tubes exceeding 1 inch in diameter. With the Fortin barometer, if the internal diameter of the cistern is 1.1 inches, then the capillary depression will be 0.008 inch, and errors on the order of 0.7 mb can be expected from this effect alone.

The *temperature error* is due to the lag of the mercury column in responding to changes in temperature. An uncertainty of 1°F in the temperature of the barometer produces an uncertainty of 0.1 mb in its indicated pressure. This is negligible in comparison with errors caused by different parts of the instrument being at different temperatures, but even so a rise in the external air temperature of 2°F per hour will result in an error of up to 0.03 mb. If the temperature stratification of the external air is nonuniform, then different parts of the barometer will be at different temperatures. The resulting error is appreciable if the rate of change of air temperature is 3°F or more per hour, and such a rate is easily exceeded if there is a stove or electric heater near the instrument.

The *vacuum error* can be very serious. If the vacuum above the mercury is not perfect, then the apparent reading of the instrument will increase; 100 cubic mm of air in the glass tube above the upper mercury meniscus will cause a Kew barometer to read 0.12 mb too high, while 0.01 milligram of water vapor will cause an error of 2.3 mb at normal pressures.

The *verticality error* is due to the effect of gravity. If a symmetric mercury barometer is mounted with the long axis of its tube inclined at $\phi°$ to the vertical, then its readings must be multiplied by $\cos\phi$ to obtain a true pressure value. Fortin barometers must be fixed so that their departure from true vertical at the lowest parts of the instruments does not exceed 0.06 inch (1.5 mm), since the fiducial point (i.e., the tip of the pointer in the cistern of this type of barometer) does not lie on the long axis of the instrument; if the fiducial point is 0.5 inch from the center line of the mercury column and the instrument is 5 minutes of arc out of true vertical, then the error will be 0.026 mb at a pressure of 1,000 mbs. For a Kew barometer, however, the error for a 5-minute departure from true vertical is only 0.0012 mb, and is therefore negligible.

With marine barometers, swinging of the instrument on its gimbal mounting causes the mercury column to both increase in length, because the instrument is inclined to the vertical most of the time, and to decrease in length, because the centrifugal force set up by the swinging causes an apparent increase in the value of g. The two errors are combined in the expression

$$CE = \frac{1}{4} H\theta^2 \left[1 - \frac{4\pi^2(h_1 - h_2)}{gt^2}\right],$$

in which θ is the angle the barometer makes with the vertical at the extreme of its swing, H is the height of the mercury column, t the time of oscillation of the barometer, and h_1, h_2, the distances of the mercury surfaces in the tube and in the cistern above and below the axis of swing. Errors due to the east-west motion of the ship are on the order of 0.05 mb, the barometer reading too high when the ship is moving east and too low when it is moving west. The error arises because the acceleration of gravity (g) is changed by an amount equal to $2\Omega v \cos\phi \sin\theta$, because of the increase or decrease in the centrifugal force of the Earth's rotation acting on a ship moving along a course that has an east-west component. Here v is the velocity of the ship, θ is the direction of its course, reckoned from true north, Ω is the angular velocity of the Earth, and ϕ the latitude.

The pressure lag is defined by the equation

$$\frac{dh}{dt} + \frac{1}{L}(h - h_0) = 0,$$

in which h is the observed height and h_0 the true height of the mercury above the cistern level, t is time, and L is the lagging time. L is defined as the time taken by the barometer to indicate 63 per cent of a sudden change in pressure, and is constant for a given instrument, varying between 6 and 9 minutes for British Meteorological Office marine barometers. When atmospheric pressure is changing at a constant rate, the recorded pressure lags behind the true pressure by an amount αL, where α equals the rate of change of pressure with time. Thus the instrument reads too high when pressure is falling, and too low when it is rising.

1.7. *Gravity correction*: The definitions of the units of pressure involve g, hence standard values must be taken to compensate for local variation in g. The correction varies with pressure, but the variation is not large in middle latitudes, so that a constant correction can be applied, corresponding to that for the mean pressure at the station. This correction equals $\dfrac{H(g_{\phi,h} - g_s)}{g_s}$ and is to be added to the observed reading, H. Here g_s is the standard value of gravity, and $g_{\phi,h}$ the value at height h in latitude ϕ as given by the empirical formula:

$$g_{\phi,h} = 980.616(1 - 0.0026373 \cos 2\phi + 0.0000059 \cos^2 2\phi) - 0.0000939h.$$

Prior to 1950, three values of "standard" gravity were in use in different parts of the world: 980.665 cm per sec^2, 980.62 cm per sec^2, and the value of gravity at mean sea level in latitude 45°. According to the International Barometer Convention, the conventional value of standard gravity, as proposed by the WMO in 1953, is 980.665 cm per sec^2, and other values were dropped after January 1, 1955.

Temperature correction: An increase in air temperature results in expansion of the mercury, its containing tube, the metal scale, and the cistern. For the Fortin barometer, the correction equals

$$-H_t \frac{(\beta - \alpha)(T - 273)}{1 + \beta(T - 273)},$$

where H_t is the observed reading in mbs corrected for index error and gravity, T is the temperature of the barometer in degrees absolute, α is the mean coefficient of linear expansion of the metal of which the scale is composed, over the appropriate temperature range, and β is the mean coefficient of volume expansion of mercury over the same temperature range. For the Kew barometer, the correction is

$$-\left[H_t \frac{(\beta-\alpha)(T-273)}{1+\beta(T-273)} + 1.33\frac{V}{A}(\beta-3\eta)\,(T-273)\right],$$

where V is the initial volume of the mercury, and η is the composite coefficient of linear expansion of the glass and the steel of which the instrument is constructed. H_t must be in mb. Before January 1, 1955, three separate standard temperatures were in use for mercury barometers. For instruments reading in millibars, a standard temperature of 12°C for the whole instrument was used. For instruments calibrated in inches, two standards were used, 62°F for the scale and 32°F for the mercury. For millimeter barometers, the standard was 0°C for the whole instrument. Since January 1, 1955, the standard temperature for all barometers is 0°C, the inch (mm) of mercury being defined as an inch (mm) of mercury at 0°C, under a gravitational attraction of 980.665 cm per sec^2.

Height correction: Pressure readings are reduced to mean sea level by adding to the observed reading the pressure due to the weight of a hypothetical vertical air column equal in length to the height of the barometer cistern above mean sea level. The pressure of this column depends on its temperature and water-vapor content, the local value of gravity, and the existing atmospheric pressure at the top of the column. For stations less than 1,000 feet above mean sea level, the effect of water vapor can be neglected. The values for mean sea-level pressure (p_0) are given by

$$\log_{10}\left(\frac{p_0}{p}\right) = H/67.4T,$$

where T is the temperature of the air column in degrees absolute (assumed to be constant and equal to the normal air temperature at station level), and H is the height of the column in meters. If H is in feet, the equation is

$$\log_{10}\left(\frac{p_0}{p}\right) = H/221.1T.$$

1.8. The *control* of the instrument is defined as the force which must be applied to the indicating mechanism at its recording point (i.e., the pen) to keep the reading constant when the value of the element being recorded changes by one unit; i.e., it is the force required to move the pen over one unit of the scale, provided the measured element remains constant. The control is arranged so that the maximum effect of friction on the pen is less than the least change that it is required to record. The greater the control, the less the effect of friction on the pen. With the ordinary aneroid barograph, one unit of the scale corresponds to one mb, and the required force equals 1,000A dynes, where A is the effective crosssectional area of the diaphragm chamber in square cm. The control thus equals $1{,}000A/M$, where M is the magnification of the lever system.

1.9. The lag coefficient is proportional to the viscosity of the oil, for a given diameter of annular gap. The viscosity of the oil varies considerably with temperature, therefore the lag coefficient also varies, from 100 seconds at 30°F to 40 seconds at 110°F (for a silicone oil).

1.10. The time-constant (λ) is equivalent to the lagging time, and is the time taken for the pressure difference between outside and inside the instrument to fall to 37 per cent (i.e., $1/e$) of its original value; λ is defined by

$$\frac{dp'}{dt} = \frac{1}{\lambda}(p - p'),$$

where p is the external atmospheric pressure, p' is the pressure inside the container, and t is time. This equation does not apply if the temperature of the air in the container is changing. If t is large compared with λ, then $p - p' = a\lambda$, where a is the rate of change of pressure with time; λ varies from about a minute to an hour or more in different instruments.

1.11. The most commonly used Standard Atmosphere is that defined in 1952 by the International Civil Aviation Organization. In its original form, the ICAO Standard Atmosphere applied to the lower 65,000 ft of the atmosphere, but was later extended to 75 km (47 miles) as a "tentative" atmosphere, and to 300 km (186 miles) as a "speculative" atmosphere. It assumed a mean sea-level pressure and temperature of 1,013.25 mb and 15°C, respectively, and a mean lapse-rate of 0.65°C per 100 m (2°C per 1,000 feet) from sea level to a height of 11 km (36,000 feet), where the mean temperature is −56.5°C. Above 11 km, the troposphere was assumed to be isothermal to 25 km (82,000 feet), with temperatures increasing above 25 km.

1.12. Using the U.S. Standard Atmosphere [prepared by the U.S. Weather Bureau in 1922, and representing the relation between pressure and temperature up to 10 km (33,000 feet), based on average conditions reported for the United States at latitude 40°N; later extended to 65,000, then to 400,000 feet], z_p equals zero at the pressure that will support a standard mercury column 760 mm high. The standard temperature, T_p, is 15°C at $z_p = 0$, decreasing upward uniformly at 6.5°C per km (1.9812°C per 1,000 feet) until it equals −55°C at $z_p = 10{,}769.2$ m (35,332 feet), above which it remains constant at this value. The air is assumed to obey the perfect-gas law and to contain no water vapor, and vertical accelerations are assumed to be absent, with g assumed constant at 980 cm per sec^2. The advantage to using this Standard Atmosphere, assuming the hydrostatic equation and the equation of state are obeyed perfectly, is that z_p and the absolute pressure units (mb or lbs per square inch) are in a one-to-one relationship.

1.13. To obtain the observed pressure at any station, subtract the reported D value from the height of the station, then convert the resulting value of z_p into inches or mm of mercury, using the relationships

$$T_p = T_0\left(\frac{p}{p_0}\right)^{0.190285} \quad \text{and} \quad z_p = \frac{1}{\alpha}(T_0 - T_p),$$

where $T_0 = 288°\text{K}$, $p_0 = 1{,}013.25$ mb, and $\alpha = 1.9812°\text{C}$ per 1,000 feet; T_p is the standard temperature corresponding to any given pressure p, defined by $T_p = T_0 - \alpha z_p$ for the standard troposphere.

1.14. According to T. E. W. Schumann, *W*,7 (1952), 370, considerably more than half the northern hemisphere shows significant ($p = .05$) autocorrelation of mean annual pressure for a one-year lag.

1.15. I. I. Schell, *JM*, 13 (1956), 592; K. S. Ramamurti, *JM*, 6 (1949), 63; C. Normand, *QJRMS*, 79 (1953), 463. H. C. Willett and F. T. Bodurtha, *BAMS*, 33 (1952),

429, describe an abbreviated form of the Southern Oscillation, defined as follows: (a) for the period June to August, Santiago pressure, plus Honolulu pressure, plus 0.7 Manila pressure, minus Batavia pressure, minus Madras temperature, minus 0.7 Darwin pressure; (b) for the period December to February, Samoa (Apia) pressure, minus Darwin pressure, minus Manila pressure, minus Batavia pressure, minus average temperature over southwest Canada, minus Apia temperature. The lag correlation between (a) and (b) is +0.74; for the complete Southern Oscillation, it is +0.83, with the secular trend removed.

1.16. The S_1 component is the pressure cycle repeating itself every 24 hours; the S_2 component repeats itself every 12 hours, the S_3 component every 8 hours, the S_4 every 6 hours, etc.

1.17. M. F. Harris, *JM*, 12 (1955), 394. According to R. Frost, *MOSP*, no. 4 (1960), the semidiurnal oscillation over Malaya is very difficult to reconcile with resonance effects. The first pressure maximum, at 10:00 A.M. local time, seems to be due to convection over the land, whereas the second maximum, at 10:00 P.M., is due to the fundamental period of free oscillation of the atmosphere of 12 solar hours.

1.18. One microbar equals 0.001 mb or 1 dyne per cm^2 or 0.00075 mm of mercury.

1.19. Each pressure surface is displaced vertically an amount δz, given by

$$\frac{\delta z}{z} = \frac{\delta T}{T},$$

where T is the mean temperature of the vertical air column between the Earth's surface and height z; δz increases from zero at the Earth's surface to a maximum at z, all higher pressure surfaces being raised by an amount equal to the value of δz at height z. Quantitatively, if pressures at the bottom and the top of the column are 1,000 and 600 mb, respectively, and if the column is heated 5°C more than the surrounding air by nonadiabatic processes, then the resulting change in pressure at the ground would be just under 9 mb.

1.20. The northern hemisphere's average mean sea-level pressure is 4 m lower in July than in January because it receives more heat than the southern hemisphere in July, less heat in January; this pressure change is nonadiabatic. In middle latitudes, mean pressures on land areas rise from summer to winter, but corresponding pressures on the oceans fall. There is a rise of 12 mb for land at latitude 40°N, and one of 11 mb for land at 50°N, whereas pressure falls 5 mb and 10 mb on oceans at latitudes 40°N and 50°N, respectively. Another nonadiabatic effect is the production of anticyclones over polar regions by surface cooling.

1.21. According to W. Schaffer, *JM*, 6 (1949), 212, the applicable equation is

$$\frac{w(\Gamma_e - \Gamma)}{T} = \frac{1}{\rho g}\frac{\partial}{\partial z}\left(V_h' \cdot \mathbf{\nabla}_p + \frac{\partial p}{\partial t}\right) - \frac{1}{p}\left(\frac{dp}{dt}\right)_H,$$

where w is the vertical velocity of the air (in cm per sec), Γ is the adiabatic lapse-rate, Γ_e the environmental lapse-rate, T is absolute temperature, ρ is density, p is pressure, V_h is the horizontal wind vector, $\partial/\partial z$ means rate of change with respect to height, dp/dt the rate of change of pressure with time, following the air particles, and $\mathbf{\nabla}_p$ the horizontal gradient of pressure.

1.22. J. Bjerknes, *Meteorol. Z.*, 54 (1937), 462, gives the pressure-tendency equation in the form

$$\left(\frac{\partial p}{\partial t}\right)_h = -\int_h^{\infty} g\left(u\frac{\partial \rho}{\partial x} + v\frac{\partial \rho}{\partial y}\right)dz - \int_h^{\infty} g\rho\left(\frac{\partial u}{\partial x} + \frac{\partial v}{\partial y}\right)dz + g(\rho w)_h,$$

where $\left(\frac{\partial p}{\partial t}\right)_h$ represents the pressure tendency at level h, ρ and p are the density and the pressure at this level, and u, v, and w are the wind components in the x, y, and z directions. The first term on the right-hand side of the equation represents the horizontal advection of mass, the second term the horizontal divergence of wind velocity, and the third term the vertical advection at level h. The equation was rewritten by H. G. Houghton and J. M. Austin, *JM*, 3 (1946), 57, in the form

$$\left(\frac{\partial p}{\partial t}\right)_h = -\int_h^{\infty} g\left(u\frac{\partial \rho}{\partial x} + v\frac{\partial \rho}{\partial y}\right)dz + \int_h^{\infty} gw\left(\frac{d\rho}{dz} - \frac{\partial \rho}{\partial z}\right)dz,$$

which shows that pressure change is decided by horizontal divergence and vertical motion, depending on the way in which the motion changes the density distribution above level h. The divergence term is large, difficult to measure, and known to only a very moderate degree of accuracy. The horizontal advection term is smaller, and easier to measure, but the vertical advection term is both large and difficult to measure. The difficulty in applying the equation is thus that the most important terms in it cannot be measured with sufficient accuracy, usually, whereas the easily measurable quantities are the least important. When observations of sufficient accuracy are available, the equation may be used in the practical computational form devised by R. Fleagle, *JM*, 5 (1948), 281,

$$\frac{1}{\rho}\left(\frac{\partial p}{\partial t}\right)_h = \frac{1-K}{p}\cdot\frac{\partial p}{\partial t} + \frac{1}{\theta}V_h\cdot\nabla_h\theta + \frac{w}{\theta}\cdot\frac{\partial \theta}{\partial z} - \frac{1}{\theta}\frac{d\theta}{dt},$$

in which $K = \frac{R}{C_p}$, R being the universal gas constant for dry air and C_p its specific heat at constant pressure, θ represents potential temperature, $\nabla_h\theta$ the horizontal gradient of potential temperature, and V_h the horizontal wind velocity vector. The pressure tendency at mean sea level is given by

$$\left(\frac{\partial p}{\partial t}\right)_0 = -\int_0^{\infty} \frac{1}{\rho}\frac{\partial p}{\partial t}\,dp.$$

1.23. This model was described by J. Bjerknes and J. Holmboe, *JM*, 1 (1944), 1. It relates the horizontal distribution of net mass divergence to a geographical pattern of high- and low-pressure areas. It shows that the anticyclones must move toward regions of mass convergence, i.e., regions of rising pressure, and cyclones must move toward regions of mass divergence (falling pressure).

APPENDIX 2

2.1. Fresnel's law states that the albedo (A) of a plane water surface under direct solar radiation is given by

$$A = 50\left[\frac{\sin^2(Z - \mathrm{r})}{\sin^2(Z + r)} + \frac{\tan^2(Z - r)}{\tan^2(Z + r)}\right],$$

where r is the angle of refraction of the beam, related to the refractive index of the water, i, by Snell's law, $i \sin r = \sin Z$, Z being the zenith angle. W. V. Burt, *JM*, 11 (1954), 283, gives the mathematics for the albedo of wind-roughened water.

2.2. Following *ITM*, p. 293, the atmospheric energy equation may be expressed as

$$\frac{dh}{dt} = \frac{d}{dt}\left(\frac{c^2}{2} + ap + gz + C_vT\right) - a\frac{\partial p}{\partial t} - (uF_x + vF_y + wF_z), \qquad \text{[A2.1]}$$

where c is the magnitude of the three-dimensional velocity vector $c^2 = u^2 + v^2 + w^2$. The equation is derived from the equations of motion and the first law of thermodynamics. If the air motion is adiabatic, and a steady state exists, then (neglecting viscous effects) the equation reduces to

$$\frac{c^2}{2} = ap + gz + C_vT = K, \qquad \text{[A2.2]}$$

where K is a constant. The left side of Equation A2.2 is constant along a given streamline, but the value of the constant may vary from one streamline to another. Equation A2.2 is a special case of Bernoulli's equation for an incompressible fluid. For a mass of air (that is, the contents of a specific volume) followed along a streamline, assuming a steady state, $\frac{c^2}{2} + ap + gz = K$, indicating that pressure must change as the elevation of the mass changes.

2.3. According to *ITM*, p. 294, the internal energy (per unit horizontal area) of an air column extending from sea level to height h is given by $C_v \int_0^h \rho T\, dz$ or, after using the hydrostatic equation, by $-\frac{C_v}{g}\int_{p_0}^{p_h} T\, dp$. The potential energy of an air layer extending from sea level to height h is given by $g \int_0^h \rho z\, dz$ or by $-\int_{p_0}^{p_h} z\, dp$. Combining the expressions for internal and potential energies gives *potential energy equals* $\frac{R}{C_v}$ *times internal energy*, where R is the universal gas constant. Consequently, if the temperature of an air column rises (i.e., if its internal energy increases), then the column must expand vertically; the expansion raises its center of mass and therefore increases its potential energy. The total kinetic energy of an atmospheric column is 53.8 U_g^2 in the friction layer, and 458 U_g^2 above gradient-wind level, i.e., 512 U_g for a complete column rising through both layers, where U is geostrophic eastward windspeed. These expressions apply to a model atmosphere with east-west trending isobars at all heights, and a pressure-gradient force per unit mass that is constant in the vertical.

2.4. M. Margules, *Meteorol. Z.* (1906), p. 243, showed that if two dry air masses of different temperature, arranged side by side, and initially at rest, are permitted to rearrange themselves within the same volume so that ultimately a minimum of potential energy is attained, then the average windspeed generated in the two masses when the maximum amount of internal and potential energy has been converted into kinetic energy is given by $\frac{1}{2}\left(\frac{gh\,\Delta T}{\bar{T}}\right)^{1/2}$, where $\bar{T}$ is the average temperature of the two air masses, h is the height of the volume of air, and ΔT is the difference in temperature between the two air masses at any height. Viscous losses of energy are ignored.

2.5. According to B. Bolin, *AGP*, 1 (1952), 102, the thermodynamic efficiency of the atmosphere is given by $\frac{(E_+ - E_1)}{E_+}$, where E_+ represents the total intensity of the energy sources and E_1 the total intensity of the energy sinks.

2.6. The classic paper is by E. N. Lorenz, *T*, 7 (1955), 157. See also J. A. Dutton and D. R. Johnson, *AGP*, 12 (1967), 334. Lorenz showed that the average available potential energy ($\bar{A}$) per unit area for the entire atmosphere equals

$$\frac{1}{2}\int_0^{\bar{p}_0} \frac{\overline{(T')^2}}{\bar{T}(\Gamma_d - \bar{\Gamma})}\,dp,$$

where the bars indicate averages over an entire isobaric surface, and T is absolute temperature. The average kinetic energy per unit area for the entire atmosphere, $\bar{K}$, is given by $\bar{K} = \frac{1}{2g}\int_0^{\bar{p}_0} \overline{V^2}\,dp$, where V represents horizontal windspeed. Following J. S. Winston and A. F. Krueger, *MRW*, 89 (1961), 307, the components of kinetic and available potential energy may be obtained as follows.

Zonal available potential energy, $\bar{A}_z$, equals

$$\frac{1}{2}\int_0^{\bar{p}_0} \frac{\overline{[T]'^2}}{\bar{T}(\Gamma_d - \bar{\Gamma})}\,dp,$$

where the square brackets indicate a latitudinal average and $[T]'$ represents the departure of the latitudinal average temperature from the over-all mean temperature.

Eddy available potential energy, $\bar{A}_E$, equals

$$\frac{1}{2}\int_0^{\bar{p}_0} \frac{\overline{(T^*)^2}}{\bar{T}(\Gamma_d - \bar{\Gamma})}\,dp,$$

in which T^* is the departure of the temperature of the atmosphere at any point from its latitudinal average.

Zonal kinetic energy, $\overline{K}_z$, equals

$$\frac{1}{2g}\int_0^{\bar{p}_0} \overline{[V]^2}\, dp,$$

where $[V]$ is the average windspeed for each circle of latitude.

Eddy kinetic energy, $\overline{K}_E$, equals

$$\frac{1}{2g}\int_0^{\bar{p}_0} \overline{(V^*)^2}\, dp,$$

where V^* is the deviation of the wind vector from its latitudinal average.

The local change in available potential energy is

$$\frac{\partial \bar{A}}{\partial t} = G - C,$$

The local change in kinetic energy is

$$\frac{\partial \bar{K}}{\partial t} = C - D.$$

Here C is the rate at which potential energy is converted into kinetic energy or vice versa, and equals

$$-\frac{R}{g}\int_0^{\bar{p}_0} \frac{\overline{T\omega}}{p}\, dp,$$

where $\omega = dp/dt$, and is the rate of vertical motion in pressure coordinates; G is the rate at which available potential energy is generated, and equals

$$\frac{1}{g}\int_0^{\bar{p}_0} \frac{\Gamma_d \overline{T'Q'}}{\bar{T}(\Gamma_d - \bar{\Gamma})}\, dp;$$

and D is the rate at which kinetic energy is dissipated by friction, and equals

$$-\frac{1}{g}\int_0^{\bar{p}_0} \overline{\mathbf{V}\cdot\mathbf{F}}\, dp,$$

where **V** is the wind vector and **F** is the frictional force vector.

2.7. Following H. R. Byers, *General Meteorology* (New York, 3d ed., 1959), pp. 274–77, for conditions on a hemispheric scale, considering a latitudinal ring of air of unit width in latitude ϕ, and assuming the effective depth of the atmosphere in this latitude to be D, the absolute angular momentum, μ_ϕ, of the ring is given by

$$\mu_\phi = M_\phi \Omega r_\phi^2 + M_\phi \bar{u} r_\phi,$$

where M_ϕ, the mass of the ring, is equal to $2\pi r_\phi \bar{\rho} D$; $\bar{\rho}$ and r_ϕ are the mean density of the atmosphere and the radius of the Earth in latitude ϕ, respectively; and $\bar{u}$ is the average relative zonal windspeed in latitude ϕ.

2.8. For details, see V. P. Starr and R. M. White, *QJRMS*, 77 (1951), 215, and *AFGP*, no. 35 (1954); J. Bjerknes, *DC*, p. 67; C. H. B. Priestley and A. J. Troup, *JAS*, 21 (1964), 459. The properties of angular momentum, energy, mass, and water are conservative, and cannot be generated or destroyed within the atmosphere. They can, however, be redistributed or (in the case of energy) converted from one form into another. Following Starr and White, the flux of any of these properties may be evaluated by means of the expression

$$\frac{1}{g}\int_0^{p_0}\int_0^t\int_0^L av\,dx\,dt\,dp = \frac{p_0 tL}{g}\,(\overline{[\alpha v]}),$$

in which α is the amount of any given property per unit mass of air, v represents the northward velocity component, the brackets indicate a mean with respect to longitude, the parentheses indicate a mean with respect to vertical pressure, and the bar indicates a mean with respect to time. The term $(\overline{[\alpha v]})$ may be expanded, following V. P. Starr and R. M. White, *QJRMS*, 78 (1952), 407, as

$$(\overline{[\alpha v]}) = ([\bar{\alpha}])\,([\bar{v}]) + (\overline{[\alpha]'}\;\overline{[v]'}) + (\overline{[\alpha]'[v]'}) + (\overline{[\alpha' v']}).$$

Here the term $([\bar{\alpha}])\,([\bar{v}])$ measures the flux of quantity α when a net shift of mass takes place across a latitude circle; $(\overline{[\alpha']}\;\overline{[v]'})$ measures the flux due to the mean meridional circulations; $(\overline{[\alpha]'\;[v]'})$ measures the flux due to the instantaneous meridional circulations; and $(\overline{[\alpha' v']})$ measures the flux due to large-scale horizontal eddies. The respective balance equations are:

for angular momentum; $(\overline{[uv]}) = ([\bar{u}]\,[\bar{v}]) + (\overline{[u]'}\;\overline{[v]'}) + (\overline{[u]'\;[v]'}) + (\overline{[u'v']})$;

for water balance, $(\overline{[qv]}) = ([\bar{q}]\,[\bar{v}]) + (\overline{[q]'}\;\overline{[v]'}) + (\overline{[q]'\;[v]'}) + (\overline{[q'v']})$;

for energy (enthalpy) balance, $(\overline{[Tv]}) = ([\bar{T}])\,([\bar{v}]) + (\overline{[T]'}\;\overline{[v]'}) + (\overline{[T]'\;[v]'}) + (\overline{[T'v']})$.

2.9. C. H. B. Priestley, *QJRMS*, 75 (1949), 28. This classic study measured the advective and eddy fluxes by $C_p \rho VT$, the amount of sensible heat carried poleward per unit area per unit time, V representing the poleward component of horizontal velocity. The mean total flux, F, was measured (for a unit area and time) by

$$F = \frac{C_p}{g}\int_0^{\bar{p}_0} \overline{VT}\,dp.$$

The expression VT was evaluated by means of simultaneous values of V and T at fixed pressure levels, i.e., $\overline{VT} = \bar{V}\bar{T} + \overline{V'T'}$, where the assumption is made that $V = \bar{V} + V'$ and $T = \bar{T} + T'$, and where $\bar{V}\bar{T}$, $\overline{V'T'}$, represent the advective and eddy fluxes, respectively.

2.10. G. S. Benton and M. A. Estoque, *JM*, 11 (1954), 462. The water-vapor transport vector, $\mathbf{w}$, is given by

$$\mathbf{w} \equiv \frac{1}{T}\int_0^T \frac{1}{g}\left(\frac{\rho_w}{\rho}\right)\mathbf{C}\,dt \approx \frac{1}{N}\sum_{i=1}^{N}\frac{1}{g}\left(\frac{\rho_w}{\rho}\right)\mathbf{C}_i,$$

where i denotes individual observations, T is the time in which N observations occur, and $\mathbf{C}$ is the horizontal wind-velocity vector; $\mathbf{w}$ is determined for each level in cm per mb per sec, and then an integrated transfer vector, $\mathbf{W}$, is found from $\mathbf{W} = \int_p^{p_o} \mathbf{w} dp$.

2.11. J. W. Hutchings, *QJRMS*, 83 (1957), 30. At any level, the advective transfer, $\mathbf{w}_M$, of water vapour is given by $\mathbf{w}_M = \frac{1}{g}(\bar{q}\bar{c})$ and the vertically integrated advective transfer is given by $\mathbf{W}_M = \int_0^{p_o} \mathbf{w}_M \, dp$. Here

$$\bar{q} = \frac{1}{T}\int q \, dt \approx \frac{1}{N}\sum_{i=1}^{N} (q_i)$$

and

$$\bar{c} = \frac{1}{T}\int c \, dt \approx \frac{1}{N}\sum_{i=1}^{N} (c_i)$$

where T denotes the time interval over which q and c are averaged, there are N observations during T, and i denotes individual observations. For any level, the eddy transfer, $\mathbf{w}_E$, of vapor is given by $\mathbf{w}_E = \mathbf{w} - \mathbf{w}_M$, where $\mathbf{w}$ is the total water-vapor transfer vector, and the vertically integrated eddy transfer, $\mathbf{W}_E$, is given by $\mathbf{W}_E = \mathbf{W} - \mathbf{W}_M$. We evaluate $\mathbf{w}$ as

$$\mathbf{w} = \frac{1}{gT}\int_0^T qc \, dt \approx \frac{1}{gN}\sum_{i=1}^{N} (q_i c_i).$$

$\mathbf{W}$ is then obtained from $\mathbf{W} = \int_0^{p_o} \mathbf{w} \, dp$.

2.12. The convective vertical heat flux is given by $\overline{\rho_1 w_1 h_1}$, where ρ_1, w_1, h_1 are simultaneous values of the fluctuating atmospheric density, vertical velocity relative to the Earth's surface, and total heat per unit mass at a point in a constant-pressure surface. The convective flux of sensible heat is $C_p p\bar{w}_1/R$.

2.13. Following Scorer, *MM*, 83 (1954), 202, the velocity of ascent of a thermal bubble is proportional to $\sqrt{gBR}$, where B is the buoyancy (i.e., the ratio of the difference in density between the bubble and its surroundings to the density of the surroundings) and R is the size (i.e., a linear dimension) of the bubble.

2.14. The Richardson number is defined as

$$Ri = \frac{\frac{g}{\theta}\frac{\partial\theta}{\partial z}}{\left(\frac{\partial u}{\partial z}\right)^2}.$$

E. L. Deacon, *CTP*, no. 4 (1955), proved that the heights of the transitional layer are given by fixed multiples of

$$\frac{C_p T \tau^{3/2}}{g\rho^{1/2} H},$$

where τ is the shearing stress and H is the convective heat flux.

2.15. See C. H. B. Priestley, *QJRMS*, 81 (1955), 139, and *UCSCM*, p. 106, for details of these procedures. During forced convection, the potential-temperature profile obeys the law $\frac{\partial\theta}{\partial z} \propto \frac{1}{z}$ and the heat flux is

$$k_\rho{}^2\, C_p z^2 \left|\frac{\partial\theta}{\partial z}\right| \frac{\partial u}{\partial z},$$

where k ($\approx$0.4) is von Kármán's constant. During free convection, the profile is given by $-\frac{\partial\theta}{\partial z} = \frac{C}{z^3}$, and the heat flux is $0.9\rho C_p \left(\frac{g}{\theta}\right)^{1/2} C^{3/2}$.

2.16. Following Panofsky, *CM*, p. 639, the *kinematic method* employs the equation

$$w_h = -\frac{1}{\rho_h}\int_s^h \rho \operatorname{div} \mathbf{V}\, dz + w_s\left(\frac{\rho_s}{\rho_h}\right),$$

w_h being the vertical velocity at level h, and w_s the vertical velocity at the Earth's surface, as estimated from the horizontal wind field and the slope of the terrain. An alternative equation is

$$w_h = \left(\frac{p_s}{p_h}\right)^{1/\gamma} w_s - \int_s^h \left(\frac{p_s}{p_h}\right)^{1/\gamma} \nabla\cdot\mathbf{V}\, dz,$$

where $\gamma = \frac{C_p}{C_v}$. In both equations the subscripts h and s refer to the arbitrary height h and the Earth's surface, respectively. The divergence ($\nabla\cdot\mathbf{V}$) is computed either from wind components,

$$\nabla . \mathbf{V} = \frac{\partial u}{\partial x} + \frac{\partial v}{\partial y} - \frac{V}{R}\tan\phi,$$

where R is the radius of the Earth in latitude ϕ, or from streamlines,

$$\nabla\cdot\mathbf{V} = \frac{\partial V_r}{\partial r} + \frac{V_r}{r},$$

where V_r is the horizontal wind speed at a distance, r, from the origin of the streamline curvature (V_r is positive if directed along the positive direction of r) and the coordinates are taken parallel and at right angles to the streamlines.

The *adiabatic method* employs the equation

$$w = -\left(\frac{\partial T}{\partial t} + \mathbf{V}\cdot\nabla T\right)\Big/(\Gamma_d - \Gamma) = -\frac{\delta T}{\delta t}\Big/\Gamma_d - \Gamma,$$

where $\frac{\delta T}{\delta t}$ is the change in temperature of the air following a horizontal trajectory, Γ is the existing lapse-rate, and w is the required vertical velocity; $\frac{\delta T}{\delta t}$ is found by subtracting the temperature at the beginning of an isobaric trajectory from the temperature at the end (say, 12 hours later) of the trajectory.

2.17. Following S. J. Smebye, *JM*, 15 (1958), 547, it may be shown, from the vorticity equations and the first law of thermodynamics, and assuming geostrophic motion, that

$$w_M\left(1.3 + 1.1 \times 10^8 \frac{\lambda Q_m}{m^2}\right) = 1.3\overline{w}_M + 1.4 \times 10^{-4} A + 4 \times 10^{-3}(\overline{H} - H),$$

with

$$A = \lambda \frac{\partial}{dp}(\mathbf{V} \cdot \nabla Q) + \nabla^2(\mathbf{V} \cdot \nabla \alpha).$$

Here A is measured in units of 50 feet per 6 hours, all other quantities in meters, tons, and seconds; Q is the absolute vorticity; V is the horizontal wind vector; λ is vertical velocity in an x, y, p, coordinate system; α is specific volume; H is the heat received per unit mass of air in a unit time (i.e., the latent heat of vaporization); m is the scale factor of the map; w is the center value of vertical velocity at the midpoint of a 600-km grid; $\overline{w}$ is the mean value at the four corners of the grid; and w_M is the mean value of vertical velocity through the layer between 1,000 and 500 mb.

The orographically-induced vertical motion (w') is given by

$$w' = w'_0\left(\frac{p}{p_0}\right)^{2 \cdot 5},$$

where w'_0 is the value of vertical velocity due to orography at the Earth's surface. The computed precipitation (N) is given by $N = N_1 + \Delta N$, where N_1 is the rainfall computed from w_m, assuming that no latent heat is released (i.e., as given in the equation of the preceding paragraph), and ΔN is the amount of additional precipitation computed from the released heat ($H - \overline{H}$).

2.18. Under normal steady-state conditions, the pressure-gradient force **PG** balances the frictional force **F** (see Figure A2.18). Separation occurs when the pressure-gradient is shifted by a topographic obstacle so that the pressure-gradient force is in the same direction as the frictional force. Then if **PG** + **F** = **V**, where **V** is the wind velocity, stagnant air results. If **PG** + **F**/ > /**V**, reversed flow will result.

2.19. The natural period of vertical oscillation of the atmosphere is given by

$$P = 2\pi \sqrt{\frac{T}{g\left(\frac{\partial T}{\partial z} + \Gamma\right)}}$$

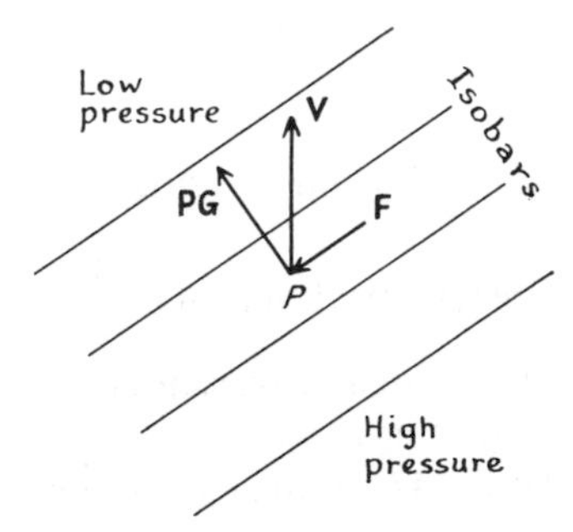

FIGURE A2.18.
The vector **V** represents the surface wind; **PG** is the vector representing the pressure-gradient force; and **F** is the vector representing the frictional force.

where Γ is the appropriate adiabatic lapse-rate. If an air particle is displaced vertically from its equilibrium level in a stable atmospheric environment, buoyancy forces will cause it to oscillate about its equilibrium level with a period of vertical oscillation given by this expression for P.

2.20. Queney's parameters are defined as

$$l_s = \sqrt{\frac{g\beta}{U}} \quad \text{and} \quad l_f = \frac{f}{U},$$

where $\beta = \frac{1}{\theta}\frac{\partial\theta}{\partial z}$, the coefficient of static stability, U is the horizontal windspeed, and f equals twice the vertical component of the Earth's angular velocity (i.e., it equals the vorticity about a vertical axis).

2.21. Scorer showed that the wavelength, λ, of the wave resulting from a combination of a horizontal windspeed with the natural period of vertical oscillation of the atmosphere is given by

$$\lambda = \frac{2\pi}{\sqrt{\frac{g}{U^2T}\left(\frac{\partial T}{\partial z} + \Gamma\right) - \frac{1}{U}\left(\frac{\partial^2 U}{\partial z^2}\right)}},$$

in which $\frac{\partial^2 U}{\partial z^2}$ represents the vertical acceleration of the air particles. Because the last term is usually very small, the equation reduces to

$$\lambda = 2\pi U \sqrt{\frac{T}{g\left(\frac{\partial T}{\partial z} + \Gamma\right)}}.$$

Real airstreams normally contain several natural wavelengths, and so it is more correct to regard λ as a parameter that describes the wind and stability characteristics of an airstream rather than as giving the true length of any orographic waves that form in it. Accordingly, λ, as defined by the second equation, is usually termed *Scorer's parameter* and is designated by the letter l. Practically, it is usually defined as

$$l^2 = \left(\frac{B}{U}\right)^2 - \frac{1}{U}\left(\frac{\partial^2 U}{\partial z^2}\right),$$

where $B^2 = \left(\frac{g}{\theta}\right)\left(\frac{\partial\theta}{\partial z}\right)$, i.e., the coefficient of static stability multiplied by the acceleration of gravity. As with the first equation, the second term in this equation can usually be ignored, because it is only important for shallow atmospheric layers. Hence $l \approx \sqrt{\frac{g\beta}{U^2}}$, where β is the coefficient of static stability. Scorer proved that orographic waves can only be noticeable if l^2 decreases with height.

2.22. The vertical displacement (ζ) of a streamline from its undisturbed level at a height z, due to the presence of a symmetric mountain ridge whose length is infinite compared to its width, is given by

$$\zeta = -Hbe^{-Kb}\,\xi \sin Kx.$$

Here H is the height of the crest of the ridge; b is a width parameter (i.e., a representative figure) for the ridge; x is the horizontal distance from the ridge crest at which ζ is measured; K equals $2\pi/W$, where W is the lee wavelength; and

$$\xi = \frac{2\pi\rho_0 f}{\rho\left(\dfrac{\partial f}{\partial K}\right)_0},$$

in which the subscript o refers to the lower boundary of a particular atmospheric layer, and f may be found by finite-difference approximations from an expression given in C. E. Wallington and V. Portnall, *QJRMS*, 84 (1958), 38, which see for details.

2.23. Following C. E. Wallington, *QJRMS*, 84 (1958), 428, a single ideal ridge, of symmetric cross section, may be described by the equation

$$h = \frac{Hb^2}{b^2 + x^2},$$

in which h is the height of the ridge at a horizontal distance x from the crest, H is the crest height, and b is a width parameter. Any orographic cross section, however complex, may be represented by a number of such single ridges, in which case

$$h = \sum_{r=1}^{n} \frac{H_r b_r^2}{b_r^2 + (x - x_r)^2},$$

where n ridges are required, r indicates any particular ridge, and x_r is the horizontal distance of the summit of this ridge from an arbitrary origin. In Figure 2.18,A (p. 160), the equation

$$h = \frac{Hb^2}{b^2 + x^2},$$

is applied to each half of the actual ridge separately, b being given the value B when $x > 0$ and the value B_1 when $x < 0$. Wallington also shows that the displacement ζ of a streamline by a composite ridge that is the algebraic synthesis of several single ideal ridges is given by

$$\zeta = -M\xi \frac{\sin 2\pi(x - X)}{W}.$$

Here $M = \sqrt{C^2 + S^2}$; $X = W \tan^{-1}(S/C)$; ξ, W, and x are defined as in Appendix 2.22; and

$$C = \sum_{r=1}^{n} H_r b_r e^{-Kb_r} \cos Kx;$$

and

$$S = \sum_{r=1}^{n} H_r b_r e^{-Kb_r} \sin Kx,$$

where $K = \dfrac{2\pi}{W}$. The mountain factor M and the phase displacement X are calculated from $H_r b_r x_r$ for various lee wavelengths for the ridge r.

2.24. Following Sheppard, *QJRMS*, 82 (1956), 528, Bernoulli's equation may be written in the form

$$-\frac{dp}{\rho} = d\left(\frac{1}{2}q^2 + gz\right)$$

for an air particle of unit mass moving in a frictionless, gravitational atmosphere, where q represents the velocity of the particle. The work done on the particle $\left(\dfrac{dp}{\rho}\right)$ by the drop in pressure along an increment of its path equals the change in the sum of its kinetic ($\frac{1}{2}q^2$) and potential (gz) energy.

2.25. See J. R. Malkus, *SP*, 43 (1955), 461; J. R. Malkus and M. Stern, *JM*, 10 (1953), 30, 105; and J. F. Black and B. L. Tarmy, *JAM*, 2 (1963), 557. In general, for a two-dimensional model in which the x direction is taken parallel to the wind and the z direction is upward, the height M of the thermal mountain at distance X from the windward edge of the island is given by $M = A\left(1 - \dfrac{1}{e}\right)$, where A, the maximum height of the mountain, is given by $A = \dfrac{T_e}{(\Gamma - \gamma)}$, where T_e is the effective temperature excess of the surface of the island. Also, X is given by $X = \dfrac{U^3}{(gsK)}$, where U is the basic windspeed, s is the static stability, and K is the coefficient of eddy conductivity.

APPENDIX 3

3.1. The coefficient of viscosity is also termed the *dynamic viscosity*, and is related to the kinematic viscosity ν by the expression $\nu = \dfrac{\mu}{\rho}$.

3.2. The mixing length, l, is defined by

$$u_* = l\left(\frac{\partial \bar{u}}{\partial z}\right),$$

where u_* is the *friction velocity*, defined by $u_*^2 = \dfrac{\tau}{\rho}$, where τ is the *eddy shearing stress*. If u' and w' denote departures from the mean of the horizontal and vertical motion respectively, then $\tau = -\rho\overline{u'w'}$.

$$\tau = A\left(\frac{\partial \bar{u}}{\partial z}\right) = K\rho\left(\frac{\partial \bar{u}}{\partial z}\right),$$

where A and K represent the interchange coefficient and the eddy viscosity, respectively.

3.3. I.e., the *exchange-coefficient hypothesis* states that

$$\text{Mean flux} = (k + A)\left(\frac{d\bar{E}}{dz}\right),$$

where k is the molecular coefficient of viscosity, conductivity, and diffusivity, and E is the amount per unit mass of any transferable conservative entity, such as heat, momentum, or matter suspended in the air.

3.4. Other important stability parameters are: the ratio $\dfrac{z}{L}$, where L is the *Obukhov scale length*, defined, following E. K. Webb, *AMM*, vol. 6, no. 28 (1965), by

$$L = -\frac{u_*^3}{\left(\dfrac{KgH}{C_p\rho\theta}\right)}.$$

where H is the heat flux ($H = C_p \rho \overline{w'T'}$); and the *Flux Richardson number*, *Rf*, defined by

$$Rf = -\frac{\dfrac{gH}{c_q \theta}}{\left[\tau\left(\dfrac{\partial U}{\partial z}\right)\right]}.$$

The Richardson number may also be expressed as

$$Ri = g\frac{\left(\dfrac{\partial T}{\partial z} + \Gamma\right)}{T\left(\dfrac{\partial \bar{u}}{\partial z}\right)^2},$$

where T is the absolute temperature of the environment.

3.5. In its simplest form (for a flat plate), Fourier's law of heat conduction states that

$$\frac{Q}{A} = \kappa\left(\frac{t_1 - t_2}{x}\right),$$

where a quantity of heat Q passes per second through an area A of a plate of conductivity κ and thickness x, the sides of which are maintained at constant temperatures t_1 and t_2.

3.6. The *scale of turbulence*, L, is defined by

$$L = \int_0^\infty R(y)dy.$$

$R(y)$ is the correlation coefficient between velocities separated by a variable distance y. Thus L represents a type of length parameter that indicates the sharpness of the rate of decrease of $R(y)$ with y.

The *microscale of turbulence*, M, is defined by

$$\frac{1}{M^2} = \lim_{y \to 0}\left\{\frac{1 - R(y)}{y^2}\right\}.$$

It is important in the decaying stages of turbulence.

3.7. The coefficient $R(y)$ in Appendix 3.6 is a *space-correlation* coefficient. A *time-correlation* coefficient, $R(\xi)$, may be defined in terms of the turbulent fluctuations measured at a fixed point at intervals of time separated by ξ. If the turbulence is uniform and steady, then the spread of a cluster of particles over a plane because of turbulence is given by

$$\overline{X}^2 = 2\overline{u'^2}\int_0^T \int_0^t R(\xi)d\xi\, dt,$$

where a particle travels a distance X in time T; $R(\xi)$ is the *autocorrelation coefficient*.

3.8. According to F. Pasquill, *Atmospheric Diffusion* (New York, 1962), Taylor proved the equations

$$R(\xi) = \int_0^\infty F(n)\cos 2\pi nt\, dn,$$

$$F(n) = 4\int_0^\infty R(\xi)\cos 2\pi nt\, dn,$$

where a plot of $F(n)$ against n forms the *power-spectrum function* of the observed turbulence, and the measured fluctuations $u'(t)$ are represented by means of the Fourier integral

$$u'(t) = 2\pi \int_0^\infty [I_1(n)\cos 2\pi nt + I_2(n)\sin 2\pi nt]\, dn,$$

where n is the frequency of the function in cycles per sec.

3.9. The values of n are: for a marked lapse-rate, 0.20; for a small or zero lapse-rate, 0.25; for a moderate inversion, 0.33; for a marked inversion, 0.50. We define n by the expression

$$\bar{u} = \bar{u}_1 \left(\frac{z}{z_1}\right)^{n/(2-n)}.$$

For the corresponding values of C for different heights, see *MAE*, p. 53. In general, for adiabatic conditions,

$$C = 0.18 - 0.422 \log_{10} z,$$

where z denotes the height above ground level for which C is required.

3.10. Sutton's equations for an instantaneous point source are as follows.

(a) For isotropic turbulence,

$$\chi = \frac{Q}{\pi^{3/2} C^3 (\bar{u}t)^{3(2-n)/2}} \exp\left[-\frac{r^2}{C^2(\bar{u}t)^{2-n}}\right],$$

where χ is the concentration of diffused matter at point x, y, z at time t; Q is the quantity of matter released instantaneously at time $t = 0$; C is defined by:

$$C^2 = \frac{4\nu^n}{(1-n)(2-n)\bar{u}^n} \left[\frac{\overline{(u')^2}}{\overline{u^2}}\right]^{1-n};$$

n is defined as in Appendix 3.9; ν is the kinematic viscosity; and

$$\bar{x}^2 = \frac{1}{2} C^2 (\bar{u}t)^{2-n}.$$

(b) For nonisotropic turbulence,

$$x = \frac{Q}{\pi^{3/2} C_x C_y C_z (\bar{u}t)^{3(2-n)/2}} \exp[-p(\bar{u}t)^{n-2}]$$

where

$$P = \left(\frac{x^2}{C_x^2}\right) + \left(\frac{y^2}{C_y^2}\right) + \left(\frac{z^2}{C_z^2}\right)$$

and the values of the virtual diffusion coefficients in the OX, OY, and OZ directions are:

$$C_x \equiv C;$$

$$C_y^2 = \frac{4\nu^n}{(1-n)(2-n)(\bar{u})^n}\left[\frac{\overline{(v')^2}}{\bar{u}^2}\right]^{1-n};$$

$$C_z^2 = \frac{4\nu^n}{(1-n)(2-n)(\bar{u})^n}\left[\frac{\overline{(w')^2}}{\bar{u}^2}\right]^{1-n}.$$

3.11. Homogeneous turbulence is defined as turbulent motion whose statistical properties $R(y)$ and u'^2 are independent of position. In such a situation, if one observing point is fixed, and the other varies in distance y from it, $R(y)$ should decrease with increasing y.

3.12. Following B. P. Harper, *JM*, 18 (1961), 487, the daily 500-mb hemispheric index of meridional circulation, V_ϕ, is defined as

$$V_\phi = \frac{-g\sum\Delta h}{R\pi\omega \sin 2\phi},$$

where Δh is the difference in height from each maximum to the corresponding minimum of 500-mb height to the east along latitude ϕ, and R, ω, are the radius and angular velocity of the Earth, respectively.

3.13. Following R. D. Elliott and T. B. Smith, *JM*, 6 (1949), 67, after Lettau, the *lateral eddy-viscosity coefficient*, A_ϕ, is defined by

$$A_\phi = \frac{\rho \sum^{N} \overline{V}_i'^2 n_i}{2N}$$

where N is the number of broad-scale "surges" of abnormal flow northward or southward during the period over which the summation $\sum$ is performed; $\overline{V}_i'$ is the mean absolute geostrophic departure from the mean flow during one "surge," as measured by the pressure differences between closely adjacent points at equal longitude distances to left and right of the data point; and $\overline{V}_i'\, n_i$ represents the "mixing length" of the "surge," where n_i is the number of days during which it was in progress.

3.14. The Prandtl number Pr may be defined as the ratio of the kinematic viscosity ν of a fluid to its thermometric conductivity κ:

$$Pr = \frac{C_p\,\nu}{\kappa}.$$

One derivative of height is the *profile contour number* α of the local wind distribution, defined by:

$$\alpha = \frac{\delta \log V}{\delta \log z} \equiv \frac{zV'}{V},$$

where V is the windspeed at height z above the ground, and V' denotes $\frac{\partial V}{\partial z}$. The α-profile determines the apparent roughness length z_0 of the wind profile, and α in effect measures the relative gradient of windspeed. The *Deacon number* of the wind profile, β, is defined by

$$\beta = \frac{-\delta \log V'}{\delta \log z} \equiv \frac{-zV''}{V'},$$

where V'' denotes $\frac{\partial^2 V}{\partial z^2}$. Combining β and α gives

$$\beta = 1 - \alpha - \frac{\delta \log \alpha}{\delta \log z},$$

and for $\beta = 1$, the logarithmic profile

$$V = k\frac{\log (z + z_0)}{z_0}$$

results, where z_0 is the roughness length.

3.15. The friction velocity, u_*, as defined in Appendix 3.2, is introduced because measurements indicate that the virtual stresses set up by turbulence are approximately proportional to the square of the mean velocity of flow. It is therefore useful to define an auxiliary reference velocity, u_*, so that virtual stresses are proportional to u_*^2.

3.16. For deeper layers,

$$\frac{\bar{u}}{u_*} = \frac{1}{k} \ln \left(\frac{z + z_0}{z_0}\right),$$

where $\bar{u} = 0$ at $z = 0$. The two equations may be combined by introducing the concept of *macroviscosity*, N, defined by $N = u_* z_0$. Then for a shallow layer,

$$\frac{\bar{u}}{u_*} = \frac{1}{k} \ln \left(\frac{u_* z}{N + \nu/9}\right),$$

and for a deep layer,

$$\frac{\bar{u}}{u_*} = \frac{1}{k} \ln \left(\frac{u_* z + N}{N + \nu/9}\right),$$

where ν is the kinematic viscosity.

If $N < 0.13\nu \approx 0.02$ cm^2 per sec, the surface is said to be *aerodynamically smooth*; if $N > 2.5\nu \approx 0.4$ cm^2 per sec, it is said to the aerodynamically *rough*.

3.17. Over surfaces that are aerodynamically rough, but relatively smooth topographically, the drag coefficient C_d is given by

$$C_d = \frac{k^2}{\log\left(\frac{z_a}{z_0}\right)^2}.$$

3.18. One parameter, e.g., is *Richardson's stability index*, obtained by writing *Ri* in the form

$$Ri = g\frac{\frac{d(\ln\theta)}{dz}}{\left(\frac{dU}{dz}\right)^2}.$$

See T. H. Ellison, pp. 400ff in G. K. Batchelor and R. M. Davies, eds., *Surveys in Mechanics* (Cambridge, Eng., 1956). This definition of *Ri* gives a "local" value of the Richardson number at height z above the Earth's surface. Near the latter, θ is very close to the actual temperature T, and consequently *Ri* will be different from *Ri* at z. The combination of these two "local" values of *Ri* gives the *bulk Richardson number* which is the usual meaning of *Ri*, e.g., as in Appendix 2.14. H. H. Lettau, *EAFM*, p. 328, discusses the connection between "local" and "bulk" values of *Ri*.

The vertical heat flux is usually written

$$\frac{H}{C_p\rho}.$$

3.19. Following H. Charnock, *SP*, 46 (1958), 470,

$$L \equiv \frac{u_*^3}{k\left(\frac{g}{T}\right)\left(\frac{H}{C_p\rho}\right)},$$

where L is the Monin-Obukhov parameter and k is von Kármán's coefficient.

3.20. According to A. K. Blackadar, *AMM*, 4, no. 22 (1960), 3, the *Deacon profile*, on which see E. L. Deacon, *QJRMS*, 75 (1949), 89, is

$$u = \frac{1}{k}\sqrt{\frac{\tau_0}{\rho}}\left[\frac{1}{1-\beta}\left\{\left(\frac{z}{z_0}\right)^{1-\beta} - 1\right\}\right],$$

where $\beta < 1$ for stable conditions, and $\beta > 1$ for unstable conditions. The profile does not apply above the lowest 5 or 10 meters of the atmosphere, because the departure of β from unity tends to increase with increasing height above the ground.

The *Monin-Obukhov profile* is given by

$$u = \frac{u_*}{k}\left(\ln\frac{z}{z_0} - \frac{\alpha z}{L}\right),$$

where L is the Monin-Obukhov parameter, defined by:

$$L = u_3^* \frac{C_p \rho T}{kgH}$$

and $\alpha \approx 0.6$. Strictly speaking, this equation is only true for small values of $\frac{z}{L}$. According to J. Neumann, *JM*, 18 (1961), 808, L may be obtained from measurements of temperature and windspeed by means of the relationship

$$Ri = \frac{z}{L}\left(1 + \alpha \frac{z}{L}\right),$$

where Ri is the Richardson number defined as

$$Ri = \frac{\frac{g}{T_0}\left(\frac{d\bar{T}}{dz}\right)}{\left(\frac{d\bar{u}}{dz}\right)^2},$$

in which T_0 is the average temperature of the air layer under consideration.

The *power-law profile* is given by

$$\frac{u}{u_1} = \left(\frac{z}{z_1}\right)^p,$$

where u_1 and z_1 are taken for some level within the layer, and p is a parameter independent of height. According to Blackadar,

$$p = \frac{u^* S}{ku},$$

where S is the Monin-Obukhov wind-shear function defined by

$$S = \frac{k}{u_*}\left(\frac{\partial u}{\partial \ln z}\right).$$

Following W. C. Swinbank, *QJRMS*, 90 (1964), 119, the *exponential wind profile* is given by

$$u_2 - u_1 = \frac{k^*}{k} \ln \frac{\left[\exp\left(\frac{z_2}{L} - 1\right)\right]}{\left[\exp\left(\frac{z_1}{L}\right) - 1\right]},$$

where u_1, u_2, are the horizontal wind speeds at heights z_1, z_2, respectively, and L is the Monin-Obukhov length. M. L. Barad, *JAM*, 2 (1963), 747, says that this Swinbank profile should be used only when there is nonneutral stability.

3.21. This equation may be represented by

$$U\frac{\partial q}{\partial x}=\frac{\partial}{\partial z}\left(K_E\frac{\partial q}{\partial z}\right),$$

for the humidity component, where x is taken to be the distance downwind from the leading edge. For the solution of this equation, see O. G. Sutton, *QJRMS*, 75 (1949), 335, J. R. Philip, *JM*, 16 (1959), 535, and D. A. De Vries, *loc. cit.*, 256.

3.22. Following O. G. Sutton, *CM*, p. 502, the generalized coefficient of diffusion C is given by

$$\sigma^2 = C^2(\bar{u}t)^{1.75},$$

where σ is the standard deviation of the distances traveled by a particle in time t.

3.23. Following E. W. Bierly and E. W. Hewson, *JAM*, 1 (1962), 383, the equation for center-line ground-level concentrations under aerodynamic downwash conditions, for example, is given by:

$$\chi_{(x,0,0)}=\frac{2Q}{\pi C_y C_z \bar{u}x_1^{2-n}},$$

where x_1 is the distance from the top of the stack to a point on the ground a horizontal distance x from the stack, or $x_1=(x^2+h^2)^{1/2}$ where h is the actual height of the stack.

3.24. The Deacon-Swinbank formula (*UCSCM*, p. 38) is

$$E=-C_c\rho\, U_c^2\left(\frac{q_j-q_c}{U_j-U_i}\right),$$

$$C_c=\left(\frac{U_a}{U_c}\right)^2 C_d,$$

where C_d is the drag coefficient. In practice, windspeed U is first measured at two heights a, b in order to determine C_d from

$$C_d=\rho k^2\frac{\left[\left(\frac{U_b}{U_a}\right)-1\right]^2}{\left[\ln\left(\frac{b}{a}\right)\right]^2}.$$

This must be carried out in neutral atmospheric conditions over the given surface. The windspeed U_c is then measured simultaneously at a lower level c in order to determine C_c. Finally, windspeeds and humidities U_i, U_j, and q_i, q_j, are measured at two new heights i, $j(j>i)$, leading to the estimation of E.

The Priestley formula (*UCSCM*, p. 106) is

$$E=1.10\times10^{-6}a^{\frac{1}{2}}(\theta_a-\theta_b)^{\frac{1}{2}}(e_a-e_b),$$

where potential temperatures (θ) and vapor pressures (e) are measured at two heights a and b, and it is assumed that $K_E=K_H$, that $b/a=4$, and that the air density is 1.2×10^{-3} g per cm^3; E is in g per cm per sec, e is mb, a is cm, and θ in °C.

3.25. Following N. Ramalingam, *N*, 185 (1960), 100, Rayleigh's limiting condition is given by

$$\frac{\rho_1 - \rho_0}{\rho_0} > \frac{27\pi^4 k\nu}{4gh^3},$$

where k is the coefficient of heat conduction in a fluid of thickness h in equilibrium, with the density ρ_1 in its upper layer being greater than the density ρ_0 in its bottom layer. If the limiting condition is satisfied, then the fluid will break down into convection. For the atmosphere, the limiting condition may be written

$$\frac{\log\left(\dfrac{p_1}{p_0}\right)}{\log\left(\dfrac{\theta_0}{\theta_1}\right)} > 19,$$

where p_0, p_1, refer to pressure at ground level and at 500 mb, respectively, and θ_0, θ_1, are the potential temperatures corresponding to the dry-bulb temperatures at these respective levels. It is assumed that $K_M \approx K_H$, and the sounding employed must be a truly vertical ascent.

APPENDIX 4

4.1. The region between the jet front and the jet axis is the thermally stable layer that slopes equatorward and downward from the core of the Polar Front jet stream and into which is concentrated the horizontal temperature and moisture gradients across the jet as well as the main horizontal and vertical wind shears.

4.2. G. Hadley, *Phil. Trans. Roy. Soc.*, 29 (1735), 58, proposed that air in the lower atmosphere flows toward the equator, where it rises and then flows poleward in the upper atmosphere.

4.3. J. P. Peixoto, T, 110 (1958), 188, and *TMA*, p. 232, derives the fundamental equations and shows that the equation of balance for water in the atmosphere may be expressed either as

$$\frac{1}{a\cos\phi}\left\{\frac{\partial Q_\lambda}{\partial\lambda} + \frac{\partial}{\partial\phi}(Q_\phi\cos\phi)\right\} = E - P$$

or, where **Q** is a vector, as

$$\frac{1}{a\cos\phi}\left\{\frac{\lambda}{\partial\lambda}\overline{\mathbf{Q}}_\lambda + \frac{\partial}{\partial\lambda}(\overline{\mathbf{Q}}_\phi\cos\phi)\right\} = \overline{E-P};$$

here λ represents longitude, E is evaporation, and P is precipitation. Hydrostatic equilibrium is assumed.

4.4. Following E. Palmén and L. A. Vuorela, *QJRMS*, 89 (1963), 131, the full equation is

$$\frac{2\pi a\cos\phi}{g}\int_0^{p_0}\overline{\widehat{gv}}\,dp = \frac{2\pi a\cos\phi}{g}\left[\int_0^{p_0}\bar{\hat{q}}\,\bar{\hat{v}}\,dp + \int_0^{p_0}\overline{\widehat{q'v'}}\,dp + \int_0^{p_0}\overline{\hat{q}''\hat{v}''}\,dp\right],$$

where q', v', represent deviations from time means, and $\hat{q}''$, $\hat{v}''$, represent local deviations from q', v'. The total mass circulation in the northern hemisphere involved in the meridional circulation is

$$\frac{2\pi a\cos\phi}{g}\int_{100\text{ mb}}^{700\text{ mb}}\bar{v}\,dp;$$

$\bar{v}$ represents the mean meridional speed at an arbitrary isobaric surface. From this, the air circulation involved in the mean Hadley cell of low latitudes is about 230×10^6 tons per sec, and that of the Ferrel cell in middle latitudes is about 3×10^6 tons per sec.

4.5. The Phillips model is described by the following equations:

$$\frac{\partial q_1}{\partial t} = -\mathbf{V}\cdot\nabla(\beta y + q_1) + A\nabla^2 q_1 + \left(\frac{2RH\lambda^2}{f_0\, C_p}\right)\left(\frac{y}{W}\right), \qquad [A4.1]$$

$$\frac{\partial q_3}{\partial t} = -\mathbf{V}_3\cdot\nabla(\beta y + q_3) + A\nabla^2 q_3 - \left(\frac{2RH\lambda^2}{f_0\, C_p}\right)\left(\frac{y}{W}\right)$$

$$-\frac{k}{2}[3q_3 - q_1 - 4\lambda^2(\psi_1 - \psi_3)], \qquad [A4.2]$$

where

$$q_1 = \zeta_1 - \lambda^2(\psi_1 - \psi_3) = \nabla^2\psi_1 - \lambda^2(\psi_1 - \psi_3), \qquad [A4.3]$$

$$q_3 = \zeta_3 - \lambda^2(\psi_1 - \psi_3) = \nabla^2\psi_3 + \lambda^2(\psi_1 - \psi_3), \qquad [A4.4]$$

in which $f = 2\Omega \sin\phi$, $\beta = \frac{df}{dy}$, $\mathbf{V}$ is the horizontal velocity vector, and ∇ is the horizontal gradient operator on an isobaric surface. The principle involved is as follows. If q and ψ are known at time t, Equations A4.1 and A4.2 give the values of $\frac{\partial q}{\partial t}$, so that values of q at time $(t + \Delta t)$ may be found by extrapolation. Equations A4.3 and A4.4 then permit ψ_1 and ψ_3 to be found at time $(t + \Delta t)$ from the values of $q(t + \Delta t)$. The prediction commences with the "atmosphere" at rest, i.e., with $\psi_1 = \psi_3 =$ zero. Heating then causes $\bar{q}_1$ and $\bar{q}_3$ to change as described by Equations A4.1 and A4.2, in turn producing changes in $\bar{\psi}_1$ and $\bar{\psi}_3$ as described by Equations A4.3 and A4.4. After $\bar{\psi}$ and $\bar{q}$ are no longer zero, the frictional terms $A\left(\frac{\partial^2 \bar{q}}{\partial y^2}\right)$ and $-K\bar{\zeta}_4$ come into operation, as well as the heating term. $A = A_V = A_T = 10^5\ m^2$ per sec, where A_V is the lateral kinematic eddy-viscosity coefficient and A_T is the lateral eddy-diffusion coefficient for heat. Thus A represents the effect of eddies smaller than the grid-size used for the computations. The method of solution requires that the boundary conditions be known for rectangular region, $0 \le x \le L, -W \le y \le W$, where x, y, are the distance coordinates to the east and north respectively. The vertical coordinate is taken to be pressure p. The subscripts 0, 1, 3, denote quantities measured at 0, 250, and 750 mb.

4.6. According to Smagorinsky, the equations of motion for a two-level model of the atmosphere mapped conformally on to a Mercator projection are given by:

$$\frac{\partial \hat{\Phi}}{\partial t} = -\frac{\partial}{\partial x}\left(\frac{\hat{\Phi}\bar{u}}{2}\right) - m^2\frac{\partial}{\partial y}\left(\frac{\hat{\Phi}\bar{v}}{2m^2}\right) - \gamma_s^2\,\hat{D} + \kappa Q + H, \qquad [A4.5]$$

$$\frac{\partial \hat{\mathbf{v}}}{\partial t} = m\hat{\mathbf{G}} - m\nabla\hat{\Phi}, \qquad [A4.6]$$

$$\nabla^2\psi^* = \operatorname{curl}\overline{\mathbf{G}}, \qquad [A4.7]$$

in which $\Phi = \phi - \{[\phi]\}$ is the deviation from the domain mean geopotential, ϕ representing geopotential; $m = \sec\theta$ is the Mercator map factor, γ_s^2 is a measure of the static stability, D is a measure of the horizontal deformation, $\kappa = R/C_p = 0.287$, Q is the rate of nonadiabatic heating of the air per unit mass, H represents the small-scale eddy heat flux divergence, **G** is the difference between the (horizontal) frictional and inertial force vectors, and $\psi^* = \dfrac{\partial\psi}{\partial t}$, where ψ is the stream function corresponding to vertically integrated flow. If **I** is the inertial force vector and **F** is the frictional force vector (both horizontal), such that $\mathbf{G} = -\mathbf{I} + \mathbf{F}$, then

$$m\mathbf{I}_1 = \mathbf{i}\left[\frac{\partial u_1^2}{\partial x} + m^4\frac{\partial}{\partial y}\left(\frac{u_1 v_1}{m^4}\right) - \frac{\hat{\mathscr{D}}\bar{u}}{4} - fv_1\right]$$

$$+\mathbf{j}\left[\frac{\partial u_1 v_1}{\partial x} + m^3\frac{\partial}{\partial y}\left(\frac{v_1^2}{m^3}\right) - \frac{\hat{\mathscr{D}}\bar{v}}{4} + \left(f + \frac{\alpha u_1}{a}\right)u_1\right],$$

$$m\mathbf{I}_3 = \mathbf{i}\left[\frac{\partial u_3^2}{\partial x} + m^4\frac{\partial}{\partial y}\left(\frac{u_3 v_3}{m^4}\right) + \frac{\hat{\mathscr{D}}\bar{u}}{4} - fv_3\right]$$

$$+\mathbf{j}\left[\frac{\partial u_3 v_3}{\partial x} + m^3\frac{\partial}{\partial y}\left(\frac{v_3^2}{m^3}\right) + \frac{\hat{\mathscr{D}}\bar{v}}{4} + \left(f + \frac{\alpha u_3}{a}\right)u_3\right],$$

where **i**, **j** represent unit vectors positive eastward and northward respectively, $\alpha = \sin\theta$, $f = 2\Omega\alpha$, $\mathscr{D}$ is the horizontal divergence, and x, y, represent the eastward and northward Mercator map coordinates.

4.7. The assumptions made in this adjustment are: (a) the criterion for saturation is a relative humidity of 100 per cent; (b) if the air is saturated and the lapse-rate is super-moist-adiabatic, free convection will be strong enough to make the equivalent potential temperature and the relative humidity uniform in the unstable layer; and (c) the sum of potential, internal, and latent energies is conserved during the convective adjustment.

4.8. According to P. A. Sheppard, *AGP*, 9 (1962), 77, the equation of continuity for any entity to be studied in terms of energy transactions may be expressed in the form:

$$\frac{\partial}{\partial t}\int_\tau \rho s\, d\tau = \int_\tau q\, d\tau - \int_\sigma \rho s\,\mathbf{V}\cdot d\delta - \int_\sigma \mathbf{F}\cdot d\delta, \qquad \text{[A4.8]}$$

where **V** represents the (unsmoothed) wind velocity, **F** is the nonconvective flux (i.e., the flux due to molecular conduction or radiation) of the (scalar) entity under consideration, q is the rate of production of this entity per unit volume of the atmosphere (i.e., the source strength of the entity), $d\tau$ is an element of volume within the atmosphere, σ represents surface area, and $d\delta$ is an element of surface on the volume τ with direction taken as that of the outward-drawn normal. There are assumed to be s units of the entity per unit mass of air. The equation simply states that the rate of change of the entity with time equals its rate of production minus the rate at which the entity is transferred outward across the surface of area σ by the air motion and by nonconvective processes, all referred to volume τ. The volume τ may be taken to be a column of air resting on the level surface of the Earth. Then the only contribution to the rate of change of the entity considered within τ made by $\int_{\sigma} \rho s\, \mathbf{V} \cdot d\delta$ will be convective transport across the vertical sides of the column and across its top surface; the term $\int_{\sigma} \mathbf{F} \cdot d\delta$ will be important only at the lower boundary of the column and (for radiative transfer) at its upper boundary. The equation may thus be written as

$$\frac{\partial(\rho s)}{\partial t} = q - \nabla \cdot \rho s\, \mathbf{V} - \nabla \cdot \mathbf{F} \qquad \text{[A4.9]}$$

by means of the divergence theorem. Since Equation A4.8 is for unsmoothed motion, the expression $\rho s\mathbf{V}$ in the convective-flux term may be regarded as consisting of a mean flux $\overline{\rho s}\overline{\mathbf{V}}$ and a turbulent flux $\overline{\rho s' \mathbf{V}'}$. The equation works best for any property for which the Earth's surface constitutes a source or sink, e.g., heat, momentum, water vapor. For example, if s is a component of momentum, then Equation A4.9 becomes the equation of motion for the coordinate in question, with q representing the sum of all forces per unit volume acting in that direction. If s refers to water vapor, then $q = 0$ if no condensation or evaporation occurs, otherwise q represents the rate of condensation (or the negative of the rate of evaporation) per unit volume of air. For full details, see Sheppard, *op. cit.* (1962).

4.9. According to E. J. Rebman, *BAMS*, 35 (1954), 498, a vector interpretation is useful. The solar-terrestrial complex operates in a three-dimensional field, and never achieves final dynamic balance. The net effect is to create a climatological vector field, stemming from the persistence of previous departures from the mean and the achievement of the new anomaly. The feedback involved in the complex generates a thermodynamic torque that changes as the complex passes from one state to another.

4.10. According to J. Bjerknes, *USCC*, p. 297, and *AGP*, 10 (1964), 1, the Ekman theory shows that

$$D = \pi \sqrt{\frac{2\mu_e}{2\rho\Omega \sin\phi}},$$

where D is the depth below which the water flows in the opposite direction to the surface current, μ_e is the eddy viscosity of the water (assumed to be independent of depth) and ρ is its density. The Ekman transport is given by

$$\rho T_y = -\frac{\tau_x}{2\Omega \sin\phi},$$

where T_y is the volume of water transported normal to the wind direction and τ_x is the wind stress. The original publication is V. K. Ekman, *Arkiv. Math. Astron. Fys.*, 2 (1905), 1.

4.11. According to Bjerknes, the theory devised by H. V. Sverdrup, *Proc. Nat. Acad. Sci. U.S.*, 33 (1947), 318, shows that steady-state geostrophic currents in a baroclinic ocean are governed by equations

$$U = \frac{1}{r}\frac{\partial\psi}{\partial\phi}$$

and

$$V = -\frac{1}{r\cos\phi}\frac{\partial\psi}{\partial\lambda} = \frac{1}{\beta}\operatorname{curl}_z \tau_s$$

if the condition of constancy of ocean level is to be satisfied. U represents the vertically integrated total zonal mass transport across a unit length of a meridian, V represents the vertically integrated total meridional mass transport across a unit length of a latitude circle, λ is latitude, τ_s is the windstress vector, ψ represents a stream function of the vertically integrated horizontal total mass transport, and

$$\beta = \frac{\partial}{\partial y}(2\Omega \sin \phi).$$

APPENDIX 5

5.1. A third scale, the tercentesimal scale, was suggested by Napier Shaw, but it has not found general acceptance; for details, see N. Shaw, *Manual of Meteorology*, III, 20. A. Celsius (1701–1744), a Swedish astronomer at Üpsala, was the first to propose dividing the thermometric scale into an interval of 100° between the boiling and freezing points of water, and in this sense was the originator of the Centigrade scale. The *SI system* (Systeme Internationale) of units is now becoming the international standard system for scientific work, but, in view of his concern with past data, the climatologist will still need to work in the c.g.s. or m.t.s. systems for convenience.

5.2. Dalton's law states that the total pressure of a mixture of gases equals the sum of the pressures of the individual gases composing the mixture (i.e., the partial pressures). It also states that each gas in the mixture occupies the whole volume of the mixture of gases, as if other gases were absent. In principle the atmosphere is assumed to be a mixture of two gases, a hypothetical gas ("clean dry air," molecular weight 28.9), and a real gas ("water vapor," molecular weight 18). Separate equations of state may be written for each of these gases, and the concept of virtual temperature is introduced to combine these two equations.

5.3. Following H. R. Byers, *General Meteorology* (New York, 1944), p. 134, the *first law of thermodynamics* may be expressed as follows. Suppose a pressure p is applied to a parcel of air in the atmosphere (see Figure A5.3); p will be exerted in all directions, and will cause the parcel to be compressed to a smaller volume by an amount dV, $-dV = Adn$, where A is the surface area of the parcel and the minus

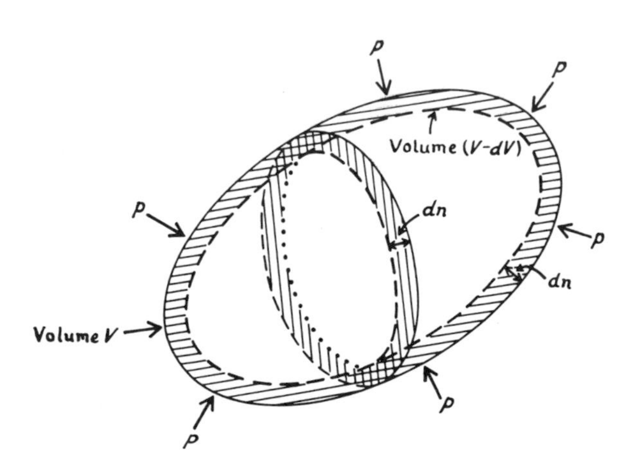

FIGURE A5.3.
Compression of an air parcel.

sign indicates that increasing pressure results in a decrease in volume. If dn is infinitesimally small, i.e., if the surface area of the parcel is assumed to be almost the same before and after the application of pressure p, then the work done, dW, on the parcel by p is given by $dW = Fdn$. But since $p = F/A$ by definition, it follows that $dW = -pdV$. The first law of thermodynamics may be stated in the form $dE = dQ + dW$, where dE is the change in internal energy of the gas, dQ is the heat added to it, and dW is the work done on the gas. Consequently

$$dE = dQ - p\ dV. \qquad [A5.1]$$

For a constant-volume process, $dV = 0$; i.e., $dE = dQ$. The heat added to the gas by changing its temperature an amount dT equals the mass of the gas times the heat required to raise the temperature of a unit mass of the gas 1°C times dT; i.e., $dQ = MC_v dT$, where M is the mass of the gas and C_v is its specific heat at constant volume. Therefore Equation A5.1 may be expressed in the form

$$dQ = MC_v dT + p\ dV.$$

Considering unit mass, and dividing by M, we obtain:

$$dq = C_v dT + p\ dv,$$

where dq is the heat added to unit mass, and dv is the change in specific volume (i.e., the change in the volume of unit mass, v). Since $C_p - C_v = R/m$, where m is the molecular weight of the gas, we obtain Equation 5.3.

5.4. By definition, an adiabatic process is one in which no exchange of heat takes place between the system under investigation and its environment.

5.5. The important thing for the nonmathematicians to note about the hydrostatic equation is that it is a differential equation, and must be integrated before it can be applied to actual measurements. For methods of numerical integration of the equation, see W. D. Duthie, *JM*, 3 (1946), 89.

5.6. Following *EDM*, p. 23, the phase-changes are described by

$$L_{1,2} = T(\phi_2 - \phi_1) = e_2 - e_1 + e_s(v_2 - v_1),$$

where L is the latent heat involved in the transformation of water from phase 1 to phase 2; ϕ_1, ϕ_2, represent the amount of heat in the system in phase 1 and phase 2, respectively (both at temperature T); e_1, e_2, represent the respective changes in internal energy, and v_1, v_2 the respective changes in specific volume, during the transformation; and e_s is the saturation vapor pressure, i.e., the vapor pressure at which both water and water vapor can exist side by side. The appropriate values of L are as follows.

Between liquid water and water vapor, $L_E = 2.500 \times 10^6$ kilojoules per ton.
Between ice and water vapor, $L_s = 2.834 \times 10^6$ kilojoules per ton.
Between ice and liquid water, $L_M = 0.334 \times 10^6$ kilojoules per ton.

Where E, S, and M denote evaporation, sublimation, and melting, respectively, the values are positive. When E denotes the condensing of water, and S and M denote the freezing of water, the values are negative. The quoted values of L are practically constant for the temperature ranges found within the atmosphere, although in general, L does vary with temperature.

5.7. Following *EDM*, p. 25, the *Clausius-Clapeyron* equation may be written in the form

$$\frac{de_s}{dT} = \frac{L_{1,2}}{T(v_2 - v_1)}.$$

In effect, the equation (first derived by Clapeyron in 1832, later rederived by Clausius) is based on entropy considerations, and gives the slope of the curve of saturation vapor pressure versus temperature as a function of latent heat, temperature, and specific volume.

5.8. A pseudoadiabatic transformation may be defined, following J. M. Van Mieghem, *CM*, p. 531, as one in which the heat received by an open system does not alter the internal transformation of which the open system is the site. An "open system" is one in which both energy and mass are exchanged with the surrounding environment: a precipitating cloud is an example.

5.9. The vertical acceleration of an air parcel not in hydrostatic equilibrium is given by *EDM*, p. 49, as

$$\frac{\partial^2 z}{\partial t^2} = -\frac{1}{\rho}\frac{\partial p}{\partial z} - g = \frac{g(\rho' - \rho)}{\rho},$$

where ρ, ρ', are the densities of the air parcel and the surrounding air, respectively. In practice, the equation is written in the form

$$\ddot{z} = gz\frac{(\gamma_e - \gamma)}{T'},$$

where γ is the actual lapse-rate of the parcel $\left(\text{i.e., } -\frac{\partial T}{\partial z}\right)$, γ_e is the environmental lapse-rate, T' is the environmental temperature at height z, and $\ddot{z} \equiv \frac{\partial^2 z}{\partial t^2}$.

5.10. A. K. Showalter, *BAMS*, 34 (1953), 250. To obtain a value of the index, lift the air parcel at 850 mb dry-adiabatically to saturation, then lift it pseudo-adiabatically to 500 mb. Then subtract the lifted 500-mb temperature from the observed 500-mb temperature algebraically.

5.11. The *Rackliff index*, on which see P. G. Rackliff, *MM*, 91 (1962), 113, ΔT, is defined by

$$\Delta T = \theta_{w9} - T_5,$$

where θ_{w9} is the wet-bulb potential temperature at 900 mb, and T_5 is the 500-mb dry-bulb temperature (both in °C). The *Jefferson index*, T_j, on which see G. J. Jefferson, *MM*, 92 (1963), 92, is defined by

$$T_j = 1.6\,\theta_{w9} - T_5 - 11.$$

The *Boyden index*, on which see C. J. Boyden, *MM*, 92 (1963), 198, I, is defined by

$$I = Z - T - 200,$$

where Z is the 1000-700-mb thickness in decameters and T is the 700-mb temperature in °C.

5.12. Potential temperature is defined as the temperature an air parcel would have if it is brought dry-adiabatically to a pressure of 1,000 mb.

5.13. To derive Equation 5.10, take logs of the equation for potential temperature, differentiate, multiply by C_p, and then substitute from the gas equation. The entropy of a body may be defined as the quantity of heat within the body divided by its temperature. The concept was introduced by Clausius and is the basis of most thermodynamic theory.

5.14. The quantity g in the hydrostatic equation, $dp = -g\rho\, dz$ varies from place to place. Therefore it is convenient to define a new variable, p_g (the *geopotential*, with dimensions L^2T^{-2}), such that $dp_g = g\, dz$; p_g is the energy imparted to a unit mass if it is lifted from sea level to a height z above sea level. Height in *geopotential meters* (i.e., in dynamic meters) equals $\frac{1}{9.81}\int_0^z g\, dz$, where z is in ordinary meters.

5.15. The *Gold Square values* for the various months, as given E. Gold, *MOPN*, no. 63 (1933), are as follows.

January, 0.4; February, 0.7; March, 1.1; April, 1.5; May, 1.9; June, 2.0; July, 1.8; August, 1.7; September, 1.3; October, 0.9; November, 0.7; December, 0.4.

5.16. That the geographical coordinate system must be an accelerating one is illustrated by the mathematical statement of circular motion. In Figure A5.16,A, which illustrates a particle of mass M moving in a horizontal circular path, it is apparent from Newton's first law of motion that a force must be acting on the particle. This is an inward force, the *centripetal force*, and its magnitude can be shown to be Mv^2/r,

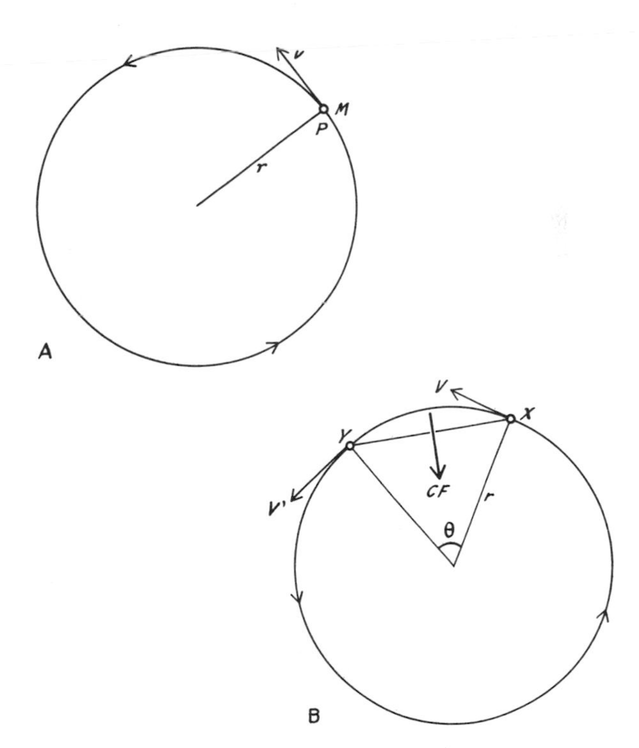

FIGURE A5.16.
In A, particle P, of mass M, is moving in a horizontal circle with linear velocity V. In B a, particle moves in a horizontal circle with constant angular velocity. CF indicates line of action of centripetal force, and θ is in radians.

where v is the linear velocity of the particle. In Figure A5.16,B, where the particle has moved from point X, where its velocity is V, to point Y, where its velocity is V', the particle may be regarded as being subject to a force V^2/r, acting in a direction perpendicular to the chord XY. This force will produce an acceleration in the *linear* velocity of the particle (i.e., $V' > V$), although its *angular* velocity θ/t, where t is the time taken for the particle to move from X to Y, remains constant. Movements of points defining the elements of a geographical coordinate system fixed to the surface of the Earth are analogous to the movements of the particle. Consequently, the system must be accelerating.

5.17. The notation devised by Leibniz used the expression dy/dx for the differential coefficient of y with respect to x, whereas Newton preferred the expression $\dot{y}$. Similarly, d^2y/dx^2 in the Leibniz notation corresponds to $\ddot{y}$ in Newton's.

5.18. Following *ITM*, p. 205, for the special case of fluid motion in which the two-dimensional divergence is zero,

$$\frac{\partial u}{\partial x} + \frac{\partial v}{\partial y} = 0.$$

This condition is satisfied by a function ψ if $u = -\dfrac{\partial \psi}{\partial y}$, $v = \dfrac{\partial \psi}{\partial x}$; ψ is defined as the *stream function*, and the equation of a streamline is

$$\frac{\partial \psi}{\partial x}\,dx + \frac{\partial \psi}{\partial y}\,dy = 0, \quad \text{or} \quad \partial \psi = 0,$$

where $d\psi$ represents an increment in ψ along a streamline at a given moment. If the flow is nondivergent, the equation of a streamline becomes ψ = constant, and therefore each streamline may be labeled with a particular value of ψ. The velocity of the fluid flow is determined by the distance apart of the streamlines. See S. V. Nemchinov, S. A. Musaeljan, and V. P. Sadotov, *T*, 15 (1963), 120, for the determination of stream functions from data for vertical motion.

5.19. From the rules of partial differentiation,

$$\frac{du}{dt} = \frac{\partial u}{\partial t}\frac{dt}{dt} + \frac{\partial u}{\partial x}\frac{dx}{dt} + \frac{\partial u}{dy}\frac{dy}{dt} + \frac{\partial u}{\partial z}\frac{dz}{dt},$$

and since $dx/dt = u$, $dy/dt = v$, $dz/dt = w$, we have

$$\frac{du}{dt} = \frac{\partial u}{\partial t} + u\frac{\partial u}{\partial x} + v\frac{\partial u}{\partial y} + w\frac{\partial u}{\partial z}.$$

Following Stokes, $\dfrac{du}{dt}$ is sometimes written as $\dfrac{Du}{Dt}$, and is called the *total* derivative of u with respect to t. Similar expressions hold for dv/dt and dw/dt.

5.20. Divergence may be estimated by means of finite differences, on which see *EDM*, p. 106. On the estimation of divergence, see: R. D. Graham, *BAMS*, 34 (1953), 68; K. Bryant, *MM*, 88 (1959), 270; D. T. Williams, *BAMS*, 41 (1960), 383. A. Wiin-Nielsen, *MWR*, 89 (1961), 67, and G. J. Haltiner, H. E. Nicholson, and W. B. Oakes, *MWR*, 91 (1963), 219, discuss divergence in numerical prediction models. An alternative method given by Byers, *op. cit.* (1944), p. 387, involves splitting each observed wind vector into west-to-east (u) and south-to-north (v) components, and then constructing maps of u and v values. The number of unit u-isopleths in a unit x-distance on the first map gives $\frac{\partial u}{\partial x}$, the number of unit v-isopleths in a unit y-distance on the second map gives $\frac{\partial v}{\partial y}$. Read off $\frac{\partial u}{\partial x}$ and $\frac{\partial v}{\partial y}$ at point P: then the divergence at P is given by $\frac{\partial u}{\partial x} + \frac{\partial v}{\partial y}$. According to A. H. Gordon, *QJRMS*, 77 (1951), 302, if N_1, N_2, N_3, N_4, and E_1, E_2, E_3, E_4, are northerly and easterly components, respectively, of the mean vector winds, using four adjacent 5° squares (see Figure A5.20), then convergence/divergence $\propto \pm[N_1 + N_2 - (N_3 + N_4) + \sec\,\phi(E_2 + E_4 - E_1 - E_3)]$, where ϕ is the mid-latitude of the squares.

5.21. This is demonstrated by means of the equation of continuity. Consider a large cube at rest on the Earth's surface (see Figure A5.21). Suppose that horizontal divergence is occurring within the cube; i.e., a deficit of mass is occurring. If the direction of the airflow through the cube is parallel to face *ABCD* as shown, and if the inflow and outflow velocities are both to be equal to *V*, then, because of continuity considerations, the decreasing mass within the cube must be compensated by air descending through face *DCQP* into the cube. Consequently, horizontal divergence is associated with subsidence and hence with adiabatic heating. On the other hand, if horizontal convergence is occurring within the cube (i.e., mass is accumulating), then if the inflow and outflow velocities are to be equal, the surplus in mass must be compensated by air rising through face *DCQP* out of the cube. Therefore horizontal convergence must be associated with ascending air and with adiabatic cooling.

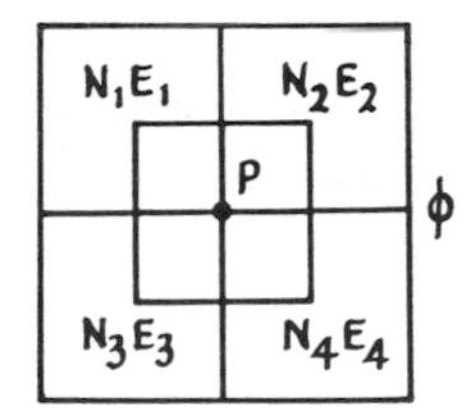

FIGURE A5.20.
Calculation of mean vector wind, refered to point *P*.

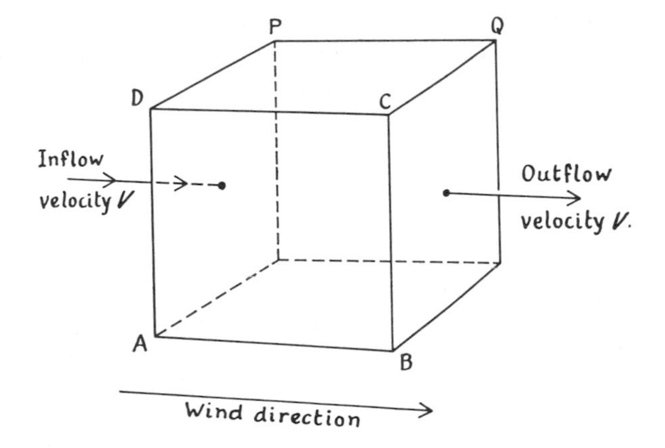

FIGURE A5.21.

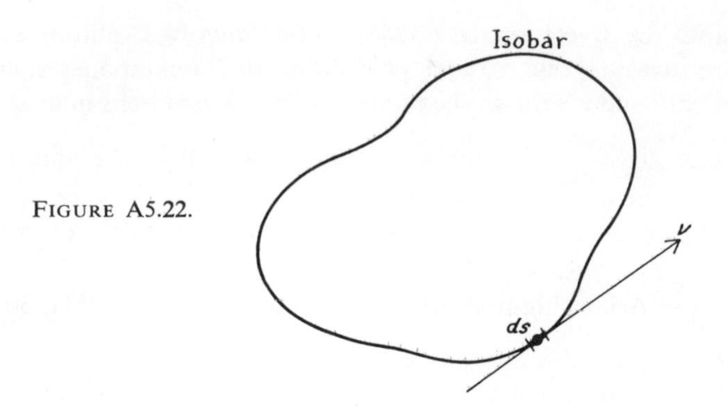

FIGURE A5.22.

5.22. In Figure A5.22, if ds is a small element of a contour of a fluid in motion (e.g., an isobar for air in geostrophic motion), then the circulation around the contour is defined to be $\oint v\, ds$, where v is the tangential component of the velocity of fluid motion at ds.

5.23. For a mass m moving in a horizontal circular path of radius r, the principle of conservation of angular momentum states that $mr\omega$ is constant, where ω is the angular velocity of the particle. In Figure A5.16,A, $\omega = v/r$.

5.24. See R. E. Petersen, *JM*, 14 (1957), 367, and 15 (1958), 335. Following B. G. Wales-Smith, *MM*, 88 (1959), 4, the vertical component of relative vorticity at O in Figure A5.24 is given by $\zeta = \dfrac{\partial v}{\partial x} - \dfrac{\partial u}{\partial y}$, or, in terms of finite differences,

$$\zeta \approx \frac{v_3 - v_1}{d} - \frac{u_4 - u_2}{d},$$

where u, v, are the wind components in the AO and BO directions, respectively. If Z_4 is the height of an isobaric surface at point 4, etc., then the *geostrophic vorticity* is given by

$$\zeta_G = \frac{4g}{\lambda d^2}\,\frac{Z_1 + Z_2 + Z_3 + Z_4 - 4Z_0}{4}.$$

A graphic method devised by Fjørtoft facilitates computations if they must be made for a large area with many data points. On the measurement of vorticity, see also G. W. Platzman, *JM*, 4 (1947), 58.

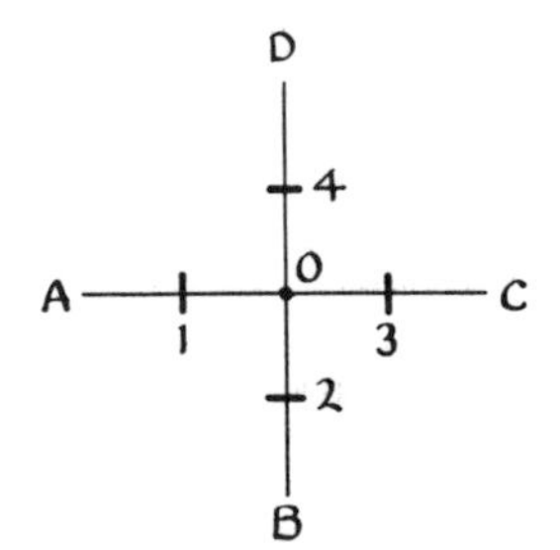

FIGURE A5.24.
Here d is the distance between point o and points A, B, C, D.

5.25. If **U** and **V** are vectors, the *vector cross product* of **U** and **V** is $\mathbf{W} = \mathbf{U} \times \mathbf{V}$, where **W** is a vector such that **W** is perpendicular to **U** and to **V** (and hence to their plane), and **W** is directed so that, when viewed from its terminal point, the rotation from **U** to **V** is positive (see *HM*, p. 220).

5.26. The *vector dot product*, $\mathbf{U} \cdot \mathbf{V}$, is a number equal to $UV \cos \theta$, where U, V, are the magnitudes of the vectors, **U**, **V**, respectively, and θ is the angle between **U** and **V**.

5.27. See *ITM*, pp. 288 and 317, for details. The finite-difference approximation to the Laplacian at point O (see Figure A5.27) is given by

$$(\nabla^2 S)_0 = \frac{\partial^2 S}{\partial x^2} + \frac{\partial^2 S}{\partial y^2} \approx \frac{S_1 + S_2 + S_3 + S_4 - 4S_0}{d^2}.$$

where S_1, etc., is the value of S at point 1, etc. If S represents the height z of a constant-pressure surface, then the Laplacian of z, which is proportional to the *geostrophic relative vorticity*, is given by

$$(\nabla^2 z)_0 \approx \frac{z_1 + z_2 + z_3 + z_4 - 4z_0}{d^2}.$$

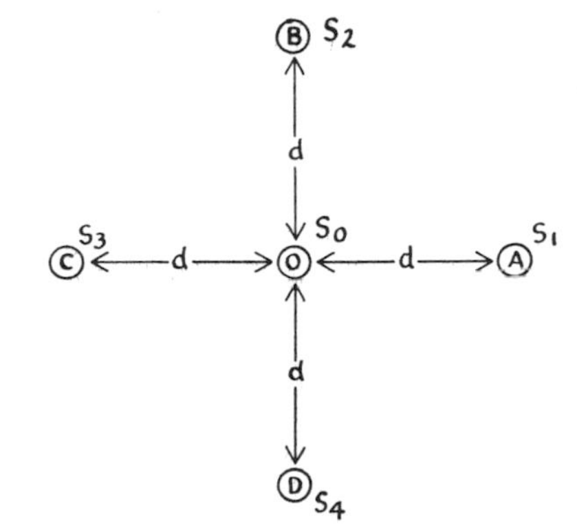

FIGURE A5.27.
Figures S_0, S_1, S_2, S_3, S_4, are the values of the element in question at points O, A, B, C, and D, respectively.

5.28. The thermal wind equations for a constant-pressure surface are

$$\frac{\partial v}{\partial z} = \frac{g}{\lambda T}\left(\frac{\partial T}{\partial x}\right)_p$$

and

$$\frac{\partial u}{\partial z} = -\frac{g}{\lambda T}\left(\frac{\partial T}{\partial y}\right)_p.$$

5.29. On these four influences, see R. C. Sutcliffe, *Proc. Toronto Meteorol. Conf.* (London, 1953), p. 139, according to whom the vorticity equation for surface circulations may be written with sufficient accuracy as

$$\frac{d}{dt}(\lambda + \zeta) = (\lambda + \zeta) \text{ div } \mathbf{V}$$

where ζ is the relative vorticity about a vertical axis. For surface cyclones, $\zeta > 0$, hence $\lambda + \zeta > \lambda$. For surface anticyclones, $\zeta < 0$, therefore $\lambda + \zeta < \lambda$.

5.30. Following Defant (also *HM*, p. 433), if the air over a coastal area is stationary, the isobars will be horizontal, but the isosteres (ρ_1 and ρ_2 in Figure A5.30) will be inclined to the horizontal, owing to the existence of differential heating at the Earth's surface. For the contour $ABCD$, circulation equals

$$\oint S\,dp = -(p_0 - p_1)(S_1 - S_0).$$

Therefore, from Bjerknes' circulation theorem, which states that

$$\frac{D}{dt}(\Gamma + 2\omega \cdot \mathbf{A}) = -\oint \frac{dp}{\rho},$$

where $\dfrac{D}{dt}$ indicates the *total* rate of change with respect to time, ω is the Earth's angular velocity, Γ is the circulation, and $\mathbf{A}$ is the vector area of the surface capping the contour $ABCD$, we have

$$\frac{D}{dt}(\Gamma + 2\omega \cdot \mathbf{A}) = (p_0 - p_1)(S_1 - S_0).$$

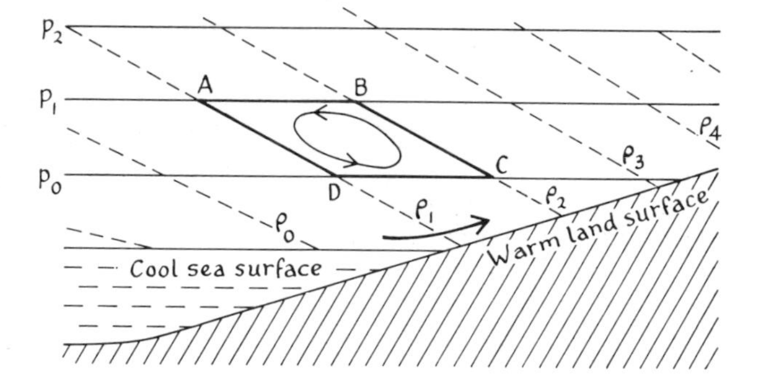

FIGURE A5.30.

APPENDIX 6

6.1. The correction for height is $-WE_h$, where E_h is the error of a value at a point on the first-guess map. The weighting factor, W, is

$$W = \frac{(N^2 - d^2)}{(N^2 + d^2)},$$

where d is the distance between the observing point and the nearest grid point, and N is the distance over which W decreases to zero. The map is smoothed by the formula

$$\bar{D} = \frac{1}{2} D + \frac{1}{2} 8 \sum_{i=1}^{4} D_i,$$

where $\bar{D}$ is the adjusted value of the central grid point, D is the height (pressure) at this point, and D_i represents values at the nearest surrounding grid points.

6.2. Following R. A. Baumgartner, *HM*, p. 759, if r_t is the radius of curvature of the trajectory and r_i is the radius of curvature of the isobar, then

$$\frac{1}{r_t} = \frac{1}{r_i}\left(1 - \frac{c}{v}\cos\psi\right),$$

where c is the speed of the pressure system, v is the speed of the gradient wind, and ψ is the angle between the direction of motion of the pressure system and the direction of the gradient wind; r_i is determined by first finding the center of curvature of the isobar in its vicinity, which is found at the intersection of the perpendicular bisectors of two chords of the isobar; and c may be found from

$$U = \frac{1}{2}\left(\frac{b_A N_A}{\Delta p_A} + \frac{b_B N_B}{\Delta p_B}\right),$$

where U is the speed of movement of the isobars (in N units every 3 hours), b is the 3-hour pressure tendency, N is the distance between two adjacent isobars (in arbitrary units), and Δp is the pressure difference between the two isobars; N is positive when pressure decreases in the positive direction of U. The subscripts A, B, indicate the points at which the axis of computation intersects the isobars, in advance of and to the rear of the symmetry line, respectively.

6.3. According to W. J. Saucier, *Principles of Meteorological Analysis* (Chicago, 1953), the following relations apply.

For a linear height scale,

$$\frac{\text{vertical scale in plotted sounding}}{\text{vertical scale in cross section}} = \frac{\text{temperature in Standard Atmosphere}}{\text{actual (absolute) temperature}}.$$

For a logarithmic pressure scale,

$$\text{vertical scale in sounding} = \frac{\text{a constant for the cross section}}{\text{actual (absolute) temperature}}.$$

6.4. The Sutcliffe equation is derived from the normal pressure-tendency equation by means of the intermediate equation

$$\lambda(\text{div}_p \mathbf{V} - \text{div}_p \mathbf{V}_0) = -\mathbf{V} \cdot \nabla_p(\zeta + \lambda) + \mathbf{V}_0 \cdot \nabla_p(\bar{\zeta} - \lambda) - \frac{1}{\lambda}\left(\frac{\partial}{\partial t}\right)_p \nabla_p^2 h', \qquad \text{[A6.1]}$$

where h' is the thickness between two isobaric surfaces and the subscript p denotes differentiation at constant pressure. Assuming quasigeostrophic motion (i.e., assuming that the wind nearly balances the pressure gradient), then

$$\zeta = \frac{1}{\lambda} \nabla_p^2 h = \frac{1}{\lambda}\left(\frac{\partial^2}{\partial x^2} + \frac{\partial^2}{\partial y^2}\right)_p h,$$

where h is the absolute height of the pressure level at which the vorticity is being measured. Substitution for ζ in Equation A6.1 then results in the basic Sutcliffe equation,

$$\lambda(\text{div}_p \mathbf{V} - \text{div}_p \mathbf{V}_0) = -\mathbf{V}' \frac{\partial}{\partial s}(\lambda + \zeta + \zeta_0),$$

in which the left-hand side represents the relative (isobaric) divergence between the two pressure levels (>0 for cyclonic development, <0 for anticyclonic development); $\mathbf{V}'$ is the relative wind (i.e., the thermal wind or wind shear vector) between these two levels; ζ, ζ_0, represent the respective vorticities, and $\frac{\partial}{\partial s}$ denotes differentiation in the direction of shear. Implicit in the theory is the assumption that no synoptic development can occur without wind shear.

6.5. If the thickness gradient is stronger in the left-hand branch of the diffluence than in the right-hand branch, then the motion of the secondary depression is to the left of the diffluence; if stronger in the right-hand branch, then the secondary moves to the right of the diffluence. In both cases, the speed of movement of the secondary is greater than with symmetric diffluence.

6.6. M. K. Miles, *MM*, 88 (1959), 193. Important weather changes also result from "relaxation," i.e., warming and weakening of a thermal trough, involving poleward movement of the thickness isopleths; surface pressures rise on the associated cold front, precipitation intensity diminishes, and any tendency to wave formation is suppressed. See M. K. Miles and G. A. Watt, *MM*, 91 (1962), 120, for details.

6.7. According to C. J. Boyden, *MM*, 87 (1958), 98, points on the power curve that relates daily mean total thicknesses at Crawley to daily mean surface temperature at London Airport (Heathrow) are as follows; the temperatures are in parentheses and in °F, thicknesses in geopotential meters.

Winter: 5,100 (26), 5,200 (31), 5,300 (37), 5,400 (44), 5,500 (52), 5,600 (62), 5,700 (74).

Summer: 5,200 (40), 5,300 (44), 5,400 (50), 5,500 (56), 5,600 (65), 5,700 (78).

An abrupt change in the "thickness season" occurs in March-April.

6.8. According to H. H. Lamb, *QJRMS*, 81 (1955), 172, the critical value varies from 17,000 to 17,900 feet, depending on (a) the temperature lapse-rate in the lowest 1 to 2 km of the atmosphere; and (b) whether the precipitation is formed in clouds within the friction layer or at higher levels in the atmosphere. The changeover from liquid to solid precipitation in cold pools occurs at thickness values of 16,900 to 17,150 feet for showers, and at 17,200 feet for other types of precipitation.

6.9. Following *WAF*, I, 258, the vorticity equation for a surface of constant θ may be written in the form

$$\frac{\partial Q_\theta}{\partial t} + u\frac{\partial \theta_\theta}{\partial x} + v\frac{\partial \theta_\theta}{\partial y} + w_\theta\frac{\partial \theta_\theta}{\partial z} = -D_\theta\, Q_\theta + qx\frac{\partial w_\theta}{\partial x} + q_y\frac{\partial w_\theta}{\partial y},$$

which may be rewritten, omitting the subscript θ, as

$$\frac{\partial Q}{\partial t} + \mathbf{V}\cdot Q\mathbf{V} + w_n\frac{\partial \theta}{\partial n} = -DQ + \boldsymbol{q}\cdot\nabla w_n + F,$$

where Q is the component of absolute vorticity normal to the constant θ surface, $\boldsymbol{q}$ is the vorticity vector tangent to the surface, and D represents divergence. Therefore

$$\mathbf{V}\cdot\nabla Q + DQ = F - w_n\frac{\partial Q}{\partial n}$$

where F is the vorticity of the frictional force. And, since $D = \nabla\cdot\mathbf{V}$, therefore

$$\nabla\cdot(Q\mathbf{V}) = F - \frac{\partial Q}{\partial \theta}\dot{\theta},$$

where

$$\dot{\theta} = \frac{\partial \theta}{\partial t} + w_n\frac{\partial \theta}{\partial n}.$$

The term $\nabla\cdot(Q\mathbf{V})$ represents the export of vorticity per unit time through the boundaries of unit area in the surface of constant θ, and thus is a measure of the intensity of the vorticity source.

6.10. For details see *WAF*, I, 190. The first space derivatives of X are its first partial derivatives with respect to x and to y. The second space derivatives of X are its second partial derivatives with respect to x and to y.

6.11. Following J. S. Sawyer, *QJRMS*, 84 (1958), 375, the critical equation is

$$V\frac{\partial \zeta}{\partial s} + w\frac{\partial \zeta}{\partial p} - \frac{\partial \omega}{\partial y}\frac{\partial u}{\partial p} + (\lambda + \zeta)\mathrm{div}_p v = 0,$$

where V is the wind velocity relative to the wind at 600 mbs, $\frac{\partial \zeta}{\partial s}$ is the rate of vorticity change along a horizontal projection of a streamline, and the coordinate system is assumed to move with the general flow of the system at the speed of the 600-mb wind; $\omega \equiv Dp/Dt$.

6.12. According to C. H. Hinkel and W. E. Saunders, *MM*, 94 (1955), 241, the relation between the mean speed of the front, U_F, and the mean geostrophic wind component perpendicular to the front, U_G, is

$$U_F = 0.75U_G + 3.7,$$

which has an rms error of 5.9 knots. If U_G' is the mean of U_G, and the mean geostrophic wind component perpendicular to the front as measured in the cold air 75 miles ahead of the front, then

$$U_F = 0.91U_G' + 2.6,$$

which has an rms error of 4.6 knots.

6.13. Following A. G. Matthewman, *MM*, 81 (1952), 266, the dynamic relationship is

$$U_F = U_G' - \frac{V_G}{\lambda}\frac{\partial V_G}{\partial y},$$

where V_G is the geostrophic wind component parallel to the front.

6.14. Following Petterssen, *Geofys. Publ.*, vol. 11, no. 6 (1936), and *WAF*, I, 169–71, the intensity of frontogenesis may be defined by

$$\mathscr{F} = \frac{d}{dt}|\nabla\theta|,$$

where $|\nabla\theta|$ is the magnitude of the gradient of potential temperature. Where the motion is such that $|\nabla\theta|$ increases more rapidly along a certain line than elsewhere, then *frontogenesis* is said to occur along that line. Assuming adiabatic motion, Petterssen showed that

$$\mathscr{F} = |\nabla\theta|\frac{1}{2}(F\cos 2\beta - D),$$

where D is the divergence $\left(D = \frac{\partial u}{\partial x} + \frac{\partial v}{\partial y}\right)$, F is the deformation $\left(F = \frac{\partial u}{\partial x} - \frac{\partial v}{\partial y}\right)$, and β is the angle between the x-axis and the isotherms of potential temperature. For any given field of motion, the sign of $\mathscr{F}$ depends entirely on the angle β. The critical angle β' at which $\mathscr{F} = 0$ is given by $\cos 2\beta' = \pm D/F$. With sufficient accuracy, $\beta' \approx \pm 45°$. If the angle between the isotherms and the axis of dilatation (i.e. the axis of stretching, OX in Figure A6.14,A) is less than 45°, the situation is frontogenetic; if the angle is more than 45°, the situation is frontolytic.

Figure A6.14,A, shows a flow situation in which pure deformation exists in the OX direction. The simplest frontogenetic field is that associated with a symmetric col (Figure A6.14,B) in which $\beta < 45°$. The temperature gradient will therefore increase, and will increase most rapidly where it is most intense. The axis of dilatation will act as a collector of isotherms, because confluence toward this axis in the col will rotate

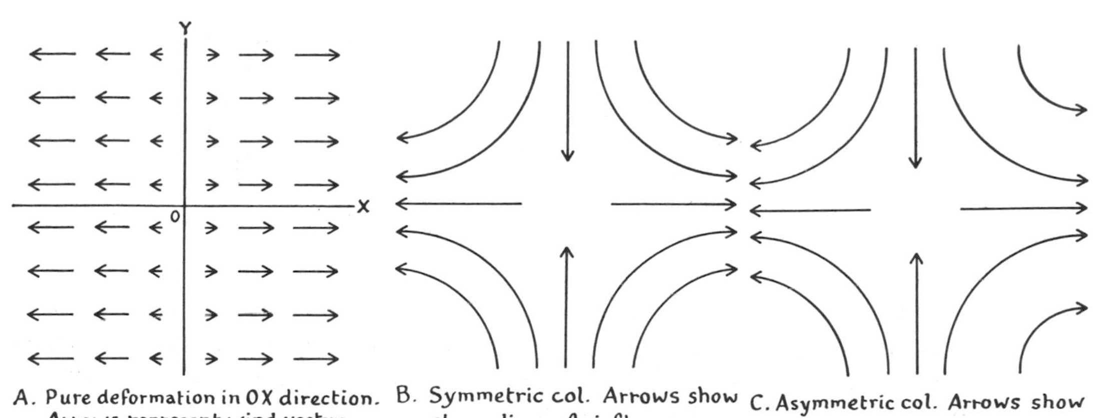
Y
X
O
A. Pure deformation in OX direction. Arrows represent wind vectors.
B. Symmetric col. Arrows show streamlines of airflow.
C. Asymmetric col. Arrows show streamlines of airflow.

FIGURE A6.14.
Deformation fields and frontogenetic cols. A, Pure deformation in OX direction; arrows represent wind vectors. B, Symmetric col; arrows show streamlines of airflow. C, Asymmetric col; arrows show stream lines of airflow.

the isotherms so that they become parallel to the dilatation axis and, since no flow can occur across the latter, isotherms on either side of the axis will crowd toward it, and a discontinuity will tend to form. In an asymmetric col (Figure A6.14,C), $\beta < 45°$, hence the temperature gradient will decrease, and will decrease most rapidly where it is largest; therefore no discontinuity can form. See also J. E. Miller, *JM*, 5 (1948), 169.

6.15. Following T. H. Kirk, *QJRMS*, 92 (1966), 374, the critical situations are defined as follows.

The *temperature discontinuity* is determined by the maximum of $\nabla(\nabla^2 v)$, where v indicates specific volume and $-\nabla^2 v = \lambda \dfrac{\partial \zeta_G}{\partial p}$, in which ζ_G is the geostrophic vorticity.

The *wind discontinuity* is defined by $\nabla^2 \mathbf{V}$, which depends on the gradient of vorticity and the gradient of divergence. The discontinuity in geostrophic wind occurs where the gradient of geostrophic vorticity is a maximum.

The *discontinuity in barometric-tendency gradient* is given by a critical value of $-\nabla^2 \dfrac{\partial \phi}{\partial t}$, where ϕ is the geopotential, i.e., by a critical value of $-\lambda \dfrac{\partial \zeta_G}{\partial t}$.

6.16. According to C. W. Newton and S. Katz, *BAMS*, 39 (1958), 129, if $\overline{V}$ is the mean wind through the in-cloud layer (see Figure A6.16), then if complete vertical mixing takes place in the storm, due to updraughts and downdraughts leading to vertical exchange of horizontal momentum within the cloud, then *relative* velocities V_l' and V_u' will be created between the in-cloud air and the environmental air at both lower and upper levels. Therefore moist air will be fed into the right flank of the storm (with respect to $\overline{V}$) in the lower levels, and moist air will be evicted from the upper levels. Cloud growth will thus be favored on the right flank of the storm, and as a consequence the latter will move at an angle to the 700-mb wind, rather than with the wind.

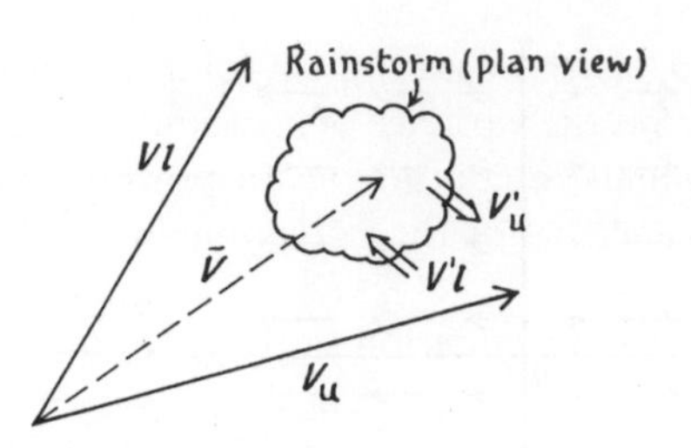

FIGURE A6.16.
Wind field in a convective rainstorm (after Newton and Katz).

6.17. For example, VR was constant at 1,000 feet above the ground, but not at lower levels. At 150 feet and 300 feet, and immediately outside the region of maximum speed of the tornado, $VR^{1.6}$ was constant; at greater radii, $VR^{0.8}$ was constant. See W. H. Hoecker, Jr., *MWR*, 88 (1960), 167.

6.18. T. Fujita, JM, 16 (1959), 38, states that the divergence at ground level was appreciable (100×10^{-5} per sec), decreasing to 30×10^{-5} per sec at 5,000 feet. The vertical velocity inside the mesosystems increased from 1 foot per sec at 1,000 feet to 3 feet per sec at 5,000 feet.

6.19. According to R. C. Miller and L. G. Starrett, *MM*, 9 (1962), 247, the threshold of thunderstorm activity in Britain is given by

$$T_8 - T_5 = 18.5 - 0.3T_5,$$

where T_8, T_5, represent the 850-mb and 500-mb temperatures in °C, respectively.

6.20. According to R. J. Donaldson, *BAMS*, 46 (1965), 185, the radar reflectivity Z_R is defined by

$$Pr = 1.1 \times 10^{-23} \frac{P_0 G^2 \theta \phi h Z_R}{\lambda^2 R^2},$$

where the notation is as in Appendix 1.7 in *Techniques*.

6.21. For example, following W. C. Jacobs, *BAMS*, 27 (1946), 306, suppose that a southeasterly weather type occurs, on the average, on 30 per cent of all days during January at 6:00 A.M., and that on 50 per cent of these days the gradient windspeed is 10–20 miles per hour. In 75 per cent of these latter cases, isobar curvature is cyclonic, with a 50 per cent frequency of mP air with mT aloft. When all these conditions hold, then on the average, stations A, B, and C, have ceilings of more than 600 feet and visibility of more than one mile. For stations D, E, and F, the corresponding frequency of such ceilings and visibility is 5 per cent. What is the probability that a situation will occur at 6:00 A.M. on a day in January such that stations A, B, C, are open (i.e., ceiling more than 600 feet, visibility more than one mile)? Then, from elementary probability algebra, probability $= 0.30 \times 0.05 \times 0.75 \times 0.50 \times 0.85 \times (1 - 0.05) = 0.045$.

6.22. Two examples of expressions involving Fisher-Tschebyscheff polynomials, taken from F. K. Hare *et al.*, *Arctic Meteorol. Res. Group, McGill Univ., Sci. Rep. no.* 3, are as follows.

1. Wadsworth's equation is

$$P_{ij} = \bar{p} + \sum_{r=1}^{a} Z_r SX_i + \sum_{r=1}^{b} Z_s SY_{sj} + \sum_{r=1}^{c} Z_{rs} SX_{ri} Y_{sj} \ldots$$

where p_{ij} is the pressure at point i, j, $\bar{p}$ is the mean pressure over all the sampling points; S is the variance of p_{ij} about $\bar{p}$; X_{ri}, Y_{sj}, are Fisher-Tschebyscheff orthonormal polynomials in coordinates of powers r and s, respectively; Z represents the normalized coefficients of the polynomials, and a,b,c, are the limiting powers of the expansion.

2. Godson's equation is

$$p_{ij} = \frac{1}{\sqrt{g_i}} \left[\sum_n \sum_m A_{nm} X_m(r_i) \cos n\theta_j + \sum_n \sum_m B_{nm} X_m(r_i) \sin n\theta_j \right]$$

in which $\sqrt{g_i}$ is a weighting factor to ensure equal-arc sampling; r_i and θ_j represent colatitude and longitude respectively; and A_{nm}, B_{nm}, are the Fourier coefficients for wave number n and power m in Fisher-Tschebyscheff orthonormal polynomials X_m.

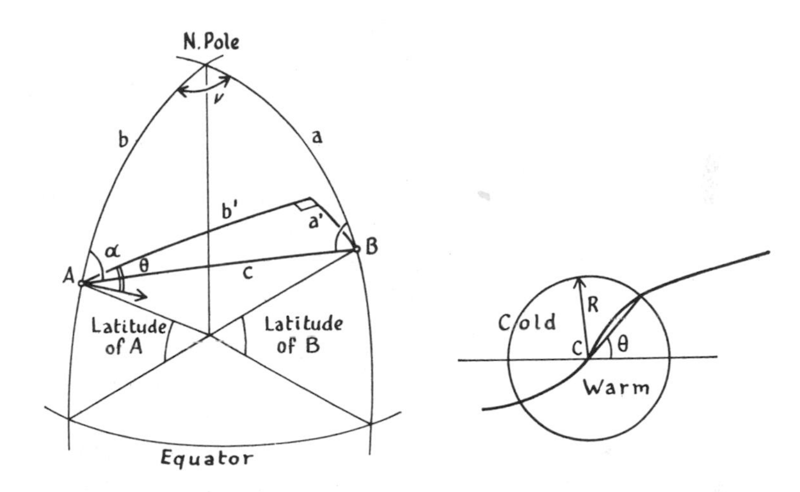

FIGURE A6.23.
Radius R is 5° of latitude; C is the center of the storm (after Jorgensen).

6.23. Following D. L. Jorgensen, *JAM*, 2 (1963), 226, the problem is as follows (see Figure A6.23). Given (1) the latitude and longitude of the sea-level low-pressure center located at A, (2) the latitude and longitude of the station at B that has reported precipitation, and (3) the amount the grid is rotated (anticlockwise) in degrees, it is required to determine the abscissae (a', b') of point B in the rotated grid. Let θ be the amount (in degrees) the grid is rotated (anticlockwise) about A to bring b to lie parallel to the thickness line through the center of the storm. Then when angle α and arc distance c are known, θ is given and therefore

$$a' = \sin^{-1}[\sin c(90 - \alpha - \theta)],$$

$$b' = \tan^{-1}[\tan c(90 - \alpha - \theta)].$$

The angle α is obtained by solving the simultaneous equations

$$\tan\frac{1}{2}(\beta-\alpha)=\frac{\sin\frac{1}{2}(b-a)}{\sin\frac{1}{2}(b+a)}\cot\frac{1}{2}v,$$

$$\tan\frac{1}{2}(\beta+\alpha)=\frac{\cos\frac{1}{2}(b-a)}{\cos\frac{1}{2}(b+a)}\cot\frac{1}{2}v,$$

where the arc distance a is 90° minus the latitude of B, the arc distance b is 90° minus the latitude of A, the angle v is the difference in longitude between A and B, and

$$c=\sin^{-1}[(\sin v\sin a)/\sin\alpha].$$

6.24. Following D. E. Martin and W. G. Leight, *MWR*, 77 (1949), 275, isobar curvature is measured at each station (S in Figure A6.24), by constructing chord BC such that its length is 10° of latitude, its endpoints are on the isobar through S, and the chord is centered on S. The perpendicular distance SD, measured in degrees of latitude, gives the curvature (positive for cyclonic curvature). Departures of 10,000-foot pressures from normal are estimated at 5° latitude and 5° longitude intersections. Air-parcel trajectory for each month is obtained by drawing the isobar through the station S, measuring the average pressure-gradient BC upstream from S, computing the geostrophic windspeed corresponding to BC, then extrapolating SD a distance upstream along the trajectory equivalent to the 48-hour travel of an air parcel moving at the speed of the geostrophic wind. See also W. H. Klein, *JAM*, 1 (1962), 154.

6.25. See S. W. C. Pack, *Weather Forecasting* (New York, 1948), Chapter 22, for examples of these empirical rules. Some illustrations are: (a) a cyclone approaching an anticyclone will become retarded and will then move in the same direction as the anticyclone; (b) the centers of warm-sector cyclones move in straight paths, whose direction lies parallel to that of the warm sector isobars; (c) elongated depressions

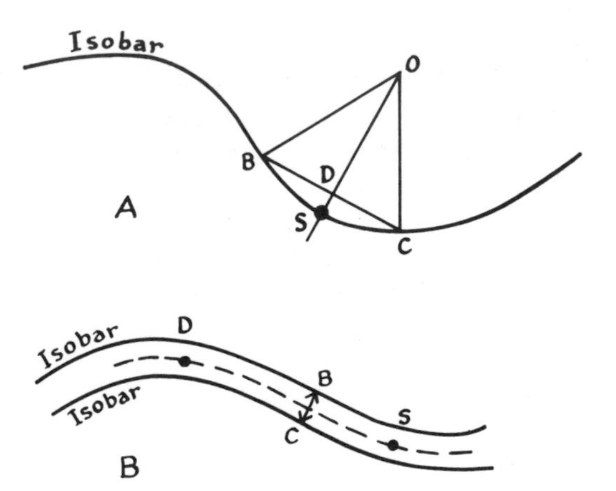

FIGURE A6.24.
A shows measurement of isobar curvature. Length of BC equals 10° of latitude; OB, OS, and OC are radii of curvature of the isobar through station S. B shows measurement of air-parcel trajectory: Broken line is the intermediate isobar; BC represents the pressure gradient.

move along their longest symmetry axis; (d) circular depressions move in the direction of the isallobaric gradient. See *WAF*, I, 44–56 and 156–88, and H. R. Byers, *General Meteorology* (New York, 1944), pp. 450–61, for an account of the kinematic methods. For Petterssen's classic forecasting rules, based on kinematics, see *WAF* (1940), pp. 398–437. Three of Petterssen's rules for predicting pressure changes are particularly useful, because they take into account both dynamic and thermal effects: (a) when the thickness isopleths descend the pressure surfaces, pressure must rise; (b) when the thickness lines ascend the pressure surfaces, pressure must fall; (c) when the thickness lines and the pressure contours are parallel, no pressure change can occur.

Abbreviations and Symbols

The following abbreviations are used in the bibliographical references in the notes at the end of each chapter and in the Appendixes.

AAAG. *Annals of the Association of American Geographers.*
AFGP. *Air Force Geophysics Research Papers.* (United States Air Force, Geophysics Research Directorate, Cambridge Center).
AFSG. *Air Force Surveys in Geophysics.*
AG. *Australian Geographer.*
AGP. *Advances in Geophysics* (American).
AGS. *Australian Geographical Studies.*
AM. *Antarctic Meteorology* (New York, 1960).
AMGB. *Archiv für Meteorologie, Geophysik, und Bioklimatologie.*
AMM. Meteorological Monographs of the American Meteorological Society.
ANYAS. *Annals of the New York Academy of Science.*
AS. *The Advancement of Science* (Journal of the British Association for the Advancement of Science).
ASM. B. Bolin, ed., *The Atmosphere and the Sea in Motion* (New York, 1959).
BAMS. *Bulletin of the American Meteorological Society.*
BWMO. *Bulletin of the World Meteorological Organization.*
CD. C. E. Anderson, ed., *Cumulus Dynamics* (New York, 1960).
CG. *The Canadian Geographer.*
CM. T. F. Malone, ed., *Compendium of Meteorology* (Boston, 1951).
CPRMS. *Centenary Proceedings of the Royal Meteorological Society* (London, 1950).
CTP. Technical Papers, Division of Meteorological Physics, Commonwealth Scientific and Industrial Research Organization (Australia).
DC. R. L. Pfeffer, ed., *Dynamics of Climate* (New York, 1960).
DMWF. C. L. Godske, T. Bergeron, J. Bjerknes, and R. C. Bundgaard, *Dynamic Meteorology and Weather Forecasting* (Washington, D.C., 1957).
EAFM. H. H. Lettau and B. Davidson, eds., *Exploring the Atmosphere's First Mile*, vol. I (New York, 1953).
EDM. A. H. Gordon, *Elements of Dynamic Meteorology* (London, 1962).
G. *Geography* (Journal of the Geographical Association of the United Kingdom).

GA.	*Geografiska Annaler* (Swedish).
GC.	*Glaciers and Climate*, special volume of *Geografiska Annaler*, vol. 31 (1949).
GJ.	*Geographical Journal* (Royal Geographical Society, London).
GPA.	*Geofisica Pura y Applicata* (Italian).
GR.	*Geographical Review* (American Geographical Society).
GRTMV.	*Geographical Reports* (Tokyo Metropolitan University).
GS.	*Geographical Studies* (British).
HBES.	M. I. Budyko, *The Heat Balance of the Earth's Surface*, trans. N. Stepanova (Washington, D.C., 1958).
HM.	F. A. Berry, E. Bollay, and N. R. Beers, eds., *Handbook of Meteorology* (New York, 1944).
HMS.	*Harvard Meteorological Studies.*
HSM.	C. E. P. Brooks and N. Carruthers, *Handbook of Statistical Methods in Meteorology* (London, 1953).
ITM.	S. L. Hess, *Introduction to Theoretical Meteorology* (New York, 1959).
JAM.	*Journal of Applied Meteorology* (American Meteorological Society).
JAS.	*Journal of the Atmospheric Sciences* (American Meteorological Society).
JGR.	*Journal of Geophysical Research.*
JM.	*Journal of Meteorology* (American Meteorological Society).
JPC.	*Japanese Progress in Climatology.*
JTG.	*Journal of Tropical Geography* (Singapore).
LSGA.	Lund Studies in Geography, Series A (Swedish).
M.	O. G. Sutton, *Micrometeorology* (New York, 1953).
MAB.	*Meteorological Abstracts and Bibliography* (American Meteorological Society).
MAE.	*Meteorology and Atomic Energy* (Washington, D.C., 1955).
MC.	V. Conrad and L. W. Pollak, *Methods in Climatology* (Cambridge, Mass., 1950).
MGAB.	*Meteorological and Geoastrophysical Abstracts and Bibliography* (American Meteorological Society).
MJTG.	*Malayan Journal of Tropical Geography.*
MM.	*Meteorological Magazine* (Meteorological Office of the United Kingdom).
MO.	Meteorological Office of the United Kingdom.
MOGM.	*Meteorological Office Geophysical Memoirs.*
MOHSI.	*Meteorological Office Handbook of Meteorological Instruments, Part I: Surface Instruments* (1956).
MOHUI.	*Meteorological Office Handbook of Meteorological Instruments, Part II: Upper-Air Instruments* (1961).
MOOH.	*Meteorological Office Observer's Handbook* (1952).
MOPN.	*Meteorological Office Professional Notes.*
MOSP.	*Meteorological Office Scientific Papers.*
MWR.	*Monthly Weather Review* (United States Weather Bureau).
N.	*Nature* (British).
NZG.	*New Zealand Geographer.*
PAS.	*Polar Atmosphere Symposium, part I* (New York, 1958).

PC.	*Publications in Climatology* (Laboratory of Climatology, C. W. Thornthwaite Associates, New Jersey).
PDM.	D. Brunt, *Physical and Dynamical Meteorology* (Cambridge, Eng., 1941).
PIGU.	*Proceedings of the International Geographical Union.*
PISRSM.	H. Wexler and J. E. Caskey, Jr., eds., *Proceedings of the First International Symposium on Rocket and Satellite Meteorology* (Amsterdam, 1963).
PM.	J. C. Johnson, *Physical Meteorology* (New York, 1954).
PPOM.	Papers in Physical Oceanography and Meteorology (M.I.T. and Woods Hole Oceanographic Institute).
PSTM.	W. Hutchings, ed., *Proceedings of the Symposium on Tropical Meteorology, Rotorua* (Wellington, N.Z., 1964).
PTMC.	*Proceedings of the Toronto Meteorological Conference* (London, 1953).
QJRMS.	*Quarterly Journal of the Royal Meteorological Society.*
RSI.	*Review of Scientific Instruments* (British).
SAGJ.	*South African Geographical Journal.*
SG.	*Soviet Geography.*
SGM.	*Scottish Geographical Magazine.*
SMC.	*Smithsonian Miscellaneous Collections* (Smithsonian Institution, Washington, D.C.).
SP.	*Science Progress* (British).
SPIAM.	*Scientific Proceedings of the International Association of Meteorology* (Rome, 1954).
T.	*Tellus* (Swedish).
TAGU.	*Transactions of the American Geophysical Union.*
TIBG.	*Transactions of the Institute of British Geographers.*
TJC.	*Tokyo Journal of Climatology.*
TMA.	D. J. Bargman, ed., *Tropical Meteorology in Africa* (Nairobi, 1960).
UCPG.	University of California Publications in Geography.
UCSCM.	*UNESCO Canberra Symposium on Climatology and Meteorology* (Paris, 1958).
USCC.	*UNESCO Symposium on Changes in Climate* (Paris, 1963).
W.	*Weather* (Royal Meteorological Society).
WAF.	S. Petterssen, *Weather Analysis and Forecasting* (New York, 2nd ed., 1956), vols. I and II.
WW.	*Weatherwise* (American Meteorological Society).

In the Appendixes, symbols are employed as defined here, unless otherwise stated.

Quantities and parameters		Basic units
c.	Horizontal windspeed	cm sec^{-1}
C_d.	Drag coefficient	dimensionless
C_p.	Specific heat of air at constant pressure	cal gm^{-1} deg^{-1}
C_v.	Specific heat of air at constant volume	cal gm^{-1} deg^{-1}
e.	Vapor pressure	dyne cm^{-2}
e_s.	Saturation vapor pressure	dyne cm^{-2}

Quantities and parameters		*Basic units*
h.	Height above mean sea level	cm
H.	Eddy heat flux	erg cm^{-2} sec^{-1}
K_E.	Eddy viscosity	cm^2 sec^{-1}
K_H.	Eddy conductivity	cm^2 sec^{-1}
K_W.	Eddy diffusivity	cm^2 sec^{-1}
L_v.	Latent heat of vaporization of water	cal gm^{-1}
p.	Actual atmospheric pressure	dyne cm^{-2}
p_0.	Pressure reduced to mean sea level	dyne cm^{-2}
p_h.	Pressure at height h above mean sea level	dyne cm^{-2}
q.	Specific humidity	dimensionless
r.	Radius of the Earth	cm
R_n.	Net radiation	erg cm^{-2} sec^{-1}
Re.	Reynolds number	dimensionless
Ri.	Richardson number	dimensionless
Ro.	Rossby number	dimensionless
u.	Horizontal windspeed component in the west-east direction	cm sec^{-1}
u_*.	Friction velocity	cm sec^{-1}
v.	Horizontal windspeed component in the south-north direction	cm sec^{-1}
V_g.	Geostrophic windspeed	cm sec^{-1}
V_{gs}.	Surface geostrophic windspeed	cm sec^{-1}
w.	Windspeed component in the vertical direction	cm sec^{-1}
x.	Distance measured in the west-east direction	cm
y.	Distance measured in the south-north direction	cm
z.	Distance measured in the vertical direction.	cm
z_0.	Roughness length	cm
Γ.	Lapse-rate	deg cm^{-1}
Γ_a.	Adiabatic lapse-rate	deg cm^{-1}
Γ_d.	Dry-adiabatic lapse-rate	deg cm^{-1}
Γ_e.	Environmental lapse-rate	deg cm^{-1}
ω.	Angular velocity	radians sec^{-1}
Ω.	Angular velocity of the Earth at the equator	radians sec^{-1}
ρ.	Density	gm cm^{-3}
ε.	Emissivity	cal cm^{-1} min^{-1} micron^{-1}
ν.	Kinematic viscosity, degrees of freedom	dimensionless
σ.	Stéfan-Boltzmann constant, standard deviation	cal cm^{-2} sec^{-1} deg^{-4}
κ.	Thermal conductivity	cal cm^{-1} sec^{-1} deg^{-1}
τ_0.	Surface tangential (shearing) stress	dynes cm^{-2}
θ.	Potential temperature	deg K
ϕ.	Latitude	degrees
λ.	Coriolis parameter	radians sec^{-1}
ψ.	Stream function	dimensionless

Further Reading

The references in the notes at the end of each chapter of *Foundations of Climatology* are intended (1) to indicate the source of the information given in the text, (2) to guide the student to useful reviews of the literature, and (3) to indicate classic or basic papers on specific topics, many of which may interest the climatologist more than the meteorologist, because the climatologist may be concerned with analysis of data from what the meteorologist may consider an outdated era of atmospheric science. Most of the topics discussed in this book are matters of current interest, and for up-to-the-minute information (in the English language) about the state of knowledge, the reader should see the following journals and periodicals.

1. For reviews of progress in specific fields: *AGP*, *AMM*, *BAMS*, *BWMO*, *Reviews of Geophysics* (Washington, D.C.), *SP*, *TAGU*, *WMO Technical Notes* (Geneva, Switz.).

2. For reviews and original research: *Agricultural Meteorology* (Amsterdam), *Atmospheric Environment* (Oxford), *Boundary-layer Meteorology* (Dordrecht, Holland), *International Journal of Biometeorology* (Leiden, Holland), *MM*, *N*, *Science* (Washington, D.C.), *The Marine Observer* (London), *W*, *WW*.

3. For original research: *AMGB*, *Atmospheric and Oceanic Physics* (Izvestiya, Academy of Sciences, U.S.S.R.; English translation by the American Geophysical Union, Washington, D.C.), *Australian Meteorological Magazine* (Melbourne), *GA*, *Geophysical Magazine* (Tokyo), *Indian Journal of Meteorology and Geophysics* (Delhi), *Japanese Journal of Geophysics* (Tokyo), *JAM*, *JAS*, *Journal of Atmospheric and Terrestrial Physics* (Oxford), *JGR*, *Journal of the Meteorological Society of Japan* (Tokyo), *MM*, *MWR*, *QJRMS*, *T*.

4. For southern hemisphere studies: *NOTOS* (South African Weather Bureau).

5. For reviews of progress and occasional original research papers in geographical climatology: *AAAG*, *AGS*, *CG*, *G*, *GA*, *GJ*, *GR*, *JTG*, *LSGA*, *NZG*, *SAGJ*, *SG*, *TIBG*.

The reader will find additional information concerning various topics covered in this book in the following books:—

ON ATMOSPHERIC PHYSICS

R. A. Craig, *The Upper Atmosphere: Meteorology and Physics* (London and New York, 1965).

R. G. Fleagle and J. A. Businger, *An Introduction to Atmospheric Physics* (London and New York, 1963).

D. L. Laikhtman, *Physics of the Boundary Layer of the Atmosphere* (Jerusalem, 1964).

H. U. Roll, *Physics of the Marine Atmosphere* (London and New York, 1965).

ON THE ATMOSPHERIC CIRCULATION

E. R. Reiter, *Jet-stream Meteorology* (Chicago and London, 1963).

W. L. Webb, *Structure of the Stratosphere and Mesosphere* (London and New York, 1966).

ON CLIMATIC CHANGE

H. H. Lamb and A. I. Johnson, *Secular Variations of the Atmospheric Circulation Since 1750*, *MOGM*, No. 110 (1966). Contains January and July surface pressure charts, 80°N–65°S, for 1750–1962.

J. M. Mitchell, Jr., ed., *Causes of Climatic Change*, *Met. Monographs*, vol. 8, no. 30 (Boston, Mass, 1968).

A. E. M. Nairn, ed., *Problems in Palaeoclimatology* (London, New York, Sydney, 1964).

ON ATMOSPHERIC TURBULENCE

J. L. Lumley and H. A. Panofsky, *The Structure of Atmospheric Turbulence* (London, New York, Sydney, 1964).

C. H. B. Priestley, *Turbulent Transfer in the Lower Atmosphere* (Chicago, 1959).

ON OCEANOGRAPHY

H. Stommel, *The Gulf Stream: A Physical and Dynamical Description* (Cambridge, Eng., 2d ed., 1965).

W. S. von Arx, *Introduction to Physical Oceanography* (Boston, 1962).

ON POLAR CLIMATOLOGY

S. S. Gaigerov, *Aerology of the Polar Regions* (Jerusalem, 1967).

M. K. Gavrilova, ed. by M. I. Bydyko, *Radiation Climate of the Arctic* (London, 1966).

M. J. Rubin, ed., *Studies in Antarctic Meteorology*, *Antarctic Research Series*, no. 9 (1966).

ON PHYSICAL CLIMATOLOGY AND MICROCLIMATE

R. Geiger, *The Climate near the Ground* (Cambridge, Mass., 4th ed., 1965).
W. D. Sellers, *Physical Climatology* (Chicago, 1965).
R. H. Shaw, ed., *Ground-Level Climatology*, Publication no. 86, American Association for the Advancement of Science (Washington, D.C., 1967).

RECENT PUBLICATIONS

Significant topics covered in publications appearing since this book went to press include the following.

Books. J. E. Caskey, Jr., *A Century of Weather Progress* (Boston, 1970); G. A. Corby, ed., *The Global Circulation of the Atmosphere* (London, 1970); A. J. Court, ed., *Eclectic Climatology* (Corvallis, 1969); H. Flohn, ed., *General Climatology* (*World Survey of Climatology*, vol. 2, New York, 1970); L. Gandin, *The Planning of Meteorological Station Networks* (Geneva: *WMO Tech. Note* no. 111, 1970); S. Orvig, ed., *Climates of the Polar Regions* (*World Survey of Climatology*, vol. 14, New York, 1970); E. Palmén and C. W. Newton, *Atmospheric Circulation Systems: Their Structure and Physical Interpretation* (New York, 1970); D. F. Rex, ed., *The Climate of the Free Atmosphere* (*World Survey of Climatology*, vol. 4, New York, 1969); W. L. Webb, ed., *Stratospheric Circulation* (*Progress in Astronautics and Aeronautics*, vol. 22, New York, 1969).

Journals. *AMGB*, series B, 17 (1969), 147, for climatic change over the Polar Ocean (Vowinckel and Orvig), and 18 (1970), 18, for analytical climatology (Thom). *BAMS*, 50 (1969), 880, for land- and sea-breeze fronts near 50 cm (Hsu). *BWMO*, 19 (1970), 223, for physical/descriptive climatology (Flohn). *JAM*, 8 (1969), 484, for short-period kinetic-energy cycles (Soong and Kung); *loc. cit.*, p. 649, for equatorial anticyclones (Fujita); *loc. cit.*, p. 799, for equatorial convectivedown drafts (Zipser); *loc. cit.*, p. 896, for atmospheric wake phenomena (Zimmerman). *JAM*, 9 (1970), 389, for the mean microclimatic wind profile over the ocean (Ruggles); *loc. cit.*, p. 417, for mesosystems within occlusions (Kreitzberg); *loc. cit.*, p. 433, for interaction of binary tropical cyclones (Brand); *loc. cit.*, p. 442, for southwest monsoon mid-troposphere cyclones (Krishnamurti and Hawkins); *loc. cit.*, p. 534, for CAT at subsidence inversions (Boucher); *loc. cit.*, p. 537, for wet-bulb zero and hail occurrence (Morgan); *loc. cit.*, p. 708, for eigenvectors of sea-level pressure (Kutzbach); *loc. cit.*, p. 760, for diurnal variation of onshore wind speed (Yu and Wagner). *JAS*, 27 (1970), 494, for surface pressure/ionospheric relations (Tolstoy and Lan). *JAS*, 28 (1971), 254, for thunderstorm/ionospheric relations (Davies and Jones). *MM*, 98 (1969), 274, for 100-mb flow/surface weather relations (Davis). *MM*, 99 (1970), 1, for large-scale equatorial disturbances (Sawyer); *loc. cit.*, p. 221, for digitized data banks in synoptic climatology (Craddock). *MM*, 100 (1971), 46, for a persistent mesoscale eddy (Findlater). *MWR*, 98 (1970), 399, for 850-mb lows and heavy snow (Browne and Younkin). *MWR*, 99 (1971), 49, for long-period variations in seasonal sea-level pressure (Wagner). *QJRMS*, 95 (1969), 710, for polar lows as baroclinic disturbances (Harrold and Browning). *QJRMS*, 96 (1970), 610, for synoptic disturbances with a time scale of a month (Sawyer); *loc. cit.*, p. 626, on stationary disturbances (Holopainen). *SG*, 10 (1969), 429, on climatic change (Budyko). *TIBG*, 50 (1970), 55, on climatic anomaly fields

(Perry). *W*, 24 (1969), 255, on mountain-sized lee eddies (Fjörchtgott). *W*, 26 (1971), 263, on human effects on climate (Sawyer). *WW*, 23 (1970), 160, on tornado suction spots (Fujita). For scale concepts, see Fiedler and Panofsky, *BAMS*, 51 (1970), 1114, and Barry, *TIBG*, 49 (1970), 61. For boundary-layer climatology, see Oke and East, *Boundary-layer Meteorology*, 1 (1971), 411; Estoque, *MWR*, 99 (1971), 193; Bourne and Ball, *JAM*, 9 (1970), 862. For mesoclimatology, see Grillo and Spar, *JAM*, 10 (1971), 56. On chinooks, see Brinckman, *W*, 26 (1971), 230, and Riehl, *loc. cit.*, p. 241. On strong down-mountain winds, see Julian, *WW*, 22 (1969), 108.

Indexes

Name Index

This index includes (a) names mentioned in the main body of the text; and (b) names given in the notes (at the end of each chapter) as authorities for statements made in the book.

Geographical Index

Numbers in **boldface** are references to figures. Except for large, famous cities, towns and cities are listed under countries, or, for the United States, under the names of the states.

Subject Index

Only the most important discussions of each subject are indexed. References to figures are indicated by boldface type.